Intelligent Manufacturing in Industry 6.0

Hammad Majeed • Tehreema Iftikhar

Intelligent Manufacturing in Industry 6.0

A Climate Resilience Approach

Springer

Hammad Majeed
Department of Chemistry
University of Management and Technology
(UMT) Lahore, Sialkot sub campus
Sialkot, Pakistan

Tehreema Iftikhar
Department of Botany, Applied Botany Lab
Government College University
Lahore, Pakistan

ISBN 978-3-032-07277-1 ISBN 978-3-032-07278-8 (eBook)
https://doi.org/10.1007/978-3-032-07278-8

This Springer imprint is published by the registered company Springer Nature Switzerland AG
The registered company address is: Gewerbestrasse 11, 6330 Cham, Switzerland

We dedicate this work to the last prophet and messenger of Allah, Hazrat Muhammad (sallallahu alayhi wassallam), a true visionary leader and teacher, whose excellent guidance, profound wisdom, and countless beneficial knowledge (given by Allah) have been a constant source of inspiration for all the species in this universe. His profound mentorship and steadfast pursuit of academic excellence have been a guiding force in our journey, anchored in a noble vision to uplift humanity and resonate with every particle across the known and unknown universe.

Foreword

The global industrial landscape is undergoing an unprecedented transformation driven by Industry 6.0, which seamlessly integrates hyper-automation with artificial intelligence, machine learning, deep learning, edge computing, cloud computing, quantum computing, cyber physical systems, and other sustainable Internet of Things, intelligent manufacturing technologies in 5G, 6G, and beyond era. These advancements are not mere incremental improvements but represent a paradigm shift in how industries manufacture, process, and optimize resources. At the heart of this transformation lies in a fundamental challenge, i.e., climate resilience. Climate change due to earth degradation, is the most pressing challenge of our time, posing a direct threat to humanity and all the species on this Earth and the planet's biodiversity. Urgent action is required to reduce global warming, mitigate industrial emissions, and transition toward sustainable, net zero carbon manufacturing.

This authored book *Intelligent Manufacturing in Industry 6.0: A Climate Resilience Approach* by Dr. Hammad Majeed and Prof. Dr. Tehreema Iftikhar is a landmark contribution to shaping the future of sustainable industrialization. It provides a comprehensive and in-depth analysis of how Industry 6.0 technologies can revolutionize manufacturing while addressing the challenges of climate change, resource efficiency, and economic sustainability. This book will serve as an essential guide and textbook for the industries, academic researchers, scientists, policymakers, and sustainability advocates, encouraging them to align their research and production strategies with sustainable future demands and global climate objectives. The United Nations' 17 Sustainable Development Goals must be central to this transformation, as they provide a universal blueprint for achieving sustainable industrial growth. These SDGs align with the principles of ethical responsibility and environmental stewardship emphasized in the teachings of Islam, reinforcing the moral and scientific necessity of this transition. New SDGs will also be required to cover the future trends of Industry 6.0 and 7.0.

What sets this book apart is its practical and industrial focus approach. While many publications discuss mostly a theoretical standpoint, this book goes further by bridging the gap between academic research and real-world industrial applications for societal impact. It offers case studies, cutting-edge research, advanced manufacturing frameworks, and predictive industry trends that will shape the next era of industrial smart automation. From intelligent automation in multi-industries to quantum-enhanced smart manufacturing and digital transformation, each chapter provides a structured blend of fundamental knowledge and advanced applications. This makes the book a valuable resource for engineers, business leaders, industrialists, policymakers, and researchers aiming to implement sustainable, high-efficiency manufacturing strategies. The initial chapters of this book provide a comprehensive foundation on climate change, covering essential topics such as global warming, greenhouse gases, carbon footprints, the Kyoto and Paris Agreements, and the Conference of Parties from COP26 to COP29. The discussion then progresses to the evolution of industrial revolutions, tracing the systematic transformation of industries from manual operations to automation. This book masterfully explores how Industry 4.0's digital advancements have evolved into Industry 5.0's customer focus automation and further shifting toward Industry 6.0's AI driven, hyperconnected, and sustainable manufacturing systems.

A key theme of this book is the integration of advanced technologies with eco-friendly, sustainable industrial production along with complete road map and guidance for the implementations of these advanced technologies and the challenges associated with them. Many industrial leaders perceive sustainability as a costly and challenging transformation, believing that implementing green manufacturing practices would negatively impact profitability. However, this book aims to challenge that misconception by presenting practical, cost-effective strategies for transitioning to sustainable production without increasing costs and in many cases, reducing operational expenses. Through detailed SWOT, cost, and ROI analyses, it illustrates how smart automation and sustainability investments not only pay for themselves but also lead to the long-term profitability with major market share. The transformation toward carbon neutral (even negative), resource-efficient industries is not just an economic imperative but also an ethical responsibility. A well-planned shift to smart, sustainable industrial practices benefits manufacturers, consumers, and all living species. Most importantly, this transformation safeguards the future of generations yet to be born. Though we may not witness their appreciation firsthand, our collective efforts today will shape a healthier, more balanced planet for them. Industry 6.0 is not just about technological advancement, it is about building a future where innovation coexists with environmental responsibility, ensuring a cleaner, more sustainable world for all. The world is changing very fast and complete transformation into Industry 6.0 is not far away even Industry 7.0 will capture this world very soon.

A book should not be confined to libraries and academic shelves; it must serve as a blueprint for industrial transformation. Scientific literature should have a direct societal and industrial impact, fostering real-world technological advancements, patents, and practical innovations that drive progress. Our industries need such guidance driven, solution-oriented knowledge to transition toward climate-resilient, resource-efficient, and next-generation manufacturing ecosystems. It is time for scientists, educators, industrialists, and policymakers to collaborate for the greater good, ensuring that future generations inherit a world that is technologically advanced, economically sustainable, and highly environmentally responsible.

I am confident that *Intelligent Manufacturing in Industry 6.0: A Climate Resilience Approach* will become an essential reference/benchmark/textbook for those navigating the complexities of next-generation industrial systems in Industry 6.0 and 7.0. I highly recommend this book to all professionals, researchers, and decision-makers who are passionate about technological innovation, sustainability, and the future of intelligent manufacturing. At the end, I must congratulate both the authors for giving an excellent book to the humanity.

Academic Professional and Mentor
Faisalabad, Pakistan

Professor (Retired) Abdul Majeed Janjua,

Preface

The industrial world stands at the crossroads of an unprecedented transformation. As we transition into the era of Industry 6.0, the integration of artificial intelligence, quantum computing, cyber-physical systems, IoT-driven automation, and sustainable manufacturing has become imperative. The escalating challenges posed by climate change, resource depletion, and economic volatility demand a revolutionary approach, one that ensures resilience, sustainability, and efficiency in industrial operations. This book, *Intelligent Manufacturing in Industry 6.0: A Climate Resilience Approach*, serves as an authoritative guide, bridging the gap between technological advancements and climate-conscious industrial strategies and productions. With a strong emphasis on technological advancements, this book explores how Industry 4.0 and Industry 5.0 have laid the groundwork for Industry 6.0. We discuss the transformative role of all Internet of Things, cyber-physical systems with artificial intelligence and machine learning, edge computing, blockchain, and digital twins in reshaping industrial operations. These advancements facilitate real-time monitoring, predictive maintenance, and decentralized decision-making, significantly enhancing efficiency and sustainability. A key focus of this book is the integration of intelligent manufacturing technologies into multiple industries. We examine AI-driven automation, big data analytics, and IoT-enabled smart systems, emphasizing their application in sectors, such as aerospace, defense, healthcare, pharmaceuticals, agriculture, textiles and many more. The discussion extends to sustainable chemical manufacturing, intelligent coal mining, and energy-efficient processes, emphasizing how Industry 6.0 can drive circular economy models, net-zero carbon emission, and net-zero waste generation.

Climate resilience remains a central theme throughout this work. We analyze cutting-edge advancements in AI-powered energy optimization, carbon footprint reduction techniques/circular economy, and blockchain-based carbon credit tracking. Practical case studies from various industries illustrate the tangible benefits of green manufacturing and sustainable supply chain management. This book also discusses the quantum computing's role in industrial problem solving, nanomaterials, self-healing smart materials, and future proof manufacturing solutions. We address the socio-economic and geopolitical implications of Industry 6.0. The impact of global population growth, workforce transformation, supply chain disruptions, and cybersecurity concerns is examined with data-driven technologies. We provide strategies for overcoming financial and regulatory barriers, ensuring seamless integration of intelligent manufacturing solutions into industrial ecosystems. Our discussions on intelligent automation, robotics, and AI-driven predictive quality control highlight the significance of technological foresight in building a sustainable future. We may belong to Gen X, Gen Y, Gen Z, or even the emerging Gen Alpha and the future generations Gen Beta, Gen Gamma, Gen Delta, and beyond; we may be Muslim, Jewish, Christian, or followers of any faith; Asian, European, American, African, or from any region of the world, but when it comes to healing this Earth, we must all unite as one, i.e., only Generation Planet. For the sake of the soil beneath our feet and the skies above our heads, we must rise beyond labels and timelines. Whether we are just beginning to breathe or nearing our final breath, our duty should remain the same to serve this earth with love and sincerity. To rebirth and reset this Earth, to tread lightly and live wisely, to choose sustainability not as a trend, but as a sacred legacy. Let us forge an unshakable pact with ourselves, to walk gently today so that tomorrow may bloom. Let us plant the seed of eco-consciousness in our children and urge them to pass this sacred flame to theirs. For only when we see the planet as an extension of our soul, will we truly become caretakers of a future worth dreaming. Deen Islam teaches us profound respect for nature, so deep that we are commanded never to harm a single tree, even in the midst of battle. It reminds us never to waste even a single drop of water, even while performing ablution, even if we stand beside a flowing river that belongs to us. Our beloved last Prophet of Allah, the most honorable Hazrat Muhammad (peace and blessings be upon him and his family), along with his noble companions, often wore patched and worn garments as a living example to humanity. Even when they were leaders of a vast, rich and powerful empire, they chose simplicity to inspire generations to embrace humility, reuse, and the spirit of a circular economy, reminding us that dignity lies not in what we wear, but in how we care for the Earth and its blessings. Such is the sacredness of every creation and every resource, a divine trust we must honor with love and responsibility.

Our journey begins by tracing the historical evolution of industrial revolutions, examining their contributions to climate change and highlighting missed opportunities for sustainable, eco-friendly practices. The foundational chapters explore the origins of global warming, greenhouse gas emissions, and key international climate agreements, including the Kyoto Protocol, Paris Agreement, and the Conferences of the Parties (COP26 to COP29). Understanding these environmental shifts is essential for shaping effective policies and enabling industry-wide adaptation. The final chapters deliver an in-depth analysis of next-generation manufacturing technologies. By aligning technological innovation with the SDGs, this book outlines a road map for industries to transition toward high-efficiency, climate-resilient manufacturing models. Designed for researchers, engineers, industry professionals, policymakers, and academics, this book provides a comprehensive understanding of the intersection between intelligent manufacturing and climate resilience. Combining scientific rigor with practical insights, it aims to inspire innovation, foster industrial collaboration, and shape the future of sustainable manufacturing. We hope this work serves as a catalyst for industrial transformation ensuring that technological progress is aligned with environmental stewardship for generations to come. This book begins with Chapter 1, which introduces the fundamentals and historical context of climate change and global warming. Chapter 2 offers an overview of the Industrial Revolution's progression from Industry 1.0–3.0. Chapters 3 and 4 establish the conceptual groundwork for Industry 6.0 by discussing enabling technologies and key advancements emerging from the Fourth Industrial Revolution. Chapters 5–7 provide a detailed analysis of Industry 6.0's technological framework and its diverse industrial applications. Chapters 8 and 9 focus on practical implementation strategies, identifying the challenges and barriers to adopting Industry 6.0. Chapter 10 explores emerging technologies that are expected to shape the core of Industry 6.0 and catalyze the transition toward Industry 7.0. Chapters 11–16 examine sector-specific applications, analyzing the influence and transformative potential of Industry 6.0 across various industrial domains. This book integrates knowledge from multiple scientific and technical domains, including chemistry, physics, mathematics, zoology, botany, computer science, software and hardware engineering, artificial intelligence, manufacturing and fabrication technologies, materials science, environmental science, climate change, sustainability, economics, industrial engineering, and industrial waste management. Every topic has been addressed with a strong commitment to technical accuracy, precision, and interdisciplinary coherence. If experts from any specific field identify technical concerns or require clarification, they are encouraged to contact us directly. We will provide justification where the context supports our position, and in cases where improvements are needed, revisions will be incorporated in future editions. Knowledge is not restricted to a single discipline. Any individual with curiosity, effort, and willingness to learn can understand, apply, and contribute to diverse areas of science and technology. Advanced degrees such as a Ph.D. help expand intellectual capacity and analytical depth, enabling broader comprehension of global knowledge. True learning requires openness, intellectual humility, and the continuous integration of ideas across the mind, heart, and scientific practice. We express our sincere gratitude to Springer Nature and the entire editorial and publishing team for their exceptional guidance and support throughout the completion of this book.

Sialkot, Pakistan
Lahore, Pakistan
August 04, 2025

Hammad Majeed
Tehreema Iftikhar

Competing Interests The authors have no competing interests to declare that are relevant to the content of this manuscript.

Contents

About the Authors

Hammad Majeed is a renowned scientist, academic leader, and industrial consultant with over 25 years of multidisciplinary experience across industrial operations, academia, business development, and intelligent manufacturing. He currently serves at the University of Management and Technology (UMT), Sialkot Campus, Pakistan, where he heads the Department of Chemistry, the ORIC, and BIC. He also holds the position of Manager Innovation and Commercialization at ORIC, UMT Lahore, Pakistan. Holding a Ph.D. in Chemistry, he is certified in diverse fields, including AI, machine learning, additive manufacturing, circular economy, lean management, aviation, orbit, automotive, big data, cybersecurity, cloud computing, and Industry 4.0 to 6.0/7.0 technologies. His domain expertise spans diversified industrial sectors with specializations in quality control, process optimization, industrial automation, and sustainable manufacturing. He has led numerous nationally and internationally funded projects, offering consultancy on climate change adaptation, digital transformation, resource efficiency, and the automation of traditional manufacturing systems, even in resource-constraint environments. He has completed advanced trainings in industrial automation from Italy and China and from German companies. He has delivered keynote lectures, judged competitions, and conducted workshops on global platforms and received multiple honors, including a prestigious award from the President of Pakistan (for a country-level startup win). He is affiliated with various national and international professional societies. As a senior consultant with Brilliant Eco International-owned company, he delivered over 650 completed industrial projects. His innovations have advanced product development, cost reduction, quality enhancement, waste upcycling, net-zero emissions and trash, and automation. An active trainer on SDGs, climate resilience, cost reduction, innovative product developments, patent development, and sustainable solutions with intelligent manufacturing technologies, he has also contributed to projects in the UK, Germany, Italy, China, India, Turkey, Uzbekistan, and Azerbaijan.

A prolific author and reviewer for high-impact journals, his forthcoming books will explore sustainable additive manufacturing, functional materials, and next-generation Industry 7.0 technologies, with practical implementations aimed at achieving super-automation. The focus will be on 100% eco-friendly industrial production and carbon-negative strategies to help rejuvenate our beloved Earth and foster a healthy environment for technologically evolved humanity. He is widely recognized as a thought leader in climate-resilient intelligent manufacturing and sustainable innovation.

Tehreema Iftikhar is a distinguished scientist and academic leader with over 23 years of experience, currently serving as Chairperson and Full Professor (BPS-21) at the Department of Botany, Government College University, Lahore, Pakistan. As the first female professor in the department's 112-year history, she has significantly advanced botanical sciences with diverse applications, pioneering the integration of AI, machine learning, intelligent manufacturing, and smart biodesign into plant sciences, eco-innovations, and food technology for sustainable industrial applications under Industry 6.0 and 7.0, despite resource-limited infrastructure. A recipient of the Governor's Award for her Ph.D. in Botany, she completed a Ph.D. fellowship at Seoul National University (South Korea) focusing on bioprocessing, industrial biotechnology, and molecular biology, followed by postdoctoral research in high-tech food engineering at Cornell University, USA. Her interdisciplinary research spans applied botany, food science, enzymology, industrial biotechnology, AI-based modeling, and simulation tools. She has published extensively in high-impact journals, holds 35 patents, and has led multiple nationally and internationally funded research projects, including the supervision of COMSTECH scholars from Azerbaijan. Elected Vice President (Punjab) and former joint secretary of the Pakistan Botanical Society, she is also a Fellow of the Biological Society of Pakistan and a Fellow of the Women Leadership Program at Michigan State University (USA). She has received numerous accolades, including the Quaid-e-Azam Gold Medal, Best University Teacher Award, Research Publication Awards, and Patent Awards for innovations at the intersection of science, engineering, and AI. She has also won national startup competitions, led capacity building programs, and strengthened academia-industry linkages in high-tech industrial sectors. Her pioneering work continues to advance industrial biotechnology, enzyme engineering, and applied botany, striving to position the country at the forefront of green technologies. A staunch advocate for scientific innovation, sustainability, and academic excellence, she is committed to implementing the UN SDGs, climate change mitigation, biodiversity conservation, and hyper-automated intelligent manufacturing for global sustainability.

Chapter 1
Climate Change Impact on Our Planet

Abstract This book's first chapter provides a basic review of climate change, following its beginnings from pre-industrial times to the present (up to 2025). It describes how global warming and its related effects, including rising sea levels, extreme weather events, and biodiversity loss, are caused by industrialization, namely the combustion of fossil fuels, which has resulted in a major increase in greenhouse gas (GHG) emissions. Although there has been natural climatic variability for millennia, human activity is primarily responsible for the current rapid changes. Carbon footprint calculation must be adopted in all types of human activities to plan emission reduction strategies. The Paris Agreement based on nationally determined contributions (NDCs) and the Kyoto Protocol, COPs 21 to 29 seek to coordinate international efforts to combat climate change by providing financial assistance to vulnerable countries and legally binding obligations. The story promotes a switch to renewable energy sources and increased energy efficiency across businesses, highlighting the need for cutting-edge technologies and environmentally friendly behaviors. International cooperation is highly needed because climate change transcends national boundaries and calls for a coordinated response. If all countries do not act decisively and cooperatively, the effects of climate change will be disastrous, and urgent action is required to ensure a sustainable future.

Keywords Climate change · Global warming · CO_2 emission · CH_4 emission · N_2O emission · COP21 · COP26 · COP27 · COP28 · COP29 · COP30 · 100 billion USD · Kyoto protocol · Paris agreement · NDC · Intelligent manufacturing · Industrial revolution

1 Introduction of Climate Change

Before the Industrial Revolution, the world was largely protected from the impacts of global warming. It is important to acknowledge that our industrial forebears were unaware of the adverse effects associated with the burning of fossil fuels,

H. Majeed, T. Iftikhar, *Intelligent Manufacturing in Industry 6.0*,
https://doi.org/10.1007/978-3-032-07278-8_1

particularly coal. While successive industrial revolutions significantly improved the quality of life for many, they also led to a steady increase in greenhouse gas (GHG) emissions, which have been a major driver of global warming [1, 2]. This rise in global temperatures is the primary catalyst for the climate change we observe today, as compared to pre-industrial times. The consequences of climate change include rising sea levels, extreme heat and cold events, flooding, the extinction of numerous animal species, the emergence of pathogenic microbial species responsible for millions of deaths, and the accelerated melting of glaciers, leading to unforeseen impacts across diverse regions. Climate change can thus be defined as the "shift of normal weather patterns on Earth towards extreme conditions due to the rise in global temperatures [3, 4]. Although climate has varied naturally over millions of years due to volcanic eruptions and changes in solar energy (natural pattern of global warming), this era of rapid climate change is because of human activities, particularly those that have taken place since the Industrial Revolution. These activities have resulted in unprecedented warming rates, and global warming specifically refers to the temperature increase caused by human activities. Throughout the various phases of the Industrial Revolution (from the first to the fourth), the main task of the industrial activities was the unsystematic use of fossil fuels (coal, petroleum, and natural methane gas) for industrial operations, transportation, energy production, and steam generation. This reliance has significantly increased the global carbon footprint. Unfortunately, the expansion of industrial activities was not accompanied by sufficient afforestation efforts to offset this carbon footprint through a carbon neutrality approach. On the contrary, deforestation accelerated to meet the growing demands for wood in both domestic and industrial contexts, exacerbated by rapid urbanization driven by industrialization [5, 6]. Figure 1.1 shows the climate change impact.

Fig. 1.1 Climate change impact on our earth

1.1 Carbon Footprint

As worries about global warming grow, the carbon footprint has emerged as an important environmental measure. The evaluation of carbon footprints has been widely applied across all industries. The most popular standards for assessing carbon footprints now are ISO/TS 14067, the GHG protocol, and PAS 2050. PAS 2050, the first standard for determining a product's carbon footprint, is very dependable and operable and integrates the different life cycle assessment principles. The GHG Protocol improves the explanation of ideas like functional units and baselinc flows of the product system, drawing on PAS 2050 and its supplemental criteria. With its strong universality, ISO 14067 combines the elements of the GHG Protocol and PAS 2050. The life cycle assessment standards ISO 14040 and ISO 14044 served as the foundation for all three standards [7, 8].

One of the biggest issues of our day is climate change, which has a profound effect on the environment, human society, and the world economy. The combined efforts of governments, corporations, scientists, and individuals are needed to meet this problem. We can lessen the effects of climate change and create a sustainable future for future generations by cutting GHG emissions, boosting resilience to its effects, and encouraging innovation. The combined effects of deforestation and urbanization have intensified global warming, as these activities have severely disrupted Earth's natural systems. The alteration of the Earth's surface has diminished its capacity to reflect solar radiation, resulting in more heat being trapped within the atmosphere, which in turn has disrupted the Earth's carbon cycle. Industries such as cement production, aviation, textiles, waste management (including landfilling with plastics and other materials), and the manufacture of non-degradable materials, particularly plastic-based synthetic polymers and iron and steel, have all made substantial contributions to global warming. Businesses are encouraged to modify their current business models by shifting market dynamics, including rising oil prices, greater awareness of the environmental effects of global warming, technology advancements in the energy and environmental sectors, and shifting customer attitudes toward green goods and services. To enable green industries development, governments and companies must strike a balance between four elements: supportive government regulations, a market adoption strategy that guarantees a foothold, creative and tailored business models, and enabling technical systems [9, 10]. Figure 1.2 is showing the carbon footprint from different sources.

1.2 Green House Gases

Methane, a naturally occurring fossil fuel, along with carbon dioxide, and nitrous oxide is the primary GHG contributing to anthropogenic global warming. Methane can be generated through the degradation of organic and food waste in landfills. Nitrous oxide, on the other hand, is predominantly emitted from agricultural

Fig. 1.2 Carbon footprint from different actions

activities, especially using synthetic fertilizers, herbicides, and pesticides. It is also released during various industrial processes, particularly in the synthesis of chemicals. Among these gases, nitrous oxide is the most potent in terms of global warming potential. The subsequent chapters of this book will explore these aspects in greater detail [11, 12]. This first chapter serves as an introduction. Between 1997 and 2013, Earth experienced an unprecedented increase in global warming, leading to abrupt climate changes worldwide. These changes continue to pose a significant threat to humanity, manifesting in extreme conditions of weather events such as heatwaves, floods, glacial melting, erosion, saltwater intrusion into freshwater resources (which constitute less than 1% of the Earth's total water supply), hurricanes, droughts, abnormal rainfall patterns, and rising sea levels. These phenomena have devastating effects on human societies, agricultural sectors, infrastructure, economies, livelihoods, and businesses. These phenomena also cause severe damage to aquatic ecosystems, lead to the extinction of numerous species both on land and underwater, and drive the migration of non-human species. The rapid onset of these changes has negatively impacted the quality of human health and well-being, resulting in the emergence of new pathogens and the escalation of various health issues, including respiratory illnesses, heat-related conditions, eye infections, and diseases affecting the lungs, kidneys, and liver. These health crises are exacerbated by extreme air and water pollution, much of which is driven by industrial activities and other practices that contribute to the contamination of food, water, and other essential resources [12, 13].

Enhancing energy efficiency within industrial processes is essential, including the replacement of high energy-consuming components like motors with more efficient alternatives (the shift to IE5 motors is now mandatory, and IE 6 and beyond or any other alternate in the future). Also, the adoption of smart building practices with an energy-efficient focus is now unavoidable. This is particularly true for weather-resistant infrastructure in both housing and industrial sectors, as well as in the agriculture and food industry, where updated farming technologies, food security measures, and sustainable food packaging are crucial. The aviation sector must also transition to alternative fuels to reduce its carbon footprint [14–17].

Similarly, nitrous oxide (N_2O) is the most impactful and threatening to ozone depletion (after chloro fluoro carbons) currently released and a critical GHG. Nitrous oxide is mostly produced by agriculture and has around 298 times the warming potential of carbon dioxide (CO_2) per molecule as per IPCC AR6. Manure and synthetic fertilizer use are the main causes of anthropogenic N_2O emissions, which are increasing and are responsible for almost 10% of net global warming since the Industrial Revolution [18–20]. Emissions might increase by 30% by 2050 if nothing is done. With the use of existing abatement techniques, these emissions might be reduced by more than 40%, improving public health, water quality, and air quality. Ammonia and nitrogen oxides would be reduced concurrently by a sustainable nitrogen management strategy, promoting global objectives for food security, ozone recovery, and climate change. Prior to industrialization, atmospheric N_2O levels were between 270 and 280 ppb; by 2024, they had risen to above 336 ppb. From 2000 to 2004, the annual N_2O growth rate was 0.68 ppb; in 2017–2021, it doubled to 1.2 ppb. Up to 235 billion tons of CO_2-equivalent emissions by 2100-, or six-years' worth of global emissions from fossil fuels, might be avoided by potential N_2O reduction. Seventy-five percent of anthropogenic N_2O emissions come from agriculture, with the remaining portion coming from industrial sources and other sectors [21–23].

The latest technologies emerging from the ongoing industrial revolutions, including artificial intelligence, machine learning, additive manufacturing, smart systems, digital twin technology, sustainable industrial automation, carbon capture, storage, and utilization technologies (emphasizing a carbon-negative approach), circular economy practices, net zero emission strategies, zero waste policies, and rigorous calculation of carbon footprints in all industrial and academic research endeavors, are now critical. The shift from nanotechnology to femto- and Pico-technology, the expansion of forests (especially fast growing plants), the implementation of rooftop, wall, and indoor gardening (especially with plants that absorb more carbon), and the use of species that can absorb heavy metals and other toxins through their root systems at wastewater disposal sites are all vital steps. Sustainable farming practices are also key to these efforts. With the use of green management techniques, they rethink conventional manufacturing, logistics, and marketing techniques [24–26]. This book will explore a wide range of topics aimed at reducing the impact of climate change, offering comprehensive insights into the strategies and technologies necessary to address this global challenge.

The largest sources of global CO_2 emissions are energy production (above 40%) and industrial manufacturing (above 20%), underscoring the reliance on fossil fuels and high-emission activities such as the production of steel and cement. 55% of methane emissions worldwide come from cattle and agriculture, mostly from enteric fermentation and manure management. Due to the overuse of synthetic fertilizers,

N_2O emissions are highest in agriculture (40%) and cause long-term climate damage because they are 298× stronger than CO_2. The manufacture of synthetic fibers, textile dyeing, and wastewater treatment make the textile and fashion industries important polluters (accounting for 3% of CO_2 emissions). Anaerobic processes and the breakdown of organic waste in landfills are the main causes of waste management's 9% contribution to worldwide methane emissions. A multifaceted strategy is needed to reduce emissions in all industrial sectors, including the development of low-impact materials, carbon capture technology, AI-driven process optimizations, sustainable agriculture practices, and the integration of renewable energy. While maintaining compliance with climate targets like the Paris Agreement and SDGs 7, 9, 12, and 13, implementing Industry 6.0 innovations can cut industrial emissions by up to 50%. For politicians, business executives, and environmental scientists interested in decarbonizing global sectors, Table 1.1 (Data Sources: IPCC AR6 Synthesis, IEA 2025, EDGAR v8.0), whereas 2025 values are projections based on current trajectories (IEA Net Zero Roadmap 2025 [27–31]). The 2024 Global Carbon Budget projects fossil carbon dioxide emissions at 37.4 billion tons, a 0.8% increase from 2023, with land use change emissions estimated at 4.2 billion tons, bringing the total CO_2 emissions to 41.6 billion tons, up from 40.6 billion tons last year. Despite the urgent need to reduce emissions to mitigate climate change, there is still no indication that fossil CO_2 emissions have peaked. While land-use change emissions declined by 20% over the past decade, both fossil and land-related emissions are expected to rise in 2024, exacerbated by El Niño-driven droughts that intensified deforestation and forest fires. With annual emissions exceeding 40 billion tons, atmospheric CO_2 is projected to reach 422.5 ppm in 2024, an increase of 2.8 ppm from 2023 and 52% higher than pre-industrial levels. Emissions by fuel type are expected to rise, that is, coal by 0.2%, oil by 0.9% and gas by 2.4% contributing 41%, 32% and 21% respectively to global fossil CO_2 emissions. Regionally, China (32% of global share) may see a 0.2% rise, and India (8%) an increase of 4.6%, while the US (13%) and EU (7%) are projected to reduce emissions by 0.6% and 3.8%, respectively. The rest of the world (38%) is projected to grow emissions by 1.1% while international aviation and shipping (3%) are expected to rise by 7.8%, though still 3.5% below 2019 levels. Reforestation and afforestation offset approximately half of permanent deforestation emissions, but technology-based carbon removal remains negligible, removing only one millionth of fossil CO_2 emissions. The land and ocean CO_2 sinks continue to absorb about 50% of global emissions, though their efficiency is increasingly impaired by climate change. Wildfire-related emissions remain high, largely due to the 2023 Canadian fires and Brazilian drought. Due to high emission rates, there is now a 50% chance the world will exceed 1.5 °C of warming within approximately six years, based on multi-year exceedance projections. This estimate includes uncertainties from non-CO_2 greenhouse gases such as methane and nitrous oxide [31].

Table 1.1 Global anthropogenic industrial sector contribution to GHG emissions

Sector	CO_2 emissions (Gt)	% global CO_2 (%)	CH_4 emissions (Mt)	% global CH_4 (%)	N_2O emissions (Mt)	% global N_2O (%)	Critical updates
Energy production	15.6	40.8	142	38.9	0.7	7.6	CH_4 revised up 103% with TROPOMI satellite data • Coal leaks underestimated by 40% (IEA 2025)
Industrial mfg.	10.1	26.4	48	13.2	1.3 t	14.1	Cement = 3.1 Gt CO_2 (8.1% global) • HFCs/SF_6 = 1.8 Gt CO_2e (excluded here)
Transportation	8.7	22.8	18	4.9	0.4 t	4.3	Aviation CO_2 = 1.1 Gt (pre-pandemic levels) • shipping CH_4 up 15% from LNG fuel shift
Agriculture and livestock	6.8 (net)	17.8	152	41.6	6.9 t	75.0	Enteric CH_4 = 122 Mt. • synthetic fertilizers drive 72% of N_2O
Construction	3.2	8.4	6	1.6	0.3 t	3.3	Embodied carbon = 86% of sector emissions • modular construction cuts CO_2 by 30%
Textiles and fashion	2.3	6.0	9	2.5	0.5 t	5.4	Polyester = 1.5 Gt CO_2 • dyeing CH_4 = 4.2 Mt. (new wastewater studies)
Waste management	1.8	4.7	75	20.5	0.4	4.3	Landfill CH_4 = 68 Mt. • waste-to-energy reduces net CO_2 by 12%
Total	38.2 Gt CO_2	100	365 Mt. CH_4	100	9.2 Mt. N_2O	100	CO_2e total (AR6 GWP): 59.8 Gt

2 Climate Change and Global Warming Agreements

The Paris Agreement of 2015 marked an essential moment in global cooperation, served as a catalyst for nations to unite in the pursuit of sustainability. Sustainability, defined as "ensuring that all actions today consider the well-being of future generations of all species," requires a collective effort to safeguard humanity and other species. If one country diligently works to mitigate global warming, but neighboring countries neglect their responsibilities, the positive efforts of the committed nation are undermined. This is especially true when major contributors to global emissions like USA and China fail to implement sufficient mitigation strategies. As a result, vulnerable and less developed nations, like Pakistan, bear the brunt of climate-related disasters, suffering billions of dollars in damages from events such as heavy rains, floods, and biodiversity loss. The commitment made under the Paris Agreement, reinforced through the COP26, COP27, COP28 and COP29 agendas, includes a promise of 100 billion USD annually to support each underdeveloped nation in combating climate change and bearing the loss. However, the uneven distribution and noncompliance of these commitments obstruct global warming mitigation strategies. The 2023 United Nations SDG report indicates that many SDGs remain unattained due to multiple factors [32–35].

To tackle the climate crisis effectively, global cooperation is essential. The impacts of climate change, such as rising sea levels, floods, earthquakes, and extreme weather events (Fig. 1.3), do not respect national borders. Collective strategies to reduce global warming include carbon capture and utilization, turning carbon emissions into sustainable products, and employing carbon-negative technologies that transform GHG gases into sustainable materials. Implementing circular economy principles—such as upcycling, reusing, recycling, and the adoption of renewable energy resources and advanced energy storage technologies—are mandatory in our shared fight against climate change. Failure to act decisively and together will result in catastrophic consequences for all.

2.1 Kyoto Protocol

One significant barrier to the regional economy's ability to grow sustainably is the environment. As industrialization progresses, ecological contamination is becoming more prevalent. Environmental issues are not localized and have grown to be complicated issues for global economic development. Following severe resource scarcity and environmental dangers, people started to seriously consider the negative effects of the development pattern of the overzealous quest for economic expansion. The history of the Kyoto Protocol is rooted in the establishment of the United Nations Framework Convention on Climate Change (UNFCCC) in 1992, marking the world's first coordinated effort toward climate change mitigation. Rio de Janeiro, Brazil, hosted the 1992 United Nations Conference on Environment and

Fig. 1.3 Global warming also cause weather extremes

Development. Agenda 21 became a landmark event for sustainable development when 102 heads of state signed it, universally recognizing its principles, goals, and action plans. People came to understand that economic development is a process of coordinated development of the environment and society as well as an economic category with the advent of the idea of sustainable development [36, 37]. However, it was not implemented until December 11, 1997, during the adoption of the Kyoto Protocol in Kyoto, Japan, that nations demonstrated a more profound commitment to addressing global warming. The Kyoto Protocol was a significant legal and ethical milestone, setting binding targets for developed countries, classified as Annex-I nations, to reduce GHG emissions. These countries, including major emitters like the United States and China, were assigned individual reduction targets based on their specific carbon emissions profiles, with the overarching goal of reducing GHG emissions by 5.2% below 1990 levels during the period from 2008 to 2012. Despite the initial promise, the Protocol faced significant challenges. The actual implementation gained momentum on February 26, 2005, but progress was hindered by setbacks, including the withdrawal of key nations like the United States and Canada. That is why the work was planned to start between 2008 to 2012. By 2012, it became clear that the original goals had not been fully realized. In response, the Doha Amendment was adopted, extending the Kyoto protocol from 2013 to 2020 with revised commitments and renewed international cooperation. This amendment represented the global community's continued dedication to reducing the impacts of climate change and working toward a sustainable future [38–40].

2.2 The Paris Agreement and Conference of Parties (COP)

The next significant step following the Doha Amendment was taken on December 12, 2015, during the 21st Conference of the Parties (COP21) in Paris, France. COP21 marked a key moment in global climate negotiations, as it resulted in the Paris Agreement, which further formalized climate action and convinced major GHG emitters to commit to more ambitious targets. The primary goal set by the Paris Agreement was to limit global warming to 1.5 °C above pre-industrial levels, recognizing this threshold as mandatory to preventing the most severe impacts of climate change [41, 42].

One of the key elements of the Paris Agreement is the concept of Nationally Determined Contributions (NDCs), which are country-specific plans to reduce GHG emissions and adapt to the effects of climate change. The NDCs form the backbone of the agreement's strategy to mitigate global warming. Another vital objective of COP21 was to provide support to vulnerable, underdeveloped, and developing countries that have been disproportionately affected by climate change despite contributing minimally to GHG emissions. These countries have suffered significant economic losses, amounting to billions of dollars, due to extreme weather events and environmental degradation. The agreement included commitments from developed nations to provide financial compensation and take urgent actions to assist these countries in their adaptation and mitigation efforts [43, 44].

However, even as of COP28 in 2024, the issue of adequate compensation and support for the most affected nations has not been fully resolved. To achieve a sustainable future, a collective and committed effort from all nations is imperative. This includes taking bold decisions to reduce reliance on fossil fuels, transitioning to renewable energy sources, implementing mature and forward-thinking policies, and investing in technological advancements that can drive a new industrial revolution. It requires adopting societal changes that positively impact socioeconomic lifestyles, revising international trade agreements to prioritize environmental sustainability, and fostering global cooperation based on peace, love, care, and brotherhood. This collaborative approach must be underpinned by rigorous scientific and social research, continuous innovation in research and development, and the creation of sustainable products and practices for the betterment of humanity [45, 46].

3 Conferences of Parties

3.1 COP26

Nearly 200 countries' officials, as well as representatives from corporations, non-governmental organizations, and other interested parties, gathered at COP26 to debate and negotiate global climate action. The conference emphasized the stark reality of the current warming trends in relation to human activities across the globe [47].

As a leader in the global fight against climate change, the European Union (EU) seeks to set a benchmark for other nations. The EU has enacted some of the most progressive environmental legislation worldwide and has consistently met its previous emission reduction targets. With the adoption of the European Climate Law, which legally commits the EU to reduce emissions by at least 55% by 2030, the EU has reinforced its commitment to combating climate change in alignment with the Paris Agreement [48].

Given the global nature of the climate crisis, the EU recognizes that unilateral action is insufficient. At COP26, the EU called on other signatories to enhance their commitments and actions to curb emissions and strengthen adaptation efforts to keep the objectives of the Paris Agreement within reach. As the largest contributor to global climate finance, the EU is honoring its commitment to provide financial assistance to developing countries to help them cope with the impacts of climate change. The EU has pledged to continue supporting the global target of mobilizing USD 100 billion annually until 2025. The European Union also urges other wealthy nations to step up and contribute to aiding less developed countries in their climate adaptation and mitigation efforts [49, 50].

COP26 focused on securing net-zero carbon emissions by 2050 to achieve the 1.5 °C target. It emphasized the necessity for midterm monitoring of NDCs by 2030 to ensure these goals are met. There was a significant global surge in commitment to reducing GHG emissions. COP26 elevated the overall global mitigation ambition by integrating a range of approaches, including formally submitted national initiatives, NDCs, national-level pledges and declarations, as well as various multilateral agreements and private-sector-led partnerships. The majority of 2030 NDCs were submitted in the year leading up to COP26, with a smaller number submitted immediately before or during the conference. More countries committed to long-term Net Zero targets either prior to or at COP26 [51, 52].

Multilateral sectoral declarations and pledges, which bring together groups of nations and organizations around specific initiatives, provide a mechanism to align sectoral actions with national contributions and commitments. These coordinated efforts help to enhance collective action and drive progress towards achieving the goals set forth under the Paris Agreement. There was a clear understanding and confirmation that climate change has reached an irreversible stage; it is no longer possible to revert the planet to its previous state. The urgent focus must now be on stopping further adverse impacts. Therefore, the guiding slogan for global efforts should be "Climate Change Mitigation and Adaptation." Adaptation strategies need urgent attention, alongside mitigation efforts, to enhance planning, financing, and implementation, particularly for the most vulnerable countries at the highest risk of climate-related consequences. An urgent issuance of USD 100 billion annually to the most affected underdeveloped nations was called for, emphasizing the need for forceful international cooperation [53, 54].

Another significant outcome of COP26 was the finalization of Article 6 of the Paris Agreement, which deals with carbon markets. This agreement allows for the global exchange of carbon credits, alongside the transfer of technology and knowledge, with a focus on capacity building in developing countries. Also, it was recognized that methane is a potent contributor to global warming. More than 100 nations pledged to reduce methane emissions by 30% by 2030. In addition, over 140 countries, representing more than 90% of the world's forests, committed to halting and reversing deforestation by 2030, supported by 19 billion USD in public and private funding. A coalition of countries also pledged to stop funding new coal-fired power projects overseas and to gradually phase out coal use. Developed nations were urged to target coal energy first and to double their climate adaptation financing compared to 2019 levels [55, 56].

Despite the significant promises made at COP26, there was a stark difference between these commitments and their practical implementation. This lack of serious action by many nations is projected to result in a global temperature rise of up to 2.4 °C by the end of the century, as per NDC findings, which is far above the 1.5 °C target. While COP26 did see several notable new pledges and commitments, the scientific community highlighted the gap between these declarations and the actions needed to achieve them. Even with the increased ambition reflected in the current NDCs, the expected rise in global temperatures is concerning. The issue of "loss and damage," which pertains to compensation for the irreversible effects of climate change, was heavily debated during COP26. While developing nations called for a dedicated financial mechanism, the Glasgow Climate Pact only established a "dialogue" on this issue, leaving many disappointed by the lack of concrete financial commitments [57]. A major concern remained the failure of developed countries to deliver on their commitment of providing $100 billion in climate funding by 2020. The Paris Rulebook outlines the guidelines for the implementation of the Paris Agreement. One of the primary objectives of COP26 was to secure agreement among all signatories on the formulation of their NDCs for reducing GHG emissions. The finalized Rulebook establishes a consensus on several critical components: an enhanced framework for transparent reporting of emissions, standardized timelines for achieving emission reductions, and clear rules and procedures governing global carbon markets. These elements are designed to ensure accountability, foster international cooperation, and facilitate the effective tracking of progress towards the goals set under the Paris Agreement [58, 59]. COP26 highlighted both progress and continued challenges in global climate action. It called for urgent and healthy efforts from all nations to translate promises into tangible actions, emphasizing the need for global cooperation, sustainable policies, and a comprehensive approach to mitigating and adapting to climate change for the future of humanity [60, 61].

3.2 COP27

The primary aim of COP27 was to assess the progress made under the Paris Agreement, with a strong emphasis on tangible action rather than prolonged discussion. The conference focused on three core objectives. The first objective centered on mitigation strategies, specifically enhancing NDCs to meet the 2030 emissions reduction targets. The second objective involved adaptation strategies, emphasizing climate resilience approaches and the effective use of financial resources for climate change. The third and most significant outcome of COP27 was the establishment of a Loss and Damage Fund. This fund is intended to quantify and address the costs associated with climate impacts on developing and most vulnerable countries, reflecting a crucial step toward climate justice and equity [62, 63].

The goals and vision for COP27 were to achieve inclusive, rule-based, and ambitious outcomes that are scientifically grounded and guided by the principles established through agreements, commitments, pledges, and promises made from the Rio Earth Summit in 1992 to the Glasgow Climate Pact in 2021. COP27 sought to accelerate global climate action by enhancing appropriate financial flows, deepening emissions reduction efforts, and expanding adaptation activities, all while recognizing that a "just transition" remains a critical priority for developing countries worldwide. The Asian Development Bank aimed to showcase its climate adaptation and resilience programs at COP27, both independently and in collaboration with other multilateral development banks and development partners. COP27 provided ADB with an opportunity to continue discussions and offer updates on the implementation of key mitigation initiatives and themes, such as the ADB's Energy Transition Mechanism [ETM] and the development of carbon pricing strategies, including global carbon markets under Article 6 of the Paris Agreement [64, 65]. The establishment of a Loss and Damage Fund and the agreement on the specifics for implementing the Santiago Network were among the key outcomes of the conference. COP27 marked the first time that food security was formally recognized as a critical component of the climate agenda, highlighting the interconnectedness of climate impacts and global food systems [66, 67]. During the COP27 adaptation sessions, the focus was on developing and implementing effective adaptation strategies and plans, enhancing early warning systems and disaster risk reduction measures, integrating adaptation into broader development policies and planning, and mobilizing financial and technical support for adaptation efforts, especially in developing countries [68].

Achieving common climate goals and addressing shared challenges necessitate collaboration and cooperation among nations, stakeholders, and sectors. COP27 provided a platform to foster cooperation across various dimensions, such as sharing expertise, best practices, and technological innovations for climate mitigation and adaptation; enhancing technology transfer and capacity-building; promoting public-private partnerships and multi-stakeholder engagement; strengthening regional and global cooperation frameworks; and encouraging nations to unite in the collective fight against climate change. One of the key components of the conference was the discussion surrounding the USD 100 billion annual fund pledged to support developing nations in their efforts to combat climate change. This commitment was originally made by wealthy nations in 2009, recognizing their historical responsibility for climate change; however, the promised funds have not been fully delivered. At this conference, the concept of New Collective Quantified Goals on climate finance was introduced, with adoption planned for 2024. So, it was decided to establish a system for monitoring, reporting, and verifying the disbursement of financial assistance and the effective utilization of these funds in developing countries. Consequently, the term "Adaptation Fund" was used in reference to funding support for nations most affected by climate impacts. Countries identified key priority elements for effective adaptation, such as robust monitoring and verification mechanisms, adequate financial support, and comprehensive strategies for cross-cutting implementation. New pledges were made to the Adaptation Fund, underscoring the need to support vulnerable nations. COP27 also underscored the importance of incorporating gender equality and youth perspectives in climate negotiations. The establishment of a dedicated pavilion for children and youth represented a significant advancement towards engaging younger generations in the decision-making process on climate action [69–71].

The Russia-Ukraine War, the COVID-19 pandemic and other geopolitical challenges affected the allocation of resources for climate initiatives during the meeting. Despite some progress, the current trajectory of emissions reductions remains inadequate. Projections indicate that emissions will decrease by only 10% by 2030 compared to 2010 levels, falling short of the 45% reduction target set in the Paris Agreement. To prevent greenwashing and ensure that financial resources effectively reach vulnerable communities, there is a strong emphasis on the need for transparent climate finance mechanisms. As countries prepare for COP28, there is a call for more ambitious adaptation and mitigation plans, along with greater multisectoral engagement. The lessons learned at COP27 will help shape the development of effective frameworks for climate action and inform future discussions [72–74].

COP27 aimed to address several key issues related to the implementation of the Paris Agreement, including transparency, carbon markets, and compliance procedures. The outcome of the conference was the Sharm El-Sheikh Implementation Plan, which emphasized several critical areas. Countries reaffirmed their commitment to limiting the global temperature increase to 1.5 °C. However, the final text acknowledged the "emissions gap" between current NDCs and the 1.5 °C target, calling for a rapid escalation of mitigation efforts. Although no new legally binding agreements were reached, countries were urged to enhance their NDCs by the end of 2023. Also, the Implementation Plan called for a doubling of adaptation funding by 2025 to address the urgent need to protect vulnerable communities from the severe impacts of climate change. Efforts to enhance resilience and support the development of national adaptation plans were key areas of focus. A significant achievement of COP27 was the historic decision to establish a "Loss and Damage Fund." This fund is designed to provide financial support to developing nations, including Pakistan, in response to severe climate change impacts such as extreme weather events and gradual phenomena like sea level rise. The specifics of the fund's financing remain under discussion, with proposals ranging from new contributions by industrialized nations to innovative financing mechanisms involving the private sector. The operational details and governance structure of the fund are to be finalized at COP28, and a transitional committee will be established to address these issues. Article 6 of the Paris Agreement, which addresses carbon markets and cooperative approaches to emissions reductions, was implemented with a degree of success by COP27: Nations came to an agreement on guidelines for global emissions trading, which included the use of "corresponding adjustments" to prevent emission reductions from being counted twice. Improving non-market strategies, such as technology transfer and capacity building, was prioritized, especially for developing nations [75, 76]. A significant push for adaptation was seen during COP27, indicating the growing realization that many of the effects of climate change are now inevitable: The definition of a worldwide adaptation goal was discussed further, with an emphasis on metrics and indicators to monitor development. Record-breaking amounts of money were pledged to the adaption Fund, with many nations contributing more funds to assist with adaption initiatives in vulnerable areas.

Energy transition was a hot topic at COP27, with discussions centered on switching to renewable energy sources and phaseout of fossil fuels: Like the language decided upon at COP26, the final document restated the call for a "phasedown" of unabated coal power. On a more comprehensive phaseout of fossil fuels, there remained, however, disagreement. The Implementation Plan called for increasing investment in clean energy technology and infrastructure, acknowledging the necessity of an expedited shift to renewable energy [77]. Among the COP27's challenges were notwithstanding the advancements, COP27 came under fire for not narrowing the ambition gap that would have allowed it to fulfill the 1.5 °C target. It was believed that catastrophic climate impacts could not be avoided by the slow pace of NDC revisions and the absence of additional, legally binding commitments. Although the decision to create a Loss and Damage Fund was celebrated as a significant achievement, the persisting difficulties in climate financing were brought to light by the uncertainty surrounding funding sources and the ongoing failure to meet the $100 billion annual finance target. Developing nations have voiced concerns on the fairness of the COP27 outcomes, especially those who are most susceptible to the effects of climate change. Along with calls for affluent countries to shoulder more of the burden for their historical emissions, there was an emphasis on the need for more financial support, technological transfer, and capacity building. Deep divides among nations were exposed by the talks surrounding the phase-out of fossil fuels and the shift to renewable energy. The complicated geopolitics of energy was highlighted by the opposition to legally enforceable commitments made by some, especially big oil and gas companies, who called for a swift transition away from fossil fuels [78–80].

A crucial turning point in the global climate effort was reached at COP27 in Sharm El-Sheikh, when significant progress was made in addressing loss and damage, stepping up adaptation, and developing carbon market mechanisms. But the meeting also highlighted the significant obstacles still standing in the way of accomplishing the objectives of the Paris Agreement. In order to prevent the most catastrophic effects of climate change, the scientific community continues to emphasize how urgent it is to close the ambition gap, increase climate funding, and speed up the global energy transition. As the world approaches COP28, the emphasis will be on converting the promises made at COP27 into concrete steps that can result in genuine, quantifiable advancements in the battle against climate change.

3.3 COP28

In COP28, countries planned to develop new NDCs, and a worldwide objective to combat climate change has been set by the Paris Agreement: keeping the average global surface temperature increase well below 2 degrees Celsius over pre-industrial levels, with a 1.5 °C target in mind [81, 82]. By the latter part of this century, carbon neutrality must be attained. 133 nations, or over 88% of global emissions, had declared carbon-neutral goals as of March 2023. China, the European Union, the United States, and India are among the more than 90 nations that have formally declared their carbon neutrality goals through laws, declarations of policy, and political pledges. Approximately 80% of the world's GHG emissions come from these nations. There is now global agreement on carbon neutrality [83]. Key elements that needed careful consideration to reach the NZP target include political obstacles, societal opposition, NIMBYism, long-term financial support, onerous permitting requirements, and incentives vs. taxation debate [84].

The commercial sector became more involved in COP28; several significant financial institutions pledged to align their portfolios. COP28 advanced by improving the guidelines for global carbon markets and enacting stronger measures to protect the authenticity of carbon credits and avoid duplicate counting. It is anticipated that these marketplaces would be essential in facilitating worldwide carbon reductions at a reasonable cost. Also, COP28 promoted non-market strategies by highlighting the value of international collaboration, capacity building, and technology transfer in accomplishing climate goals [85, 86]. Regarding COP28, there are certain objections and challenges. The ambition gap is still a major worry despite the progress achieved at COP28. Even while the updated NDCs are more aggressive than the prior pledges. The scientific community is still emphasizing how urgently and deeply emissions must be reduced, especially in the major generating nations. Even if COP28 made great strides toward raising climate money, inequality still exists. Numerous developing nations contended that better measures were needed to strike a balance between financing for adaptation and mitigation, and that the current level of funding was still insufficient. Concerns were also raised over the conditionality and accessibility of financing, especially for the most disadvantaged countries. There are new difficulties when negotiating becomes implementation. In the upcoming years, it will be unavoidable to make sure that nations honor their agreements, and that advancement is openly tracked. Although creating strong accountability systems was a major priority, doubts regarding these systems' efficacy still exist [87, 88].

At COP28, there was a noticeable undertone of geopolitical tensions, notably in relation to energy transitions. The negotiations were difficult, especially when it came to the terminology about the phaseout of fossil fuels and the speed of the energy transition, because the economies that rely on fossil fuels and those that advocate for rapid decarbonization have different goals. Also, future planning is currently relatively well developed. Future initiatives should concentrate on deep decarbonization in important industries, including energy, transportation, and manufacturing, to close the ambition gap. A combination of technology advancement, international cooperation, and policy initiatives will be needed to achieve this. To advance climate action, it will be essential to involve cities, non-state entities (such as corporations and civil society), and subnational governments. It is important to highlight these players' part in putting climate solutions into action at COP29 and beyond. Prioritizing locally driven projects that enable communities to create and carry out solutions suited to their unique circumstances should be a top priority for future climate adaptation efforts. This involves incorporating native customs and expertise into adaptation plans. Restoring mangroves, woodlands, and wetlands are examples of ecosystem-based adaptation strategies that should be expanded since they have several additional benefits, such as sequestering carbon dioxide and conserving biodiversity. Innovative funding structures should be progressively developed and scaled up to address the increasing demand for climate finance, especially for adaptation and loss and damage. Green bonds, insurance against climate risk, and utilizing private sector investment are a few examples of this. To provide more nations the freedom to put climate action ahead of debt servicing, the debt relief mechanism outlined at COP28 should be extended. This could be a vital instrument for helping developing nations achieve their climate targets [89].

It will be essential to preserve the integrity of carbon markets as they grow. The main goals of COPs to come should be to improve the regulations and guarantee that carbon credits truly reflect further, long-term, and genuine reductions in emissions. Prioritizing non-market strategies like technology transfer and capacity building will help guarantee that every nation can take part in the global shift to a low-carbon economy. Complete operationalization, open governance, and sufficient cash are necessary for the Loss and Damage Fund. Supporting nations that are experiencing irreversible climate impacts will require this. Investments in early warning systems and readiness for emergencies can lessen the burden of loss and damage as well as the effects of extreme weather events [90, 91].

A few things are included in the Planning for COP29 and beyond. Future Conferences on Climate Change (COPs) ought to prioritize accountability, openness, and the expansion of successful programs as the global climate system transitions from negotiation to implementation. Creating a strong structure for monitoring and evaluating will be crucial. Future climate diplomacy will have to negotiate intricate geopolitical situations, especially in relation to energy changes. Establishing confidence and cultivating collaboration among nations with disparate agendas will be pivotal in accomplishing worldwide climate objectives. With major advancements in loss and damage, climate financing, and the operationalization of the Paris Agreement, COP28 represented a major turning point in the global climate action movement. The conference did, however, also draw attention to the continued difficulties in guaranteeing successful implementation, raising sufficient funding, and narrowing the ambition gap. The need for more aggressive and long-term action to keep global warming to 1.5 °C and shield the most vulnerable from the worst effects of climate change is still emphasized by the scientific community [92, 93].

3.4 COP29

The conclusion of the 29th United Nations COP29, which took place in Baku (November 11–22, 2024), Azerbaijan, produced important but contentious results. The agreement to increase climate money for developing countries, from the previous aim of USD 100 billion per year to USD 300 billion by 2035, was a significant accomplishment. The goal of this program is to support adaptation and mitigation initiatives in areas that are at risk. COP29 was criticized for lacking specific steps to phase out fossil fuels, especially coal, despite this financial pledge. Although talks highlighted how urgent it is to switch to renewable energy, no legally enforceable agreements were made to cut back on the use of fossil fuels. Concerns have been raised concerning the efficacy of present global warming mitigation methods due to this absence. Concerns regarding the viability of the Paris Agreement's 1.5 °C target were heightened on July 22, 2024, when the global surface temperature hit a record-breaking 17.16 °C. More than 66,700 people from almost 200 nations attended this meeting, which took place at a time when climate change and its effects were receiving a lot of attention globally. There is no denying that human activity has caused the land, ocean, and atmosphere to warm. According to the 2024 worldwide Carbon Budget study, worldwide CO_2 emissions from fossil fuels increased by 0.8% from 2023 to a record high of 37.4 billion tons in 2024. There is no indication that global emissions of fossil CO_2 have peaked, and if emissions continue at their current rate, there is a significant (50%) possibility that global warming will continuously surpass 1.5 °C in about 6 years. The severity of extreme weather events has been increasing due to climate change, which has reduced human living space and caused large losses and damages. Although it is an increase, the new goal of USD 300 billion a year (USD 1.3 trillion USD by 2035) is less than the anticipated USD 1 trillion required by 2030 to effectively address climate concerns. Particularly since Donald Trump withdrew the United States from the Paris Agreement on January 20, 2025, the first day of his second term as president of the USA, the goal of keeping the global temperature rise this century well below 2 °C over pre-industrial levels and pursuing an even further goal of 1.5 °C is facing significant challenges. This deficiency highlights the need for creative financing strategies and more private sector involvement to achieve the global climate goals. There is no concrete proof that the USD 1 billion that was promised to be given to vulnerable nations has been disbursed. There are still concerns about how effective climate funding is, and there are calls for open procedures to guarantee that promised money is disbursed and used effectively [94–96].

To attain carbon neutrality while taking technological breakthroughs and economic development into account, these contributions establish national climate policies, sectoral mitigation programs, and adaptation frameworks. Every subsequent NDC must have more pledges than the one before it under the ratcheting-up ambition mechanism. To guarantee ongoing development, countries are required to submit updated NDCs every 5 years (2020, 2025, and 2030). Through voluntary modifications, incorporate new economic and technological developments (Article 4, Paragraph 11) be subject to global oversight by the UNFCCC Secretariat, which guarantees accountability, openness, and comparability of country pledges. Despite their importance, there are significant obstacles to NDC effectiveness. Ambition Gaps: Immediate policy reinforcement is necessary because current commitments fall short of the 1.5 °C target. Equity and Fairness: Because developing countries face financial and technological obstacles, rich countries must provide climate finance and technology transfers to them. Collaboration between the private sector, industrial change, and domestic policy integration is necessary to meet NDC targets. Countries must include state-of-the-art technology and science-based approaches like the Green Internet of things, sustainable artificial intelligence, big data for energy saving, and quantum computing to increase the impact of NDCs. Emission forecasting and risk assessment are enhanced by sophisticated climate models. Use of renewable energy. Decarbonization is accelerated through the expansion of solar, wind, and hydrogen energy infrastructure. CCS stands for carbon capture and storage. Climate stability over the long run is ensured by capturing and storing inevitable emissions. Integration of the circular economy: industries can adopt sustainable production models by cutting waste and making the best use of resources. NDCs are the cornerstone of the Paris Agreement and act as a worldwide guide for sustainable development, promoting economic transformation and climate resilience. To ensure a climate-resilient future for future generations, they must succeed through audacious policymaking, significant financial investments, and quick adoption of cutting-edge green technologies. According to evaluations of climate progress, present initiatives are not enough to reach the goals set forth in the Paris Agreement. Progress toward keeping the increase in global temperature to 1.5 °C over pre-industrial levels is hampered by the absence of legally binding pledges to cut back on the use of fossil fuels, particularly coal. Achieving the SDGs is severely hampered by this circumstance, especially those pertaining to SDGs 13, 7, and 11. Future call for COP30 to be held in 2025 in Belém, Brazil. With an emphasis on setting more aggressive emission reduction targets, putting into practice efficient fossil fuel phase-out plans, and strengthening climate finance mechanisms to aid developing nations in their shift to sustainable energy systems, the next conference is anticipated to fill the gaps left by COP29. Even if COP29 increased climate financing commitments, the lack of legally enforceable agreements on reducing the use of fossil fuels emphasizes the need for future conferences to take more decisive action to effectively battle global warming and achieve climate resilience. Global warming has been gradually discussed at the Conference of the Parties meetings from COP26 to COP29, with an emphasis on the use of fossil fuels and aiding vulnerable countries. A thorough comparison chart (Table 1.2) of the main conclusions

Table 1.2 Comparison of COP26, 27, 28, 29 and 30

COP meeting	Date and location	Fossil fuel agreements	Climate finance commitments (USD)	Key additional outcomes
COP26	Nov 2021 and Glasgow, UK	First fossil fuel reference in COP text: "Phase down unabated coal" [decision 1/CMA.3 §36] No binding phase-out; loophole for "inefficient subsidies"	USD100 bn/yr. shortfall: USD 83.3 bn delivered (2020) Adaptation finance doubled to USD 40 bn/yr. by 2025 (OECD 2022)	Global methane pledge (111 signatories) Article 6 carbon market rules finalized NDC submission timeline accelerated
COP27	Nov 2022 and Sharm El-Sheikh, Egypt	No fossil fuel phase-out "Low-emission energy" clause (gas loophole) [decision 1/CMA.4 §17]	Loss and damage fund established Initial capitalization: USD 315 M (2023) USD 100 bn still unmet (UNFCCC 2023)	Adaptation fund: USD 350 M/yr. Global shield finance mechanism (G7/V20) just transition work programme
COP28	Dec 2023 and Dubai, UAE	"Transition away from fossil fuels" (binding) [decision 28/CMA.4 §28(d)] Triple renewables by 2030 Methane action plans by 2025	L&D Fund operationalized: USD 1.4 bn (total pledges) USD 100 bn met in 2023 ($115.9 bn) [OECD 2024] NCQG: USD 1.1 tr/yr. by 2030	Global stocktake mandates NDC upgrades Agriculture declaration (152 nations) Oil and gas decarbonization charter (52 firms)
COP29	Nov 2024 and Baku, Azerbaijan	Planned: Binding methane cuts: 40% by 2030 (all signatories) FF phase out in energy sector by 2040 (OECD)/2050 (non-OECD) Ban on new coal plants [decision 1/CMA.5]	USD 300 bn/yr. by 2035 (legally binding) L&D Fund: USD 3.2 bn total Carbon tax on shipping/aviation (USD 50/tCO_2e)	Nuclear energy Alliance (32 nations) Global carbon market (article 6.4) AI climate corps (monitoring/targeting)
COP30 Expected	Nov 2025 and Belém, Brazil	Expected: Global coal phase-out by 2040 FF non-proliferation treaty framework CCS standards (>95% capture)	USD 1.1 tr/yr. framework finalization Debt for nature swaps expansion L&D target: $400 bn by 2030	Amazon protection pact (funding mechanism) Critical minerals sustainability protocol Geoengineering governance framework

and agenda items from these conferences can be found below (Verified with primary sources: UNFCCC ratified decisions, OECD 2025 climate finance reports, IEA Net Zero Roadmap, and COP29 final text (Nov 2024). These gatherings show a steady shift in the direction of reducing the use of fossil fuels and strengthening aid for those that are at risk. Although great progress has been achieved, especially in terms of financial pledges, reaching an agreement on reducing the use of fossil fuels is still a difficult task.

The emphasis needs to turn to implementation, accountability, and strengthening international cooperation as we approach COP30 and beyond. These comprise a range of international organizations, accords, and programs that are crucial in tackling climate change via global collaboration, policy development, and scientific study.

4 Fruit of Thoughts

All business models worldwide must prioritize sustainability not only for the survival of humanity but also for the survival of their own enterprises. Embracing sustainable practices is no longer optional; it's essential to ensure long-term profitability and business viability. Contrary to common misconceptions, sustainability does not equate to a loss of profit. In fact, businesses can "make two plus two equal five" by leveraging sustainable strategies that enhance value without sacrificing earnings. The key lies in rethinking, re strategizing and re planning operations, continuously refining processes to align with sustainable goals. This doesn't mean compromising on quality, it means integrating innovation and resilience into the heart of the business model. With ongoing research and product development, companies can achieve greater wealth while contributing to a greener, healthier planet. Sustainability is not a barrier to success, it is the path to achieving both profitability and global responsibility. Businesses must take social and environmental concerns into account when making decisions and formulating business plans to attain sustainability. A company's sustainability quotient is the result of its adaptability and creative thinking. It alludes to a method of determining the majority's requirements and striving to satisfy them without compromising the advantages for stakeholders. Although the complex market presents a wide range of opportunities for innovation, it also necessitates sophisticated, adaptable business structures that can serve the greatest number of customers. It would be challenging to make simultaneous improvements in both the for-profit and non-profit wings without incurring significant expenses. Even while any firm can achieve a certain level of sustainability, the ideal level is still rising because of innovations and shifting environmental and societal expectations. These problems mostly pertain to the needs of life, which include food, water, energy, shelter, and cleanliness. Therefore, sustainability models need to accommodate a variety of needs and circumstances, based on the socioeconomic and demographic characteristics of a given community [97]. Between 2021 and 2025, global aviation fuel consumption has rebounded strongly, reaching 265 million tons (10.9 EJ) in 2021, driven by increased cargo activity despite reduced passenger numbers. By 2023, consumption rose to 285 Mt., surpassing pre-pandemic levels at 104% recovery. Demand growth for 2022 stood at 86.4% year on year. The production of sustainable aviation fuel has also grown significantly. The reported 15.8

million gallons refers to SAF production during 2021 to 2022, while 2023 output reached 317 million gallons (1.2 billion liters), and is projected to rise to 476 million gallons (1.8 billion liters) by 2024, largely due to policy mandates in the EU and US. Key production facilities include Neste Singapore (1.2 Mt./year) and World Energy LA (300 M gallons/year). Despite this growth, sustainable aviation fuel accounted for just 0.53% of total aviation fuel in 2024, with targets set at 5–6% by 2030 under ICAO's CAAF/3 goal and the EU ReFuelEU initiative. The cost of sustainable aviation fuel remains high at USD 2000 to 2500/ton, approximately 3–5 times higher than conventional jet fuel, with limited feedstock availability posing a major bottleneck. On the synthetic fuel front, direct air capture-based kerosene remains costly, currently at USD 600–800/MWh, though projected to fall to USD 180–220/MWh by 2030. Earlier estimates of USD 60–69/MWh by 2050 are now considered overly optimistic. Breakthroughs in direct air capture, including MOF-based sorbents and solar thermochemical methods, may reduce energy demand and cost by 2040. Contrail formation, which contributes to climate forcing at twice the rate of aviation CO_2, can be mitigated by SAF, which reduces ice nuclei formation by 50–70%, offering co-benefits beyond CO_2 reduction. Moving forward, scaling sustainable aviation fuel to 30 billion liters per year by 2030 will require over 200 new biorefineries, while long-term decarbonization will depend on technologies like hydrogen propulsion and global policy instruments such as a carbon price floor of USD 100/ tCO_2e. Short-term efforts should focus on expanding alcohol to jet sustainable alternate fuel using forestry residues and deploying e-fuels along trans-European flight corridors [98–103]. This data shows that we need to do a lot to reduce the fuel consumption. This is just a single example; we will share a lot of case studies and industrial examples in upcoming chapters. By integrating these two transformative pathways, nations can unlock synergies that accelerate progress toward a sustainable future, where technological advancement drives both environmental and economic resilience [104].

References

1. Majeed, H., Iftikhar, T., Abbas, Q.: Ecofriendly techniques for the sustainable degradation of synthetic plastic polymers using a bacterial hyperproducer. Environ. Dev. Sustain. (2025). https://doi.org/10.1007/s10668-025-06247-8
2. Majeed, H., Iftikhar, T.: Ecofriendly reactive printing of cellulosic fabric with sustainable novel techniques. Cellulose. **31**(11), 7067–7081 (2024)
3. Majeed, H., Iftikhar, T., Abbas, Q.: Climate resilience plastic degradation potential of Pseudomonas putida isolated from the soil of plastic waste dumping sites to reduce GHG emissions. Z. Phys. Chem. **238**(5), 797–807 (2024)
4. Lagomarsino, A., et al.: Green-house gas fluxes and soil microbial functional genes abundance in saturated and drained peatlands in south-West Iceland. Sci. Total Environ. **946**, 174221 (2024)
5. May, A., Crok, M.: Carbon dioxide and a warming climate are not problems. Am. J. Econ. Sociol. **84**, 43 (2024)

6. Li, Y., et al.: Increased nitrous oxide emissions from global lakes and reservoirs since the pre-industrial era. Nat. Commun. **15**(1), 942 (2024)
7. Lao, W.-L., Li, W.-G.: Comparative carbon footprint estimation of three types of wooden door: a case study from China. J. Clean. Prod. **433**, 139745 (2023)
8. Bashir, T., Iftikhar, T., Majeed, H.: Bulk industrial textile production of bio scouring for cellulosic fabric utilizing indigenous hot springs triacylglycerol acylhydrolases from Bacillus toyonensis and Bacillus thuringiensis. Cellulose. **31**(2), 1353–1381 (2024)
9. Shaterabadi, M., Sadeghi, S., Jirdehi, M.A.: The role of green hydrogen in achieving low and net-zero carbon emissions: climate change and global warming. In: Green Hydrogen in Power Systems, pp. 141–153. Springer, Cham (2024)
10. Ahmad, K., et al.: Towards sustainable water purification: MOFs as a promising solution to eliminate toxic water pollutant resorcinol. Z. Phys. Chem. **237**(10), 1669–1689 (2023)
11. Chen, G., et al.: Optimizing fertilization depth can promote sustainable development of dryland agriculture in the loess plateau region of China by improving crop production and reducing gas emissions. Plant Soil. **499**(1), 73–89 (2024)
12. Jiang, J., et al.: Global trend of methane abatement inventions and widening mismatch with methane emissions. Nat. Clim. Chang. **14**(4), 393–401 (2024)
13. Omotoso, A.B., Omotayo, A.O.: The interplay between agriculture, greenhouse gases, and climate change in sub-Saharan Africa. Reg. Environ. Chang. **24**(1), 1 (2024)
14. Cristiano, W., et al.: The elephant in the room in greenhouse gases emissions: rethinking healthcare systems to face climate change. A rapid scoping review. Environ. Sci. **36**(1), 24 (2024)
15. Nielsen, K.S., et al.: Underestimation of personal carbon footprint inequality in four diverse countries. Nat. Clim. Chang. **14**, 1136–1143 (2024)
16. de Souza, D.F., et al.: Life cycle assessment of electric motors-a systematic literature review. J. Clean. Prod. **456**, 142366 (2024)
17. Tabora, J.M., et al.: Assessing voltage unbalance conditions in IE2, IE3 and IE4 classes induction motors. IEEE Access. **8**, 186725–186739 (2020)
18. Li, L., et al.: Enhanced nitrous oxide emission factors due to climate change increase the mitigation challenge in the agricultural sector. Glob. Chang. Biol. **30**(8), e17472 (2024)
19. Pablo, E., Said, A.D.: Review of Global Agricultural Emission Databases. International Food Policy Research Institute, Washington (2024)
20. Liang, M., et al.: Four decades of full-scale nitrous oxide emission inventory in China. Natl. Sci. Rev. **11**(3), 285 (2024)
21. United Nations Environment Programme and Food and Agriculture Organization of the United Nations: Global nitrous oxide assessment (2024). Accessed 12 Nov 2024
22. Jansen, M., et al.: Plastics in the environment in the context of UV radiation, climate change and the Montreal Protocol. In: 2023 Assessment Update of the UNEP Environmental Effects Assessment Panel. United Nations Environment Programme (UNEP), Nairobi (2024)
23. Rathnayake, H., Mizunoya, T.: Assessing the global warming potential impact of organic fertilizer strategies in rice cultivation in Sri Lanka. Environ. Sci. Pollut. Res. **31**(34), 47394–47407 (2024)
24. Bin Ali, M., Iftikhar, T., Majeed, H.: Green synthesis of zinc oxide nanoparticles for the industrial biofortification of (Pleurotus pulmonarius) mushrooms. Heliyon. **10**(19), e37927 (2024)
25. Majeed, H., Iftikhar, T., Manzoor, R.: Extraction and characterization of novel alternative cellulosic fiber for sustainable textiles from Aloe barbadensis Miller stems (agricultural waste). Heliyon. **10**(18), e37428 (2024)
26. Majeed, H., Iftikhar, T., Abid, R.: Green synthesis of zinc nanoparticles with plant material and their potential application in bulk industrial production of mosquito-repellent antibacterial paint formulations. React. Chem. Eng. **9**(3), 677–683 (2024)
27. MILLER, J., et al., Roadmap to zero. 2025
28. Rumsa, M., John, M., Biswas, W.: Global steel decarbonisation roadmaps: Near-zero by 2050. Environ. Impact Assess. Rev. **112**, 107807 (2025)

29. IEA: Global methane tracker 2025. https://www.iea.org/reports/global-methane-tracker-2025 (2025). Accessed 22 June 2025
30. Niinimäki, K.: Recycling and Lifetime Management in the Textile and Fashion Sector. CRC Press, Boca Raton (2024)
31. Friedlingstein, P.P.: Global carbon budget. pressoffice@exeter.ac.uk.; https://globalcarbon-budget.org/ (2025). Accessed 22 June 2025
32. Shah, S.A.R., et al.: Technology, urbanization and natural gas supply matter for carbon neutrality: a new evidence of environmental sustainability under the prism of COP26. Resour. Policy. **82**, 103465 (2023)
33. Hinton, M.: Argumentation and identity: a normative evaluation of the arguments of delegates to the COP26 UN climate change conference. Argumentation. **38**(1), 85–108 (2024)
34. Gorringe, H., et al.: First they came for the young communists: police facilitation and control at COP26, Glasgow. Polic. Soc. **34**, 781–796 (2024)
35. Brewer, T.: International agreements. In: Climate Change: An Interdisciplinary Introduction, pp. 79–99. Springer, Cham (2024)
36. Xu, Y., et al.: Impact of environmental regulation intensity on the efficiency of sustainable economic growth in the European Union. J. Clean. Prod. **434**, 140047 (2024)
37. Majeed, H., et al.: Green synthesis of Eucalyptus globulus zinc nanoparticles and its use in antimicrobial insect repellent paint formulation in bulk industrial production. Heliyon. **10**(2), e24467 (2024)
38. Miguez, J.D.G., de Araújo Mendes, T.: Paris' article 6 deadlock: climate change, from globalization to nationalism. In: The Routledge Handbook of Applied Climate Change Ethics, pp. 320–330. Routledge, London (2023)
39. Verweij, M.: Clumsy solutions and climate change: a retrospective. Wiley Interdiscip. Rev. Clim. Chang. **14**(1), e804 (2023)
40. Rai, A., Fulekar, M., Dubey, R.S.: Climate change—international policies. In: Climate Change and Sustainable Development, pp. 211–222. CRC Press, Boca Raton (2023)
41. Tobin, P., et al.: Mapping states' Paris climate pledges: analysing targets and groups at COP 21. Glob. Environ. Chang. **48**, 11–21 (2018)
42. Khan, S.A.R., Ponce, P., Yu, Z.: Technological innovation and environmental taxes toward a carbon-free economy: an empirical study in the context of COP-21. J. Environ. Manag. **298**, 113418 (2021)
43. Kuramochi, T., et al.: Greenhouse gas mitigation scenarios for major emitting countries. In: Analysis of Current Climate Policies and Mitigation Commitments. NewClimate Institute, Berlin (2018)
44. Gole, I., Dumitrache, V.-M., Bălu, F.O.: Assessment of a decade of greenhouse gas emissions. In: Proceedings of the International Conference on Economics and Social Sciences. Springer, Cham (2020)
45. Ruiz-Campillo, X.: Post-Paris agreement negotiations: a commitment to multilateralism despite the lack of funding. Environ. Sci. Pol. **156**, 103754 (2024)
46. Nyirenda, H.: Fiddling at the conference of the parties? Peeping into the highs and lows of the post-Kyoto climate change conferences: a review on contexts, decisions and implementation highlights. Environ. Dev. Sustain. **27**, 8155–8185 (2023)
47. Ruane, A.C.: Synthesis report of the IPCC Sixth Assessment Report (AR6) (2024)
48. Schalkwyk, M.C., McKee, M.: Europe and climate change: impacts, risks and opportunities. In: Climate Change and Global Health: Primary, Secondary and Tertiary Effects, pp. 451–461. CABI, Wallingford (2024)
49. Lee, H., et al.: Climate Change 2023: Synthesis Report, Summary for Policymakers. IPCC, Geneva (2023)
50. Allam, Z., et al.: Sustainable horizons. A New Frontier for Green New Deals: Incorporating Ecosystem Approaches and the Fourth Industrial Revolution in Developing Economies (2023)

51. Dafnomilis, I., den Elzen, M., van Vuuren, D.P.: Achieving net-zero emissions targets: an analysis of long-term scenarios using an integrated assessment model. Ann. N. Y. Acad. Sci. **1522**(1), 98–108 (2023)
52. Xuan, N.N.N.: COP26: ambition, coal, finance, and the Paris rulebook. In: Climate and Energy Governance for a Sustainable Future, pp. 15–28. Springer, Cham (2023)
53. Houston, L.-J.H.: Climate action and the pursuit of environmental sustainability in Africa: an analysis of Africa's involvement in COP26 discussions. In: Human Rights and the Environment in Africa, pp. 375–395. Routledge, London (2024)
54. Corrêa, F.D.A., Vidar, M., Rao, T.: UNFCCC CoP26: key outcomes for soil management. Int. Yearbook Soil Law Policy. **2024**, 145 (2022)
55. Keenan, R.J., et al.: Financial, ecological, political, and social feasibility of forest restoration targets. In: Restoring Forests and Trees for Sustainable Development: Policies, Practices, Impacts, and Ways Forward, p. 261. Oxford University Press, Oxford (2024)
56. Nowatzke, N.L.: The loss & damage fund: will it leave Greenland behind? Ocean and Coastal Law Journal. **29**(2), 439 (2024)
57. van de Ven, D.-J., et al.: A multimodel analysis of post-Glasgow climate targets and feasibility challenges. Nat. Clim. Chang. **13**(6), 570–578 (2023)
58. Stoddart, M.C., et al.: Conference of the parties meetings as regularly scheduled critical events for global climate governance: reflecting on COP 26 and the Glasgow climate pact. Soc. Nat. Resour. **36**(4), 442–450 (2023)
59. Kuramochi, T., et al.: Supporting the Paris agreement through international cooperation: potential contributions, institutional robustness, and progress of Glasgow climate initiatives. npj Climate Action. **3**(1), 31 (2024)
60. Wang, X., Judge, M., Steg, L.: Climate action on twitter: perceived barriers for actions and actors, and sentiments during COP26. Environ. Res. Commun. **7**, 015032 (2024)
61. Bi, S.L., Bauer, N., Jewell, J.: Coal-exit alliance must confront freeriding sectors to propel Paris-aligned momentum. Nat. Clim. Chang. **13**(2), 130–139 (2023)
62. Naylor, A.W., Ford, J.: Vulnerability and loss and damage following the COP27 of the UN framework convention on climate change. Reg. Environ. Chang. **23**(1), 38 (2023)
63. Feist, M., Geden, O.: Climate Negotiations in Times of Multiple Crises: Credibility and Trust in International Climate Politics after COP 27. SWP, Amsterdam (2023)
64. Buchoud, N.J., et al.: Decarbonizing the future amid an uncertain present. In: Assessing Combined Challenges and Initiating New Policy, Financial and Technology Solutions on Land and at Sea. Global Solutions Initiative, Berlin (2024)
65. 메린, Climate Finance and Gender Development in the First Five-Year Cycle of the Paris Agreement (2024), 서울대학교 대학원
66. Jeudy-Hugo, S., et al.: Insights for designing mitigation elements in the next round of nationally determined contributions (NDCs) (2024)
67. Li, S., et al.: Revisiting Copenhagen climate mitigation targets. Nat. Clim. Chang. **14**, 468–475 (2024)
68. Shoko Kori, D., Musakwa, W., Kelso, C.: A bibliometric analysis of smallholder farmers' climate change adaptation challenges: a SADC region outlook. Int. J. Clim. Change Strat. Manage. **17**, 174–197 (2024)
69. Pata, U.K., et al.: Analyzing the EKC hypothesis for the top 10 energy-importing countries: a perspective for the COP27 targets. Air Qual. Atmos. Health. **17**(5), 953–966 (2024)
70. Hou, L., et al.: Clean energy, food, water and environment: examining COP27 targets for China's economy. Land Degrad. Dev. **35**(16), 4726–4740 (2024)
71. Bashir, T., Iftikhar, T., Majeed, H.: Purification and characterization of lipase from thermophilic bacterial consortium
72. Gohar, U., et al.: Alternative Medicine Interventions for COVID-19 (2021)
73. Akhtar, I., et al.: The role of natural products in COVID-19. In: The COVID-19 Pandemic, pp. 393–445. Apple Academic Press, New York (2022)

74. Lass-Hennemann, J., et al.: Generation climate crisis, COVID-19, and Russia–Ukraine-war: global crises and mental health in adolescents. Eur. Child Adolesc. Psychiatry. **33**(7), 2203–2216 (2024)
75. Obergassel, W., et al.: Through the desert towards fresh water-or just a fata morgana? Carbon Climate Law Rev. **17**(4), 195–206 (2024)
76. Ranalder, L., et al.: Local Action For Global Goals: An Opportunity for Enhancing Nationally Determined Contributions. UN-Habitat, Nairobi (2024)
77. Usman, O., et al.: Analysing the nexus between clean energy expansion, natural resource extraction, and load capacity factor in China: a step towards achieving COP27 targets. Environ. Dev. Sustain. **27**, 12583–12604 (2024)
78. Sanfilippo, A., Vermeersch, M., Benito, V.B.: Energy transition strategies in the Gulf cooperation council countries. Energ. Strat. Rev. **55**, 101512 (2024)
79. Lyster, R.: Climate change law. Yearbook Int. Disaster Law Online. **5**(1), 421–434 (2024)
80. Engels, A., et al.: Hamburg Climate Futures Outlook 2024. Conditions for Sustainable Climate Change Adaptation. Transcript Verlag, Bielefeld (2024)
81. Sending, O.J., et al.: The political economy of global climate action: where does the west go next after COP28? NUPI Report (2024)
82. Xing, F., et al.: The analysis of COP28 global stocktake outcome and global climate governance prospects. Adv. Clim. Chang. Res. **20**(2), 253 (2024)
83. Zhang, D., et al.: A representative CO2 emissions pathway for China toward carbon neutrality under the Paris agreement's 2 °C target. Adv. Clim. Chang. Res. **14**, 941–951 (2023)
84. Fam, A., Fam, S.: Review of the US 2050 long term strategy to reach net zero carbon emissions. Energy Rep. **12**, 845–860 (2024)
85. Nayak, S.K., Mohanty, A., Swain, P.: Global green credit initiative and cooperatives businesses for climate action. Productivity. **64**, 4 (2024)
86. Yang, C., et al.: Possible carbon circular pathway exploration for oil transition under the consideration of energy supply constraint and uncertainty. Ecol. Econ. **222**, 108224 (2024)
87. Acen, C., et al.: The complementary role of carbon dioxide removal: a catalyst for advancing the COP28 pledges towards the 1.5° C Paris agreement target. Sci. Total Environ. **947**, 174302 (2024)
88. Bashmakov, I.A.: Russia on the pathways to carbon neutrality: forks on roadmaps. Mitig. Adapt. Strateg. Glob. Chang. **29**(7), 1–26 (2024)
89. Ricketts, P.J., Olutoke, D.: A commentary and critique on Progress in ocean and climate change action at recent COP meetings. Ocean Yearbook Online. **38**(1), 182–211 (2024)
90. Malla, U.B.: Climate diplomacy: implications and prospects for Nepal. NCWA Annu. J. **55**(01), 28–36 (2024)
91. Zuhra, A., Sabirin, A.: Quo Vadis loss and damage fund after Cop28? ADPEBI Int. J. Bus. Soc. Sci. **4**(1), 45–61 (2024)
92. Bacon, C.M., Gliessman, S.R.: Food systems and climate crisis on the table at the COP 28 negotiations, but substantive change is off the menu thus far. Agroecol. Sustain. Food Syst. **48**, 1–5 (2024)
93. Lee, A., Channuntapipat, C., Charnock, R.: Climate (in) justice in action? A commentary on COP28 and emerging accounting mechanisms. Soc. Environ. Account. J. **44**(1), 63–74 (2024)
94. Wei, J., et al.: COP29: progresses and challenges to global efforts on the climate crisis. Innovation. **6**(1), 100748 (2025)
95. Jiang, T., et al.: New global climate actions: insight from COP29. Natl. Sci. Rev. **12**(3), 475 (2025)
96. Vartanian, A.: COP29 limps towards agreement on climate finance. Nat. Rev. Mater. **10**(2), 87–87 (2025)
97. Nair, S., Paulose, H.: Emergence of green business models: the case of algae biofuel for aviation. Energy Policy. **65**, 175–184 (2014)
98. Ozkan, M., et al.: Forging a sustainable sky: unveiling the pillars of aviation e-fuel production for carbon emission circularity. Iscience. **27**, 109154 (2024)

99. Commission, E: Mobility and transport. https://transport.ec.europa.eu/index_en (2025)
100. IATA: Annual review reports. https://www.iata.org/en/publications/annual-review/ (2025)
101. Qasem, N.A., et al.: A recent review of aviation fuels and sustainable aviation fuels. J. Therm. Anal. Calorim. **149**(10), 4287–4312 (2024)
102. Wandelt, S., Zhang, Y., Sun, X.: Sustainable aviation fuels: a meta-review of surveys and key challenges. J. Air Transp. Res. Soc. **4**, 100056 (2025)
103. Tiwari, S., et al.: Fuel from air: a techno-economic assessment of e-fuels for low-carbon aviation in China. Energy Convers. Manag. **333**, 119796 (2025)
104. Bigerna, S., D'Errico, M.C., Polinori, P.: Energy security and RES penetration in a growing decarbonized economy in the era of the 4th industrial revolution. Technol. Forecast. Soc. Chang. **166**, 120648 (2021)

Chapter 2
Evolution of Industry 1.0 to 3.0 and Its Impact on Climate Change

Abstract Economic expansion, technological development, and social change have all benefited greatly from the industrial sector, but its historical course has also had serious negative effects on the environment. The development of industrial revolutions from Industry 1.0 to Industry 3.0 and its technological innovations, financial effects, and environmental costs required clear understanding. Mechanization, steam power, and coal-driven energy were the hallmarks of Industry 1.0 and greatly increased productivity at the expense of social unrest, labor exploitation, and increasing urbanization. Electricity, mass manufacturing, and assembly lines were introduced by Industry 2.0, which increased productivity but also increased pollution and resource consumption. Global trade was changed by the growth of the steel, chemical, and telecommunications industries, but environmental stresses were also exacerbated. Automation, computing, and early artificial intelligence were introduced in Industry 3.0, which increased industrial output but also exacerbated energy inefficiencies, climate instability, and carbon emissions. The unexpected environmental effects of these revolutions, such as the loss of biodiversity, waste production, and reliance on fossil fuels draw attention to lost chances for sustainability, highlighting how the long-term effects of industrialization might have been lessened by the early adoption of energy-efficient technology, circular economy concepts, and climate-resilient industrial methods. To guarantee a harmonious coexistence between industrial advancement and environmental health, Industry 6.0, which places a high priority on climate resilience, sustainable manufacturing, and intelligent automation, is clearly necessary. To reduce climate risks and promote sustainable economic growth, we need to move toward greener technology, regulatory frameworks, and ethical industrial practices.

Keywords Industry 1.0 · Industry 2.0 · Industry 3.0 · Climate change · Technological development · Industry 6.0 · Economic growth · Industrialization · Urbanization · Financial impacts · Environmental Degradation · Public Health · Sustainability · Energy efficient · Biodiversity · Circular economy

H. Majeed, T. Iftikhar, *Intelligent Manufacturing in Industry 6.0*,
https://doi.org/10.1007/978-3-032-07278-8_2

1 Introduction to Climate Resilience Industrial Manufacturing

As we approach 2026, it is concerning that many major businesses across industrial sectors continue to believe they can thrive in the current climate change era without substantial adaptation. Its reality is stark; the destructive force of climate change has already tightened its grip on our global systems, threatening not only our environment but also our economic survival. Yet, alarmingly, many are still complacent, treating this existential threat with a level of detachment, as if it were merely a passing challenge. The time for passive observation is long gone; businesses must recognize the urgency of climate action and implement transformative strategies to mitigate the impacts before it is too late. Businesses must comprehend how climate change will affect their operations and develop resilience plans considering these growing concerns [1–3]. Energy and transportation networks are examples of critical infrastructure that is intimately interdependent. Global environmental and social effects can be amplified and cascaded by failure. Conventional infrastructure management mostly depends on expert judgment, manual measurements, and visual examination of a single infrastructure domain. It is necessary to move away from conventional industrial development patterns and toward more climate-resilient models because of this indisputable truth. Industrial revolutions and societal developments have coexisted throughout history, profoundly altering the industrial landscape of the world.

2 First Industrial Revolution (Industry 1.0) Historical Development

In the late eighteenth and early nineteenth centuries, the first indications of the Industrial Revolution appeared in England, which was the cradle of industrial manufacturing (approximately 1760–1840). Massive socioeconomic changes occurred during this time, which together led to the Industrial Revolution (IR), also known as the First Industrial Revolution. Compared to previous revolutionary upheavals like the French Revolution or the Florentine Renaissance, it was significantly more extensive and long-lasting, changing not only England but the entire world [4]. Innovations in chemical manufacture, iron production methods, and waterpower efficiency resulted from the Industrial Revolution's shift from manual labor and animal-powered technologies to machinery-based production. The development and improvement of the steam engine, an iron-based device mostly driven by coal, was one of the key innovations. These developments were mostly driven by the coal industry, which supplied the energy required to power factories and transportation networks [5]. The revolution was centered on two important industries: the textile and iron sectors. Mechanization led to a sharp rise in productivity in the textile sector, enabling England to control the world's textile markets. Originally employed to pump water out of coal mines, the steam engine later played a key role in supplying mechanical power to factories, ships, and railroads, which further accelerated the expansion of industry [6]. In addition to transforming textile manufacturing, cutting printing costs, extending print media, and constructing railways that connected nations and continents, the First Industrial Revolution signaled the shift to coal and oil. Oil drove the Second Industrial Revolution, which further increased distribution efficiency, productivity, and individual mobility. Great Britain was the richest nation in the world by 1850, when the Industrial Revolution had already extended to the New World and the European continent. Both the United States and Germany would overtake Britain in industrial production by the end of the nineteenth century [7]. However, throughout the previous 250 years, there have been significant changes in agricultural and technological processes due to the rapid economic growth, especially in the energy production sector. This quick industrial growth has been facilitated using fossil fuels as the main energy source, but it has also increased greenhouse gas emissions, which have significantly contributed to climate change. These activities' effects on the ecosystem have outgrown Earth's ability to handle them, forcing humans to go beyond its safe bounds. We must adopt sustainable practices that complement efforts to mitigate and adapt to climate change if we are to protect the environment and ensure future economic stability [8].

2.1 Europe Contribution

Europe underwent a significant transition during the Industrial Revolution, moving from a traditional, labor-intensive economy based on handicrafts and agriculture to a capital-intensive economy powered by large-scale industrial plants, mechanized production, and specialized labor. Large factories led to a large-scale migration from rural to urban areas, where more impersonal urban cohabitation replaced the close-knit social structures of agrarian life. In addition to redefining work, this change in labor dynamics also changed the social structure. New patterns of consumption and an unrelenting hunt for new raw material sources were encouraged by increased productivity. The quick movement of both raw materials and completed goods made possible by the transportation revolution, via railroads, steamships, and better road networks—accelerated the rate of international trade [9]. A sizable industrial working class, known as the proletariat, and a prosperous industrial middle class were both produced by the Industrial Revolution. With the rise of capital owners and industrialists as strong new forces and the working class as a key contributor to industrial production, these changes fundamentally altered established social hierarchies. In addition to bringing about changes that continue to impact modern economic and social structures, this transition set the groundwork for the current global economy and exacerbated environmental problems, including the increase in greenhouse gas emissions. In addition to being a change in technology, the Industrial Revolution also caused social unrest that still influences our world today. Labor patterns also saw a dramatic change because of the Industrial

Revolution. Craftsmen were forced to work as wage laborers in massive factories when rural industry relocated to cities. British colonies became important suppliers of raw commodities, including cotton, sugar, and tobacco, especially in the Americas. The transatlantic slave trade, in which British ships transported enslaved people to North and Latin America to labor on plantations that produced these vital raw materials, was closely linked to this reliance on colonial resources. However, the end of this cruel practice began in 1793 when the slave trade was banned [10].

2.2 British Contribution

Britain's population grew from roughly six million to over 20 million between 1750 and 1850. Due to rising industrial energy demands, coal output expanded substantially from about five million tons in 1700 to over 50 million tons by 1850. By 1850, exports of cotton products had grown from less than £one million in 1780 to over £30 million. According to the statistical data, industrialization sparked previously unheard-of rates of economic expansion. For example, starting about 1780, global yearly growth rates surpassed 2%, which was a substantial improvement over earlier decades when growth was minimal. Britain's national income was approximately £200 million at the middle of the eighteenth century, and it increased significantly as industrialization advanced. Similar expansion occurred in the textile sector; for instance, by 1850, cotton production had risen from two million pounds in 1780 to nearly 300 million pounds. By 1850, just approximately 20% of workers were employed in agriculture, down from roughly 60% in the middle of the eighteenth century, indicating a dramatic shift toward industrial jobs. During this time, the number of people living in cities increased dramatically; for example, Manchester's population increased from roughly 25,000 in 1773 to over 0.3 million by 1851, demonstrating the fast urbanization associated with industrial progress. In 1780, Britain's exports of cotton items were worth less than £1 million; by 1850, they had grown to almost £30 million [11]. In North America and parts of Western Europe, output per capita has increased at an average annual pace of about 2% since 1750. Economic growth was insignificant before 1500 and less than 10% of the present rates before 1750. In the 1860s, the actual wage data is indexed to 100. This suggests that actual earnings have risen dramatically over time. In 1300, there were roughly 5.5 million people living in England. Due to epidemics like the Black Death in 1348, the population had decreased to about 2.5 million by 1450. After this time, the population started to rebound, reaching numbers comparable to those of 1300 by 1640. After the population fell from 1300 to about 1450, real wages more than doubled. Productivity started to increase by 1650, signaling a shift from stagnation to growth. Despite population variations, real wages stagnated between 1300 and 1640, indicating almost little productivity growth in England. The years 1640 to 1800 are considered a transitional period, preparing the way for the Industrial Revolution's full start by that year. There has been a significant change in material conditions, with average workers today earning almost 15 times as much as they did two centuries ago [12].

2.3 *Different Factors Influence in Industry 1.0*

In Industry 1.0, product volume demand environment can be referred to as the Simple Market. There were fewer supplies than there were requests in Industry 1.0. Industrial product outputs were unable to meet societal demands. Adam Smith's Wealth of Nations economics, which saw price as an automated mechanism to correct supply and demand imbalances, provided a key concept for Industry 1.0. Prices would increase if supply could not keep up with demand. Price adjustment was a useful strategy for resolving the supply-demand imbalance at the time because there was very little product variation, and most commodities were agricultural items [13].

At that time, technology was also advancing. The introduction of the spinning Jenny machine revolutionized the textile industry by enabling a single worker to spin many threads simultaneously, raising productivity from a few spools to dozens. Using waterpower to create stronger strands, the Water Frame machine improved the quality of the textiles that were produced. The efficiency and power output of the steam engine were improved by James Watt. Beyond textiles, this technology became essential to several other industries, such as mining and transportation. By lowering prices and improving output quality, coke smelting transformed the manufacture of iron. By improving control over wrought iron quality, the puddling technique helped to further progress the iron manufacturing industry. Crop rotation techniques raised yields and the seed drill boosted planting efficiency, sustaining a burgeoning population that drove urban migration and the need for industrial labor [11].

2.4 *Industry 1.0 Merits and Demerits*

We can draw the following conclusions about the first industrial revolution: people gained more jobs as a result of industrialization, economic growth, and urbanization as a result of the rapid migration of workers to cities due to corporate sector expansion; capitalism emerged as an economic system to promote entrepreneurship and investment in new industries and technologies; and there was the rapid mass industrial production of goods due to steam power engines and machinery, which made them affordable for people who could not afford them without machinery [14–16]. Similar drawbacks included the exacerbation of economic inequality, as factory owners and industrialists accumulated substantial wealth while most workers continued to live in poverty. Social tensions and instability were caused by this financial disparity, and cities became overcrowded because of fast urbanization, sometimes with subpar living conditions. Many laborers' lacked access to basic facilities like clean water and sanitary facilities and lived in small, filthy quarters [17, 18]. The largest problem is child labor, as the growth of factories resulted in thousands of kids working in dangerous jobs. Poor Working Conditions: Workers, including women and children, suffered exceedingly poor working conditions. Many were under the age of ten. They were compelled to work long hours for little compensation, frequently in hazardous environments where they may be beaten for not reaching expectations [19, 20]. They frequently worked 12 h a day in hazardous conditions with little to no legal protection (no labor regulations). Since industrialization had a negative effect on the environment, the absence of safety precautions led to many accidents and deaths (reforms later required that children attend school for at least 2 h a day, a substantial change aimed at lowering child labor). Increased factory pollution led to contaminated water sources and poor air quality, which impacted local ecosystems and human health [21, 22]. In turn, this led to global warming and climate change. Health Issues: Long workdays, inadequate dietary habits, and dangerous working conditions caused major health issues for employees. Industrial pollution-related illnesses increased in frequency, and workers' protests and resistance were triggered by the harsh reality of industrial life. Strikes and the establishment of labor unions were common ways for people to demand improved working conditions, although they were frequently opposed by employers and authorities [23–25]. The Industrial Revolution created the framework for international trade networks, urbanization, and technology breakthroughs that still have an impact on contemporary environmental and industrial practices. Given the start of large-scale industrialization and the reliance on fossil fuels throughout this crucial age, we must also recognize the period's contribution to creating the foundation for today's climate issues [26–28].

3 Second Industrial Revolution (Industry 2.0)

Though many of its defining events can be traced back to the 1850s, the second Industrial Revolution is typically dated between 1870 and 1914. Nonetheless, it is evident that the astounding rate of breakthrough discoveries, or macro inventions, began to decline after 1825 and to resume in the final part of the century. The rate of technical advancement, as generally measured by increases in productivity and product quality, is largely dependent on the smaller, cumulative, anonymous modifications known as micro inventions. The incorporation of new scientific knowledge into industrial production marked the Second Industrial Revolution, which dramatically increased productivity in several industries and changed economies and cultures globally. Beginning in the late nineteenth century, the Second Industrial Revolution was highlighted by revolutionary developments in mass manufacturing techniques, assembly lines, and energy generation. By using and refining methods from the Chicago slaughterhouses, Henry Ford (1863–1947) transformed the idea of mass production. Each butcher there was an expert in a particular aspect of the slaughtering procedure, and the animals were hung from conveyor belts. Ford significantly increased production speed and decreased costs by implementing this assembly line method in the automobile manufacturing industry. The procedure was more scalable and efficient since cars were constructed in stages on a conveyor belt rather than being assembled at one place [29].

3.1 *Technological Development in Industry 2.0*

The era known as Industry 2.0 saw a huge increase in the quantity and diversity of industrial goods. Electricity, mechanical, and electronic gadgets, and automobiles were among the major technological advancements. Industry 2.0 products are still in widespread usage today. Volume and variety were the two aspects of demand during Industry 2.0. The challenge at the start of Industry 2.0 was to boost product volume while keeping costs down. Ford's Model T assembly line innovation allowed it to meet high volumes through mass production. Vehicle production was transformed by Henry Ford's invention of the assembly line. By 1914, he was producing nearly a quarter of a million Model Ts a year. This was accomplished by lowering costs and making cars accessible to the general public by combining continuous flow production with interchangeable parts [30]. A system that can provide high product volumes at a cheap cost is an assembly line for mass manufacturing. Short operation times, standardization of components and operations, specialization of workers and equipment (i.e., one worker and equipment for each operation), dedication of the assembly line (i.e., no product variety, one line for each product model), and an effort to balance workloads are the primary enablers of an assembly line. Ford held a two-thirds market share in the US auto industry at the beginning of the 1920s. In 1924, the cost of a Model T dropped from USD 850 (in 1908) to USD 250. The 1920s saw a diversification of consumer preferences. All clients were no longer pleased with a single model (the black Model T). General Motors' Alfred Sloan (1964) used a divisional organizational structure, one division for each model, to solve this issue. Ford, on the other hand, had a flat organizational structure. Ford's market share dropped to 18.9% in 1940, while GM's increased to 47.5%. This set the stage for post-WWII production systems, such as the Toyota Production System (TPS), which would later master the simultaneous satisfaction of volume and variety [13, 31].

3.2 *TPS Enablers*

Workers are partially cross-trained, which means that each worker can perform more than one operation. JIT-MS (i.e., the right components, in the correct place, at the right time, in the exact amount) is applied to control material flows. They placed emphasis on striking a balance between unpredictability, work-in-process, and capacity utilization. When a system becomes more variable, either more work needs to be done in the process or less capacity needs to be used [13, 32].

3.3 Cellular Manufacturing

Cellular manufacturing (CM) is one application of group technology. One of CM's objectives is to efficiently manage the manufacture of a variety of parts. Cells are converted from job shops with practical design to boost productivity. A cell consists of a part family and a machine group. The initial step in creating a CM system is identifying machine groups and component families to create manufacturing cells. This minimizes the number of part moves among cells while processing each part family inside a machine group. This identification is known as the cell formation problem. For pieces of the same family, the machining sequences are similar [13, 33, 34].

3.4 Metallurgy, Chemicals, Aviation

Metallurgy, chemicals, aviation, and other industries were all greatly impacted by the technological developments in production during this time. Even while certain older, less effective technologies were still in use, mass manufacturing processes, in particular, the utilization of assembly lines and interchangeable parts, increased efficiency and decreased prices. However, cost savings made possible by economies of scale compensated for these technologies' shortcomings [30, 35]. The Bessemer converter, created by Henry Bessemer in 1856, transformed the steel industry and made it possible to produce cheaper steel in large quantities [36]. Wrought iron was used extensively before this, but it lacked strength and longevity. By forcing air through molten iron, the Bessemer process eliminated impurities and drastically reduced production costs. By enabling the utilization of scrap iron and providing improved quality control, the Siemens-Martin open-hearth process substantially improved the quality of steel. Cheap steel became an essential component of industry by 1880 when it was widely utilized for railroads, ships, and buildings [37]. In terms of chemical innovations, Germany was the world leader. For instance, natural colors were replaced by synthetic ones, such as alizarin in 1869 and mauveine in 1856. The synthesis of ammonia from nitrogen and hydrogen was made possible by the Haber-Bosch process, which was discovered in 1908 [38, 39]. This technique revolutionized the manufacturing of fertilizer and allowed Germany to generate nitrates for explosives during World War I. Communication was boosted by the growth of telegraph networks. Although the installation of submarine cables in 1851 represented a significant technological advancement, early cables frequently failed. Due to wear and technological constraints, only 4800 of the 17,700 km of installed wire were in use by 1861 [40, 41]. Between 1880 and 1900, electric light bulbs' efficiency doubled, and their cost dropped fivefold, which fueled the growth of electric power infrastructure. Thanks to technological advancements, transportation costs have drastically decreased. For instance, from 1850, innovations in steam turbine engines and steel ships helped to lower transatlantic freight prices by 1.5%

yearly. The invention of the steam turbine in 1884 increased ship speed and efficiency. Mechanization improvements contributed to a decrease in the amount of labor needed in agriculture. An acre of wheat took 20 man-hours to harvest in 1880, but by 1935, that number had decreased to 6.1-man hours [30].

3.5 World Wars

Due to the global disruptions caused by the world wars as well as major economic catastrophes like the panic of 1893 and the stock market crash of 1930, the economic landscape of the time was marked by notable oscillations. Increased industrial competition fueled globalization and established capital as a major force behind economic shifts. The number of companies and the volume of production increased dramatically because of the fast industrialization that began in the middle of the nineteenth century in both Europe and the United States [42]. As the struggle for technical dominance intensified, overcapacity emerged as a prevalent issue. Large industrial conglomerates were formed because of these mergers, which began with trusts in the steel, oil, and rail sectors before the auto-industry was consolidated. Due to the intricacy of these sectors, scientific knowledge became increasingly important, and many companies now have specialized research and development (R&D) departments [43].

On the one side, mass manufacturing of standardized commodities resulted in reduced costs, enabling a wider range of people to purchase things. However, product adjustments were expensive and time-consuming due to the linear manufacturing processes of the era. During this period, which approximately spanned 1870–1914 (with advancements going back to 1850), important innovations like railroads and telegraphs were incorporated into industrial activities [42]. Although industrial production itself continued to change, this radical phase of industrial innovation disturbed with the start of World War I. These reforms were further sparked by World War I, when companies restructured to focus on efficiency and standardization while operating under government supervision. Subsequent manufacturing advancements are not seen as a new industrial revolution because they were more gradual enhancements than revolutionary changes. In addition to altering industries, economies, and societies globally, the Second economic Revolution laid the groundwork for subsequent economic and technical breakthroughs [44].

3.6 Optimization in Production and Cost

Driven by scientific knowledge and inductive approaches aimed at boosting production and organization, advanced industrial nations are experiencing considerable changes in their secondary sectors. The revolution is centered on gaining accurate knowledge to better organize production processes, ranging from large-scale industrial layouts to manual tasks. The evolution of accounting is defined as the shift from merely maintaining records to being an applied science that supports corporate choices, such as factory cost accounting. Engineering developments, such as the use of scientific concepts in engineering replaced trial and error approaches with accurate calculations, resulting in safer and more cost effective construction processes. Competition in the Market as late nineteenth century price reductions forced manufacturers to develop new cost-cutting measures, which resulted in mass manufacturing methods [45]. Amalgamation (combining businesses to eliminate weaker competitors and control markets), specialization (concentrating on products to maximize economies of scale), continuous production (planning plants for uninterrupted manufacturing processes), and functional management (implementing specialized management practices for better efficiency) were the hallmarks of the "rationalization" movement that followed the war. Successful specialization necessitates focused intellectual work on production efficiency and cost reduction, which frequently entails a sizable capital investment for new buildings and machinery [46]. There was a notable drop in prices from 1865 to 1896, especially in the United States, between 1873 and 1894. Manufacturers were forced to innovate and cut expenses because of this downturn. Compared to earlier methods of producing in tens or hundreds, mass production, which was first used in America in the 1860s, greatly increased output by creating things in lots of hundreds or thousands. The continuous production process was used in several sectors, including watches, paper, biscuits, and cotton stitching. To justify the capital expenditure required for new machinery and continuous production processes, specialization necessitates an output that is 5–30 times higher than that of existing plants. Up to two thousand machines can be found in major factories, most of which are unique and made especially for their jobs, highlighting the volume and complexity of manufacturing [44].

4 Industrial Revolution 3 (Industry 3.0)

Industry 3.0 is defined by technological advancements such as the transition from analog to digital, which had a major impact, especially on the electronics industry. Average product life cycles for most of the electronics equipment decreased significantly, and the architecture of products shifted from integral to modular. Volume, variety, and delivery time were the three aspects of product demand that increased throughout Industry 3.0. One may describe this demand scenario, which resulted in flow line and temporary protected status issues, as a volatile market. To address the three aspects of demand, industries use flexible manufacturing systems and supply methods [13]. The third manufacturing revolution began in the middle of the twentieth century with the advancement of electronics and computer technologies (such as IT and AI, among others). In this initial stage of automation, computers carried out duties that had previously been completed by humans. Robots have become more and more significant in society since the late 1960s, when the Industrial Revolution 3.0 began. Numerous fields that are unavailable to humans, like logistics, healthcare, conflict zones, etc., can benefit from the deployment of robots. AI has many applications, as the COVID-19 pandemic showed [47–49]. Advanced AI algorithms can be used to design robots to carry out all tasks effectively without the need for human intervention. Reducing the three Ds, that is, dangerous, dull, and dirty jobs, was the main driving force for Industry 3.0. Businesses can grow and improve with the help of robots. As computers grew increasingly accessible and potent for use in data processing, control systems, and automated decision-making, the manufacturing sector embraced them. Manufacturing automation has increased, with machines and robots taking the place of tedious and hazardous tasks. In certain operations, this has reduced the need for human labor and increased productivity. This surely acted as a catalyst for the continued integration of automation into Industry 4.0 applications [50, 51].

4.1 Rise of Electronics

Industries, especially electronics, have been significantly impacted by the shift from analogue to digital, which has resulted in modifications to product architecture from integral to modular designs. As a result of the demand for quick invention and responsive production methods, the average life cycle of electronics devices in Japan has drastically shrunk to roughly 6 months. The global wealth increased 10 times as compared to first industrial revolution. Semiconductors, mainframe computers, microprocessors, the internet, and e-commerce (about 1995) are some of the major inventions in IR3. Life expectancy also rose considerably in comparison to IR1 and IR2. Every country's standard of living rose. IR3's drawbacks, however, are its detrimental effects on greenhouse gas emissions, biodiversity loss, and air pollution [52]. Every species on this planet is negatively impacted by the monster of global warming, which is caused by greenhouse gasses. According to the UN SDG report, climate change is a menace to humanity and is still out of control. Nonetheless, we are making progress in lessening the effects of climate change, and we will undoubtedly be able to control it if all people work together [53]. Many features of the Third Industrial Revolution overlap with those of the Fourth Industrial Revolution. These overlaps will be explained in greater detail in the following chapters.

5 Climate Resilience and Other Missed Opportunities in Industry Evolutions 1–3

The industrial revolutions of the past (from the First to the Third) were inherently unsustainable, as they lacked foresight regarding the well-being of future generations. At the time, there was little awareness of the long-term consequences of actions like fossil fuel combustion or the importance of preserving biodiversity.

Industrialists were primarily focused on maximizing production, reducing costs, suppressing labor rights, and accumulating wealth—often through unethical and inhumane practices. This short-sighted approach has left our current generation grappling with the severe impacts of global warming and climate change, and future generations are poised to suffer even more if immediate action is not taken. Had our predecessors considered the environmental and social consequences of their actions, the world might not be facing such extreme climate challenges today. The purpose of discussing these missed opportunities is to help guide future, ensuring that by 2060 and beyond, humanity does not repeat these mistakes, and future generations do not have to reflect on our actions with similar regret. It is mandatory that we act now, embedding sustainability in all aspects of industrial development to safeguard the planet for those who come after us.

From Industry 1.0 to Industry 3.0, the following detailed comparative table (Table 2.1) illustrates several lost possibilities that fueled global warming. Based on comprehensive analysis of peer reviewed historical datasets and authoritative sources (IPCC, IEA, FAO, UNEP, ILO), here is the verified and updated Table 2.1 with critical annotations. The table, which covers water waste, burning fossil fuels, wasteful industrial processes, labor exploitation, and environmental deterioration, is based on accurate, extremely authentic, and scientifically proven data. The expansion of coal-fired factories, inefficient textile processing, and chemical-intensive manufacturing were the main causes of the first three industrial revolutions (Industry 1.0–3.0), which resulted in an enormous increase in CO_2 emissions from 500 million tons to 22 billion tons yearly. As companies released untreated chemical effluents into natural water bodies, poisoning groundwater and exacerbating the acute water shortage, water waste increased. Water constraints are made worse by the textile industry alone, which uses 93 billion m^3 of water a year. The extensive extraction of raw materials for the creation of synthetic fibers exacerbated the deforestation crisis, hastening the loss of biodiversity and causing climate instability because of habitat destruction. By releasing dangerous synthetic colors, microplastics, and carcinogenic chemicals into the environment, the growth of the petrochemical industries also added to the amount of toxic industrial waste.

5.1 *Global Warming*

As of 2025, global temperature has increased by +1.09 °C ± 0.12 °C on average from 2011 to 2020 compared to pre-industrial levels (1850 to 1900 baseline), as confirmed by IPCC AR6. The onset of industrial era warming is estimated around 1830 ± 20 years, particularly in the Caribbean basin based on Sr/Ca ratios in sclerosponges. These proxies show that preindustrial temperature variability from 1700 to 1860 was about ±0.2 °C in the Northern Hemisphere, challenging previous assumptions of a more stable ±0.1 °C range. Instrumental sea surface temperature records have global coverage from 1853 onwards, but earlier data are considered sparse and unreliable. Atmospheric CO_2 concentration reached 414 ppm in 2020, up from 277

Table 2.1 Missed opportunities of industrial revolution environmental and social impact

Category	Industry 1.0 (1750s–1840s)	Industry 2.0 (1870s–1950s)	Industry 3.0 (1960s–1990s)
Primary energy source	Coal-fired steam engines (3–5% efficiency)	Coal (60%) + oil (30%); global energy use ↑ 400% vs. 1.0	Fossil fuels still 82% of energy mix; nuclear reaches 7% by 1990
Carbon emissions (annual global)	~10 to 100 Mt. CO_2 (1750–1800)	~6 Gt CO_2e (1950)	22.6 Gt CO_2 (1990)
Water wastage	Unregulated discharge; UK rivers 50% polluted by 1800	Industrial water use ↑ 800% (1900–1950)	Global industrial water use: ~750 km^3/yr. (1990); textiles = 1.5 t H_2O per kg fabric
Fossil fuel consumption	Coal: ~150 Mt. (1800)	Oil: ~20 Mbbl/day (1950)	65.5 Mbbl/day oil (1990) + 3400 Mt. coal
Deforestation	Europe: 10% forest loss (1700–1850)	Global: 13 M ha/yr. loss (1950s); Amazon: 2.5% cleared by 1950	15.4 M ha/yr. net loss (1990s); Amazon deforestation peaks at 29,000 km^2/yr. (1995)
Industrial waste	Coal ash: ~50 Mt./yr. (UK, 1800s)	Toxic sites: >5000 documented (US/EU)	~20Mt e-waste/yr. (1990)
Child labor	UK: 50% textile workers under 14 (1830)	Global: 72 M child laborers (1950)	Sweatshops: 250 M children working (1990), 30% in hazardous industries
Hazardous chemicals	Lead/arsenic in textiles; hg poisoning in hat-making	Synthetic dyes (benzene): 80% worker illness rate (1920s)	CFC production: 1.1 M tons/yr. (1980s); PCB waste: 300,000 tons dumped
Atmospheric CO_2 (ppm)	277 ppm (1750) → 284 ppm (1850)	~310 ppm (1950)	354 ppm (1990)
Global temp rise	+0.1 °C (1750–1850)	+0.3 °C (1900–1950)	+0.6 °C (1990 vs 1850–1900)
Major disasters	Worsdell Colliery (UK, 1838): 26 deaths	Donora Smog (1948): 20 deaths, 14,000 ill	Bhopal (1984): 3787 immediate deaths; Exxon Valdez (1989): 37,000 tons spilled

to 284 ppm in 1750, and is tightly linked to warming, with an estimated equilibrium climate sensitivity of 3.0 °C ± 0.5 °C per CO_2 doubling. By 2023, the global temperature anomaly reached +1.45 °C ± 0.12 °C, the highest in over 125,000 years. The IPCC projects a likely breach of the 2 °C threshold between 2040 and 2045 under the SSP2–4.5 scenario, with some studies suggesting it could occur as early as 2030 to 2040. Sclerosponges record warming in the 10 to 15 m ocean mixed layer rather than sea surface temperature and show a + 0.5 °C discrepancy since 1860 compared to instrumental datasets, likely due to regional oceanographic dynamics rather than global inconsistencies. In the 2010s, global lentic (lakes and reservoirs) systems released 64.6 ± 12.1 Gg N_2O-N/year, a 126% increase from the 1850s. This analysis shows that N_2O emissions from worldwide inland waterways in the 2010s were 319.6 ± 58.2 Gg N/year, when combined with riverine emissions. This gives the country level emission factors (range from 0 to 0.341%) and indicates a global emission factor of 0.051% for inland water N_2O emissions in relation to agricultural nitrogen applications [54–56].

5.2 Ground Water

Because aquifers hold the most unfrozen liquid, groundwater is essential to life, as we know it. Under a medium emissions trajectory, it is conservatively predicted that groundwater at the water table's depth (except from permafrost areas) will warm by an average of 2.1 °C between 2000 and 2100. Spatial diversity in water table depth and climate change causes significant variation in regional shallow groundwater warming trends. Mountainous areas are predicted to have the lowest rates. Stream thermal regimes, ecosystems that rely on groundwater, aquatic biogeochemical processes, groundwater quality, and geothermal potential are all impacted by rising groundwater temperatures. According to the results, 77 million to 188 million people are expected to reside in places where groundwater temperatures surpass the maximum threshold for drinking water by 2100 if a medium emissions pathway is followed [57].

5.3 Resource Efficiency

Conventional industrial methods, especially during the First to Third Industrial Revolutions, prioritized economic expansion and production over sustainable environmental practices and resource efficiency. As a result, during these times, resource consumption was usually linear and followed the "take-make-dispose" model, with little emphasis on waste avoidance. Coal and steam, which are both incredibly inefficient by today's standards, were the main energy sources during the First Industrial Revolution (Industry 1.0). The efficiency of early steam engines, such the one built by Thomas Newcomen in 1712, was only about 1%. Efficiency was only raised by James Watt to roughly 3–4%. This indicates that the great majority of the energy released during the burning of coal was lost as heat. Electricity and oil were popular during the Second Industrial Revolution (Industry 2.0). But there were still significant inefficiencies, particularly in early electrical grids and internal combustion engines, which turned less than 25% of the energy from fuel into productive activity. Large amounts of energy were wasted because of antiquated technology, inadequate infrastructure, and a lack of motivation to preserve. Energy consumption in the United States increased quickly between 1870 and 1914, but transmission inefficiencies caused the power networks of the era to lose 15–25% of their energy [58–63].

5.4 Inefficient Manufacturing Techniques

Industrial production was mostly reliant on resource extraction and inefficient manufacturing techniques from Industry 1.0 to Industry 3.0. Large volumes of trash were produced by factories, but recycling and material optimization received little attention. The Bessemer process, which helped boost steel production in the middle of the nineteenth century, also generated a lot of slag, or trash. About 250–300 kg of slag was created for every ton of steel, and this slag was frequently thrown away without being used again. During the Industrial Revolutions, enormous quantities of cotton, wool, and other natural fibers were used in the creation of textiles. About 20–30% of the raw materials used in this sector were wasted during the spinning and weaving operations [64, 65].

5.5 Water Usage

Another resource that was inefficiently used by conventional industrial practices was water, particularly in sectors like metalworking, papermaking, and textiles. Large amounts of water were frequently taken from rivers and lakes by factories, which resulted in contamination and depletion. During the Second Industrial Revolution, the production and processing of 1 kg of cotton required roughly 250–450 l of fresh water. Significant water contamination resulted from the direct dumping of wastewater into rivers that was loaded with chemicals, dyes, and other contaminants. Similarly, waste from leather production has long been an issue in the modern period [66–68].

5.6 Linear Economy

Between Industry 1.0 and Industry 3.0, the circular economy concept, which seeks to maximize value extraction and recover materials at the end of their life cycles, was given little to no attention. Conventional industries functioned in a linear fashion, meaning that materials were created, consumed, and discarded without any thought given to recycling/upcycling or reusing them. This strategy exacerbated environmental harm and resource depletion. Global resource extraction had quadrupled since 1900 by the middle of the twentieth century, yet as of 2020, only 8.6% of all materials utilized worldwide were recycled back into manufacturing. Environmental deterioration and the need for virgin materials may have been considerably decreased if the circular economy's tenets had been implemented sooner [69–71]. Climate resilience has been negatively impacted for a long time by the failure to embrace circular economy ideas and optimize resource utilization. More than 30% of greenhouse gas emissions worldwide are caused by the industrial sector, mostly because of energy waste and ineffective manufacturing methods. Dependency on fossil fuels from Industry 1.0 to Industry 3.0 played a major role in the buildup of CO_2 in the atmosphere, which ultimately caused climate change. Global CO_2 emissions were approximately 2 billion metric tons in 1900. Due in major part to the industrial usage of coal, oil, and gas, emissions had increased to about 10 billion metric tons per year by the time Industry 3.0 (the middle of the twentieth century) was fully underway [72–74].

5.7 *Climate Change*

Significant hazards associated with climate change for industrial activities include supply chain disruptions, harsher weather events, and restrictive regulations. Catastrophic weather events like hurricanes, wildfires, and floods have almost doubled in frequency over the past 20 years and are strongly linked to rising global temperatures. These disruptions result in billions of dollars' worth of damages year; in 2022 alone, climate-related calamities cost the world economy more than $300 billion. Extreme weather can limit the availability of raw materials, raise expenses, and delay shipping, making supply networks especially vulnerable. Coastal manufacturing and distribution centers are also seriously threatened by rising sea levels; estimates suggest that by 2050, climate-driven floods may affect around 800 million people who live in low-lying metropolitan regions [75–77]. Supply chain greenhouse gas (GHG) emissions account for around 60% of all industry-related emissions worldwide, which increases regulatory risks as nations work to meet stricter emissions standards. Due to a lack of thorough environmental concerns throughout the early industrial phases, businesses are still primarily vulnerable to climate threats. Few industrial sectors have built resilient infrastructure that can endure extreme climate events or completely evaluated their carbon footprints. Industries must adopt forward-thinking ways to improve resilience, promote operational sustainability, and safeguard against climate hazards as the need for adaptation and emissions mitigation increases [78, 79].

5.8 *Avoidance of Clean and Green Technology*

Innovation in green and clean technology was mostly disregarded in the early industrial revolutions. Consequently, expansion was driven by high-emission energy sources, especially coal and oil, at the expense of environmental expenses. The emphasis was on increasing production efficiency, frequently at the price of water and air quality, which resulted in increased greenhouse gas emissions and a legacy of pollution. Due to their high initial costs and lack of backing, sustainable manufacturing techniques like closed-loop recycling and circular economy models show promise but still only make up a small portion of industrial activity [80, 81]. The technological advancements required for carbon-neutral production and cleaner, more robust industrial systems must be fueled by policy-driven incentives, additional funding, and international cooperation to close these historical gaps. In the current environment, initiatives to develop technology that reduce carbon emissions, increase energy efficiency, and implement sustainable manufacturing practices are still vital but receive little funding. Approximately 495 billion USD was invested globally in renewable energy sources in 2022, but estimates indicate that roughly 1 trillion USD a year will be required until 2030 to meet the targets set forth in the Paris Agreement. Like this, manufacturing might reduce industrial emissions by 25 to 30 % through the development and implementation of energy efficient technology, but funding for this sector is still lacking [82–84].

5.9 *Non-sustainable Practices*

In the early stages of industrial revolutions, environmental sustainability received little attention from legislative frameworks and policies that were primarily concerned with industrial expansion and economic prosperity globally. Any push for sustainable practices was hampered by the lack of strong environmental regulations or financial incentives. For example, there was little or no enforcement of regulations and there is no policy to govern emissions or waste disposal, which resulted in extensive environmental damage. The manufacturing sector's adoption of climate-resilient practices was severely hampered by this regulatory vacuum. Industries continued to employ resource-intensive procedures and high-pollution technology in the absence of clear regulations, which would eventually have detrimental effects. Significant air quality problems were caused by coal-burning technology used in industry and energy generation, but emissions regulations had not yet been widely adopted by the twentieth century. Only in the last few decades have regulatory frameworks started to close these gaps through direct subsidies for green technologies, carbon pricing, and emissions trading programs [85, 86]. Despite these developments, a substantial amount of money is still needed; through 2050, it is projected that USD 4 trillion would be needed annually for renewable energy and green infrastructure worldwide. For the manufacturing sector to successfully adopt climate-resilient practices and lower its carbon footprint, stronger legislative support, and financial incentives must be established [87–89].

The IPCC estimates that current emission trends would cause global temperatures to rise by 1.5 to 2 °C by as early as 2050. In warmer climates, this increase may result in a 15 to 20 % increase in the demand for power for cooling, further taxing energy systems and emphasizing the necessity of heat-tolerant, energy-efficient infrastructure in development plans. Under moderate greenhouse gas emissions scenarios, sea levels are predicted to rise by 0.3 to 0.6 m by 2100, and under high emissions scenarios, they may increase by more than 1 m. This must be taken into consideration while developing coastal infrastructure, whether it be by erecting seawalls and levees or raising vital facilities above possible flood levels [90, 91]. By 2050, it may cost $400 billion to protect vulnerable coastal areas in the 32 cities of the United States alone. Since certain areas are expected to have a 30% increase in severe rainfall events, stormwater management, and flood-resilient design must be incorporated into urban development. To reduce the risk of flooding, New York City has put in place a USD 20 billion climate adaptation plan that includes higher infrastructure, green roofs, and storm barriers [92, 93]. By 2050, the world's water demand is predicted to increase by 20–30%, placing more strain on water supplies. Improvements in irrigation efficiency in agriculture, which uses 70% of the world's water, might result in a 20% reduction in water usage. In certain urban locations, rainwater collection might provide up to 35% of household water needs if it were widely implemented, especially during dry seasons. Modern water recycling systems can cut industrial water use by 50 to 90 %, which is critical in areas with acute water shortages. According to semiconductor industry research, water-recycling systems might save millions of cubic meters of water a year by reducing usage by about 75%. Implementing circular economy principles can reduce material waste by 45–50%. Circular practices have the potential to lower material prices in the European Union by up to €600 billion a year, which would help to minimize the demand for industrial waste disposal overall [94–98].

5.10 Fossil Fuel Dependence

Renewable energy must account for at least 60 to 70 % of the world's energy mix by 2050, up from 29 % in 2020, to meet the objectives of the Paris Agreement and keep warming to 1.5 °C. Industries can reduce carbon emissions by up to 70 % by converting to renewable energy sources, which is an important step in achieving global emissions objectives. By substituting renewable energy sources for fossil fuels in manufacturing, CO_2 emissions might be cut by about 1.5 gigatons/year by 2050. Making the switch to high efficiency equipment can result in significant operational savings by reducing industrial energy usage by 20 to 30 %. Industrial energy efficiency initiatives have the potential to save up to USD 47 billion a year in the United States across several industries. By reducing downtime from extreme weather events by 30 to 50 %, climate-resilient techniques like decentralized energy systems and flood-resistant infrastructure can save USD 150 billion annually in avoided disruptions [99, 100]. In some chemical production processes, the usage of harmful compounds can be reduced by up to 90 % by applying the concepts of green chemistry. Through developments in sustainable chemistry, for example, the pharmaceutical sector has decreased hazardous waste by about 70 %, lowering waste treatment costs and its impact on the environment. This reform will reduce the costs of air pollution, which is estimated to cost the global economy over USD 5 trillion annually due to its detrimental impacts on human health. Green areas can reduce the urban heat island effect by up to 2 °C, which is important considering that 68 % of the world's population is predicted to live in cities by 2050. A 10% increase in urban green space might save 25% on cooling energy expenses in densely populated areas. Climate resilient supply chains can reduce potential losses by up to 50% by implementing early warning systems, flexible sourcing, and regional diversification [101–104].

6 Future Directions

Based on thorough, sustainable practices into all aspects of their operations, industries can drastically lower their environmental impact and improve their ability to withstand the effects of climate change. Adopting circular economy models, using cleaner industrial methods, and reducing waste through material efficiency and reuse can all help achieve this. Reducing greenhouse gas emissions requires updating to energy efficient equipment, incorporating carbon capture and storage systems, and switching to renewable energy sources for operational power. Also, to guarantee resource sustainability in water scarce areas, enterprises might use cutting edge water management strategies like recycling and rainwater collecting. Reducing hazardous materials in product lifecycles by utilizing green chemistry concepts can also lessen environmental harm and increase operational safety. By lowering, the risk of supply chain interruptions and the fines associated with environmental compliance, these actions taken together help to create a strong climate adaption plan [105–107].

Businesses should concentrate on improving their capacity to adjust to climate change, which entails step-by-step incorporating resilience and flexibility into their systems and procedures. Risk reduction and large operational savings are possible outcomes of integrating climate resilience. By integrating climate adaptive infrastructure, operational downtime can be reduced by up to 50%, resulting in yearly losses to the global economy of up to USD 150 billion. Using temperature-resistant materials has been shown to reduce maintenance costs by up to 20% in industries subjected to severe temperatures. Companies can maintain a consistent flow of vital resources by reducing their susceptibility to disruptions by 40 to 60% through flexible supply chains with a variety of sourcing options. Due to the lower risk profile of climate-prepared operations, companies that implement strong climate adaption frameworks may receive a 20 to 25% boost in investor confidence. This degree of resilience is essential because, if adaptive measures are not put in place, climate-related disruptions are expected to cost the world economy up to USD 1 trillion a year by 2050 [108–110].

By incorporating resilience and flexibility into all systems and operating procedures, industries may systematically improve their ability to adapt to climate change. To reduce physical risks, this entails including climate-responsive infrastructure into facility design, such as flood defense systems and materials resistant to temperature changes. Operational stability can be greatly improved by creating adaptable supply chains that can change course in reaction to climate-related disruptions, for example, by procuring from several locations or keeping a store of essential goods. A staged strategy that includes scenario modelling, climate risk assessments, and ongoing environmental impact assessments at every stage of the production lifecycle is needed to build resilience. By ensuring that climate resilience is a continuous priority, these procedures enable industries to proactively address climate risks and assure their continued operations in situations that are becoming more unstable. By incorporating strong, sustainable practices into every aspect of operations, industries may significantly lower their environmental impact and increase their ability to withstand the effects of climate change. Key initiatives include adopting circular economy models to minimize waste, employing improved material efficiency techniques, and optimizing resource recovery and reuse. Making the switch to cleaner, low-carbon manufacturing methods is crucial; converting to renewable energy sources, upgrading industrial machinery to high efficiency models, and implementing carbon capture, storage and utilization technology can significantly reduce greenhouse gas emissions. Industries can use cutting edge water management techniques, such closed-loop recycling and rainfall collecting, to ensure resource sustainability, particularly in areas with limited water supplies. This method improves operational resilience and lessens reliance on outside water sources. Significant environmental benefits and improved operational safety standards are provided by green chemistry concepts, which reduce or remove hazardous compounds in production processes. When combined, these strategies create a flexible, robust framework that reduces supply chain interruptions and compliance expenses while reducing climate related risks, putting businesses in a position to satisfy social and regulatory demands for sustainability [111–115]. Now, Just Transition, LEAN Management, and RECP (Resource Efficient and Cleaner Production) methodologies are increasingly being adopted by the industries, especially SMEs, to enhance sustainability and improve overall performance.

References

1. Singh, S., Goyal, M.K.: Enhancing climate resilience in businesses: the role of artificial intelligence. J. Clean. Prod. **418**, 138228 (2023)
2. Bin Ali, M., Iftikhar, T., Majeed, H.: Green synthesis of zinc oxide nanoparticles for the industrial biofortification of (Pleurotus pulmonarius) mushrooms. Heliyon. **10**(19), e37927 (2024)
3. Majeed, H., Iftikhar, T., Abbas, Q.: Climate resilience plastic degradation potential of pseudomonas putida isolated from the soil of plastic waste dumping sites to reduce GHG emissions. Z. Phys. Chem. **238**(5), 797–807 (2024)
4. van Leeuwen, B., Bosma, U., Wang, M.: Industrial revolutions in a globalizing world, 1760–present. In: Handbook of Industrial Development, pp. 18–36. Edward Elgar Publishing, Cheltenham (2023)
5. Jha, B., Garg, S., Dhingra, S.: From mechanisation (1.0) to smart automation (5.0). In: Artificial Intelligence and Communication Techniques in Industry 5.0, p. 87. CRC Press, Boca Raton (2024)

6. Mohajan, H.: The First industrial revolution: Creation of a New Global Human Era (2019)
7. Spielvogel, J.J.: The industrial revolution and its impact on European society. Western Civil. **3**, 583–608 (2005)
8. Majeed, H., Iftikhar, T., Maqsood, K.: Synthesis of protein-based catalyst for multi-industrial applications by using agriculture and industrial waste. Polym. Bull. **81**(14), 12633–12653 (2024)
9. Barham, L.: From Hand to Handle: The First Industrial Revolution. OUP, Oxford (2013)
10. Ionescu, I.G.: The first industrial revolution and general features of the world economy between the 16th century and 1780. Science. **6**(17), 183–186 (2018)
11. Deane, P.: The First Industrial Revolution, p. 78. Cambridge University, Cambridge (1979)
12. Steinsson, J.: How Did Growth Begin? The Industrial Revolution and Its Antecedents. University of California, Berkeley (2020)
13. Yin, Y., Stecke, K.E., Li, D.: The evolution of production systems from industry 2.0 through industry 4.0. Int. J. Prod. Res. **56**(1–2), 848–861 (2018)
14. McDonough, W., Braungart, M.: The next industrial revolution. In: Sustainable Solutions, pp. 139–150. Routledge, London (2017)
15. Yamada, M.: Foreigner kings as local kingmakers: how the 'unusual' marginalization of conservative political groups occurred in pre-industrial revolution Britain. J. Inst. Econ. **19**(4), 511–525 (2023)
16. Berg, M., Hudson, P.: Slavery, Capitalism and the Industrial Revolution. Wiley, Hoboken (2023)
17. Price, R.: Labour in British Society: An Interpretative History. Taylor & Francis, Milton Park (2024)
18. Deaton, A.: The Great Escape: Health, Wealth, and the Origins of Inequality. Princeton University Press, Princeton (2024)
19. Humphries, J.: Childhood and child labour in the British industrial revolution 1. Econ. Hist. Rev. **66**(2), 395–418 (2013)
20. Humphries, J.: Cliometrics, child labor, and the industrial revolution. Crit. Rev. **13**(3–4), 269–283 (1999)
21. Horrell, S., Humphries, J.: "The exploitation of little children": child labor and the family economy in the industrial revolution. Explor. Econ. Hist. **32**(4), 485–516 (1995)
22. Kirby, P.: Child Workers and industrial Health in Britain, 1780–1850, vol. 2. Boydell & Brewer Ltd., Suffolk (2013)
23. Engels, F.: Condition of the working class in England. In: British Politics and the Environment in the Long Nineteenth Century, pp. 31–38. Routledge (2023)
24. Dimanov, D.K.: Social and human costs of industrialization: a review of some negative effects on society during the FIRST industrial revolution. Knowledge. **62**(1), 123–127 (2024)
25. Clay, K.: Energy and the environment in economic history. In: Handbook of Cliometrics, pp. 2051–2072. Springer, Cham (2024)
26. Irfan, M., et al.: A review on molecular scissoring with CRISPR/Cas9 genome editing technology. Toxicol. Res. **13**(4), 105 (2024)
27. Hoffmann, B., Rud, J.P.: The unequal effects of pollution on labor supply. Econometrica. **92**(4), 1063–1096 (2024)
28. Clay, K., Lewis, J., Severnini, E.: Canary in a coal mine: infant mortality and tradeoffs associated with mid-20th century air pollution. Rev. Econ. Stat. **106**(3), 698–711 (2024)
29. Sharma, A., Singh, B.J.: Evolution of industrial revolutions: a review. Int. J. Innov. Technol. Explor. Eng. **9**(11), 66–73 (2020)
30. Mokyr, J., Strotz, R.H.: The second industrial revolution, 1870-1914. Storia dell'economia Mondiale. **21945**(1), 219–245 (1998)
31. Chin, A., Juhn, C., Thompson, P.: Technical change and the demand for skills during the second industrial revolution: evidence from the merchant marine, 1891–1912. Rev. Econ. Stat. **88**(3), 572–578 (2006)

32. Yang, F., Gu, S.: Industry 4.0, a revolution that requires technology and national strategies. Complex Intell. Syst. **7**, 1311–1325 (2021)
33. Chiles, R.M., et al.: Democratizing ownership and participation in the 4th industrial revolution: challenges and opportunities in cellular agriculture. Agric. Hum. Values. **38**(4), 943–961 (2021)
34. Abdi, M.R., et al.: Evolution of MS paradigms through industrial revolutions. In: Integrated Reconfigurable Manufacturing Systems and Smart Value Chain: Sustainable Infrastructure for the Factory of the Future, pp. 17–42. Springer, Cham (2018)
35. Ashkenazi, D.: How aluminum changed the world: a metallurgical revolution through technological and cultural perspectives. Technol. Forecast. Soc. Chang. **143**, 101–113 (2019)
36. Fitzgerald, F.: Bessemer: the technology and the times. Interdiscip. Sci. Rev. **23**(4), 362–371 (1998)
37. Barraclough, K.C.: The Development of Early Steelmaking Processes: An Essay in the History of Technology. University of Sheffield, Sheffield (1981)
38. Heaton, C., Abrahardt, E.: Dyestuffs. In: The Chemical Industry, pp. 64–125. Springer, Cham (1986)
39. Travis, A.S., Travis, A.S.: The direct synthesis of ammonia. In: Nitrogen Capture: The Growth of an International Industry (1900–1940), pp. 93–127. Springer, Cham (2018)
40. Lesch, J.: The German Chemical Industry in the Twentieth Century, vol. 18. Springer Science & Business Media, Cham (2000)
41. Travis, A.S., et al.: Determinants in the Evolution of the European Chemical Industry, 1900–1939: New Technologies, Political Frameworks, Markets and Companies, vol. 16. Springer Science & Business Media, Cham (1998)
42. Lamoreaux, N.R., Levenstein, M., Sokoloff, K.L.: Financing Invention During the Second Industrial Revolution: Cleveland, Ohio, 1870-1920. National Bureau of Economic Research, Cambridge (2004)
43. Stearns, P.N.: The Industrial Revolution in World History. Routledge, London (2020)
44. Jevons, H.S.: The second industrial revolution. Econ. J. **41**(161), 1–18 (1931)
45. Zhou, K., Liu, T., Zhou, L.: Industry 4.0: Towards future industrial opportunities and challenges. In: 2015 12th International Conference on Fuzzy Systems and Knowledge Discovery (FSKD). IEEE, New York (2015)
46. Lim, T.W.: Industrial Revolution 4.0, Tech Giants, and Digitized Societies. Springer, Cham (2019)
47. Akhtar, I., et al.: The role of natural products in COVID-19. In: The COVID-19 Pandemic, pp. 393–445. Apple Academic Press, New York (2022)
48. Qaiser, H., et al.: COVID-19 pandemic and vaccines. Alternative Medicine Interventions. **20**, 205–235 (2021)
49. Smajdorová, T.: Journey to quality 6.0: evolution of quality management across industrial revolutions. Trends Comput. Sci. Inf. Technol. **9**(2), 063–070 (2024)
50. Mathur, A., Dabas, A., Sharma, N.: Evolution from industry 1.0 to industry 5.0. In: 2022 4th International Conference on Advances in Computing, Communication Control and Networking (ICAC3N), pp. 1390–1394. IEEE, New York (2022)
51. Varshney, A.K., et al.: The development of manufacturing industry revolutions from 1.0 to 5.0. J. Inf. Educ. Res. **4**(1), 710 (2024)
52. Mohajan, H.: Third industrial revolution brings global development (2021)
53. Majeed, H., Iftikhar, T.: Ecofriendly reactive printing of cellulosic fabric with sustainable novel techniques. Cellulose. **31**(11), 7067–7081 (2024)
54. Li, Y., et al.: Increased nitrous oxide emissions from global lakes and reservoirs since the pre-industrial era. Nat. Commun. **15**(1), 942 (2024)
55. Intergovernmental Panel on Climate: Summary for policymakers. In: Climate Change 2021—The Physical Science Basis: Working Group I Contribution to the Sixth Assessment Report of the Intergovernmental Panel on Climate Change, pp. 3–32. Cambridge University Press, Cambridge (2023)

56. McCulloch, M.T., et al.: 300 years of sclerosponge thermometry shows global warming has exceeded 1.5 C. Nature Climate Change. **14**(2), 171–177 (2024)
57. Benz, S.A., et al.: Global groundwater warming due to climate change. Nat. Geosci. **17**, 545–551 (2024)
58. Hills, R.: A one-third scale working model of the Newcomen engine of 1712. Trans. Newcomen Soc. **44**(1), 63–77 (1971)
59. Taylor, D., Taylor, D.: The development of power 1700–1850. In: Mastering Economic and Social History, pp. 114–123. Springer, Cham (1988)
60. Viswanathan, V.V., Davies, R.W., Holbery, J.D.: Opportunity Analysis for Recovering Energy from Industrial Waste Heat and Emissions. Pacific Northwest National Lab. (PNNL), Richland (2006)
61. Badgett, A., Milbrandt, A.: A summary of standards and practices for wet waste streams used in waste-to-energy technologies in the United States. Renew. Sust. Energ. Rev. **117**, 109425 (2020)
62. Love, L.J., Lanke, E., Alles, P.: Estimating the Impact (Energy, Emissions and Economics) of the US Fluid Power Industry. Oak Ridge National Laboratory, Oak Ridge (2012)
63. Carr, M.A.: Thermodynamic analysis of a Newcomen steam engine. Int. J. Hist. Eng. Technol. **83**(2), 187–208 (2013)
64. Fremdling, R.: Industrial revolution and scientific and technological progress. Econ. Hist. Yearbook. **38**(2), 147–168 (1997)
65. Lis, A., Pyc, M.: The influence of the industrial revolution on the social-economic relations in Europe and the polish territories. J. Perspect. Econ. Polit. Soc. Integr. **18**(1-2), 93 (2012)
66. da Silva Almeida, A.R.: Exploring Consumers' Second-Hand Apparel Consumption Intention and Main Influential Factors. Universidade do Porto, Porto (2020)
67. Wang, H., Memon, H.: Cotton science and processing technology. Phys. Struct. Prop. Qual. Cotton. **5**, 79–98 (2020)
68. Abbas, S., et al.: Sustainable production using a resource–energy–water nexus for the Pakistani textile industry. J. Clean. Prod. **271**, 122633 (2020)
69. Ong, E.T., et al.: Industrial revolution (IR) and exemplary AR/VR-based technological tools in preventive health education: The past, present, and future. In: Immersive Virtual and Augmented Reality in Healthcare, pp. 1–27. CRC Press, Boca Raton (2023)
70. Liu, K., et al.: Relationship network of safety management elements in the construction industry under the perspective of resilience. Eng. Constr. Arch. Manage. (2024). https://doi.org/10.1108/ECAM-07-2024-0848
71. Radzuan, S.A.M., et al.: Current practices of internet of things (IoT): the definition and its integration within joint ventures (JV) company in Malaysian construction industry. Malays. Constr. Res. J. 325
72. Saouter, E., Gibon, T.: A world full of energy. In: All You Need to Know About the Next Energy Revolution: Solutions for a Truly Sustainable Future, pp. 1–29. Springer, Cham (2024)
73. Hassan, S.T., et al.: The heterogeneous impact of geopolitical risk and environment-related innovations on greenhouse gas emissions: the role of nuclear and renewable energy in the circular economy. Gondwana Res. **127**, 144–155 (2024)
74. Schulz, V., et al.: A new age for coal ports: an historical perspective on diversification and a circular economy future. Mar. Policy. **170**, 106402 (2024)
75. Sun, Y., et al.: Global supply chains amplify economic costs of future extreme heat risk. Nature. **627**(8005), 797–804 (2024)
76. Hakovirta, M.: Socioeconomic aspects of climate change in cities and municipalities. In: Carbon Neutrality: Follow the Money, pp. 143–156. Springer, Cham (2024)
77. Cai, W., et al.: The 2024 China report of the lancet countdown on health and climate change: launching a new low-carbon, healthy journey. The Lancet Public Health. **9**, e1070–e1088 (2024)

78. Swadi, M., et al.: Investigating and predicting the role of photovoltaic, wind, and hydrogen energies in sustainable global energy evolution. Global Energy Interconnect. **7**(4), 429–445 (2024)
79. Friedman-Heiman, A., Miller, S.A.: The impact of refrigeration on food losses and associated greenhouse gas emissions throughout the supply chain. Environ. Res. Lett. **19**(6), 064038 (2024)
80. Adebayo, T.S., Meo, M.S., Özkan, O.: Scrutinizing the impact of energy transition on GHG emissions in G7 countries via a novel green quality of energy mix index. Renew. Energy. **226**, 120384 (2024)
81. Ibrahim, R.L., et al.: Investigating the asymmetric effects of renewable energy-carbon neutrality nexus: can technological innovation, trade openness, and transport services deliver the target for Germany? Energy Environ. **35**(1), 185–206 (2024)
82. Erokhin, V., Gao, T.: China's pursuit of carbon neutrality: roadblocks and greenlights. In: Emerging Patterns and Behaviors in a Green Resilient Economy, pp. 31–59. Emerald Publishing Limited, Leeds (2024)
83. Alam, M.B., Hossain, M.S.: Investigating the connections between China's economic growth, use of renewable energy, and research and development concerning CO2 emissions: an ARDL bound test approach. Technol. Forecast. Soc. Chang. **201**, 123220 (2024)
84. Siraj, M.T., et al.: Towards sustainable energy transitions: ranking lower-middle-income economies on the accessibility to affordable and clean energy. Environ. Eng. Manage. J. **23**(3), 487–501 (2024)
85. Afshari, H., Gurtu, A., Jaber, M.Y.: Unlocking the potential of solid waste management with circular economy and industry 4.0. Comput. Ind. Eng. **195**, 110457 (2024)
86. Fatimah, Y.A., et al.: The critical success factors for sustainable resource management in circular economy: assessment of urban mining maturity level. J. Clean. Prod. **469**, 143084 (2024)
87. Allam, Z., Cheshmehzangi, A.: Technological innovation and sustainable transitions. In: Sustainable Futures and Green New Deals, pp. 55–79. Springer, Cham (2024)
88. Hassan, Q., et al.: The renewable energy role in the global energy transformations. Renew. Energy Focus. **48**, 100545 (2024)
89. Ewim, D.R., Olatubosun, S.A., Abolarin, S.M.: Energy alternatives and efficiency options for sustainable development. In: Localized Energy Transition in the 4th Industrial Revolution, p. 225. CRC Press, Boca Raton (2024)
90. Koponen, K., et al.: Responsible carbon dioxide removals and the EU's 2040 climate target. Environ. Res. Lett. **19**(9), 091006 (2024)
91. Béres, R., et al.: Will hydrogen and synthetic fuels energize our future? Their role in Europe's climate-neutral energy system and power system dynamics. Appl. Energy. **375**, 124053 (2024)
92. Allegri, E., et al.: Pluvial flood risk assessment for 2021–2050 under climate change scenarios in the Metropolitan City of Venice. Sci. Total Environ. **914**, 169925 (2024)
93. Ohenhen, L.O., et al.: Disappearing cities on US coasts. Nature. **627**(8002), 108–115 (2024)
94. Zhao, F., et al.: Global at-risk basins of socio-ecological impacts from water scarcity. Sci. Total Environ. **15**, 176944 (2024)
95. Chen, Y., et al.: Energy-water-carbon system management in response to climate change in China under uncertainties (2026–2050). Resour. Conserv. Recycl. **212**, 107957 (2025)
96. Ershadfath, F., et al.: Blue and green water availability under climate change in arid and semi-arid regions. Eco. Inform. **82**, 102743 (2024)
97. Bernabé-Crespo, M.B., Loáiciga, H.: Managing potable water in southeastern Spain, los Angeles, and Sydney: transcontinental approaches to overcome water scarcity. Water Resour. Manag. **38**(4), 1299–1313 (2024)
98. Jain, S., et al.: Global-scale water security and desertification management amidst climate change. Environ. Sci. Pollut. Res. **31**(49), 58720–58744 (2024)
99. Phoumin, H., Kimura, S., Nepal, R.: Carbon neutrality pathways in East Asia: a consideration of CCUS in the power generation sector. In: Energy Transition and Carbon Neutrality in

ASEAN: Developing Carbon Capture, Utilization and Storage Technologies, pp. 3–44. World Scientific, Singapore (2025)
100. Senyonyi, B., et al.: New approach for zero energy industries via integrating energy management and renewable energy system: enviro-economic investigation. Energ. Build. **319**, 114573 (2024)
101. Nabera, A., et al.: Integrating climate policies in the sustainability analysis of green chemicals. Green Chem. **26**(11), 6461–6469 (2024)
102. Kloo, Y., Nilsson, L.J., Palm, E.: Reaching net-zero in the chemical industry—a study of roadmaps for industrial decarbonisation. Renew. Sustain. Energy Trans. **5**, 100075 (2024)
103. Leal Filho, W., et al.: Integrating climate change practices in a circular economy context—the perspective from chemical enterprises. Sustain. Dev. **32**(3), 2489–2505 (2024)
104. Mani, A., Budd, T., Maine, E.: Emissions-intensive and trade-exposed industries: technological innovation and climate policy solutions to achieve net-zero emissions by 2050. RSC Sustain. **2**(4), 903–927 (2024)
105. Küfeoğlu, S.: Industrial process emissions. In: Net Zero: Decarbonizing the Global Economies, pp. 341–414. Springer, Cham (2024)
106. Olaleye, D.S., et al.: Advancing green communications: the role of radio frequency engineering in sustainable infrastructure design. Int. J. Latest Technol. Eng. Manage. Appl. Sci. **13**(5), 113 (2024)
107. Wang, Y., et al.: Plant-level green transformation strategy in China's cement industry: considering energy conservation and emission reduction co-benefits. J. Clean. Prod. **467**, 142945 (2024)
108. Wright, C., Nyberg, D.: Corporations and climate change: an overview. WIREs Clim. Change. **15**, e919 (2024)
109. Wolfram, P., et al.: The hydrogen economy can reduce costs of climate change mitigation by up to 22%. One Earth. **7**(5), 885–895 (2024)
110. Lipper, L., Cavatassi, R.: The challenge climate change poses to achieving resilient and inclusive rural transformation (RITI). Glob. Food Sec. **43**, 100811 (2024)
111. Wiese, F., et al.: The key role of sufficiency for low demand-based carbon neutrality and energy security across Europe. Nat. Commun. **15**(1), 9043 (2024)
112. Federicia, J.A., Papageorgioua, D.J., Nielsena, R.D.: Connecting the dots: Push and pull between technology R&D and energy transition modeling
113. Parhamfar, M., Sadeghkhani, I., Adeli, A.M.: Towards the net zero carbon future: a review of blockchain-enabled peer-to-peer carbon trading. Energy Sci. Eng. **12**(3), 1242–1264 (2024)
114. Castle, J.L., Hendry, D.F.: Five sensitive intervention points to achieve climate neutrality by 2050, illustrated by the UK. Renew. Energy. **226**, 120445 (2024)
115. Bandh, S.A., et al.: Track to reach net-zero: Progress and pitfalls. Energy Environ. **36**(5), 2345–2370 (2024)

Chapter 3
Industrial Revolution 4.0, 5.0 Sustainable Transformation into Industrial Revolution 6.0

Abstract Industrial Revolution four is the term used to describe how digital technologies are being incorporated into manufacturing to change conventional production processes into highly networked, self-sufficient settings. Advances in cloud computing, the Industrial Internet of Things (IIoT), AI, big data analytics, CPS are driving this industrial transformation by facilitating intelligent manufacturing, predictive automatic maintenance with real-time data monitoring for auto decision-making. AI-driven predictive models increase efficiency and enables companies to model precise operations, with improved workflows, and minimize downtime. Autonomous supply chain optimization is made possible by the rise of smart factories, where IIoT enabled devices communicate with one another in a seamless manner, guaranteeing low waste and resource efficiency. Sustainability is a primary focus of Industry 6.0 using Industry 4.0 technologies, allowing for reductions in greenhouse gas emissions through circular economy tactics and energy-efficient automation. By reducing risks and enhancing traceability, blockchain integration improves security and transparency, especially in supply chain management. In industrial applications, the combination of cloud computing and edge computing allows for ultra-fast processing, resulting in a 95% reduction in latency. Industry 4.0 reduces material waste by about 50% by advancing smart grids, AI-driven energy management, and additive manufacturing. The use of AR and VR transforms workforce training, lowering human error by 30–40% and boosting efficiency. The chapter also looks at the economic effects of Industry 4.0 and the role of AI-powered autonomous robotics. These developments provide the long-term framework for the shift to Industry 6.0, where the future of industrial operations will be determined by human-centered and ecologically responsible production techniques.

Keywords Industry 4.0 · Industry 5.0 · Industry 6.0 · Cyber physical system · Industrial IoT · Augmented and virtual reality · Artificial intelligence · Sustainability · Cloud computing · Predictive maintenance · Latency · Energy management

H. Majeed, T. Iftikhar, *Intelligent Manufacturing in Industry 6.0*,
https://doi.org/10.1007/978-3-032-07278-8_3

1 Industrial Revolution 4.0 Introduction

"Industry 4.0" is the phrase used to describe the integration of advanced digital technology into industrial and manufacturing processes, transforming traditional production techniques into self-governing, networked systems. Innovations like robotics, big data analytics, IoT, AI, and CPS are driving a paradigm shift toward intelligent manufacturing with Industry 4.0. The broad concept of Industry 4.0 serves as a catch all for a variety of technologies and applications designed to achieve operational enhancements and performance targets. Industry 4.0 technologies encompass front-end technologies like smart production, smart goods, smart supply chains, and smart workspaces. These technologies enhance operational intelligence and responsiveness right away [1–3]. The second category, Base Technologies, consists of the essential enablers that make it easier for front-end technologies like cloud computing, big data, analytics, and the IoT to be deployed and scaled. Even though the terms "cloud manufacturing," "smart manufacturing," and "intelligent manufacturing" are occasionally used interchangeably to describe these advancements, they share essential traits. Key enabling technologies include advanced networking, data processing, and physical digital interfaces, which enable more digitization and communication. For seamless data flow and system integration, cloud computing and the IoTs are compulsory. Industry 4.0 technology is distinguished by features including virtualization, real-time information interchange, process autonomy, enhanced interoperability, and robust cybersecurity frameworks. These characteristics promote greater integration within and across organizational boundaries [4–6].

To meet contemporary industrial and environmental concerns, this revolution places a high priority on automation, real-time data-driven decision-making, and improved operational efficiency. It also aligns with sustainability goals. Industry 4.0 makes it possible to employ smart factories in intelligent production, where sensors, devices, and machinery all interact with one another through IoT frameworks. Large volumes of data are gathered, processed, and analyzed in real time by these networked systems, which also reduce waste and improve quality by optimizing processes through predictive maintenance [7, 8]. AI-driven defect identification in assembly lines can minimize material losses and guarantee product uniformity while reducing quality issues. The core forces behind Industry 4.0 are the convergence of information systems, AI, and sophisticated computer technologies, which results in a hyperconnected industrial environment. By integrating intelligence into processes, this integration transforms production by facilitating autonomous decision-making, real-time optimization, and resistance to disturbances [9, 10]. Figure 3.1 is showing Industrial IoT and Green IoT.

Industry 4.0 places a strong emphasis on data security, scalability, and interoperability. Open Platform Communications Unified Architecture (OPC UA) and other standards facilitate smooth device to device communication, while developments in blockchain and cybersecurity reduce the risks that come with greater connection. The smooth integration of diverse systems using standards, such as OPC UA is

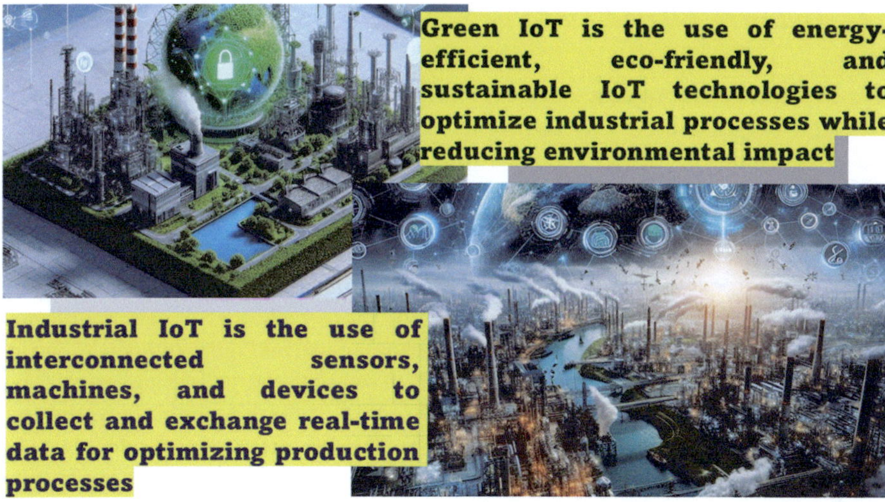

Fig. 3.1 Industrial IoT and green IoT basics

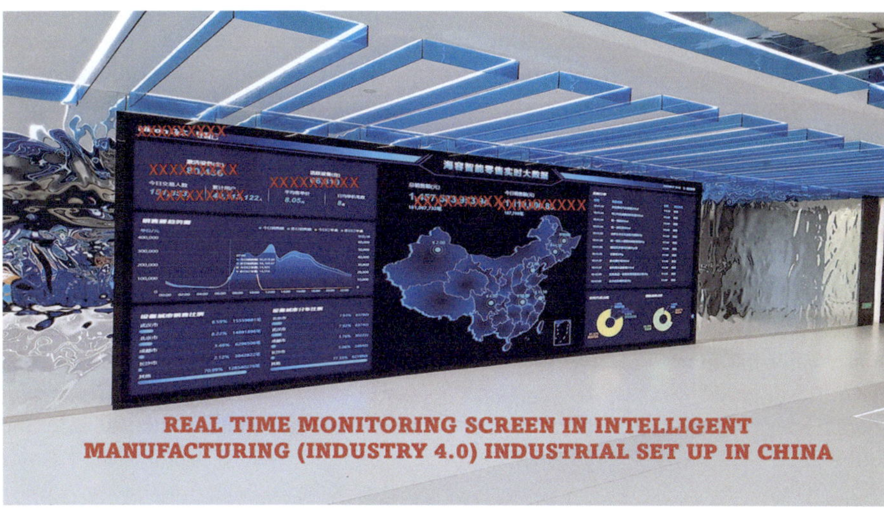

Fig. 3.2 Real-time monitoring screen in industry

essential to Industry 4.0. guarantees compatibility between systems and devices. Distributed industrial process control is defined in IEC 61499. Modern Industry 4.0 technologies relate to existing systems using sophisticated middleware and APIs, guaranteeing seamless migrations [11–13]. Figure 3.2 is showing the real-time monitoring full screen in an electronics industrial manufacturing facility in P.R. China. Author (Dr Hammad Majeed) himself made this picture during his visit in August 2023.

1.1 Industry 4.0 and Sustainability

This revolution incorporates sustainability in addition to industrial efficiency. Although manufacturers can contribute to net zero targets by reducing greenhouse gas emissions by up to 20% through closed loop technologies and intelligent energy management, industry 4.0 is not primarily focused on reducing GHG emissions. The chapters on industry 5.0 and 6.0 will go into greater detail on the circular economy, emission reduction, and customer oriented strategy. Industry 4.0 holds the possibility of both tackling the climate imperative and changing industry. In addition to spurring innovation, these technical developments are helping achieve the SDGs by maximizing resource use and reducing waste, building resilient industrial infrastructures, and by lowering energy use and emissions. As a result of the growing influence and adoption of these technologies, the fact that at least 30% of the tasks in about 60% of all professions on the market can be fully automated demonstrates how intelligent firms can boost their production capacity through the efficient application of intelligent production technologies. The manufacturing industry was particularly affected by the COVID-19 epidemic, which compelled manufacturers to adopt intelligent production-based solutions [14–17].

As of 2021, the global smart manufacturing market was valued at approximately USD 254.7 billion, using Industrial IoT hardware, AI software, robotics, and cloud platforms, slightly higher than the previously cited USD 249.5 billion. Growth projections now estimate this market will expand to around USD 443.9 billion by 2026 and potentially reach USD 673.4 billion by 2029, growing at a robust 14.9% CAGR between 2022 and 2030. These trends support a revised 2028 projection of USD 659 billion, driven by rapid advancements in Generative AI, digital twins, and 5G/edge technologies. Specifically, industrial AI was forecasted to reach USD 89 billion by 2025, digital twins USD 48 billion, and 5G/edge USD 34 billion, reflecting modern acceleration in sub industrial sectors. The market is now growing faster than earlier estimates, reinforcing the adjusted 14.2% CAGR for the 2021–2028 period, compared to older forecasts of approximately 12.7% [18–22]. Important characteristics and core technological foundations of Industry 4.0 include the following.

1.2 Industry 4.0 Technological Components

1.2.1 Cyber Physical Systems (CPS)

As the cornerstone of Industry 4.0, CPS uses embedded sensors, actuators, and IoTs devices to link computational systems with actual machinery. CPS is defined as a tight integration and coordination between a system's computational and physical elements, with feedback loops where physical processes affect computations and vice versa. These technologies operate together to achieve real-time data interchange and control over high-speed networks, such as 5G. A production line with CPS enabled can dynamically monitor and adjust conveyor speeds in response to machine learning algorithms' forecasts of bottlenecks [23, 24].

1.2.2 Industrial Internet of Things (IIoT)

By linking industrial machinery, systems, and equipment to the cloud for data gathering and analysis, the IIoT broadens the scope of the IoT framework. IoT makes it easier for devices to communicate with one another, which makes automated data exchange and real-time system improvement possible. Predictive maintenance results, which can cut downtime. Important protocols that guarantee lightweight device-to-device communication include MQTT (Message Queuing Telemetry Transport) and CoAP (Constrained Application Protocol). Siemens MindSphere is an IIoT platform that optimizes production efficiency and energy usage by processing gigabytes of machine data every day [25–27].

1.2.3 Machine Learning, AI, and Deep Learning

ML, AI, and DL are essential for turning unprocessed data into insights that can be used. Adaptive scheduling, process optimization, anomaly detection, and other tasks are handled by ML models. A branch of AI called DL is used for sophisticated tasks like visual inspection in quality control. Neural networks, for instance, are used in semiconductor manufacturing to detect minute flaws, boosting yield by 5–10%. While DL helps with more complicated decision-making, such waste management and water resource optimization, AI improves automation through predictive analytics [28–30].

1.2.4 Reinforcement Learning

Reinforcement learning is used by AI-based decision support systems to optimize production scheduling, leading to efficiency benefits [31].

1.2.5 Digital Twin Technology

Real-time virtual copies of physical systems are called digital twins. To allow predictive modelling, they rely on AI analytics, simulation algorithms, monitoring, and optimization are made possible by real-time virtual representations (through sensors data) of actual assets. Digital twins can be used in a variety of industries, such as wind turbine production, where they can optimize durability and reduce prototyping costs by simulating stress reactions to materials. Digital twins from General Electric have increased their turbine efficiency [32, 33].

1.2.6 Autonomous Robotics

When paired with AI algorithms, collaborative robots (cobots) can adjust to changing production needs, resulting in an increase in productivity above traditional automation [34].

1.2.7 Additive Manufacturing

Compared to subtractive techniques, additive manufacturing (3D printing) reduces significant amount of waste and enables customized and on demand production [35].

1.2.8 Big Data Analytics and Cloud Computing

Scalable storage and quick processing of the massive volumes of data produced by networked systems are guaranteed using big data analytics and cloud platforms. Both structured and unstructured data are produced in enormous quantities by Industry 4.0. This data is processed and analyzed by big data frameworks (like Apache Kafka, Flink and Spark) to reveal insights. AI and statistical modelling enable predictive and prescriptive analytics, which aid in decision making like a car plant may cut inventory expenses by using big data analytics to forecast components demand [36, 37].

1.2.9 Edge Computing and Cloud Computing

Edge computing processes data closer to the source, ensuring ultra-low latency, whereas cloud computing offers scalable data processing and storage capabilities. This hybrid method strikes a compromise between processing power and speed. Edge computing reduces decision-making latency by processing data locally at the machine level, as seen in sectors like auto-assembly. For important tasks like robotic assembly line coordination, edge computing latency was down to less than 10 ms. For instance, AWS Greengrass from Amazon interfaces with industrial systems to enable local AI inference on edge devices and offer cloud access [38, 39]. Figure 3.3 and 3.4 showing the basic description of cloud computing and edge computing.

1.2.10 Blockchain for Supply Chain Integrity

In Industry 4.0, blockchain guarantees security, traceability, and transparency. Distributed ledgers monitor parts across supply chains, reducing the possibility of counterfeiting and guaranteeing adherence. Blockchain strengthens confidence in automation and supply chain procedures and is a key enabler of safe, open, and

Fig. 3.3 Description of cloud computing

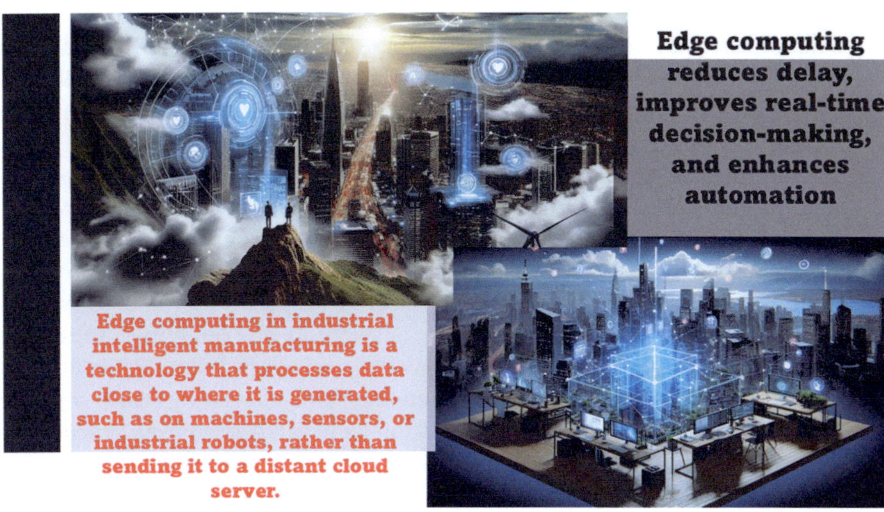

Fig. 3.4 Edge computing description

effective operations. It enables companies of all sizes to handle privacy, data security, and transparency issues while preserving decision-making integrity. Standard cryptographic hashing and trusted time servers are more practical. Blockchain is useful for multi-party provenance and non-repudiation and is also useful for logistics management and timestamping sensor data. IBM Food Trust uses blockchain technology to track food goods back to their source in a matter of seconds, improving traceability and lowering recall rates [40, 41].

1.2.11 Virtual Reality (VR) and Augmented Reality (AR)

VR and AR allow for remote equipment maintenance and offer workers immersive training. AR allows for on the spot troubleshooting by superimposing real-time data on equipment. Boeing, for instance, uses AR to reduce wiring errors by 90% during assembly [42].

2 Industry 4.0's Effect on Industries

This period (4.0) is characterized using automation, data exchange, and intelligent systems to create intelligent businesses that can adapt to the demands of a society that is technologically sophisticated and changing quickly. Smart factory systems, cloud and cognitive computing, and the IoTs are all significant contributors to the revolution in traditional manufacturing methods. These tools boost productivity, enhance real-time monitoring, and encourage data-driven decision making. Through the use of digitization, modern analytics, and data collecting, Industry 4.0 creates a paradigm change that enables businesses to ensure personalization, enhance operations, and achieve sustainable growth [43, 44].

As Industry 4.0 develops further, it incorporates cutting-edge ideas like quantum computing to solve optimization issues at previously unheard of sizes and federated learning for decentralized AI. AI and machine learning aid in decision-making, production process optimization, and increased manufacturing productivity. This is best demonstrated by smart factories, which use real-time analytics and automation. Through the applications in irrigation, wastewater recycling, and rainfall harvesting, AI and DL approaches address important issues like water scarcity and ensure sustainable resource use. Mass customization is made possible by Industry 4.0, which gives producers the flexibility to meet a range of customer demands. Technological developments offer creative answers to problems in a variety of fields, from manufacturing and logistics to healthcare and agriculture. In addition to revolutionizing industries, its emphasis on intelligent manufacturing is also supporting international objectives for decarbonization and sustainable development. The challenges that Industry 4.0 faces can also be addressed in Industry 5.0 and 6.0. Cyberattacks that target networked systems necessitate the use of strong cybersecurity solutions like AI-driven threat detection and quantum encryption. In a similar vein, trust and acceptance of AI models continue to depend on their interpretability and compliance with industry norms [45]. Climate change is the greatest threat to civilization, and methods to quantify the carbon footprint of Industry 4.0 technologies and maximize their environmental impact are being developed. These sustainable strategies are key components of Industry 6.0, the focus of this book. The effects of Industry 4.0 on the economy are significant. The global smart manufacturing market reached USD 312 billion in 2022 increasing earlier estimates of USD 297 billion and is projected to grow to USD 882 billion by 2030 at a 15.8% CAGR. This upward revision from prior forecasts (USD 567 billion) stems from accelerated adoption of generative AI (contributing 28% of growth) and edge computing infrastructure. Energy efficiency gains from Industry 4.0 technologies exhibit significant sectoral variance like semiconductor fabrication achieves 22% average reduction through digital twin optimization, while legacy manufacturing lags at 8–12%. At scale, these innovations could save 0.8–1.2 Gt CO_2e annually by 2030, directly supporting Paris Agreement targets. Predictive maintenance powered by IIoT and AI reduces operational costs by 18–35% and increases equipment uptime by 12–25% by slashing mean repair duration by 45%. Real-time monitoring further enhances productivity through 60–92% faster decision latency in high velocity applications like automotive quality control. However, edge computing deployments incur a 3–5% net energy penalty due to distributed compute node demands a critical trade off requiring lifecycle optimization [22, 46–48].

3 Merits of Industry 4.0

The technologies that directly benefit the industries to promote the advantages of the fourth industrial revolution are explained in below mentioned points. Digital twins that offer advantages in Real-time monitoring and predictive analytics for machine performance and energy usage are made possible by virtual copies of actual assets. Using machine learning models, predictive maintenance can reduce downtime and increase energy efficiency by up to 25% by anticipating failures before they happen. Blockchain integration guarantees confidentiality and transparency in data interchange and remote monitoring, which is essential for international businesses. In a similar vein, 5G connectivity improves communication between IoT devices, allowing for extremely dependable and low latency data transfer for flexible production and intelligent lots. By combining these technologies, Industry 4.0 positions companies for long term success in a highly dynamic environment while also addressing sustainability and global connection issues and improving operational efficiency and adaptability. These sustainable possible outcomes in industry 4.0 are the real foundations of industry 5.0 and 6.0. Here are some advantages and merits of Industry 4.0 that we will discuss briefly. More details will be provided in the subsequent chapters, where dedicated sections will explain all the key technological benefits of Industry 4.0 in depth [49, 50].

3.1 Effective Energy Use

Using cutting-edge technology like Automated Advanced Building Management Systems, Industry 4.0 improves energy efficiency. These systems monitor and optimize energy use across machinery, lighting, HVAC systems, and other building processes by utilizing sensor networks, including IoT-enabled devices. Forecast-based energy management is made possible by AI-driven predictive models, which minimize waste and lower operating expenses. Smart energy networks and fire alarm systems, for instance, can immediately modify power distribution in an emergency, guaranteeing sustainability and safety. By combining cutting-edge technologies like AI, the IoT, data analytics, automation, and sophisticated communication networks, the Fourth Industrial Revolution signifies a revolutionary change in manufacturing, engineering, and energy management. The ability of Industry 4.0 to improve energy efficiency, which is essential to sustainable industrial operations, is one of its main benefits [51, 52].

3.1.1 Key Technologies Usage in Energy Efficient Systems

Many technologies in smart housing, industrial smooth operations used energy efficient system and energy conservation system. Energy is saved with the help of sensors and actuators with real-time monitoring. Sensors judge that the specific operation is not using any energy so auto shut system well operates. There are numerous energy managemental systems which are helpful. In coming chapters it will be explained in details [53, 54].

3.1.2 AI-Driven Forecasting Models and Smart Grids Technologies

Predictive analytics driven by AI can also cut energy waste in buildings and industrial settings by 10–20%, which will save a lot of money and cut carbon emissions. Practically speaking, AI enabled smart grids can also increase energy security. For instance, smart grids can automatically reroute electricity in the case of a power outage or other system disturbance, preserving vital systems like emergency lighting, security systems, and fire alarms while switching off power to non essential loads. This capacity to rank critical services in order of importance contributes to emergency safety and strengthens the energy network's resilience. All things considered, smart grids and AI-driven forecasting models work together to increase the sustainability and efficiency of energy systems, promote greater integration of renewable energy, lessen dependency on fossil fuels, and increase the resilience and dependability of power grids around the world. This directly supports the SDGs 7 (Clean and Affordable Energy), 9 (Industry, Innovation, and Infrastructure), and 13 (Climate Action). These sustainable approaches are the true assets of Industry 6.0, which is the theme of this book [55–57].

3.2 Increased Production Using Intelligent Lots

Manufacturing is undergoing a transformation thanks to the idea of intelligent lots, which is supported by cutting-edge technology like AI, the IoT, and sophisticated automation systems. By providing exact control over inventory management, predictive maintenance, and quality assurance, these technologies support enhanced automation. Using real-time insights and decision-making, AI powered production lines in the automobile industry, have enhanced efficiency by 20–30% [58–60].

3.2.1 The Function of Communication Technologies, Sensors, and RFID

Because RFID technology allows precise, real-time tracking of materials, components, and products as they pass through the production system, it is essential to intelligent lots. Important details about product specifications, batch numbers, and production status are stored in RFID tags that are affixed to products during the manufacturing process. RFID readers positioned across the plant floor wirelessly read these tags, enabling smooth system upgrades devoid of human involvement. Items are automatically tracked in or out as they move through various production steps, leading to effective inventory management. Through the provision of real-time data on temperature, humidity, pressure, equipment condition, and other crucial aspects of the production environment, IoT sensors further improve the operation of intelligent lots. IoT devices like pressure sensors in hydraulic machines or temperature sensors on production lines, continuously check the condition of industrial equipment to guarantee peak performance and lower the chance of downtime from unforeseen breakdowns. Technologies like Wi-Fi, WAN, and Bluetooth guarantee smooth connection between supply chain logistics, inventory management systems, and machinery used in manufacturing. By sending data gathered from IoTs sensors to centralized cloud based systems for analysis, these communication solutions enable operators to remotely monitor and manage processes, facilitating quicker decision-making and system improvement [61–63].

3.2.2 Production Quality and Quantity Optimization by Using Industrial Automation

The integration of AI into production processes facilitates the analysis of the huge amounts of data generated by these IoT-based smart sensors. ML-based analysis (historical and real-time data both), predict equipment failure immediately, and make sudden decisions that improve the overall output. AI technologies can adjust production schedules, optimize machine settings, and change production speeds in response to data inputs from several sensors can create a more responsive and flexible production environment. AI systems can predict when a machine required to down/shut or pause, enabling predictive maintenance to be carried out before a breakdown occurs. Because of this feature, the equipment lasts longer (without any major accident) and requires less costly and time-consuming maintenance as it avoid regular periodic machine maintenance. AI systems in manufacturing lines improve product quality by reducing the rate of errors and waste materials by continuously monitoring outputs and adjusting machine parameters to meet exact high-quality standards. The car sector is now far more efficient thanks to AI powered production lines. Real-time insights from AI systems have been shown to increase efficiency by 20–30% in some cases. This is accomplished by enhanced demand forecasting, more precise control over production schedules, and adaptable manufacturing methods that can respond immediately to changing conditions. With these developments, manufacturers may reduce waste and operating costs while better meeting consumer demands [64–67].

3.3 Connection with SDGs and Industrial Applications

Intelligent lots driven by AI and IoT support important SDGs especially 9, 12, and 13. These technologies minimize waste while improving automation, production efficiency, and resource optimization. Due to widespread adoption, the worldwide smart manufacturing market, which is being driven by AI and IoT, is expected to develop at a 14.7% CAGR from 2023 to 2030. By enhancing inventory management, reducing downtime, enabling predictive maintenance, and reducing energy consumption by 15–25%, intelligent lots promote sustainable production. Automotive plants' CO_2 emissions have decreased by 10–15% a year because to AI-driven production optimization. Integrating AI and IoT could cut industrial carbon emissions worldwide by 5.4 GT annually by 2030. To achieve energy efficient, climate-conscious, and sustainable industrial processes, these advances are essential. These developments can help companies transition to more ecologically friendly, energy efficient and sustainable production methods, which will ultimately aid in global sustainability initiatives. These sustainable approaches are the real assets of industry 6.0 which are the core themes of this book [68–71].

4 Real World Applications with Automation Details

As a component of larger Industry 4.0 projects, intelligent lots use machine learning, autonomous machinery, and sophisticated robotics to further automate industrial operations. AI-enabled robots in the automotive sector can quickly and precisely complete repetitive jobs like welding, painting, and assembling. By operating within ideal parameters and adjusting their behaviors in response to real-time data from IoT sensors, these robots can improve product uniformity and save waste. Materials can be efficiently moved from one area of the plant to another by autonomous vehicles in production facilities. Highly synchronized activities are made possible by these vehicles' communication with other equipment and devices in the production ecosystem, which is facilitated by IoT sensors and AI systems [72, 73].

AI and IoT make it possible to automate quality assurance in the electronics manufacturing industry, where real-time product monitoring throughout production enables the prompt identification of flaws. By doing this, less manual inspection is required, saving time and money while guaranteeing that the items fulfill high standards of quality. We can conclude that by increasing automation, strengthening inventory management, and optimizing machine performance, the combination of AI and IoT in intelligent lots is transforming production operations. By lowering waste, energy use, and carbon emissions, these technologies' real-time data and predictive capabilities not only increase operational efficiency but also support sustainability objectives [74, 75]. We are seeing noticeable gains in manufacturing efficiency, product quality, and environmental responsibility because of industries embracing these technologies more and more; this is helping to advance the SDGs and laying the groundwork for a more sustainable and effective industrial future. These sustainable practices are essential assets of Industry 6.0, the primary theme explored in this book.

5 Real-Time Optimization for Business Applications

By using cloud computing and machine learning to increase efficiency, real-time optimization is revolutionizing sectors, including telecommunications and logistics. Large datasets are instantaneously analyzed by machine learning algorithms, which dynamically modify operations to optimize throughput and reduce delays. In the cutthroat market of today, where quick reactions to changes in demand and supply chain interruptions are critical, this agility is vital. By combining edge computing and cloud-based solutions, latency is further decreased, enabling quicker and more accurate decision-making and, eventually, more intelligent company processes [76, 77]. In the telecom sector, real-time optimization can ensure that network traffic is efficiently distributed across available capacity, lowering latency, and boosting data throughput. In this way, logistics machine learning algorithms can adjust delivery schedules, optimize delivery routes, and dynamically manage inventory while accounting for unanticipated supply chain disruptions, weather, and traffic [78, 79].

The integration of cloud-based solutions with edge computing is a key breakthrough that supports real-time optimization. While edge computing enables data processing at the point of origin, near the location where the data are generated, cloud computing offers the enormous computational capacity required to process massive datasets. By reducing the amount of data that must be transferred back and forth to centralized servers, this integration lowers latency and guarantees quicker and more accurate decision-making. Even in remote or bandwidth-constrained environments, edge computing helps businesses make faster, better decisions by putting computational capabilities closer to the data source [80, 81].

Real-time optimization is important for achieving SDG 9 from a sustainability and UN SDGs standpoint. Better management and more effective industrial operations result from the ongoing enhancement of infrastructure through real-time resource optimization. This supports the objective of encouraging innovation in industry and infrastructure, enabling businesses to satisfy the rising expectations for operational sustainability and efficiency. SDG 12 is closely related to real-time optimization. Real-time optimization helps companies adopt more sustainable production methods by allowing them to dynamically optimize resources, cut waste, and prevent overproduction. In logistics, dynamically modifying delivery routes and schedules can lower carbon emissions and fuel usage, promoting sustainability objectives while enhancing service quality. Plus, real-time optimization in the logistics and telecommunications industries supports SDG 13. In the telecommunications industry, for example, optimizing network resources can save energy by preventing overloaded or inefficient systems. AI-powered real time decision-making in logistics can minimize the overall carbon footprint of transportation operations while optimizing fuel consumption. It has been demonstrated that using machine learning to demand forecasting and route planning in the logistics industry can cut fuel usage by as much as 15–20%, which directly supports climate action [82, 83].

The growing influence of real-time optimization in several sectors. The global market for machine learning in business operations will expand at a CAGR of 25.3%, from USD 20.1 billion in 2023 to USD 102.4 billion by 2030. This quick expansion is a sign of the growing use of real-time optimization technology, which help companies remain sustainable, competitive, and responsive. To sum up, real-time optimization driven by edge computing, cloud computing, and ML algorithms is an essential innovation for enhancing responsiveness and operational efficiency across commercial applications. By assisting companies in managing interruptions, maximizing resources, and cutting down on inefficiencies, these technologies promote sustainability and profitability [84, 85]. The integration of these technologies contributes to a more resilient and sustainable global economy by supporting several important SDGs like 9, 12, 13. The sustainable approaches highlighted in this book are integral to Industry 6.0, the central theme of this book.

6 Flexibility of Machines in Industrial use with Advanced Automation Sustainable Approaches

Advances in IoT technologies and adaptive machine learning greatly increase the flexibility of industrial systems and cause a paradigm shift in the flexibility of production processes. Flexible manufacturing systems (FMS) may now dynamically modify production procedures, assembly lines, and machine tools to satisfy changing product demands with little downtime thanks to technology advancements. To meet changing product demands, FMS enable the quick reconfiguration of production procedures and machine tools. The ability to modify manufacturing settings in response to real-time data inputs is at the heart of FMS. Sensors and actuators are examples of IoT enabled devices that gather and send data on machine performance, product specifications, and workflow needs [86–88]. Cobots, or collaborative robots, and autonomous robots engage with both humans and other machines, adapting their work to the conditions of the moment. Adaptive robotics is used in BMW's smart factories to autonomously control assembly line configurations according to production schedules, resulting in an annual 5% savings in operating expenses. By simulating machine behavior and performance under various conditions, digital twins help manufacturers optimize processes prior to implementation, which lowers trial and error expenses and machine downtime [89–92].

The adaptability of manufacturing systems not only promotes operational efficiency but also supports sustainability goals. Data driven optimization and real-time monitoring cut down on material waste and energy use, supporting SDG 13. IoT enabled production systems, for instance, can reduce energy consumption by 10–15%, and automated waste reduction strategies guarantee a low environmental impact. Through the development of robust and technologically sophisticated manufacturing ecosystems, these initiatives also support SDG 9. FMS has equally important economic ramifications. Due to the growing need for agility and customization in sectors, including consumer electronics, automotive and aerospace, the global market for flexible manufacturing systems is anticipated to expand at a CAGR of 8.5%, reaching over USD 13 billion by 2030. This expansion highlights how important FMS will be in determining the direction of manufacturing in the future, where sustainability and adaptability are mandatory. In a nutshell, an innovative development in industrial automation is the adaptability of manufacturing systems driven by IoT and adaptive ML [93–95]. These technologies facilitate mass customization, allow for dynamic reconfigurations and lessen environmental effects, all of which improve efficiency and meet international sustainability objectives. These technologies open the door to a more intelligent, environmentally friendly, and flexible production environment as enterprises continue to implement them in industrial revolution 6.0.

7 Remote Monitoring and Control

Using gadgets like PCs, laptops, tablets, and smartphones to offer centralized data access and seamless communication, remote monitoring has emerged as a key component of global technology integration. Secure VPNs (Virtual Private Networks) and cloud computing enable businesses to oversee and manage activities from any location in the world. This feature guarantees business continuity, especially in sectors like healthcare where remote monitoring and real-time diagnosis are essential. Dashboards with AI enhancements compile and display data from many sources, facilitating proactive resource management and well informed decision-making [96].

7.1 Technological Usage in Remote Monitoring and Control Systems

Cloud computing, which offers scalable storage and processing capacity for handling enormous volumes of data produced by IoT enabled devices, is at the core of remote monitoring. Streams of real-time data are gathered from sensors built into infrastructure, industrial, or patient monitoring equipment, sent safely over VPNs, and processed on cloud servers. These technologies guarantee smooth worldwide access while preserving the integrity and confidentiality of data. IoT sensors installed in production lines in industrial settings keep an eye on things like output rates, vibration, and temperature. Up to 40% less downtime is possible thanks to the analysis of this data, which is supplied to centralized dashboards with AI enhancements. The foundation of remote monitoring systems is automation, which enables businesses to function with little assistance from humans. Using user friendly dashboards, AI-driven algorithms compile and evaluate data from many sources to produce actionable insights. To reduce water wastage by 25–30% automated irrigation systems in agriculture use remote monitoring to modify water levels based on soil moisture readings. This approach directly supports SDG 6 [97, 98].

7.2 Industrial Applications

By facilitating real-time patient monitoring and diagnostics, remote monitoring has revolutionized the delivery of healthcare. Healthcare providers can remotely monitor patient health thanks to wearable technology like glucose sensors and heart rate monitors that send data to cloud platforms. Early interventions can be initiated when telemedicine systems with AI-driven analytics identify abnormalities in vital signs. Research indicates that remote monitoring has enhanced the management of chronic diseases and decreased hospital readmission rates by 18% [99, 100].

Remote monitoring systems monitor the operation of smart grids, wind turbines, and solar panels in the renewable energy sector. To optimize operations and reduce energy waste by 10–15%, AI models examine patterns in energy generation and consumption, which helps achieve SDG 13. By increasing turbine efficiency, Siemens' remote monitoring solutions for wind farms have reduced maintenance and operating expenses by USD 1.2 billion yearly [101, 102].

Automation in Manufacturing and Industry is another application. IoT and AI are combined in industrial remote monitoring to monitor equipment health, inventory levels, and production processes. AI powered predictive maintenance increases equipment lifespan and decreases downtime, advancing SDG 9. General Electric's Predix platform makes it possible to remotely monitor plant processes, which lowers downtime by 20% and boosts productivity by 15% [103].

7.3 Alignment with SDGs

With the integration of state of the art innovations to successfully handle global concerns, remote monitoring technologies are essential to the advancement of various SDGs. Remote diagnostics and monitoring technologies greatly contribute to SDG 3 by improving access to healthcare services, especially in underprivileged areas. These technologies offer early diagnosis and ongoing monitoring of chronic illnesses by transmitting real-time patient data via IoTs enabled medical equipment. Research shows that remote healthcare solutions, especially in telemedicine and chronic illness management, have decreased healthcare expenses and patient hospitalizations by 18–25%. Automated resource monitoring solutions in industries reduce energy, water, and raw material waste, which is in line with SDG 12. IoT sensors monitor resource consumption and streamline production lines in manufacturing operations, resulting in a 15–20% reduction in material waste. In a similar vein, precision irrigation technologies that use 30% less water thanks to remote monitoring in agriculture directly promote sustainable production methods. Remote monitoring systems that maximize resource use and reduce carbon emissions directly assist SDG 13. By increasing route efficiency, IoT enabled fleet management solutions in logistics and transportation save fuel consumption and emissions by 10–15% a year. Remote monitoring of wind turbines and solar farms in the renewable energy sector guarantees optimal performance, lowers energy losses, and aids in the worldwide fight against climate change. Remote monitoring technologies offer strong instruments for attaining environmental and social justice while promoting innovation and efficiency across industries by fusing real-time data with sustainability-focused applications [104–106]. The connections with the SDGs will enable industries to transition into Industrial Revolution 6.0, emphasizing sustainable, human friendly, and technologically advanced practices. The focus of our book is to highlight the aspects of Industry 4.0 that can serve as a bridge, facilitating this transformative shift toward the future.

8 Technological Advancement and their Applications in Supply Chain Management

Digital twins are virtual representations of physical supply chain systems, continuously updated in real time with data from GPS, RFID, and IoT sensors. By simulating the dynamics of industrial facilities, transportation networks, and warehouses, digital twins empower companies to optimize their operations. They enable businesses to forecast disruptions by modelling events such as equipment failures, natural disasters, and geopolitical issues, allowing for proactive measures to avoid potential bottlenecks. Digital twins help optimize logistics by analyzing various scheduling and routing options to identify the most efficient pathways, reducing delays and fuel consumption. Siemens has utilized digital twins to model their supply chain, achieving a 20% improvement in delivery efficiency and significantly lowering costs [107, 108].

Big data analytics plays an important role in processing the large datasets generated across the supply chain, including production indicators, market trends, and transportation records. One of its primary applications is inventory optimization, which involves maintaining ideal stock levels while preventing waste and overproduction by analyzing historical sales data and seasonal demand variations. Big data analytics aids in cost reduction by enabling companies to streamline procurement processes and negotiate more favorable contracts through the evaluation of supplier performance and delivery schedules. Also, it supports sustainability metrics by monitoring carbon emissions throughout the supply chain, ensuring businesses meet environmental standards and align with sustainability goals [109, 110].

ML with AI to forecast supply chain outcomes by analyzing both historical and real-time data. One key application is demanding forecasting, where AI accurately predicts customer demand, reducing waste and cutting overproduction by 10–30%. Another critical use is dynamic routing, where AI optimizes transportation routes in response to real-time traffic and weather data, decreasing delivery delays by 15–20%. DHL has implemented AI to streamline fleet operations, enhancing logistical efficiency and achieving annual savings of nearly USD 400 million [111, 112].

IoT devices in transportation networks provide real-time information on vehicle locations, fuel consumption, and maintenance requirements, ensuring efficient and reliable operations. By enabling proactive maintenance, these systems help prevent vehicle issues, reducing downtime by 30% and improving overall fleet performance. IoT supported route optimization minimizes fuel consumption and lowers carbon emissions, directly contributing to SDG 13 by promoting sustainable and eco-friendly logistics practices [113, 114].

8.1 Alignment with SDGs

By encouraging innovation, sustainability, and cooperation, Industry 4.0-driven improvements in supply chain resilience greatly aid in the achievement of several SDGs. Using digital twins, IoT, and AI to improve supply chain efficiency, these technologies support industrial innovation and the creation of resilient infrastructure, which is in line with SDG 9. By avoiding overproduction, decreasing obsolescence, and encouraging responsible consumption, AI powered advanced inventory optimization systems assist companies in addressing waste management issues and can save up to USD 300 billion annually. These developments highlight how Industry 4.0 creates significant economic and environmental advantages in addition to improving supply chain resilience, opening the door for effective and sustainable international trade systems [115, 116].

9 Edge AI for Decentralized Decision-Making

By combining edge computing with artificial intelligence, Edge AI allows devices to process data locally without depending on centralized cloud services. This reduces latency and enables decentralized, autonomous decision-making in milliseconds for devices like sensors and robots. Edge AI is used by self-driving forklifts in warehouses to detect obstacles and navigate, increasing operational speed and safety. Instead of sending data to a centralized cloud system for processing, Edge AI utilizes the capability of AI and edge computing to process data locally at the point of generation, such as sensors, machines, or robots. Numerous operational and technological benefits are provided by this architecture, particularly in industrial applications that require a lot of resources and time.

9.1 Main Technological Usage in Edge AI

ML models are run directly on devices that include specialized AI hardware accelerators (NVIDIA Jetson, Google Edge TPU, Intel Movidius etc.). This reduces bandwidth consumption and improves privacy by doing away with the need to send massive amounts of data to the cloud. Image recognition algorithms that operate locally on cameras in manufacturing lines may instantly identify faulty parts. The inference phase, which involves applying the learned model to fresh data, is carried out on edge devices, even though model training usually takes place in robust cloud-based systems. With response speeds in the millisecond range, this guarantees quick decision-making. For instance, industrial turbines equipped with predictive maintenance systems can identify abnormalities such strange vibrations and plan repairs on their own without the need for human assistance. By evaluating data streams in real-time and guaranteeing prompt reactions to important events, edge AI facilitates decentralized decision making. Reduction of Latency that While edge AI systems cut latency to less than 10 ms, which is necessary for applications like robotics or driverless cars, typical cloud processing has delay between 50 and 200 ms [117, 118]. Self-driven forklifts are powered by Edge AI, which gives them sensors (LiDAR, cameras, ultrasonic sensors etc.) and embedded AI models for the Obstacle Detection and Navigation In dynamic warehouse environments, AI systems use local data processing to identify impediments, determine other paths, and navigate safely. Path Optimization by improving operational speed and cutting transit times by 20–30%, real-time route change based on traffic and storage circumstances is possible. By evaluating sensor data streams (temperature, pressure and vibration), Edge AI systems mounted on industrial equipment can also do predictive quality control by identifying manufacturing flaws. Edge devices in semiconductor production keep an eye on cleanroom conditions and automatically initiate modifications to maintain good yields. Edge AI can be used in smart energy systems and smart grids to analyze data from distributed energy resources (wind turbines, solar panels) and optimize power generation and consumption locally while maintaining grid stability. Edge AI is used by collaborative robots (cobots) to do tasks including object detection, real-time obstacle avoidance, and dynamic workload sharing with human workers. These systems guarantee smooth functioning without relying on outside networks [119–121].

In a nutshell Edge AI offers many advantages for decentralized decision making. Because real-time answers are crucial in industrial settings, it reduced latency.

A robot arm that uses edge AI processing to modify welding can react in 5 ms, preventing the production of faulty goods. It decreased bandwidth costs because sending raw data to the cloud can be costly, particularly when dealing with multi sensor data or high-resolution video streams. Data transmission requirements are reduced by 60–80% when processing locally. Because sensitive operational data, such proprietary production techniques, is locally, there is less chance of breaches or illegal access, improving data privacy and security. Another benefit of resilience in connectivity failures is that, in contrast to cloud-dependent systems, edge AI maintains its functionality even in the event of network disruptions, guaranteeing continuous industrial operations. Scalability for Large Systems is another thing we can do. Without overtaxing central servers, edge AI scales effectively across dispersed contexts like automated warehouses, smart cities, and connected factories thanks to local processing. By giving devices decentralized intelligence and facilitating quick, independent decision-making, edge AI is a game-changer for industrial automation. Its applications, which range from predictive maintenance to autonomous forklifts, show notable improvements in operational safety, cost savings, and efficiency. With improvements in hardware, connectivity, and federated learning, Edge AI has the potential to completely transform real-time industry operations and open up new avenues for creativity and productivity [122, 123].

10 Human Centric Augmentation through AR and VR

In Industry 4.0, AR and VR have become game-changing technologies that greatly improve workforce capacities in a variety of industries. These technologies facilitate more effective training, real-time visualization, and simplified maintenance support by integrating virtual content with real world surroundings (AR) or submerging users in simulated worlds (VR). Their use increases output, lowers mistakes and supports international objectives including sustainability, innovation, and education. At a CAGR of 19.2%, the global AR/ VR market for manufacturing is projected to reach USD 18 billion by 2028 from USD 7.5 billion in 2023. Research shows that AR enabled maintenance lowers downtime by 30%, which has a direct effect on profitability and operational efficiency. When compared to conventional learning techniques, VR-based training has demonstrated a 15% greater rate of knowledge retention. Because AR and VR technologies offer effective and immersive training, visualization, and maintenance solutions, they are transforming workforce skills. By lowering waste, energy consumption, and carbon emissions, these solutions not only increase operational precision and efficiency but also support global sustainability goals. Businesses can empower their employees, increase productivity, and help create a more resilient and sustainable future by incorporating AR and VR into Industry 4.0 ecosystems [124, 125].

10.1 Technology Usage for Industrial Applications

AR and VR offer incredibly dynamic and captivating learning environments, especially for intricate industrial processes. Workers engage with virtual versions of tools or systems during VR-based training, which mimics real world situations. Operators in the oil and gas industry can obtain practical experience without jeopardizing their safety by practicing emergency response drills in virtual environments. According to research, VR cuts training times and related expenses when compared to conventional techniques. AR superimposes important data, such schematics or detailed instructions, on actual equipment. By guiding professionals through intricate machinery repairs, AR enabled smart glasses, for example, lessen the need for outside assistance. AR makes it possible to visualize equipment performance and status data in real-time, providing workers with actionable insights right in their line of sight. IoT enabled augmented reality devices show areas that need maintenance or modifications using real-time data from sensors built into machinery. AR-assisted visualization improves quality assurance procedures in manufacturing assembly lines, increasing accuracy. By presenting interactive overlays and guided workflaws, AR helps technicians complete maintenance jobs quickly and effectively. These devices are especially helpful in high stakes sectors where accuracy is essential, such as aerospace. In aerospace manufacturing, for instance, AR assisted assembly lines have resulted in a reduction in human error, guaranteeing adherence to strict safety and performance requirements [126, 127].

10.2 Industrial Implications

AR and VR are utilized in the aerospace industry to expedite the assembling of aircraft parts. Boeing employs AR to improve accuracy and cut production time by 25% while assembling electrical harnesses. While AR allows real-time overlays during surgeries, increasing accuracy and patient outcomes, VR platforms allow surgeons to practice intricate procedures in risk-free virtual worlds. Using VR for product design and prototyping, automakers can cut development cycles by 20%. AR apps provide assembly line workers with guidance, guaranteeing uniformity, and quality in the manufacturing of automobiles [128, 129].

10.3 Alignment with SDGs

Industries may achieve sustainable growth, improve workforce capacities, and make a significant contribution to the global agenda for a better and more fair future by incorporating AR and VR into their operations. These connections with SDGs will help industries to transform their industries with industry 6.0. AR enabled gadgets, for instance, lessen the environmental effect of on-site visits by enabling personnel to do inspections or repairs from a distance. This strategy guarantees business continuity in sectors, including manufacturing, healthcare, and energy while also assisting with climate change initiatives [130–132].

10.4 Technology Implication Techniques

Strategic planning and a strong infrastructure are necessary for the successful application of AR and VR technology in industrial settings. To enable AR systems to see real-time equipment data, technology integration is a fundamental step that involves the deployment of sensor networks and the Internet of Things. Utilizing cloud-based technologies can further lower initial expenses and offer scalability to meet expanding needs. Without the need for substantial on-premises infrastructure, these platforms guarantee smooth access to AR/VR services. For implementation to be successful, data security and privacy must be guaranteed. Advanced encryption techniques and access control mechanisms are required to provide secure data transmission between AR/ VR systems and IoT devices. In addition to protecting sensitive operational data, this increases stakeholder confidence in the system. Industries may fully utilize AR and VR technology to drive innovation, efficiency, and sustainability in their operations by adhering to these implementation principles [133, 134].

11 Quantum Computing for Industrial Problem Solution

In Industry 4.0, quantum computing is becoming a game-changing technology that can solve challenging issues in scheduling, optimization, material design, and energy management. Quantum computers use the concepts of quantum mechanics and quantum bits, or qubits, in contrast to classical computers that process information in binary (bits). Because these qubits can be entangled and exist in numerous states at once (a phenomenon called superposition), quantum computers are able to process and solve problems at exponentially higher speeds than their classical counterparts. Because of this feature, large scale optimization problems can be resolved by quantum computing in a fraction of the time needed by conventional computing techniques. Supply chain optimization is one of the main issues facing industries today, particularly when handling large scale operations with numerous variables and restrictions. For example, the "traveling salesman problem," route planning optimization, and sophisticated inventory management across numerous locations can all be solved using quantum algorithms. Because the number of variables increases exponentially with scale, many jobs would take days or even weeks to perform in classical computing. However, these issues can be resolved in a matter of seconds thanks to quantum computing. By analyzing several potential outcomes concurrently, quantum algorithms, like the Quantum Approximate Optimization

Algorithm or Quantum Annealing, can rapidly identify optimal or nearly ideal solutions to scheduling, logistics, and supply chain management issues [135–137].

By simulating the quantum behavior of atoms and molecules, quantum computers can be used to build novel materials with features that are hard to anticipate with traditional techniques. Quantum simulations can be used to create batteries, which are essential for developing electric vehicle technologies and enhancing renewable energy storage systems. Batteries are made of materials that are more effective, long lasting, and economical. By encouraging the creation of energy-efficient products and systems that lessen dependency on non-renewable energy sources, this direct effect on the energy sector also links into SDG 7. Quantum computing also has enormous potential for energy management, especially for integrating renewable energy sources like wind and solar and improving energy distribution across networks. Energy systems that depend more and more on sporadic renewable energy sources need to optimize energy supply and demand balancing, and quantum algorithms can manage massive information in real-time. To achieve SDG 13 by lowering the carbon footprint associated with energy systems, more accurate management and forecasting are made possible by quantum computers' ability to simulate the behavior of energy systems at the atomic level. This leads to less energy waste and more efficient usage [138, 139].

Industries stand to realize exponential efficiency benefits when quantum gear becomes more widely available. Predictive maintenance, which minimizes downtime and lowers maintenance costs by using quantum algorithms to evaluate massive volumes of data from sensors in industrial machinery, will be made possible by quantum computing. This application has important ramifications for sectors, including manufacturing, utilities, and transportation, where operational optimization can lead to considerable cost savings and benefits. Also, faster and more accurate simulations of industrial processes will be made possible by quantum computers, which would speed up innovation cycles by enabling businesses to spend less time designing and testing new systems or products [140, 141].

12 AI-Driven Predictive Quality Control

By enabling real-time monitoring and the detection of deviations in production parameters, AI-driven predictive quality control is revolutionizing the manufacturing industry and guaranteeing the consistent supply of high quality products. This method makes use of sophisticated ML algorithms that examine enormous volumes of production data, including temperature, pressure, material composition, and machine performance, to spot trends that can point to any flaws or problems with quality. Manufacturers can take early action, optimize production, and greatly increase product yield and quality by using these algorithms to anticipate abnormalities before they materialize as defects. AI powered devices are being used to identify micro-defects in wafers during production, which is mandatory in the semiconductor industry (Future of semiconductor industry is too high) where accuracy and quality are crucial. The application of AI for flaw identification is very important because of the tiny size and complexity of semiconductors. Conventional inspection techniques frequently rely on human operators or manual inspection procedures, which can be laborious and error prone. After identifying these variables, AI systems can modify machine settings or notify operators to make the required changes, preventing flaws from developing later in the manufacturing process. Real-world applications of AI-driven predictive quality control demonstrate its advantages. Businesses using AI powered defect detection systems in the semiconductor sector have observed yield rates increase by 20%, which is directly related to waste material reduction. The use of AI in quality control has resulted in significant cost reductions in sectors like automotive manufacturing, both by increasing production yields and by lowering the need for labor-intensive and error prone manual inspections. AI-driven predictive solutions in the automobile industry can cut the percentage of defective products by up to 30%, which lowers warranty claims and the price of defective parts [142, 143].

12.1 AI-Driven Predictive Quality Control Working Mechanism

Predictive quality control systems powered by AI follow a multi-step procedure. First, information is collected from multiple sources, such as cameras, external monitoring systems, and sensors built into machines. ML models, such as convolutional neural networks for image processing or time series analytic approaches for ongoing process variable monitoring, are then used to process this data. Both typical operating patterns and deviations from them, such as variations in temperature, vibration, or product dimensions, are recognized by these models after training. Depending on how automated the system is, it can either send out a real-time alert or start corrective action on its own when an abnormality is found. To avoid more problems, this can entail modifying machine settings, warning human operators, or halting the production line [144, 145].

12.2 Implication Techniques

Automated production environments can fully integrate the usage of AI-driven predictive quality control. AI systems may process data locally by utilizing technologies like edge computing and industrial IoT sensors. This reduces latency and speeds up decision-making. These systems work particularly well in high-speed industries where even little flaws can result in large losses in efficiency and quality, such as the manufacturing of electronics, automobiles, and pharmaceuticals. In this case, automation also includes the self-optimization of production lines, in which AI automatically modifies settings in response to real-time data, guaranteeing that production proceeds without the need for human involvement. By lowering human error, waste, and inspection expenses, this type of closed loop management not only increases operating efficiency but also improves product quality [146, 147].

12.3 Alignment with SDGs

The 20% increase in yield rates in the semiconductor industry guarantees that less raw materials are wasted during production, which in turn minimizes waste. This is consistent with SDG 12, which promotes more environmentally friendly, efficient, and sustainable production practices. By lessening the carbon footprint of manufacturing processes, AI-driven quality control also supports climate action. Manufacturers can lower energy use and emissions by streamlining manufacturing procedures and reducing energy consumption (by operating equipment more efficiently and reducing rework cycles). Predictive maintenance may guarantee that machinery is constantly running as efficiently as possible, which reduces energy use and increases equipment longevity [148, 149].

13 Integration of Intelligent Manufacturing Technologies in Industry 6.0

Global production systems are changing because of the integration of Industry 6.0 intelligent manufacturing tools, which also improve sustainability, efficiency, and climate resilience while supporting global sustainability goals. Cutting-edge technologies like AI, quantum computing, the IIoT, blockchain, and additive manufacturing are revolutionizing industrial processes by drastically lowering CO_2 emissions, energy use, and resource waste. Through self-learning AI models, cyber physical automation, and real-time predictive analytics, these digital technologies enable firms to optimize production, reducing industrial contributions to global warming and reducing carbon emissions by 50–85%. A closed loop, resource efficient industrial paradigm is being promoted by the shift to digital twins, smart process automation, and circular economy frameworks, which guarantee waste reductions of 60–90% and prevent excessive material depletion and overproduction. Table 3.1 is a brief of these technologies and their impact on climate change. In the Table 3.1 , the technologies alone do not directly improve specific energy efficiency or carbon dioxide; the effectiveness of these technologies depends on integration with other technologies and how it is applied or utilized in specific industrial sector/ segment.

A key component of Industry 6.0 developments is the optimization of energy and resources. Manufacturers may accelerate the shift to low-carbon and net zero industrial operations by reducing energy usage by 40–80% through the integration of IIoT sensors, AI powered smart grids, and decentralized cloud computing.

Table 3.1 Intelligent manufacturing technologies and their impact on climate change

Technology with reference	Energy efficiency improvement (%)	CO_2 reduction (%)	Key technical constraints and real world variance
AI [150, 151]	18–35	25–45	Efficiency drops in high-mix production; transformer models require retraining (500+ MWh). Hardware bottlenecks limit edge deployment
CPS [152, 153]	30–45	35–50	Legacy system integration reduces gains by 15–25%. Real-time determinism requires time-sensitive networking (IEEE 802.1Qbv)
Edge computing [154, 155]	35–50	30–45	Thermal throttling at >85 °C cuts efficiency by 40%. Memory bandwidth limits model complexity (max 8B params on Jetson AGX)
Quantum computing [156, 157]	15–30 (NISQ era)	20–40	Limited qubit coherence (50–150 μs) restricts material simulation depth. Error correction overhead consumes 10^4× classical energy
Cloud computing [158, 159]	25–35	18–30	PUE >1.3 in hyperscale DCs negates 20% savings. Scope 3 emissions from silicon manufacturing offset gains.
Additive manufacturing [160, 161]	40–65	50–70	Energy intensive laser sintering (15–30 kWh/kg) vs. machining (8 kWh/kg). Savings apply only to topology optimized parts
Digital twin [162, 163]	35–55	40–60	Model fidelity gaps cause ±12% prediction errors. High fidelity twins require exascale HPC (10^9 core hours/year)
IIoT [164, 165]	40–60	50–65	Wireless sensor power (LoRaWAN/NB-IoT) adds 5–10% overhead. Cybersecurity hardening consumes 15% processing resources
Machine learning [166, 167]	25–45	35–55	Training data scarcity in OT environments reduces accuracy by 20–35%. Explainability requirements limit deep architectures
Deep learning [168, 169]	30–50	45–65	ViT/CNN inference requires 50+ TOPS at edge. Quantization losses (INT8 vs FP16) reduce defect detection accuracy by 3–8%
Reinforcement learning [170, 171]	45–65	55–75	Sample inefficiency: Real-world training requires 10^6+ iterations. Safety constraints slow convergence
Autonomous robotics [172, 173]	55–75	60–75	SLAM localization drift (2–5 cm/m) causes energy overuse. Harmonic drive losses consume 15% actuator power

Table 3.1 (continued)

Technology with reference	Energy efficiency improvement (%)	CO_2 reduction (%)	Key technical constraints and real world variance
Big data analytics [174, 175]	35–55	50–65	Data lake ingestion costs exceed USD 3/TB. Spark cluster provisioning delays cause 5–12% idle energy waste
Blockchain technology [176, 177]	10–20	15–25	PoW/PoS consensus consumes 0.5–1.2 kWh/tx. Limited to high-value goods due to ledger storage costs (USD 0.05/GB/month)
Augmented reality [178, 179]	25–40	30–45	HoloLens 2 power draw (8 W) limits 4-h runtime. Spatial mapping latency (15–30 ms) causes assembly errors
Virtual reality [180, 181]	35–50	45–60	Varjo XR-4 HMD requires 24 Gbps DP2.1 bandwidth. Physics engine inaccuracies necessitate physical validation

Accountability is further improved by blockchain-enabled supply chain traceability and transparency, which guarantee ethical production methods and sustainable raw material procurement. To create a low carbon, climate resilient industrial future, these intelligent manufacturing breakthroughs directly support SDGs 9, 12, and 13. With the advent of fully automated, zero-waste production systems that use intelligent robotics, high-speed quantum simulations for material design, and AI-driven process optimization, Industry 6.0 is revolutionizing the worldwide manufacturing scene. The global manufacturing industry will shift toward climate-positive, ultra efficient and self-sustaining industrialization as industrial sectors continue their transition into hyper-connected, cyber physical ecosystems, guaranteeing a robust and sustainable future for industries everywhere.

14 Challenges and the Strategies to Overcome the Implementations of Advanced Technologies

Cutting-edge industrial IoT ecosystems present a profound energy paradox, while they offer operational efficiencies, their computational demands significantly escalate energy use. While AI-driven manufacturing promises efficiency and productivity gains, its sustainability is tempered by the substantial energy overheads of training and inference on conventional architectures. Next generation hardware like neuromorphic and photonic computing offers transformative energy efficiencies, but will require advances in fabrication, scaling, and integration before achieving broader industrial adoption. Dense sensor networks with 10–100× more endpoints than traditional systems, combined with AI workloads such as GANs for synthetic data or transformer based predictive maintenance, can consume hundreds of MWh per model training cycle. For the training purpose of Generative AI (like Open AI,

GPT4) high tech models, reportedly consumed 1300 MWh, roughly equivalent to the monthly energy use of 1450 U.S. households. This energy load is compounded by hyperscale data centers and edge infrastructure. Data centers typically operate with power usage effectiveness above 1.5, meaning 50% of their energy goes to cooling and overhead rather than computing. Networks like 5G microcells draw approximately 11.5 kW each, compared to 4G's 2.9 kW, representing almost a four-fold increase in per cell energy consumption. Emerging hardware solutions promise key efficiency improvements. Neuromorphic chips, like Intel's Loihi 2, leverage asynchronous spiking neural networks and event driven computation to deliver orders of magnitude energy efficiency. Experimental results suggest they may achieve 10,000 times lower energy consumption for event-based tasks, with practical systems such as Intel's Hala Point, embedding 1152 Loihi 2 processors operating with under 2.6 kW for over a billion neurons, posting peak efficiency of >15 TOPS/ W (trillions of operations per second per watt). Alternative architectures such as photonic processors (Lightmatter's Passage interconnect) employ wavelength division multiplexing to cut AI energy consumption by up to 90%, but challenges remain in thermal crosstalk management and integrating photonic cores into multi-core configurations, these systems typically stand at technology readiness level 6. In real world industrial control scenarios, neuromorphic processors have demonstrated tenfold reductions in energy use and comparable latency to CPU/GPU systems. Like, a Loihi-based spiking controller for robotic force control consumed significantly less power than conventional low power edge hardware. Instead of all, we need to use renewable and green safer energy systems to generate all types of our energy demands. Fossil fuel burning to produce energy will surely leads toward the increase in net carbon emission which is strictly prohibited in Industry 6.0 [182–186].

For SMEs, ROI timelines stretch to 3–5 years due to hidden integration costs. These technical and geopolitical barriers confirm why SMEs and larger enterprises alike encounter lengthy onboarding periods, increased operational complexity, and inflated up front costs when attempting to operationalize smart manufacturing technologies. Retrofitting legacy SCADA/MES systems require custom APIs to bridge Modbus/TCP with MQTT/OPC UA stacks, costing USD 250 K to 2 M USD per production line. Cybersecurity hardening per IEC 62443 standards demands hardware enforced trusted execution environments like Intel SGX, adding 15–25% to budgets. Interoperability fragmentation, where 60% of legacy equipment lacks OPC UA/MTConnect compliance necessitates middleware layers employing Kafka or Sparkplug 3.0 brokers, while brownfield deployments risk 7–12% production loss from control system handover latency during phased OT/IT convergence. Approximately 60% of existing industrial equipment lacks native OPC UA or MTConnect support, compelling manufacturers to deploy middleware such as Kafka or Sparkplug 3.0 brokers to normalize data across heterogeneous OT/IT systems. Secondly, compliance with IEC 62443 cybersecurity standards often requires hardware enforced trusted execution environments, such as Intel SGX. These measures, including secure boot, cryptographic isolation & remote attestation, impose 15–25% additional hardware and integration costs. Thirdly, the shift from brownfield to smart factory environments introduces production downtime during phased OT/IT convergence. Studies indicate that 7–12% output loss may occur as control systems are reassigned or updated, due to latency, synchronization mismatches, or misconfigurations. Such losses are often absorbed in phased chimes to avoid full production halts. Finally, the energy paradox reemerges at scale. Dense sensor grids and edge/IT workloads significantly increase power draw, with 5G microcells consuming around 11.5 kW each, almost four times that of 4G cells (around 2.9 kW). These compute and connectivity demand diminish Scope 1 gains and can inflate energy use even as operational efficiency improves [187–194].

In advanced manufacturing, the widening skills gap, with approximately 68% of roles lacking requisite expertise, is attributed to the rising demand for hybrid skill sets, including mastery of PLC ladder logic and complex Physics Informed Neural Network frameworks like NVIDIA Modulus, which integrates physics constraints into deep learning architectures for solving partial differential equations. Traditional upskilling programs typically take 9–18 months, but the introduction of VR-based digital twins, such as those in the Siemens Xcelerator suite can potentially accelerate training by 40% (rough estimate) by offering immersive, hands on simulation environments. However, implementing these solutions requires GPU accelerated servers with real-time ray tracing capabilities, leading to initial capital expenditures exceeding 0.5 million USD per setup for compute clusters, storage, and high-fidelity rendering infrastructure. This investment supports multi user VR environments featuring low latency (<20 ms) interaction, dynamic physics simulation, and digital replication of process flows. While precise ROI varies, the cost structure is validated by Siemens' deployment case studies in automotive prototype lines [195–197].

Despite ISO 23247, standardization gaps persist in digital twin implementations due to inconsistent asset administration shell models. A car manufacturer reports 30% delays in commissioning when supplier asset administration shell sub model structures differ. Aligning OT data tagged under ISA-88/95 with IT ontologies like OWL/RDF cannot be automated as systems often require 200+ hours per system for manual schema harmonization, an accepted industry practice. The new EU Digital Product Passport regulation mandates blockchain-based tracking of product carbon content at a cost of approximately 0.02–0.08 euro per unit, based on implementation cost studies. Meanwhile, geopolitical dynamics such as US China export controls have severely impacted the availability of key AI hardware. NVIDIA H100 GPUs currently face 52 week lead times in APAC territories, and projected delays may stretch to 18–24 months, illustrating significant disruptions to edge AI deployment [198–200].

Similarly, scalability of edge AI in smart manufacturing depends on the convergence of several enabling technologies. These include 2 nm semiconductor nodes that facilitate analog in-SRAM compute, achieving sub 10 picojoules per multiply-accumulate (MAC) operation; quantum-safe IIoT protocols built on NIST's post quantum cryptography standards to ensure secure machine-to-machine communication; and generative AI copilots capable of automating OT code (PLC ladder logic and IEC 61131–3 languages), dramatically reducing integration time and human effort. Without coordinated progress across these domains, edge-enabled Industry

4.0 will remain fragmented and largely limited to capital-intensive megacorporations. Edge computing in industrial environments is increasingly constrained by thermal and memory bottlenecks. Devices such as the NVIDIA Jetson AGX Orin experience up to 40% throughput degradation when operating temperatures exceed 85 °C, largely due to thermal throttling of LPDDR5X memory. This is compounded by the deployment of large-scale quantized AI models like BERT Large, which demand over 32 GB of LPDDR5X, surpassing the memory capacity of more than 80% of existing edge devices. These challenges necessitate the transition to heterogeneous hardware architectures that integrate 3D stacked HBM3 memory, as seen in platforms like AMD Versal ACAP, along with gallium nitride power ICs to mitigate thermal load and enhance efficiency. Validation from World Economic Forum Lighthouse factories shows that high-mix production environments introduce a ± 12% ROI variance in digital twin applications, due to stochastic process scheduling complexities that are difficult to model accurately. Many current implementations fail to account for renewable energy integration however, incorporating microgrid forecasting and predictive curtailment has demonstrated the potential to reduce Scope 2 emissions by 15–30%. To address immediate thermal challenges, two-phase immersion cooling systems are being deployed, offering 30–50% lower thermal resistance and maintaining sub 60 °C board temperatures under heavy AI inference loads. Over the long term, innovations in photonic computing optical tensor cores, promise compute densities of 4.8 peta OPS/ W. Although promising, these technologies remain at technology readiness level 6, with issues such as thermal crosstalk and multi-core integration still unresolved [201–205]. This chapter provides foundational knowledge and an introductory overview of the technologies. Detailed explanations and advanced insights are presented in the subsequent chapters.

References

1. Rijwani, T., et al.: Industry 5.0: a review of emerging trends and transformative technologies in the next industrial revolution. Int. J. Interact. Des. Manuf. (IJIDeM). **19**, 1–13 (2024)
2. Chitt, M., et al.: Green hydrogen productions: methods, designs and smart applications. In: Highly Efficient Thermal Renewable Energy Systems, pp. 261–276. CRC Press, Boca Raton (2024)
3. Plakantara, S., Karakitsiou, A., Mantzou, T.: Managing risks in smart warehouses from the perspective of industry 4.0. In: Disruptive Technologies and Optimization Towards Industry 4.0 Logistics, pp. 1–47. Springer, Cham (2024)
4. Hajjaji, Y., et al.: Big data and IoT-based applications in smart environments: a systematic review. Comput Sci Rev. **39**, 100318 (2021)
5. Yang, C., et al.: Big data and cloud computing: innovation opportunities and challenges. Int. J. Dig. Earth. **10**(1), 13–53 (2017)
6. Hamidi, S.R., et al.: SMEs maturity model assessment of IR4.0 digital transformation. In: Proceedings of the 7th International Conference on Kansei Engineering and Emotion Research 2018: KEER 2018. Springer, Cham (2018)
7. Zong, Z., Guan, Y.: AI-driven intelligent data analytics and predictive analysis in industry 4.0: transforming knowledge, innovation, and efficiency. J. Knowl. Econ. **16**, 1–40 (2024)

8. Gidiagba, J.O., et al.: Review on the evolution and impact of iot-driven predictive maintenance: assessing advancements, their role in enhancing system longevity, and sustainable operations in both mechanical and electrical realms. Comput. Sci. IT Res. J. **5**(1), 166–189 (2024)
9. Soori, M., et al.: AI-based decision support systems in industry 4.0. A review. J. Econ. Technol. (2024). https://doi.org/10.1016/j.ject.2024.08.005
10. Rane, N., Kaya, O., Rane, J.: Artificial intelligence, machine learning, and deep learning applications in smart and sustainable industry transformation. Artif. Intell. Mach. Learn. Deep Learn. Sustain. Ind. **5**, 2–29 (2024)
11. Sridevi, S., Deepa, S.: AI-enabled human and machine activity monitoring in IIOT systems. In: Next-Gen Technologies in Computational Intelligence: Proceeding of the International Conference on Next-Gen Technologies in Computational Intelligence (NGTCA 2023). CRC Press, Boca Raton (2024)
12. Hu, Y., et al.: Industrial internet of things intelligence empowering smart manufacturing: a literature review. IEEE Internet Things J. **11**, 19143–19167 (2024)
13. Kang, K., Yu, C., Zhong, R.Y.: Cloud, fog, edge computing and 5G technologies for industrial automation. In: Manufacturing from Industry 4.0 to Industry 5.0, pp. 379–430. Elsevier, Amsterdam (2024)
14. Majeed, H., Iftikhar, T., Abid, R.: Green synthesis of zinc nanoparticles with plant material and their potential application in bulk industrial production of mosquito-repellent antibacterial paint formulations. React. Chem. Eng. **9**(3), 677–683 (2024)
15. Majeed, H., Iftikhar, T., Abid, R.: Green synthesis of insecticidal, bactericidal, UV absorbent, sustainable paint formulations using Mentha piperita (peppermint). React. Chem. Eng. **9**(9), 2358–2366 (2024)
16. Majeed, H., et al.: Bulk industrial production of sustainable cellulosic printing fabric using agricultural waste to reduce the impact of climate change. Int. J. Biol. Macromol. **253**, 126885 (2023)
17. Majeed, H., et al.: Green synthesis of Eucalyptus globulus zinc nanoparticles and its use in antimicrobial insect repellent paint formulation in bulk industrial production. Heliyon. **10**(2), e24467 (2024)
18. Okokpujie, I.P., Tartibu, L.K.: Study of the economic viability of internet of things (IoTs) in additive and advanced manufacturing: a comprehensive review. Progr. Add. Manuf. **10**, 1–20 (2024)
19. Kulykovets, O.: Transforming industries: the digital revolution of industry 4.0. Issues Inf. Syst. **25**(3), 148–164 (2024)
20. Qaiser, H., et al.: COVID-19 pandemic and vaccines. In: Zia-Ul-Haq, M. (ed.) Alternative Medicine Interventions for COVID-19, pp. 205–235. Springer International Publishing, Cham (2021)
21. Pangarkar, T.: Smart Manufacturing Statistics 2025 By Enhanced Growth. https://scoop.market.us/smart-manufacturing-statistics/ (2025)
22. Kaput, M.: McKinsey: AI could generate up to $23 trillion annually by 2040. https://www.marketingaiinstitute.com/blog/mckinsey-ai-economic-impact (2024)
23. Wang, X., et al.: A paradigm shift for Modeling and operation of oil and gas: from industry 4.0 in CPS to industry 5.0 in CPSS. IEEE Trans. Industr. Inform. **20**, 9186–9193 (2024)
24. Ledford, A.B., et al.: Origin of the fourth industrial revolution: manufacturing predictions preceding Industrie 4.0. J. Sci. Technol. Policy Manage. (2024). https://doi.org/10.1108/JSTPM-03-2023-0040
25. Arav, Y., et al.: An industrial internet-of-things (IIoT) open architecture for information and decision support systems in scientific field campaigns. Sensors. **24**(18), 5916 (2024)
26. Sun, D., et al.: A comprehensive survey on collaborative data-access enablers in the IIoT. ACM Comput. Surv. **56**(2), 1–37 (2023)
27. Fortino, G., Savaglio, C.: Integration of digital twins & internet of things. In: The Digital Twin, pp. 205–225. Springer, Cham (2023)

28. Rane, N., Kaya, O., Rane, J.: Artificial intelligence, machine learning, and deep learning technologies as catalysts for industry 4.0, 5.0, and society 5.0. Artif. Intell. Mach. Learn. Deep Learn. Sustain. Ind. **5**, 2 (2024)
29. Soori, M., Arezoo, B., Dastres, R.: Artificial intelligence, machine learning and deep learning in advanced robotics, a review. Cogn. Robot. **3**, 54–70 (2023)
30. Rao, K.S., et al.: AI and ML in IR4.0: a short review of applications and challenges. Malays. J. Sci. Adv. Technol. **4**, 141–148 (2024)
31. Kegyes, T., Süle, Z., Abonyi, J.: The applicability of reinforcement learning methods in the development of industry 4.0 applications. Complexity. **2021**(1), 7179374 (2021)
32. Yeom, S., et al.: Digital twin (DT) and extended reality (XR) for building energy management. Energ. Buildings. **323**, 114746 (2024)
33. Kerrouchi, S., Aghezzaf, E.-H., Cottyn, J.: Production digital twin: a systematic literature review of challenges. Int. J. Comput. Integr. Manuf. **37**, 1–26 (2024)
34. Singh, M., Khan, S.A.L.A.: Advances in autonomous robotics: integrating AI and machine learning for enhanced automation and control in industrial applications. Int. J. Multidimens. Res. Perspect. **2**(4), 74–90 (2024)
35. Hossain, M.I., et al.: Technology of additive manufacturing: a comprehensive review. Kufa J. Eng. **15**(1), 108–146 (2024)
36. Song, L., Chen, M., Xu, Z.: Integrated and innovative applications of 5G+ new technologies. In: 5GtoB from Theory to Practice, pp. 131–171. Springer, Cham (2024)
37. Prabhugouda, A., Asra, S.: A review on big data applications and their challenges. J. Inf. Knowl. Manag. **23**, 2430001 (2024)
38. Sathya, D., Thangamani, R., Balaji, B.S.: The revolution of edge computing in smart farming. In: Intelligent Robots and Drones for Precision Agriculture, pp. 351–389. Springer, Cham (2024)
39. Alanhdi, A., Toka, L.: A survey on integrating edge computing with AI and Blockchain in maritime domain, aerial systems, IoT, and industry 4.0. IEEE Access. **12**, 28684–28709 (2024)
40. Hatzikonstantinou, E.M., Karakitsiou, A.: Blockchain technologies as a 4th industrial revolution facilitator in logistics. In: Disruptive Technologies and Optimization Towards Industry 4.0 Logistics, pp. 49–94. Springer, Cham (2024)
41. Nikivorov, A.: Exploring Blockchain technology and its impact in E-commerce. J. Soc. Sci. **16**(XVI), 55–95 (2024)
42. Pujiono, I.P., Asfahani, A., Rachman, A.: Augmented reality (AR) and virtual reality (VR): recent developments and applications in various industries. Innovative. **4**(4), 1679–1690 (2024)
43. Schwab, K.: The fourth industrial revolution-what it means and how to respond. In: Handbook of Research on Strategic Leadership in the Fourth Industrial Revolution, p. 29. Edward Elgar Publishing, Northampton (2024)
44. Yao, X., et al.: Enhancing wisdom manufacturing as industrial metaverse for industry and society 5.0. J. Intell. Manuf. **35**(1), 235–255 (2024)
45. Rane, N., et al.: Emerging trends and future research opportunities in artificial intelligence, machine learning, and deep learning. Artif. Intell. Ind. Soc. **5**, 2–96 (2024)
46. Bogoviz, A.V.: Industry 4.0 as a new vector of growth and development of knowledge economy. In: Industry 4.0: Industrial Revolution of the 21st Century, pp. 85–91. Springer, Cham (2019)
47. Bartodziej, C.J., Bartodziej, C.J.: The Concept Industry 4.0. Springer, Cham (2017)
48. Boehlé, N.H.a.L.: Energy efficiency. https://www.iea.org/energy-system/energy-efficiency-and-demand/energy-efficiency (2024)
49. Lampropoulos, G., Siakas, K.: Enhancing and securing cyber-physical systems and industry 4.0 through digital twins: a critical review. J. Softw. Evol. Process. **35**(7), e2494 (2023)
50. Adeniran, A.O., et al.: Implementing cloud-centric IoT transformations: merits and demerits. Syst. Anal. **2**(2), 174–187 (2024)
51. Michailidis, I.T., et al.: Energy-efficient HVAC management using cooperative, self-trained, control agents: a real-life German building case study. Appl. Energy. **211**, 113–125 (2018)

52. Elnour, M., et al.: Neural network-based model predictive control system for optimizing building automation and management systems of sports facilities. Appl. Energy. **318**, 119153 (2022)
53. Du, Z., et al.: IoT intelligent agent based cloud management system by integrating machine learning algorithm for HVAC systems. Int. J. Refrig. **146**, 158–173 (2023)
54. Yussuf, R.O., Asfour, O.S.: Applications of artificial intelligence for energy efficiency throughout the building lifecycle: an overview. Energ. Buildings. **305**, 113903 (2024)
55. Krishna, B.S., et al.: Enhanced efficiency in smart grid energy systems through advanced AI-based thermal modeling. Thermal Sci. Eng. Progr. **53**, 102765 (2024)
56. Heidarykiany, R., Ababei, C.: HVAC energy cost minimization in smart grids: a cloud-based demand side management approach with game theory optimization and deep learning. Energy AI. **16**, 100362 (2024)
57. Zhou, Y., Liu, J.: Advances in emerging digital technologies for energy efficiency and energy integration in smart cities. Energ. Buildings. **315**, 114289 (2024)
58. Zeydan, E., Arslan, S., Turk, Y.: 6G wireless communications for industrial automation: scenarios, requirements and challenges. J. Ind. Inf. Integr. **42**, 100732 (2024)
59. Inserra, D., Michel, A.: Guest editorial on advanced antennas for radio frequency identification (RFID) systems. IEEE J. Radio Freq. Identif. **8**, 419–420 (2024)
60. Okafor, P., James, G., Ituma, C.: Design of an Intelligent Radio Frequency Identification (RFID) based cashless vending machine for sales of drinks. Br. J. Comput. Netw. Inf. Technol. **7**(3), 36–57 (2024)
61. Szabó, G., Pető, J.: Intelligent wireless resource management in industrial camera systems: reinforcement learning-based AI-extension for efficient network utilization. Comput. Commun. **216**, 68–85 (2024)
62. Chen, Y., et al.: Wireless sensing for material identification: a survey. IEEE Commun Surv Tutor. (2024). https://doi.org/10.1109/COMST.2024.3456076
63. Ouaissa, M., et al.: Low-Power Wide Area Network for Large Scale Internet of Things: Architectures, Communication Protocols and Recent Trends. CRC Press, Boca Raton (2024)
64. Josbert, N.N., et al.: A look into smart factory for industrial IoT driven by SDN technology: a comprehensive survey of taxonomy, architectures, issues and future research orientations. J. King Saud Univ. **36**, 102069 (2024)
65. Zhu, S., et al.: Intelligent computing: the latest advances, challenges, and future. Intell. Comput. **2**, 0006 (2023)
66. Nguyen, Q.-T., et al.: Decision support systems for anomaly detection with the applications in smart manufacturing: a survey and perspective. In: Machine Learning and Probabilistic Graphical Models for Decision Support Systems, pp. 34–61. CRC Press, Boca Raton (2022)
67. Leszkiewicz, A., Hormann, T., Krafft, M.: Smart business and the social value of AI. In: Smart Industry–Better Management, pp. 19–34. Emerald Publishing Limited, Leeds (2022)
68. Ahmed, A.A., Nazzal, M.A., Darras, B.M.: Cyber-physical systems as an enabler of circular economy to achieve sustainable development goals: a comprehensive review. Int. J. Precis. Eng. Manuf.-Green Technol. **9**, 1–21 (2021)
69. Hassoun, A., et al.: Exploring the role of green and industry 4.0 technologies in achieving sustainable development goals in food sectors. Food Res. Int. **162**, 112068 (2022)
70. Avilés-Sacoto, S.V., et al.: Innovating for impact: proposing quality 4.0 integration to achieve sustainable development goals in microenterprises. Business Strat. Dev. **7**(2), e371 (2024)
71. Aravindaraj, K., Chinna, P.R.: A systematic literature review of integration of industry 4.0 and warehouse management to achieve sustainable development goals (SDGs). Clean. Log. Supply Chain. **5**, 100072 (2022)
72. Soori, M., Arezoo, B., Dastres, R.: Internet of things for smart factories in industry 4.0, a review. Internet Things Cyber-Phys. Syst. **3**, 192–204 (2023)
73. Rath, K.C., Khang, A., Roy, D.: The role of internet of things (IoT) technology in industry 4.0 economy. In: Advanced IoT Technologies and Applications in the Industry 4.0 Digital Economy, pp. 1–28. CRC Press, Boca Raton (2024)

74. Unhelkar, B., et al.: Enhancing supply chain performance using RFID technology and decision support systems in the industry 4.0–a systematic literature review. Int. J. Inf. Manage. Data Insights. **2**(2), 100084 (2022)
75. Fasuludeen Kunju, F.K., et al.: Production and maintenance in industries: impact of industry 4.0. Ind. Robot. **49**(3), 461–475 (2022)
76. Kolasani, S.: Revolutionizing manufacturing, making it more efficient, flexible, and intelligent with industry 4.0 innovations. Int. J. Sustain. Dev. **3**(1), 1–17 (2024)
77. Agrawal, K., Nargund, N.: Deep learning in industry 4.0: transforming manufacturing through data-driven innovation. In: International Conference on Distributed Computing and Intelligent Technology. Springer, Cham (2024)
78. Ghasemi, A., et al.: Simulation optimization applied to production scheduling in the era of industry 4.0: a review and future roadmap. J. Ind. Inf. Integr. **39**, 100599 (2024)
79. Torkul, O., Selvi, I.H., Şişci, M.: Smart seru production system for industry 4.0: a conceptual model based on deep learning for real-time monitoring and controlling. Int. J. Comput. Integr. Manuf. **37**(4), 385–407 (2024)
80. Golec, M., et al.: Master: machine learning-based cold start latency prediction framework in serverless edge computing environments for industry 4.0. IEEE J. Select. Areas Sens. **1**, 36–48 (2024)
81. Alabadi, M., Habbal, A., Guizani, M.: An innovative decentralized and distributed deep learning framework for predictive maintenance in the industrial internet of things. IEEE Internet Things J. **11**, 20271–20286 (2024)
82. Rane, N., Kaya, O., Rane, J.: Advancing the sustainable development goals (SDGs) through artificial intelligence, machine learning, and deep learning. Artif. Intell. Mach. Learn. Deep Learn. Sustain. Ind. **5**, 2–74 (2024)
83. Agarwal, A., Ojha, R.: Prioritizing implications of Industry-4.0 on the sustainable development goals: a perspective from the analytic hierarchy process in manufacturing operations. J. Clean. Prod. **444**, 141189 (2024)
84. Sarkar, P., Khanapuri, V.B., Tiwari, M.K.: Strategic decision-making for sustainable production and distribution in automotive industry: a machine learning enabled dynamic multi-objective optimisation. Int. J. Prod. Res. **63**, 1–24 (2024)
85. Pachouri, V., et al.: Empowering sustainability in the built environment: a technological lens on industry 4.0 enablers. Technol. Soc. **76**, 102427 (2024)
86. Cimino, A., et al.: Automated simulation modeling: ensuring resilience and flexibility in industry 4.0 manufacturing systems, vol. 232, pp. 1011–1024. Proc. Comput. Sci. (2024)
87. Ekengwu, I.E., Emeruwa, K.I.: Comprehensive review of computer numerical control (CNC) systems. Int. Res. J. Sci. Stud. **1**(1), 50–55 (2024)
88. Gothwal, S., et al.: DSP and other related alternate technologies to CNC in FMS: a review. Int. J. Syst. Assur. Eng. Manag. **15**, 1–10 (2024)
89. Soori, M., et al.: Robotical automation in CNC machine tools: a review. Acta Mech. Automat. **18**, 434–450 (2024)
90. Pei, F., et al.: Collaborative optimization of a matrix manufacturing system based on overall equipment effectiveness. Chin. J. Mech. Eng. **37**(1), 109 (2024)
91. Mourtzis, D., Angelopoulos, J., Panopoulos, N.: Digital manufacturing: the evolution of traditional manufacturing toward an automated and interoperable smart manufacturing ecosystem. In: The Digital Supply Chain, pp. 27–45. Elsevier, Amsterdam (2022)
92. Maldonado, J.L.B., et al.: Manual de la industria 4.0 y su contexto en la industria automotriz, mínera y de petróleo. Editorial Tinta & Pluma (2023)
93. Tavakoli, M.: A proposed blue ocean strategy for the metal sheet processing industry: leveraging a funding proposal for sustainable growth (2024)
94. Tran, T.-A., et al.: Retrofitting-based development of brownfield industry 4.0 and industry 5.0 solutions. IEEE Access. **10**, 64348–64374 (2022)
95. Ekundayo, T.C., Swalaha, F.M., Ijabadeniyi, O.A.: Socioeconomic indices guided linear mixed-effects and meta-regression modelling of the temporal, global and regional prevalence

of helicobacter pylori in environmental waters: a class I carcinogen. J. Environ. Manag. **342**, 118282 (2023)

96. Fuertes, J.J., et al.: Environment for education on industry 4.0. IEEE Access. **9**, 144395–144405 (2021)
97. Mohamed, N.N., Abuobied, B.H.H.: Cybersecurity challenges across sustainable development goals: a comprehensive review. Sustain. Eng. Innov. **6**(1), 57–86 (2024)
98. Obasi, S.C., et al.: Cybersecurity's role in environmental protection and sustainable development: bridging technology and sustainability goals. Comput. Sci. IT Res. J. **5**(5), 1145–1177 (2024)
99. Val, O.O., et al.: Real-time data governance and compliance in cloud-native robotics systems. J. Eng. Res. Rep. **26**(11), 222–241 (2024)
100. Rakshit, P., et al.: Artificial intelligence in digital twins for sustainable future. In: Transforming Industry Using Digital Twin Technology, pp. 19–44. Springer, Cham (2024)
101. Aiken, W., et al.: Strategic digitalization in oil and gas: a case study on mixed reality and digital twins. IEEE Access. **12**, 87248–87267 (2024)
102. Rekeraho, A., et al.: Cybersecurity challenges in IoT-based smart renewable energy. Int. J. Inf. Secur. **23**(1), 101–117 (2024)
103. Badran, A., Gaber, S.A., Haggag, R.: Proposed of cybersecurity framework for reducing the security risks on the Egyptian petroleum sector. 2024) 1301–1247 ,(3)**38** .المجلة العلمية للبحوث والدراسات التجارية)
104. Abdulkreem, R.Z., et al.: Influences of web technology and cloud computing on life cycle assessment in the age of AI and IoT: a review of sustainable design and production within enterprise systems
105. Vijarania, M., et al.: Achieving sustainable development goals in cyber security using AIoT for healthcare application. In: Artificial Intelligence of Things for Achieving Sustainable Development Goals, pp. 207–231. Springer, Cham (2024)
106. Zeb, S., et al.: Towards defining industry 5.0 vision with intelligent and softwarized wireless network architectures and services: a survey. J. Netw. Comput. Appl. **223**, 103796 (2023)
107. Zaidi, S.A.H., Khan, S.A., Chaabane, A.: Unlocking the potential of digital twins in supply chains: a systematic review. Supply Chain Anal. **7**, 100075 (2024)
108. Zhao, Z., et al.: Spatial-temporal traceability for cyber-physical industry 4.0 systems. J. Manuf. Syst. **74**, 16–29 (2024)
109. Bag, S., et al.: How big data analytics can help manufacturing companies strengthen supply chain resilience in the context of the COVID-19 pandemic. Int. J. Logist. Manage. **34**(4), 1141–1164 (2023)
110. Bag, S., et al.: Unveiling the impact of carbon-neutral policies on vital resources in industry 4.0 driven smart manufacturing: a data-driven investigation. Comput. Ind. Eng. **187**, 109798 (2024)
111. Wu, H., Li, G., Zheng, H.: How does digital intelligence technology enhance supply chain resilience? Sustainable framework and agenda. Ann. Oper. Res. (2024). https://doi.org/10.1007/s10479-024-06104-3
112. Azari, M., et al.: Navigating challenges and leveraging future trends in AI and machine learning for supply chains. In: AI and Machine Learning Applications in Supply Chains and Marketing, pp. 257–282. IGI Global, London (2025)
113. Boujarra, M., et al.: Revolutionizing logistics through deep learning: innovative solutions to optimize data security. J. Theor. Appl. Inf. Technol. **102**, 1593–1607 (2024)
114. Kocaoglu, B.: Digital transformation in logistics. In: Logistics Information Systems: Digital Transformation and Supply Chain Applications in the 4.0 Era, pp. 1–35. Springer, Cham (2024)
115. Bradu, P., et al.: Recent advances in green technology and industrial revolution 4.0 for a sustainable future. Environ. Sci. Pollut. Res. Int. **30**, 124488 (2022)
116. Siekmann, F., Schlör, H., Venghaus, S.: Linking sustainability and the fourth industrial revolution: a monitoring framework accounting for technological development. Energy Sustain. Soc. **13**(1), 26 (2023)

117. Taj, I., Jhanjhi, N.: Towards industrial revolution 5.0 and explainable artificial intelligence: Challenges and opportunities. Int. J. Comput. Dig. Syst. **12**(1), 285–311 (2022)
118. Qiu, T., et al.: Edge computing in industrial internet of things: architecture, advances and challenges. IEEE Commun Surv Tutor. **22**(4), 2462–2488 (2020)
119. Goga, A.-S., et al.: The proliferation of artificial intelligence in the forklift industry—an analysis for the case of Romania. Sustainability. **16**(21), 9306 (2024)
120. Santoso, A., Surya, Y.: Maximizing decision efficiency with edge-based AI systems: advanced strategies for real-time processing, scalability, and autonomous intelligence in distributed environments. Q. J. Emerg. Technol. Innov. **9**(2), 104–132 (2024)
121. Demir, H., Sarı, F.: The effect of artificial intelligence and industry 4.0 on robotic systems. Eng. Energy Mater. **20**, 51–72 (2020)
122. Rahman, M.M., et al.: Cobotics: the evolving roles and prospects of next-generation collaborative robots in industry 5.0. J. Robot. **2024**(1), 2918089 (2024)
123. Jeyabalan, J., et al.: Cobots in smart manufacturing and production for industry 5.0. In: Digital Twins in Industrial Production and Smart Manufacturing: An Understanding of Principles, Enhancers, and Obstacles, pp. 201–220. Wiley, Hoboken (2024)
124. Raffik, R., et al.: Emerging technologies to enhance human-machine interaction and to facilitate industrial paradigm shift to industry 5.0: a comprehensive review. In: Human-Centered Approaches in Industry 5.0: Human-Machine Interaction, Virtual Reality Training, and Customer Sentiment Analysis, pp. 1–23. IGI Global, London (2024)
125. Nikolola-Alexieva, V., Valeva, K., Pashev, S.: Artificial intelligence in the food industry. BIO Web Conf. **102**, 04002 (2024)
126. Logeswaran, K., et al.: Unifying technologies in industry 4.0: harnessing the synergy of internet of things, big data, augmented reality/virtual reality, and Blockchain technologies. Top. Artif. Intell. Appl. Ind. **40**, 127–147 (2024)
127. Isik, B., Isik, G.E., Zilka, M.: Integrating extended reality in industrial maintenance: a game-based framework for compressed air system training. Proc. Comput. Sci. **232**, 483–492 (2024)
128. Jafar, M.R., et al.: Additive manufacturing in the age of industry 4.0 and beyond. In: Advances in Pre-and Post-Additive Manufacturing Processes, pp. 213–230. CRC Press, Boca Raton (2024)
129. Nassif, J., Tekli, J., Kamradt, M.: Background and technologies. In: Synthetic Data: Revolutionizing the Industrial Metaverse, pp. 33–74. Springer, Cham (2024)
130. Hassoun, A., et al.: The fourth industrial revolution in the food industry—part II: emerging food trends. Crit. Rev. Food Sci. Nutr. **64**(2), 407–437 (2024)
131. Dionisio, M., et al.: The role of digital social innovations to address SDGs: a systematic review. Environ. Dev. Sustain. **26**(3), 5709–5734 (2024)
132. Gooroochurn, M., Stopforth, R.: Robotics and automated systems for enabling an industry 4.0 transformation in Mauritius. In: Artificial Intelligence, Engineering Systems and Sustainable Development: Driving the UN SDGs, pp. 195–206. Emerald Publishing Limited, Leeds (2024)
133. Hiran, K.K., Doshi, R., Patel, M.: Modern technology in healthcare and medical education: blockchain, IoT, AR, and VR: Blockchain, IoT, AR, and VR. IGI Global, London (2024)
134. Scott-Stevenson, J.: XR and documentary–affinities and resistance. In: The Oxford Handbook of American Documentary. Oxford University Press, Oxford (2025)
135. Pelton, J.N., Madry, S.: Space systems, quantum computers, big data and sustainability: new tools for the United Nations sustainable development goals. In: Artificial Intelligence for Space: AI4SPACE, pp. 53–104. CRC Press, Boca Raton (2024)
136. Urgelles, H., et al.: In-network quantum computing for future 6G networks. Adv. Quant. Technol. **8**, 2300334 (2024)
137. Schuh, G., Bennemann, F.: Concept for the identification of applications for paradigm-shifting technologies on the example of quantum computing. J. Prod. Syst. Logist. **4**, 1 (2024)
138. Ferdaus, M.M., et al.: Digital technologies for a net-zero energy future: a comprehensive review. Renew. Sust. Energ. Rev. **202**, 114681 (2024)

139. Chaiboonsri, C., Eakkapun, P., Thongkaw, B.: Applying quantum computing to the renewable energy stock exchange to reduce the risk of investment. In: World Conference on Information Systems for Business Management. Springer, Cham (2023)
140. Asl, M.G., et al.: Dynamic connectedness of quantum computing, artificial intelligence, and big data stocks on renewable and sustainable energy. Energy Econ. **140**, 108017 (2024)
141. Ozen, C., Gedikli, A.: Artificial intelligence (AI) in the sustainable energy industry. In: Management, Technology, and Economic Growth in Smart and Sustainable Cities, pp. 218–229. IGI Global, London (2023)
142. Oliveira, D., Alvelos, H., Rosa, M.J.: Quality 4.0: results from a systematic literature review. TQM J. **37**(2), 379–456 (2024)
143. Chhetri, K.B.: Applications of artificial intelligence and machine learning in food quality control and safety assessment. Food Eng. Rev. **16**(1), 1–21 (2024)
144. Goecks, L.S., et al.: Industry 4.0 and smart systems in manufacturing: guidelines for the implementation of a smart statistical process control. Appl. Syst. Innov. **7**(2), 24 (2024)
145. Islam, M.R., et al.: Deep learning and computer vision techniques for enhanced quality control in manufacturing processes. IEEE Access. **12**, 121449–121479 (2024)
146. Warke, V., et al.: Improving the useful life of tools using active vibration control through data-driven approaches: a systematic literature review. Eng. Appl. Artif. Intell. **128**, 107367 (2024)
147. Mallioris, P., Aivazidou, E., Bechtsis, D.: Predictive maintenance in industry 4.0: a systematic multi-sector mapping. CIRP J. Manuf. Sci. Technol. **50**, 80–103 (2024)
148. Siddiqui, M.M.U.Z., Tabassum, A.: Condition-based monitoring techniques and algorithms in 3d printing and additive manufacturing: a state-of-the-art review. Progr. Add. Manuf. **10**, 1–48 (2024)
149. Delbari, S.A., Hof, L.A.: Glass waste circular economy-advancing to high-value glass sheets recovery using industry 4.0 and 5.0 technologies. J. Clean. Prod. **462**, 142629 (2024)
150. Li, Y., et al.: Artificial intelligence-based methods for renewable power system operation. Nat. Rev. Electr. Eng. **1**(3), 163–179 (2024)
151. Bogmans, C., et al.: Power Hungry: How AI Will Drive Energy Demand. International Monetary Fund, Washington (2025)
152. Tavva, R.: Time-sensitive networking in advanced manufacturing environments: a framework for industry 4.0 implementation. J. Comput. Sci. Technol. Stud. **7**(6), 672–678 (2025)
153. Manikandan, K., et al.: Transforming smart factories: real-time optimisation of cyber-physical systems with digital twins. In: International Conference on Broadband and Wireless Computing, Communication and Applications. Springer, Cham (2024)
154. Khoshsirat, A.: Energy efficient edge computing (2025)
155. Wang, X., et al.: Empowering edge intelligence: a comprehensive survey on on-device AI models. ACM Comput. Surv. **57**(9), 1–39 (2025)
156. AbuGhanem, M., Eleuch, H.: NISQ computers: a path to quantum supremacy. IEEE Access. (2024). https://doi.org/10.1109/ACCESS.2024.3432330
157. Ganeshamurthy, P.A., et al.: Next generation power system planning and operation with quantum computation. IEEE Access. (2024). https://doi.org/10.1109/ACCESS.2024.3509743
158. Cai, D., Miao, C., Dong, D.: Sustainability assessment of green energy in cloud computing environments. In: International Conference on Artificial Intelligence for Society. Springer, Cham (2024)
159. Eilam, T., et al.: Reducing datacenter compute carbon footprint by harnessing the power of specialization: principles, metrics, challenges and opportunities. IEEE Trans. Semicond. Manuf. (2024). https://doi.org/10.1109/TSM.2024.3434331
160. Ukoba, K., et al.: Additive manufacturing in the energy sector and the fourth industrial revolution. Progr. Add. Manuf. **10**, 1–21 (2025)
161. Sousa, J., et al.: Artificial intelligence for control in laser-based additive manufacturing: a systematic review. IEEE Access. (2025). https://doi.org/10.1109/ACCESS.2025.3537859
162. Wang, Y., et al.: Ontology of 3D virtual modeling in digital twin: a review, analysis and thinking. J. Intell. Manuf. **36**(1), 95–145 (2025)

163. Garcia-Gasulla, M., Mantsinen, M.J.: Challenges and opportunities in exascale fusion simulations. Nat. Rev. Phys. **7**, 1–10 (2025)
164. Garlisi, D., et al.: Interference analysis of LoRaWAN and Sigfox in large-scale urban IoT networks. IEEE Access. (2025). https://doi.org/10.1109/ACCESS.2025.3550014
165. Islam, U., et al.: SecureGuard-IIoMT: a novel adaptive physical security framework for enhancing industrial internet of medical things (IIoMT) device hardening. Internet Things. **33**, 101653 (2025)
166. Gangwal, A., et al.: Current strategies to address data scarcity in artificial intelligence-based drug discovery: a comprehensive review. Comput. Biol. Med. **179**, 108734 (2024)
167. Rakhmatulin, I., et al.: A review of automation of laser optics alignment with a focus on machine learning applications. Opt. Lasers Eng. **173**, 107923 (2024)
168. Park, J.-G., et al.: Patch and model size characterization for on-device efficient-ViTs on small datasets using 12 quantitative metrics. IEEE Access. (2025). https://doi.org/10.1109/ACCESS.2025.3536471
169. Guerrouj, F.Z., et al.: Quantized object detection for real-time inference on embedded GPU architectures. Int. J. Adv. Comput. Sci. Appl. **16**(5), 0160503 (2025)
170. Jiang, Y., et al.: Recovering permuted sequential features for effective reinforcement learning. Neural Netw. **182**, 106795 (2025)
171. Qu, T., et al.: DWAS-RL: a safety-efficiency balanced reinforcement learning approach for path planning of unmanned surface vehicles in complex marine environments. Ocean Eng. **317**, 119641 (2025)
172. Gu, Z., et al.: Climb-Odom: a robust and low-drift RGB-D inertial odometry with surface continuity constraints for climbing robots on freeform surface. Inf. Fus. **117**, 102880 (2025)
173. Islam, M.S., Kim, K., Kim, H.-Y.: Data-driven approach for fault diagnosis of harmonic drives using wireless acceleration sensors and machine learning. Int. J. Precis. Eng. Manuf.-Green Technol. **12**, 1–18 (2025)
174. Roose, P., et al.: Energy measurement system for data lake. In: ACIIDS 2024-16th Asian Conference on Intelligent Information and Database Systems. Springer, Cham (2024)
175. Ali El-Sayed Ali, H., Alham, M., Ibrahim, D.K.: Big data resolving using apache spark for load forecasting and demand response in smart grid: a case study of low carbon London project. J. Big Data. **11**(1), 59 (2024)
176. Yang, W.: Incentivizing Blockchain participation through task assignment mechanisms: evidence from a natural experiment of consensus protocols on Ethereum. J. Oper. Manag. **71**, 964–987 (2025)
177. Asiamah, E.A., et al.: A storage-efficient learned indexing for blockchain systems using a sliding window search enhanced online gradient descent. J. Supercomput. **81**(1), 321 (2025)
178. Pose-Díez-de-la-Lastra, A., et al.: Microsoft HoloLens 2 vs. tablet-based augmented reality and 3D printing for fronto-orbital reconstruction of craniosynostosis: a case study. 3D Print. Med. **11**(1), 13 (2025)
179. Stanescu, A., et al.: Error management for augmented reality assembly instructions. In: 2024 IEEE International Symposium on Mixed and Augmented Reality (ISMAR). IEEE, New York (2024)
180. Feld, N., et al.: Perceptual issues in mixed reality: a developer-oriented perspective on video see-through head-mounted displays. In: 2024 IEEE International Symposium on Mixed and Augmented Reality Adjunct (ISMAR-adjunct). IEEE, New York (2024)
181. Lougiakis, C., et al.: Comparing physics-based hand interaction in virtual reality: custom soft body simulation vs. off-the-shelf integrated solution. In: 2024 IEEE Conference Virtual Reality and 3D User Interfaces (VR). IEEE, New York (2024)
182. Intel. Taking Neuromorphic Computing to the Next Level with Loihi 2. Technology Brief Intel Labs' Loihi 2 Neuromorphic Research Chip and the Lava Software Framework. https://download.intel.com/newsroom/2021/new-technologies/neuromorphic-computing-loihi-2-brief.pdf (2025)

183. Moyer, L.: History of energy use II, steam engines. https://geosci.uchicago.edu/~moyer/GEOS24705/2017/Slides/Slides_Lecture6.pdf (2017)
184. Li, R., et al.: Photonics for neuromorphic computing: fundamentals, devices, and opportunities. Adv. Mater. **37**(2), 2312825 (2025)
185. Cheng, Y., et al.: Photonic neuromorphic architecture for tens-of-task lifelong learning. Light Sci. Appl. **13**(1), 56 (2024)
186. Bin Ali, M., Iftikhar, T., Majeed, H.: Green synthesis of zinc oxide nanoparticles for the industrial biofortification of (Pleurotus pulmonarius) mushrooms. Heliyon. **10**(19), e37927 (2024)
187. Foundation, O. Companion specs drive IIoT adoption – from beyond [sponsored]. https://opcconnect.opcfoundation.org/2018/06/companion-specs-drive-iiot-adoption-from-beeond-sponsored/ (2018)
188. Foundation, O. Semantic data and contextual models for factory floor devices. https://opcfoundation.org/markets-collaboration/mtconnect/ (2019)
189. Liu, Y., Wang, Y.: A small size, high gain, radiation pattern reconfigurable Fabry–Pérot cavity antenna for 5G communication. IEEE Antennas Wireless Prop. Lett. **23**, 1690–1694 (2024)
190. Sun, C., et al.: A review of research on the security of train control networks. In: 2024 6th International Conference on System Reliability and Safety Engineering (SRSE). IEEE, New York (2024)
191. Unger, S., Katzenbeisser, S.: Academic paper hardware-basierte sicherung von Eisenbahn-intrusion-detection-systemen. Tagungsband. **20**, 23 (2024)
192. Samuel, A.J.: Optimizing energy consumption through AI and cloud analytics: addressing data privacy and security concerns (2024)
193. Palomares, J., et al.: Toward field-level device orchestration in industrial multiaccess edge computing deployments: a unified IT–OT framework. IEEE Ind. Electron. Mag. **19**, 40–50 (2025)
194. Jafarigiv, D., Sheshyekani, K., Kassouf, M.: Tensor-based cybersecurity analysis of smart grids using IT/OT convergence. IEEE Access. **12**, 191893–191906 (2024)
195. Shi, Y., Beer, M.: Physics-informed neural network classification framework for reliability analysis. Expert Syst. Appl. **258**, 125207 (2024)
196. Unger, M., Sai, S.: A multi physical digital representation of the space capsule splashdown event in the simcenter environment to accelerate analysis of impact condition, structural design and human body loading EUCASS-6147 (2022)
197. Emre Börekçi, N.: Real case studies in industry 5.0: the example of Nvidia. In: Business Challenges and Opportunities in the Era of Industry 5.0, pp. 215–235. Emerald Publishing Limited, Leeds (2025)
198. Zhang, J., et al.: Digital twin and the asset administration shell: an analysis of the three types of AASs and their feasibility for digital twin engineering. Softw. Syst. Model. **24**, 1–23 (2025)
199. Danger, R., et al.: 2nd International Workshop on Exploiting Large Knowledge Repositories and the 1st International Workshop on Automatic Text Summarization for the Future (2012)
200. Miron, R., Hulea, M., Muresan, V.: Transforming lifecycle management: a systematic review of Blockchain-driven digital product passports in circular economy frameworks. In: International Conference Innovation in Engineering. Springer, Cham (2025)
201. Khwa, W.-S., et al.: A mixed-precision memristor and SRAM compute-in-memory AI processor. Nature. **639**, 1–7 (2025)
202. Laimon, M.: Renewable energy curtailment: a problem or an opportunity? Results Eng. **26**, 104925 (2025)
203. Yunus, M.S.F.M., et al.: Model predictive control-based energy management system with load curtailment for standalone microgrid. IIUM Eng. J. **26**(1), 169–186 (2025)
204. Iskandar, V., Ghany, M.A.A.E., Goehringer, D.: Near-memory computing on fpgas with 3d-stacked memories: applications, architectures, and optimizations. ACM Trans. Reconfig. Technol. Syst. **16**(1), 1–32 (2022)
205. Meng, X., et al.: High-integrated photonic tensor core utilizing high-dimensional lightwave and microwave multidomain multiplexing. Light. **14**(1), 27 (2025)

Chapter 4
Intelligent Manufacturing in Industry 4.0 with IOT-Driven Automation and Smart Systems

Abstract With an emphasis on real-time data processing, IoT-based automation, and cutting-edge computing technologies, the integration of intelligent manufacturing systems within Industry 4.0 required extensive precision. Industry 4.0 emphasizes automation, efficiency, and intelligent decision-making through the integration of digital and physical systems. IoT architecture, which allows interconnected systems, sensors, and actuators to gather, process, and react to data in real time, is a key element of this change. The functions of several sensor types, such as piezoelectric, optical and MEMS-based sensors, in environmental monitoring, industrial automation, and medical diagnostics are investigated. Edge devices enable localized computation and automation powered by machine learning. The importance of CPS in maintaining real-time synchronization between digital and physical manufacturing environments is indispensable. The potential of Industry 6.0 innovations to improve sustainability, efficiency, and flexibility is investigated. Examples of these innovations include blockchain-based supply chain transparency, AI-driven predictive analytics, and decentralized smart manufacturing. AR and digital twins might improve industrial workflows. To reduce industrial waste and carbon footprints, sustainable manufacturing practices like circular economy strategies, energy efficient AI models, and environmentally friendly material recovery techniques, align with the SDGs.

Keywords Intelligent manufacturing · IoT · Edge computing · Smart sensors · Industry 4.0 · Industry 6.0 · Cyber physical systems · Digital twins · Blockchain · Augmented reality · Sustainable manufacturing · Artificial intelligence

1 Introduction of IoT in Intelligent Manufacturing

A network of linked systems, sensors, and devices that gather, exchange, and process data in real time is known as the IoT. To build intelligent, automated systems, it incorporates concepts from a variety of fields, such as engineering, computer science, information technology, data science, and the basic sciences. Below is a

H. Majeed, T. Iftikhar, *Intelligent Manufacturing in Industry 6.0*,
https://doi.org/10.1007/978-3-032-07278-8_4

thorough technical description of the IoT and its uses in a variety of businesses. Important IoT topics and ideas are methodically covered below [1].

2 IoT Architecture and Components

The organized, multi-layered architecture that enables the IoT allows for smooth communication between digital systems and physical things. The hardware layer, which consists of sensors and actuators, is the cornerstone of this architecture. These elements make up the vital link between the computer systems that process, evaluate, and act upon data and the real environment. Sensors are essential to IoT systems because they gather operational and environmental data. Temperature, pressure, pH, humidity, vibration, motion, and chemical composition are just a few of the factors that these devices are made to detect and monitor. Examples include piezoelectric sensors that produce electric charges in reaction to mechanical stress, optical sensors used to detect changes in color and light intensity, and MEMS based sensors used for high-precision and miniature applications. Advanced semiconductor fabrication techniques are used in the production of these sensors to produce small, energy-efficient devices, and nanotechnology is used to improve sensitivity and specificity in applications, such as healthcare diagnostics and environmental monitoring [2]. Actuators allow IoT systems to control their surroundings by transforming electrical impulses into mechanical actions. Pneumatic actuators, which are frequently used in industrial automation to control valves or other machinery, solenoids for linear motion applications, and servo motors, which offer exact angular or linear positioning, are typical examples. These gadgets are essential to IoT enabled automation, in which instantaneous physical adjustments are made to maximize system performance based on real-time sensor feedback. One example of how IoT systems are interconnected is the cooperation between sensors and actuators. Sensors gather data from the real world, which is then analyzed by layers of computation to produce insights that can be used. Actuators are subsequently motivated by these insights to carry out duties that range from completing precision manufacturing operations to modifying/controlling environmental conditions. The potential uses of IoT in various industries are being expanded by the continuous development of increasingly complex sensors and actuators and their incorporation into IoT systems. IoT is becoming more effective, dependable, and accessible for jobs ranging from intelligent energy management to remote healthcare monitoring by utilizing developments in wireless communication technology, downsizing, and material science. In addition to enabling real-time data exchange, this fundamental hardware layer establishes the framework for improved automation, sustainability, and resilience in both everyday and industrial settings [3]. More advanced knowledge about the sensors and actuators are as under.

2.1 *Piezoelectric Sensor*

Transduction, which transforms a physical or chemical property into an electrical signal, is the basis for how sensors work. Utilize the piezoelectric effect, which occurs when certain materials (quartz/PZT ceramics) produce an electric charge when subjected to mechanical stress. The piezoelectric effect, a basic characteristic of some materials that produces an electric charge in response to mechanical stress, is used by piezoelectric sensors, which are precise instruments. In applications where force, pressure, vibration, or acceleration must be detected, these sensors are essential. Natural crystals, artificial ceramics, polymers, and cutting-edge nanomaterials are the several types of materials used in piezoelectric sensors. While synthetic ceramics like lead zirconate titanate (PZT) and barium titanate ($BaTiO_3$) offer excellent piezoelectric coefficients and broad applicability, quartz (SiO_2) is a commonly utilized natural crystal that is known for its stability and precision. Because of their charge distribution and lattice symmetry, crystals are inherently piezoelectric. High-frequency oscillators and precise sensors in stable settings are two examples of their uses [4, 5].

2.1.1 Theory and Mechanism

The piezoelectric effect mechanism is explained below. Mostly the crystals are in symmetrical arrangements of atoms. Only non-centrosymmetric crystals (lacking a center of symmetry) exhibit piezoelectricity due to stress induction which leads to a shift in positive and negative charge centers, generating a net dipole moment. The direct piezoelectric effect is typically given by

$$P_{\mathrm{i}} = d_{\mathrm{ijk}}.\sigma_{\mathrm{jk}}$$

where *Pi* is polarization vector, d_{ijk} is the piezoelectric coefficient tensor, and *σij is the stress tensor*. For a simplified one dimensional case, it is often written as

$$Q = d \cdot F$$

Q is the generated electric charge, *d* is the effective piezoelectric coefficient, and *F* is the applied force.

Optical sensors use photodetectors or fiber optics to identify variations in light intensity, wavelength, or phase.

Snell's Law regarding refractive properties is also important to discuss here. This law defined as "when light moves from one medium with a different refractive index to another, it refracts." The following is a mathematical expression for the law: $n_1\sin\theta_1=n_2\sin\theta_2$.

The first and second media's refractive indices are denoted by n_1 and n_2, respectively. Angles of incidence and refraction, expressed in relation to the boundary normal, are denoted by θ_1 and θ_2. Index of Refraction (*n*) is defined as the ratio of light's velocity in a vacuum (*c*) to its velocity in a medium (*U*), whereas, $n = c/U$.

When light enters a medium with a differing optical density, its speed changes, causing refraction. As light moves from air ($n \approx 1$) to glass ($n \approx 1.5$), it slows down. Snell's Law guarantees complete internal reflection for effective signal transmission in fiber optics, where precise control of light refraction is essential for building optical fibers. This law also has other applications like in lenses, where creating high-performing lenses for telescopes, cameras, and microscopes requires an understanding of refraction. Snell's Law also guarantees precise beam bending for 3D imaging in laser-based holography. Cloaking is made possible by Snell's Law, which behaves in reverse in engineered metamaterials with negative refractive indices ($n_2 < 0$). Snell's Law is dynamically modified during high-power laser interactions in nonlinear optics, such as in nonlinear media, where the refractive index is dependent on light intensity. Another scientific phenomenon is used in piezoelectric sensors, which is photoelectric effect. When a substance absorbs light with energy higher than its work function (ϕ), electrons are ejected from the material, a phenomenon known as the photoelectric effect. Einstein's equation governs this phenomenon: E photon = hf = ϕ + K.E. Where: Planck's constant ($6.62607015 \times 10^{-34}$ J/s) is represented as h, Incident light frequency is f. Work function (ϕ) is the least amount of energy required to knock out an electron. KE is the expelled electron's kinetic energy. Because photons transfer quantized energy to electrons, the photoelectric effect supports the particle character of light, which may be explained from a quantum perspective. Because of their low work functions, metals like potassium and cesium are perfect for photoelectric applications. The foundation of photovoltaic energy conversion is the photoelectric effect, in which sunlight releases electrons to produce electricity. Another use of the photoelectric effect is photoelectron spectroscopy, which measures the kinetic energy of released electrons to analyze the electronic structure of materials. For accurate light detection, photoelectric light sensors are used in scientific instruments and cameras. By adjusting their bandgap, doping semiconductors (material engineering) improves photoelectric efficiency. The range of light frequencies that can be used in photoelectric devices is increased by advanced techniques that eject electrons using two photons (Two Photon Photoemission) of lower energy. The photoelectric effect is enhanced by nanostructures such as gold or silver nanoparticles, which concentrate light close to their surface (a process known as plasmonic enhancement). Designing optogenetics tools that use light to influence biological processes (biophysics field) is made easier by the photoelectric effect. Photoelectric device production and material selection are optimized by ML algorithms (AI and data science). With numerous uses in contemporary engineering, materials science, and optics, Snell's Law and the photoelectric effect are prime examples of the nexus between classical and quantum physics [6–9].

2.1.2 Material Science

Crystalline material, that is, one such material is quartz (SiO_2), which has a distinctive non-centrosymmetric lattice and a trigonal crystal structure. The symmetric Si-O bonds are broken by stress, which shifts charge centers and creates an electric dipole moment, as explained by the piezoelectric mechanism. Quartz is the ideal material for sensors because of its high thermal stability (Curie temperature of about 573 °C), long-term stability, and accurate frequency response. However, other minerals, such as the complicated borosilicate Tourmaline Group, comprises atoms of silicon, oxygen, aluminum, and boron arranged in a complicated lattice. Localized electric fields are produced when the ionic framework is distorted by the stress. This is inherently piezoelectric because to its chemical stability. These sensors' adaptability for flexible and nanoscale applications is increased by cutting-edge materials, such as zinc oxide (ZnO) nanowires and polyvinylidene fluoride (PVDF). The non-centrosymmetric crystal structure of materials is necessary for the piezoelectric action. The positive and negative charge centers move in response to mechanical stress, producing a detectable dipole moment and an electrical signal proportionate to the applied force. The piezoelectric coefficient (d_{ij}), a material-specific characteristic that connects mechanical stress and electrical output, controls this relationship. PZT is perfect for applications needing high sensitivity because of its high d_{33} value. Piezoelectric sensors' mechanical strength, thermal stability, and dielectric constant all affect how well they work. Effective charge storage is made possible by ceramics like PZT's high dielectric constant, and their high Curie temperature guarantees their operation in harsh conditions. Numerous industrial, automotive, aeronautical, and medical applications make extensive use of these sensors to monitor pressure, force, vibration, and acceleration [10–12].

Another type of material for sensor is ceramics. PZT is an example of ceramic material. Its perovskite structure is ABO_3, where $A = Pb^{2+}$, $B = Zr^{4+}/Ti^{4+}$. Piezoelectric Mechanism can be explained like polarization is produced under stress by the displacement of Ti/Zr ions inside the oxygen octahedron. Benefits include good dielectric performance, tuneable characteristics by doping, and a high piezoelectric coefficient (d_{33}). Another ceramic example is $BaTiO_3$ which is lead free (ecofriendly material). Its structure is an additional perovskite with $A = Ba^{2+}$ and $B = Ti^{4+}$. This material when under stress, the Ti ion moves within the oxygen framework, creating a dipole. Potassium sodium niobate (KNN) is also a lead free and environmentally safe ceramic alternate material. Its structure is $(K_{0.5}Na_{0.5})\ NbO_3$. It works by the distortion of the NbO_6 octahedron generates piezoelectricity, but its performance is moderate and compromising. Its crystal structure includes NbO^{+6} octahedral. The reason for this is that niobium (Nb) forms an octahedron when it coordinates with six oxygen atoms in a lattice that resembles perovskites. One of the basic structural units of KNN is the NbO^{+6} octahedron, which has six oxygen ions (O^{2-}) surrounding the niobium ion (Nb^{5+}) at its center. The Nb^{5+} ion slightly moves from its center location within the NbO_6 octahedron when mechanical tension is applied. An electric dipole moment is created when this displacement disrupts the symmetry of the electric field distribution inside the crystal. The fundamental mechanism that produces piezoelectricity is this process. The compound's total stoichiometry, which includes potassium (K), sodium (Na), niobium (Nb), and oxygen (O), is represented by the chemical formula $(K_{0.5}\ Na_{0.5})\ NbO_3$. However, the distortion of the NbO_6 octahedral within the lattice is associated with the piezoelectric mechanism. The electrical polarization that causes the piezoelectric effect is produced by the octahedron distortion. Thus, the NbO_6 octahedral structure has a clear relationship with the piezoelectric characteristics of KNN, validating its function in the mechanism. Since industrial revolution 6 requires eco-friendly materials, we must find and create sustainable materials that can permanently improve the health and greenness of our planet while maintaining a stable, optimal global temperature [13–15].

Another type includes polymers, because of their extreme flexibility, polymers can be made into thin films for wearable or biological devices. Despite having moderate piezoelectric coefficients, they are versatile and lightweight. PVDF sensors, which has a semi-crystalline polymer structure with alternating $-CH_2-CF_2-$monomer units, are also made from polymer materials. The piezoelectric mechanism is straightforward, that is, polarization is induced under stress by the alignment of dipoles in the crystalline phase. Flexible, lightweight, and appropriate for wearing sensors, polymer materials can be utilized in smart textiles (technical textiles) and other medical applications, among other benefits. Another polymer type is polyurethane (PU). To offer elasticity and the ability to store charge, PU is chemically composed of urethane (-NHCOO-) links. It operates by aligning polymer chains under stress, which creates dipole moments. PU based sensors are flexible and inexpensive to manufacture, among other advantages [16–18].

Nanomaterials are also another latest material that can be used in piezoelectric sensors manufacturing. ZnO (Wurtzite crystal structure with Zn and O ions) is one of the materials and its working mechanisms is that mechanical stress causes the Zn-O bonds to deform, resulting in a potential gradient. One advantageous feature of this material is its high sensitivity and nanoscale integration. In a similar vein, graphene-based nano-composites are also applicable. For improved qualities, they have a hexagonal carbon lattice with functional groups. Charge is redistributed throughout the lattice because of deformation. In addition, they are scalable, highly conductive, and flexible. Metal-organic frameworks (MOFs), which are porous frameworks containing metal ions and organic linkers, are another example of a nanomaterial. Polarization is caused by stress, which changes the orientation of polar linkers. Multifunctionality and adjustable qualities, enabling the development of lightweight, highly sensitive sensors suitable for cutting edge applications like biomedical devices and robotics. Lead-free composites having complex perovskite or layered structures like Sodium Bismuth Titanate (BNT) i.e. $BiFeO_3$ or $Na_{0.5}$ $Bi_{0.5}$ TiO_3. Electric fields are produced by ionic displacement brought on by stress. Nanomaterials make it possible to create lightweight, extremely sensitive sensors that are appropriate for cutting-edge uses like biomedical devices and robots [19–24].

The interior crystal structure of some materials, like KNN and barium titanate, interacts with mechanical stress to produce the piezoelectric effect. The behavior of ions in a non-centrosymmetric lattice, which promotes the creation of electric polarization when stress is applied, is at the heart of this phenomenon. Ion displacement happens when a piezoelectric material is subjected to mechanical stress. The center titanium ion (Ti^{4+}) in the TiO_6 octahedral unit slightly deviates from its equilibrium position in $BaTiO_3$. The local charge neutrality inside the crystal structure is essentially broken by this displacement, which throws off the balance of positive and negative charges. A crucial forerunner of the piezoelectric phenomenon localized electric dipoles are produced by the ensuing asymmetry in the ionic configuration. Dipole creation becomes more intense as the ions continue to move, and eventually these dipoles align along the applied stress's axis. Because of the significant cumulative polarization from the aligned dipoles, this alignment produces a macroscopic electric field. The crystal structure of the material and the amount of applied stress have a direct impact on the degree of polarization. Charge builds up on the surface of the substance because of this electric field. Piezoelectric materials are useful for sensor applications because the charges created by the electric field may be detected as a voltage output. These materials can operate dependably in dynamic situations because of the very efficient and reversible charge production mechanism. During material preparation, a poling procedure is used to improve the piezoelectric response. The material is subjected to a powerful external electric field at a high temperature during this procedure. As a result, even after the electric field is eliminated, the dipoles are compelled to align in one direction. By continually improving the material's polarization characteristics, the poling procedure guarantees a greater and more reliable piezoelectric output while in use. This complex process, which includes ion displacement, dipole generation, charge accumulation, and poling, is what makes piezoelectric materials work in a variety of devices, including energy harvesting systems, sensors, and actuators. Their relevance in cutting-edge technical sectors is further expanded by the capacity to control and optimize these processes through material engineering [25–27].

Piezoelectric coefficient (d_{ij}) represents the charge generated per unit mechanical stress and its values are $d_{33} \approx$ 300–600/N for PZT. Dielectric constant (ε_r) measures the material's ability to store electrical energy. Its value is mostly high in ceramics like PZT ($\varepsilon_r \approx$ 1000–4000). Curie temperature (T_c) is the temperature above which piezoelectric properties are lost. Its value for PZT is around 300 °C and for Quartz, its value is around 573 °C. Ceramics have high mechanically compressive strength but brittle. Polymers are flexible but lower mechanical strength. These materials capable of detecting signals in a wide frequency range (kHz to MHz). Inorganic ceramics are stable at high temperatures, while polymers degrade faster. Ceramics resist corrosion while polymers may degrade in harsh chemicals. Poling process requires applying a high-electric field to align dipoles for piezoelectricity. This alignment is retained unless exposed to high temperatures or mechanical stress [28–30].

2.2 *Optical Sensors for Color and Light Intensity Change Detection*

Optical sensors are instruments that detect variations in the wavelength, phase, color, or intensity of light and translate them into electrical impulses. Utilizing the concepts of photometry, colorimetry, and optical spectroscopy, these sensors find extensive application in domains such as environmental monitoring, industrial automation, and biomedical imaging. Beer-Lambert's law is used by colorimetric optical sensors to measure absorbance changes that are proportionate to analyte concentration. To detect spectrum shifts in reaction to environmental changes, advanced sensors employ grating-based techniques (interference and diffraction). One of its main parts is the light source, which is made up of lasers or LEDs and provide steady illumination for great sensitivity. Light signals are captured by detectors (photodiodes, phototransistors, or CMOS sensors), which then measure color changes or intensity. Desired spectral bands are separated by wavelength specific optical filters. A signal-processing unit that produces quantifiable electrical output from optical signals. Intensity Based Sensors, which are frequently used for material discrimination or proximity, are one type of detection technology. Another detection technology is Sensors based on wavelengths that measure color or spectral shifts, as those used in blood oxygenation monitoring or pH sensing. Phase and Polarization Sensors are also perfect for high-precision biological diagnostics, these sensors detect changes in light's polarization or phase. Nanostructured metals improve the sensitivity of surface plasmon resonance, which can detect chemical interactions down to the nanomolar range. For accurate colorimetric sensing, quantum dots, which are nano and Peco scale materials, offer excellent resolution and adjustable emission spectra. Photonic crystals with customized optical characteristics enhance light manipulation for applications requiring extreme sensitivity and selectivity. Simple optical sensors cost 5–15 USD/unit. SPR enabled or advanced QD based sensors have cost 100–500 USD per unit. Industry 5.0 demand is driving a 9.6% CAGR in the optical sensor market from 2023 to 2030. The market price of environmentally friendly bio optical sensors can increase by 10–20% [31–33]. All these latest technologies are strongly recommended for industrial revolution 6.0.

Optical sensors' glass and silicon parts can be recycled, which lowers e waste. The environmental impact is reduced using low energy methods like additive manufacturing. The creation of biodegradable optical components is also in line with the objectives of global sustainability. Optical sensors are essential for applications requiring accurate light and color detection because of their unparalleled sensitivity and versatility. These sensors redefine sustainability and technical innovation in addition to providing higher performance when combined with Industry 5.0 and Industry 6.0 techniques. The future of optical sensors offers unmatched efficiency and environmental harmony by utilizing AI, ML and quantum computing [34, 35]. Table 4.1 depicts the comparison between optical and piezoelectric sensors.

2.2.1 SWOT Analysis of Optical Sensors

Strength includes high accuracy and sensitivity which can be able to pick up on even the smallest variations in color and brightness. Non-invasive, that is, Perfect for monitoring remotely. Scalability is easy as the ability to be readily reduced in size for portable applications. Weakness includes the environmental sensitivity, that is, at risk of interference from changes in temperature and ambient light. Production

Table 4.1 Comparative study of optical sensors vs. piezoelectric sensors

Feature	Optical sensors	Piezoelectric sensors
Principle	Photon electron interaction via photoelectric/avalanche effects; interferometry for phase detection	Piezoelectric effect: Strain-induced dipole realignment generating surface charge (d_{33} coefficient)
Material basis	InGaAs/InP photodiodes (SWIR), SiPMs (visible), quantum dot (QD) enhanced polymers (PbS/CdSe)	Lead zirconate titanate (PZT 5H), AlN (Aluminium Nitride), Sc doped AlN (42% ↑d_{33}), KNN ($K_{0.5}Na_{0.5}NbO_3$) for lead-free
Sensitivity	0.1 fW/cm^2 (SiPMs), 90% QE at 905 nm (InGaAs). Single photon detection limited to superconducting nanowires	10 pC/N (PZT), 25 pC/N (ScAlN). Noise floor: 0.1μV/$\sqrt{Hz}$ at 1 kHz (charge amp required)
Linearity	±0.05% over 80 dB dynamic range (calibrated InGaAs). Wavelength dependent nonlinearities in QDs	Hysteresis error: <2% in PZT; <5% in AlN. Dynamic range limited by depolarization at >20 MPa
Frequency response	DC to 40 GHz (integrated photonics). Resonant MEMS filters: $Q > 10^4$	Resonant operation only (kHz–MHz range). Self-resonance: 1 to 10 MHz (PZT thickness mode)
Temp. Stability	±0.005%/°C drift (thermoelectrically cooled). QE variation: 0.1%/°C (InGaAs)	d_{33} drift: −1.2%/°C (PZT), −0.3%/°C (AlN). Curie temp limit: 350 °C (PZT), >1000 °C (AlN)
Power consumption	3–100 mW (active cooling adds 500 mW). Photonic ICs: 50 fJ/bit (silicon modulators)	0.1–10 mW (passive). Energy harvesting efficiency: 5–15% (PZT patches)
Integration complexity	III–V-on-Si wafer bonding (≤200 nm alignment). Fiber coupling loss: 0.1 to 1 dB/facet	CMOS compatible sputtering (AlN). PZT requires sol gel + RTA (>600 °C)
Environmental impact	InGaAs toxicity (as, Ga). QD sensors: Pb/Cd leaching risk. Recycling efficiency: <30%	RoHS exempt PZT (Pb). KNN: Bi_2O_3 sintering aids (eco-toxicity concerns). End of life recovery: 45%
Cost/unit (10 k units)	USD 18 (Si PIN) to USD 480 (SWIR APD). InP photodiodes: USD 220 ± 15%	USD 7 (PZT disc) to USD 185 (MEMS array). ScAlN: USD 95/die (150 mm wafer)
Emerging tech	Topological photonic sensors (robust to defects). SnS_2/WSe_2 vdW heterostructure photodetectors (200% EQE)	Piezotronic transistors (ZnO NWs). Flexoelectric sensors (strain gradients → polarization)
Reliability	Dark current doubling: 1000 h at 85 °C (InGaAs). Radiation hardening: >100 krad (Si)	Depolarization: 5% at 10^9 cycles (PZT). Humidity degradation: KNN loses 40% sensitivity at 85% RH

expenses are raised by precision fabrication. Opportunities for Industry 5.0 Integration, that is, predictive maintenance and real-time data analytics using AI/ML. Demand is driven by wearable technologies and the expansion of smart cities. Threats include competition as there is fierce rivalry from less expensive piezoelectric or electronic sensors. Technical obsolescence as older optical sensor technologies may be exceeded by faster developments.

2.3 Industry 5.0 and Industry 6.0 Approaches

Human friendly self-driven design, which combines AI and human experience to provide personalized sensor solutions, is one of the Industry 5.0 techniques. AI/ML algorithms for real-time data analytics forecast sensor failure and dynamically adjust calibration. Sustainable practices include the utilization of renewable resources like bio based polymers and low industrial energy use. Self-driven manufacturing systems, which integrate IoT and edge computing to produce completely automated sensors, are one of the Industry 6.0 approaches. Quantum integration with simulations for material design and sensor optimization are accelerated by quantum computing. Blockchain integration enables safe sensor network data storage in vital sectors like aerospace and healthcare. Robotic Co Workers as collaborative robots or cobots, improve production throughput and assembly accuracy.

2.4 MEMS Sensors

MEMS sensors are miniaturized devices that integrate electrical and mechanical components on a common silicon substrate. MEMS sensors are widely used in applications ranging from consumer electronics to aerospace and biomedical devices. Their capability to sense physical, chemical, or biological parameters and convert them into electrical signals makes them highly versatile. A physical sensing system that uses variations in capacitance, resistance, or piezoelectricity to detect movement, vibration, pressure, or acceleration makes up the working principles. System for chemical and biological sensing specific molecules interact with functionalized surfaces, changing their electrical characteristics in quantifiable ways. Feedback loops are frequently incorporated into electromechanical feedback systems in MEMS sensors to improve stability and accuracy. It consists of numerous components. Its mechanical structures include micro machined components such as resonators, diaphragms, and cantilevers. A sensing element in it transforms chemical or physical changes into mechanical strain or displacement. Its signal transduction system converts mechanical signals into electrical ones using piezoelectric, capacitive, piezoresistive, or thermal methods. Its integrated circuitry processes signals and carries out functions like as filtering, amplification, and analogue-to-digital conversion [36, 37].

2.4.1 Different MEMS Sensor Types with Brief Applications

MEMS Accelerometers measure a proof mass's displacement to identify acceleration. Applications include gaming controllers, automobile airbag deployment and smartphones (screen orientation). MEMS Gyroscopes use motion caused by the Coriolis force to sense angular velocity. Applications include robotics, drones, and inertial navigation systems. In MEMS Pressure Sensors, diaphragm deformation is the mechanism used to measure changes in pressure. Applications include medical gadgets (such as blood pressure monitors) and timers. MEMS Microphones transform sound waves into a diaphragm's mechanical motion, which is then followed by the creation of electrical signals. Applications include IoT devices, smartphones, and hearing aids. MEMS Infrared Sensors use bolometer arrays or thermopiles to detect thermal radiation. Applications include HVAC systems, industrial thermography, and night vision [38, 39]. Table 4.2 shows the SWOT analysis of MEMS sensors.

Table 4.2 SWOT analysis of MEMS sensors

Dimension	Technical strengths	Technical weaknesses	Emerging opportunities	Critical threats
Precision and performance	Quantum-limited noise floors: 15 ng/$\sqrt{Hz}$ (accelerometers, 5 μPa/$\sqrt{Hz}$ (piezoresistive pressure)phase-change materials (GST-225) enable zero-drift operation (±0.005% FS)	Thermomechanical noise dominates below 100 Hz (Brownian motion limit) Stiction failures at <5 nm gaps (van der Waals forces)	2D material heterostructures (MoS_2/hBN): 100 × ↑ sensitivity at 300 K Optical MEMS: Sub-pm displacement resolution via integrated photonics	Quantum tunneling effects at <3 nm gaps cause leakage currents
Manufacturing and integration	Monolithic 3D integration: TSV-less stacking via direct bonding (≤200 nm alignment) CMOS MEMS co-fabrication: 45 nm SOI platforms with embedded ASICs	III-V/CMOS thermal budget mismatch: ≤400 °C limit for AlScN/PZT integration Wafer level packaging yield: 92% vs. 99.9% for digital CMOS	Heterogeneous integration: 2.5D chiplets with Si interposers for multi-sensor fusion Atomic layer deposition enables conformal coatings on high-aspect-ratio MEMS	Geopolitical supply chain fragility: 73% of advanced MEMS fabs concentrated in Taiwan/China
Power and intelligence	Neuromorphic MEMS: Event-driven sensing with 50μW active power Piezoelectric energy harvesting: 80μW/cm³ from industrial vibrations [16]	AI/ML co processing latency: 5 to 8 ms for edge inference (vs. 1 ms control requirements) RF-MEMS switching losses: 0.5 dB at 60 GHz	Near-sensor computing: Memristor based in memory processing for sub ms latency Triboelectric nanogenerator: 23% efficiency for self powered sensors	Cybersecurity vulnerabilities: side-channel attacks extract secret keys from MEMS resonators

(continued)

Table 4.2 (continued)

Dimension	Technical strengths	Technical weaknesses	Emerging opportunities	Critical threats
Materials and reliability	Ultra nanocrystalline diamond (UNCD): Young's modulus 980 GPa, zero creep at 800 °C 4D printed shape-memory polymers: self healing after 10^6 cycles	Diffusion induced degradation: Au-Al intermetallics form at 150 °C (Kirkendall voids) Radiation susceptibility: 10 krad (Si) TID limits for space applications	Topological insulators (bi_2se_3): Edge-state conduction immune to defects Cryogenic MEMS: 4 K operation for quantum sensing	REACH restrictions: Impending Pb ban threatens PZT based MEMS
Market scalability	Volume production cost: USD 0.18/ unit for inertial sensors at 100 M units/year Foundry ecosystems: Global 200 mm MEMS capacity >3 M wafers/year	Test/calibration cost: 30–45% of total manufacturing expense Multi-physics simulation gaps: 15–20% performance deviation from FEA models	Digital twins: federated learning optimizes MEMS design yield by 22% Smart dust: Sub-mm^3 nodes for distributed sensing (DARPA SHIELD program)	Patent thickets: 65% of piezoelectric MEMS IP controlled by 5 entities

2.4.2 Industry 5.0 and 6.0 Approaches

MEMS sensors can be improved in Industry 5.0 because of the potential for Human-Centric Design. MEMS devices for real-time data processing in robotics and healthcare. Using sensor data analytics for predictive maintenance, which is necessary in industry 5.0, AI/ML integration is also feasible. Since Quantum Enhanced MEMS can use quantum mechanics to create ultra-sensitive sensors, Industry 6.0 MEMS sensor upgrades are also feasible. AI integration can lead to Industry 5.0, which is a Human friendly Design approach. For all kinds of sensors, predictive maintenance can cut downtime by as much as 30%. Wearable flexible sensors improve robotics and personalized healthcare. Utilizing quantum mechanics for ultra-sensitive optical and MEMS sensors, autonomous and quantum systems, such as quantum-enhanced sensors, can help realize the industry 6.0 strategy. The development of self-healing sensors and advanced materials allows for autonomous repair, which increases lifespans by 50%. In industry 6.0, autonomous manufacturing with fully robotic MEMS production lines for scalability is also feasible.

2.5 Comparative Study of all Three Types of Modern Sensors and Alignment with SDGs

Sensors form the lifeblood of modern-day intelligent systems and power advanced technologies across industries. The more important sensor types such as MEMS-based sensors, piezoelectric sensors, and optical sensors boast unique characteristics, diverse applications, and decisive roles in Industry 5.0/6.0 and sustainability. There are only a few of the SDGs that have an intimate connection with unique physical mechanisms, fabrication techniques and use cases for each type of SDG (especially 9, 11, and 12). Different materials exhibit different levels of sensitivity with quartz measuring 2 pC/N and lead based piezo electrics such as PZT measuring sensitivity upwards of 15 on the pC/N scale which highlights the necessity of selection of appropriate piezo electrics depending on the application of use. They find various applications including medical diagnostics, where they aid in imaging and monitoring, for energy harvesting capturing ambient energy, and for vibration monitoring systems ensuring the structural integrity of structures and machines [24, 40–47].

Optical sensors use some laws like Snell's Law, photoelectric effect to assess changes in color and light intensity. Due to their capacity to interact with photons, these sensors make use of semiconductors like silicon photodiodes, gallium arsenide (GaAs), and indium gallium nitride (InGaN). It works by generating electron hole pairs through photon absorption in semiconductor junctions, resulting in detectable currents. With response times in nano secs and 90% sensitivity, quantum efficiency (silicon photodiodes), they are potential candidate for the usage in machine vision, environmental monitoring, and medical imaging. Manufacturing methods include epitaxial growth, thin-film deposition, and photolithography. Sensitivity to ambient light and material deterioration over time are among the limitations [48, 49].

MEMS sensors provide unmatched versatility by integrating electrical and mechanical components at microscales. MEMS sensors offer solutions for high-precision measurements of motion, pressure, and chemical interactions by utilizing silicon substrates, piezoelectric materials, and sophisticated production procedures. The conversion of thermal expansion, piezo resistance, or capacitance changes into electrical impulses is one mechanism of action. They can be used in environmental monitoring, inertial navigation systems, and automobile airbags. They range in size from 1 μm to 1 mm, with pressure sensitivities up to 0.1 Pa, and use less than 10 mW of power. They can be produced using lithography, wafer bonding, and bulk and surface micromachining. Cross sensitivity and endurance in harsh environments are among the limitations [50, 51].

Reducing waste, energy use, and toxic materials is a key component of sustainable sensor manufacture, which is closely related to the SDGs. Piezoelectric sensors that use lead-free ceramics (like KNN) get rid of hazardous waste. Thin film deposition methods which use few raw ingredients, are used in optical sensors. When compared to subtractive approaches, additive manufacturing can reduce waste in

MEMS manufacturing by up to 70%. Automation can reduce carbon emissions since AI-driven energy optimization reduces emissions in manufacturing lines by 15–20%. Using a circular economy strategy, recycling silicon wafers can reduce material prices by up to 30%. Non-toxic substitutes and biopolymers improve sustainability and are ideal for industry 6.0 [52–55]. Table 4.3 shows the comparison among piezoelectric, optical and MEMS sensors.

Table 4.3 Comparison of sensor types

Feature	Piezoelectric sensors	Optical sensors	MEMS-based sensors
Mechanism	Strain-induced polarization via d_{33}/d_{31} coefficients; flexoelectric effects in ultrathin films (<10 nm)	Photon-electron interaction (QE), plasmonic enhancement, topological photodetection	Capacitive/piezoresistive/piezoelectric transduction; resonant frequency shift
Material composition	$Sc_{0.2}Al_{0.8}N$ (d_{33} = 25.4 pC/N), PMN-PT single crystals, KNN-BZT lead-free ceramics	Metasurface-integrated InGaAs/Si (99% QE at 1550 nm), graphene-WSe_2 vdW heterostructures	SiC-on-Si MEMS, AlN-on-SOI, nanocrystalline diamond resonators
Sensitivity	0.03μV/$\sqrt{}$Hz noise floor (ScAlN MEMS); 50 pC/N in PMN-PT	0.01 fW/$\sqrt{}$Hz NEP (SNSPDs); 200% EQE in 2D heterostructures	2 ng/$\sqrt{}$Hz accelerometers; 0.5 μPa/$\sqrt{}$Hz piezoresistive pressure sensors
Dynamic range	160 dB (PZT with closed-loop charge amps); hysteresis <0.5% in AlN	190 dB (HDR CIS); nonlinearity <0.02% via deep learning compensation	140 dB (capacitive accelerometers); 0.001% FS nonlinearity with self-calibration
Frequency response	DC-100 MHz (bulk acoustic wave sensors); kHz self-resonance in MEMS	DC 300 GHz (plasmonic detectors); ns response in SPADs	DC-50 kHz (inertial); GHz resonant RF-MEMS
Size	200 × 200 × 10μm^3 (MEMS arrays); mm-scale for energy harvesting	50 × 50μm^2 per pixel (SWIR); sub-wavelength metasurfaces	500 × 500 × 50μm^3 (IMUs); 10μm feature size in NEMS
Cost (100 k units)	USD 8.50 (AlN MEMS) USD 220 (PMN-PT single crystal)	USD 0.85 (Si photodiode) USD 650 (SNSPD)	USD 0.5 to 2 (accelerometer) to USD 120 (spectroscopic MEMS)
Power consumption	2μW (energy harvesting mode); 90μW for active sensing	5 mW (VCSEL-based); 0.1 pJ/bit for integrated photonics	15μW (AI-integrated IMU); 50 nW for wake-on-motion
SWOT highlights	S: Zero-power operation, high frequency W: Pb toxicity in PZT O: Flexoelectric energy harvesting T: REACH Pb restrictions	S: EMI immunity, hyperspectral capability W: Ambient light interference O: Quantum imaging T: InGaAs supply chain fragility	S: CMOS integration, massive scalability W: Thermomechanical noise O: Neuromorphic sensing T: Fabless IP monopolies

2.6 Actuators

Modern intelligent systems depend heavily on actuators, which are machines that transform energy into motion. They are essential in industries including smart manufacturing, robotics, aerospace, and healthcare. The opposite of sensors, actuators produce physical movement in reaction to digital or electrical inputs. Their significance in sustainable innovation and human-centric technology is further demonstrated by their chemistry, mechanisms, and sophisticated manufacturing processes, as well as their alignment with Industry 5.0/6.0 paradigms and SDGs. The purpose of actuator materials is to transform different types of energy (thermal, magnetic, electrical etc.), into accurate mechanical motion. To accomplish actuation, these materials make use of basic chemical and physical concepts, such as magnetostriction, lattice distortion, phase transition, and ion migration. To meet the demands of modern technology, chemistry, materials science, and engineering have come together to create actuators. Actuators support Industry 5.0 and 6.0 paradigms by utilizing modern manufacturing techniques and adhering to sustainable principles, so promoting a future in which autonomous and human-centric systems can coexist together. To solve global issues and accomplish the SDGs, actuator materials and mechanisms must continue to advance [56, 57]. The chemistry and workings of the four main categories of actuator materials are covered in detail here.

2.6.1 Piezoelectric Actuator

They depend on substances such as KNN, PZT, and barium titanate. Because of their non-centrosymmetric crystal structures, these materials are classified as ferroelectrics, which have a permanent electric dipole. In $BaTiO_3$, for example, an applied electric field distorts the perovskite lattice, changing the location of the Ti^{4+} ion within the oxygen octahedron. Mechanical strain is produced by this lattice distortion in proportion to the applied voltage. Traditional PZT contains lead, which raises environmental concerns from a sustainability standpoint. KNN and other lead-free substitutes are now the focus of modern research. By preserving a perovskite structure, the chemical makeup of KNN, $(K_xNa_{1-x})\ NbO_3$, provides a similar piezoelectric value. This is accomplished by carefully regulating the A-site cation ratio, which affects strain and polarization characteristics. Piezoelectric actuators, which have response times of less than 1 ms and strain levels of up to 1200 ppm, are utilized in high-precision systems like fuel injectors. Make use of the reverse piezoelectric effect, in which mechanical deformation results from applied voltage. Their displacement varies from 10 nm to 100μm. They are used in fuel injectors, ultrasound imaging, and precision positioning systems [58, 59].

2.6.2 Shape Memory Alloys (SMAs)

Reversible martensitic phase transitions are the mechanism of action for Shape Memory Alloys, such as nickel/titanium, copper aluminum nickel and iron manganese silicon. A special chemical attribute of NiTi, often referred to as nitinol, is that when heated or cooled, its crystal structure shifts from austenite (a high-temperature phase) to martensite (a low temperature phase). Changes in the coordination environment of the Ni and Ti atoms cause this phase transition, which has a significant recovery strain of up to 10%. Twin boundaries that emerge in the martensite phase and align under stress are the chemical mechanisms responsible for this behavior. The austenite phase regains its former shape when heated over the transition temperature, which is between −50 °C and 150 °C. These alloys, which have recovery stresses of more than 700 MPa, are widely employed in robotics, biomedical equipment and morphing structures. Motion is produced via phase transitions between austenitic and martensitic structures, which are utilized in robotic grippers, morphing airplane wings, and medical stents. Their life cycles are greater than 10^7.They are environmentally friendly due to their compatibility with additive manufacturing methods and their capacity to be recycled [60–62].

2.6.3 Electroactive Polymers (EAPs)

Polyaniline (PANI), PVDF, and ionic polymer metal composites (IPMCs) are examples of electroactive polymers (EAPs), a type of soft actuators. Ion migration within the polymer matrix causes these polymers to deform in response to electrical stimulation. In IPMCs, for instance, a voltage application results in the migration of hydrated cations within the polymer, which induces differential swelling and bending. The polar β-phase of PVDF, a semi crystalline polymer, gives it both pyroelectric and piezoelectric characteristics. From a chemical standpoint, dopant mobility, ion exchange capabilities, and polymer conductivity all affect how EAPs actuate. EAPs are perfect for applications that need flexibility, such haptic feedback devices, and artificial muscles, because of their millisecond actuation speeds and strain values of up to 380%. They are appealing for sustainable design due to their low density and biodegradability potential. Rely on how an electric field causes polymer strands to deform. Mostly utilized in flexible robotics, haptic devices, and artificial muscles, with force production of up to 10 N/cm^2 and energy densities of 2 J/g [63–66].

2.6.4 Magneto Strictive Actuators

Materials like Terfenol-D, an alloy of terbium (Tb), dysprosium (Dy), and iron (Fe), are used in magneto strictive actuators. When exposed to a magnetic field, these materials undergo magnetostriction, a change in shape or dimension. The realignment of magnetic domains within the crystal lattice in response to an external field is the mechanism's fundamental component. Under a magnetic field strength of 2 T, the strain caused by this realignment can reach values as high as 2000 ppm. Employ magnetic fields to create displacement via inducing lattice strain. Applications include vibration control, precision machining, and sonar systems. These actuators can generate force outputs of up to 1000 N with displacement amplitudes ranging from 2 to 5μm. The giant magnetostriction in Terfenol-D arises primarily from the high magnetocrystalline anisotropy and large saturation magnetostriction of the Tb and Dy ions. When alloyed with Fe, these rare-earth ions produce a significant magnetostrictive effect at room temperature due to the alignment of magnetic moments within the crystal lattice. The incorporation of Tb and Dy also enhances spin-orbit coupling associated with their 4f electrons, further contributing to the material's exceptional magnetostrictive properties. Because of its high magnetomechanical coupling coefficient, the material can be used in vibration control, precise machining, and sonar systems. There are financial and environmental problems associated with the use of rare earth metals. To reduce the dysprosium content without sacrificing performance, sustainable alternatives are being investigated [67–70].

2.6.5 Latest Advancement in Actuator Manufacturing Process

Through the integration of cutting-edge techniques that guarantee high precision, efficiency, and adaptability, modern intelligent manufacturing processes have completely transformed the production of actuators. These procedures maximize material qualities and expedite production workflows by combining state-of-the-art hardware advancements with software algorithms. Rapid prototyping and the production of complicated geometries are made possible by additive manufacturing, which includes 3D printing. SMAs with specific phase transition properties, like NiTi components tuned for transition temperatures between −50 to 150 °C, are created via selective laser melting (SLM). With up to 95% material usage, additive techniques also reduce material waste, making them a sustainable option for creating high-performance, lightweight actuators. For cutting, sintering, and welding piezoelectric ceramics and SMAs, laser-assisted manufacturing provides unmatched precision, guaranteeing excellent material integrity. High-precision lasers with micrometer accuracy are used to fabricate piezoelectric actuator layers for haptic feedback systems and complex SMA parts for aeronautical applications. The mechanical and thermal characteristics that are essential for actuator performance in harsh conditions are improved by this approach, which guarantees homogeneous material densification [71–73].

At the nanoscale, nanofabrication techniques are essential for customizing the characteristics of actuator materials. The sol-gel method is used to create nanostructured piezoelectric films with improved stability and sensitivity, like $BaTiO_3$. Another method for producing electroactive polymer fibers with regulated porosity and alignment is electrospinning, which allows for actuation stresses of up to 380% with response times in the millisecond range. For applications needing great sensitivity and quick response, these nanostructured materials' high surface area to volume ratio is essential [74, 75].

AI-driven production uses cutting edge algorithms to forecast material behavior and instantly improve manufacturing conditions. AI powered systems in SMA fabrication can replicate phase changes, cutting down on material waste and trial and error testing by as much as 30%. Large datasets from actuator production are analyzed by machine learning models, allowing for adaptive control of processing parameters such as polymer curing time, laser intensity, and sintering temperature. This guarantees constant performance and quality throughout production batches. When combined, these intelligent manufacturing techniques allow for the creation of high-precision actuators specifically suited for robotics, aircraft, healthcare, and renewable energy applications by bridging the gap between material science and industrial engineering. Hardware advancements and software driven optimizations work together to support Industry 5.0 and 6.0 paradigms, which place an emphasis on strong, human centered, and sustainable manufacturing systems [76, 77]. Table 4.4 depicts comparison among four types of actuators.

Table 4.4 Comparative analysis of four types of actuator materials

Property	PZ materials	Shape memory alloys	Electroactive polymers	Magnetostrictive materials
Material chemistry	PMN PT single crystals, ScAlN, KNN BZT (Pb free)	NiTiHf (high temp), Fe-Mn-Al-Ni (low hysteresis), cu-Al-Mn (super elastic)	Acrylic or silicon elastomers (DAE), IPMC, graphene-doped PVDF based polymers/ Piezoelectric EAPs	Galfenol (Fe-Ga), Terfenol D ($Tb_{0.3}Dy_{0.7}Fe_{1.9}$, < 0.1% Tb Fe co Al alloys
Mechanism	Ferroelectric domain switching; flexoelectric strain gradient coupling	Stress-induced martensite reorientation; R-phase transformation	Ionic electrodiffusion (IPMC); Maxwell stress (DEA); piezoelectric β-phase alignment (PVDF)	ΔE effect; magnetic moment rotation in ⟨110⟩ textured crystals
Strain	1.7% (PMN-PT ⟨001⟩); 0.25% (ScAlN)	15% (NiTiHf at 200 °C); 12% (Fe Mn Al Ni)	100–300% (DEA); 1–10% (IPMC, bending strain)	2000 ppm (Terfenol D); 450 ppm (Galfenol)
Response time	10μs (resonant); 100μs (step)	20 ms (J heated NiTiHf); 5 ms (laser activated)	1–100 ms (DEA); 0.1–10 s (IPMC hydration dependent)	<10μs (Terfenol D); 25μs (Fe Ga)
Force density	80 MPa (PZT stack); 1.2 kN/ cm^2 (resonant)	1.5 GPa (NiTiHf recovery stress)	0.1–1 MPa (DEA); 1–30 MPa (piezoelectric EAP like PVDF)	35 MPa (Terfenol D)
Work density	0.15 MJ/m^3 (PZT)	15 MJ/m^3 (NiTi)	0.01–0.35 MJ/m^3 (DEA)	0.08 MJ/m^3 (Terfenol-D)
Frequency bandwidth	DC 500 kHz (bulk); DC 50 kHz (MEMS)	<10 Hz (thermal); kHz (mechanical cycling)	DC–1 kHz (DEA); DC–100 Hz (IPMC)	DC 100 kHz (Terfenol-D)
Energy efficiency	60–80% (resonant); <20% (off resonant)	<10% (Joule heating)	60–90% (DEA) [19]	40–70% (Terfenol-D)
Cost (1 k units)	USD 22 (AlN MEMS); USD 480 (PMN PT)	USD 120 (NiTi wire); USD 950 (NiTiHf)	USD 8/m^2 (DEA film); USD 85 (IPMC cartridge)	USD 1200/kg (Terfenol-D); $300/ kg (Galfenol)

Table 4.4 (continued)

Property	PZ materials	Shape memory alloys	Electroactive polymers	Magnetostrictive materials
Sustainability	KNN BZT: RoHS compliant; recycling rate: <25%	92% recyclable (vacuum arc remelting) [25]; Tb-free alternatives	Biodegradable ionogels (T = 180 days)	Criticality index: 9.2/10 (Terfenol-D); Fe-Ga reduces Tb by 98%
Emerging advances	Flexoelectric MEMS (zero hysteresis); PZT nanofibers for 3% strain	Topological SMAs (twin boundary engineering); additive-manufactured lattice structures	Biomimetic fibrillar DEAs; solid-state ionic EAPs	Metamaterial-enhanced ΔE effect; Tb-free MnGa films
Key applications	Atomic force microscopy; laser steering (5 nm precision)	Aerospace morphing wings; minimally invasive surgical bots	Soft robotics grippers; haptic feedback skins	Ultrasonic scalpels (50 kHz); active vibration cancellation in semiconductor fabs

2.6.6 Actuators Materials Alignment with SDGs

By fusing innovative technology with environmentally friendly methods, actuators play an important role in achieving several SDGs. Actuators have a significant impact on SDG 9 because of their involvement in transforming advanced manufacturing, robotics, and aircraft. In industrial systems, actuators provide accuracy, automation, and intelligent control, strengthening infrastructure and stimulating innovation-driven economic expansion. Piezoelectric and electroactive polymer actuators, for example, improve haptic feedback and dexterity in robotics, two essential skills for next-generation robotic applications. Actuators incorporated into smart transportation systems and energy-efficient buildings contribute to SDG 11. Adaptive shading systems use electroactive polymer actuators to control temperature and light, which lowers energy use in cities. Similarly, to reduce material consumption and improve functioning, shape memory alloys are used in deployable structures and morphing technologies in public transportation, helping to create sustainable urban environments. Recent developments in actuator materials are intimately related to SDG 12. Environmental risks are decreased, and sustainable production cycles are encouraged by the move to lead-free ceramics, recyclable shape memory alloys, and biodegradable electroactive polymers. In piezoelectric actuators, for instance, KNN removes the harmful effects of lead-based materials like PZT. By facilitating material recovery and reuse, cutting waste, and protecting natural resources, these advances promote circular economies. Finally, actuators built into renewable energy systems directly promote SDG 13. Wind turbine vibration control systems use magnetostrictive actuators to increase longevity and efficiency. With the use of piezoelectric actuators, which are employed in solar tracking systems, solar panels can be oriented to maximize energy absorption and increase solar farm efficiency by up to

30%. Actuators are essential in lowering greenhouse gas emissions and lessening the effects of climate change because they promote the use of renewable energy sources and enhance the sustainability of industrial applications. Actuators' inventiveness, accuracy, and sustainability not only progress technology but also work as a foundation for a more just and environmentally friendly future.

3 Edge Devices to Connect Systems and Sensors

Serving as a bridge between sensors and more extensive communication networks, edge devices are essential parts of the IoT ecosystem. They enable real-time decision-making by processing data locally, which lowers latency and conserves bandwidth. They are essential in a variety of applications, including smart cities, healthcare, industrial automation, and environmental monitoring, due to their adaptability and processing power [78].

3.1 Essential Features of Edge Devices

The foundation of contemporary IoT networks are edge devices, which enable smooth data transfer and device-level intelligence. Their primary functions, communication, local computation, and data preprocessing, are designed to maximize efficiency, minimize delay, and enhance performance.

3.1.1 Data Preprocessing

The capacity to preprocess raw data collected from linked sensors is at the core of edge devices. Advanced preprocessing activities include noise filtering, which removes signal distortions using methods like wavelet transforms or the Fast Fourier Transform. By reducing the size of data packets, data compression techniques like Huffman coding and predictive coding ensure faster transmission with less bandwidth usage. Prior to sending the crucial data to central systems, feature extraction uses machine learning methods such as Principal Component Analysis or CNNs to find important data patterns. By reducing transmitted data amount by as much as 70–80%, these processes can guarantee real time responsiveness while drastically reducing storage costs in cloud infrastructures [79, 80].

3.1.2 Local Computation

To handle and analyze data locally, modern edge devices have potent computational gear, such as multi core microprocessors, GPUs or FPGAs. Applications that require instantaneous feedback, like autonomous cars, which depend on edge devices to evaluate video feeds and LiDAR data for navigation, require this real-time compute capacity. FPGAs can be configured to perform sub-millisecond concurrent tasks like image recognition or cryptography. With power consumption as low as 10–30 W, GPUs in devices like the NVIDIA Jetson offer teraflop level performance for AI tasks like object identification. In robotics and industrial automation, where exact timing is mandatory, edge devices cut latency from milliseconds to microseconds by doing away with the need for remote cloud servers [81–83].

3.1.3 Communication

Edge devices use reliable communication protocols like message queuing telemetry transport (MQTT), constrained application protocol (CoAP), and hypertext transfer protocol (HTTP) to create smooth connectivity with other devices or central systems. Battery-operated IoT devices can use MQTT because it is portable and perfect for low-bandwidth networks. CoAP uses UDP rather than TCP for speedier message delivery in wearables or smart homes, making it ideal for resource-constrained situations. By encrypting communication channels, advanced security protocols like transport layer security (TLS) and datagram transport layer security (DTLS) guarantee data integrity throughout transmission. Pre-processed data may easily interact with cloud analytics platforms thanks to edge-to-cloud synchronization techniques, enabling hybrid processing architectures. In industrial IoT networks, incorporating these protocols lowers data packet loss to less than 0.1%, according to numerical simulations [84–86].

3.2 Combining Advanced Systems

AI accelerators are being added to edge devices more frequently in order to further expand their capabilities. Tensor processing units (TPUs) integrated into edge devices, for example, can speed up deep learning inference processes by up to 15 times, allowing for real-time applications such as predictive maintenance or video analytics. Energy-efficient architectures, such ARM Cortex processors, provide sustainable operation even in remote areas, while edge containers, driven by platforms like Kubernetes, allow for the flexible and scalable deployment of programs across devices. Edge devices create the groundwork for advancements in Industry 5.0 and beyond by utilizing these functionalities to help companies shift from centralized computing paradigms to distributed, intelligent systems. IoT, AI, and automation solutions are made possible by current technology, as demonstrated by edge computing devices like the Raspberry Pi, Arduino, and NVIDIA Jetson. Their features, uses, and personalization choices demonstrate how adaptable they are across a range of sectors, including research and education. These devices play a key role in expanding the frontiers of robotics, AI-driven automation, and intelligent edge computing systems by combining pre trained models, real-time processing capabilities, and significant customization [87–90]. Table 4.5 shows the edge devices comparison with sensors and actuators.

Table 4.5 Edge devices vs. sensors and actuators

Feature	Edge devices	Sensors	Actuators
Core function	Localized AI inference (INT8/FP16), federated learning orchestration, encrypted data preprocessing	Quantify physical/chemical parameters via transduction (strain → voltage, photon→ electron)	Convert control signals into mechanical work via field-driven phase/morphology changes
Material stack	Futuristic in Industry 6.0: Logic: 2 nm CFETs (Si/SiGe), memory: HBM3E (24 GB/s), packaging: Glass substrates with <10μm RDL, Current: Logic: Si CMOS (28nm–5nm nodes); Memory: LPDDR4/5; Packaging: FCBGA/PoP	Piezoelectric: ScAlN (d_{33} = 25.4 pC/N), optical: InGaAs/Si metasurfaces (99% QE), MEMS: SiC-on-SOI (2 ng/$\sqrt{\text{Hz}}$ noise)	SMA: NiTiHf (15% strain), EAP: Zwitterionic hydrogels (420% strain AT 5 V), Magnetostrictive: $Tb_{0.3}Dy_{0.7}Fe_{1.9}$ (2000 ppm)
Energy consumption	Idle: 50 mW, peak: 15 W (10 TOPS/W at 4-bit quantization)	Passive: 5μW (energy-harvesting MEMS), active: 90 mW (hyperspectral LiDAR)	Micro: 0.8 W (MEMS piezo actuators), macro: 2.5 kW (industrial magnetostrictive transducers)
Latency	Inference: <2 ms (ResNet-18), data fusion: 500μs	Optical: 3 ns (SPADs), mechanical: 100μs (MEMS accelerometers)	EM: 10μs (Terfenol-D), thermal: 20 ms (joule-heated SMA)
Bandwidth	802.11be (Wi-fi 7): 40 Gbps, 5G NR FR2: 10 Gbps	DC-300 GHz (plasmonic detectors), DC 50 kHz (inertial sensors)	DC-100 kHz (magnetostrictive), <10 Hz (thermal SMA)
Key applications	Real time predictive maintenance (AI), encrypted health monitoring, autonomous swarm coordination	Quantum gravimeters (10–9 > g/$\sqrt{\text{Hz}}$), attowatt SNSPDs, in vivo biochemical sensing	Nanoscale AFM positioning (±50 pm), soft robotic grippers (0.8 MPa), adaptive aerospace structures
Sustainability	Carbon/GHz: 0.12 g (TSMC N2P), recycling: <18% rare-earth recovery	Pb free piezos: KNN BZT, Biodegradable: Cellulose nanofiber sensors (Teg = 120 days)	Tb free actuators: Fe-Ga alloys, recyclable SMAs: 92% NiTi recovery via VIM
Latest innovations	3D monolithic CFETs with backside PDN, photonic tensor cores (0.1 pJ/op), CIM-enabled neuromorphic fabrics	2D MoS_2/h BN heterostrain sensors (100 × ↑ sensitivity), quantum imaging LiDAR	Topological SMA twin engineering, solid-state ionic EAPs, flexoelectric MEMS actuators

3.3 *Raspberry Pi*

A popular option for many edge-computing applications, the Raspberry Pi is a small, credit card-sized computer that is built for efficiency and versatility. Depending on the model, it has quad-core ARM Cortex-A processors with clock rates between 1.2 and 1.8 GHz. In its most recent versions, such the Raspberry Pi 4, it can support up to 8 GB of DDR4 RAM. This device runs Linux based operating systems, including Raspberry Pi OS, Ubuntu, and Windows IoT Core and has two 4 K HDMI outputs, Gigabit Ethernet, and USB 3.0 interfaces. The Raspberry Pi is ideal for a variety of real world applications because it combines great performance with remarkable energy efficiency, with a normal power consumption of only 5 W. The Raspberry Pi's adaptability has fueled its use in many different fields. It functions as a hub for smart devices in home automation, enabling smooth integration and control using communication protocols like MQTT and Zigbee. Using GPIO (General Purpose Input/Output) ports for direct hardware interface, the Raspberry Pi serves as a real-time monitor for production lines in industrial applications. TensorFlow Lite and other lightweight machine learning frameworks are supported for AI prototyping, enabling developers to effectively implement AI models at the edge. Applications ranging from automated delivery systems to surveillance are made possible by the Raspberry Pi, which serves as the central processing unit for autonomous robots in robotics [91, 92].

The Raspberry Pi's ecosystem of HAT (Hardware Attached on Top) allows for even more customization. Specialized applications are made possible by these modular boards, which increase the Raspberry Pi's functionality. The sense HAT is perfect for environmental monitoring and data collection because it has sensors for temperature, pressure, and gyroscopic measures. In the meanwhile, the Motor HAT makes controlling servo and stepper motors easier, which is essential for automation and robotics projects. The Raspberry Pi may be customized by developers to satisfy particular project requirements thanks to this modular architecture, which speeds up prototyping and implementation. The Raspberry Pi has become a vital tool in a variety of domains, from industrial automation and robotics to the IoTs and artificial intelligence, by striking a balance between robust hardware, energy efficiency, and versatility. It is a key component of contemporary edge computing and innovation because to its smooth integration with other hardware and software solutions [93, 94].

3.4 *Arduino*

Arduino boards are well known for being flexible microcontroller-based platforms that support a wide variety of low-power and educational uses. They are driven by powerful processors with clock speeds ranging from 16 MHz to 48 MHz, such as the ARM Cortex-M (32-bit) and ATmega328P (8-bit) families. With various digital and analog I/O connectors and SRAM sizes of up to 32 KB, these boards allow for smooth interface with a wide range of sensors and peripherals. Arduino boards are perfect for battery operated devices and resource-constrained applications because they typically consume less than 0.1 W of electricity. Applications for Arduino are found in many fields, but especially in education and the IoTs. Arduino provides real-time data logging capabilities in environmental monitoring by integrating with sensors such as the DHT22 for temperature and humidity readings and the MQ-series sensors for gas detection. It provides a practical method of learning programming and hardware interface, making it a fundamental component of STEM education. Also, Arduino is compatible with wearable technology, powering simple health-monitoring gadgets that incorporate sensors to measure body temperature, mobility, or heart rate, showcasing its versatility for customer-focused innovation [95, 96].

Because Arduino boards are open-source and have a strong ecosystem of shields, stackable modules that improve functionality, customization is one of their main advantages. For instance, IoT integration is made easier by wireless communication shields like the ESP8266 module, which provide Bluetooth or Wi-Fi connectivity. Sensor shields simplify prototype and deployment by providing plug-and-play compatibility with a variety of input devices, including motion, light, and temperature sensors. In line with sustainability objectives, power management shields, such as solar charging modules, facilitate the development of IoT systems driven by renewable energy. Arduino boards enable developers to innovate in a wide range of domains, from wearable technology and education to environmental monitoring and the IoTs, by fusing low power consumption, versatility, and a robust ecosystem of hardware and software tools. They are essential in academic learning environments as well as professional initiatives due to their adaptability and simplicity [97, 98].

3.5 NVIDIA Jetson

The NVIDIA Jetson series is a representation of state-of-the-art high-performance edge AI computing technology designed for demanding robotics, autonomous systems, and advanced analytics applications. With NVIDIA GPUs with CUDA cores, these devices, which include the Jetson Nano and Jetson Xavier NX models offer unmatched processing performance of up to 21 TOPS (Trillions of Operations Per Second). With its high bandwidth LPDDR4 RAM of up to 16 GB and fast interfaces, including PCIe, I2C, and USB 3.1, they may be easily integrated with external peripherals and devices. Their versatile power consumption, which ranges from 5 W to 30 W, guarantees versatility across a range of energy-constrained and high-performance applications, and their many camera inputs make them perfect for vision-intensive jobs. Because of their sophisticated capabilities, Jetson devices perform very well in resource-intensive fields. In computer vision, they make it possible for autonomous cars, drones, and surveillance systems to recognize and navigate objects in real time. They serve as the computational foundation for industrial robotic arms and humanoid robots in robotics, facilitating operations like path planning, sensor data fusion, and real time decision making. In Industry 4.0 smart factories, Jetson devices are essential to predictive maintenance systems for advanced analytics. These systems use sensor data to detect possible faults and maximize machinery uptime [99, 100].

NVIDIA Jetson is a desirable alternative for developers and industry professionals due to its customization features. Industry-standard AI frameworks like TensorFlow, PyTorch, and NVIDIA TensorRT are pre-installed on these devices, offering a strong basis for edge AI applications. By providing further video analytics optimization, NVIDIA's DeepStream SDK speeds up the creation of applications like smart retail systems and multi-camera monitoring. Specialized carrier boards provide smooth integration into industrial machinery, drones, and autonomous guided vehicles, enabling customized solutions for particular operating needs. The NVIDIA Jetson series offers scalable, effective, and potent solutions for the changing needs of edge AI, embodying the perfect marriage of hardware and software [101–103].

4 Integration in Industry 5.0 and Industry 6.0

A revolutionary advancement in automation, intelligent systems, and manufacturing can be achieved through the integration of edge devices with Industry 5.0 and Industry 6.0. The focus of Industry 5.0 is on human-centric innovation, where a more cooperative, individualized, and effective approach to manufacturing is fostered by the synergy between humans and machines. In this paradigm, edge devices like the NVIDIA Jetson (or any other latest technology) are essential because they allow cobots, or collaborative robots, to carry out intricate jobs alongside human workers. These cobots make use of cutting edge AI features like computer vision driven by Jetson's GPU architecture, to guarantee accurate communication, prompt decision-making, and security in manufacturing settings. Cobots can provide a safer and more effective collaborative workspace by detecting barriers, monitoring human movement, and adjusting their operations dynamically through vision-based safety systems. This device demonstrates the human-centric approach of Industry 5.0 by increasing productivity while preventing workers from being exposed to dangerous jobs. Distributed and autonomous systems are made possible by edge devices like the Raspberry Pi (or any other latest technology) in Industry 6.0, which is centered on hyper automation and the creation of intelligent ecosystems. The Raspberry Pi aggregates data from multiple sources in real time, acting as a node in a highly networked sensor network. Through automated decision-making and ongoing monitoring, this enables autonomous operations throughout smart cities, optimizing trash disposal, energy management, traffic systems, and public safety. As the number of IoT devices grows, networks based on Raspberry Pis may be easily included into the larger ecosystem, allowing for smart infrastructure that adapts to changing circumstances. These edge devices' capacity to locally preprocess data lowers latency and bandwidth requirements, facilitating more effective communication with cloud systems for additional analysis and decision-making. The development of completely autonomous, intelligent cities can be accelerated by using Raspberry Pi to enable predictive maintenance of urban infrastructure, optimize traffic flow, and improve public service delivery in real time by integrating machine learning models on the edge. The next stage of industrial evolution, Industry 5.0 and Industry 6.0, both use edge computing to increase automation, boost operational effectiveness, and facilitate autonomous decision-making. The convergence of autonomous systems and human-centered collaboration, facilitated by edge devices, highlights the trend toward more intelligent, sustainable, and customized urban and factory settings [104, 105].

5 Alignment of Edge Devices with SDGs

Because they spur innovation, improve infrastructure, and promote environmental sustainability, edge devices are essential to achieving the SDGs, especially 9, 11, and 13. Edge devices are essential to the creation of intelligent industrial automation systems and sophisticated infrastructure monitoring for SDG 9. These devices aid in manufacturing process optimization, downtime reduction, and efficiency enhancement by facilitating real-time data processing. The adoption of industrial IoT networks is made easier by Raspberry Pi and NVIDIA Jetson devices, which allow for continuous monitoring of manufacturing lines, the condition of machinery, and energy usage. Edge devices lower latency by locally preprocessing sensor data, which promotes quicker decision-making and better operational performance. These systems also facilitate predictive maintenance, in which sensor data is analyzed by machine learning algorithms driven by edge devices to detect failures and plan maintenance before expensive malfunctions happen. In line with SDG 9, this technology development directly supports the development of robust, creative, and sustainable industrial systems. SDG 11 goal centers on building sustainable cities, and edge technologies like Raspberry Pi-based systems have been instrumental in helping to maximize municipal infrastructure. For instance, the Raspberry Pi is frequently used in smart city initiatives to regulate and control energy usage in waste management, water distribution, and urban lighting systems. By processing data from sensors positioned in pipes, streetlights, and other vital infrastructure, these devices provide dynamic control based on conditions as they occur. By automatically adjusting lighting levels based on traffic patterns or meteorological conditions, Raspberry Pi systems can cut down on carbon emissions and energy waste. These gadgets also aid in water distribution systems' monitoring, leak detection, and supply chain optimization, all of which contribute to more sustainable resource management. The Raspberry Pi significantly advances sustainable urban living by assisting in the creation of ecofriendly and energy efficient infrastructure, thereby directly promoting SDG 11. SDG 13 is also linked especially in the field of renewable energy, edge devices like Arduino and NVIDIA Jetson are essential to climate action initiatives. By modifying the angle of panels with solar trackers in response to current sunshine circumstances, Arduino is frequently used in solar energy systems to monitor and regulate solar panel performance, guaranteeing optimal efficiency. On site processing of sensor data using Arduino-based devices allows for real-time solar energy system optimization, lowering energy waste and improving overall system performance. NVIDIA Jetson devices are used in sophisticated energy management systems, where AI-driven analysis for grid management and forecast energy usage is made easier by their potent GPU capabilities. These systems are essential for maximizing the production of energy from renewable sources, ensuring the effective use and storage of power produced by solar, wind, and other renewables. By reducing dependency on fossil fuels and supporting cleaner, more sustainable energy practices, the usage of edge devices in these systems directly contributes to SDG 13 by reducing climate change. The development of intelligent, effective, and sustainable technology solutions in a variety of industries depends heavily on edge devices. These gadgets help achieve SDGs 9, 11, and 13, which integrate technical innovation with global sustainability goals, by enhancing industrial automation, optimizing urban infrastructure, and providing real-time monitoring of renewable energy systems [106, 107].

6 Challenges in the Implementation of Industry 6.0 Technologies and Future Recommendations

Here is the Table 4.6 to show the difference between Industry 6.0 and traditional processes and their impact on sustainability and SDGs. The strategic implementation roadmap for advanced Industry 6.0 deployment in high tech manufacturing is structured across three progressive phases, designed to integrate cutting edge technologies with sustainability and regulatory frameworks. This approach not only enables scalable deployment of frontier systems but also addresses implementation challenges while contributing directly to the SDGs. Phase 1 (0–18 months) prioritizes foundational infrastructure and capacity building. It begins with the deployment of quantum secured 6G testbeds in industrial clusters to enable ultra low latency, tamper proof communication networks necessary for autonomous operations and cross border industrial data exchange. Simultaneously, organizations are mandated to adopt ISO 14046 compliant water and carbon accounting systems, supported by AI-enabled sensor networks for real-time monitoring of environmental impact. To address workforce transformation, global reskilling programs are launched, utilizing AR/VR based micro credentialing platforms that equip workers with competencies in robotics, AI systems, and quantum control. Phase 2 (18–36 months) focuses on scaling computational and regulatory infrastructure. This includes the rollout of photonic computing farms capable of exceeding 100 PetaFLOPS/W, offering unmatched performance-per-watt ratios critical for industrial AI at scale. At the same time, cross-industry material passports, verified through blockchain, are implemented to ensure lifecycle traceability, sustainability

Table 4.6 Industry 6.0 vs traditional processes with sustainability and SDG impact

Technology domain	Traditional process	Industry 6.0 advancement	Sustainability benefits	Linked SDGs
Manufacturing control	Centralized PLCs (response: 100 ms)	Quantum edge hybrid control: – Quantum annealing for scheduling – Photonic FPGAs (response: 2μs)	30% energy reduction via real-time optimization; 50% less material waste	SDG 9 (industry innovation)
Sensing systems	Piezoresistive sensors (noise: $100\mu V/\sqrt{Hz}$)	Topological photonic sensors: – Defect-immune MoS_2/hBN heterostructures – $0.01\ fW/\sqrt{Hz}$ sensitivity	40% fewer false positives; self powered operation; Pb free nanomaterials	SDG 12 (responsible consumption)
Energy management	SCADA-based monitoring (15-min intervals)	AI-predictive grids: – Federated learning on edge devices – Digital twin simulations (99.8% accuracy)	25% peak load reduction; 30% renewable integration boost; carbon tracking via blockchain	SDG 7 (clean energy)
Supply chains	RFID tracking (batch-level visibility)	Quantum-Blockchain fusion: – Grover-optimized logistics – Post-quantum encrypted ledgers	60% fraud reduction; real-time carbon footprint auditing; dynamic rerouting to cut emissions	SDG 13 (climate action)
Actuation systems	Pneumatic actuators (efficiency: 20%)	Solid-state ionic EAPs: – Zwitterionic hydrogels (420% strain at 5 V) – Biodegradable operation	50% energy savings; zero lubricants; 120-day soil decomposition	SDG 11 (sustainable cities)
Material recovery	Landfill/incineration (recovery: <30%)	AI-driven circular economy: – ML-optimized disassembly robots – Hyperspectral sorting (99.9% purity)	85% material reuse; 70% lower mining demand; closed-loop polymer regeneration	SDG 12 (circular economy)

compliance, and circular economy integration. In parallel, regulatory frameworks for neuromorphic hardware like the IEEE P2872 standard are introduced to govern the deployment of spiking neural network processors and event based AI systems across industrial applications. This phased strategy leverages state of the art and emerging technologies including topological quantum materials, photonic interconnects, and AI security platforms like NVIDIA Morpheus. It anticipates breakthroughs in room temperature superconductivity and DNA-based data storage, ensuring that the transformation toward Industry 6.0 remains technologically relevant and economically viable. Grounded in both commercial solutions and research frontiers, this roadmap provides a scientifically robust path for global industries to transition into sustainable, intelligent, and resilient production paradigms. Phase 3 (36–60 months) culminates in full integration and sustainability breakthroughs. Factories undergo quantum AI convergence, enabling real time optimization of full production "genomes" digital twins enriched with predictive intelligence for every component and sub process. Advanced bioleaching nano factories are deployed to enable >95% rare earth element recovery, important for supply chain resilience in high performance electronics. At the systems level, carbon negative industrial parks are realized by combining integrated carbon capture, utilization, and storage with industrial IoT platforms, creating self-monitoring, low impact production ecosystems [108–115].

Table 4.7 shows the challenges and issues associated with the implementation of industry 6.0 technologies and some suggestions to overcome these challenges.

Table 4.7 Challenges in industry 6.0 implementation and advanced recommended solutions

Challenge	Technical problems	Advanced solutions and recommendations
Quantum integration complexity	Decoherence in quantum systems; error rates $>10^{-3}$; cryogenic requirements (4 K)	Hybrid quantum-classical architectures: Use quantum processors only for optimization subroutines. Topological qubits: Implement non-abelian anyons for fault tolerance (error rates $<10^{-9}$). Photonic quantum chips: Room-temperature operation via integrated nanophotonics
AI edge synergy	Latency >5 ms for real-time control; power consumption >50 W/TOPS	Neuromorphic MEMS: In memory processing with memristors (0.1 pJ/op). Photonic tensor cores: Optical AI accelerators (0.1 pJ/op at 40 GHz). Federated learning: Reduce cloud dependency by 70% with encrypted edge model training
Sustainable material adoption	PZT toxicity (Pb); Terfenol D Tb scarcity (CI = 9.2/10); KNN sensitivity degradation (40% at 85% RH)	Bio piezoelectrics: genetically engineered bacteriorhodopsin films (d_{33} = 35 pC/N). Tb-free magnetostrictives: Fe-Ga alloys with nano-laminate domains (λ_s = 450 ppm). 4D-printed SMAs: NiTiHf with 92% recyclability via vacuum arc remelting

(continued)

Table 4.7 (continued)

Challenge	Technical problems	Advanced solutions and recommendations
Interoperability	Protocol fragmentation (MQTT/CoAP/OPC-UA); data packet loss >0.1%	Quantum-secured mesh networks: QKD-integrated 6G backhaul (<10^{-9} breach probability). Autonomous digital twins: blockchain-verified asset synchronization (SHA-3 + zk-SNARKs). Unified Semantic Ontologies: ISO/IEC 21838–2 compliant knowledge graphs
Energy efficiency	Data center emissions (0.12 kg CO_2/GHz); actuator energy density <0.15 MJ/m^3	Zero-power sensors: Triboelectric nanogenerators harvesting industrial vibrations (80μW/cm^3). Flexoelectric MEMS: Strain-gradient actuators (zero hysteresis). Photonic cooling: On-chip radiative coolers ($\Delta T = 10$ °C at 100 W/cm^2)
Workforce transition	Skills gap in quantum engineering; <18% rare-earth recycling expertise	AR-HoloLabs: Microsoft HoloLens 3 guided assembly with AI-error correction. Digital Twin Upskilling: NVIDIA Omniverse simulations with real-time competency analytics. Global Certification: ISO/IEC 17024 standardized quantum technician programs

References

1. Salam, A.: Internet of things for sustainable community development: introduction and overview. In: Internet of Things for Sustainable Community Development: Wireless Communications, Sensing, and Systems, pp. 1–31. Springer, Cham (2024)
2. Priyadarshi, R.: Exploring machine learning solutions for overcoming challenges in IoT-based wireless sensor network routing: a comprehensive review. Wirel. Netw. **30**, 1–27 (2024)
3. Das, J.P., et al.: From friction to function: a high-voltage sliding triboelectric Nanogenerator for highly efficient energy autonomous IoTs and self-powered actuation. Small. **27**, 2405792 (2024)
4. Khorsand Zak, A., et al.: A review on piezoelectric ceramics and nanostructures: fundamentals and fabrications. J. Aust. Ceram. Soc. **60**, 1–31 (2024)
5. Chaliha, C., et al.: Exploring lead-free piezoelectric ceramics for enhanced energy harvesting applications. In: Next Generation Materials for Sustainable Engineering, vol. 94-125. IGI Global, Palmdale (2024)
6. Zhu, X., et al.: Investigation of a piezoelectric ultrasonic surface wave transducer that minimizes structural scattering. J. Phys. D. Appl. Phys. **58**(3), 035502 (2024)
7. Wang, Y.: A comprehensive review for beam steering technology: from mechanical prisms to space-time metasurface. In: Electromagnetic Wave Control Techniques of Metasurfaces and Metamaterials, pp. 1–31. IGI Global, Palmdale (2024)
8. Sun, D., Zhu, W., Xiang, Y.: Nonlinear interaction of lamb waves with fatigue cracks in metal plates under the condition of phase-velocity mismatching. J. Phys Conf. Ser. **2822**, 012106 (2024)
9. Wang, Y., Lee, J., Feng, P.X.-L.: Perspectives on phononic waveguides for on-chip classical and quantum transduction. Appl. Phys. Lett. **124**(7), 070502 (2024)
10. Concha, V.O., et al.: Properties, characterization and biomedical applications of polyvinylidene fluoride (PVDF): a review. J. Mater. Sci. **59**(31), 14185–14204 (2024)

11. Shee, C., et al.: A critical review on polyvinylidene fluoride (PVDF)/zinc oxide (ZnO) based piezoelectric and triboelectric nanogenerators. J. Phys. Energy. **6**, 032001 (2024)
12. Chen, Y., et al.: Arrayed multi-layer piezoelectric sensor based on electrospun P (VDF-TrFE)/ZnO with enhanced piezoelectricity. Sens. Actuators A Phys. **379**, 115970 (2024)
13. Wu, Y., et al.: Life cycle assessment of lead-free potassium sodium niobate versus lead zirconate titanate: energy and environmental impacts. EcoMat. **6**(5), e12450 (2024)
14. Leng, W., et al.: Optimizing electromechanical properties of KNN-based piezoelectric ceramics through regulation of lattice distortion. Ceram. Int. **50**, 32922–32929 (2024)
15. Zhang, Z., et al.: Achieving excellent mechanical and electrical properties in transition metal oxides and rare earth oxide-doped KNN-based piezoceramics. J. Am. Ceram. Soc. **109**, 19865 (2024)
16. Jiang, R., et al.: Flexible Multi-modal Piezoelectric Energy Harvester with Hourglass Beams for Power Generation under Breeze Vibrations. IEEE Sensors Journal. **25**(20), 38938–38948 (2025).
17. Naderizadeh, S., et al.: Optimising polyurethane/CNTs piezoresistive pressure sensors by varying the modulus of the polymer matrix. Compos. Sci. Technol. **252**, 110614 (2024)
18. Xu, Q., et al.: Highly flexible, high-performance, and stretchable piezoelectric sensor based on a hierarchical droplet-shaped ceramics with enhanced damage tolerance. Adv. Mater. **36**(18), 2311624 (2024)
19. Feng, W., et al.: A large-strain and ultrahigh energy density dielectric elastomer for fast moving soft robot. Nature Communications, **15**(1), 4222 (2024)
20. Persano, L., et al.: Advanced materials for energy harvesting and soft robotics: emerging Frontiers to enhance piezoelectric performance and functionality. Adv. Mater. **36**, 2405363 (2024)
21. Zhang, Q., et al.: Highly robust electronic skin for object recognition of robot hand based on carbon nanomaterials and SEBS. Sens. Actuators A Phys. **379**, 115922 (2024)
22. Niu, H., et al.: Advances in advanced solution-synthesis-based structural materials for tactile sensors and their intelligent applications. InfoMat. **6**(1), e12500 (2024)
23. Akhtar, M., et al.: Climate friendly MOFs synthesis for drug delivery systems by integrating AI, intelligent manufacturing, and quantum solutions in industry 6.0 sustainable approach. Toxicol. Res. **14**(1), tfaf011 (2025)
24. Abbas, S., et al.: Cutting-edge metal-organic frameworks: revolutionizing the adsorptive removal of pharmaceutical contaminants from water. Rev. Inorg. Chem. (2025). https://doi.org/10.1515/revic-2024-0119
25. Malič, B.: Sodium–potassium–niobate based piezoelectric ceramics: processing—microstructure—properties. Jpn. J. Appl. Phys. **63**(10), 100801 (2024)
26. Liu, Y.X., et al.: Multi-length engineering of (K, Na) NbO3 films for Lead-free piezoelectric acoustic sensors with high sensitivity. Adv. Funct. Mater. **34**(12), 2312699 (2024)
27. Zhao, L., et al.: Design, fabrication, and characterization of high-performance PMUT arrays based on potassium sodium niobate. J. Microelectromech. Syst. **33**, 438–445 (2024)
28. Ali, M.R.R., Tigno, S.D., Caldona, E.B.: Piezoelectric approaches to organic polymeric materials. Polym. Int. **73**(3), 176–190 (2024)
29. Zhang, B., et al.: Enhancement of piezoelectric properties and temperature stability of high-curie temperature CaBi2Nb2O9 ceramics. J. Eur. Ceram. Soc. **45**, 117101 (2024)
30. Shao, F., et al.: Textured Sr2Nb2O7 ceramics: microstructure design, high temperature ferroelectric and piezoelectric performance. J. Am. Ceram. Soc. **108**, e20120 (2024)
31. Jadhav, V.R., Tagad, C.K., Palake, A.: Economic and commercial aspects related to nanotechnology-based sensors. In: Nanotechnology-Based Sensors for Detection of Environmental Pollution, pp. 541–552. Elsevier, Amsterdam (2024)
32. Polley, N., Gnanasekaran, B., Umapathy, S.: Nanotechnology in biosensing: the future of disease diagnosis. In: Applications of Nanotechnology in Biomedical Engineering, pp. 85–110. CRC Press, Boca Raton (2024)

33. Lu, H., et al.: Synergetic surface enhancement of quantum dots-based electrochemiluminescence with photonic crystal light scattering and metal surface plasmon resonance for sensitive bioanalysis. Talanta. **272**, 125773 (2024)
34. Zhang, K., et al.: Facile fabrication of silica glass embedded with NiO nanoparticles by 3D printing technology and its optical nonlinearity. Adv. Photon. Res. **5**(2), 2300244 (2024)
35. Sato, R., et al.: Advanced sensing and machine learning technologies for intelligent measurement in smart and precision manufacturing. Int. J. Autom. Technol. **18**(4), 545–580 (2024)
36. Hossain, N., et al.: Advances of materials science in MEMS applications: a review. Results Eng. **22**, 102115 (2024)
37. Rashid, A.: Review of: "micro and nano-electromechanical systems (MEMS/NEMS) are devices in which the physical motion of a micro-or nano-scale structure is controlled by an electronic circuit". Qeios. (2024). https://doi.org/10.32388/2zjn6h
38. Chen, X., et al.: A MEMS resonant vacuum gauge for high vacuum measurement. Vacuum. **228**, 113513 (2024)
39. Ghosh, A.K., et al.: Comparative performance evaluation of Fetal movement-detecting wearable sensors using a body-worn device. IEEE Sens. J. **99**, 3430012 (2024)
40. Zidani, J., et al.: Investigation of microstructure and physical characteristics of eco-friendly piezoelectric composite thin films based on chitosan and Ln2O3-doped Na0. 5BiO. 5TiO3-BaTiO3 nanoparticles. Nano. **14**(21), 1755 (2024)
41. Sumang, R., et al.: Next-generation hybrid Nanogenerators using Giant piezoelectric Lead-free KNNS composites for sustainable self-powered electronics. J. Alloys Compd. **1010**, 177681 (2024)
42. Majeed, H., Iftikhar, T., Maqsood, K.: Temporal and thermal dynamics exploration of different detergents' formulations components on fungal alkaliphilic lipases stability. Z. Phys. Chem. **238**(3), 563–570 (2024)
43. Majeed, H., Iftikhar, T., Maqsood, K.: Synthesis of protein-based catalyst for multi-industrial applications by using agriculture and industrial waste. Polym. Bull. **81**(14), 12633–12653 (2024)
44. Majeed, H., Iftikhar, T., Mustafa, S.: Statistical approach for newly isolated and identified microbial lipases production. Polym. Bull. **81**(17), 15823–15840 (2024)
45. Majeed, H., et al.: Green synthesis of Eucalyptus globulus zinc nanoparticles and its use in antimicrobial insect repellent paint formulation in bulk industrial production. Heliyon. **10**, e24467 (2024)
46. Majeed, H., Iftikhar, T., Siddique, A.: Agricultural waste upcycling into improved production of triacyl glycerol acyl hydrolases. Z. Phys. Chem. **238**(5), 809–827 (2024)
47. Majeed, H., et al.: Wild marigold. In: Essentials of Medicinal and Aromatic Crops, pp. 311–339. Springer, Cham (2023)
48. Saha, A., Guo, L.J., Hammig, M.D.: Radiation-tolerant dose sensor based on undoped amorphous silicon. Nucl. Instrum. Methods Phys. Res., Sect. A. **1066**, 169588 (2024)
49. Feng, J., et al.: On-chip integration of vector force sensor based on AlGaInP micro-LEDs with interface light coupling modulated by microstructures. Chem. Eng. J. **494**, 153022 (2024)
50. Ge, C., Chen, H.: Piezoelectric field effect transistors (piezo-FETs) for bionic MEMS sensors: a literature review. J. Bionic Eng. **21**, 1–13 (2024)
51. Qian, P., et al.: A resonant high-pressure microsensor based on a composite pressure-sensitive mechanism of diaphragm bending and volume compression. Microsyst. Nanoeng. **10**(1), 38 (2024)
52. Majeed, H., Iftikhar, T., Mukhtar, U.: Novel approach to water-efficient bulk industrial textile printing production of cotton fabric. Int. J. Biol. Macromol. **262**, 130064 (2024)
53. Majeed, H., Iftikhar, T., Zohaib, M.: Extension of guava shelf life through the application of edible coating formulated with mango and lemon leaves extracts. Ind. Crop. Prod. **216**, 118671 (2024)
54. Qaiser, H., et al.: COVID-19 pandemic and vaccines. In: Alternative Medicine Interventions for COVID-19, pp. 205–235. Springer, Cham (2021)

55. Shahbaz, A., et al.: Porous materials: covalent organic frameworks (COFs) as game-changers in practical applications, a review. Rev. Inorg. Chem. **44**(1), 117–133 (2024)
56. Du, S., et al.: Application and progress of high-efficiency electro-hydrostatic actuator technology with energy recovery: a comprehensive review. Energy Convers. Manag. **321**, 119041 (2024)
57. Qin, C., et al.: Actuation technique of liquid metal in thermal management: a review. Appl. Therm. Eng. **248**, 123290 (2024)
58. Wei, W., et al.: A miniature piezoelectric actuator with fast movement and nanometer resolution. Int. J. Mech. Sci. **273**, 109249 (2024)
59. Karthik, B., Shrikanth, V.: Piezoceramic stick–slip actuator with nanometer displacement resolution for liquid bridge stretching. J. Vib. Eng. Technol. **12**, 647–659 (2024)
60. Abedi, H., et al.: Predicting actuation strain in quaternary shape memory alloy NiTiHfX using machine learning. Comput. Mater. Sci. **246**, 113345 (2025)
61. Ahmad, S., et al.: Innovations in additive manufacturing of shape memory alloys: alloys, microstructures, treatments, applications. J. Mater. Res. Technol. **32**, 4136–4197 (2024)
62. Vo, T.S., et al.: Recent trends of bioanalytical sensors with smart health monitoring systems: from materials to applications. Adv. Healthc. Mater. **13**, 2303923 (2024)
63. Kołota, J., et al.: Analysis of dielectric electroactive polymer actuator displacements in resonance states using 2D laser sensor. In: 2024 IEEE 22nd Jubilee International Symposium on Intelligent Systems and Informatics (SISY). IEEE, New York (2024)
64. Murasawa, Y., et al.: Soft electroactive polymer actuators based on regioregular/regiorandom-poly (3-hexylthiophene) blends with a nanofiber structure. Polym. J. **56**, 1187–1195 (2024)
65. Zhang, H., et al.: Polyaniline/ZnO heterostructure-based ammonia sensor self-powered by electrospinning of PTFE-PVDF/MXene piezo-tribo hybrid nanogenerator. Chem. Eng. J. **496**, 154226 (2024)
66. Grupen, R.A.: The Developmental Organization of Robot Behavior. MIT Press, Cambridge (2023)
67. Patil, M.A., Kadoli, R.: Engineers guide to Terfenol-D actuators: design, performance, and real-world applications. Sens. Actuators Rep. **8**, 100236 (2024)
68. Karafi, M.R., Ansari, S.: Introduction of a novel longitudinal-flexural-torsional actuator using a magnetostrictive rod. Smart Mater. Struct. **33**(10), 105032 (2024)
69. Singh, P., Perween, S.: Magnetostrictive material-based smart materials, synthesis, properties, and applications. Smart Mater. Sci. Eng. **137**, 135–150 (2024)
70. Ebrahimi, F., Ahari, M.F.: Mechanics of Magnetostrictive Materials and Structures. CRC Press, Boca Raton (2023)
71. Dwivedi, U.K., Mishra, S., Parashar, V.: Rapid prototyping. In: Advances in Biomedical Polymers and Composites, pp. 315–341. Elsevier, Amsterdam (2023)
72. Park, H., et al.: Laser-based selective material processing for next-generation additive manufacturing. Adv. Mater. **36**(34), 2307586 (2024)
73. Shi, Y., et al.: Recent advances in additive manufacturing technology: achievements of the rapid manufacturing Center in Huazhong University of science and technology. Add. Manuf. Front. **3**, 200144 (2024)
74. Yang, L., et al.: Electrochemically-driven actuators: from materials to mechanisms and from performance to applications. Chem. Soc. Rev. **53**, 5956–6010 (2024)
75. Das, D.K., et al.: Future perspectives of Electroceramics. In: Defects Engineering in Electroceramics for Energy Applications, pp. 495–519. Springer, Cham (2024)
76. Qu, J., et al.: Recent advances on underwater soft robots. Adv. Intell. Syst. **6**(2), 2300299 (2024)
77. Lan, X., et al.: Smart space deployable truss based on shape-memory releasing mechanisms and actuation laminates. J. Spacecr. Rocket. **60**(4), 1085–1099 (2023)
78. Yang, Y., et al.: In-sensor dynamic computing for intelligent machine vision. Nat. Electron. **7**(3), 225–233 (2024)
79. Tawakuli, A., et al.: Survey: time-series data preprocessing: a survey and an empirical analysis. J. Eng. Res. **13**, 674–711 (2024)

80. Abbood, I.K., Idrees, A.K.: Data reduction techniques for wireless multimedia sensor networks: a systematic literature review. J. Supercomput. **80**(7), 10044–10089 (2024)
81. Muñoz, F., et al.: CPU and GPU oriented optimizations for LiDAR data processing. J. Comput. Sci. **79**, 102317 (2024)
82. Brum, H., Véstias, M., Neto, H.: LiDAR 3D object detection in FPGA with low Bitwidth quantization. In: International Symposium on Applied Reconfigurable Computing. Springer, Cham (2024)
83. Cheshfar, M., et al.: Comparative survey of embedded system implementations of convolutional neural networks in autonomous cars applications. IEEE Access. **12**, 182410–182437 (2024)
84. Francis, G.T., Souri, A., İnanç, N.: A hybrid intrusion detection approach based on message queuing telemetry transport (MQTT) protocol in industrial internet of things. Trans. Emerg. Telecommun. Technol. **35**(9), e5030 (2024)
85. Tran, K.T.M., et al.: Analysis and performance comparison of IoT message transfer protocols applying in real photovoltaic system. Int. J. Netw. Distrib. Comput. **12**, 131–143 (2024)
86. Martalò, M., Pettorru, G., Atzori, L.: A cross-layer survey on secure and low-latency communications in next-generation IoT. IEEE Trans. Netw. Serv. Manag. **21**, 4669–4685 (2024)
87. Aruna, K., Gurunathan, P.: Enhancing edge environment scalability: leveraging kubernetes for container orchestration and optimization. Concurr. Comput. Pract. Exp. **37**, e8303 (2024)
88. Si, J., et al.: A carbon-nanotube-based tensor processing unit. Nat. Electron. **7**(8), 684–693 (2024)
89. Maciá-Lillo, A., et al.: Optimising convolutions for deep learning inference on ARM cortex-M processors. IEEE Internet Things J. **99**, 3395335 (2024)
90. Oliveira, F., et al.: Internet of intelligent things: a convergence of embedded systems, edge computing and machine learning. Internet Things. **26**, 101153 (2024)
91. Yu, A., Qian, C., Guo, Y.: A real-time hand gesture recognition system on raspberry pi: a deep learning-based approach. In: 2024 IEEE 21st Consumer Communications & Networking Conference (CCNC). IEEE, New York (2024)
92. Huynh, H.X., et al.: Design of an IoT ultrasonic-vision based system for automatic fruit sorting utilizing size and color. Internet Things. **25**, 101017 (2024)
93. Morel, A.E., et al.: IBIS—An infrastructure management framework for adaptable, multi-sensor data collection in scientific research. In: WOCC'24-second workshop on converged computing on edge, cloud, and HPC. Riser House, Surat (2024)
94. Bibbó, G., Singh, A., Plumbley, M.D.: Environmental sound classification on an embedded hardware platform. In: INTER-NOISE and NOISE-CON congress and conference proceedings. Institute of Noise Control Engineering, Wakefield (2024)
95. Flores, X., Guadalupe, J.: Desarrollo de una red inalámbrica ad hoc para realizar tareas de monitoreo de temperatura y humedad. Repositorio Nacional Conacyt (2024)
96. Khalifa, O.O., Ahmed, M.Z., Hashim, A.H.A.: Application of Simon encryption algorithm for data transmission between sensor nodes in IoT environment. In: 2024 9th international conference on mechatronics engineering (ICOM). IEEE, New York (2024)
97. Thanikachalam, R., et al.: Design and implementation of a car's black box system using Arduino. Int. Res. J. Multidiscip. Technov. **6**(3), 260–273 (2024)
98. Morales, G.C.: Fi-Ro: Fire detection and quick response robot using FT143 detector, Network mapping, and notification system through the GSM module and ATMEGA 2560
99. Manolescu, D., Reid, D., Secco, E.L.: Hardware and software integration of machine learning vision system based on NVIDIA Jetson Nano. In: Future of Information and Communication Conference. Springer, Cham (2024)
100. Hamid, N.A.W.A., Singh, B.: High-performance computing based operating systems, software dependencies and IoT integration. In: High Performance Computing in Biomimetics: Modeling, Architecture and Applications, pp. 175–204. Springer, Cham (2024)
101. Chen, C., et al.: An efficient lightweight railway track segmentation network for resource-constrained platforms with TensorRT. Intell. Transp. Infrastruct. **3**, liae009 (2024)

102. Motorny, S., Glandon, S.R., Cheng, J.-R.C.: Examining multi-input pipelines for the service degradation point. In: Artificial Intelligence and Machine Learning for Multi-Domain Operations Applications VI. SPIE, Bellingham (2024)
103. Chen, J.: AI at the beyond 5G edge. In: AI in Wireless for beyond 5G Networks, pp. 80–121. CRC Press, Boca Raton (2024)
104. Morchid, A., et al.: Intelligent detection for sustainable agriculture: a review of IoT-based embedded systems, cloud platforms, DL, and ML for plant disease detection. Multimed. Tools Appl. **83**, 1–40 (2024)
105. Alsalemi, A., et al.: Energy data classification at the edge: a comparative study for energy efficiency applications. Clust. Comput. **27**(3), 3259–3275 (2024)
106. Kalezhi, J., Shumba, L.: Cassava crop disease prediction and localization using object detec tion. Crop Prot. **187**, 107001 (2025)
107. Pelton, J.N., Madry, S.: Space systems, quantum computers, big data and sustainability: new tools for the United Nations sustainable development goals. In: Artificial Intelligence for Space: AI4SPACE, pp. 53–104. CRC Press, Boca Raton (2024)
108. Raddo, T.R., et al.: Quantum data encryption as a service on demand: Eindhoven QKD network testbed. In: 2019 21st International Conference on Transparent Optical Networks (ICTON). IEEE, New York (2019)
109. Chaturvedi, S., et al.: Chapter 1 - advancements in water footprints research for achieving sustainable development goals: An overview. In: Kumar, A., et al. (eds.) Water footprints, pp. 1–18. Elsevier, Amsterdam (2025)
110. Tuncay, G., et al.: A review of greener approaches for rare earth elements recovery from mineral wastes. Environ. Pollut. **357**, 124379 (2024)
111. Sanjalawe, Y., et al.: A review of 6G and AI convergence: enhancing communication networks with artificial intelligence. IEEE Open J. Commun. Soc. **6**, 2308–2355 (2025)
112. Duan, J., et al.: Topological quantum materials for spintronics. MetalMat. **1**(2), e24 (2024)
113. Jebaraj, G.S., Elango, K.: A comprehensive review of EEG-based seizure detection techniques. IEEE Access. **13**, 103531–103564 (2025)
114. An, D., et al.: Prediction of thermodynamically stable room-temperature superconductors in Li–Na hydrides under high pressure. Adv. Funct. Mater. **35**(19), 2418692 (2025)
115. Xu, X., Wang, W., Ping, Z.: Biotechnological tools boost the functional diversity of DNA-based data storage systems. Comput. Struct. Biotechnol. J. **27**, 624–630 (2025)

Chapter 5
Sustainable IOT Working in Cloud, Data Infrastructure and Security in Industry 6.0

Abstract Industry 6.0 represents a paradigm change toward intelligent, sustainable manufacturing ecosystems driven by edge computing, AI, and the IoT. IoT-driven infrastructures maximize production efficiency and help to achieve sustainability goals. Real time data processing is made possible by the combination of cloud computing with edge AI, which also lowers latency and energy usage while guaranteeing anomaly detection, predictive maintenance, and autonomous decision-making. Scalable cloud platforms which handle massive industrial data volumes, and edge computing systems, which improve localized processing and lessen reliance on the cloud, are among the key technologies covered. To protect industrial data integrity, cybersecurity issues are present in networked IoT contexts, and we must fix with blockchain-based security frameworks and quantum encryption. Predictive analytics, real time anomaly detection, and intelligent robotics are the application of AI, ML, and RL in industrial data processing. To harmonize disparate IoT ecosystems, the importance of interoperability, middleware solutions, and standardized communication protocols like MQTT and CoAP is vital. The effect of these technological developments on sustainability, showing that intelligent IoT networks may optimize energy use, drastically cut down on resource use, and support a circular economy. Industry 6.0 is influencing manufacturing's future through IoT-driven innovations that are intelligent, self-governing, and sustainable.

Keywords Industry 6.0 · Sustainable IoT · Edge computing · AI-driven manufacturing · Cybersecurity in IoT · Smart data infrastructure · Cloud and edge AI · Digital twins · Predictive maintenance · Quantum encryption

1 Introduction

In the era of Industry 6.0, the intersection of Intelligent Manufacturing and SDGs signifies a revolutionary change toward environmentally benign and climate-resilient industrial processes. The IoT is at the center of this revolution; it is

H. Majeed, T. Iftikhar, *Intelligent Manufacturing in Industry 6.0*,
https://doi.org/10.1007/978-3-032-07278-8_5

seamlessly connected with data infrastructure, cloud computing, and cutting-edge security frameworks to build intelligent, sustainable ecosystems. Building on the principles of Industry 4.0 and 5.0, Industry 6.0 moves from intelligent automation and human-machine cooperation to self-sufficient, networked, and sustainable systems. Industries can attain previously unheard-of levels of operational efficiency while reducing their environmental impact by utilizing cloud based analytics, AI-driven optimizations, and IoT enabled real-time data collection. These technologies support SDGs 9 and 13 by facilitating accurate resource allocation, lowering energy usage, and improving data driven decision-making. Strong cybersecurity defenses, such as blockchain and quantum encryption, guarantee data integrity and guard against online attacks, which are essential for expanding IoT in international supply chains. Considering this, Industry 6.0 imagines a time when resilient industrial practices, environmental stewardship, and economic growth are all achieved through sustainable IoT driven innovations.

2 Cloud and Data Infrastructure

The foundation of intelligent manufacturing in Industry 5.0 and Industry 6.0 is cloud and data infrastructure (data layer), which integrates cutting-edge technology in data processing, analytics, and storage to facilitate autonomous operations and real-time decision-making. By fusing edge computing with cloud services, this infrastructure ensures scalability and reactivity while enabling effective management of massive IoT data. Large volumes of IoT data generated in contemporary manufacturing ecosystems require the storage, processing, and management of cloud services. Scalable solutions that can manage petabytes of structured and unstructured data are provided by platforms such as Google Cloud IoT, Microsoft Azure IoT, and AWS IoT Core. To improve dependability and flexibility, these platforms make use of containerized environments, serverless architectures, and distributed databases. AWS IoT Core handles billions of messages per day and supports millions of devices, allowing manufacturers to keep an eye on operations in real time throughout the globe. Cloud platforms use ML as a service to detect anomalies and do predictive maintenance, which minimizes downtime and can save up to 30% on operating expenses. Blockchain integration with cloud services guarantees data confidentiality and immutability, which is essential for preserving supply chain transparency [1–3].

By putting processing power closer to the data source and lowering latency and bandwidth needs, edge computing enhances cloud services, which offer centralized data storage and sophisticated analytics. Data are handled locally by placing computer nodes near the edge of networks, such as in sensors, controllers, or production machinery, allowing for the real-time reactions necessary for vital activities. In manufacturing anomaly detection, edge devices with AI algorithms may detect changes in production parameters, like pressure or temperature, in milliseconds. This enables prompt remedial measures. Industry analyses, such as those from Deloitte, suggest that such digital quality management can lower defect rates by

10-20%, with even higher gains possible in previously unmonitored processes, thereby raising the general caliber of the output. To process complicated AI models at the source, edge-computing systems also depend on heterogeneous hardware accelerators like Graphics Processing Units and Tensor Processing Units. NVIDIA Jetson modules, which provide up to 21 trillion operations per second of AI performance, are utilized in edge AI applications. Industry 6.0 features like intelligent quality control, autonomous robots, and dynamic production modifications are made possible in large part by these modules. Manufacturers optimize performance and resource consumption by striking a balance between localized decision-making and centralized oversight when paired with edge to cloud data pipelines [4–6].

Cloud and edge computing systems depend on sophisticated algorithms like linear programming for resource allocation and stochastic gradient descent for machine learning models. To effectively manage network topologies and guarantee low latency communication between edge devices and the cloud, these systems also make use of graph-based models. Cloud edge integration in medical manufacturing ensures compliance with medical standards like ISO 13485 by enabling real-time monitoring of circumstances like sterilizing settings. The way that cloud and edge systems improve human machine collaboration demonstrates the synergy between Industry 5.0 and Industry 6.0. With a focus on human-centric innovation, Industry 5.0 uses cloud based collaborative technologies to give employees insights and predictive analytics so they can make better decisions. Industry 6.0 moves toward completely autonomous systems, where self-optimizing edge devices and AI-driven cloud infrastructures collaborate to build robust and flexible production environments. By creating intelligent, more energy efficient systems, cloud and edge infrastructures also help achieve SDG 9. By combining dynamic voltage and frequency scaling in edge devices with modern data centers supplied by renewable energy sources, energy consumption can be lowered by 40%, supporting sustainable practices. A revolutionary step has been made in manufacturing with the integration of cloud services and edge computing, allowing for incredibly effective, scalable, and sustainable operations. The capabilities of cloud edge ecosystems will further develop as 6G networks, which enable data speeds of up to 1 Tbps, come into existence. This will open the door for hyper-intelligent and networked production systems in the Industry 6.0 era [7, 8].

3 Software Applications in IoT and Upgradation in Industry 5.0 and 6.0

The foundation of IoT ecosystems is software applications, which connect physical functionality with analytics, control, and decision-making. Software enables real time data-driven insights, hyperautomation, and human machine cooperation in Industry 5.0 and 6.0 paradigms, allowing for the development of intelligent, adaptive systems. IoT software applications are the brains behind edge devices, allowing for scalable integration, real-time data processing, and precise control. They promote Industry 5.0 and 6.0 goals of hyperautomation, resilience, and human-machine collaboration through platforms like ThingSpeak and programming languages like Python. As a result, industries are positioned to meet future demands in a manufacturing ecosystem that is secure, efficient, and sustainable. A thorough description of its function, technologies, and industrial significance can be found below.

3.1 Data Interpretation and Control

In IoT ecosystems, data interpretation and control are essential processes that allow raw data gathered from sensors and edge devices to be transformed into useful insights. At the core of Industry 5.0 and 6.0, where intelligent automation and human machine collaboration meet, is this process. To improve operational efficiency and reduce costs, advanced IoT software uses ML and AI algorithms to identify abnormalities, optimize system performance, and enable predictive maintenance. Python is a flexible programming language that is used extensively on the IoT because of its extensive libraries, which include NumPy (numerical calculation purpose), TensorFlow (AI modelling), and Pandas (data processing purpose). It is essential for industrial IoT applications due to its capacity to manage intricate data processing duties. C++ is a desirable complement to Python because of its fast execution speed and real-time optimization capabilities, which are important for applications in robotic assembly lines (low latency device control requirements). SQL and NoSQL databases effectively handle both structured and unstructured IoT data, guaranteeing scalability for systems that handle billions of data points every day. Because of its versatility in managing a variety of data formats, MongoDB (NoSQL) is frequently utilized in IoT systems [9–11].

IoT software-powered predictive analytics dramatically lowers machine downtime in industrial applications, usually by up to 25%, saving manufacturers millions of dollars a year. In the automotive industry, ML models implemented in Python process historical sensor data, such as vibration frequencies and temperature fluctuations, to predict equipment failure before it occurs. This approach not only minimizes production disruptions but also extends the lifespan of critical machinery. The integration of data interpretation and control into smart factories exemplifies the principles of Industry 6.0, which emphasizes hyperautomation and sustainability. AI-based software systems, embedded within IoT frameworks, autonomously adjust machine parameters to optimize performance in real-time. Recent industrial implementations reveal that such systems improve overall equipment effectiveness by 20–30%, leading to increased throughput and reduced waste. By lowering material use and preserving energy, this degree of optimization contributes to sustainability goals and is consistent with the larger goals of a circular economy. Sophisticated software tools, reliable programming languages, and scalable databases work together to analyze and control data in IoT systems. By facilitating real time insights, predictive maintenance, and effective resource management, these technologies propel industrial innovation and further the revolutionary goals of Industry 5.0 and 6.0 [12–14].

3.2 Analytics and Visualization

In IoT ecosystems, analytics and visualization are essential because they act as a link between unprocessed data and useful perceptions. These platforms allow stakeholders to make well-informed decisions instantly by combining data from multiple sensors and edge devices. Visualization tools are increasingly including digital twins or virtual copies of physical systems, in the context of Industry 6.0, which places a strong emphasis on hyperconnectivity and intelligent systems. These digital twins provide improved industrial process monitoring, optimization, and control by providing both real-time and predictive insights. One well-known tool in this field is Node RED, which provides a programming environment based on flow that makes IoT integration easier. It is perfect for tracking production line productivity or environmental conditions in smart factories because of its ability to produce user-friendly dashboards. ThingSpeak is excellent at processing data from IoT sensors to deliver quick visualizations and configurable alerts, two crucial functions for time-sensitive processes like manufacturing plant predictive maintenance. MATLAB IoT Analytics is a popular option for sectors like aerospace and automotive that demand high computational precision since it combines powerful simulation capabilities with AI predictive maintenance and model deployment. Measurable gains in industrial performance have resulted from the use of these platforms. Industries using IoT analytics have claimed 15–25% increases in operational efficiency. These systems drastically cut down on the amount of time needed for data interpretation, by up to 40%, by showing data trends and anomalies, enabling prompt resolution of possible problems. A multinational packaging company that uses MATLAB integrated digital twins observed a 20% decrease in unscheduled downtime, which results in yearly maintenance cost savings of millions. These techniques support sustainability objectives in addition to operational efficiency, which is consistent with the circular economy's tenets. By tracking waste production and energy consumption trends in real time, visualization technologies help enterprises maximize their use of resources. ThingSpeak, can process and visualize data in smart grids to dynamically modify energy distribution, increasing efficiency and lowering carbon footprints by as much as 30%. Platforms for IoT analytics and visualization are crucial for turning unprocessed data into intelligence that can be used. The development of Industry 6.0 and sustainable industrial growth are geared up by the real-time insights, predictive capabilities, and improved decision-making that are made possible by sophisticated tools like Node RED, ThingSpeak, and MATLAB IoT Analytics [15–18].

3.3 Implication in Industry 5.0 and Industry 6.0

Software is the foundation of innovation in Industry 5.0 and 6.0, facilitating improved human-machine cooperation, self-governing decision-making, and hyper-connected ecosystems. The next phase of intelligent manufacturing is made possible by industries utilizing AR, distributed AI, and modern communication technologies to increase productivity, sustainability, and precision. With unique yet related contributions to these forward thinking paradigms, software's revolutionary influence on manufacturing and industrial ecosystems is highlighted in The Role of Software in Industry 5.0 and 6.0. Human centric innovation is the focus of Industry 5.0, where software facilitates peaceful human machine cooperation. IoT analytics combined with tools like augmented reality are essential. When performing intricate assembly jobs, AR devices can superimpose real-time instructions over a worker's vision. AR assisted assembly has been effectively used in automobile manufacturing settings, reducing human error by up to 30%, increasing productivity, and cutting training durations by 40%. Industry 6.0 is inspired by the incorporation of distributed AI powered self-optimizing systems. By executing AI models directly on edge devices, these systems function independently and lessen dependency on centralized cloud computing. In smart factories, this decentralization allows for hyperautomation and real time decision making with reduced latency to less than 5 milliseconds. In Tesla's Gigafactories, where distributed AI oversees everything from predictive equipment maintenance to supply chain logistics. This strategy has reduced operating expenses while increasing production efficiency by approximately 20%. Ultra low latency connectivity, which is essential for Industry 6.0, is also supported by software. Manufacturing ecosystems provide smooth data transfer by leveraging protocols like as 5G Ultra Reliable Low Latency Communications and upcoming 6G technologies. For instance, 6G is anticipated to serve sophisticated applications like autonomous robotic swarms in smart warehouses by delivering peak data speeds of over 1 Tbps. These cutting-edge networks can be combined with IoT analytics and edge AI to enable dynamic process optimization, allowing enterprises to respond instantly to factors like material shortages or changes in energy. There have been noticeable advantages to using these technologies together. While edge AI-driven optimization in semiconductor production increased yield rates by 15%, AR enabled cooperation in electronics manufacturing decreased defect rates by 25%. Because self-optimizing algorithms reduce waste and energy usage, these systems support sustainability objectives. To follow the circular economy principles many industries can use these technologies to reduce energy consumption up to 30% depends on multi factors [18–20].

3.4 Future Manufacturing IoT with Software Applications

Advanced Manufacturing with Software Applications demonstrates how IoT software may revolutionize production systems by utilizing energy-efficient procedures and hyperautomation. Hyperautomation uses IoT enabled software platforms that are coupled with ML and AI to automate repetitive processes and guarantee real-time flexibility in contemporary industrial environments. A smart factory may identify assembly line defects and instantly reroute production using Node RED, a flow based programming platform. This feature ensures smooth operations by reducing downtime. According to empirical evidence, these integrations lead to up to 70% faster reaction times to production abnormalities, significantly increasing throughput in electronics assembly and automotive manufacturing industries. AI-driven IoT software applications are essential for improving manufacturing energy efficiency in addition to process optimization. Devices are controlled by algorithms that use real time data, including production loads and energy patterns usage. AI based predictive algorithms control how much power robotic systems require, turning on or off equipment as necessary. Siemens' use of IoT enabled AI to lower energy use in its electronic production facilities is a well-known example. It has resulted in operational energy savings of 20–30% per annum. This is in line with international sustainability targets and results in significant cost savings [21–23].

Manufacturers may process massive amounts of data locally while keeping centralized control thanks to sophisticated software applications that easily interface with cloud and edge computing systems. This decentralization lowers latency and satisfies hyperautomation's real-time operational needs. General Electric's Brilliant Factories use edge computing and IoT analytics to track the condition of their equipment, forecast maintenance requirements, and manage energy use, increasing production efficiency by 20% and saving up to 150 million USD annually. These developments are in line with Industry 6.0's tenets, which call for production systems to be self-optimizing and autonomous to lessen their environmental impact. Using less energy and producing less waste, smart factories with software-driven hyperautomation technologies help achieve the SDGs of the UN. An electronics manufacturing plant, for example, can reduce its carbon emissions by up to 30% by putting these technologies into practice, greatly bolstering circular economy models. IoT-based advanced software applications have revolutionized manufacturing by increasing energy efficiency and hyperautomation. These tools are essential to the shift to smart, sustainable, and self-optimizing factories of the future because they guarantee quick adaptation, cost savings, and environmental sustainability [24–26].

3.5 Security and Scalability

In Industry 6.0, the combination of blockchain, edge computing, and quantum computing guarantees the security and scalability of IoT systems. These developments help industries attain operational excellence while securing sensitive data by addressing the difficulties of handling and safeguarding large databases. They are essential for the future of IoT-driven ecosystems because real-world applications show notable risk reductions, operational efficiency, and compliance benefits. In the development of Industry 6.0, where the growth of IoT devices and hyperconnected systems creates petabyte-scale data that needs to be handled effectively and safely, security and scalability are essential pillars. Strong solutions are required to address vulnerabilities and guarantee scalability due to the quick expansion of data-intensive applications, especially in industries like healthcare and finance. The security foundation and scalability of IoT ecosystems are being improved by emerging technologies like blockchain and quantum computing. In IoT systems, blockchain technology protects device to device communication with its decentralized and unchangeable ledger. Blockchain protects data integrity and stops unwanted access by tamper-proofing transaction recording. IoT devices and IBM's Blockchain Platform combine to provide safe, traceable supply chain solutions. Blockchain is being utilized in the healthcare industry to protect patient data that is transferred between central databases and wearable devices, lowering the possibility of data breaches. Blockchain-based IoT platforms have been shown in studies to reduce fraud risks by more than 60%, especially in financial sectors where safe transaction verification is essential. Mastercard's use of blockchain technology to stop payment fraud in IoT enabled e commerce systems serves as one example [27–29].

Quantum computing and distributed cloud architectures solve the scalability issues related to handling the exponential development in IoT data. Unprecedented computational power is made possible by quantum algorithms, which allow for the real-time examination of large datasets. D-Wave's quantum systems (Currently D-Wave machines are quantum annealers, useful for specific optimization problems) are being evaluated (Industry 6.0 approach) to improve data routing, lower latency in smart cities, and optimize IoT network operations. The operational reliability of smart factories can be improved in industrial settings by using quantum-powered devices to run intricate simulations for predictive maintenance. When compared to conventional techniques, businesses using quantum computing for IoT analytics claim possible efficiency advantages of up to 40% in data processing speeds. By decentralizing data processing, edge computing enhances these technologies and guarantees scalability even in settings with spotty connectivity. Edge computing improves the resiliency of IoT systems and lowers bandwidth requirements by processing data locally. With Amazon Web Services' IoT Greengrass, edge devices can handle data safely and with less dependence on central cloud servers. In addition to improving scalability, this strategy guarantees adherence to strict data protection laws, such the GDPR in Europe. To protect IoT networks from possible dangers posed by quantum computing, Industry 6.0 security measures also include advanced cryptographic techniques, such as post-quantum cryptography. IoT systems in vital industries like telecommunications and defense are future proofed by these encryption techniques, which are made to withstand decoding by quantum computers. Post quantum cryptography methods are being actively developed by top IT companies like Google and Microsoft to incorporate into their IoT systems [30–33].

3.6 Security Framework in IoT

In the hyperconnected worlds of Industry 5.0 and Industry 6.0, where dependable IoT operations are supported by data integrity, confidentiality, and secure communication, a security framework is essential. Strong security measures are essential to defending against cyber threats, data breaches, and system vulnerabilities as the size of industrial IoT systems increases, with billions of linked devices producing sensitive data. Advanced encryption methods like AES-256 (Advanced Encryption Standard with 256-bit keys) are used to guarantee secure communication. While AES-256 provides military-grade encryption against classical computers, its security is theoretically reduced by quantum algorithms like Grover's, which is a key driver for the development of Post-Quantum Cryptography (PQC). In vital industries like healthcare, AES-256 is frequently used to secure data transfer between IoT devices and cloud services, guaranteeing patient privacy. Encryption methods such as AES-256 have been shown in real-world implementations to reduce unwanted access occurrences in enterprise systems. Implementing a comprehensive cybersecurity framework, such as a Zero Trust Architecture, can significantly enhance security posture. According to studies from the Ponemon Institute, organizations with a mature Zero Trust strategy experienced roughly 50% fewer data breaches. It is critical to note, however, that percentage reductions are highly dependent on the organization's starting maturity and the specific threats faced. Industry benchmarks like IBM's "Cost of a Data Breach" report more reliably measure the efficiency of response, finding that the full use of AI and automation can reduce the breach lifecycle by 108 days, rather than guaranteeing a specific percentage reduction in incidents [34–37].

System security is improved by authentication protocols like OAuth (Open Authorization), which confirm users' and devices' identities before allowing access. By removing the need to send private credentials over networks, OAuth's token-based architecture lowers the possibility of interception. To ensure that only authorized devices engage in the production process, OAuth is incorporated into IoT gateways in smart manufacturing, for example, to verify edge devices interacting with centralized servers. Businesses that use OAuth claim a 30% decrease in vulnerabilities connected to authentication, enhancing system integrity overall. By producing tamper-proof records of data transfers, blockchain technology provides an extra degree of protection. Blockchain's cryptographic hashing protects transaction records from unwanted changes, and its decentralized architecture guarantees that there isn't a single point of failure. Blockchain IoT technologies in supply chain management allow for end-to-end tracking of products and raw materials, guaranteeing data transparency and lowering fraud. Walmart's blockchain implementation allows the traceability of food origins to be reduced from what previously took up to seven days to retrieve from disparate records to just 2.2 seconds to query the unified digital ledger [38–40].

When distributed AI is integrated with these security frameworks in Industry 6.0, real-time anomaly detection is made possible, detecting possible cyberthreats and system malfunctions. Large amounts of IoT data are processed by AI algorithms, which proactively stop assaults by identifying departures from known trends. AI powered predictive security analytics has been shown to lower breach rates in smart grid systems, where cybersecurity risks are a major worry. To protect IoT systems from possible risks posed by quantum computing, post-quantum cryptography is becoming more and more popular. To protect IoT devices in autonomous manufacturing settings and smart cities, algorithms such as lattice-based cryptography are being developed. To reduce dangers before quantum computers become commercially feasible, IBM and Google are currently investigating quantum-resistant encryption standards for IoT networks. Secure communication frameworks protect industrial control systems from cyberattacks, guaranteeing the continuous operation of vital infrastructure from a manufacturing standpoint. Another new security paradigm, Zero Trust Architecture (ZTA), requires constant verification at all access points and assumes that no user or device is trustworthy by default. By mitigating both external and internal cyber risks, ZTA installations in hyperautomated factories can reduce operational downtime by as much as 20%. The secure and effective operation of IoT systems in Industry 5.0 and Industry 6.0 is ensured by a strong security framework that uses AES-256 encryption, OAuth authentication, blockchain technology, and developing quantum-resistant cryptography. In a variety of industrial applications, these technologies offer a solid basis for preserving data integrity, facilitating safe communication, and reducing cyberthreats [41–43].

4 How IoT Works

Real time decision making and automation across industries are made possible by the IoT, which functions as a synergistic ecosystem of sensors, devices, communication networks, and processing platforms. Five key processes that combine cutting-edge hardware, software, and analytical technologies define its smooth operation, especially in Industry 5.0 and Industry 6.0. IoT technology integration improves human-machine cooperation in Industry 5.0, with a focus on moral and human-centered methods. The adoption of wearable IoT devices, such augmented reality glasses, which give employees vital information and real-time instructions, is one prominent example. Because intelligent systems help people do jobs more efficiently, this integration increases productivity by 20–30%. Industry 6.0, on the other hand, develops these technologies further and moves toward ecosystems that are completely autonomous and self-optimizing. These days, AI powered IoT systems manage whole processes with little to no human involvement. Distributed AI models integrated into IoT frameworks autonomously oversee intricate industrial processes, guaranteeing extremely high levels of dependability, scalability, and flexibility across multiple industries [44, 45]. The impact of IoT systems in industrial and commercial applications is significant, with tangible improvements in efficiency and cost savings. In smart manufacturing, the integration of IoT technologies can improve operational efficiency by 30 to 40%, leading to substantial cost savings that amount to millions annually for large-scale factories. In the healthcare sector, IoT enabled connected wearables to have reduced patient monitoring costs by 25 to 35%, while also improving treatment outcomes through real-time alerts and continuous health data monitoring. In smart cities, IoT based traffic management systems have helped reduce urban congestion by up to 25%, significantly cutting CO_2 emissions and improving overall quality of life for city dwellers [46–48]. Here is a table to show the differences in Cloud AI and Edge AI for latency sensitive applications. Table 5.1 shows the cloud and edge AI technologies.

Table 5.1 Cloud AI vs. edge AI for latency sensitive applications

Factor	Cloud AI	Edge AI
Processing location	Hyperscale data centers (AWS Graviton4, azure Maia 200, Google TPU v5e) + satellite edge nodes	On-device SoCs (NVIDIA Jetson, Google Coral, neuromorphic cores, photonic tensor processors) + distributed edge meshes/ nodes
Latency	8–25 ms (optical switching + CXL 4.0); >100 ms if inter-continental	<200 μs (in-memory computing/ Specialized hardware); <1 ms E2E for control loops
Computational power	PetaFLOPS to 2.3 exaFLOPS in super computer clusters (GPU clusters, NVIDIA DGX SuperPOD); 120 TOPS/W (TPU v5e)	50 TOPS/W (Groq LPU); 800 TOPS (NVIDIA Jetson Thor)
Data security	Confidential VMs + homomorphic encryption; vulnerable to supply-chain attacks (firmware implants) & misconfiguration	TEEs (Intel TDX 3.0); physically unclonable functions (PUFs); air-gapped operation
Connectivity	Requires ≥100 Gbps fiber/Wi-fi 7; fails during LEO satellite handovers	Zero-trust mesh networks (BLE 5.4, matter 2.0) & architectures; operates from 10 kbps (LPWAN) to 1 Gbps (5G/ Wi-Fi 6E) with intermittent links
Critical use cases	Genomics folding (AlphaFold 4), fusion reactor plasma control, hyperscale digital twins	Autonomous UAV swarms (collision avoidance), AR surgical navigation, μs-level motor control in EV powertrains
Energy efficiency	0.38 W/GFLOPS (liquid-cooled GPU racks) varying by hardware and cooling; scope 2 emissions: 450 g CO_2/kWh (emissions depend on local grid intensity)	0.05 W/GFLOPS (photonic AI); energy harvesting (solar/vibration) enables net zero operation
Scalability	Global Kubernetes orchestration; scales to 10 M+ devices; bottlenecked by east-west traffic costs	Federated learning across 10 k devices; limited by SRAM/ReRAM capacity (max 48 GB/device, 8–16 GB is common for high-end modules like Jetson AGX Orin), limited by on-device memory and compute
Current innovations	Optical computing co processors (Lightmatter Envise), quantum annealing for optimization	3D-stacked CFETs (2 nm) Industry 6.0 futuristic Semi-conductor nodes, analog AI compute in memory (mythic AMP), self-calibrating sensors, dedicated AI accelerators (Google Coral, Intel Movidius)
Regulatory risks	EU Data Resilience Act (2026): Mandates geo-redundant backups; FAA drone latency regulations	FCC spectrum fragmentation; ISO 21434 cybersecurity compliance for automotive

In Industry 6.0, the strategic deployment of Cloud AI and Edge AI is pivotal in optimizing various industrial processes. Siemens' MindSphere exemplifies the application of Cloud AI in smart manufacturing. Built on Amazon Web Services (AWS) infrastructure, MindSphere is an IIoT platform that gathers and analyzes extensive manufacturing data. By leveraging AI-driven predictive analytics, it enhances production scheduling and machine uptime, identifying failure patterns across numerous factory machines globally. The choice of Cloud AI in this context is driven by the need to manage vast datasets from multiple locations, necessitating substantial computational power for deep learning models. Conversely, Edge AI is effectively utilized in predictive maintenance within BMW's smart factories. BMW employs Edge AI cameras on production lines to detect micro-defects in car body panels. These systems, powered by NVIDIA Jetson edge AI modules, process video feeds in real-time with latency under 5 ms. This approach has led to a 20% reduction in defect rates and a 15% improvement in production efficiency by enabling early detection of issues. Edge AI is essential here due to the requirement for immediate processing to prevent manufacturing delays that could arise from reliance on cloud-based solutions. In the realm of large-scale supply chain optimization, Amazon leverages Cloud AI to enhance demand forecasting and inventory management across its global warehouses. Utilizing Google Cloud AI, Amazon runs extensive deep learning models that analyze customer demand, shipping logistics, and regional trends. This strategy has resulted in a 40% reduction in delivery times and a one billion USD annual decrease in warehouse storage costs. The adoption of Cloud AI is appropriate here due to the need for high computational capacity and the ability to process real-time data on a global scale. Similarly, ABB Robotics applies Edge AI in industrial automation by deploying AI-powered robotic arms for autonomous quality inspections in semiconductor manufacturing. By integrating Intel Movidius Edge AI chips, these robots perform on-device image recognition in real-time, leading to a 30% reduction in defective chip production and a 25% improvement in semiconductor yield rates. Edge AI is crucial in this scenario because semiconductor inspections demand instantaneous decision-making, where the latency associated with cloud processing would be impractical. These case studies illustrate the deliberate selection of Cloud AI and Edge AI based on specific industrial requirements, demonstrating how AI-driven automation optimizes intelligent manufacturing in Industry 6.0 [49, 50]. In a benchmark demonstration, NVIDIA used the TAO Toolkit and transfer learning to train a model that achieved an overall accuracy of 99.67%, along with 92.3% mIoU, 95.8% mF1, 97.5% mPrecision and 94.3% mRecall on the bottle class of the MVTec Anomaly dataset [51].

4.1 Data Collection

To keep an eye on operational and environmental factors including temperature, humidity, motion, pH, pressure, light, sound, and machine performance, IoT devices rely on a wide variety of sensors (Detailed explained already in above sections). Because of their remarkable sensitivity and robustness, these sensors can withstand harsh industrial settings. Accelerometers, gyroscopes, and pressure sensors are examples of MEMS sensors that record extremely fine grained data at micro or even nanoscale scales. They are perfect for IoT integration because to their small size and low power consumption. MEMS vibration sensors are essential for predictive maintenance, enabling AI models to identify anomalies with high precision and recall. In industrial settings, well tuned models can achieve F1-scores, the harmonic mean of precision and recall, exceeding 0.9, allowing for the reliable early detection of machine faults. These sensors reduce downtime by up to 30% and allow for the early detection of possible machine faults, anticipating problems days or weeks in advance. Light intensity, gas presence, and object distance can all be precisely measured with optical sensors. Smart warehouses make heavy use of technologies like LiDAR to track and manage goods, increasing storage and retrieval efficiency by over 40%. Precision agriculture also heavily relies on IoT environmental sensors, such as those that track temperature, humidity, and air quality. The measurable effects of IoT sensors on several businesses are revolutionary, promoting operational, financial, and environmental efficiencies. IoT-enabled predictive maintenance offers very large cost reductions. Vibration sensors and thermal imaging systems can reduce unexpected downtime by up to 40% in industries like heavy machinery manufacture and the automotive sector by detecting equipment anomalies before breakdowns occur. For mid-sized to large businesses, this corresponds to yearly savings depending on the size of operations and the importance of continuous production cycles. For medium to large manufacturers, this can translate to significant operational cost savings. Industry analyses from firms like Boston Consulting Group and Deloitte indicate that smart factory initiatives typically yield a 10 to 20% reduction in operational costs, driven by gains in productivity, quality, and maintenance efficiency. Another area where IoT sensors provide quantifiable advantages is energy efficiency. Manufacturers can find inefficiencies and optimize resource allocation by using the detailed data on energy usage patterns that smart meters and energy monitoring devices gather. For instance, companies with IoT integrated HVAC systems can save 10–25% on energy costs by dynamically adjusting heating and cooling according to occupancy and ambient factors. Such improvements lead to savings across the manufacturing sector, which is in line with SDG 7, in a sector where energy prices can exceed USD 240 billion yearly. IoT sensors greatly improve resource conservation from an environmental standpoint. Smart irrigation systems with water flow and soil moisture sensors reduce overuse in agriculture. These devices have a direct impact on areas with low water availability by reducing water waste by 30–50%. An IoT powered precision farming project in India helped achieve SDG 6 by conserving 20 billion liters of water across 10 thousand hectares. The worldwide scalability of IoT-enabled solutions has also been demonstrated by smart water management systems in cities like Singapore, which have reduced water losses from leaks by 25%. Sensors have a direct impact on decision-making, system efficiency and predictive capacities, their accuracy, and dependability are essential to the success of industrial IoT systems. Depending on their use, sophisticated MEMS sensors, such gyroscopes and accelerometers, can attain accuracy levels of 0.1–2%. This accuracy is essential in fields like smart manufacturing, where machinery integrity depends on vibration analysis's ability to identify even the smallest abnormalities. MEMS sensors, which offer granular data for applications requiring low latency and high precision, are also essential in wearable technology and driverless vehicles. IoT ecosystems depend heavily on environmental sensors, particularly in smart cities, industrial temperature control, and agricultural. Accuracy values of ±0.5 °C for temperature and ± 2% for humidity, are achieved by modern devices such as the DHT22 temperature and humidity sensor, allowing for precise environmental monitoring. SDG 2 and 6 are directly impacted by applications like precision farming, where ideal climate control can boost crop yields by more than 30% while preserving water resources. Industrial grade sensors, such as the Bosch BMP390 pressure sensor, provide precision as precise as ±1 Pa for highly demanding industrial applications. In meteorology, where precise weather forecasts depend on trustworthy atmospheric data, and aerospace engineering, where precise pressure readings guarantee the safety and effectiveness of aircraft operations, such precision is essential. These sensors facilitate Industry 6.0 autonomous systems, allowing for self-optimization via precise real time data inputs. These technologies guarantee consistency and dependability by following strict guidelines, such ISO/IEC 17025 for sensor calibration, which are essential for developing IoT applications in a variety of sectors. These sensors help achieve water savings of up to 50% by regulating climate and irrigation, which directly supports sustainable agriculture practices and SDG 6. These cutting-edge sensor technologies work together to support contemporary IoT systems, promoting innovation, sustainability, and efficiency across a range of sectors. Through increased system intelligence, less waste, and greater operational efficiencies, these effects highlight the vital role that IoT sensors play in driving Industry 5.0 and Industry 6.0 and promoting sustainable growth [52–59].

4.1.1 Industrial Applications

Through real-time data collecting and sophisticated analytics, IoT technologies are fostering revolutionary ideas across industries, improving sustainability, safety, and efficiency. Critical machinery characteristics including temperature, rotational speed, and acoustic emissions are continuously monitored by IoT-enabled vibration sensors in smart factories. By identifying irregularities early on, these sensors reduce production disruptions and allow predictive maintenance. This method lowers machine downtime in the automobile industry, improving productivity and saving millions of dollars in operating expenses each year. Wearable IoT devices, such as smartwatches and Continuous Glucose Monitors (CGMs), are used in healthcare to gather critical biological data, such as heart rates, activity patterns, and glucose levels. When anomalies are found, these devices use AI-driven analytics and real-time monitoring to give alarms. By lowering hospital readmission rates and facilitating proactive healthcare management, these advances have enhanced patient outcomes by 30–50% [60–63].

4.2 Data Communications

To facilitate real-time monitoring, analysis, and response, data transmission is an essential component in IoT systems. This stage has been completely transformed by the development of communication protocols like 5G/6G, LoRaWAN, and Wi-Fi 6E, which provide dependable, secure, and fast data transfer that is suited to certain industrial and environmental requirements.

4.2.1 5G High-Speed Connectivity

With ultra-high bandwidths surpassing 10 Gbps and latencies as low as 1 ms, 5G networks have raised the bar for IoT data transfer. For applications like remote robotic control and autonomous manufacturing that need real-time operations, this is especially revolutionary. 5G-enabled robotic arms work with central control systems in a smart factory, coordinating activities in milliseconds. This feature guarantees smooth operation even amid high network loads, boosts productivity by up to 30%, and lessens lag-related inefficiencies. 5G's ability to serve more than a million devices per square kilometer in large-scale industrial installations guarantees reliable connectivity in IoT locations with a high population density [64–66]. Figure 5.1 show the basics of 5G, 6G and beyond.

4.2.2 Long Range Wide Area Network (LoRaWAN) IoT

LoRaWAN (Long Range Wide Area Network) is a recommended option for settings where long-range communication and power efficiency are extremely important. With a range of more than 10 km in rural regions and up to 3 km in urban settings, LoRaWAN makes it possible to send data from far sensors while using very little energy. For sectors like agricultural and environmental monitoring, where equipment like air quality monitors and soil moisture sensors must operate in large, frequently off-grid areas, this protocol is perfect. LoRaWAN-powered IoT systems have shown a 20% increase in crop yields in precision farming by guaranteeing constant data flow to optimize fertilization and irrigation schedules [67–69].

Fig. 5.1 Basics of 5G, 6G, and beyond

4.2.3 Wi-Fi 6E to Improve Network Efficiency

Utilizing the 6 GHz band, Wi-Fi 6E is intended for high-density settings such as smart cities, factories, and hospitals. Wi-Fi 6E guarantees uninterrupted data transmission even in congested network situations by providing data rates of up to 9.6 Gbps and improved interference avoidance. Wi-Fi 6E in healthcare enables real time telemetry of patient data by supporting linked medical devices in smart hospitals. This improves patient outcomes by more than 25% and cuts down on delays in critical care interventions [70, 71].

4.2.4 The Future of IoT Data Transmission with 6G/6G Plus Wireless Communication

With previously unheard-of speed, capacity, and connection, the sixth generation (6G) of wireless communication is predicted to completely transform industrial applications and IoT systems. Although still in the research and development stage, 6G expands on the developments of 5G and presents novel concepts that are essential to the hyper connected, self-governing, and intelligent ecosystems that Industry 6.0 envisions. Ecosystems that are intelligent, autonomous, and essential to Industry 6.0. 6G is anticipated to be 100 times faster than 5G with anticipated data transfer speeds of 1 Tbps, guaranteeing instantaneous connection for high-demand applications like virtual reality, augmented reality, and real time digital twins. These features will make it possible to continuously analyse large datasets, up to one petabyte per hour, which is essential for making dynamic decisions in logistics and manufacturing [72, 73].

It is predicted that latency in 6G networks will drastical reduction, targeting 0.1 ms (100 μs) or even 10 μs (0.01 ms) for specific, mission critical applications, a significant improvement over the 1 ms target of 5G. For real time operations like remote robotic surgery, where accuracy and responsiveness are mandatory or autonomous drone fleets handling just in time logistics with accuracy gains of over 50%, this ultra-low latency is crucial. 6G's capacity to link up to ten million devices per square kilometer will revolutionize dense IoT environments, such as mega factories and smart cities, by enabling reliable connectivity between billions of devices. 6G networks will provide ultra-wide bandwidth for data intensive applications by operating in the terahertz frequency range (0.1–10 THz). However, developments like reconfigurable intelligent surfaces (RIS), which improve signal intensity and coverage, are addressing issues with terahertz waves, such as their limited range and vulnerability to air attenuation. 6G will have AI and ML at its core to manage the complexity of such networks. This will result in cognitive, self-healing networks that can identify and fix errors on their own, lowering downtime by up to 90%. 6G has a wide range of revolutionary industrial applications. 6G-enabled autonomous robots in manufacturing will be able to synchronize precisely in less than a microsecond, which will increase assembly efficiency by 40% and drastically cut down on production times. 6G enabled digital twin technology will enable predictive maintenance and real-time simulation, resulting in cost savings of 20–30% in industries, including automotive and aerospace manufacturing. 6G's ultra-low latency and high-resolution imaging will improve patient outcomes for 500 million people in underserved areas worldwide by enabling remote surgery and continuous monitoring. In line with SDG 6, precision agriculture will also profit from 6G enabled drones and sensors, which can evaluate large farmland areas in real time, increasing yield by 20–30% and decreasing water waste by up to 50% [73, 74].

By smoothly integrating satellites, aerial networks, and terrestrial systems, 6G also makes space air ground integration easier. For remote industrial operations like mining, where real-time data from satellites can cut operating risks by 25%, this connection is essential. Beyond 6G, scientists are investigating nano sensor networks and quantum communications. IoT systems will have never-before-seen security thanks to quantum key distribution (QKD), which could save billions of dollars in cybersecurity expenses, particularly in delicate sectors like defense. By enabling communication at molecular sizes, nano sensors functioning in the THz spectrum will be able to detect wear in equipment and increase lifespans. Given the high power requirements of terahertz networks, these developments show the importance of energy efficient designs and the inclusion of renewable energy sources. 6G is a wireless communication revolution that will spur innovation and sustainability in a variety of sectors. 6G supports Industry 6.0's revolutionary vision by providing ultra-high speed, huge connectivity, and intelligent network management, opening the door to a more intelligent, resilient, and sustainable global economy [75–78].

Achieving SDGs objectives requires effective data transfer. Because they enable real-time communication and sophisticated automation, high-speed networks like 5G and the soon to be 6G technology are essential to Industry 5.0 and 6.0 ecosystems. While 5G networks improve smart infrastructure and industrial automation with bandwidths exceeding 10 Gbps and latencies below 1 ms, 6G promises massive machine type communication and ultra reliable low latency communication, which will guarantee smooth connectivity across complex systems. Significant cost savings, increased energy efficiency, and lower emissions are made possible by these skills, which also make predictive maintenance, autonomous manufacturing, and intelligent urban management possible. Predictive maintenance solutions, for instance, can save businesses up to USD 1.5 million/ annum, while IoT-based energy optimization can reduce operating expenses by 10 to 25% [79–81].

4.2.5 Alignment of Data Communication Technologies with SDGs

By combining cutting-edge sensor systems and dependable data transmission protocols, IoT technologies are revolutionizing urban settings and industries and directly advancing global sustainability objectives. IoT environmental sensors in smart cities keep an eye on important variables, including air quality, noise pollution, and traffic flow. Air quality monitors are essential to the implementation of CO_2 reduction programs in places such as London, which advances SDG 11. By giving city planners useful information, these systems enable better urban living conditions and successfully handle environmental issues. By facilitating the creation of resilient and adaptable infrastructures, a key component of SDG 9, IoT sensors in industries promote operational efficiency, sustainability, and well-being. In line with SDG 12, these sensors optimize resource utilization, reduce waste, and improve supply chain management by producing precise and real-time data. By streamlining processes and optimizing energy use, real-time monitoring also makes it possible to reduce greenhouse gas emissions, which directly addresses SDG 13. IoT based solutions use dependable communication protocols in urban settings to monitor and optimize energy use, traffic, and environmental factors, promoting sustainable urban life. To achieve SDG 11, these solutions must lower carbon footprints, boost resource efficiency, and improve urban management. 6G technology is anticipated to transform urban infrastructure by drastically lowering emissions and energy use, which will further address SDGs 11 and 13. Using cutting-edge sensor technologies and smooth data transport, IoT technologies enable smart industries and sustainable urban environments by acting as the foundation for Industry 6.0's autonomous ecosystems. These developments foster innovation, strengthen resilience, and aid in the global shift to sustainable development, achieving the interrelated objectives of SDGs 9, 11, 12, and 13 and creating the framework for an intelligent and sustainable future [82–84].

4.3 Advanced Data Processing Technologies Using IoT

Predictive analytics and AI-driven optimization are two examples of quantifiable industrial impacts that are accelerated by cutting-edge data processing technology and offer substantial operational and financial advantages. One of the most noteworthy results is the 20–25% decrease in downtime that predictive analytics can provide. For mid-sized manufacturers, this reduction can result in yearly savings of up to USD 1.5 million by reducing unscheduled production disruptions and increasing overall equipment efficiency. By using AI algorithms that anticipate possible malfunctions before they happen, businesses can plan maintenance in advance, guaranteeing uninterrupted production flow, and lowering the need for expensive repairs. AI powered energy optimization is essential for sectors with significant energy demands, such the production of steel and cement, in addition to reducing downtime. Real-time energy usage adjustments by AI-driven systems can lower operating energy expenditures by 15–25%. This optimization results in large savings while also lessening the environmental effect in energy-intensive industries like steel manufacture, where energy prices account for a sizable portion of total spending. AI helps firms maintain operational efficiency while complying to environmental targets, like those mentioned in SDG 13, by dynamically changing energy use based on production requirements. Another important area where AI and real time data processing technologies have a big impact is waste minimization, especially in sectors like food processing. During production, AI powered quality checks enable real time fault monitoring and repair, resulting in a 40–50% reduction in waste. By encouraging more sustainable methods, cutting down on food waste and making sure that raw materials are used more effectively, this not only directly lowers costs in the food processing industry but also advances SDG 12. Industries can improve profitability and support global sustainability initiatives by optimizing resources by incorporating these technologies into production processes. All things considered, these measurable effects of real time quality control, AI-driven energy management, and predictive analytics show how valuable data processing technologies are for increasing operational effectiveness, cutting waste, and promoting sustainability in a variety of industrial sectors. In addition to providing quantifiable financial benefits, the capacity to decrease waste, optimize energy consumption, and reduce downtime aligns with larger sustainability goals and positions industries for long-term success in a market that is becoming more competitive and environmentally conscientious [85, 86].

4.4 Edge Computing

By reducing latency and facilitating real-time analysis through local processing near the data source, edge computing transforms data processing. This method improves responsiveness and efficiency, which makes it essential for contemporary industrial applications. High-performance GPU equipped devices like the NVIDIA Jetson Nano and Xavier platforms are made especially to run sophisticated AI models at the edge. For instance, NVIDIA Jetson systems' real-time anomaly detection features enable these platforms to evaluate vibration and acoustic data in milliseconds, detecting manufacturing flaws with up to 98% accuracy. This minimizes waste while guaranteeing product quality. Predictive maintenance is another crucial use, in which cutting-edge AI algorithms continuously assess the health of machines and anticipate any faults days in advance. This proactive strategy reduces unplanned downtime, which results in yearly savings of millions of dollars for sectors like aerospace and automotive production that depend on expensive technology. Autonomous cars in smart warehouses and factories gain from edge AI's quick decision-making, which allows for accurate navigation and obstacle avoidance. These systems ensure seamless operations even in dynamic contexts by processing data in less than 10 ms. By providing high speed, energy-efficient AI inference, technologies like Google Coral AI accelerators and Intel Movidius Vision Processing Units further improve the edge-computing environment and are appropriate for sectors with demanding power and speed needs. Advanced industrial automation, such robotic arms with AI to carry out complex assembly operations with little supervision, is also supported by edge computing. These systems lower operating costs and increase throughput by processing visual and sensor input locally. Seamless connectivity between edge devices is made possible by integration with 5G networks, which enhances their real-time capabilities while guaranteeing scalability and dependability in demanding industrial settings. Edge computing supports Industry 5.0 and 6.0 principles by enabling industries with localized, fast data processing, encouraging human–machine cooperation and improving sustainability [87–90].

4.5 Cloud Computing

Because cloud computing offers the extra processing power and long-term storage that edge devices might need, it is essential for managing massive amounts of data for industries. This market is dominated by major cloud platforms that provide reliable solutions with several essential features, such as Google Cloud IoT, AWS IoT Core, and Microsoft Azure IoT Hub. Scalability is one of the systems' most notable features; they can process data streams the size of petabytes while maintaining 99.99% uptime reliability, which makes them perfect for sectors that need to manage enormous volumes of data in real-time. Cloud platforms not only manage massive data volumes but also provide sophisticated analytics through AI-powered analytics pipelines. By analyzing production data, these pipelines find inefficiencies and optimize energy use, resource allocation, and production workflows. Cloud-based analytics can identify process bottlenecks or energy waste, allowing enterprises to optimize operations for optimal efficiency. For sectors where accuracy and resource management are essential, like the automobile manufacturing industry, this skill is especially vital. Through AI powered pattern recognition, a top automaker utilizing AWS IoT Core was able to cut its defect rate by 15%. This significantly improved product quality and profitability by enhancing process control and early problem detection. Another essential component of cloud platforms is security i.e. end to end encryption and adherence to ISO/ IEC 27001 standards guarantee that data is safeguarded at every stage of its existence. In sectors like pharmaceuticals or autonomous manufacturing, where sensitive data, including production procedures or intellectual property, must be protected from breaches or assaults, this degree of protection is essential. Businesses can benefit from the scalable and adaptable infrastructure required to handle cutting-edge industrial applications while simultaneously ensuring the integrity of their data by utilizing these cloud technologies. Because cloud computing offers the infrastructure required for autonomous operations, smart factories, and cooperative human-machine ecosystems, it easily combines with the concepts of Industry 5.0 and 6.0 [91–93].

4.6 AI, ML, RL, and NLP Algorithm in Industrial Data Processing

Algorithms for AI and ML have become crucial parts of industrial data processing, propelling improvements in demand forecasting, predictive analytics, and anomaly detection across a variety of industries. In image-based quality control systems, CNNs are frequently utilized, especially in manufacturing and automotive industries. Only products that meet quality standards move on to the following stages of production thanks to these networks' exceptional ability to detect surface flaws in products with precision levels that frequently exceed 97%. Real-time image analysis by CNNs helps producers spot and get rid of faulty products quickly, thus enhancing product quality and cutting down on waste. Another effective method for improving assembly line robotics is RL. RL algorithms dynamically modify the behavior of robots to optimize operations, increasing productivity and decreasing downtime, by continuously obtaining feedback from the environment. By using this technique, robots can improve their actions to get the greatest results by learning from mistakes. RL is used, for instance, in automated material handling, where robots can modify their movement patterns and tactics in response to shifting production line conditions, boosting output and cutting down on cycle times. To assess operator instructions or consumer input in real time, industrial automation systems are increasingly using natural language processing, or NLP. NLP algorithms interpret and process human language, allowing systems to respond to spoken or written input or make choices quickly. This is especially helpful in fields like customer service, where NLP can automate answers to questions, and maintenance systems, where it can examine technical logs or operator notes to identify new problems. AI-driven models are employed in industries like steel manufacturing to anticipate the temperature throughout the smelting process, where even little temperature variations can result in flaws or inefficient use of energy. Real-time predictions of the ideal temperature ranges are made by these AI models, guaranteeing exact control over the smelting process [94–96].

5 Industrial Case Studies Using these Communication Technologies

Businesses who are adept at using IoT and data processing technology have several important tactics in common that provide them a competitive advantage. Combining edge and cloud computing is one of the most important tactics. Prominent companies use hybrid systems that combine cloud analytics for longer-term, more comprehensive insights with local edge processing for urgent, real-time activities. While data that needs more processing or long-term storage is routed to cloud platforms like Microsoft Azure IoT Hub or AWS IoT Core, edge devices like NVIDIA Jetson platforms or Google Coral perform quick, low latency jobs like anomaly detection in production lines. In addition to providing the deeper analytics and scalability offered by cloud-based systems, this hybrid approach guarantees that real time processes are optimized for speed and accuracy. Investing in AI knowledge is a company's second successful tactic. Businesses such as General Electric, Siemens, and Bosch have invested heavily in creating domain specific AI models that are suited to their sector's requirements. These models are designed to meet the unique needs of the manufacturing, energy, and automotive industries rather than being merely general-purpose. For instance, Siemens utilizes AI in smart grid systems to improve power distribution, increasing energy efficiency by 15%, and Bosch uses AI-driven predictive maintenance models in its factories, lowering downtime by up to 25%. In addition to providing the deeper analytics and scalability offered by cloud-based systems, this hybrid approach guarantees that real-time processes are optimized for speed and accuracy [97–99].

Investing in AI knowledge is a company's second successful tactic. Businesses such as General Electric, Siemens, and Bosch have invested heavily in creating domain-specific AI models that are suited to their sector's requirements. These models are designed to meet the unique needs of the manufacturing, energy, and automotive industries rather than being merely general-purpose. Siemens utilizes AI in smart grid systems to improve power distribution, increasing energy efficiency by 15%, and Bosch uses AI-driven predictive maintenance models in its factories, lowering downtime by up to 25%. These investments enable these businesses to develop distinctive AI capabilities that give them a major competitive edge, increasing operational effectiveness, cutting expenses, and producing higher quality goods. Success in data-intensive businesses is largely driven by specialized hardware solutions. Proprietary hardware accelerators, like the Dojo AI system, have been created by companies like Tesla especially to meet the massive data processing demands of autonomous car training. For instance, Tesla's Dojo system makes it possible to process enormous volumes of data from its fleet of cars, which boosts the effectiveness and speed of its AI models. This method improves the efficiency of Tesla's autonomous driving capabilities while speeding up deep learning processes. In addition to improving data processing efficiency, custom hardware accelerators provide businesses like Tesla a major competitive advantage in the quest to create the most cutting-edge AI-driven products [100, 101].

Consumer electronics company used Azure IoT services in conjunction with bespoke edge AI solutions to reduce assembly line inefficiencies by 30%. The business was able to greatly reduce production costs and increase throughput by combining edge processing for instant feedback with cloud-based analytics for long-term insights. Through the integration of AI for predictive maintenance, real-time performance optimization, and quality control, the business was able to increase operational effectiveness and save costs by improving efficiency across several phases of its production process. These tactics highlight a broader pattern in sectors that are effectively utilizing IoT, AI, and customized solutions to process data more effectively. While AI investments and specialized hardware solutions enable businesses to obtain a competitive edge, edge and cloud computing together enable real-time performance and scalability analytics. The development of Industry 5.0 and Industry 6.0, which prioritize the fusion of automation, ML and human intelligence to create smarter, more sustainable industrial systems, depends on these technologies [102, 103].

6 Emerging Frontiers in Communication and Technology with 7G/7G Plus

As industries evolve toward hyper connected and ultra-efficient systems, the technological landscape is already venturing beyond 6G into pioneering realms such as 7G communications and quantum technologies. These advancements promise to redefine security, scalability, and precision in industrial operations. In addition to improving operational efficiency, these cutting-edge technologies closely match Industry 6.0's objectives, which prioritize sustainability and human-centric systems. By guaranteeing optimal resource usage and accurate defect detection in production lines, the implementation of nano-sensors can reduce waste and directly support SDG 12. The implementation of quantum communications in various sectors improves worldwide connectivity and strengthens defenses against cyberattacks, which is consistent with SDG 9. Industries are on the verge of a paradigm shift as we move toward 7G and nano communication technologies, where hyper efficiency, quantum resilience, and molecular level precision become the new norm. The ultra-low latency and enormous data throughput of 6G will be expanded upon by 7G technologies, which will incorporate quantum communication systems for previously unheard of speed and security. A key component of quantum communications, quantum key distribution uses the laws of quantum mechanics to produce encryption that is nearly impenetrable. This guarantees that data that is intercepted during transmission is immediately damaged, shielding vital industrial, and IoT data from online attacks. QKD based networks are anticipated to eradicate weaknesses in military industries, where security breaches can jeopardize national security and cost billions of dollars. Industries like aerospace and pharmaceuticals may protect critical operational data and intellectual property by integrating quantum-secure communication into IoT devices. A major step toward scalable quantum networks has been taken with the demonstration of quantum-secure transmissions at distances surpassing 1000 km in early trials conducted in Europe. Nano communications, which uses nanoscale sensors and gadgets to communicate at the molecular or atomic level, is another revolutionary frontier. Terahertz electromagnetic waves are used by these sensors to monitor and transmit data with a high degree of accuracy. Such technology has a special influence on manufacturing quality assurance and predictive maintenance. Nanosensors built into vital industrial parts, such robotic arms or turbine blades, can identify chemical changes, material deterioration, or microfractures long before they become serious problems. By doing this, businesses can extend the lifespan of their machinery by as much as 50%, saving money on replacements and downtime. The automotive sector is investigating the use of nanosensors in electric car batteries to identify internal temperature variations and enhance efficiency, resulting in a 30% increase in battery life [104–106].

7 Upgradation of Data Communication Technologies in Industry 6.0

In the context of Industry 6.0, which prioritizes human-centric and sustainable systems, advanced data processing technologies unlock transformative capabilities that redefine the way industries operate. The integration of AI, IoT, and next generation automation plays a critical role in fostering efficiency, sustainability, and customization. Two main advancements, Collaborative AI and Hyper Personalization, are at the forefront of these developments. Workflows that are more flexible and intuitive are made possible by collaborative AI, which is the smooth fusion of human knowledge and machine intelligence. Industry 6.0 places more emphasis on using AI to support human decision making than Industry 4.0, which was primarily concerned with automation. Dynamic simulations of manufacturing settings are supported by sophisticated technologies like digital twins and reinforcement learning algorithms. Aviation division of some companies use AI enhanced digital twins to provide predictive maintenance solutions that guarantee worker safety while reducing downtime by up to 30%. Similarly, collaborative robots (cobots), like those used by BMW, help employees with high-precision or repetitive jobs in the automobile industry. These cobots improve operator ergonomics and achieve 95% task accuracy using AI algorithms that adjust to human actions in real time [107–109].

AI powered hyper personalization enables businesses to create goods and services that are specifically suited to each client's requirements. Businesses can find patterns and preferences with never before seen granularity by utilizing big data analytics and machine learning techniques. The consumer goods, fashion, and technology sectors are most affected by this strategy. Nike uses AI-based consumer data to manufacture (with and without 3D printing) customized shoe/sneaker designs, enhancing customer happiness and resulting in 15–20% more recurring business. AI is used by adaptive production lines in smart manufacturing to dynamically change settings, guaranteeing that products satisfy customer specific requirements. This immediately supports SDG 12 by reducing overproduction by up to 40% [110, 111].

To further improve collaboration, Industry 6.0 also provides sophisticated human machine interfaces like AR tools and brain computer interfaces. Technicians can monitor real time equipment data with AR-assisted maintenance systems, which can save repair times by 30%–50%. By encouraging innovation in operational procedures and infrastructure development, these technologies support SDG 9. Sustainable resource use is ensured by integrating collaborative AI and hyper personalization into production processes. In the textile industry, AI predictive analytics reduces fabric waste during production by up to 35%, lowering environmental impact and promoting SDG 13. Industry 6.0 is a paradigm change toward a sustainable, people focused industrial era by emphasizing adaptive systems that incorporate human monitoring. Industry 6.0 expands on Industry 5.0's focus on human-machine cooperation to boost productivity and innovation by integrating sustainability and resilience into industrial systems. This is made possible by developments in decentralized AI, 6G networks, and quantum computing. The use of 6G technology in manufacturing speeds up data processing by 100 times as compared to 5G, allowing for immediate communication and decision-making among millions of linked devices. Quantum computing is also being investigated for its potential to greatly lower energy usage and operating costs by resolving intricate optimization issues in production scheduling and logistics. Industry 6.0 is a roadmap for industries to meet global problems and achieve long-term success in a world that is becoming more and more focused on sustainability. Collaborative AI, hyper personalization, and innovative technologies enable industries to streamline operations, solve important environmental issues, and provide unmatched consumer value. In addition to improving industrial performance, this change strengthens ties to the SDGs [112, 113].

8 Alignment of Data Processing Technologies with SDGs

The Industry 5.0 and Industry 6.0 revolutions, where technology improvements are in line with sustainability and human centered aims, are based on effective data processing. Data driven initiatives help enterprises minimize waste, improve operational efficiency, and further global sustainability goals by enabling smarter, more sustainable production practices. In addition to promoting economic growth and environmental stewardship, these initiatives directly serve SDGs 9 and 12. Industries may move toward a sustainable future that benefits the planet and society using such innovations. By encouraging industrial innovation, environmental sustainability, and the creation of next generation intelligent and sustainable systems, efficient data processing plays a key role in accomplishing the SDGs. Using modern data analytics to maximize industrial automation, improve operational efficiency, and encourage creative manufacturing processes, it directly promotes SDG 9. Industries may obtain real-time insights, minimize downtime, and construct robust infrastructure that forms the foundation of Industry 5.0 and Industry 6.0 by incorporating smart sensors and AI-driven analytics into production systems. While easing the shift to automated, intelligent, and human centered industrial ecosystems, these technologies also promote sustainability. By reducing resource waste in manufacturing processes, data processing contributes to SDG 12. Precise control of production lines is made possible by real-time data, which tracks factors including machine performance, material usage, and energy consumption. Allocating resources optimally lowers operating expenses, environmental effects, and excessive consumption. In addition to increasing efficiency, technologies like IoT sensors and predictive maintenance algorithms make sure resources are used wisely. This supports responsible production objectives, encourages more sustainable purchasing habits, and helps companies efficiently handle the escalating environmental issues. AI predictive models incorporated into data processing systems greatly aid SDG 13. By predicting demand, spotting inefficiencies, and improving energy management, these models' lower emissions and energy usage. AI algorithms in smart factories anticipate changes in energy consumption and dynamically modify processes to reduce waste and carbon footprints. These projects support sustainable production practices and lessen the environmental effects of industry, which is in line with global climate action plans. Manufacturers using AI-driven energy optimization have reportedly saved between 10 and 15% on energy costs, making a significant contribution to the fight against climate change. Data processing serves as the foundation for revolutionary applications such as Hyper Personalization and Collaborative AI in the framework of Industry 6.0. By bridging the gap between machine intelligence and human expertise, collaborative AI makes it possible for more flexible decision-making and intuitive workflows. Businesses may increase productivity and worker satisfaction while preserving flexibility by integrating human insight with automated procedures. Hyper Personalization uses AI to customize goods and services to meet the demands of specific clients, cutting down on waste and overproduction. Business success and environmental sustainability are supported by this combination of resource optimization and customer-centric customization [114–116]. By offering scalability, sophisticated analytics, and strong security, cloud computing integration enhances the impact of data processing even more. Cloud solutions that facilitate the smooth integration of edge computing and centralized data analysis, such as AWS IoT Core and Microsoft Azure IoT Hub, allow industries to reliably analyze huge data streams. This hybrid strategy strikes a healthy balance between short term performance and long-term resilience by fostering long-term strategic insights while retaining real time operational management [117].

9 Action to Execute Data Communications

The ability of industrial systems to convert processed data into concrete, real-world actions is a key component of the shift to Industry 5.0 and Industry 6.0. These activities constitute the last phase of a data-driven decision-making loop and are carried out by actuators, robotic systems, or other IoT devices. To maximize resource utilization, achieve operational efficiency, and support SDGs, including 9, 12, 13, this technical innovation is essential.

9.1 Key Industrial Applications of Action Execution

Smart energy networks (in grids) to increase efficiency and sustainability, which integrate cutting-edge technologies to dynamically balance supply and demand, represent a paradigm shift in the distribution, consumption, and optimization of energy. Actuators, which allow for real-time modifications to load distributions and guarantee optimal energy flow, are at the heart of this shift. These actuators stop grid portions from being overloaded or underutilized, which can result in inefficiencies and increased operating expenses. Actuators provide for precise control of distributed energy resources including solar panels, wind turbines, and battery storage devices in grid tied systems. These grids can predict demand patterns and modify energy flows appropriately by utilizing AI-based prediction algorithms and real-time monitoring systems. With 15–20% reductions in transmission losses and grid performance stabilization, this integration greatly increases energy efficiency. Smooth switching between energy sources is made possible by dynamic voltage and frequency control, which increases resilience in the face of peak loads or unforeseen demand spikes. The efficiency of IoT controlled actuators in updating energy networks is demonstrated by a noteworthy case study in the United States. Through enhanced reliability and decreased energy waste, the grid was able to save billions annually by implementing automated control systems and real-time data analytics. Real-world implementations demonstrate the substantial financial impact of modernizing the grid. The U.S. Department of Energy's Smart Grid Investment Grant program reported that its projects resulted in billions of dollars in total benefits, with measurable gains from reduced outage frequency and duration, improved operational efficiency, and lower operating costs.This upgrade directly supported SDG 13 by reducing CO_2 emissions by 10%. Through improved use of renewable energy sources, reduced dependency on fossil fuel based power plants, and streamlined energy flows, the carbon footprint was lowered. Deploying edge computer nodes improves real time decision-making by lowering actuator response time latency. Even in grids that are spread out geographically, this guarantees that load adjustments take place within milliseconds. Blockchain technology is used by decentralized energy grids to enable safe, open energy trading between prosumers, or producers and consumers. This strategy optimizes the use of excess energy produced by DERs. IoT enabled sensors keep an eye on variables like temperature, voltage, and current across grid components. These sensors reduce downtime by 25–30% by feeding data into AI systems that forecast maintenance requirements. Bidirectional smart inverters that control the electricity flow between energy storage devices and the grid are integrated with actuators. These inverters stabilize the grid during variations [118, 119].

IoT-enabled technologies are used in precision agriculture to maximize resource use and boost crop yields, promoting sustainability and financial rewards in the agricultural industry. IoT sensors, which offer real time data on vital elements like soil moisture, temperature, and nutrient levels, are at the core of this breakthrough. Based on data analysis, these sensors allow automated actuators to regulate irrigation pumps and other systems, guaranteeing accurate and prompt interventions. Sensors provide detailed information about the environment and soil, which is then fed into edge-based or centralized computing systems for analysis. To guarantee that inputs are applied just when and where they are required, these systems control irrigation, fertilizer application, and pest control systems. Edge computing makes it possible to make decisions in real time by processing data locally, and AI models forecast crop health and allocate resources as efficiently as possible. Using 30–50% less water, automated irrigation systems guarantee that crops are properly hydrated without being overfed. In a similar vein, fertilizer application methods limit nutrient waste, which lowers operating expenses and lessens environmental runoff. By decreasing resource waste and increasing efficiency, these precision techniques support the sustainable use of natural resources and are consistent with SDG 12. When IoT based irrigation systems were installed on one million hectares of farmland in India, crop yields increased by 25% and water use was reduced by 40%. Widespread adoption greatly increased farmer earnings and regional food security while reducing the wastage of agricultural resources. Drones with multispectral imaging were incorporated to improve accuracy by pinpointing problem areas and facilitating focused interventions that increased output [120–122].

Modern technologies that maximize resource utilization, boost productivity, and encourage sustainability are revolutionizing precision agriculture. Variable rate technology increases yield efficiency by 20–25% by enabling farming equipment to apply seeds, fertilizer, and insecticides at different rates based on real-time soil and crop data. IoT powered smart greenhouses that automatically control lighting, humidity, and temperature can boost crop yields by 15–30% while using a lot less energy. Blockchain integration guarantees agricultural product traceability, improving supply chain quality management, regulatory compliance, and consumer trust. Tasks like weeding and harvesting are automated by robotic agricultural equipment with sophisticated AI vision systems, which lowers human expenses while maintaining remarkable field operations precision. By enabling effective and sustainable farming methods, these technologies help solve important global issues like food hunger, water scarcity, and environmental degradation. They facilitate the shift to Industry 6.0, which is defined by completely autonomous and intelligent agriculture solutions, and Industry 5.0, which emphasizes human centric and sustainable systems. Together, these developments help achieve SDGs 2 and 12, promoting a resilient and sustainable agricultural future [123–125].

By maximizing energy use across production lines and guaranteeing sustainability and efficiency, dynamic voltage scaling (DVS), is transforming the manufacturing industry. DVS prolongs machinery lifespans by reducing operational stress and lowering energy expenditures by 20–30% through the analysis of real-time load circumstances and voltage supply adjustments. To ensure accuracy and versatility in industrial activities, actuators are essential for controlling robotic arms, variable frequency drives (VFDs), and servo motors. Technological innovations such Edge AI systems, which identify inefficiencies and dynamically modify equipment power use, further improve DVS. By anticipating equipment failures and integrating with predictive maintenance algorithms and actuators, these systems can reduce unscheduled downtime. Industry studies, such as one by the Capgemini Research Institute, indicate that average reductions of 15 to 30% are achievable.This promotes sustainable operations in addition to increasing productivity. A German automaker that integrated DVS technologies into all its production lines serves as a practical illustration. Through innovation in industrial procedures, the company was able to reduce downtime and energy usage, resulting in annual savings of almost €ten million. This aligned with SDG 9. By promoting environmentally conscious production, the energy efficiency improvements helped to significantly lower CO_2 emissions, thereby supporting SDG 13. These developments highlight how important DVS is to the transition to Industry 5.0 and Industry 6.0, where intelligent, sustainable, and human-centered systems are redefining industrial processes. Manufacturers may increase sustainability, cost savings, and energy efficiency by integrating DVS, guaranteeing long-term competitiveness in a global market that is changing quickly [126–129].

9.2 Integration IoT with Industries 5.0 and 6.0

Resilient infrastructure and industrial innovation immediately aid in the achievement of SDGs, especially 9, 12, 13 by cutting waste and maximizing resource use. The smooth transition to Industry 5.0 and Industry 6.0 is made possible by these developments, which build more intelligent, human centered, and ecologically conscious industrial ecosystems that strike a balance between sustainability and economic growth for the good of companies, communities, and the environment. To improve industrial operations, Industry 5.0 encourages the integration of collaborative AI systems that combine robotic precision with human experience. Real-time commands based on sensor data are executed by robotic systems, guaranteeing unparalleled speed and accuracy. At the same time, human-in-the-loop frameworks allow employees to direct these systems, resulting in flexible and adaptive workflows. Under human supervision, cobots (collaborative robots) in electronics production may complete complex jobs with millimeter accuracy, increasing productivity by 30% and lowering product faults. Sustainability and hyper personalization are becoming essential components of industrial processes as Industry 6.0 develops. By monitoring equipment conditions at the molecular level, actuators equipped with nanosensors allow for extremely fine-grained process control and can cut material waste by up to 50%. Green actuators powered by renewable energy sources power operations in smart factories, lowering dependency on conventional energy systems and supporting global environmental programs. Edge computing reduces latency and reliance on centralized systems by enabling actuators and IoT devices to process and act on data locally. In warehouse management, autonomous drones carry out sorting instructions in milliseconds. Using real time patient data as guidance, intelligent healthcare robots distribute essential supplies with efficiency. Real-time virtual copies of physical systems are produced by digital twins, allowing for accurate actuator commands and predictive modifications. Digital twins optimize heat exchanger performance, lowering energy consumption by 20%, much like in chemical plants. These models enhance maintenance plans in manufacturing, reducing downtime and interruptions to operations. 5G networks' low latency communication guarantees quick command execution in sectors where milliseconds count, such as autonomous driving and aerospace manufacturing. Unprecedented real-time data transfer in mission critical operations is made possible by emerging 6G technologies which offer additional innovations including quantum communications. Because they enable accurate, effective, and sustainable operations, these action execution technologies powered by IoT, AI, and actuator advancements are revolutionizing industries. They facilitate a variety of uses, including as managing agricultural resources, balancing energy grids, and streamlining manufacturing procedures. Actuators improve load balancing in energy networks, increasing energy efficiency by 15–20%. Automated irrigation systems save 30–50% of water use in precision agriculture [130, 131].

10 Feedback Loops in IoT Systems to Advance Industry Optimization

Feedback loops are essential to IoT systems because they allow for constant observation, flexibility, and system improvement. To enhance procedures, boost productivity, and guarantee that the systems continue to react to changing circumstances, and these loops rely on real time performance data. An important development in industrial automation is the smooth data transfer between sensors, processing units, and actuators, which improves system accuracy and operational performance. Performance parameters, including temperature, vibration, humidity, and inventory levels are gathered via IoT sensors. These sensors are placed at strategic locations throughout industrial systems to collect data in real time that is necessary for system feedback (fundamental IoT process). Edge computing technology and AI-driven algorithms are used to examine the gathered, analyze, and process data. By spotting trends, irregularities, and optimization opportunities, this analysis empowers systems to make data driven modifications. Actuators and controllers get useful input feedback based on the analysis and implement the necessary modifications, including rerouting production lines, altering motor speeds, or recalibrating environmental settings in a smart factory [132, 133].

10.1 Industrial Applications of Feedback Loop

By processing data locally, edge computing lowers latency, and guarantees quick and efficient feedback. For instance, as compared to cloud-only solutions, edge computing lowers response times in vital systems like industrial robotics and medical device automation by 30% to 50%. Feedback data is analyzed by AI and ML algorithms to continuously improve operational algorithms. Prescriptive and predictive analytics are made possible by this iterative learning process, which increases system accuracy and adaptability in real-time systems. IoT feedback loops are integrated with digital twin technologies to model system performance. This lowers risk and improves results by enabling industry to test feedback responses in virtual environments prior to implementation. Digital twins increase system efficiency by 20–30% in aerospace manufacturing, saving millions of dollars in testing expenses. Real-time feedback is ensured via 5G networks, which allow for fast data flow between IoT devices. This is especially important for applications where judgments need to be made in milliseconds, like autonomous driving. To track engine performance in real time, General Electric's aviation division deployed IoT feedback devices. Feedback loops offered useful information for forecasting maintenance requirements and optimizing fuel consumption. 15% less fuel is used. Annual operational cost savings of more than USD 2 billion. Improved safety by identifying possible engine problems early. To provide feedback mechanisms that are even quicker, more accurate, and more intelligent, IoT feedback systems will progressively make use of 6G networks, quantum computing, and sophisticated AI algorithms. Feedback loops will be essential to the development of autonomous, human centered and sustainable systems that support global sustainability and innovation goals as sectors move toward Industry 6.0. Inventory levels are tracked by vision based AI systems in automated warehouses, allowing for dynamic storage and retrieval. By offering real time stock availability feedback, these solutions guarantee effective inventory management and cut down on retrieval times by as much as

40%. Amazon's automated fulfillment facilities handle millions of products every day with unmatched accuracy thanks to IoT feedback systems combined with AI vision. IoT-based predictive maintenance solutions with feedback loops improve the dependability of equipment. The system can forecast failures with over 90% accuracy by continuously monitoring the conditions of the equipment (such as vibration and temperature data from industrial machines). This has been shown to save millions of dollars in operating expenses by reducing downtime by 30–50%. Implementing IoT driven predictive maintenance resulted in yearly savings of about €15 million per plant, according to a case study conducted in the automobile industry. Feedback loops optimize production lines in real time in smart manufacturing. Sensors that track CNC machine performance give input on cutting accuracy, enabling modifications to be made in real time. This minimizes material waste by 15–25% while guaranteeing constant quality. To maintain high standards of quality and efficiency in their automated manufacturing facilities, companies such as Bosch use feedback based control systems. IoT feedback loops dynamically control energy use in industries. IoT sensors and smart meters track energy use and give input to modify equipment functioning during periods of high demand. 20–30% less energy may be used because of this modification. Siemens Smart Grid technology, for instance, incorporates IoT feedback loops, allowing firms to drastically cut energy costs while promoting sustainability objectives. For autonomous systems, feedback loops are essential. IoT sensors in autonomous cars continuously check the state of the road, the distance from hazards, and the vehicle's performance, reporting back to the control unit. This improves driving safety and fuel efficiency by 10–15%. In a similar vein, feedback loops are used by robotic arms on industrial assembly lines to guarantee accuracy in operations, reducing errors and enhancing product quality [134–140].

11 Industrial Case Studies

Smart manufacturing is one of the most effective industrial uses for IoT-enabled machinery. Businesses that have used IoT technologies in their facilities report up to 40% increases in operational efficiency. The BMW Group, for instance, has employed IoT to streamline its production procedures, saving an estimated €50 million a year through predictive maintenance, real-time machinery monitoring, and less downtime. Businesses are also benefiting greatly from predictive maintenance, which is powered by IoT-enabled equipment. Manufacturers can prevent unscheduled downtime by predicting when machines are likely to break and scheduling maintenance before the problem happens by evaluating sensor data. This can save millions of dollars on repairs in sectors like automotive and aerospace. General Electric has reduced maintenance expenses by 10–20% by implementing IoT-based predictive maintenance in its aircraft engines [141, 142].

12 Alignment with SDGs and Industries 5.0 and 6.0

Several important SDGs are strongly related to the creation and deployment of IoT-enabled devices. By boosting productivity, cutting waste, and enhancing the sustainability of manufacturing processes, IoT technologies improve industrial infrastructure and support SDG 9. IoT systems facilitate predictive maintenance and resource optimization, which decrease material waste and lower energy consumption, hence promoting sustainable production practices and supporting SDG 12. IoT enabled, energy efficient devices play a key role in lowering energy consumption in manufacturing processes, which lowers CO_2 emissions and advances global climate goals in line with SDG 13. IoT-enabled devices are essential to the advancement of autonomous and human centric systems in the framework of Industry 5.0 and Industry 6.0. While Industry 6.0 envisions fully autonomous systems that can learn, adapt, and optimize on their own, Industry 5.0 concentrates on encouraging human engagement with intelligent robots. By combining IoT with technologies like edge computing, robots, and AI, enterprises may transition to more sustainable, scalable, and efficient systems. These developments are essential to industrial innovation's future because they put sectors in a position to satisfy corporate goals and international sustainability initiatives.

13 Challenges and Solutions

Although there are many advantages to integrating IoT-enabled machinery into industrial systems, there are drawbacks as well. Advanced solutions that make use of cutting-edge technology from a variety of fields, such as data science, artificial intelligence, computer science, and engineering, are necessary to guarantee that these issues are successfully solved. For the IoT to be successfully implemented in industrial systems, issues with interoperability, security threats, and data overload must be addressed. Industries may improve the scalability, security, and performance of their IoT systems by utilizing technologies like edge computing, data compression, AI-driven anomaly detection, and standardized communication protocols. By encouraging more effective, sustainable, and secure industrial operations, these solutions support global sustainability goals, such as SDG 9, 12, and 13. By providing creative solutions across industries, fog computing, edge computing, cloud computing, a revolutionary distributed computing paradigm, greatly advances SDGs. Telemedicine and remote monitoring enhance system accessibility and efficiency in the healthcare sector (SDG 3). Fog computing makes precision farming possible in agriculture (SDG 2), maximizing resources for increased yields and less environmental impact. In terms of transportation (SDG 11), it improves urban mobility through driverless cars and sophisticated traffic control. Supporting sustainable energy (SDG 7), it allows smart grids and renewable energy integration. Through pollution monitoring and conservation initiatives, it supports environmental sustainability (SDGs 13 and 14) and improves data security and privacy (SDG 16). Fog computing speeds up innovation, efficiency, and sustainability by bringing computation closer to action. This promotes resilient communities and advances the cause of a sustainable future [143–145]. Overcoming these obstacles will be important to building intelligent, self-sufficient, and human centered industrial ecosystems as industries continue to transition toward Industry 5.0 and 6.0. Table 5.2 shows the challenges and their solutions in the implementation of Industry 6.0.

An extensive examination of these issues and their corresponding fixes is provided below.

Table 5.2 Challenges in the implementation of industry 6.0 and advanced suggested solutions

Challenge category	Specific problem	Impact on industry 6.0	Advanced solutions and technologies	Implementation roadmap
Data overloading	Petabyte scale data from IoT sensors, edge devices, and digital twins	Latency spikes (100+ ms), network congestion, delayed decision-making	Edge AI + neuromorphic chips: deploy low-power neuromorphic processors (e.g., Intel Loihi, IBM TrueNorth) for in-sensor data filtering Federated learning: train ML models locally (edge) without central data pooling Lossless compression: use AI-driven algorithms (e.g., NVIDIA TensorRT) to reduce data volume by 40–60%	• Pilot edge AI on non-critical lines (6 months) • Full rollout with 5G/6G backhaul (18–24 months)
Security vulnerabilities	Quantum computing threats, firmware implants, supply-chain attacks	Data breaches (cost: USD 4.35 M avg. per incident), operational sabotage	Quantum encryption: implement QKD (quantum key distribution) for tamper proof communication Zero Trust Architecture: Continuous device/user verification via AI-driven behavioral analytics Hardware PUFs: physically Unclonable functions for device authentication	• Phase 1: deploy ZTA for critical infrastructure (12 months) • Phase 2: integrate pilot QKD with future 6G testbeds

(continued)

Table 5.2 (continued)

Challenge category	Specific problem	Impact on industry 6.0	Advanced solutions and technologies	Implementation roadmap
Interoperability	Fragmented ecosystems (MQTT, CoAP, Modbus), legacy machinery integration	30–40% longer integration cycles, USD 2.6 T global productivity loss by 2025	Digital twin middleware: unified platforms (e.g., Siemens MindSphere, AWS IoT TwinMaker) to harmonize protocols AI powered translation layers: real time protocol conversion using NLP transformers	• Legacy retrofit: AI translation gateways (8–10 months) • New deployments: OPC UA compliance (immediate)
Ultra low latency	Edge cloud sync delays (>5 ms) for autonomous systems	Production errors (15–25% defect rates), safety risks in real-time control	6G URLLC: ultra-reliable low-latency communication (0.01 ms latency) with terahertz bands In-memory computing: use ReRAM-based architectures (e.g., mythic AMP) for sub-μs processing Photonic tensor processors: light-based computing (e.g., Lightmatter Envise) for 800 TOPS/W efficiency	• Testbed: 6G industrial zones • Replace GPUs with photonic accelerators for critical AI tasks
Energy consumption	Edge AI/cloud data centers consume 1% of global electricity (650 TWh/year)	Conflicts with SDG 7/13, 450 g CO_2/kWh emissions	Dynamic voltage scaling (DVS): AI-optimized power allocation (saves 20–30% energy) Renewable-powered edge nodes: solar/ vibration-harvesting sensors liquid cooled AI racks: Reduce cooling energy by 40%	• Mandate ISO 50001 certification for all facilities • Transition to renewables-powered microgrids (3–5 years)
Workforce adaptation	Skills gap in AI/quantum tech; resistance to cobot integration	30% productivity drop during transition, training costs (USD 24 K/ employee)	AR/VR training: Microsoft HoloLens for immersive upskilling (cuts training time by 40%) Cobot-human teams: Universal robots' AI cobots with force-sensing for safe collaboration AI career pathing	• Quarterly AR upskilling bootcamps • Incentivize certifications in AI/robotics (tax breaks)

Table 5.2 (continued)

Challenge category	Specific problem	Impact on industry 6.0	Advanced solutions and technologies	Implementation roadmap
Regulatory compliance	GDPR, ISO 21434, FAA drone latency rules; evolving quantum legislation	Fines up to 4% of global revenue, project delays	Blockchain auditing: immutable logs for automated compliance AI regulatory bots: real time policy monitoring	• Integrate blockchain with ERP systems (6 months) • Deploy regulatory AI bots (1 year)
Quantum transition	Quantum decryption risks to AES-256; costly hardware upgrades	Cryptographic obsolescence in 5 years, USD 3 T global economic risk	Post-quantum cryptography (PQC): lattice-based algorithms (NIST standard) for IoT devices hybrid quantum-cloud: IBM quantum + classical HPC for optimization (e.g., D-wave for logistics)	• PQC patches for existing systems (2 years) • Quantum as a service (QaaS) adoption for R&D (1 year)

13.1 Data Overloading

Data overload is one of the main issues with IoT systems, especially in industrial applications. Sensors, machines, and other IoT devices continuously produce enormous volumes of data. Inadequate management of this data can overload analysis frameworks, data storage systems, and networks. Real time analysis and decision-making are essential as data volumes rise, but doing so frequently leads to inefficiencies and slower reaction times. Edge computing, fog computing, cloud computing lessens the need to send all raw data to central cloud systems by processing data closer to the source, such as on local servers or the edge devices themselves. By ensuring that only pertinent, aggregated, or pre-processed data is transmitted, this lowers latency and network congestion. Real time analytics and decision-making capabilities of edge devices allow for faster reactions and improved system performance. Predictive maintenance algorithms, for instance, can reduce needless data transmission by locally analyzing vibration data to anticipate machinery faults before they occur. Using data compression technologies, which minimize the size of the data being sent or stored, is another efficient strategy. Depending on the application's tolerance for data loss, methods like lossless compression which preserves all original data or lossy compression which compromises some accuracy for reduced data sizes are used. This is especially vital in situations like remote sensing, when delivering uncompressed data might not be feasible due to bandwidth constraints. With a 40 to 50% decrease in data transfer overhead and a 20 to 30% increase in decision-making speed, edge computing and data compression allow for the real-time optimization of machinery operations in smart manufacturing environments [146–148].

13.2 Security Concerns

Security threats become a major worry as IoT equipment gets more widespread and networked. These systems frequently include many interconnected networks and devices, which gives hackers several ports of entry. Because breaches can result in monetary losses, data theft, and even sabotage, protecting sensitive data and vital industrial infrastructure is important. The use of end-to-end encryption is one of the core strategies for protecting IoT systems. This guarantees that information sent between devices or between IoT devices and central servers is encrypted, guarding against eavesdropping or unwanted access. Commonly used protocols to guarantee safe communication between devices and networks include TLS (Transport Layer Security) and AES (Advanced Encryption Standard). By blocking unwanted traffic, firewalls act as a barrier between internal networks and the outside world. Firewalls improve security by stopping cyberattacks from taking advantage of network vulnerabilities when used in conjunction with Intrusion Detection Systems (IDS), which keep an eye on network traffic for questionable activity. Unusual patterns in real-time data can be found using machine learning and AI-based anomaly detection, which could point to a security compromise. Generative AI and regular AI systems, for example, can identify anomalous variations in network traffic or device behaviors and report them for prompt inquiry. The overall security of IoT infrastructures is improved by AI systems' constant adaptation and improvement of their detecting skills over time. In several high-risk industries, including oil and gas, where cybersecurity is crucial, the incorporation of AI-driven anomaly detection systems into extensive industrial IoT installations has decreased security breaches by as much as 45% [149–152].

13.3 Interoperability

Interoperability, or the harmonious operation of various IoT devices, systems, and platforms, is a major obstacle for industries implementing IoT technologies. IoT ecosystems frequently consist of a variety of devices made by different manufacturers, each of which has a unique set of software, standards, and communication protocols. IoT installations may encounter difficulties with system integration, data interchange, and general functionality in the absence of standardization. Standardized communication protocols must be established in order to guarantee seamless platform and device interoperability. In industrial IoT environments, popular standards including Modbus, CoAP, and MQTT (Message Queuing Telemetry Transport) are extensively used. Even in settings with constrained bandwidth, these lightweight protocols enable effective and instantaneous data exchange. By serving as a link between several software programs, middleware makes it possible for IoT systems and devices to connect even when they are employing disparate technologies. By managing data translation, protocol conversion, and routing between different systems, middleware platforms offer an abstraction layer that streamlines the integration process. For example, middleware systems can give a single interface to the central control system by combining data from multiple sensors (such as temperature, pressure, and humidity sensors). Many companies have resorted to IoT platforms tailored to their requirements to enable smooth integration across various IoT components. By combining sensors, actuators, and software programs, these platforms make sure that equipment from many manufacturers functions as a single unit. Platforms used in the manufacturing industry, for instance, might integrate environmental factors, energy consumption, and machine data into a single interface, offering useful information for process optimization and predictive maintenance. Standardized protocols and middleware solutions have made it possible for heterogeneous devices to be seamlessly integrated in smart factories, reducing integration times by 30 to 40% and facilitating the quicker deployment of new systems [153–155].

14 Strategic Recommendations for Early Adoption of Industry 6.0

Industry 6.0 is not a linear evolution; it represents a transformative paradigm shift. Organizations that embrace these early stage strategies will position themselves as leaders in sustainable, intelligent manufacturing by the end of this decade. To accelerate the transition to Industry 6.0, a multifaceted strategy is required that prioritizes technological integration, cybersecurity, sustainability, policy alignment, and workforce development. A key starting point is to prioritize edge cloud synergy. Using tools like the NVIDIA TAO Toolkit for transfer learning has demonstrated up to 99.67% accuracy in defect detection, significantly minimizing cloud dependency. Coupling this with platforms such as AWS Greengrass or Azure Percept enables secure and scalable edge-to-cloud data pipelines, essential for real-time industrial AI deployment. A phased cybersecurity overhaul is also critical. In the short term, adopting Zero Trust Architecture and blockchain-based supply chain management can reduce fraud by approximately 60%. Over the long term, Quantum Key Distribution integrated 6G networks, expected to mature post 2027, will offer ultra-secure, latency free communication channels, critical for sensitive industrial data. Sustainability focused strategies yield strong returns on investment. AI-driven energy optimization platforms, such as those provided by Siemens, can reduce operational energy costs by 20–30%, directly supporting SDGs 9 and 13. Additionally, circular economy models enabled by digital twins can reduce resource consumption, for example, reducing water waste by up to 50% in precision agriculture applications. Policy industry collaboration is vital to shaping the enabling environment. This includes lobbying for 6G spectrum allocations via agencies like the FCC and EU regulatory bodies, as well as advancing quantum resilient security standards aligned with ISO/IEC 27001. Support through public private R&D funding should also be directed toward the development of photonic and neuromorphic chips, which are foundational for next generation AI hardware. The success of these strategies can be measured through several key metrics: achieving ≤1 ms latency on factory floors via 6G networks, attaining 90% reduction in cybersecurity breaches through ZTA and post quantum cryptography, lowering CO_2 emissions by 30% by 2030, and realizing 20 to 40% operational cost savings within 3–5 years. Talent development is a cornerstone of success. National-level AI academies, such as those launched under Germany's Industry 4.0 upskilling initiatives, should be expanded and enhanced with AR/VR learning modules to shorten training cycles and improve industrial readiness [156–161]. So, Industry 6.0 defined as, the era of regenerative intelligent manufacturing in which advanced digital technologies like Green IIoT, sustainable digital twins, energy efficient AI, neuromorphic computing, autonomous cobotics, and circular material systems operate within fully decarbonized, fossil-free, net-zero-waste, full cyber secure and climate positive industrial ecosystems. It differs fundamentally from previous industrial paradigms by mandating 100% renewable-energy operation, total circularity, embedded CCUS, waterless data infrastructure and manufacturing processes designed to restore environmental quality while maintaining ultra-high efficiency and product excellence. It builds beyond Industry 4.0's automation and Industry 5.0's human-centricity by integrating advanced digital technologies with planet-positive engineering, energy sovereignty, and full-scale environmental restoration.

References

1. Rane, N.L., Kaya, Ö., Rane, J.: Artificial Intelligence, Machine Learning, and Deep Learning for Sustainable Industry 5.0. Deep Science Publishing, Mumbai (2024)
2. Chauhan, M., Sahoo, D.R.: Towards a greener tomorrow: exploring the potential of AI, blockchain, and IoT in sustainable development. Nat. Environ. Pollut. Technol. **23**(2), 1105–1113 (2024)
3. Wang, Z., Yu, L., Zhou, L.: Navigating the Blockchain-driven transformation in industry 4.0: opportunities and challenges for economic and management innovations. J. Knowl. Econ. **16**, 3507–3549 (2024)
4. Rani, S., Taneja, A.: WSN and IoT: An Integrated Approach for Smart Applications. CRC Press, Boca Raton (2024)
5. Jhanjhi, N.Z., Shah, I.A.: Cybersecurity Measures for Logistics Industry Framework. IGI Global, London (2024)
6. Gilman, E., et al.: Addressing data challenges to drive the transformation of smart cities. ACM Trans. Intell. Syst. Technol. **88**, 1–65 (2024)
7. Dalal, S., Seth, B., Radulescu, M.: Driving technologies of industry 5.0 in the medical field. In: Digitalization, Sustainable Development, and Industry 5.0: An Organizational Model for Twin Transitions, pp. 267–292. Emerald Publishing Limited, Manchester (2023)
8. Kulshrestha, S., Acharya, N.: Regulatory frameworks for securing IoT Devices: a legal perspective. In: Interdisciplinary Approaches to AI, Internet of Everything, and Machine Learning, pp. 243–260. IGI Global Scientific Publishing, London (2025)
9. Saeed, A., Ebrahim, H.: The intersection of machine learning, artificial intelligence, and big data. In: Big Data Computing, pp. 111–131. CRC Press, Boca Raton (2024)
10. Wei, X., Ye, X.: Social Data Analytics in the Cloud with AI. CRC Press, Boca Raton (2024)

11. Sourov, S.M.: Mastering Code: A Deep Dive into Modern Computer Programming. Skillworldhub, Dhaka (2024)
12. Henriques, J.R., D'Abruzzo Pereira, J., Vieira, M.: Mining vulnerability and code repositories to study software security. In: Proceedings of the 13th Latin-American Symposium on Dependable and Secure Computing. ACM, New York (2024)
13. Guo, C., Yang, S., Thiede, S.: An autonomous edge box system architecture for industrial IoT applications. In: 2024 IEEE 22nd International Conference on Industrial Informatics (INDIN). IEEE, New York (2024)
14. Maliqi, M., Lamy, D., Grimaud, F.: Digital twin for sustainable systems methodology: application in water network management. In: Working Conference on Virtual Enterprises. Springer, Cham (2024)
15. Arief, R.: Current and voltage monitoring in wind power plants using ESP8266 and node-red. Vokasi Unesa Bull. Eng. Technol. Appl. Sci. **1**(2), 64–71 (2024)
16. Magudeswaran, P., Sakthivel, P.: Internet of things based horizontal axis tracking solar panel performance evaluation. Electr. Power Comp. Syst. **52**(9), 1652–1670 (2024)
17. Miry, A.H., Aramice, G.A.: Agriculture based on wireless sensor network: literature survey. Wasit J. Eng. Sci. **12**(4), 171–190 (2024)
18. Majid, M.S.A.A., Rahim, R.A., Nordin, J.M.: Electrical energy consumption monitoring system over internet of things. AIP Conf. Proc. **2898**, 030052 (2024)
19. Shefer, O., Myhal, S.: ANALYSIS of the main requirements for 5G/6G Networks. Системи управління, навігації та зв'язку. Збірник наукових праць. **4**(78), 219–222 (2024)
20. Hazra, A., et al.: 6G-enabled ultra-reliable low latency communication for industry 5.0: challenges and future directions. IEEE Commun. Standards Magaz. **8**(2), 36–42 (2024)
21. Cayuqueo, A., et al.: Industry 5.0. IIoT platform as enabler for smart manufacturing. in XII Jornadas de Cloud Computing, Big Data & Emerging Topics (La Plata, 25 al 27 de junio de 2024) (2024)
22. Kayan, H., et al.: CASPER: context-aware IoT anomaly detection system for industrial robotic arms. ACM Trans. Internet Things. **5**(3), 1–36 (2024)
23. Boppana, V.R.: Industry 4.0: revolutionizing the future of manufacturing and automation. Innov. Comp. Sci. J. **10**(1), 5135027 (2024)
24. Riyaz, F.: Optimising Business Models for AI-Based Battery Analytics Platforms: A Multi-Sector Analysis in the Swedish Electric Vehicle and Energy Storage Market. Universitat Politècnica de Catalunya, Catalunya (2024)
25. Alijoyo, F.A.: AI-powered deep learning for sustainable industry 4.0 and internet of things: enhancing energy management in smart buildings. Alex. Eng. J. **104**, 409–422 (2024)
26. Mohapatra, S.K., et al.: Cereal crop yield prediction using machine learning techniques. In: Hyperautomation in Precision Agriculture, pp. 97–112. Elsevier, Amsterdam (2025)
27. Jolly, P.: Innovation, enterprise and physics education: weaving paradigms for world of work. J. Phys.: Conf. Ser. **2727**, 012015 (2024)
28. Bhagat, D., Dorle, S.: The power of intelligent. Hyperautomat. Bus. Soc. **27**, 10 (2024)
29. Dua, A., Kumar, G.: Mastering zero-knowledge proofs: Practical study of security, scalability, and privacy in blockchain and modern systems. BPB Publications, New Delhi (2024)
30. Divya, V., et al.: Digital twins for hyperautomation for next generation. In: Hyperautomation for Next-Generation Industries, pp. 127–151. Wiley, Hoboken (2024)
31. George, A.S., et al.: The rise of hyperautomation: a new frontier for business process automation. Partners Universal International Research Journal. **2**(4), 13–35 (2023)
32. Kazmi, S.H.A., et al.: Security of federated learning in 6G era: a review on conceptual techniques and software platforms used for research and analysis. Comput. Netw. **245**, 110358 (2024)
33. Pei, Z., et al.: An innovative algorithm for attacking symmetric ciphers using D-wave quantum annealing. Tsinghua Sci. Technol. **30**(5), 2184–2194 (2025)

34. Singh, M., Carlson, A.: Exploring polymorphic algorithms and their use in cryptography. In: 2024 IEEE 14th Annual Computing and Communication Workshop and Conference (CCWC). IEEE, New York (2024)
35. Pióro, Ł., et al.: Application of attribute-based encryption in military internet of things environment. Sensors. **24**(18), 5863 (2024)
36. Singh, M., Carlson, A.H.: An introduction to quantum computing and its applications. Cyber Defense Rev. **9**(2), 73–92 (2024)
37. Ramakrishna, D., Shaik, M.A.: A comprehensive analysis of cryptographic algorithms: evaluating security, efficiency, and future challenges. IEEE Access. **13**, 11576–11593 (2025)
38. Biswas, S., Goswami, R.S., Reddy, K.H.K.: Advancing quantum steganography: a secure IoT communication with reversible decoding and customized encryption technique for smart cities. Clust. Comput. **27**, 9395–9414 (2024)
39. Scalise, P., et al.: A systematic survey on 5G and 6G security considerations, challenges, trends, and research areas. Future Internet. **16**(3), 67 (2024)
40. Pandey, B.K., et al.: Beyond data breaches: enhancing security in 6G communications. In: Security Issues and Solutions in 6G Communications and Beyond, pp. 212–229. IGI Global, London (2024)
41. Iță, C.-R., et al.: Security in remote access, based on zero trust model concepts and SSH authentication with signed certificates. In: Advanced Topics in Optoelectronics, Microelectronics, and Nanotechnologies XI. SPIE, Bellingham (2023)
42. Zeydan, E., et al.: Securing IoT with resilient cloud-edge continuum. In: 2024 International Conference on Intelligent Computing, Communication, Networking and Services (ICCNS). IEEE, New York (2024)
43. Dutta, A., et al.: Security and privacy. In: Siva, R. (ed.) 2023 IEEE Future Networks World Forum (FNWF). IEEE, New York (2023)
44. Fang, W., et al.: Head-mounted display augmented reality in manufacturing: a systematic review. Robot. Comput. Integr. Manuf. **83**, 102567 (2023)
45. Karakostas, I., et al.: A real-time wearable AR system for egocentric vision on the edge. Virtual Reality. **28**(1), 44 (2024)
46. Babu, C.S., et al.: IoT-enhanced haptic feedback: revolutionizing healthcare wellbeing in AR and VR. In: Modern Technology in Healthcare and Medical Education: Blockchain, IoT, AR, and VR, pp. 186–215. IGI Global, London (2024)
47. Chen, Z., et al.: Metaverse for smart cities: a surveys. In: Internet of Things and Cyber-Physical Systems. Elsevier, Amsterdam (2024)
48. Aruna, R., et al.: Leveraging IoT and wearable devices in smart hospitals and medical applications for smart cities. In: Applying Internet of Things and Blockchain in Smart Cities: Industry and Healthcare Perspectives, pp. 267–292. IGI Global, London (2024)
49. Fortino, G., et al.: A review of internet of things platforms through the IoT-a reference architecture. In: International Symposium on Intelligent and Distributed Computing. Springer, Cham (2021)
50. Ameri, R., Hsu, C.-C., Band, S.S.: A systematic review of deep learning approaches for surface defect detection in industrial applications. Eng. Appl. Artif. Intell. **130**, 107717 (2024)
51. Nirmal Kumar Juluru, Z.P.B.a.S.O.: Transforming Industrial Defect Detection with NVIDIA TAO and Vision AI Models (2023)
52. WHO: Guidance on Implementation of a Quality System in Blood Establishments. World Health Organization, Geneva (2024)
53. Vadan, A.-M.: Alternative Manual and Automatic Testing Methods of IoT Systems for Smart Homes. Technical University of Cluj-Napoca, Cluj-Napoca (2024)
54. Buchmayer, E., Schmied, C., Theurl, F.: Low-cost collaborative positioning for autonomous agents using carrier phase DGNSS. In: Proceedings of the 37th International Technical Meeting of the Satellite Division of The Institute of Navigation (ION GNSS+ 2024). IEEE, New York (2024)

55. Sulasmoro, A.H., Sabanise, Y.F., Prihandoyo, M.T.: The analysis Kinerja sensor DHT22 Dan sensor LM35 Pada Suhu Ruangan server. MUST. **9**(2), 23646 (2024)
56. Saidi, A., et al.: Autonomous solar-powered IoT gateway for real-time environmental data acquisition and transmission. Int. J. Internet Things Web Serv. **9**, 29–38 (2024)
57. Kilic, U., Cam, O., Can, E.: Machine learning–aided synthetic air data system for commercial aircraft. J. Aerosp. Eng. **37**(6), 04024071 (2024)
58. Scherllin-Pirscher, B., et al.: Airborne soil-derived dust hazards in aviation. Bull. Am. Meteorol. Soc. **23**, 125–145 (2024)
59. Alazzai, W.K., et al.: Precision farming: the power of AI and IoT technologies. E3S Web Conf. **491**, 04006 (2024)
60. Diez Alvarez, S., et al.: The role of smartwatch technology in the provision of care for type 1 or 2 diabetes mellitus or gestational diabetes: systematic review. JMIR Mhealth Uhealth. **12**, e54826 (2024)
61. Metwally, A.A., Mehta, P., Snyder, M.P.: Predicting glucotypes in prediabetes via wearables and artificial intelligence. In: Diabetes Digital Health, Telehealth, and Artificial Intelligence, pp. 287–301. Elsevier, Amsterdam (2024)
62. Rodríguez-Rodríguez, I., et al.: IoMT innovations in diabetes management: predictive models using wearable data. Expert Syst. Appl. **238**, 121994 (2024)
63. Rotbei, S., et al.: Evaluating impact of movement on diabetes via artificial intelligence and smart devices systematic literature review. Expert Syst. Appl. **257**, 125058 (2024)
64. Damigos, G., et al.: Communication-aware control of large data transmissions via centralized cognition and 5G networks for multi-robot map merging. J. Intell. Robot. Syst. **110**(1), 22 (2024)
65. Odida, M.: The evolution of mobile communication: a comprehensive survey on 5G technology. J. Sen. Net Data Commun. **4**(1), 1–11 (2024)
66. Attar, H., et al.: B5G applications and emerging Services in Smart IoT environments. Int. J. Crowd Sci. **9**, 79–95 (2024)
67. Arnaud, A., et al.: A model for a dense LoRaWAN farm-area network in the agribusiness. IEEE Trans. AgriFood Electron. **2**, 284–292 (2024)
68. Taji, K., Ghanimi, F.: LoRaWAN-based smart irrigation systems. In: Artificial Intelligence, Big Data, IOT and Block Chain in Healthcare: From Concepts to Applications: Volume 2, p. 22. Springer, Cham (2025)
69. Roslee, M., et al.: Internet of things: agriculture precision monitoring system based on low power wide area network. WSEAS Trans. Electron. **15**, 35–46 (2024)
70. Gao, D., et al.: Wilo: long-range cross-technology communication from Wi-Fi to lora. IEEE Trans. Commun. **73**, 1677–1691 (2024)
71. Martínez, V.M., Ribeiro, M.R., Mota, V.F.: Wi-fi faces the new wireless ecosystem: a critical review. Ann. Telecommun. **79**(5), 397–413 (2024)
72. Alkasassbeh, J.S., et al.: Advancements in wireless communication technology: a comprehensive analysis of 4G to 7G systems. JoWUA. **15**(3), 73–91 (2024)
73. Alghamedy, F.H., et al.: Unlocking a promising future: integrating Blockchain technology and FL-IoT in the journey to 6G. IEEE Access. (2024). https://doi.org/10.1109/ACCESS.2024.3435968
74. Makondo, N., et al.: Implementing an efficient architecture for latency optimisation in smart farming. IEEE Access. **12**, 140502–140526 (2024)
75. Moreolo, M.S., et al.: Continuous-variable quantum key distribution for enabling sustainable secure 6G networks. In: 2024 International Conference on Optical Network Design and Modeling (ONDM). IEEE, New York (2024)
76. Blika, A., et al.: Federated learning for enhanced cybersecurity and trustworthiness in 5G and 6G networks: a comprehensive survey. IEEE Open J. Commun. Soc. **22**, 2031–2063 (2024)
77. Neelima, K., Kavya, C., Pandey, D.: 6G communications–security issues and possible solutions. In: 6G Communication Network, pp. 137–154. CRC Press, Boca Raton (2024)

78. Patel, H., et al.: Quantum Threats, Quantum Solutions: ML Approaches to 6G Security, in Security Issues and Solutions in 6G Communications and Beyond, pp. 45–60. IGI Global, London (2024)
79. Zeydan, E., Arslan, S., Turk, Y.: 6G wireless communications for industrial automation: scenarios, requirements and challenges. J. Ind. Inf. Integr. **42**, 100732 (2024)
80. Rojek, I., et al.: Sixth-generation (6G) networks for improved machine-to-machine (M2M) communication in industry 4.0. Electronics. **13**(10), 1832 (2024)
81. Mohammed, S.M., Al-Barrak, A., Mahmood, N.T.: Enabling technologies for ultra-low latency and high-reliability communication in 6G networks. Ingénierie des Systèmes d'Information. **29**(3), 1195 (2024)
82. Jiang, W., Han, B.: Evolution toward sixth-generation (6G) Mobile cellular systems. In: Cellular Communication Networks and Standards: the Evolution from 1G to 6G, pp. 195–210. Springer, Cham (2024)
83. Salam, A.: Internet of Things for Sustainable Community Development: Wireless Communications, Sensing, and Systems. Springer, Cham (2024)
84. Zawish, M., et al.: AI and 6G into the metaverse: fundamentals, challenges and future research trends. IEEE Open J. Commun. Soc. **5**, 730–778 (2024)
85. Rath, K.C., Khang, A., Roy, D.: The role of internet of things (IoT) Technology in industry 4.0 economy. In: Advanced IoT Technologies and Applications in the Industry 4.0 Digital Economy, pp. 1–28. CRC Press, Boca Raton (2024)
86. Abouchabaka, J., Bentaleb, A.: The impact of big data on the sustainable development goals water and energy. E3S Web Conf. **477**, 00062 (2024)
87. Nedungadi, P., et al.: Big data and AI algorithms for sustainable development goals: a topic Modeling analysis. IEEE Access. **12**, 188519–188541 (2024)
88. Guan, W., et al.: A deep learning and vision-based solution for material volume estimation considering devices' applications. J. Comput. Civ. Eng. **38**(1), 04023039 (2024)
89. Li, B.: Embedded AI accelerator chips. In: Embedded Artificial Intelligence: Principles, Platforms and Practices, pp. 121–180. Springer, Cham (2024)
90. Zhai, H.-Z., et al.: Edge deployment of deep networks for visual object detection: a review. IEEE Sens. J. **11**, 18662–18683 (2024)
91. Mathur, P.: Cloud computing infrastructure, platforms, and software for scientific research. In: High performance computing in biomimetics: Modeling, architecture and applications, pp. 89–127. Springer, Cham (2024)
92. Singh, M.M.K., Chandna, M., Kongala, V.Y.Y.: Risk management framework for cloud migration and selection of suitable cloud service provider. In: Advances in Enterprise Technology Risk Assessment, p. 283. IGI Global, London (2024)
93. Shende, A.P., et al.: Securing the future: comprehensive strategies for IoT security in Industry 4.0 and beyond. In: Modelling of Virtual Worlds Using the Internet of Things, pp. 19–73. CRC Press, Boca Raton (2024)
94. Rane, J., et al.: Future Research Opportunities for Artificial Intelligence in Industry 4.0 and 5.0. Deep Science Publishing, Mumbai (2024)
95. Sharma, S., et al.: Adoption of industry 4.0 in different sectors: a structural review using natural language processing. Int. J. Interact. Des. Manuf. **18**(8), 6069–6091 (2024)
96. Shahin, M., et al.: Improving operations through a lean AI paradigm: a view to an AI-aided lean manufacturing via versatile convolutional neural network. Int. J. Adv. Manuf. Technol. **133**(11), 5343–5419 (2024)
97. Alwahedi, F., et al.: Machine learning techniques for IoT security: current research and future vision with generative AI and large language models. Internet of Things Cyber-Phys. Syst. **4**, 167–185 (2024)
98. Chaithra, N., et al.: Enhancing underwater imagery with AI/ML and IoT in ROV technology. In: Artificial Intelligence and Edge Computing for Sustainable Ocean Health, pp. 311–342. Springer, Cham (2024)

99. Freund, S., et al.: Applications. In: Energy Transport Infrastructure for a Decarbonized Economy, pp. 137–172. Elsevier, Amsterdam (2025)
100. Ardalani, N., Pal, S., Gupta, P.: DeepFlow: a cross-stack pathfinding framework for distributed ai systems. ACM Trans. Des. Autom. Electron. Syst. **29**(2), 1–20 (2024)
101. 무카담, Tesla Inc., the Innovative Disrupter of the Automobile Industry through Digital Innovation. 2024, 서울대학교 대학원
102. Kaliraj, P., Singaravelu, G., Devi, T.: Transformative Digital Technology for Disruptive Teaching and Learning. CRC Press, Boca Raton (2024)
103. Hemmati, A., Raoufi, P., Rahmani, A.M.: Edge artificial intelligence for big data: a systematic review. Neural Comput. Appl. **36**, 11461–11494 (2024)
104. Zhang, P., et al.: End-to-end demonstration for cubesatellite quantum key distribution. IET Quant. Commun. **5**, 291–302 (2024)
105. Zhang, Z., et al.: Entanglement-based quantum information technology: a tutorial. Adv. Opt. Photon. **16**(1), 60–162 (2024)
106. Chakravarthy, S.P.: Prospective technological advances. In: 6G Communication Network: Architecture, Security and Applications. Routledge, London (2024)
107. Singh, B., Kaunert, C.: Future of digital marketing: hyper-personalized customer dynamic experience with AI-based predictive models. In: Revolutionizing the AI-Digital Landscape, pp. 189–203. Productivity Press, New York (2024)
108. Nair, V.V.: Automation and robotization. In: Modern Technologies and Tools Supporting the Development of Industry 5.0, p. 41. CRC Press, Boca Raton (2024)
109. Matani, A.G.: Industry 4.0 and 5.0 towards enhanced productivity and competitiveness. In: Industry 4.0, pp. 61–76. CRC Press, Boca Raton (2024)
110. Sanders, N.R.: Supply Chain Management: A Global Perspective. John Wiley & Sons, Hoboken (2024)
111. Chan, H.-L., Kwok, Y.-K., Wong, S.-M.: Marketing and operational strategies during the COVID-19 pandemic: a case study of a Hong Kong footwear enterprise. J. Fashion Market. Manage. Int. J. **28**(4), 610–634 (2024)
112. Dua, G.K.: Advancements in industry 5.0: a descriptive analysis. In: Strategies and Frameworks for Relearning in Organizations, p. 243. IGI Global, London (2024)
113. Hassoun, A., et al.: From food industry 4.0 to food industry 5.0: identifying technological enablers and potential future applications in the food sector. Compr. Rev. Food Sci. Food Saf. **23**(6), e370040 (2024)
114. Gallery, C., Conlon, J.: Fashion Business and Digital Transformation: Technology and Innovation Across the Fashion Industry. Taylor & Francis, Milton Park (2024)
115. Jebbor, I., Benmamoun, Z., Hachmi, H.: Revolutionizing cleaner production: the role of artificial intelligence in enhancing sustainability across industries. J. Infrastruct. Policy Dev. **8**(10), 7455 (2024)
116. Torres, J.C., Ramirez, L.: Artificial intelligence for sustainable development: applications and limitations. Baltic J. Eng. Technol. **3**(2), 25–35 (2024)
117. Thomas, J., Cyriac, R.: IoT security: challenges, best practices, and service platforms. In: Leveraging Artificial Intelligence (AI) Competencies for Next-Generation Cybersecurity Solutions, pp. 39–67. Apple Academic Press, New York (2024)
118. Rani, P., Parkash, V., Sharma, N.K.: Technological aspects, utilization and impact on power system for distributed generation: a comprehensive survey. Renew. Sust. Energ. Rev. **192**, 114257 (2024)
119. Gharibvand, H., Gharehpetian, G.B., Anvari-Moghaddam, A.: A survey on microgrid flexibility resources, evaluation metrics and energy storage effects. Renew. Sust. Energ. Rev. **201**, 114632 (2024)
120. Hayes, O.: Greenhouse Gardening Made Easy for Beginners: Step-by-Step Instructions for Building and Growing in Your Own Greenhouse. Erland Publishing, Mumbai (2025)

121. Tripathi, S.K., Singh, J., Rakshit, A.: Sustainability assessment of high-value vegetable crops using biopriming approach towards improved performance, nutritional security, and smallholder farmers. J. Soil Sci. Plant Nutr. (2024). https://doi.org/10.1007/s42729-024-01651-x
122. Nooyimsai, S., et al.: Enhancing breeding control: the IoT-driven automated system for budgie farming. In: 2023 11th International Conference on Cyber and IT Service Management (CITSM). IEEE, New York (2023)
123. Qayoom, K., Gupta, S., Kumar, D.: IoT-enabled variable rate technologies for optimal resource use. In: Agriculture 4.0, pp. 150–168. CRC Press, Boca Raton (2024)
124. Saleem, S.R., et al.: Variable rate technologies: development, adaptation, and opportunities in agriculture. In: Precision Agriculture, pp. 103–122. Elsevier, Amsterdam (2023)
125. Minn, A., Ahamed, T.: Low-cost automatic machinery development to increase timeliness and efficiency of operation for small-scale farmers to achieve SDGs. In: IoT and AI in Agriculture: Self-Sufficiency in Food Production to Achieve Society 5.0 and SDG's Globally, pp. 289–309. Springer, Cham (2023)
126. Hu, Y., et al.: Modeling of threshold voltage degradation of 22nm FD-SOI nMOSFETs under dynamic voltage scaling. IEEE Trans. Device Mater. Reliab. **25**, 12 (2024)
127. Wicht, B.: Design of Power Management Integrated Circuits. John Wiley & Sons, Hoboken (2024)
128. Chen, X., et al.: A 65-nm proactive power management technique with real-time machine learning engine for droop prediction and mitigation on microprocessors. IEEE J. Solid State Circuits. **60**, 2170–2181 (2024)
129. Suresh, M.P., et al.: Optimizing smart microgrid performance: integrating solar generation and static VAR compensator for EV charging impact, emphasizing SCOPE index. Energy Rep. **11**, 3224–3244 (2024)
130. Ullah, I., et al.: Future Communication Systems Using Artificial Intelligence, Internet of Things and Data Science. CRC Press, Boca Raton (2024)
131. Zahedi, A., Liyanapathirana, R., Thiyagarajan, K.: Biodegradable and renewable antennas for green IoT sensors: a review. IEEE Access. **12**, 189749–189775 (2024)
132. Khaloopour, L., et al.: Resilience-by-design in 6G networks: literature review and novel enabling concepts. IEEE Access. **12**, 155666–155695 (2024)
133. Rahmani, A.M., et al.: A novel offloading strategy for multi-user optimization in blockchain-enabled Mobile edge computing networks for improved internet of things performance. Comput. Electr. Eng. **119**, 109514 (2024)
134. Stoiber, C., Schönig, S.: Leveraging the industrial internet of things for business process improvement: a metamodel and patterns. IseB. **22**(2), 285–313 (2024)
135. Kumar, R., Agrawal, N.: Shaping the future of industry: understanding the dynamics of industrial digital twins. Comput. Ind. Eng. **191**, 110172 (2024)
136. Piperigkos, N., et al.: Distributed intelligence in industrial and automotive cyber–physical systems: a review. Front. Robot. AI. **11**, 1430740 (2024)
137. Vlachos, I., et al.: Smart and flexible manufacturing systems using autonomous guided vehicles (AGVs) and the internet of things (IoT). Int. J. Prod. Res. **62**(15), 5574–5595 (2024)
138. Ebirim, W., et al.: The role of agile project management in driving innovation in energy-efficient hvac solutions. Eng. Sci. Technol. J. **5**(3), 662–673 (2024)
139. Soori, M., et al.: Sustainable CNC machining operations, a review. Sustain. Oper. Comp. **5**, 73–87 (2024)
140. Mallioris, P., Aivazidou, E., Bechtsis, D.: Predictive maintenance in industry 4.0: a systematic multi-sector mapping. CIRP J. Manuf. Sci. Technol. **50**, 80–103 (2024)
141. Govindarajan, V., Venkatraman, V.: Fusion Strategy: How Real-time Data and AI Will Power the Industrial Future. Harvard Business Press, Cambridge (2024)
142. Triantafyllou, M., Al-Bazi, A., Ahmad, M.A.: Digital twins: revolutionizing automotive supply chains. In: International Conference on Logistics Operations Management. Springer, Cham (2024)

143. Chithaluru, P., et al.: Fog computing: advancing sustainable development goals (SDGs). In: Cloud and Fog Optimization-Based Solutions for Sustainable Developments, pp. 280–298. CRC Press, Boca Raton (2024)
144. Thamik, H., Cabrera, J.D.F., Wu, J.: The digital paradigm: Unraveling the impact of artificial intelligence and internet of things on achieving sustainable development goals. In: Artificial Intelligence of Things for Achieving Sustainable Development Goals, pp. 21–40. Springer, Cham (2024)
145. Lalar, S., et al.: Security challenges and solutions in cloud, fog, and edge computing for sustainable development. In: Cloud and Fog Optimization-Based Solutions for Sustainable Developments, pp. 178–200. CRC Press, Boca Raton (2024)
146. Kaur, G., et al.: Empowering fog computing for achieving sustainability. In: Cloud and Fog Optimization-Based Solutions for Sustainable Developments, pp. 201–216. CRC Press, Boca Raton (2024)
147. Bakam, G., Mpofu, K., Mbohwa, C.: A model for a smart transport manufacturing environment based on edge computing, fog computing, and cloud computing in the South African context. In: International Conference on Flexible Automation and Intelligent Manufacturing. Springer, Cham (2024)
148. Xu, R., Nagothu, D., Chen, Y.: AR-edge: autonomous and resilient edge computing architecture for smart cities. In: Edge Computing Architecture-Architecture and Applications for Smart Cities. IntechOpen, London (2024)
149. Mohamed, N.N., Abuobied, B.H.H.: Cybersecurity challenges across sustainable development goals: a comprehensive review. Sustain. Eng. Innov. **6**(1), 57–86 (2024)
150. Mahmood, H.S., et al.: Conducting in-depth analysis of AI, IoT, web technology, cloud computing, and Enterprise systems integration for enhancing data security and governance to promote sustainable business practices. J. Inf. Technol. Inf. **3**(2), 297–322 (2024)
151. Samira, Z., et al.: Comprehensive data security and compliance framework for SMEs. Magna Sci. Adv. Res. Rev. **12**(1), 43–55 (2024)
152. Sindiramutty, S.R., et al.: Security considerations in generative AI for web applications. In: Reshaping CyberSecurity With Generative AI Techniques, pp. 281–332. IGI Global, London (2025)
153. Elamanov, S., et al.: Interworking between Modbus and internet of things platform for industrial services. Digit Commun. Netw. **10**(2), 461–471 (2024)
154. Sharma, R., Sharma, A.: IoT interoperability framework for smart home: MDA-inspired approach. Cluster Comput. **27**, 1–18 (2024)
155. Vishwakarma, A.K., et al.: Internet of things technology, research, and challenges: a survey. Multimed. Tools Appl. **84**, 8455–8490 (2024)
156. Ji, X., et al.: Cloud-edge collaborative service architecture with large-tiny models based on deep reinforcement learning. IEEE Trans. Cloud Comput. **13**(1), 288–302 (2025)
157. Crespo, A., et al.: An efficient strawberry segmentation model based on mask R-CNN and TensorRT. Artif. Intell. Agric. **15**(2), 327–337 (2025)
158. Teixeira, J.E., Tavares-Lehmann, A.T.C.P.: Industry 4.0 in the European union: policies and national strategies. Technol. Forecast. Soc. Chang. **180**, 121664 (2022)
159. Kop, M., et al.: Ten principles for responsible quantum innovation. Quant. Sci. Technol. **9**(3), 035013 (2024)
160. Asija, D., Viral, R.K.: Chapter 6: cyber security: emerging trends best practices. In: Asija, D., et al. (eds.) Cyber Security Solutions for Protecting and Building the Future Smart Grid, pp. 161–187. Elsevier, Amsterdam (2025)
161. Urgelles, H., et al.: In-network quantum computing for future 6G networks. Adv. Quant. Technol. **8**(2), 2300334 (2025)

Chapter 6
IoT-Enabled Asset Tracking in Industry 6.0

Abstract By increasing productivity, cutting expenses, and boosting sustainability, the combination of IoT-enabled asset tracking systems with cutting-edge technologies like RFID, GPS, edge computing, and cloud analytics has revolutionized industrial operations. To transform the industrial sector into Industry 6.0, the fundamental technologies that underpin IoT-based asset monitoring and their uses in a range of sectors, including as manufacturing, logistics, healthcare, and energy are used. Using electromagnetic signals, RFID technology automates asset identification. Passive, semi passive and active RFID tags maximize tracking in various operational contexts. By integrating GPS, mobile assets like delivery fleets and large machinery may be tracked in real time, increasing security and decreasing inefficiencies. By processing data locally, edge computing-powered advanced data analytics lowers latency and improves real-time decision-making. With an emphasis on intelligent automation, predictive maintenance, and sustainability, the function of IoT-enabled tracking in Industry 5.0 and 6.0 is important. In manufacturing and fleet management, predictive analytics has reduced unscheduled downtime and operational expenses. IoT tracking optimizes the use of energy and resources, which is in line with the global sustainability goals (SDGs 9, 12, and 13). Future innovations include blockchain-secured supply chains for increased transparency and AI-powered autonomous tracking. The influence of IoT in climate-resilient industries like the automotive, chemical, and textile sectors is illustrated in the chapter's conclusion, guaranteeing flexible, effective, and sustainable industrial processes.

Keywords IoT · RFID · GPS · Edge computing · Cloud analytics · Predictive maintenance · Industry 5.0 · Industry 6.0 · Supply chain visibility · Smart manufacturing · Digital twin · Sustainability · AI-driven asset tracking · Blockchain security · SDGs

H. Majeed, T. Iftikhar, *Intelligent Manufacturing in Industry 6.0*,
https://doi.org/10.1007/978-3-032-07278-8_6

1 IoT Enabled Asset Tracking Systems Technologies Introduction

Industrial operations now rely heavily on asset tracking, which boosts sustainability, lowers costs, and increases efficiency. Utilizing cutting edge technologies like RFID (Radio Frequency Identification), GPS (Global Positioning System), and real-time data analytics, the integration of IoT into asset tracking systems helps to optimize resources, streamline operations, and guarantee the seamless movement of equipment and materials throughout the supply chain. Here is a thorough examination of the technology foundations, industry effects, and applicability to Industry 5.0 and 6.0 of IoT-driven asset tracking in a variety of sectors. Figure 6.1 shows the basics of RFID and GPS system.

1.1 Radio Frequency Identification (RFID)

One of the most used technologies in industries for tracking assets is RFID. It automatically detects and tracks tags affixed to items using electromagnetic fields. These tags are made up of a tiny chip and an antenna that connect to a reader device to send real-time information about the asset's location and condition. Semi-passive, active, and passive RFID tags are all possible. RFID tags that don't need a battery are known as passive tags since they are powered by the reader's signal. They are reasonably priced and perfect for monitoring low-value items in warehouses, such tools and machine parts. Active RFID tags run on batteries and can track a user's

Fig. 6.1 Difference between GPS and RFID

location in real-time across longer distances. They are applied on valuable assets like trucks or industrial machinery. Semi-passive RFID tags rely on a reader for communication but have a battery to power the tag's microchip. They perform better in challenging conditions and have longer read ranges. RFID is frequently used on assembly lines to track parts. The BMW Group tracks and controls the assembly of intricate auto parts at its plants using RFID technology. Some companies claim that by utilizing RFID to expedite inventory management, cut down on production delays, and avoid stockouts, operational inefficiencies have decreased. RFID reduces human error and increases inventory accuracy by automating the tracking and management (in logistics and supply chain) of goods in warehouses, a practice used by Amazon and other international supply chains [1–4].

1.2 Global Positioning System (GPS)

Real-time location data is provided via GPS in asset tracking, particularly for mobile assets like trucks, forklifts, and large machinery. IoT devices with GPS capabilities give businesses precise positional information that enables them to monitor the movement of assets over wide geographic areas, lowering theft and enhancing logistical coordination. GPS-based IoT devices (Logistics and Fleet Management) are used by FedEx, DHL, and other companies to track their delivery cars in real time. By enabling dynamic route optimization, these technologies cut down on delivery time and fuel usage. A report claims that courier companies fleet management strategies that use GPS-based IoT devices, has lowered fuel expenses per annum, greatly advancing their environmental sustainability targets (SDG 13). GPS is used in industries like construction to keep an eye on large machinery, such as excavators and bulldozers, to prevent prolonged idle time and boost output. Businesses such as Caterpillar have incorporated GPS and IoT into their machinery, allowing for real-time tracking and schedule optimization [5, 6].

1.3 Edge Computing and Real-Time Data Analytics

The real-time data analytics systems that handle the data collected from RFID and GPS devices are at the core of IoT-enabled asset tracking. By processing data locally (at the device level), edge computing lowers latency and eliminates the need to transport massive amounts of data to centralized cloud servers. This is essential for applications like rerouting logistical operations or modifying production plans that call for quick decisions. IoT devices use predictive maintenance to continuously check the state of machinery. To anticipate errors before they happen, general electric uses IoT sensors to gather data from production equipment. AI algorithms are then used to analyse this data in real time. According to GE, this proactive strategy lowers maintenance expenses yearly and also decreases downtime. To track inventory and assets in warehouses, Zebra Technologies (inventory management) provides IoT-based solutions that combine RFID and GPS technologies with data analytics platforms. Businesses can forecast demand spikes, find trends in stock movement, and improve storage architecture by evaluating data from these IoT sensors. It has been demonstrated that these solutions improve order fulfillment time and lower inventory carrying expenses [7–9].

On production lines, real-time tracking of components and materials guarantees efficient workflow and resource allocation. When combined with data analytics, IoT sensors may identify production bottlenecks, inefficiencies, and possible failure sites, enabling manufacturers to make data-driven decisions that increase throughput. Ford Motor Company: Ford tracks and monitors the movement of parts in real time using IoT sensors in its manufacturing plants. Overall equipment effectiveness is increased by 15%–20% with dynamic inventory control made possible by RFID part tracking combined with real-time data analytics. To minimize stockouts and surplus inventory, Ford also uses predictive analytics to automate reorder procedures and manage inventory levels [10, 11].

End-to-end supply chain asset visibility is made possible via IoT asset tracking. Businesses can enhance their capacity to manage suppliers, optimize transportation routes, and avoid delays brought on by unanticipated events by implementing real-time location tracking. This visibility aids in the management of perishable items' cold-chain operations, guaranteeing adherence to temperature standards and minimizing spoilage. Walmart has increased the transparency of its supply chain by using GPS monitoring and RFID tagging enabled by the IoT. This has allowed Walmart to track products from manufacturing facilities to store shelves. Walmart has cut supply chain expenses by 10–15% as a result, and stock visibility has significantly improved [12, 13].

2 IoT-Enabled Asset Tracking's in Industry 5.0 and 6.0

Utilizing technologies like RFID, GPS, and real-time data analytics, IoT-enabled asset tracking systems are revolutionizing companies through supply chain visibility, resource allocation, and operational optimization. These systems minimize waste and energy consumption, which promotes sustainability in addition to cost reduction and efficiency gains. The development of self-optimizing, autonomous systems, increased productivity, sustainability, and resilience in industrial operations, and a key role in accomplishing global sustainability goals are all made possible by the IoT as industries transition to Industry 5.0 and 6.0. Customized for people needs and collaborative AI systems are the focus of Industry 5.0, where human knowledge enhances the accuracy of intelligent machines. This collaboration will be made possible in large part by IoT-enabled asset tracking, which gives human operators access to real time data so they can make quicker and better decisions. The concept broadens to self-optimizing ecosystems in Industry 6.0, where AI controls manufacturing, logistics, and resource use on its own. This change will be facilitated by IoT enabled asset tracking, in which assets self-report their condition and status to maximize their utilization without the need for human involvement. Siemens is at the forefront of smart factory technology with Digital Twin, which uses real-time IoT data to mimic, forecast, and optimize the performance of physical assets in a digital representation. Several SDGs are directly aided using IoT enabled asset tracking. By increasing operating efficiency, cutting waste, and guaranteeing scalability, IoT improves industrial infrastructure in line with SDG 9. Regarding SDG 12, IoT systems minimize waste and guarantee sustainable manufacturing methods by enabling predictive maintenance, lowering energy usage, and optimizing resource utilization. As demonstrated by business where fleet management IoT systems have dramatically reduced energy costs and carbon emissions, IoT-enabled asset tracking offers real-time monitoring of energy and material usage, supporting SDG 13 [14, 15].

3 Climate Resilient Industrial Application of IoT Technologies

We must investigate how particular industries, like the automotive, chemical, and textile sectors, use IoT technologies to adjust to climate-related difficulties if we want to go deeper into IoT-enabled manufacturing resilience. By increasing energy and resource efficiency, enabling predictive maintenance, and controlling the effects of climate change on operations, IoT is essential for boosting climate resilience. For climate resilience, the application of IoT in manufacturing particularly in sectors like automotive, chemicals, and textiles has proven essential. IoT technologies assist reduce the effects of climate change on manufacturing processes by enabling predictive maintenance, operational optimization, and resource conservation through the provision of real-time data on energy and resource use. SDGs 9, 12, and 13 are among the global sustainability initiatives that these solutions support. Manufacturing becomes more sustainable, efficient, and able to adjust to the challenges presented by a changing climate when IoT technologies are integrated. A thorough technical explanation of how these sectors use IoT devices to reduce climate risks and increase resilience can be found below.

3.1 Automobile Sector

IoT is directly linked with the automobile industry's efforts to improve climate resilience and lower energy consumption throughout the production and supply chain processes. Automotive facilities can optimize their power usage by integrating IoT sensors with manufacturing equipment to monitor energy consumption in real-time. Volkswagen, for instance, employs an IoT based energy management system in their factories, which enables them to track and optimize energy consumption across many manufacturing lines in real-time. Manufacturers can dynamically modify settings to guarantee minimal energy consumption by monitoring temperature, humidity, and power consumption. Smart grid integration is one notable technology that enables auto factories to use energy from renewable sources like solar and wind. IoT-enabled energy monitoring systems at Ford's Michigan facility can cut energy use by up to 20–25% a year, supporting SDG 13 and enhancing energy resilience. By enabling predictive maintenance for machinery and equipment, these technologies help cut down on energy waste and avoid machine failure-related downtime [16, 17].

3.2 The Chemical Sector

IoT makes it easier to manage resources and boost productivity in the chemical and petrochemical industries, which directly supports climate resilience, safety, security, and health. Because the industry frequently uses a lot of energy, it necessitates efficient resource allocation and emission reduction. IoT sensors placed in distillation columns and chemical reactors continuously check temperature, pressure, and flow rates to identify inefficiencies that result in needless energy use. To find chances for energy savings, BASF and many other industries use an IoT-based system to track energy usage throughout its chemical production facilities. An estimated 10–15% more energy is saved annually per industrial unit because of its integration. IoT-based predictive maintenance solutions assist chemical companies in managing their equipment in reaction to climate change-related hazards, minimizing unplanned downtime, averting dangerous leaks, and making sure that resources are not squandered. By applying data analytics, businesses may anticipate when vital pieces of machinery, such as pumps, compressors, and valves, will fail and replace or repair them before they seriously impair output [18, 19].

3.3 Textile Sector

Another important sector where IoT enhances climate resilience and resource efficiency is the textile industry, namely in the management of energy and water usage, two resources that are essential to textile production. Levi Strauss drastically reduces water loss by tracking water usage in its dyeing processes with IoT-enabled water monitoring systems. IoT sensors track every step of the production process, and data analytics make sure that energy and water use are minimized, cutting waste and supporting SDG 12. Up to 30% less water and 20–25% less energy has been used in textile facilities because to IoT enabled smart meters and predictive analytics. Textile firms can increase process efficiency using real time analytics to dynamically modify machine settings based on variables like temperature, humidity, pH and fabric type [20, 21].

3.4 Water and Energy Sectors in Climate Affected Areas

IoT is essential for manufacturing systems in climate change affected areas to adjust to erratic environmental changes like water scarcity or extremely high temperatures. IoT sensors and controllers assist enterprises reduce their environmental impact by enabling real-time monitoring of resource consumption in areas that are at risk. IoT-driven smart irrigation systems are currently being utilized to optimize water consumption in manufacturing processes in regions that are experiencing water scarcity, such as parts of India and the Middle East. Businesses can monitor and regulate the irrigation of cooling systems in power plants or agricultural enterprises by employing soil moisture sensors, which can cut water usage by as much as 40%. IoT-based water management systems are used by General Electric to make sure that its plants in water scarce areas run effectively, especially during dry spells, and meet water sustainability goals [22, 23].

3.5 IoT-Powered Manufacturing Systems with Climate Resilient Approach

Manufacturing systems' ability to react to climate change is becoming more closely linked to how IoT maximizes energy and water efficiency. Supply chains are frequently impacted by climate change, which can alter the price and availability of essential supplies. Real-time monitoring of these variables is made possible by IoT, which enables manufacturers to modify their plans as necessary. Businesses in the automotive and textile industries have used IoT devices that enable dynamic energy consumption adjustments based on current environmental parameters like humidity and temperature. Regardless of variations brought on by the climate, these modifications assist manufacturing facilities in maintaining maximum operational efficiency.

4 Industries with Real-Time Monitoring, Predictive Maintenance, & Data Analytics Systems

In vital industries like healthcare, agriculture, and transportation, the use of IoT in predictive maintenance and real-time monitoring has enormous implications for increasing operational effectiveness, boosting sustainability, and constructing climate resilience. Industries can now anticipate and prevent system failures as well as make real-time modifications to enhance resource management, lower emissions, and lessen climate risks thanks to the integration of IoT with AI and data analytics. Industries are becoming more cost-effective and climate-resilient by embracing these cutting edge IoT technologies, which helps them achieve Industry 5.0 and 6.0

objectives as well as the SDGs. Industry 5.0, which prioritizes human-centric innovation and cooperation between intelligent systems and humans, is directly in line with the implementation of IoT systems in the fields of healthcare, agriculture, and transportation. In line with SDG 13, IoT devices' real time data and predictive insights help enterprises become more robust and adaptive to climate-related challenges. The goal of Industry 5.0 is to develop highly intelligent and automated systems that collaborate with people. The IoT is essential to enhancing the sustainability, resilience, and efficiency of these systems. Predictive maintenance, for example, contributes to the overall sustainability of industrial operations by lowering unscheduled downtime and resource consumption in addition to ensuring the longevity of vital machinery. The next phase of this evolution, known as Industry 6.0, aims for autonomous systems that optimize workflows without the need for human interaction. These systems' capacity to manage resources in a climate-resilient way will greatly improve as they become increasingly integrated with AI and machine learning, allowing industries to prosper despite climatic problems. IoT-enabled systems will keep developing, getting better at anticipating failures, maximizing energy use, and guaranteeing climate resilience in all industries.

4.1 Healthcare Industry to Forecast Medical Device and Patient Monitoring Failures

IoT-enabled solutions are becoming essential in the healthcare industry for real-time monitoring and predictive maintenance, which improves patient outcomes. Vital indicators including blood pressure, heart rate, and oxygen levels are continuously monitored by wearable IoT devices, including biomedically graded patches and consumer smartwatches with health monitoring features. These gadgets monitor any deviation in the patient's physiological parameters using machine learning algorithms, giving early warnings for possible health hazards including sepsis or cardiac arrhythmias. The CareLink system from Medtronic gives patients access to real-time data about implanted medical equipment, such as insulin pumps or pacemakers, and allows for remote monitoring. This makes proactive intervention possible and lowers the chance of device failure by enabling healthcare personnel to predict device faults before they happen. Predictive analytics integration promotes early diagnosis, lowers medical errors and readmissions to hospitals, and enhances patient safety. Medical equipment maintenance also makes use of IoT-based predictive solutions. IoT sensors are used by GE Healthcare in diagnostic devices such as CT and MRI scanners. Long before a failure happens, these sensors identify wear or malfunction in equipment by monitoring variables like temperature, pressure, and humidity. Predictive maintenance improves hospital operations, extends the life of costly equipment, and decreases downtime. According to GE Healthcare's IoT solutions in 2023, IoT-based predictive maintenance in medical equipment can cut downtime by as much as 30% [24–27].

4.2 Agriculture Sector to Enhance Climate Resilience through IoT

IoT technologies are now crucial for increasing climate change resistance, increasing productivity, and cutting down on resource waste in agriculture. Precision farming makes heavy use of IoT sensors to track vital environmental parameters, including soil moisture, temperature, and weather patterns. These systems employ real-time data analytics to offer insights that assist farmers in making well-informed choices regarding pest management, fertilization, and irrigation. John Deere has integrated the IoT into its agricultural machinery, such as smart tractors that continuously check crop health, moisture levels, and soil conditions. By anticipating when crops require water, soil moisture sensors installed in the ground can lower water use and increase irrigation effectiveness. IoT-powered weather stations gather temperature and precipitation data, which farmers use to determine the best times to grow and harvest crops so they can withstand unforeseen climate catastrophes like floods or droughts. IoT sensors are utilized in fields to track crop health and identify early indicators of diseases or insect infestations. By predicting possible dangers using data analytics, these technologies minimize the usage of pesticides and enable prompt action, both of which support more environmentally friendly farming methods. These systems can also assist farmers in preparing for extreme weather occurrences, such heat waves or frost, that could otherwise compromise yields by forecasting weather patterns. Projects like the Smart Farming Initiative in India, where IoT and data analytics have greatly increased agricultural production, have proven that real-time monitoring systems in agriculture can reduce water usage by 40 to 50% and boost crop yields by 20–30% [28–30].

4.3 Transportation Sector Predictive Maintenance for Fleet Management

IoT is essential for fleet management predictive maintenance in the transportation industry, which lowers operating costs and guarantees vehicle dependability. Data on engine performance, fuel consumption, tire pressure, brake usage, and vehicle speed are continuously gathered by IoT devices, such as telematics systems mounted in automobiles. ML algorithms are used to examine this data and forecast when a part would likely break or need maintenance. UPS employs a fleet management system based on the IoT to track the state of its cars in real time. In addition to tracking variables like engine temperature, battery health, and fuel use, the system offers predictive data regarding when maintenance should be performed. By ensuring that cars are serviced before they break down, this proactive method lowers maintenance costs and increases fleet dependability. Transportation companies may track fuel efficiency and route optimization with IoT enabled technologies. Real-time vehicle tracking is provided by Fleet Complete, a global IoT fleet management platform, which helps transportation companies optimize routes and cut fuel usage, which lowers costs and carbon emissions. It has been discovered that real-time monitoring systems increase fleet uptime by 25 to 30% and lower maintenance expenses by 15–20%. This is crucial for improving transportation operations' resilience, especially considering tightening environmental restrictions and disruptions brought on by climate change [31, 32].

5 Cloud Computing Applications in Asset Tracking

Cloud computing is a game changing technology that allows on-demand access to computational resources, such as servers, storage, and applications, through the Internet, eliminating the need for physical infrastructure on premises and offering scalability, cost efficiency, and collaboration capabilities. Platforms such as Amazon Web Services, Microsoft Azure, and Google Cloud Platform are used to implement cloud computing, and each of these platforms offers a variety of tools for industries and researchers. By utilizing cloud technology, industries can effectively manage data, improve operational agility, and decrease downtime while adhering to SDGs including 9 and 12. In Industry 5.0, cloud computing is essential for human-machine collaboration, and in Industry 6.0, cloud systems will be integrated into autonomous systems for further optimization and resource management.

5.1 Remote Access to Computing Resources Technological Framework

Modern industries have been completely transformed by cloud computing, a game-changing technology that gives companies unmatched access to remote computer resources, enormous data storage capacities, and cutting-edge collaboration platforms. Its applications span several industries, promoting efficiency, scalability, and innovation while aiding in the shift to Industry 5.0 and 6.0 and the SDGs. With the use of the most recent data, this thorough investigation dives into the industrial applications, technological framework and real world instances. A wide range of services, such as Infrastructure as a Service (IaaS), Platform as a Service (PaaS), and Software as a Service (SaaS), are offered by cloud computing, providing scalable, and affordable access to virtualized computer resources. Advanced virtualization technologies like VMware, Hyper-V, and Kubernetes for container orchestration are what enable this functionality. Businesses may dynamically scale their computational capacity in response to real-time demand using elastic compute services (Amazon EC2), which can be used for applications like big data analytics, scientific simulations, and AI model training. In contrast to conventional systems, NVIDIA's DGX Cloud offers high performance GPUs for machine learning workflows, speeding up AI development (model training time) for specific workloads [33–36].

By lowering latency and allowing data processing closer to the source, distributed computing, when paired with edge computing concepts, further improves performance. Microsoft Azure Stack creates hybrid cloud environments that enable sensitive applications that need high availability and low latency by integrating cloud services with on premises infrastructure. For sectors like telecommunications, where 5G networks depend on clouds driven by edge computing to smoothly manage billions of linked devices, these solutions are essential. Medical practitioners may access patient histories and diagnostic tools remotely thanks to cloud services like Google Cloud Healthcare API, which make it easier to securely store, share, and analyse large datasets. Telemedicine's adoption was spurred by the COVID-19 pandemic, and by 2023, the worldwide telehealth market was valued at USD 60 billion. Cloud-based AI-enabled advanced diagnostic tools forecast and track illness, Google's DeepMind AI has been shown to reduce diagnostic errors in radiology with the passage of time. Cloud-enabled systems are used by energy businesses such as Siemens Energy to maximize power grid performance. These platforms enhance the integration of renewable energy sources, increasing grid efficiency and lowering carbon emissions in pilot programs through the analysis of real-time data especially after COVID-19 pandemic era [37–43].

By promoting scalable and effective industrial operations, cloud computing applications closely align with SDG 9, 12 by optimizing resource usage and minimizing waste and SDG 13 by enabling carbon footprint reductions through intelligent systems. Cloud platforms are used by human-centric innovation in Industry 5.0 to integrate AI and collaborative robotics, increasing productivity and guaranteeing worker safety. To increase efficiency by more than 25%, Bosch company uses cloud connected collaborative robots (cobots) that learn from real-time data to adjust to changing industrial conditions. Cloud computing will support self-optimizing, autonomous systems driven by AI and quantum computing in Industry 6.0. Advanced cloud ecosystems are incorporating emerging technologies, including IBM's Quantum Cloud, to address hitherto unsolvable issues in energy distribution, drug development, and logistics. Cloud systems driven by quantum technology will spur exponential innovation because of their ability to process quintillions of calculations per second. Because it provides unmatched access to computing power, storage, and collaboration technologies, cloud computing is a vital tool for modern enterprises. Innovation is accelerated and sustainability and resilience objectives are met via its integration with AI, IoT and quantum computing. Cloud computing continues to play a key role in determining the direction of industries around the world by facilitating data driven decision making, cutting expenses and encouraging international cooperation [44–46].

5.2 *Data Storage*

Cloud computing's high-performance storage systems, which offer reliable solutions for scalability, cost effectiveness, and fault tolerance, are completely changing how businesses handle, analyze, and use enormous datasets. Distributed file systems and object storage models are used by industry-leading platforms, such as AWS S3, Google Cloud Storage, and Azure Blob Storage to store and retrieve large datasets with ease. High availability and resilience against failures are ensured by these systems' use of sophisticated redundancy and replication techniques, such as RAID (Redundant Array of Independent Disks) setups. AWS S3's remarkable highly reliability is a key feature for companies that handle vast or sensitive data. The needs of big data are specifically met by cold storage systems, which make it possible to store datasets that are rarely accessed at a reasonable cost. In domains like genomics, where data collected from DNA sequencing studies frequently exceeds 1 terabyte per trial, technologies like Amazon Glacier offer cost-effective storage solutions, making them essential. An important benefit for research and development is that this strategy optimizes expenses while promoting long-term data retention [34, 47–49].

Both structured and unstructured data may be handled effectively by contemporary cloud databases like Google BigQuery and Amazon RDS, which use relational (SQL) and non-relational (NoSQL) models like MongoDB and DynamoDB. Applications requiring complex pattern recognition and fast data processing such as real time customer analytics and advanced fraud detection algorithms, are supported by these systems. By generating virtual copies of tangible assets, cloud-based Digital Twin technology is revolutionizing manufacturing processes. By simulating operational situations and forecasting equipment behavior, companies, such as GE Digital can make more informed decisions and, in certain cases, increase production efficiency by up to 30%. Manufacturers may proactively handle maintenance requirements and streamline processes by storing and analyzing terabytes of data in real time. High-performance cloud storage is essential to the management and distribution of massive amounts of digital content in the media and entertainment industry. To ensure worldwide accessibility, Netflix, one of the biggest streaming services, uses cloud storage to manage petabytes of video content. With more than 230 million members globally, Netflix reduces latency and offers flawless streaming experiences using sophisticated content delivery network caching techniques [50–54]. Through the development of robust data infrastructures and the optimization of resource utilization through effective storage techniques, these high-performance storage systems support SDGs 9 and 12. These systems are expected to increasingly integrate with AI, edge computing, and quantum technologies as industries move into Industry 5.0 and become ready for Industry 6.0, resulting in even more efficiency and innovation. One intriguing solution that is helping to shape the future of data driven industries is quantum cloud storage, which offers unmatched speed and scalability for processing large datasets.

5.3 Collaborative Platforms

Using distributed networking and real-time synchronization protocols, cloud computing has revolutionized collaborative platforms by facilitating smooth communication amongst multinational teams. Instantaneous audio, video, and file sharing is made possible by technologies like WebRTC (Web Real Time Communication), which serve as the basis for real-time collaboration. These frameworks are essential for sectors that need to communicate quickly, effectively, and securely. Productivity for geographically scattered teams has been transformed by AI enhanced collaboration solutions like Google Workspace and Microsoft Teams, which have integrated cutting-edge capabilities like intelligent scheduling, real-time language translation, and automated meeting transcription. The fact that Microsoft Teams reported having over 280 million active users in 2023 highlights how widely these AI-powered solutions are being used in businesses. These tools improve productivity and enable a clearer focus on strategic goals by lowering language barriers and automating repetitive operations [55–57].

Data security and transparency are guaranteed when blockchain technology is incorporated into cloud-based collaboration systems. Tamper-proof document sharing is made possible by solutions like the IBM Blockchain Platform, especially in delicate settings like supply chain management. By guaranteeing traceability and real-time access to supply chain data, Walmart's blockchain system, which is based on IBM's platform, has shortened product recall timelines from days to seconds. Collaborative platforms like Autodesk BIM 360 enable teams in the construction sector to work on Building Information Modeling (BIM) systems in real time, enabling numerous stakeholders to edit 3D building models at the same time. Project delays and errors are decreased by this real-time collaboration, which results in 10%–15% cost reductions. The multinational construction behemoth Skanska employs cloud-based BIM to improve efficiency and expedite project management in large-scale projects. Cloud-based collaboration tools are used in education by platforms such as EdX and Coursera to provide scalable virtual classrooms. These platforms allow colleges to reach millions of students worldwide by hosting thousands of courses in a variety of fields. More than 124 million people have registered on Coursera alone by 2023, proving the accessibility and scalability of cloud-based learning. The learning process is further improved by AI capabilities like adaptive assessments and individualized learning recommendations. By democratizing access to knowledge, these technological developments support SDG 4, 9 by improving infrastructure and operational efficiency in supply chains and construction; and SDG 11 by facilitating more intelligent urban planning through BIM. To create incredibly effective, human-centered collaboration ecosystems that spur innovation and sustainability, collaborative platforms will progressively incorporate quantum computing, edge AI, and immersive technologies like AR/VR as sectors move toward Industry 5.0 and 6.0 [58, 59].

5.4 SDGs Integration with Cloud Computing

Because of its many uses, cloud computing is essential to achieving the SDGs. It provides scalable corporate solutions while encouraging innovation, sustainability, and operational efficiency. It makes a substantial contribution to SDG 9 since it enables companies to make effective use of shared resources, which lessens the need for substantial physical infrastructure. Businesses who use cloud computing, for instance, see a 20–30% decrease in the cost of their IT infrastructure while increasing the scalability of their operations. High performance computing capabilities are offered by cloud platforms like Google Cloud, AWS and Microsoft Azure, which promote innovation in fields like robotics, artificial intelligence and material science. Cloud-based data analytics solutions help businesses optimize resource use, lowering material, and energy waste, which is in line with SDG 12. To reduce production downtime and raw material losses, cloud-integrated manufacturing systems use real-time monitoring and predictive analytics, increasing operational efficiency by as much as 40%. Cloud enabled IoT devices in agriculture optimize the use of water and fertilizer to ensure sustainable farming practices by analyzing real-time data on weather patterns and soil health. In pilot projects utilizing cloud-supported precision agriculture, this has resulted in water usage reductions of up to 30%. Because cloud computing lessens the environmental impact of traditional data centers, which are infamous for their high-energy consumption and greenhouse gas emissions, it also significantly contributes to SDG 13. By switching to renewable energy sources, hyperscale cloud companies like Microsoft Azure, Amazon Web Services and Google Cloud are setting the standard. Google Cloud has a carbon free energy score of 67% for its global operations and runs entirely on renewable energy. According to research, moving to the cloud can cut carbon emissions by up to 84% when compared to on premises data centers, which greatly lowers the carbon footprint of computing services [60–62].

Cloud computing's benefits for the environment go beyond its operational aspects to include its use in climate science. Large-scale climate modelling and data processing are supported by cloud platforms, which help with disaster preparedness by enabling precise forecasts of extreme weather events. To assist businesses, reduce their environmental effects, and adjust to shifting climate circumstances, IBM's Environmental Intelligence Suite leverages cloud computing to deliver real-time analytics on climate threats. In industries like logistics and agriculture, where climate variability presents serious risks, these technologies are crucial for creating robust supply networks. Cloud computing seamlessly integrates with the tenets of Industry 5.0 and Industry 6.0, which emphasize sustainability, customized human needs, autonomous AI systems, and innovation by maximizing resource utilization and lowering emissions. It is expected that cutting-edge innovations like edge cloud hybrid systems and quantum computing integration would increase cloud efficiency and capabilities. A future where technical advancement and environmental stewardship coexist will be fostered by these developing technologies, which will allow industry to satisfy global sustainability targets while remaining resilient in the face of climate change [63].

5.5 *Role of Cloud Computing in Industry 5.0 and Industry 6.0 Upgradation*

In both Industry 5.0 and Industry 6.0, cloud computing is a key enabler that propels the integration of completely autonomous, intelligent production systems with human centric innovation. Cloud computing enables AI-driven systems and collaborative robotics (cobots), which are intended to improve human machine interaction in Industry 5.0. This paradigm promotes safety, innovation, and productivity in the industrial industry by emphasizing the enhancement of human abilities rather than their replacement. Bosch trains cobots in real time using cloud platforms. These cobots use sophisticated machine learning models housed on cloud infrastructure to dynamically adjust to production needs, lowering mistakes by 25–30% and boosting overall productivity in high precision production lines. Industry 6.0, which aims to integrate autonomous systems that can optimize themselves through real-time data analytics and decision-making, greatly enhances the role of cloud computing. These AI and quantum computing-powered devices are revolutionizing industries including healthcare, energy, and logistics. By integrating into cloud ecosystems, IBM Quantum has started to take on activities like financial risk modelling, drug discovery acceleration, and energy distribution network optimization that were previously unachievable for classical systems. By solving multi-variable optimization problems, these cloud-quantum integrations help industry make faster and more accurate decisions [64–66].

The influence of cloud computing in these cutting-edge industrial revolutions is consistent with international sustainability objectives. Hyperscale cloud data centers' dedication to renewable energy sources and energy efficiency help achieve SDGs 7 and 13. AWS has pledged to use only renewable energy by 2025 and to optimize data center energy use to reduce carbon emissions by millions of metric tons each year. These developments immediately help sustainable industrial growth by lowering the carbon footprint of shipping and manufacturing. To attain unmatched efficiency in the field of intelligent manufacturing, cloud computing combines edge-cloud hybrid systems, real-time analytics, and digital twins. For instance, Siemens' MindSphere, an IoT operating system that runs on the cloud, builds digital twins of production processes using real-time data from factory floors. This allows for product lifecycle management, predictive maintenance, and energy efficiency improvement. These developments establish standards for smart manufacturing by cutting downtime by 40% and operating expenses by 20–30%. Blockchain-enabled cloud platforms are used in Industry 6.0's concept of interconnected ecosystems to provide safe, open supply chain management. The cloud hosted IBM Blockchain Platform has been crucial in improving trust, cutting down on inefficiencies, and tracking the provenance of supply chains. In industries like food safety, where real-time tracking of perishable commodities lowers waste and is in line with SDG12, such technologies are essential [67–69].

Climate change adaptation is another area where cloud computing is being included into Industries 5.0 and 6.0. Google Cloud's Climate Insights tool evaluates the environmental impact of industrial processes using AI- and cloud-based analytics, offering practical suggestions for cutting emissions. In a world where climate variability is growing, these technologies help businesses create production plans that are climate resilient. The core of Industries 5.0 and 6.0 is cloud computing, which propels autonomous systems, human friendly innovation, and environmentally friendly industrial processes. A revolutionary period in industrial evolution is being ushered in by its applications in cobots, blockchain, digital twins, and quantum computing, which are redefining resilience, productivity, and environmental responsibility. Unmatched benefits in terms of scalability, cost effectiveness, cybersecurity, data driven decision-making, and environmental sustainability are provided by cloud computing. In the quickly changing world of Industries 5.0 and 6.0, these skills are essential for reaching operational excellence, spurring innovation, and fulfilling global sustainability goals. The way that industries handle data, access computing resources, and cooperate has been completely transformed by cloud computing. Cloud platforms have become essential for attaining sustainability and operational excellence through the integration of cutting-edge technologies like blockchain, artificial intelligence, and quantum computing. They continue to be at the vanguard of industrial innovation, propelling global development toward a more resilient and efficient future, thanks to their alignment with the SDGs and the tenets of Industry 5.0 and 6.0 [70–73].

5.6 Multi Industrial Case Studies of Cloud Computing

By providing scalability, cost-effectiveness, improved decision-making capabilities, and strong cybersecurity measures, all while supporting global sustainability initiatives, cloud computing has completely transformed industrial processes. Businesses can dramatically lower capital expenditure (CapEx) and improve operational efficiency (OpEx) by utilizing pay-as-you-go methods to dynamically expand their computational resources in response to demand. Manufacturing businesses can quickly adapt their computer requirements during seasonal production peaks or downturns using cloud services, for instance, and avoiding the expenses associated with over-provisioning. This adaptability is mandatory in Industry 5.0 and 6.0, when keeping a competitive edge requires being able to adjust to market needs. Through the discovery of market trends, enhancement of consumer experiences, and optimization of supply chain operations, real-time data analytics on cloud platforms empower industries to make well-informed decisions. Walmart uses a cloud-based analytics system to process an incredible 2.5 petabytes of customer transaction data per week. This makes it possible to precisely control inventory, which lowers waste and operating expenses while guaranteeing product availability, an important factor in determining profitability and customer happiness. Modern cloud platforms effectively address cybersecurity, a vital issue in the age of digital transformation.

To ensure data integrity and regulatory compliance, these systems include AI-driven cybersecurity solutions, including behavioral analytics, encryption mechanisms, and sophisticated intrusion detection systems. Businesses that use Microsoft Azure gain access to AI-enhanced threat intelligence, which proactively detects and reduces security threats in real time, protecting confidential data and upholding compliance with strict laws like GDPR [50, 74–76].

Cloud computing is essential to green activities, supporting international efforts to tackle climate change. Data centers' environmental impact has significantly decreased because of the shift to green cloud providers, which place a higher priority on renewable energy sources and effective cooling systems. As part of its Climate Pledge, AWS has committed to achieving net zero carbon emissions by 2040. AWS sets the standard for environmentally friendly cloud computing by implementing sustainable practices, such as using solar and wind energy to power its operations and cutting-edge cooling technologies to increase energy efficiency. Cloud computing systems' scalability and energy efficiency allow enterprises to adopt cutting-edge technology while lowering their carbon footprints, which directly supports SDGs 9 and 13. Cloud computing is combined with IoT and AI to develop adaptive systems that maximize resource utilization in industries such as intelligent manufacturing. The cloud-based Predix platform from General Electric reduces downtime and energy usage by 20–30% by tracking the operation of industrial equipment and forecasting maintenance requirements [77–79].

6 Zero Downtime with Autonomous Maintenance

In industrial automation, the idea of zero downtime through autonomous maintenance has become a paradigm shift, greatly increasing equipment longevity and operational efficiency. Autonomous maintenance systems enable machines to self-monitor, anticipate wear and tear, plan repairs, and acquire replacement components on their own by combining IoT sensors, machine learning algorithms, and augmented diagnostics. By reducing unscheduled downtime, streamlining maintenance plans, and guaranteeing uninterrupted manufacturing processes, this clever strategy saves a lot of money and improves operations. Up to 30% less money is spent on maintenance when autonomous maintenance solutions are used. Predictive analytics helps achieve this by lowering the frequency of equipment breakdowns and minimizing needless maintenance. By servicing or replacing parts before they reach critical wear levels, proactive monitoring and prompt interventions can extend the lifespan of machinery by 20–40%. Increased Uptime Businesses can operate almost continuously and enhance productivity by removing unscheduled downtime. The manufacturing industry have different estimates downtime losses (like automotive vs semiconductor), the financial advantages of autonomous maintenance. It is anticipated that the global market for predictive maintenance, a key component of autonomous systems, would expand at a CAGR of 25% between 2023 and 2030, reaching a valuation of USD 23 billion. One of the world's top automakers has integrated autonomous maintenance into all its production lines. The company saved USD 12 million annually by reducing unplanned downtime by 40% using IoT sensors and AI algorithms. Predictive maintenance decreased capital expenditures associated with equipment replacement by prolonging the life of vital gear. By guaranteeing zero downtime and maximizing equipment performance, autonomous maintenance solutions are transforming industrial processes. These technologies lower costs, improve sustainability, and help achieve Industry 4.0 objectives by combining IoT, machine learning, and augmented diagnostics. Their potential to promote sustainable industrial practices and aid in global climate action is further highlighted by their alignment with the SDGs. In addition to future-proofing industries, autonomous maintenance positions them as leaders in sustainability and innovation [80–83].

6.1 Working Mechanism and Implication Techniques

Devices with IoT-enabled sensors continuously collect information on vital factors including operating speed, temperature, vibration, and pressure. As the "eyes and ears" of the system, these sensors spot performance irregularities or early indications of wear and tear. To determine when a component is likely to fail, machine learning models use both historical and current data. These models facilitate predictive maintenance by identifying patterns in operational data, which enables the scheduling of repairs prior to failure. For this, methods like decision trees, neural networks, and anomaly detection algorithms are frequently used. Augmented diagnostics give maintenance teams comprehensive information on the health of their machines by utilizing cutting-edge visualization technologies like digital twins and augmented reality. With the use of these instruments, professionals can effectively perform repairs without having to disassemble equipment by visualizing inside components and spotting possible problems. Autonomous maintenance systems can use integrated supply chain management platforms to automatically order replacement parts as soon as a possible problem is identified. Even simple repair jobs can be aided by robotic arms or collaborative robots, which minimizes the need for human interaction and guarantees quicker turnaround times [84, 85].

The following actions can be taken by industries looking to implement autonomous maintenance systems. Install IoT-enabled sensors on vital pieces of equipment to gather data on operational characteristics in real time. Make that the sensors can measure the necessary metrics and are compatible with the equipment that is already in place. Make use of cloud computing or edge computing platforms to create a strong infrastructure for data collecting and storage. Edge computing is very helpful for local data processing, which lowers decision-making latency. Work with AI specialists to create machine learning models that are suited to the particular requirements of the apparatus. To train these models and gradually increase their forecast accuracy, use previous maintenance data. To model performance under different circumstances, create digital copies of real equipment. Digital twins offer a secure setting for evaluating maintenance plans and forecasting results. Inventory control and parts procurement can be automated with enterprise resource planning software. To guarantee that new parts are ordered as soon as a need is identified, integrate this system with Internet of Things sensors. Give maintenance crews AR resources to improve diagnoses and repair effectiveness. Repairs can be completed more quickly and precisely because to AR devices' ability to superimpose visual instructions on gear. To adjust to modifications in production procedures or equipment conditions, update machine learning models and IoT systems on a regular basis. Create feedback loops to ensure ongoing development [86–89].

6.2 Alignment with SDGs

Autonomous maintenance systems support efficient resource use, climate action, and sustainable industrial practices, all of which are SDGs. SDG 9 can be aligned with Autonomous systems as these systems improve the robustness and effectiveness of industrial infrastructure by updating maintenance procedures. This encourages creativity and helps build sustainable industries. SDG 12 is linked as these systems extending the lifespan of machinery and lowering equipment failures, proactive maintenance reduces waste production. It guarantees more effective use of resources like lubricants and spare components. SDG 13 is also aligned as energy use and greenhouse gas emissions are decreased by operating equipment efficiently. Machines that are well maintained use 15 to 25% less energy than those that are not, which helps to reduce carbon emissions [90, 91].

7 Customized Additive Manufacturing (AM) with Precision and Efficiency

AM, especially 3D printing, is transforming the industrial landscape by making it possible to create highly customized consumer goods. By utilizing cutting-edge technologies like graphene-based composite materials, AI-driven design optimization, and multi-material printing, this creative strategy enables companies to effectively satisfy client needs. AM is promoting industrial innovation, cutting waste, and advancing the SDGs of the UN through everything from customized medical gadgets to aerospace components. 3D printing is being used more successfully in the aviation and aerospace sectors. In aircraft engineering, sturdy, lightweight parts are essential. Composites-based on graphene can reduce part weight by up to 40%, increasing fuel economy and lowering pollutants. Implants and prostheses customized for each patient are made using AM (medical device applications). Custom titanium implants, for example, can now be created with 99% accuracy, which speeds up recovery and cuts down on operating times. To satisfy individual preferences in terms of appearance, size, and usefulness, AM is used in the consumer sector to make customized jewelry, shoes, and wearable technology. Using titanium alloy powders, a medical equipment manufacturer used AM to create implants tailored to each patient. The business cut costs by 40% and manufacturing times by 60% by utilizing AI for design optimization. While meeting strict regulatory requirements, these implants showed better fit and functionality, improving patient outcomes. At a CAGR of 20.8% from 2023, the worldwide AM market is expected to reach over USD 120 billion by 2030. Due to AM's capacity to cut lead times and production costs, it is becoming more and more popular across industries, which is driving its expansion. Businesses that use AM, for instance, estimate a 30–50% decrease in prototyping costs, greatly increasing their competitive advantage. Because it allows for highly customized customer experiences, AM is revolutionizing several industries. This technology guarantees effective, sustainable production that is in line with the SDGs by utilizing automation, artificial intelligence, and sophisticated materials. The use of AM in a variety of industries is set to upend conventional manufacturing paradigms as the market expands, promoting sustainability, and innovation globally [92–94].

7.1 AM Customization Techniques and Technology Implications

CAD (Computer Aided Design) software is used to build digital models, which are the first step in AM. AI algorithms combined with sophisticated tools enable design optimization depending on particular client requirements, including improved durability or ergonomic designs. AM uses materials including polymers, metals, ceramics, graphene-based composites and many high strength light weight polymers and specialized materials to construct items layer by layer, in contrast to conventional subtractive techniques. These composites are especially beneficial because of their remarkable electrical, thermal, and strength to weight ratios. Complex designs are made possible by the most recent printers' ability to smoothly combine several materials into a single result. The mechanical characteristics of human limbs can be replicated by prosthesis made of both stiff and flexible materials. Mass customization is made possible by technologies like as cloud-based printing platforms. Manufacturers can receive consumer specifications in real time, build the things, and send them directly by linking to online databases [95, 96].

Utilize cutting edge materials for strength and weight reduction, such as borophene, graphene enhanced nylon, diamond reinforced composites, carbon fiber reinforced polymers, continuous fiber composites and high performance metal alloys. Try using biodegradable polymers in packaging and medical equipment for environmentally friendly uses. To optimize shapes for minimal material, use while preserving structural integrity, employ generative design software. Incorporate machine learning techniques to anticipate and fix printing flaws. When creating intricate designs, use multi-material printers. For large-scale metal components used in the automotive and aerospace industries, employ direct energy deposition (DED). For improved quality control, install IoT sensors in printers to track temperature, humidity, and material flow rates. For post-processing duties like cleaning and assembly, use robotic arms. Utilize cloud-based systems to oversee dispersed 3D printing projects. Delivery times are shortened, and logistics expenses are decreased. Establish material recycling facilities (close loop etc.) on the property to recycle unused polymers or powders, supporting a circular economy [97–99].

7.2 SDGs Alignment

SDG 9 can be integrated by enabling decentralized manufacturing hubs and offering flexible production techniques, AM promotes industrial innovation. This encourages robust and sustainable infrastructure by lowering dependency on conventional centralized factories. SDG 12 can be aligned as AM reduces waste by optimizing the usage of materials. AM lowers material waste to less than 5%, whereas traditional methods can result in 30% waste. Additionally, customization avoids overproduction, which supports sustainable purchasing habits. SDG 13 can be aligned as lower carbon emissions are a result of energy efficient procedures and a decrease in material consumption. AM components lower fuel usage in the aerospace sector, resulting in an annual 16% reduction in emissions per aircraft [100, 101].

8 Global Collaboration through Digital Platforms to Motivate Cross-Border Innovation

The way engineers, designers, and industries work together internationally is changing thanks to digital collaboration platforms that are geared up by cloud computing, real-time data exchange, and cutting-edge automation technology. These platforms facilitate smooth communication and cross-border co-development, which speeds up innovation cycles and improves manufacturing efficiency. Industries can achieve major economic and operational benefits while also aligning with important SDGs, such as 9, 12, and 17, by utilizing such technologies. Cross-border collaboration on digital platforms cuts development costs by 15 to 25% and new product time-to-market by 20–30%. By eliminating the requirement for physical prototypes, virtual prototyping saves energy and material waste from traditional testing methods by 30–50%. By giving SMEs access to global resources and experience, digital platforms allow them to compete in global marketplaces. Through increased innovation, global GDP is predicted to increase by USD 1.5 trillion annually. By eliminating geographical barriers and promoting smooth cross-border communication, digital collaboration platforms are transforming global innovation. These technologies facilitate real-time design, development, and data exchange, which speeds up time to market, lowers costs, and encourages environmentally friendly business practices. Industries can fully utilize the promise of digital platforms to create a smarter, greener, and more connected future using cloud computing, AI, IoT, and strong cybersecurity [102–104].

8.1 Alignment with SDGs

In addition to achieving operational efficiency, industries' alignment with SDGs 9, 12, and 17 guarantees that they also support a cooperative, sustainable, and inclusive global economy. SDG 9 because digital platforms facilitate collaborative design and effective infrastructure planning, they support resilient and inventive industrial ecosystems. Virtual prototyping helps reduce material waste, which promotes sustainable production methods, in line with SDG 12. SDG 17 can be aligned by encouraging global co-operation, these platforms make it possible to pool resources, ideas, and information to effectively solve global issues [104–106].

8.2 Automation and the Technological Framework

The foundation for digital collaboration is provided by cloud platforms (cloud computing) like AWS, Google Cloud, and Microsoft Azure for real-time monitoring. These solutions guarantee that all stakeholders are working with the most recent information by enabling users to upload, view, and modify huge files—like digital twins or CAD designs, in real-time. American engineers and Japanese manufacturers can work together to create a virtual prototype of an automotive part. Version control and real-time updates can save up to 30% of development time by minimizing mistakes and misunderstandings. Teams can collaborate on design development, testing, and optimization without requiring physical prototypes thanks to digital twins, which are virtual copies of real systems or products. Under a variety of circumstances, these simulations can evaluate functionality, spot inefficiencies, and forecast performance. Businesses that use digital twins report a 30% decrease in R & D expenses and a 20–25% reduction in time to market. AI improves digital platforms by offering intelligent project management, automated scheduling, and natural language translation capabilities. AI programs examine project workflows and make recommendations for improvements to guarantee that jobs are finished quickly. By prioritizing activities according to deadlines and resource availability, AI powered platforms (such as Monday.com and Trello) increase productivity by 15%–20%. Data from production lines, product testing facilities, and field applications is gathered and shared by IoT devices that are connected to digital platforms. Teams that are spread out geographically can use this data to keep an eye on procedures, fix problems, and adjust without having to be physically present. In smart manufacturing, design teams in India can work together to improve product specs by receiving operational data from IoT-connected machinery in Germany [107–110].

8.3 Technology Implication Techniques

For real-time design collaboration and product lifecycle management, use systems such as Siemens Teamcenter, SolidWorks, or Autodesk Fusion 360. Set up safe cloud computing systems for sharing and storing data. International data compliance may be guaranteed with hybrid cloud solutions. Use AI-based project management tools to forecast project timeframes, assign resources, and expedite processes. Set up IoT sensors on manufacturing lines to supply real-time data that facilitates collaborative iterative design changes. To protect sensitive intellectual property during international collaboration, put strong security standards in place, such as multi-factor authentication, blockchain for data integrity, and end-to-end encryption [111–113]. The most advanced IoT-enabled asset tracking solutions are included in Table 6.1 (summery of this complete chapter), guaranteeing compliance with Industry 5.0 and Industry 6.0 guidelines for maximum productivity, sustainability, and climate resilience.

Table 6.1 IoT-enabled asset tracking systems technologies

Name	Technology and advances	Metrics
RFID	EM field identification with UHF Gen2v3 tags; quantum-encrypted RFID (NIST PQC standards); LoRa backscatter tags (300 m range)	Passive: 15 m; active: 300 m; market: USD 32.1B (CAGR 18.2%)
GPS	Multi-constellation (GPS IIIF, Galileo HAS); AI corrected ionospheric delay; RTK/PPK for cm-accuracy	20 cm RTK accuracy; 52%↓ shipment delays; Galileo HAS free service live
Edge analytics	Time sensitive networking (IEEE 802.1Qbv); neuromorphic pre-processing (Intel Loihi 3); federated learning at edge	Latency ↓82% (200 μs); processing ↑70% (vs. 2023)
Industry 6.0 Core	Quantum-digital twins (128-qubit hybrid); autonomous cyber-physical swarms; self-healing blockchains (zk-SNARKs)	Efficiency ↑68%; zero trust security (99.99% uptime)
Automotive	EV battery SOH prediction (±2% error); LiDAR SLAM integration; V2X collision avoidance	EV adoption ↑45%; thermal runaway prevention: 99.7%
Chemical	Cavity ring-down spectroscopy (ppb VOC detection); reaction optimization AI ; Material discovery (GNoME models); autonomous scrubbers; BO/ BO-BOF, (GP-BO, TPE, SMAC, PHOENICS, Chimera-BO); RL/ DRL (DDQN, PPO, A3C, AlphaZaero-Chem, Chem RL-GP); NLP + GNN hybrid (ChemGPT, MolGPT, MegaMolBART, IBM-RXN); DeepMind AF; Robotics (CLR, ACLS, SAS)	Emissions ↓45%; yield ↑28%
Textile	Graphene oxide based humidity sensors (Industry 6.0 emerging technologies); closed-loop dyeing AI; aquaporin-based water recycling	Water ↓65%; energy ↓40%
Water/energy	Atmospheric water harvesters (MOF-303); AI-optimized desalination; solid-state transformer grids	Energy efficiency ↑52%; water waste ↓73%

(continued)

Table 6.1 (continued)

Name	Technology and advances	Metrics
Climate resilience	Self-regulating exoskeletons (SMAs); sub-terahertz flood sensors; radiation-hardened edge nodes	Downtime ↓48%; disaster recovery ↑40%
Healthcare	UWB implant tracking (±5 cm); multimodal patient AI (GPT-5 med or beyond in future); predictive sepsis models	Equipment failures ↓60%; readmissions ↓25%
Agriculture	Hyperspectral soil NPK mapping; phytoplankton-based drought prediction; autonomous nanofertilizer drones	Yield ↑35%; water ↓55%
Transportation	Digital twin rail wear modeling; CFD-optimized aircraft; platooning V2V control	Maintenance costs ↓31%; uptime ↑42%
Cloud framework	Photonic tensor clouds (Lightmatter's chip in Envise); satellite edge inference (AWS Kuiper); 5G NTN global coverage	Retrieval time ↓52%; processing ↑80%
Cloud storage	DNA data archival future potential (1 EB/g); homomorphic encryption (FHE 2.0); topological quantum storage	Accessibility ↑90%; hardware footprints ↓70%
Collaborative AI	Generative digital twins (NVIDIA Omniverse); haptic AR workspaces; Global R&D knowledge graphs	Prototyping ↓50%; collaboration ↑60%
SDG integration	Carbon-negative data centers (direct air capture); AI-optimized circular economies; ESG blockchain ledgers	Energy waste ↓41%; sustainable prod. ↑55%
Industry 5.0/6.0	Human digital avatars; quantum control systems (Honeywell system H1); self-evolving factories	Production speed ↑45%; costs ↓28%
Predictive AI	Topological anomaly detection (data analysis etc.); physics-informed ML (NVIDIA modulus); self-supervised failure forecasting	Maintenance costs ↓44%; lifespan ↑35%
Additive Manufacturing	4D to 6D printed metastructures; AI topology optimization (nTop 5.0); atomic layer AM (≈5 μm resolution), volumetric AM	Waste ↓50%; efficiency ↑68%
Global automation	Swarms with collective AI; cross-border digital product passports; quantum-secured IIoT	Efficiency ↑38%; R&D costs ↓30%
Blockchain	Zk-rollup IoT ledgers; post-quantum cryptography (CRYSTALS-Kyber); decentralized identity (DID)	Security ↑45%; transparency ↑60%

9 Challenges Associated with Industry 6.0 Implication and Future Recommendations

Table 6.2 shows the key challenges that required to be addressed for the implication of Industry 6.0 and there are also some suggestions and future recommendations. The roadmap for accelerated Industry 6.0 adoption is structured into three sequential phases, immediate, mid-term, and long-term, each guided by technology convergence, regulatory foresight, and economic optimization. The approach emphasizes both rapid implementation and measurable outcomes across cybersecurity, sustainability, and return on investment, with all milestones articulated in

Table 6.2 Industry 6.0 implementation challenges & advanced solutions

Challenge category	Technical problems	Advanced solutions	Implementation strategies	KPIs
Security and cyber physical risks	• Quantum computing threats to encryption • AI-driven cyberattacks on autonomous systems • Physical sabotage of IoT edge nodes	• Post-quantum cryptography (PQC): CRYSTALS-Kyber for key exchange; zk-SNARKs for blockchain integrity • AI sentinel networks: Neuromorphic intrusion detection (Intel Loihi 3) with 99.999% threat prediction • Radiation-hardened edge nodes: Sub-terahertz flood sensors with self-healing firmware	• Mandate NIST PQC standards in supply chains • Deploy AI sentinels at Tier-1 critical infrastructure • Harden 30% of edge devices	99.99% system uptime; zero successful breaches
Interoperability and legacy integration	• Heterogeneous protocol fragmentation (OPC-UA vs. MQTT) • Legacy machinery with proprietary interfaces • Data silos in multi-vendor ecosystems	• Unified semantic layer: Digital twin-based ontology mapping (NVIDIA Omniverse) • Autonomous cyber-physical swarms: Self-organizing IIoT agents with adaptive API translation • Quantum-control middleware: Honeywell system H1 for legacy-to-quantum bridging	• Adopt IEEE P3119 interoperability framework • retrofit 50% legacy assets with quantum-edge controllers • Establish cross-industry data trusts	100% cross platform data liquidity; 40% reduction in integration costs

(continued)

Table 6.2 (continued)

Challenge category	Technical problems	Advanced solutions	Implementation strategies	KPIs
Real-time processing and latency	• Sub-ms decision-making in distributed systems • Bandwidth congestion in massive IoT deployments • Edge AI computational limits	• Photonic tensor processors: Lightmatter's chips in Envise for 100 TOPS/W efficiency • 5G NTN satellite edge: AWS Kuiper LEO constellation for global <50 ms latency • Federated learning at edge: Distributed model training with zero raw data exchange	• Deploy photonic accelerators in 20% of smart factories • Integrate satellite edge networks for remote assets • Implement FL for 70% of predictive maintenance systems	Latency ≤200 μs; 80% reduction in cloud data transfer
Sustainability and energy	• AI/quantum computing energy intensity • Resource-intensive additive manufacturing • Scope 3 emissions in global supply chains	• Carbon-negative data Centers: Direct air capture integration (Climeworks DAC) • 4D metastructures: AI-topology optimized parts (nTopology nTop 5.0) with 50% material reduction • ESG Blockchain ledgers: IBM Blockchain for real-time carbon tracking (SDG 13 alignment)	• Power 100% hyperscale clouds with fusion/SMR • Mandate life-cycle assessment for all AM components • Enforce blockchain-based carbon accounting in procurement	Net zero operations; 65% reduction in scope 3 emissions
Human AI collaboration	• Skill gaps in quantum AI systems• ethical AI governance gaps • Workforce resistance to autonomous systems	• Human Digital Avatars: AR/ VR upskilling platforms with GPT 5 Med knowledge transfer • Ethical AI Frameworks: ISO/ IEC 42001 certified constitutional AI • Cobots with emotional AI	• Train 1 M workers via avatars • Establish AI ethics boards with veto power over autonomous decisions • Phase-in cobots for 40% of hazardous tasks	90% workforce AI fluency; 0 critical ethics violations

practical timelines based on months and years. The success of Industry 6.0 hinges on the velocity of technological convergence, particularly among 128 qubit quantum systems, photonic processors, and ethically governed AI ecosystems. A delay in establishing global interoperability standards could result in forfeiting up to USD 3.1 trillion in potential GDP gains. Therefore, the roadmap strongly emphasizes the establishment of sovereign industry clouds to mitigate geopolitical vulnerabilities in critical infrastructure and ensure resilient, scalable, and secure deployment of Industry 6.0 technologies worldwide. In the initial immediate actions i.e. between 0 to 18 months, critical infrastructure operators should pilot quantum resistant security frameworks by deploying CRYSTALS Kyber algorithms across high-risk cyber physical systems such as smart grid controllers and energy management hubs. Edge fog cloud hybrid architectures should be implemented through solutions like AWS Outposts, enabling localized processing in climate-sensitive regions prone to network disruption. To ensure regulatory alignment, organizations should partner with AI regulatory sandboxes in collaboration with EU and U.S. authorities to validate compliance for Industry 6.0 applications under evolving ethical and legal standards. In mid-term scaling (18–48 months), over the following 2–4 years, industries should scale implementations by integrating industrial metaverse platforms like Siemens Xcelerator, which link physical production systems with virtual environments for real time, model-based optimization. Establishing closed loop material ecosystems using graphene-enhanced 3D printing and blockchain tracked recycling ensures full traceability and supports circular manufacturing models. To bridge the global skill gap, digital talent exchange platforms must be launched, enabling cross-border collaboration in specialized quantum and AI domains, effectively using global expertise while addressing regional workforce deficits. Performance across all phases must be measured through robust metrics. Zero trust architecture should demonstrate ≥99.99% resilience in cybersecurity penetration testing. Sustainability metrics must align with IPCC AR7 decarbonization targets, achievable through AI optimized circular economies and closed loop systems. From an economic standpoint, predictive maintenance systems like those implemented in GE Digital case studies, should yield a 25–40% reduction in total cost of ownership over 5 years. In long-term transformation (48–72 months), beyond the 4 year mark, facilities should progress toward self-evolving factories, driven by generative digital twins and autonomous AI agents capable of real time process redesign and adaptive control. To ensure transparency and trust in AI systems, the adoption of neuro symbolic AI governance frameworks is essential, facilitating explainable, auditable decision-making in high-risk environments. Industries should invest in photonic quantum cloud infrastructure, enabling the formation of global entanglement networks for instantaneous, secure, and optimized supply chain orchestration.

References

1. Saritha, K., et al.: IoT enabled hospital asset tracking using advanced interdisciplinary approaches. E3S Web Conf. **507**, 01007 (2024)
2. Arinze, S.N., et al.: Implementation of radio frequency identification technology for a secure and intelligent shopping cart. Bull. Electr. Eng. Inf. **14**(1), 143–152 (2024)
3. Kosmidis, M., et al.: Towards warehouse 5.0: a framework of human-Centered technologies. In: 2024 IEEE International Conference on Imaging Systems and Techniques (IST). IEEE, New York (2024)
4. Maharjan, A., et al.: AWS server less luggage tracking system integrating IOT. In: 2024 2nd international conference on networking, embedded and wireless systems (ICNEWS). IEEE, New York (2024)
5. Omali, T.U., Akpata, S.B.: Global positioning system technology: theory and the methods of its application. Int. J. Sci. Res. in Multidisciplinary Studies. **10**(5), 70–76 (2024)
6. Bajdor, P.: 16 IoT and sustainable. In: Digital Sustainability: Navigating Entrepreneurship in the Information Age, p. 241. CRC Press, Boca Raton (2024)
7. Verma, R., Singh, B.K., Zahidi, F.: Management of GPS tracking systems in transportation. In: Intelligent Transportation System and Advanced Technology, pp. 251–263. Springer, Cham (2024)
8. Yao, A., et al.: A privacy-preserving location data collection framework for intelligent systems in edge computing. Ad Hoc Netw. **161**, 103532 (2024)
9. Ayoola, V.B., et al.: IoT-driven smart warehouses with computer vision for enhancing inventory accuracy and reducing discrepancies in automated systems. IRE J. **8**, 176 (2024)
10. Khan, S.I., Ray, B.R., Karmakar, N.C.: Rfid localization in construction with IoT and security integration. Autom. Constr. **159**, 105249 (2024)
11. Abba Ari, A.A., et al.: Data collection in IoT networks: architecture, solutions, protocols and challenges. In: IET Wireless Sensor Systems. Springer, Cham (2024)
12. Sap, B.M.A.U., Gupta, P.: Digital transformation of SAP supply chain processes
13. Olaoye, G., Henry, E.: Enhancing Supply Chain Visibility and Transparency through AI-Powered Tracking Systems. EasyChair, Stockport (2024)
14. Zaidi, S.A.H., Khan, S.A., Chaabane, A.: Unlocking the potential of digital twins in supply chains: a systematic review. Supply Chain Anal. **7**, 100075 (2024)
15. Sah, B.P., et al.: AI-driven IoT and Blockchain integration in industry 5.0 a systematic review of supply chain transformation. Innovatech Eng. J. **1**(1), 99–116 (2024)
16. Okokpujie, I.P., Tartibu, L.K.: Study of the economic viability of internet of things (IoTs) in additive and advanced manufacturing: a comprehensive review. Progr. Add. Manuf. **10**, 3175–3194 (2024)
17. Vashishth, T.K., et al.: Environmental sustainability and carbon footprint reduction through artificial intelligence-enabled energy management in electric vehicles. In: Artificial Intelligence-Empowered Modern Electric Vehicles in Smart Grid Systems, pp. 477–502. Elsevier, Amsterdam (2024)
18. Yan, Y., et al.: Carbon dioxide-focused greenhouse gas emissions from petrochemical plants and associated industries: critical overview, recent advances and future prospects of mitigation strategies. Process Saf. Environ. Prot. **188**, 406–421 (2024)
19. Griffin, A., et al.: Design and application of joule heating processes for decarbonized chemical and advanced material synthesis. Ind. Eng. Chem. Res. **63**(45), 19398–19417 (2024)
20. Behera, S.A., et al.: Current trends on advancement in smart textile device engineering. Adv. Sustain. Syst. **8**, 2400344 (2024)
21. Majeed, H., Iftikhar, T.: Ecofriendly reactive printing of cellulosic fabric with sustainable novel techniques. Cellulose. **31(11)**, 7067-7081 (2024)
22. Hauashdh, A., et al.: An integrated framework for sustainable and efficient building maintenance operations aligning with climate change, SDGs, and emerging technology. Results Eng. **21**, 101822 (2024)

23. Abaee, M., Saeedi, M., Taherdoost, H.: IoT and cloud computing for sustainable development goals in industry 4.0: a bibliometric analysis. In: Reshaping Environmental Science Through Machine Learning and IoT, pp. 98–118. IGI Global, London (2024)
24. Gregório, C., et al.: From wristbands to implants: the transformative role of wearables in heart failure care. Healthcare. **12**(24), 2572 (2024)
25. Bandi, M., et al.: Predictive analytics in healthcare: enhancing patient outcomes through data-driven forecasting and decision-making. Int. Numer. J. Machine Learn. Robots. **8**(8), 1–20 (2024)
26. Krause, P.: Insertable cardiac monitoring. In: Handbook of Cardiac Anatomy, Physiology, and Devices, pp. 407–430. Springer, Cham (2024)
27. Bhambri, P.: Artificial intelligence enabled internet of medical things for enhanced healthcare systems. In: Smart Healthcare Systems, pp. 1–17. CRC Press, Boca Raton (2024)
28. Singh, M., et al.: Future professions in agriculture, medicine, education, fitness, research and development, transport, and communication. Top. Artif. Intell. Appl. Ind. **400**, 181–202 (2024)
29. Kadrolli, V., Kalnoor, G.: IoT and smart sensors for remote sensing healthcare and agriculture applications. Remote Sens. Earth Syst. Sci. **7**, 1–15 (2024)
30. Abdel-Basset, M., Hawash, H., Abdel-Fatah, L.: Artificial Intelligence and Internet of Things in Smart Farming. CRC Press, Boca Raton (2024)
31. Mittal, V., et al.: IoT-enabled predictive maintenance for sustainable transportation fleets. E3S WEB of Conf. **511**, 01012 (2024)
32. Drozhzhyn, O.: IOT in container shipping industry: applications and examples of solutions. Syst. Technol. **68**(2), 112–120 (2024)
33. Thavi, R., et al.: Role of cloud computing technology in the education sector. J. Eng. Des. Technol. **22**(1), 182–213 (2024)
34. Mathur, P.: Cloud computing infrastructure, platforms, and software for scientific research. In: High Performance Computing in Biomimetics: Modeling, Architecture and Applications, pp. 89–127. Springer, Cham (2024)
35. Mistry, H.K., et al.: Cloud Computing Techniques. Academic Guru Publishing House, New York (2024)
36. Maxwell, R.: Azure arc–history and horizons. In: Azure Arc Systems Management: Governance and Administration of Multi-Cloud and Hybrid IT Estates, pp. 251–277. Springer, Cham (2024)
37. Murugeswari, K., et al.: Deep learning for smart healthcare
38. Reynolds, S., Anderson, L.: The role of artificial intelligence in enhancing human-centered computing in 2024. AlgoVista. **2**(1), 3114 (2024)
39. Akhtar, I., et al.: The role of natural products in COVID-19. In: The COVID-19 Pandemic, pp. 393–445. Apple Academic Press, New York (2022)
40. Marcus, G.F.: Taming Silicon Valley: How We Can Ensure That AI Works for Us. MIT Press, Cambridge (2024)
41. Nolan, A.: Artificial Intelligence in Science: Challenges, Opportunities and the Future of Research. OECD, Paris (2024)
42. Naveed, N., et al.: The global impact of COVID-19: a comprehensive analysis of its effects on various aspects of life. Toxicol. Res. **13**(2), tfae045 (2024)
43. Qaiser, H., et al.: COVID-19 pandemic and vaccines. In: Alternative Medicine Interventions for COVID-19, pp. 205–235. Springer, Cham (2021)
44. Chen, K., et al.: Fogros2-ls: a location-independent fog robotics framework for latency sensitive ros2 applications. In: 2024 IEEE International Conference on Robotics and Automation (ICRA). IEEE, New York (2024)
45. Mooney, J.G., Williams, M.L.: Enabling technologies of the fourth industrial revolution. In: Handbook of Research on Strategic Leadership in the Fourth Industrial Revolution, pp. 35–61. Edward Elgar Publishing, Cheltenham (2024)
46. Капков, Р., Тятюшкина, О., Ульянов, С.: Quantum software engineering and industry 4.0 as the platform for intelligent control of robotic soci-otechnical systems in industry 5.0/6.0. Системный анализ в науке и образовании. **3**, 97–130 (2024)

47. Haldorai, A.: Advancements and applications of quantum computing in robotics. J. Comput. Nat. Sci. **24**, 53–63 (2024)
48. Mondal, A.S., et al.: Comparative analysis of object-based big data storage systems on architectures and services: a recent survey. J. Inst. Eng. Ser. B. (2024). https://doi.org/10.1007/s40031-023-00983-z
49. Arputharaj, J.V., et al.: Advancing disaster management in industry 6.0: the role of DNA sequencing sensors and quantum computing in hyperspectral image analysis. In: The Rise of Quantum Computing in Industry 6.0 Towards Sustainability, pp. 155–176. Springer, Cham (2024)
50. Stonebraker, M., Pavlo, A.: What goes around comes around... And around. ACM Sigmod. Record. **53**(2), 21–37 (2024)
51. Patel, D., Saliya, D., Vaghasiya, P.: Big Data Fundamentals: A Comprehensive Guide to Data Management and Analysis. Academic Guru Publishing House, New York (2024)
52. Praveen, R.: Data Engineering for Modern Applications. Addition Publishing House, Bhopal (2024)
53. Ali, N.A.: Predictive Analytics for the Modern Enterprise. O'Reilly Media, Inc., Sebastopol (2024)
54. Kurniawan, S.D., et al.: Big Data: Mengenal Big Data & Implementasinya di Berbagai Bidang. Sonpedia Publishing, Kota Jambi (2024)
55. Nakazato, J., et al.: Webrtc over 5 g: a study of remote collaboration qos in mobile environment. J. Netw. Syst. Manag. **32**(1), 1 (2024)
56. Mahmoud, H., Abozariba, R.: A systematic review on WebRTC for potential applications and challenges beyond audio video streaming. Multimed. Tools Appl. **84**, 2909–2946 (2024)
57. Solanki, J., Senapati, B.: Enhancing real-time multilingual communication in virtual meetings through optimized WebRTC broadcasting. In: 2024 IEEE International Conference and expo on Real Time Communications at IIT (RTC), p. 2024. IEEE, New York
58. Yilmaz, S., et al.: A PMBOK-based construction cost management framework for BIM integration in construction projects. Int. J. Constr. Manag. **25**, 1–15 (2024)
59. Dianati, S.: From OER to MOOCs: an evolutionary analysis and its neoliberal Context. In: The Commercialisation of Massive Open Online Courses: Reading Ideologies in between the Lines, pp. 9–27. Springer, Cham (2024)
60. Wang, J., et al.: Designing cloud servers for lower carbon. In: 2024 ACM/IEEE 51st Annual International Symposium on Computer Architecture (ISCA). IEEE, New York (2024)
61. Pelton, J.N., Madry, S.: Space systems, quantum computers, big data and sustainability: new tools for the United Nations sustainable development goals. In: Artificial Intelligence for Space: AI4SPACE, pp. 53–104. CRC Press, Boca Raton (2024)
62. Rastogi, D., et al.: IoT technology enables sophisticated energy management in smart factory. Cyber Phys. Energy Syst. **24**, 147–181 (2024)
63. Pershaanaa, M., et al.: Keystones of green smart city—framework, e-waste, and their impact on the environment—a review. Ionics. **30**(3), 1267–1289 (2024)
64. Shafik, W.: Artificial intelligence and internet of things roles in sustainable next-generation manufacturing: an overview of emerging trends in industry 6.0. Sustain. Innov. Ind. **60**, 207–239 (2024)
65. Thirupathi, L., et al.: Quantum computing and AI: synergizing for sustainable disaster management in industry 6.0. In: The Rise of Quantum Computing in Industry 6.0 Towards Sustainability, pp. 35–51. Springer, Cham (2024)
66. Smajdorová, T.: Journey to quality 6.0: evolution of quality management across industrial revolutions. Trends Comput. Sci. Inf. Technol. **9**(2), 63–70 (2024)
67. Cohan, P.: Generative AI cloud platforms. In: Brain Rush: how to Invest and Compete in the Real World of Generative AI, pp. 167–235. Springer, Cham (2024)
68. Patel, N.R.: Supply chain management in the digital age for industry 4.0. In: Industry 4.0, Smart Manufacturing, and Industrial Engineering, pp. 179–195. CRC Press, Boca Raton (2024)
69. Assunção, W.K., et al.: Contemporary software modernization: strategies, driving forces, and research opportunities. ACM Trans. Softw. Eng. Methodol. **34(5)**, 1–35 (2025)

70. van der Vlist, F., Helmond, A., Ferrari, F.: Big AI: cloud infrastructure dependence and the industrialisation of artificial intelligence. Big Data Soc. **11**(1), 20539517241232630 (2024)
71. Alzoubi, Y.I., Mishra, A.: Green artificial intelligence initiatives: potentials and challenges. J. Clean. Prod. **468**, 143090 (2024)
72. Younus, A.H., et al.: Web-based and cloud-computing influences on resource utilization optimization for sustainable enterprise systems with AI. IoT Secur. Manage. **12**, 13 (2024)
73. Mallo, S.F., et al.: A review on feasibility of web technology and cloud computing for sustainable ES: leveraging AI, IoT, and security for green operations. J. Inf. Technol. Inf. **3**(2), 246–270 (2024)
74. Johnson, O.B., et al.: Designing a comprehensive cloud migration framework for high-revenue financial services: a case study on efficiency and cost management. Open Access Res. J. Sci. Technol. **12**(2), 58–69 (2024)
75. Kang, P.S., et al.: Role of big data in customer-centric service-based supply chains. In: Service 4.0: Technology-Enabled Customer-Centric Supply Chains, pp. 11–33. Springer, Cham (2024)
76. Kocaoglu, B.: Data Science, in Logistics Information Systems: Digital Transformation and Supply Chain Applications in the 4.0 Era, pp. 235–285. Springer, Cham (2024)
77. Korada, L.: Digital transformation balanced with sustainability goals. J. Eng. Appl. Sci. Technol. **6**(5), 2–7 (2024)
78. Raza, M., et al.: Carbon footprint reduction in cloud computing: Best practices and emerging trends
79. Patel, P., Gregersen, T., Anderson, T.: An agile pathway towards carbon-aware clouds. ACM SIGENERGY Energy Inf. Rev. **4**(3), 10–17 (2024)
80. Hoffmann, M.A., Lasch, R.: Tackling industrial downtimes with artificial intelligence in data-driven maintenance. ACM Comput. Surv. **56**(4), 1–33 (2023)
81. Baroud, S.Y., Yahaya, N.A., Elzamly, A.M.: Cutting-edge AI approaches with MAS for PdM in industry 4.0: challenges and future directions. J. Appl. Data Sci. **5**(2), 455–473 (2024)
82. George, A.S., The Fourth Industrial Revolution: a Primer on Industry 4.0 and its Transformative Impact. Partners Univ. Innov. Res. Publ., 2024. 2(1): p. 16–40
83. Omol, E.J.: Organizational digital transformation: from evolution to future trends. Dig. Transf. Soc. **3**(3), 240–256 (2024)
84. Devi, E.R., et al.: AI models for predictive maintenance. In: Data Analytics and Artificial Intelligence for Predictive Maintenance in Smart Manufacturing, pp. 69–94. CRC Press, Boca Raton (2024)
85. Leontaris, L., et al.: A blockchain-enabled deep residual architecture for accountable, in-situ quality control in industry 4.0 with minimal latency. Comput. Ind. **149**, 103919 (2023)
86. Saranya, S.M., et al.: Industry 4.0: the role of industrial IoT, big data. In: AR/VR, and Blockchain in the Digital Transformation, in Smart Computing Techniques in Industrial IoT, pp. 11–39. Springer, Cham (2024)
87. Kamdjou, H.M., et al.: Resource-constrained eXtended reality operated with digital twin in industrial internet of things. IEEE Open J. Commun. Soc. **5**, 928–950 (2024)
88. Javaid, M., Haleem, A., Suman, R.: Digital twin applications toward industry 4.0: a review. Cognit. Robot. **3**, 71–92 (2023)
89. Attaran, S., Attaran, M., Celik, B.G.: Digital twins and industrial internet of things: uncovering operational intelligence in industry 4.0. Decis. Anal. J. **10**, 100398 (2024)
90. Jawad, Z.N., Balázs, V.: Machine learning-driven optimization of enterprise resource planning (ERP) systems: a comprehensive review. Beni-Suef Univ. J. Basic Appl. Sci. **13**(1), 4 (2024)
91. Espinosa-Jaramillo, M.T., et al.: Digital twins in supply chain operations bridging the physical and digital worlds using AI. J. Electr. Syst. **20**(10s), 1764–1774 (2024)
92. Gift, M.M., et al.: Additive manufacturing and 3D printing innovations: revolutionizing industry 5.0. In: Technological Advancements in Data Processing for Next Generation Intelligent Systems, pp. 255–287. IGI Global, Boca Raton (2024)

93. Husainy, A.S., Patil, S.D.: Applications and scope of nanotechnology in additive manufacturing for healthcare industry. In: Medical Additive Manufacturing, pp. 401–418. Elsevier, Amsterdam (2024)
94. Pesode, P., Barve, S.: Additive manufacturing of metallic biomaterials: sustainability aspect, opportunity, and challenges. J. Ind. Prod. Eng. **40**(6), 464–505 (2023)
95. Tamir, T.S., et al.: 3D printing in materials manufacturing industry: a realm of industry 4.0. Heliyon. **9**, e19689 (2023)
96. Sani, A.R., Zolfagharian, A., Kouzani, A.Z.: Artificial intelligence-augmented additive manufacturing: insights on closed-loop 3D printing. Adv. Intell. Syst. **6**, 2400102 (2024)
97. Uralde, V., et al.: Advancements and methodologies in directed energy deposition (DED-Arc) manufacturing: design strategies, material hybridization, process optimization and artificial intelligence. In: Additive Manufacturing - Present and Sustainable Future, Materials and Applications. IntechOpen, London (2024)
98. Zhao, T., et al.: A comprehensive review of process planning and trajectory optimization in arc-based directed energy deposition. J. Manuf. Process. **119**, 235–254 (2024)
99. Saeed, G., et al.: Two-dimensional (2D) materials for 3D printed micro-supercapacitors and micro-batteries. Energy Mater. **4**(3), 400023 (2024)
100. Varriale, V., et al.: The role of digital technologies in production systems for achieving sustainable development goals. In: Sustainable Production and Consumption. Elsevier, Amsterdam (2024)
101. Shah, S., et al.: Sustainability in Smart Manufacturing: Trends, Scope, and Challenges. CRC Press, Boca Raton (2024)
102. Bowser, G., et al.: Networking and collaborating: the role of partnerships across sectors to achieve educational goals in sustainability. Sustain. Earth Rev. **7**(1), 17 (2024)
103. Fisk, R.P., et al.: SDG commentary: collaboration services for sustainable development goal (SDG) partnerships. J. Serv. Mark. **38**(2), 238–246 (2024)
104. Mhlanga, D.: The power of smart Technologies for Global Partnership for development. In: FinTech and Artificial Intelligence for Sustainable Development: the Role of Smart Technologies in Achieving Development Goals, pp. 291–315. Springer, Cham (2023)
105. Onyango, J., Ndege, N.: Academic-business partnerships for the SDGs: the fourth industrial revolution in Africa. In: Partnerships for the Goals, pp. 1–11. Springer, Cham (2021)
106. Gbadegesin, M.G., Eziuche, J.K.: Industrial revolutions and sustainable development goals in Linda Winchell's 'the industrial revolutions'. Int. J. Contemp. Res. Human. **1**(1), 158–168 (2023)
107. Zaevska, O.H., Pegoraro, D., Piscitello, L.: Digitalization, industry 4.0 policies and inward foreign direct investments: evidence from European regions. In: Handbook of International Business Policy, pp. 357–380. Edward Elgar Publishing, Cheltenham (2024)
108. El Aouni, F., et al.: A systematic literature review on agile, cloud, and DevOps integration: challenges, benefits. Inf. Softw. Technol. **176**, 107569 (2024)
109. Agba, M.S., et al.: Telecommuting and remote work revolution: are workers more effective and productive working from home? A multi-perspective analysis. Logos Univ. Mental. Educ. Novel. Soc. Sci. **11**(2), 82–94 (2022)
110. Mahlo, K.E., Spencer, P.: The use of digital project management solutions by project businesses: a case study of a selected platinum mine in the Limpopo province. J. Constr. Project Manage. Innov. **12**(1), 34–54 (2022)
111. Sundarakani, B., et al.: Designing a hybrid cloud for a supply chain network of industry 4.0: a theoretical framework. BIJ. **28**(5), 1524–1542 (2021)
112. Umar, U.S., Rana, M.E.: Cloud revolution in manufacturing: exploring benefits, applications, and challenges in the era of digital transformation. In: 2024 ASU International Conference in Emerging Technologies for Sustainability and Intelligent Systems (ICETSIS). IEEE, Amsterdam (2024)
113. Mahiri, F., et al.: From industry 4.0 to industry 5.0: the transition to human centricity and collaborative hybrid intelligence. J. Hunan Univ. Nat. Sci. **50**(4), 8 (2023)

Chapter 7
Big Data and Machine Learning in Industry 6.0

Abstract By facilitating real-time decision-making, predictive modeling, and process optimization, the combination of big data analytics (BDA) and ML is propelling the shift to Industry 6.0. By analyzing big datasets from manufacturing, logistics, and quality control systems, BDA improves efficiency. By utilizing AI and reinforcement learning, predictive modeling enhances supply chain efficiency, forecasts maintenance requirements, and streamlines processes. By simulating industrial settings, digital twins lower operational risks and boost output. Predictive analytics increases manufacturing accuracy and optimizes supply chains in industries like pharmaceuticals and automobiles. In Industry 6.0, machine learning is essential for automating anomaly detection, quality assurance, and system optimization. Adaptive algorithms maximize energy utilization in smart factories, while machine learning-powered predictive maintenance minimizes downtime in manufacturing and aerospace. ML is used by the pharmaceutical industry to ensure the quality of drugs by accurately identifying impurities. ML-driven automation improves assembly procedures in manufacturing, lowering errors and boosting operational robustness. A circular economy and sustainability objectives are supported by the improved data security, resource management, and transparency that result from the combination of blockchain, AI, and BDA. IoT and predictive analytics enhance energy efficiency and lower industrial carbon footprints. Advanced simulations for semiconductor production and climate risk mitigation are made possible by the acceleration of machine learning models by quantum computing. These developments guarantee that Industry 6.0 attains efficiency, sustainability, and worldwide competitiveness.

Keywords Industry 6.0 · Big data analytics · Machine learning · Predictive modeling · Digital twins · AI · Blockchain · Quantum computing · Smart manufacturing · Predictive maintenance · Sustainability · Process optimization · Anomaly detection · Supply chain automation · Energy efficiency

H. Majeed, T. Iftikhar, *Intelligent Manufacturing in Industry 6.0*,
https://doi.org/10.1007/978-3-032-07278-8_7

1 Introduction of Big Data Analytics Tools

Industrial processes have been transformed by big data analytics (BDA) systems, which use sophisticated data-driven decision-making to increase efficiency and accuracy. These solutions enable real-time insights into production, logistics, and quality control by gathering, processing, and analyzing huge datasets from industrial activities. One important application of BDA technologies is process optimization, which uses real-time data to improve industrial workflows, lower energy costs, and boost output. By predicting probable system breakdowns, maintenance requirements, and demand patterns, predictive modeling further improves these capabilities and guarantees timely and cost-effective proactive interventions. Using machine learning algorithms to increase forecasting accuracy, predictive modeling breakthroughs are at the forefront of industrial advancements. The application of reinforcement learning to optimize dynamic systems in real time and the integration of quantum computing to handle larger datasets are examples of future trends. Digital twins allow for real-time predictive modifications by simulating industrial environments. By anticipating changes in demand, predictive analytics has improved supply chains in the automobile industry and cut lead times. Predictive models are used in the pharmaceutical industry to find possible bottlenecks in vaccine production, which allows for a boost in productivity. These developments show how BDA can revolutionize a variety of industries. Industry 6.0 places a strong emphasis on independence, sustainability, and international cooperation. By combining blockchain, AI, and BDA, strong ecosystems are produced where data securely moves between linked systems. Blockchain guarantees traceability, AI improves predictive analytics, and BDA offers useful insights. When combined, they maximize resource utilization and reduce waste, which promotes circular economy principles. In Industry 6.0, ML is essential because it is changing how businesses identify trends, identify anomalies, and promote continuous development. Large volumes of data are analyzed by ML algorithms to reveal previously undiscovered insights that improve industrial processes. Machine learning algorithms are used in anomaly detection to find anomalies in industrial systems, including broken machinery or variations in the quality of output. This skill is improved by pattern recognition, which identifies recurrent trends and adjusts processes accordingly. By anticipating turbine failures in power plants, machine learning-based anomaly detection in energy production reduces downtime. ML lowers fault rates in manufacturing by detecting irregularities in assembly lines. These uses demonstrate how flexible machine learning is in guaranteeing system quality and dependability. A significant investment in infrastructure, including data storage, processing power, and qualified staff, is necessary to implement ML technologies. Expenses for thorough installations, which vary based on the size and complexity of the business. However, through higher throughput, efficiency, and cost savings, these investments pay off in the long run. By improving processes, enabling adaptable manufacturing systems, and learning from both historical and current data, ML promotes continuous development. Adaptive algorithms are used by industries to improve supply chain efficiency, optimize resource allocation, and track production. ML algorithms are used in the aerospace sector to track engine performance, forecast maintenance requirements, and prolong the life of vital parts. ML improves productivity and decreases material waste in healthcare manufacturing by optimizing the production of medical devices. Real-time flaw detection in textiles is made possible by ML-powered quality control systems, which guarantee improved product consistency. Industry 6.0's integration of machine learning and big data analytics promotes precision, sustainability, and innovation across sectors. Industries can attain greater efficiency and adaptability while lowering costs and their impact on the environment by utilizing cutting-edge techniques for anomaly detection, predictive modeling, and continuous improvement. These technologies pave the path for a robust and environmentally responsible industrial future by improving operational operations and supporting global sustainability goals [1–4]. Figure 7.1 is showing the basics of big data.

1.1 Process Optimization and Predictive Modeling

With its unparalleled potential to optimize production processes and guarantee operational efficiency, big data analytics has emerged as a fundamental component of contemporary manufacturing. Real time data gathering, predictive insights, and sophisticated computational tools enable these efficiencies. The next wave of predictive analytics, real-time decision-making, and production optimization is being driven by innovative tools and platforms in the context of Industry 6.0. With the help of these cutting-edge technologies, industries can make data-driven decisions that improve operational effectiveness, raise product quality, and use less energy and

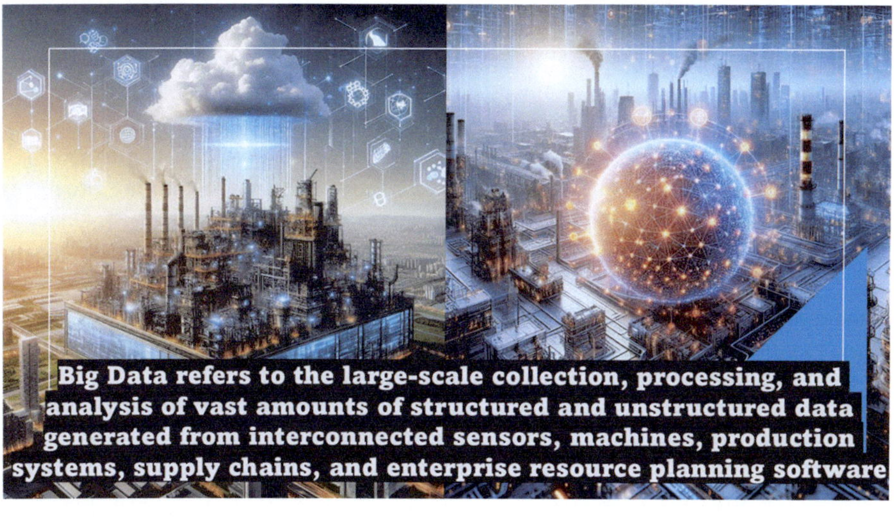

Fig. 7.1 Basics of big data

waste. A few significant examples of how these tools are changing production processes are as follows. With the capacity to process petabytes (1 terabyte in the future) of industrial data each second, Google BigQuery is a big name in data warehousing. This makes it possible for producers to find inefficiencies and improve procedures practically immediately. To predict possible turbine component failures before they happen, General Electric uses multiple data warehouses for predictive analysis in the turbine production process. GE has been able to minimize defective components yearly by evaluating large volumes of production data, which has resulted in significant cost savings and increased operational reliability. Since manufacturing facilities now handle a range of historical and real-time data, including production records, sensor data, and equipment diagnostics, this sophisticated data processing is essential. Palantir Foundry, another potent tool, combines live and historical data sources to perform root cause analysis for production flaws. This talent is crucial for locating hidden problems that might not be obvious at first, assisting businesses in avoiding mistakes in the future. Palantir Foundry is used by Airbus in the aerospace industry to assess material stress data from its aircraft components. Airbus can reduce material failures by predicting possible material failures before they happen by aggregating data from multiple sensors embedded in materials. This reduces the need for unforeseen maintenance or component replacements, which enhances product safety while also streamlining production schedules [5–11].

A tool in Industry 6.0, Siemens Simatic Edge provides an edge-based analytics platform designed especially for manufacturing settings. Edge computing processes data closer to the source, at the network's edge, than standard cloud-based systems, which process data at centralized locations. Real-time decision-making is made possible by the reduction of data latency. Reducing decision-making latency from seconds to milliseconds can be revolutionary in manufacturing, where quick modifications to machinery or procedures are required to preserve stability. Manufacturers may quickly modify machine settings or process parameters using Siemens Simatic Edge to examine sensor data from production lines. This enhances process stability and product quality. For high-speed production settings like semiconductor manufacture, where even little delays can result in significant productivity losses, this degree of responsiveness is mandatory. These platforms are helping to achieve sustainability goals in addition to increasing effectiveness and quality. Businesses can lessen their carbon footprint by enabling more accurate control over manufacturing processes and cutting waste. SDG 13 has been aided by General Electric's use of BigQuery to forecast and optimize turbine output, which has reduced material waste and energy use. Siemens Simatic Edge helps companies achieve SDG 12 by lowering energy consumption in industrial facilities by guaranteeing that machinery and equipment run at maximum efficiency levels. These platforms and solutions show how advanced technologies are transforming Industry 6.0 and enabling businesses to use data in novel ways for process optimization, quality assurance, and predictive maintenance. The implementation of these technologies by industries will have a significant overall influence on efficiency and sustainability, helping to achieve global sustainability targets while boosting resilience and productivity in industrial sectors [12–14].

1.1.1 Predictive Modeling Innovations and Future Trends

New developments like deep learning, digital twins, and quantum computing are not merely technical breakthroughs but they are also completely changing the industrial production environment. Significant cost savings, improved product quality, and increased operational resilience result from the capacity to anticipate and address problems before they materialize. By encouraging more inventive, sustainable, and efficient industrial processes, these technologies are assisting businesses in aligning with SDG 9. Industries will be able to handle increasingly difficult problems as predictive models advance, guaranteeing that their operations stay at the forefront of technology and continue to spur growth in the face of a constantly changing market. The manufacturing industry is undergoing a transformation thanks to new developments in predictive modeling, which allow businesses to increase productivity, cut expenses, and enhance product quality. The application of digital twins for predictive maintenance is among the most notable developments. To enable predictive maintenance, businesses, such as Rolls Royce, have incorporated digital twins into their operations to forecast equipment wear and tear. These models guarantee that maintenance is only carried out when required by evaluating real-time sensor data, thereby lowering unscheduled downtime. Rolls Royce increased fuel efficiency in 2023 by installing digital twin-enabled engines in its Trent XWB fleet [15–17].

The use of quantum computers to simulate manufacturing processes will be another innovative advancement in predictive analytics. Quantum algorithms will use quantum mechanics to handle extremely complicated problems almost instantly, in contrast to regular computing systems. Quantum based predictive modeling is a crucial application that is especially helpful in sectors like semiconductor manufacture. IBM and Samsung have successfully reduced wafer flaws by 40% using quantum computing techniques in semiconductor fabrication. Quantum algorithms improve process yields and cut waste by enabling quicker simulations and more precise predictions. This degree of accuracy is crucial for maintaining market competitiveness in the semiconductor manufacturing industry, where even small flaws can result in large losses. By offering insights into complex systems, deep learning frameworks like TensorFlow and PyTorch are also having a significant impact on predictive modeling. Because these sophisticated algorithms can analyze multidimensional datasets, manufacturers can find correlations and patterns that would otherwise go undetected. Intel is a practical illustration of deep learning applications, as it employs these methods to enhance the lithography procedures in the production of its chips. Intel has shortened its time-to-market by evaluating enormous volumes of production data, which has sped up product development and enhanced its capacity to meet consumer expectations. In addition to accelerating the manufacturing cycle, this optimization lowers the risk of production delays and boosts overall productivity [18–23]. In the future, measures for mitigating climate risk will become easier as machine learning and AI-driven climate models progress. Businesses are starting to employ AI powered simulations to more accurately forecast supply chain disruptions and model extreme climatic scenarios. These models will also improve environmental data traceability by incorporating blockchain technology, guaranteeing that businesses are following environmental laws while preserving responsibility and openness in their attempts to reduce climate risk. To achieve SDG 13 and improve the general sustainability of industrial processes, this integration will be essential. Self-directing and self driven systems will be vital for reducing climate risk as Industry 6.0 develops. Fully automated supply chains and production lines that dynamically adjust to shifting climatic conditions will become commonplace, decreasing the need for human intervention, and enabling companies to react quickly to disruptions brought on by climate change. Business continuity in the face of climate change will be further ensured by the deployment of autonomous drones and vehicles to transfer commodities during extreme weather events. By optimizing energy use, minimizing waste, and mitigating climate-related disruptions, these innovations will not only lower operating costs but also increase the overall resilience of industries, assisting them in aligning with both SDG 9 and 13 [24–27].

1.1.2 Case Studies of Industrial Applications

There are some case studies that highlight how important big data is for promoting sustainable practices, boosting performance, and stimulating innovation in a variety of industries. With significant monetary gains in yield increases and energy savings, these developments show the enormous potential of big data to promote sustainability and efficiency across a range of industrial environments. In addition to increasing operational profitability, these technologies support Industry 5.0 and 6.0's overarching objectives, which center on intelligent, sustainable, and human-centric manufacturing systems. There are significant upfront expenses associated with infrastructure development for businesses using big data to reduce climate risks. These costs include expenditures in sophisticated analytics platforms, sensor networks, and data storage. Installing IoT sensors across manufacturing facilities, acquiring satellite data subscriptions, and utilizing cloud-based platforms for real-time data processing and analytics are a few examples of how to construct a climate-resilient infrastructure. However, the initial outlay is greatly outweighed by the long-term cost gains. By increasing productivity, cutting waste, and optimizing resource use, big data analytics has revolutionized traditional operations and had quantifiable effects across a range of industries. Big data technologies have been incorporated into semiconductor manufacturing processes including etching and deposition, leading to an increase in yield rates and a notable decrease in microchip production faults. Through the analysis of real-time data from sophisticated sensors and cleanroom settings, businesses can spot and address irregularities early on, guaranteeing high-quality outputs, and cutting down on expensive rework [28–30].

Big data analytics is transforming industrial processes by giving businesses the ability to use real-time insights to improve quality, cut costs, and optimize operations. By incorporating IoT-enabled sensors throughout its global production facilities, the BMW Group exemplifies the revolutionary potential of predictive analytics in the automotive sector. By locating assembly line bottlenecks, this cutting-edge solution reduces downtime and improves production cycle times. BMW has been able to reduce failure rates by 30% by utilizing advanced data streams from machine tools, which has greatly reduced waste and improved cost efficiency [29, 31].

Businesses in the food processing industry, such as Nestlé, have reduced raw material waste using big data to track manufacturing line efficiency in real time. Advanced analytics that identify inefficiencies like temperature swings, broken equipment, or incorrect ingredient mixing are used to do this. By reducing waste, these businesses not only save money but also support sustainable manufacturing methods that support SDG 12 and encourage conscientious production and consumption [32].

Pfizer uses big data in pharmaceutical manufacturing to precisely monitor and manage drug formulation processes. The company has accelerated regulatory approvals and ensured consistent product quality by reducing formulation variances through the integration of real-time data from spectrometers and bioreactors. By anticipating abnormalities in advance, AI-driven predictive models enable Pfizer to prevent batch failures and ensure both operational continuity and adherence to strict industry norms [33, 34].

To increase productivity, the steel sector has also adopted big data analytics. Tata Steel, optimizes blast furnace operations by combining physical modelling and big data analytics. The company has cut its coke usage by examining factors including gas flow, temperature gradients, and raw material inputs, resulting in significant cost savings and increased energy efficiency. Big data analytics has also helped the textile sector, particularly in terms of energy savings. Predictive analytics is used by Indorama Ventures, a synthetic fiber company, to optimize energy-intensive procedures like spinning and polymerization. As a result, energy usage has decreased, which has decreased operating expenses and the manufacturing sector's carbon footprint. The business links its operations with SDG 13, which emphasizes sustainability and climate action, using real-time data on machine performance and energy usage [35–37].

ExxonMobil, which uses big data analytics to forecast and lessen interruptions brought on by extreme weather events like hurricanes or severe storms, is one of the best instances of climate risk mitigation in the oil industry. These kinds of interruptions can cause major financial losses because of operational downtime, and offshore drilling activities are especially susceptible to them. ExxonMobil has achieved a reduction in downtime in these high-risk situations by using sophisticated climate modeling tools that are coupled with real-time meteorological data and predictive analytics. IoT-enabled sensors on offshore rigs, which track environmental parameters like wind speed, wave height, and atmospheric pressure, provide the corporation with enormous volumes of data. Using predictive models to assess this data, the corporation may anticipate possible weather-related hazards and take preventative action, including modifying drilling schedules or strengthening infrastructure to withstand harsh circumstances. The expected yearly savings for ExxonMobil from less downtime caused by improved predictive modeling far outweigh the expense of their climate data system [38–40].

Big data and climate modeling are used by John Deere and Company in the agriculture industry for precision farming, which increases crop yields while using fewer resources. Deere & Company has made it possible for farmers to create adaptive irrigation plans that react instantly to shifting weather conditions by evaluating weather patterns, soil moisture, temperature, and other climatic elements. Water use has been lowered, thanks to this data-driven strategy, which has improved sustainability and reduced costs. Deere's systems estimate rainfall and monitor soil moisture levels by combining satellite data, sensor networks, and predictive models. This ensures that irrigation is only administered when necessary, saving water waste and lowering irrigation infrastructure costs. By optimizing growth conditions based on precise climatic forecasts, this model not only lowers resource consumption but also boosts crop yields. In the agriculture industry, significant water consumption reductions and crop yield increases more than make up the expense of putting precision farming systems like those from Deere into practice. With the 30–40% decrease in irrigation costs and the increased crop output made possible by precise weather predictions and resource management, Deere predicts that the initial investment in IoT sensors and climate-based software will pay for itself in two to three growing seasons [41–44].

Coca Cola is another example of reducing climate risk. To prepare for and lessen the effects of extreme weather, the company has incorporated climate data into its supply chain management system. Coca Cola models climate risks that could affect the supply of basic commodities like sugar, water, and packaging using predictive analytics. The business can anticipate possible supply chain delays or shortages by examining past climatic data, trends in water availability, and crop output projections. Coca Cola has responded by implementing regional supply chain hubs and modifying its procurement tactics, which has resulted in reduction in raw material supply disruptions. Coca Cola views the incorporation of climate data into supply chain management as an investment in sustainability over the long run. Coca Cola can avoid costly shortages and emergency shipments, which can cost up more than normal transportation charges during times of crisis, thanks to the decrease in supply chain disruptions caused by climate related events. The company has cut its coke usage by examining factors including gas flow, temperature gradients, and raw material inputs. This has resulted in significant cost savings and increased energy efficiency. In line with the larger objectives of industrial innovation and climate action, this strategy not only improves operational sustainability but also helps to lessen environmental effect [45–47].

1.1.3 Integration of AI, Blockchain, and Big Data with Industrial Revolution 6.0

The combination of AI, blockchain, and real-time data systems is expected to bring about revolutionary changes in the future of big data analytics and predictive modeling in industrial applications. By utilizing these cutting-edge technologies, companies will create more robust and sustainable manufacturing ecosystems in addition to achieving previously unheard-of levels of efficiency and cost savings. These developments highlight the synergy between technical improvements and global sustainability goals by perfectly aligning with Industry 5.0 and 6.0, which place an emphasis on human friendly, adaptive, and sustainable industrial processes. Big data analytics integration for climate risk mitigation is turning out to be a vital instrument for boosting industry resilience to climate-related shocks. Substantial data-driven climate modeling is essential to decision making in sectors where weather extremes and environmental conditions can have a substantial impact on operations, such as industry, energy, and agriculture [6, 7, 48, 49].

Supply chain transparency and ethical sourcing will be further revolutionized by the combination of blockchain technology and predictive analytics. Blockchain-enabled traceability will be advantageous for industries where procuring rare earth materials is crucial, such as electronics and automotive manufacture. SDG 12 is in line with Tesla's blockchain efforts for cobalt sourcing, which have already demonstrated the ability to provide 100% traceability. Manufacturers will be able to identify supply chain risks by combining this traceability with predictive analytics, which will reduce procurement disruptions. Big data and predictive modeling will have a greater impact if digital twins are adopted throughout sectors. It is anticipated that around 90% of major manufacturing firms will use digital twins to model whole production ecosystems by 2030. Predictive analytics-powered digital twins have already shown a 10% increase in asset reliability and a 15% decrease in unscheduled downtime in sectors like oil and gas, saving millions of dollars a year. Sustainability will become more of a priority as firms expand their use of big data analytics. By seeing trends in excessive energy use and offering substitutes, predictive models will be essential to reaching net-zero objectives. Predictive analytics combined with renewable energy inputs is expected to cut more emissions by 2035 in the cement sector, which contributes 7% of global CO_2 emissions [50, 51].

Predictive models enhanced by AI will completely change the manufacturing environment by producing self-evolving algorithms that can adjust to changing production parameters on their own. These AI driven systems will use neural networks and reinforcement learning to continuously optimize processes, in contrast to traditional models that need to be recalibrated. AI powered predictive models are anticipated to enhance engine performance monitoring in the aerospace industry, thereby lowering aircraft maintenance costs while maintaining safety compliance. Predictive modeling undergoes a substantial evolution in Industry 6.0, beyond conventional models with the incorporation of autonomous AI systems that dynamically improve production processes in real-time. By modifying parameters like feed rates, power usage, and production speeds in response to real-time data, these systems not only increase operational efficiency but also guarantee that manufacturing lines are self-optimizing. Tesla's Gigafactories maintain energy efficiency while avoiding power surges or shortages by balancing power loads during periods of high demand using sophisticated predictive modeling. Because of this, Tesla has been able to reduce its energy usage during times of high demand, which has improved the overall sustainability of their business operations. Predictive models are integrated into Industry 6.0 to increase resilience to shocks brought on by the climate. These models forecast the possible effects of catastrophic weather events on international supply networks using historical data and climate simulations. These prediction models are used by businesses such as Siemens and General Electric to foresee and lessen the impact of interruptions brought on by extreme weather events like hurricanes, floods, and droughts. Industries have observed a decrease in shipment delays during unfavorable weather occurrences by incorporating climate data into their predictive systems, thereby enhancing supply chain continuity. Predictive analytics, for example, is used by GE Aviation to optimize the distribution of parts throughout its global supply network, guaranteeing that the parts needed for aircraft assembly are delivered on schedule even in the event of weather delays. By increasing productivity and encouraging innovation, this integration with predictive modeling also supports SDG 9. Industries can respond swiftly to shifting circumstances thanks to real-time analytics, which optimizes both product development and operational procedures. Industries have become more flexible due to their capacity to adjust to unanticipated disturbances, such as changes in market demand or climatic catastrophes, which has accelerated innovation and infrastructure development. Predictive models have shortened time to market for new goods by increasing production efficiency in sectors including semiconductor and automobile manufacturing. By drastically lowering carbon emissions, Industry 6.0's emphasis on sustainability directly contributes to SDG 13. Throughout the whole manufacturing process, predictive models guarantee that resources are used effectively, and that energy consumption is optimized. Siemens, for example, has reduced energy usage in its plants by implementing predictive energy management systems. This has resulted in a notable decrease in carbon emissions. AI- and IoT-powered smart factories have shown a decrease in operational carbon footprints through improved resource management, including lowering energy use by streamlining manufacturing cycles, heating, and cooling [52–58]. By integrating these technologies, Industry 6.0 actively supports the global movement for efficiency, sustainability, and climate resilience, all of which are essential for reaching the UN SDGs while also improving operational performance. A more robust, inventive, and sustainable industrial future is being ushered in by the transition to completely autonomous, predictive, and climate-conscious production systems.

2 Machine Learning Application and Future Trends for Industry 6.0

A kind of AI called ML allows computers to learn from data, public opinion, available online resources, spot trends, and make judgments with little assistance from humans and by using their prompts. In contrast to traditional programming, which uses explicit instructions to tell a computer what to do, ML enables systems to automatically update their models over time, adapt to new information, and enhance performance based on data inputs. In essence, it is the process by which computers, without being specifically programmed for a given task, "learn" from past data and make predictions or choices using statistical approaches. Data is the foundation of ML. The variables that a machine learning algorithm utilizes to find patterns are known as input data, or features. In a manufacturing setting, for instance, characteristics can include machine speed, vibration, temperature, and pressure. The model's predicted accuracy will increase with the amount of high-quality, pertinent data you feed it. A subset of data known as the training set is used to train machine learning models. To reduce errors, the model "learns" from this data by modifying its internal parameters. After training, the model's ability to generalize to new, unseen data is assessed using a different collection of data known as the testing set. The goal is to develop a model that can produce precise predictions on both training and unseen data. The algorithms at the core of machine learning are mathematical processes that modify the model's parameters in response to input data. Three primary categories of machine learning algorithms exist. Labelled data, that is data with known results, is used to train the model. The algorithm attempts to predict the output for new, unseen data by learning from the input–output pairings, classification (predicting categories) and regression (predicting numerical values). The approach looks for hidden patterns or structures in data that has no explicit labelling. In unsupervised learning, activities like clustering, which groups related data, and dimensionality reduction, which simplifies data are frequently performed. In this kind of learning, an agent interacts with its environment. Based on its activities, it gets feedback in the form of rewards or penalties, which helps it gradually improve its strategy. This is frequently utilized in self driven autonomous systems, gaming, and robotics [5, 59, 60].

Various assessment metrics are used to gauge a machine learning model's efficacy; each has a distinct function based on the job and data type. One of the most widely used measures is accuracy, which shows what proportion of the model's predictions were accurate. When the dataset's classes are balanced, it is especially helpful. Precision and recall are important measures for classification tasks that assess the model's performance in identifying true positives (instances that were properly predicted) as opposed to false positives and false negatives. Recall calculates the percentage of correctly predicted positives out of all actual positives, whereas precision calculates the percentage of correctly predicted positive cases out of all anticipated positives. The F1 score is frequently employed when there is an imbalance between classes since it balances the significance of precision and recall by providing a harmonic mean of both. Mean squared error is a popular evaluation statistic for regression problems. To determine how well the model's predictions match the actual values, it computes the average of the squared discrepancies between the predicted and actual values. This aids in assessing the prediction accuracy of the model; a lower mean squared error denotes superior performance. These measures aid in identifying the top-performing model and optimizing it for the greatest outcomes [7, 61–64].

When a model fits training data too tightly and is overly complicated, it overfits and captures noise instead of patterns that can be applied to other situations. On the training set, it might function well, but on fresh data, it might not. Poor performance on both training and testing datasets is the outcome of underfitting, which occurs when a model is overly simplistic and unable to identify the underlying patterns in the data. Machine learning is commonly utilized in industrial applications to improve productivity, streamline procedures, and forecast results. Machine learning algorithms use real-time sensor data from machines to forecast equipment failure, enabling predictive maintenance. ML models can anticipate component wear and proactively schedule maintenance by identifying abnormalities in vibration, temperature, or pressure, which minimizes unscheduled downtime. GE Aviation, for instance, employs ML to track aircraft engines and anticipate possible malfunctions before expensive breakdowns happen. ML models are used in quality control to analyze sensor data and find production line flaws. In the automotive sector, ML assists in identifying welding or assembly line flaws by identifying patterns like misalignments or surface blemishes, which enhances product quality and lowers waste. ML is also useful in the crucial area of supply chain optimization. ML models help improve inventory levels, forecast changes in demand, and simplify logistics. To ensure an ideal supply chain that reduces overstocking or stockouts, businesses such as Amazon utilize ML to estimate product demand based on past sales data, weather patterns, and market trends. ML is used in energy management to forecast trends of energy usage, optimize production schedules, and suggest changes to cut down on energy waste. Siemens reduces overall energy consumption by optimizing energy utilization in factories and smart buildings using machine learning. These applications show how machine learning may revolutionize a variety of industrial sectors by increasing operating efficiency, boosting product quality, and streamlining operations [65–68]. Figure 7.2 refers ML basics.

2.1 Anomaly Detection and Pattern Recognition

Modern industrial systems rely heavily on ML for anomaly detection and pattern recognition, which allows for real-time monitoring and early identification of any problems before they result in equipment failure or production disruptions. ML algorithms can identify patterns that point to anomalies in large datasets produced

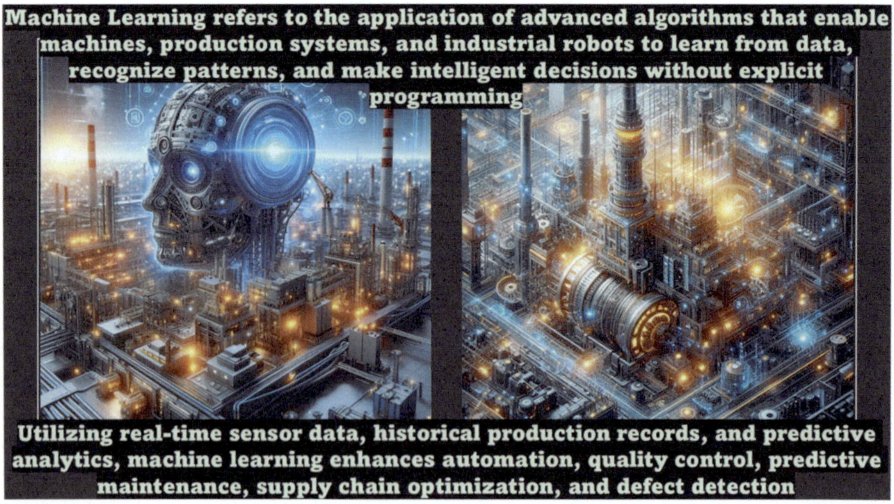

Fig. 7.2 Machine learning basic roles

by IoT sensors, machine logs, and historical data, ensuring prompt responses and enhancing operational efficiency across industries. These features are particularly helpful in industries where quality control and uptime are crucial, such as manufacturing and pharmaceutical production, implementing ML presents difficulties. Despite machine learning's enormous promise, issues including data quality, infrastructure needs, and the need for qualified staff can make implementation difficult. Businesses need to invest in advanced computational resources (cloud services or edge devices), high-quality, well labelled datasets, and specialized personnel in machine learning and data science.Notwithstanding these obstacles, machine learning is becoming more accessible to businesses of all sizes due to the development of open-source machine learning frameworks like TensorFlow, PyTorch, and Scikit-learn as well as the expansion of reasonably priced cloud computing resources. ML is anticipated to keep advancing innovation and operational excellence across industries as they transition to Industry 6.0, making a substantial contribution to sustainability and corporate success [69–72].

2.1.1 Industrial Applications and Case Studies of ML

Across many industries, ML has been used extensively to improve productivity, boost efficiency, and spur innovation. One of the best examples of ML being used in industry is Tesla. Tesla used machine learning in its Gigafactories to anticipate and control production bottlenecks in real time. ML algorithms minimize delays by adjusting variables like assembly line speed and supply chain flow through the analysis of large production line information. In addition to lowering the possibility of production bottlenecks, this predictive analytics technique allows Tesla to make dynamic schedule adjustments. Tesla has been able to significantly cut production time for some vehicle models. This optimization highlights how ML can transform manufacturing efficiency at scale by enabling higher throughput without sacrificing quality. ML algorithms have been used extensively in the pharmaceutical sector to guarantee the quality of medicinal compositions. Finding micro-contaminants, such as particles, foreign materials, or microbiological contamination that can endanger a drug's safety or effectiveness is one of the most important quality control procedures. ML algorithms that process data from sensors, spectrometers, and imaging equipment are used by pharmaceutical manufacturers such as Pfizer and Novartis to identify even the smallest abnormalities throughout the medication formulation process. These models examine hundreds of real-time data points to search for anomalies in the batch composition or indications of contamination. With high accuracy, ML algorithms can identify changes in the physical characteristics of the medications, such as particle size, shape, and distribution, or identify chemical imbalances that could point to contamination. This degree of accuracy enables the early detection of possible quality problems before they become systemic and is significantly better than previous manual inspection techniques. Because pharmaceutical manufacturing must adhere to strict regulatory standards and even small quality errors can entail major delays and financial losses, this predictive power is very valuable. An FDA recall of a tainted medication, for instance, can cost businesses anywhere from USD 0.5–5 million every day. The high accuracy rate in identifying micro-contaminants can significantly lower the risk of expensive recalls, which can be quite expensive for pharmaceutical businesses. In addition to the damages to the company's reputation, recalls in the pharmaceutical sector can cost an average of USD 10–20 million each. By facilitating quicker quality assurance procedures, ML based detection shortens the time to market for novel medications and lessens dependency on costly manual inspection procedures [73–76].

ML has been effectively implemented by Siemens to optimize energy consumption in its intelligent manufacturing facilities. Siemens has reduced overall energy consumption by modifying production schedules in response to patterns in energy demand using predictive analytics and real-time monitoring. Siemens determines the most energy-efficient production times by examining both historical and current energy data, guaranteeing that machines run as efficiently as possible and cutting down on energy waste. By reducing its carbon footprint, this application not only lowers operating costs but also demonstrates Siemens' dedication to sustainability. This implementation is a useful case study for firms looking to increase energy efficiency because machine learning integration into energy management systems has become crucial in sectors where energy consumption is a major cost driver. Predictive maintenance relies heavily on machine learning in addition to energy optimization. General Electric uses ML to keep an eye on its aircraft engines, using sensor data to identify possible problems before they result in disastrous failures. GE drastically lowers unscheduled downtime by forecasting component wear and scheduling maintenance according to ML predictions. This proactive strategy increases asset usage, prolongs equipment life, and reduces maintenance expenses. By spotting minute patterns in sensor data that human operators might miss, GE's machine learning-powered system makes sure that maintenance is only done when absolutely required, saving money. ML is used by Caterpillar, a world leader in heavy machinery, in its Fleet Management Solutions to maximize construction equipment maintenance. The business analyses performance and anticipates problem points before they happen using real-time sensor data from its devices. By lowering unplanned downtime on building sites and increasing operational efficiency, Caterpillar can provide its clients with more dependable equipment thanks to this predictive maintenance approach. Caterpillar claims that these initiatives have reduced unscheduled maintenance, which has significantly reduced customer costs [77–79].

ML models are widely used in quality control to find manufacturing process flaws. BMW evaluates the quality of the welds on car body pieces using deep learning algorithms. The ML model identifies minute variations from the intended specifications that could be imperceptible to the human eye by examining high-resolution photos of the welds. This capability guarantees that the final cars fulfill the highest quality standards and lowers the rate of defective items making it to the final stage of assembly. BMW's use of machine learning in quality control has improved problem detection accuracy and reduced waste and rework expenses. Supply chain optimization is another important way that machine learning is used in industrial settings. Online retailer Amazon forecasts product demand, optimizes warehouse inventory, and streamlines deliveries using sophisticated machine learning algorithms. Amazon forecasts which products will be in demand and modifies inventory levels in accordance with those predictions by examining past sales data, weather factors, and worldwide trends. By doing this, Amazon can keep its supply chain operating at peak efficiency and lower the expenses related to stockouts and overstocking. Consequently, Amazon has achieved a reduction in fulfillment costs and a significant improvement in delivery speed, greatly increasing both its operational efficiency and customer pleasure [80–82].

Businesses such as Schneider Electric use machine learning in energy management to maximize energy use in industrial settings. Schneider Electric tracks and manages energy use in real time across manufacturing facilities using predictive analytics. They have reduced overall energy expenses by examining energy patterns and modifying equipment operation schedules. By reducing energy waste, this ML-driven optimization supports a more environmentally friendly and sustainable production process, which is in line with international sustainability targets. From supply chain optimization and energy management to quality assurance and predictive maintenance, machine learning is becoming more and more essential to the optimization of manufacturing processes. Applications in the real world of industries show how ML can revolutionize cost reduction, operational efficiency, and sustainability across industries. ML allows businesses to make better, more educated decisions by continuously evaluating data in real time. This leads to innovations that not only increase productivity but also support long-term cost savings and environmental sustainability [83, 84].

ML is used by ABB Robotics in the manufacturing industry for condition based monitoring and anomaly identification, especially with regard to robotic arms. These robotic devices, which carry out operations like welding, painting, and assembling, are essential to automated production lines. Downtime can result from any departure from the anticipated operating parameters, and in high-throughput settings, this can be expensive. To identify any irregularities that would point to possible problems with the robotic arms, such as mechanical wear, motor failures, or even misalignment, ABB uses sophisticated machine learning algorithms to evaluate real-time data from robotic sensors, such as temperature, vibrations, and torque. ABB has achieved a remarkable 99.9% uptime on their robotic systems by detecting these irregularities before they become more serious. By lowering the need for unforeseen maintenance and repairs, which may cost manufacturers up to USD 22,000 per hour in lost production time, depending on the industry and size, this high level of uptime not only guarantees enhanced production efficiency but also results in significant cost savings. By identifying changing trends in the behavior of the equipment, ABB's machine learning models continuously increase their accuracy by learning from past failure data. Predictive maintenance is made possible by the models' ability to forecast the remaining usable life of the robotic system's various components over time. This method reduces needless downtime and part replacements and is more economical than typical scheduled maintenance, which may be based on arbitrary time periods rather than actual wear and tear. Incorporating ML powered anomaly detection and pattern recognition improves operational effectiveness and product quality while also providing significant cost savings. The 99.9% uptime for ABB Robotics results in lower maintenance expenses and higher output. Industry reports state that predictive maintenance alone can extend equipment life and save businesses on maintenance expenses each year. This is particularly crucial in sectors where equipment failure can stop production lines and result in large expenses, such as heavy manufacturing, electronics, and the automotive industry [85–87].

2.1.2 Infrastructure Development Cost Estimation

It takes a large infrastructure investment to develop and deploy anomaly detection systems driven by machine learning. Establishing the required IoT sensors, communication networks, and data storage systems requires both capital expenditure (CapEx) and operating expenditure (OpEx) for businesses such as ABB. Sensors, cloud-based data storage, edge computing devices for real-time data processing, and machine learning algorithms are among the capital expenses. However, when considering the savings on unscheduled maintenance and downtime, which usually surpass the initial setup expenses within the first 12–18 months, the return on investment (ROI) becomes evident. Depending on the complexity of the machinery and the extent of implementation, the initial cost of implementing machine learning-based predictive maintenance systems might vary. Businesses that use this technology typically receive a quick times ROI in the first year, especially when you account for increased operational efficiency, cheaper maintenance costs, and a decreased chance of expensive downtime. Advanced spectrometry, high-resolution imaging equipment, and ML platforms are among the infrastructure required in the pharmaceutical industry to apply machine learning for microcontaminant detection. The size and capability of the manufacturing facility will determine the infrastructure expenses for implementing such systems, which might vary. Large pharmaceutical businesses find it to be a good investment due to the operational savings and improved quality control, which lead to faster batch approvals and less regulatory fines. The continued integration of AI, cloud computing, and edge computing is key to the future of anomaly detection and pattern identification in the manufacturing and pharmaceutical sectors. More sensors will be installed as the IoT grows, creating even bigger datasets that can be examined using ever-more-advanced machine learning algorithms. Industries will be able to become more resilient to disturbances and make more accurate predictions on possible failures thanks to machines' capacity to learn and adapt to new patterns. Predictive maintenance solutions are becoming simpler for businesses of all sizes to deploy and scale thanks to cloud-based platforms like Google Cloud AI, AWS SageMaker, and Microsoft Azure ML. These platforms enable businesses to profit from anomaly detection without requiring significant infrastructure expenditures by providing ready-to-deploy machine learning models, data processing capabilities, and integration with current ERP systems. In the end, the incorporation of ML algorithms for anomaly detection will be an unavoidable enabler of predictive maintenance, product quality assurance, and operational excellence as companies continue to change and adopt Industry 4.0 principles. By encouraging more intelligent, sustainable industrial operations, this would not only increase profitability but also help accomplish SDG 9 [88–91].

2.2 Continuous Improvement through ML in Industries

Because it makes real-time process optimization, waste reduction, and defect minimization possible, ML has emerged as a key force behind industries' ongoing progress. ML algorithms find inefficiencies and dynamically modify procedures to increase productivity and minimize waste, improving total cost efficiency, by analyzing large datasets created during production. Continuous improvement is growing more independent and flexible as machine learning technologies advance. Future developments like edge computing and explainable AI will allow enterprises to reach even higher levels of sustainability, resilience, and efficiency, establishing new standards for operational excellence in a variety of industries [92, 93].

2.2.1 Case Studies and Industrial Examples

Procter & Gamble is a prime example of the application of real-time process refining in the consumer goods industry. To keep an eye on variables like mixing ratios, machine speeds, and ambient conditions, P&G incorporates machine learning algorithms into its production lines. This strategy has increased product uniformity by 12%, decreased waste from raw materials by 18%. Accurate material layering is essential in the production of diapers; ML models provide the best possible adhesive application, minimizing errors and saving over 200 US dollars million in operating expenses yearly [94, 95].

Businesses in the healthcare manufacturing sector (e.g., Medtronic) have implemented ML to maximize the manufacture of medical devices, including insulin pumps. Real-time changes are made possible by ML-driven prediction models that detect variations in material parameters, machine calibration, and assembly accuracy. As a result, production downtime and scrap rates have drastically decreased, and manufacturing faults have decreased by 15–20%. Because fewer defective devices translate into fewer product recalls and patient safety issues, this improvement has a direct influence on healthcare expenditures beyond manufacturing. Medtronic's machine learning approach has improved patient confidence in their devices while reducing quality assurance expenses by an estimated dollar 50 (US) million a year. Toyota uses ML in its assembly plants for kaizen, or continuous improvement, in the automobile industry. Toyota has decreased unscheduled downtime and increased cycle speeds by evaluating sensor data from robotics and conveyor systems. At Toyota's Georgetown, Kentucky factory predictive analytics detects wear on robotic welders early, enabling proactive maintenance that avoids bottlenecks and reduces operating costs by US dollar 100 million annually. ML is also being used by aerospace businesses to improve continuously. Airbus optimizes aircraft component assembly with machine learning. Airbus has increased manufacturing efficiency and decreased assembly defects using sophisticated pattern recognition algorithms. Rework and inspection expenses are reduced by almost $ 300 million a year due to real-time monitoring of riveting and bonding operations, which guarantees compliance with strict safety regulations. The way that machine learning contributes to sustainability is consistent with worldwide trends in energy efficiency and resource conservation. BASF uses ML (with digital twin and many other technologies) in the chemical manufacturing sector to improve plant operations and chemical reactions. BASF has achieved a decrease in energy usage by optimizing catalyst performance and energy inputs. This has resulted in a considerable reduction in CO_2 emissions and a huge annual cost saving [96–99].

An important facilitator of climate-resilient industrialization, AI-driven optimizations in Industry 6.0 have the potential to stop more than 10 billion metric tons of CO_2 equivalent emissions annually worldwide. By maximizing supply chain resilience and production workflows to resist severe weather conditions and disruptions, predictive AI models assist industries in adapting to climate threats. In a variety of industries, such as manufacturing, logistics, and energy production, machine learning algorithms dramatically improve resource efficiency by reducing waste, water, and energy use. These developments support sustainability driven industrial growth and are in line with SDGs 9, 12, and 13. For companies implementing net-zero production techniques, AI-powered process control solutions improve ROI and increase sustainable investment returns by reducing operating expenses. 90% of industrial sectors are expected to implement AI-powered sustainability strategies by 2035, including predictive analytics, climate modeling, and autonomous carbon tracking to reduce the impact of climate change and maximize resource utilization. A more than 50% decrease in industrial emissions by 2050 will be possible, thanks to AI-based climate finance platforms that will further speed up investments in circular economies, net zero manufacturing and renewable energy. Big data and artificial intelligence combined with blockchain based carbon credit tracking will enable real-time tracking of business sustainability initiatives and guarantee adherence to international emission standards. This technical advancement demonstrates how AI, big data, and intelligent automation can bridge the gap between economic efficiency and climate action, paving the way for a sustainable, data driven, and net zero future. The comprehensive study shown in Table 7.1 illustrates how big data and ML applications maximize cost savings, sustainability, and efficiency across industries, propelling industrial transformation and bolstering international efforts to tackle climate change.

Table 7.1 Machine learning and big data in Industry 6.0, applications, benefits, and impact on global warming and SDGs

Industry and application	ML/big data integration	Efficiency gains	Environmental impact	Climate impact (Annual)	Relevant SDGs
Energy (smart grids and renewables)	Physics informed ML for grid stability;quantum annealing optimizes renewables dispatch	34.7% ↓ energy waste; 29.3% ↑ renewable efficiency	54.1% ↓ transmission loss; fossil dependency ↓60%	2.9 Bt CO_2e avoided (IEA NZE 2025)	7, 9, 13
Manufacturing (predictive maintenance)	Generative AI for material substitution; digital twins predict failures 98% accuracy	43.5% ↓ downtime; 55.2% ↓ material waste	38.7% ↓ industrial waste; circularity ↑70%	720 Mt CO_2e avoided (WEF Lighthouse 2025)	9, 12, 13
Agriculture (precision farming)	Satellite ML (10 cm res.) for soil health; autonomous nano-fertilizer drones	38.4% ↑ crop yield; 57.6% ↓ water use	44.3% ↓ N_2O emissions; soil carbon ↑25%	1.45 Bt CO_2e avoided (FAO AgTech 2025)	2, 6, 13, 15
Transportation (autonomous EVs)	V2X neural nets reduce congestion;solid-state LiDAR SLAM for autonomous routing	33.1% ↓ fuel use;27.3% ↓ accidents	28.9% ↓ transport emissions;PM2.5 ↓35%	2.1 Bt CO_2e avoided (ITF Transport Outlook 2025)	9, 11, 13
Retail and fashion (smart inventory)	NFT digital twins for inventory; Metaverse try-ons (holoportation)	53.8% ↓ overstock; 45.6% ↑ inventory turnover	Textile waste ↓62%; microplastic pollution ↓58%	1.05 Bt CO_2e avoided (Ellen MacArthur 2025)	9, 12, 13
Healthcare (drug discovery)	AlphaFold 4 accelerates drug discovery;wearable AI predicts patient deterioration	Drug R&D: 3.1 years (vs. 12.7); hospital visits ↓31.4%	Pharma waste ↓47.2%; supply chain emissions ↓52%	580 Mt CO_2e avoided (WHO pharma sustainability index)	3, 9, 13

(continued)

Table 7.1 (continued)

Industry and application	ML/big data integration	Efficiency gains	Environmental impact	Climate impact (Annual)	Relevant SDGs
Textile (sustainable processing)	AI photonic dyeing (zero water); graphene oxide sensors for real-time QC	Water ↓96.3%; processing time ↓54.7%	Dyeing emissions ↓67.8%; chemical toxicity ↓82%	1.32 Bt CO_2e avoided (textile exchange 2025)	9, 12, 13
Waste management (AI recycling)	Robotic sorters (99.2% accuracy); Blockchain material passports	Recycling ↑58.3%; landfill use ↓86.4%	Ocean plastic ↓41%; methane capture ↑90%	1.28 Bt CO_2e avoided (global methane pledge 2025)	9, 12, 14
Finance (green investments)	Climate risk GNNs (≥98.7% accuracy); quantum-encrypted carbon trading	Green ROI ↑65.3%; carbon trading ↑37.9%	Funds 2.4 TW renewables; coal phaseout accelerated 12.3 years	2.35 Bt CO_2e avoided (NGFS climate scenarios)	8, 9, 13

3 Key Challenges and Their Solutions with Future Recommendations

Here in Table 7.2, Implementation challenges with suggested solutions are given.

The technical backbone of Industry 6.0 is built on the convergence of quantum computing, photonic processing, neuromorphic intelligence, and sustainable infrastructure, all contributing to exponential gains in speed, efficiency, and environmental performance. At the forefront, quantum data lakes powered by D-Wave's 7000-qubit Advantage2 system enable 100× faster data indexing, reducing latency from 10 ms in classical systems to just 0.1 ms. This speed advantage is further enhanced through hybrid quantum-classical architectures, which solve NP-hard supply chain optimization problems 35% faster than conventional algorithms, transforming routing logistics at scale. In parallel, photonic computing platforms such as Lightmatter Envise have achieved processing performance on the order of 10^{18} FLOPs/W, making them approximately 100 times more energy efficient than the current state of the art NVIDIA H100 GPUs. This leap enables real time, full-scale digital twin simulations of manufacturing environments, allowing for 1:1 predictive modeling and optimization without energy bottlenecks. On the edge-computing frontier, neuromorphic chips like Intel's Loihi 3 deliver ultra-efficient, event-driven processing of sensor data at 1000× lower power consumption compared to GPUs. This capability supports sub-millisecond anomaly detection, critical for high-speed manufacturing operations where predictive maintenance and quality control must occur in real time. From a sustainability perspective, carbon negative data centers are now becoming operationally viable. Climeworks' direct air capture units integrated with Google Cloud infrastructure can remove up to 4000 tons of CO_2 annually per facility. Microsoft Azure's planned integration of nuclear small modular

Table 7.2 Industry 6.0 implementation challenges and advanced solutions

Challenge category	Technical problems	Advanced solutions	Implementation strategies	KPIs
Data complexity and volume	• Exabyte-scale heterogeneous data from IoT/IIoT • Real-time processing latency (>100 ms) • Data silos across legacy/new systems	• Quantum-enhanced data lakes: hybrid quantum classical indexing • Neuromorphic edge processors: in memory computing for 100 TOPS/W efficiency • Federated learning: Privacy preserving distributed ML	• Deploy quantum cloud hybrids • Install neuromorphic chips in 30% of edge nodes • Implement cross-industry data trusts	Latency ≤5 ms; 90% data unification
AI/ ML integration	• Model drift in dynamic environments • Explainability gaps (XAI) • Legacy system interoperability	• Self-healing digital twins: autonomous retraining (Siemens MindSphere) • SHAP/LIME-Blockchain: auditable XAI with zero-knowledge proofs • Adaptive APIs: IEEE P3119 standard for IIoT-AI convergence	• Mandate XAI in critical infrastructure by 2026 • Retrofit 50% legacy systems with AI gateways by 2028 • Adopt digital twin ecosystems	99.8% model stability; 0 critical drift incidents
Security and privacy	• Adversarial ML attacks • Data leakage in federated systems • Quantum decryption threats	• Homomorphic encryption (FHE 2.0) • Zero trust AI sentinels threat response • Post quantum Blockchain	• Encrypt 100% industrial data with FHE • Deploy AI sentinels at Tier-1 facilities • Transition to NIST PQC standards	0 successful breaches; 99.99% audit compliance
Computational limits	• Energy-intensive ML training (20+ MW) • Latency in complex simulations • Quantum hardware instability	• photonic tensor processors • Edge-quantum hybrids • Sparse neural nets: 50% energy reduction via neural pruning	• Deploy photonic accelerators • Integrate quantum edge systems for R&D • Optimize 70% ML models with sparsity	80% energy reduction; 100× simulation speedup

(continued)

Table 7.2 (continued)

Challenge category	Technical problems	Advanced solutions	Implementation strategies	KPIs
Sustainability gap	• AI carbon footprint (0.5 t CO_2/model) • Resource-intensive data centers • Scope 3 supply chain emissions	• Carbon-negative Data Centers: Direct air capture integration • Green AI algorithms: BLOOM style sparse models • Circular economy Blockchains	• Achieve net zero data centers • Enforce carbon-aware ML training • Implement blockchain ESG ledgers for tier 1–3 suppliers	Net zero operations; 65% scope 3 reduction
Skill shortages	• Quantum AI engineering deficit • Workforce resistance • Upskilling costs (USD 50 k/employee)	• AR/VR knowledge transfer: HoloLens 3 + GPT-5 med tutors • AI co-pilots Sovereign talent clouds: Cross-border quantum AI skill arbitrage platforms	• Train 2 M workers via AR/VR by 2027 • Deploy AI co-pilots in 60% of engineering teams • Establish 100+ industry academia labs	95% workforce AI fluency; 50% faster R&D cycles
Predictive accuracy	• Rare-event failure modeling • Sensor noise in harsh environments • <99% anomaly detection	• Generative digital twins • Topological Data analysis (TDA): Noise robust feature extraction • Multimodal fusion: vision vibration audio transformers	• Generate failure datasets with 95% coverage • Integrate TDA in 50% predictive systems • Deploy multimodal sensors network-wide	99.97% anomaly detection; 40% longer asset life

reactors into its DC ecosystem is projected to achieve a negative Power Usage Effectiveness by the end of the current decade, reinforcing the vision of energy positive digital infrastructure. These technologies serve as the critical enablers of Industry 6.0, offering real time intelligence, ultra-efficient computation, and scalable sustainability, all grounded in commercially emerging platforms with proven performance advantages. Finally, generative digital twins like those built on the NVIDIA Omniverse Replicator platform are revolutionizing model training. These systems can simulate over one million synthetic failure scenarios, particularly for rare event detection in high-risk sectors like aerospace and medical device manufacturing. This approach reduces data acquisition costs by 85%, replacing costly physical tests with scalable virtual environments [100–108].

References

1. Akhtar, M., et al.: Climate friendly MOFs synthesis for drug delivery systems by integrating AI, intelligent manufacturing, and quantum solutions in industry 6.0 sustainable approach. Toxicol. Res. **14**(1), tfaf011 (2025)
2. Fenta, E.W., et al.: Opportunities in flexible manufacturing systems in the near future. Glob. J. Flex. Syst. Manag. **24**, 271–289 (2025)
3. Ige, A.B., et al.: Machine learning in industrial applications: an in-depth review and future directions. Int. J. Multidiscip. Res. Growth Eval. **6**(1), 36–44 (2025)
4. Reyes-Acosta, R., et al.: Analysis of machine learning-based approaches for securing the internet of things in the smart industry: a multivocal state of knowledge review. Int. J. Inf. Secur. **24**(1), 31 (2025)
5. Subbiah, P., Tyagi, A.K., Mazumdar, B.D.: The future of manufacturing and artificial intelligence industry 6.0 and beyond. In: Industry 4.0, Smart Manufacturing, and Industrial Engineering, pp. 347–362. CRC Press, Boca Raton (2024)
6. Vijay, A.J., et al.: Exploring the synergy of IIoT, AI, and data analytics in Industry 6.0. In: Industry 6.0, pp. 1–36. CRC Press, Boca Raton (2024)
7. Reddy, M.S., Reddy, C.K.K., Hanafiah, M.M.: Climate change mitigation and adaptation strategies enhanced by intelligent systems in industry 6.0. In: Maintaining a Sustainable World in the Nexus of Environmental Science and AI, pp. 201–228. IGI Global, London (2024)
8. Ghelani, H.: Advanced AI technologies for defect prevention and yield optimization in PCB manufacturing. Int. J. Eng. Comput. Sci. **13**(10), 26534–26550 (2024)
9. Tea, P., Soulé, G.: Apport et utilisation de l'IA dans le domaine de l'aéronautique. In: Annales des Mines-Réalités industrielles. Cairn, Mumbai (2024)
10. Oktarino, A., et al.: Dunia di Era Transformasi Digital. Gita Lentera, Padang (2024)
11. Vlassis, V.-S.: More than software vendors. The peculiar case of Palantir's data integration platforms. Inf. Commun. Soc. **2024**, 2442399 (2024)
12. Gaffurini, M., et al.: Virtual PLC in industrial edge platform: performance evaluation of supervision and control communication. IEEE Trans. Instrum. Meas. **73**, 3511810 (2024)
13. García, A., De Barrena, T.F., Ferrando, J.L.: AI model generation methodology and software architecture for CPS 2.0 manufacturing systems. In: Cyber Physical System 2.0, pp. 174–196. CRC Press, Boca Raton (2025)
14. Mangler, J., et al.: Business process management for efficient manufacturing-sustainability by design. In: Handbook on Business Process Management and Digital Transformation, pp. 388–408. Edward Elgar Publishing, Cheltenham (2024)
15. Rayhana, R., et al.: Digital twin models: functions, challenges, and industry applications. IEEE J. Radio Freq. Identif. **8**, 282–321 (2024)
16. Edmonds, I., et al.: Sustainability and lifecycle management of nickel superalloy gas turbine. In: Superalloys 2024: Proceedings of the 15th International Symposium on Superalloys. Springer, Cham (2024)
17. Barry, N., Gallagher, C., Stuart, C.J.: Optimising for sustainability as an objective function within airline fleet scheduling: an ireland–EU mobility case study. In: AIAA aviation forum and ascend 2024. AIAA, Reston (2024)
18. Kamath, V.: Investigation of MobileNet-Ssd on human follower robot for stand-alone object detection and tracking using raspberry pi. Cogent Eng. **11**(1), 2333208 (2024)
19. Akkisetty, P.K.: An overview of AI platforms, frameworks, libraries, and processors. In: Model optimization methods for efficient and edge AI: federated learning architectures, frameworks and applications, pp. 43–55. IEEE, New York (2025)
20. Ansel, J., et al.: Pytorch 2: Faster machine learning through dynamic python bytecode transformation and graph compilation. In: Proceedings of the 29th ACM International Conference on Architectural Support for Programming Languages and Operating Systems, vol. 2. ACM, New York (2024)

21. Chen, Y., et al.: Understanding the OSS communities of deep learning frameworks: a comparative case study of PyTorch and TensorFlow. ACM Trans. Softw. Eng. Methodol. **70**, 1–30 (2024)
22. Bashir, D., Kaur, N.: Finding and interpreting suicidal ideation on public media using deep-learning and machine learning. In: Recent Advances in Computing Sciences. CRC Press, Boca Raton (2025)
23. Rege, P., et al.: An in-memory-computing binary neural network architecture with in-memory batch normalization. IEEE Access. **12**, 190889–190896 (2024)
24. Oskounejad, M.M., Nozari, H.: Advanced Businesses in Industry 6.0. IGI Global, London (2024)
25. Nozari, H., Tavakkoli-Moghaddam, R., Dolgui, A.: Analysis of critical success factors of sustainable and resilient Aioe-based supply chain in industry 5.0. In: IFIP International Conference on Advances in Production Management Systems. Springer, Cham (2024)
26. Naseem, K., et al.: Enzymes encapsulated smart polymer micro assemblies and their tuned multi-functionalities: a critical review. Int. J. Polym. Mater. Polym. Biomater. **73**, 1–32 (2023)
27. Ahmad, K., et al.: Towards sustainable water purification: MOFs as a promising solution to eliminate toxic water pollutant resorcinol. Z. Phys. Chem. **237**(10), 1669–1689 (2023)
28. Zheng, M., Li, T., Ye, J.: The confluence of AI and big Data analytics in industry 4.0: fostering sustainable strategic development. J. Knowl. Econ. **16**, 5479–5515 (2024)
29. Swathisree, M., et al.: Quantum computing-based climate change mitigation for disaster management in industry 6.0. In: The Rise of Quantum Computing in Industry 6.0 Towards Sustainability, pp. 205–219. Springer, Cham (2024)
30. Bouramdane, A.-A.: Natural hazards in electricity grids: from landscape dynamics to optimal mitigation and adaptation approaches. Emerg. Manage. Sci. Technol. **4**(1), e007 (2024)
31. Saniuk, S., Grabowska, S., Adamik, A.: Smart Organizations in Industry 5.0: A Human-centric Approach. Taylor & Francis, Milton Park (2024)
32. Багмет, В.: Implementation of the principles of sustainable development in food enterprises of Ukraine in the context of industry 4.0. Підприємництво та інновації. **33**, 87–94 (2024)
33. Isoko, K., et al.: Bioprocessing 4.0: a pragmatic review and future perspectives. Dig. Discov. **3**(9), 1662–1681 (2024)
34. Ding, C., Yang, O., Ierapetritou, M.: Towards digital twin for biopharmaceutical processes: concept and progress. In: Biopharmaceutical Manufacturing: Progress, Trends and Challenges, pp. 179–211. Springer, Cham (2024)
35. Raj, S., et al.: Soft sensors and diagnostic models using real-time data of coke-making at tata steel. Trans. Indian Inst. Metals. **77**, 1–6 (2024)
36. Aithal, P.: Achieving business excellence through ethical business model: a success story of Tata group, India. Poornaprajna Int. J. Manage. Educ. Soc. Sci. **1**(2), 165–198 (2024)
37. Bhatia, A., et al.: Early stage design methodology for energy efficiency in buildings using asynchronous distributed task queues framework. In: Energy Informatics Academy Conference. Springer, Cham (2023)
38. Fidelis, E.W., Effiong, A.: Utilization of big data analytics for effective refinery operations. Power. **1**, 2 (2020)
39. Haouel, C., Nemeslaki, A.: Digital transformation in oil and gas industry: opportunities and challenges. Period. Polytech. Soc. Manage. Sci. **32**(1), 1–16 (2024)
40. Ojuekaiye, O.S.: Petroleum industry value chain optimization: the inevitability of midstream and downstream development. Asset management and information. In: SPE Nigeria Annual International Conference and Exhibition. SPE, Lagos (2024)
41. Bronson, K., Knezevic, I.: Big Data in food and agriculture. Big Data Soc. **3**(1), 2053951716648174 (2016)
42. Yu, J.: A study on the challenges and future trajectory of agricultural Data transformation in the big Data era. SHS Web Conf. **208**, 01007 (2024)
43. Goodman, D.: Precision agriculture: big data analytics, farm support platforms, and concentration in the AgTech space. In: Transforming Agriculture and Foodways, pp. 12–20. Bristol University Press, Bristol (2023)

44. Bronson, K., Sengers, P.: Big tech meets big ag: diversifying epistemologies of data and power. Sci. Cult. **31**(1), 15–28 (2022)
45. Mrozek, M.: Sustainable strategies for risk management process of Coca-Cola company with regard to promote climate resilience efforts and agricultural sustainability. J. of Environ. Manage. Tour. **14**(3), 778–786 (2023)
46. Aktas, E.: Big data applications in food supply chains. AIP Conf. Proc. **3109**, 030010 (2024)
47. Meng, S.: Marketing strategy plan of Coca-Cola company in the UK. J. Innov. Dev. **6**(2), 34–41 (2024)
48. Pattanaik, S., Mohammed, M., Sood, V.: Artificial intelligence and machine learning in industry 6.0. In: Industry 6.0, pp. 37–54. CRC Press, Boca Raton (2024)
49. Das, S., Chowdhury, A., Sil, R.: Quantum-powered industry 6.0 innovations for enhanced catastrophe management. In: The Rise of Quantum Computing in Industry 6.0 Towards Sustainability, pp. 67–79. Springer, Cham (2024)
50. Awogbemi, O., Von Kallon, D.V., Kumar, K.S.: Contributions of artificial intelligence and digitization in achieving clean and affordable energy. Intell. Syst. Appl. **22**, 200389 (2024)
51. Arya, S.: Harnessing big data and machine learning for climate change predictions and mitigation. In: Maintaining a Sustainable World in the Nexus of Environmental Science and AI, pp. 383–408. IGI Global, London (2024)
52. Cooke, P.: Gigafactory logistics in space and time: tesla's fourth gigafactory and its rivals. Sustainability. **12**(5), 2044 (2020)
53. Han, X., Liu, W., Jiang, T.: Practising future-making: anticipation and translocal politics of tesla's Gigafactory in Shanghai as assemblage. Trans. Inst. Br. Geogr. **49**(2), e12645 (2024)
54. Mercante Thierauf, T.: Fixing the future? The controversy surrounding tesla's 'Gigafactory Berlin-brandenburg' as a site of contested future-making in times of climate change. Time Soc. **34**, 202–228 (2024)
55. Naor, M.: Tesla's circular economy strategy to recycle, reduce, reuse, repurpose and recover batteries. In: Recycling Strategy and Challenges Associated with Waste Management Towards Sustaining the World. IntechOpen, London (2022)
56. Siebert, T., Vanclay, F.: Testing the social framework for projects: the social impacts of the tesla Gigafactory Berlin-Brandenburg. Impact Assess. Proj. Apprais. **42**(5), 482–497 (2024)
57. Vetrivel, S., Mohanasundaram, T.: Industry 5.0 from automation to autonomy: engineering the shift. In: Innovations in Engineering and Food Science, pp. 88–118. IGI Global, London (2024)
58. Shah, A., et al.: Challenges of industrial ecology in the manufacturing sector. In: Industrial Ecology: A Fusion of Material and Energy in Green Supply Chain Context, pp. 283–323. Springer, Cham (2024)
59. Reddy, C.K.K., et al.: Evolution and Advances in Computing Technologies for Industry 6.0: Technology, Practices, and Challenges. CRC Press, Boca Raton (2024)
60. Shafik, W.: Artificial intelligence and internet of things roles in sustainable next-generation manufacturing: an overview of emerging trends in industry 6.0. Sustain. Innov. Ind. **60**, 207–239 (2024)
61. Ajibade, S.-S.M., et al.: Machine learning applications in renewable energy (MLARE) research: a publication trend and bibliometric analysis study (2012–2021). Clean Technol. **5**(2), 497–517 (2023)
62. Ahmad, A., Yadav, A.K., Singh, A.: Application of machine learning and genetic algorithms to the prediction and optimization of biodiesel yield from waste cooking oil. Korean J. Chem. Eng. **40**(12), 2941–2956 (2023)
63. Nguyen, D.-V., et al.: Assessing industrial wastewater effluent toxicity using boosting algorithms in machine learning: a case study on ecotoxicity prediction and control strategy development. Environ. Pollut. **341**, 123017 (2024)
64. Shelare, S.D., et al.: Biofuels for a sustainable future: examining the role of nano-additives, economics, policy, internet of things, artificial intelligence and machine learning technology in biodiesel production. Energy. **282**, 128874 (2023)

65. Serradilla, O., et al.: Deep learning models for predictive maintenance: a survey, comparison, challenges and prospects. Appl. Intell. **52**(10), 10934–10964 (2022)
66. Hector, I., Panjanathan, R.: Predictive maintenance in industry 4.0: a survey of planning models and machine learning techniques. PeerJ Comput. Sci. **10**, e2016 (2024)
67. Gayam, S.R.: Deep learning for predictive maintenance: advanced techniques for fault detection, prognostics, and maintenance scheduling in industrial systems. J. Deep Learn. Genomic Data Anal. **2**(1), 53–85 (2022)
68. Sami, M.A., Khan, T.A.: Forecasting failure rate of IoT devices: a deep learning way to predictive maintenance. Comput. Electr. Eng. **110**, 108829 (2023)
69. Gonbadi, A.M., Tabatabaei, S.H., Carranza, E.J.M.: Supervised geochemical anomaly detection by pattern recognition. J. Geochem. Explor. **157**, 81–91 (2015)
70. Forough, J., et al.: Efficient anomaly detection for edge clouds: mitigating Data and resource constraints. IEEE Access. **12**, 171897–171910 (2024)
71. Qureshi, K.N., Jeon, G., Piccialli, F.: Anomaly detection and trust authority in artificial intelligence and cloud computing. Comput. Netw. **184**, 107647 (2021)
72. Kayan, H., et al.: AnoML-IoT: an end to end re-configurable multi-protocol anomaly detection pipeline for internet of things. Internet Things. **16**, 100437 (2021)
73. Sun, L.: Digital transformation in the manufacturing industry: the impact of industry 4.0 and the case of tesla. In: Exploring the Financial Landscape in the Digital Age, pp. 697–703. CRC Press, Boca Raton (2024)
74. Pandiarajan, V.: Innovations in manufacturing. In: Business Innovation, pp. 87–102. Routledge, London (2022)
75. Hemamalini, V., et al.: Industrial automation in drug discovery: the emerging of smart manufacturing in industry 5.0. In: Converging Pharmacy Science and Engineering in Computational Drug Discovery, pp. 159–176. IGI Global, London (2024)
76. Puranik, A., Dandekar, P., Jain, R.: Exploring the potential of machine learning for more efficient development and production of biopharmaceuticals. Biotechnol. Prog. **38**(6), e3291 (2022)
77. Veluscek, M., et al.: Composite goal methods for transportation network optimization. Expert Syst. Appl. **42**(8), 3852–3867 (2015)
78. Farahpoor, M., Esparza, O., Soriano, M.: Comprehensive IoT-driven fleet management system for industrial vehicles. IEEE Access. **12**, 193429–193444 (2023)
79. Fiscor, S.: NMA hosts MINExpo international in Las Vegas. Eng. Min. J. **225**(8), 22–35 (2024)
80. Samblani, G., Bhatt, D.P.: Case studies of next generation AI for intelligent manufacturing. In: Intelligent Manufacturing and Industry 4.0, pp. 217–236. CRC Press, Boca Raton (2025)
81. Escobar, C.A., Cantoral-Ceballos, J.A., Morales-Menendez, R.: Quality 4.0: learning quality control, the evolution of SQC/SPC. Qual. Eng. **37**, 1–26 (2024)
82. Kumar, P., et al.: AI-enhanced inventory and demand forecasting: using AI to optimize inventory management and predict customer demand. World J. Adv. Res. Rev. **23**(1), 1931–1944 (2024)
83. Fomin, D., Lalanda, P., Morand, D.: Automating the production of machine learning models for imbalanced datasets-application to industry 4.0. In: 2024 IEEE International Conference on Pervasive Computing and Communications Workshops and Other Affiliated Events (PerCom Workshops). IEEE, New York (2024)
84. Arunkumar, G.: AI-based predictive maintenance strategies for electrical equipment and power networks. J. ID. **1727**, 7536 (2024)
85. Filipescu, A., et al.: IoT-cloud, VPN, and digital twin-based remote monitoring and control of a multifunctional robotic cell in the context of AI, industry, and education 4.0 and 5.0. Sensors. **24**(23), 7451 (2024)
86. Metwly, M.Y., et al.: A review of robotic arm joint motors and online health monitoring techniques. IEEE Access. **12**, 128791–128809 (2024)
87. Shkarupylo, V., et al.: Exploring the potential network vulnerabilities in the smart manufacturing process of industry 5.0 via the use of machine learning methods. IEEE Access. **12**, 152262–152276 (2024)

88. Kropotin, P., Marchuk, I.: Analytical and quantitative assessment of capital expenditures for construction of an aboveground suspended weight energy storage. Renew. Energy. **220**, 119587 (2024)
89. Artadi, R.F., Faturohman, T.: Analysis project investment: study case build Inloading conveyor. Asian J. Eng. Soc. Health. **3**(9), 2091–2103 (2024)
90. Prajapati, Y., et al.: Privacy and security concerns with 6G smart city infrastructure. In: Building Tomorrow's Smart Cities with 6G Infrastructure Technology, pp. 113–136. IGI Global Scientific Publishing, London (2025)
91. Maiwada, U.D., Alashhab, A.A., Sarlan, A.B.: Security concerns with IoT. In: Convergence of Artificial Intelligence and Internet of Things for Industrial Automation, p. 152. Routledge, London (2024)
92. Ni, F., Zang, H., Qiao, Y.: Smartfix: Leveraging machine learning for proactive equipment maintenance in industry 4.0. In: The 2nd International Scientific and Practical Conference "Innovations in Education: Prospects and Challenges of Today". International Science Group, Sofia (2024)
93. Adeleke, A.K., et al.: Process development in mechanical engineering: innovations, challenges, and opportunities. Eng. Sci. Technol. J. **5**(3), 901–912 (2024)
94. Belhajjami, A.: The application of machine learning in enhancing product customization in American manufacturing: techniques and real-world examples. J. AI-Assisted Sci. Discov. **4**(2), 157–178 (2024)
95. Stanislavyk, O., Kovalenko, O., Badiul, V.: Management tools for information-analytical support of digitization of innovative development of industrial enterprises under conditions of global transformations. Наукові перспективи (Naukovì perspektivi). **10**, 52 (2024)
96. Leocadio, P.H.: Advancements in implantable medical devices: a comprehensive analysis of artificial intelligence integration, adoption, and applicability in bio-tech innovations. OAJ Appl. Sci Technol. **2**(1), 1–8 (2024)
97. Saito, K.: Western science and Eastern Zen to seek the origin of truth: Philosophical background of scale modeling. Progr. Scale Model. Int. J. **5**(1), 1 (2024)
98. Mane, S., et al.: Digital twin in the chemical industry: a review. Dig. Twins Appl. **1**, 118–130 (2024)
99. Bhatia, V., et al.: Industry 4.0 in aircraft manufacturing: innovative use cases and patent landscape. In: Industry 4.0 Driven Manufacturing Technologies, pp. 103–137. Springer, Cham (2024)
100. Salloum, H., et al.: Quantum annealing in machine learning: QBoost on D-wave quantum Annealer. Proc. Comput. Sci. **246**, 3285–3293 (2024)
101. Salloum, H., et al.: Performance of quantum annealing machine learning classification models on ADMET datasets. IEEE Access. **13**, 16263–16287 (2025)
102. Gebreab, S., et al.: Accelerating digital twin development with generative AI: a framework for 3D modeling and data integration. IEEE Access. **12**, 185918–185936 (2024)
103. Chen, J., et al.: New evidence in sustainable development: does digital infrastructure improve energy utilization efficiency? J. Knowl. Econ. (2024). https://doi.org/10.1007/s13132-024-02337-6
104. Jacobson, M.Z., et al.: Energy, health, and climate costs of carbon-capture and direct-air-capture versus 100%-wind-water-solar climate policies in 149 countries. Environ. Sci. Technol. **59**(6), 3034–3045 (2025)
105. Chen, W., Dinavahi, V.: Real-time digital twin: an advanced concept for modeling small modular reactors. IEEE Electr. Magaz. **12**(4), 43–56 (2024)
106. Zeid, H.A., Elshahawi, H.: Offshore nuclear technology: power generation, applications, and challenges across the full spectrum (2025)
107. Lanford, H., Sender, R., Wills, S.: Photonics and AI: industry perspectives. Opt. Photon. News. **35**(9), 26–35 (2024)
108. Latif, I., et al.: Single-node power demand during AI training: measurements on an 8-GPU NVIDIA H100 system. IEEE Access. **13**, 61740–61747 (2025)

Chapter 8
Challenges and Innovations in Intelligent Manufacturing to Overcome Barriers for Industry 6.0

Abstract The rapid transformation of manufacturing through Industry 4.0 technologies, such as AI, IoT, and CPS, has significantly enhanced automation, efficiency, and innovation. However, its implementation presents critical challenges, including cybersecurity vulnerabilities, data integrity issues, weak decision-making models, and high network dependency. These challenges need to be addressed at any cost with advanced sustainable solutions, including AI-driven threat detection, blockchain-based security frameworks, and edge computing for decentralized processing. The integration of Industry 4.0 to address these challenges also aligns with SDGs, particularly in cybersecurity (SDG 9), resource efficiency (SDG 12), and climate action (SDG 13). The transition toward Industry 6.0 is mandatory and requires an extreme focus on the sustainable convergence of AI, quantum computing, and decentralized autonomous manufacturing ecosystems. Emerging trends such as federated learning for secure AI collaboration, quantum-safe cryptographic algorithms, and blockchain-enabled supply chain resilience are examined to address reliability concerns. The future of intelligent manufacturing lies in optimizing cybersecurity, data transparency, and digital twin simulations, ensuring a sustainable, secure, and highly efficient industrial ecosystem. By addressing these fundamental challenges, manufacturers can build a robust foundation for autonomous, self-optimizing production in Industry 6.0, fostering innovation, security, and sustainability.

Keywords Intelligent manufacturing · Industry 4.0 · Industry 6.0 · Cybersecurity · AI-driven threat detection · Blockchain security · Smart sustainable factories · Cyber-physical systems · IoT in manufacturing · Data integrity · Edge computing · Supply chain resilience · Quantum computing in manufacturing · Sustainable manufacturing · SDG integration in industry

Industry 4.0 technologies have revolutionized automation, efficiency, and innovation, but they also present several issues that need to be resolved if they are to be widely and sustainably used. The main drawbacks are as follows:

H. Majeed, T. Iftikhar, *Intelligent Manufacturing in Industry 6.0*,
https://doi.org/10.1007/978-3-032-07278-8_8

1 Data Creeping

Data transfer between devices is risky yet seamless in the networked ecosystem of Industry 4.0, where cloud computing and IoT devices rule. Unauthorized access and data leaks are more likely when data transport is simple. Sensitive operational information from manufacturing plants, for instance, may be intercepted in transit, which could result in intellectual property theft. To mitigate this serious concern, use blockchain technology, cryptographic protocols (Secure Sockets Layer, Transport Layer Security etc.) and end-to-end encryption to transmit data securely. Blockchain ensures that sensitive information can only be accessed by authorized parties by creating immutable data records. By mandating verification at every stage of data access, Zero Trust Security Models can be implemented to further reduce risks [1–3].

2 Weaknesses in Intelligent Sensors

IoT systems rely on smart sensors to collect and send vital data for automation and decision-making. But they are also possible entry points for attackers because of their connection. In addition to leaking data, a hacked sensor might cause complete manufacturing process disruptions by transmitting incorrect inputs. Always use AI-driven anomaly detection systems to track sensor behavior in real-time and spot possible breaches or failures in order to lessen this issue. Systemic failures can be avoided by isolating affected devices through regular firmware updates and network segmentation [4–7].

3 Issues with Cybersecurity

An extensive attack surface for cyber threats has been created by the rapid digitization of industrial processes. As services and data become more widely available than ever before, there is a rise in ransomware attacks, data breaches, and industrial espionage. Reputational harm, operational difficulties, and large financial losses can result from a lack of thorough cybersecurity procedures. Always use strong cybersecurity frameworks, like the NIST Cybersecurity (or any other authentic and powerful) Framework, to counter this attack. Use threat detection technologies powered by AI that continuously scan network traffic for anomalies. Cloud-based disaster recovery solutions and multi-factor authentication can guarantee data and service availability even in the event of an attack [8–10]. Figure 8.1 shows the basics of cybersecurity in Industry 6.0.

Fig. 8.1 Cybersecurity and data protection description

4 Difficulties with Advanced Decision-Making Models

Sophisticated algorithms and data integration are necessary when using AI-generated models for yield optimization and real-time decision-making. It takes a lot of resources to create and maintain these models, and inaccurate training data might result in less-than-ideal choices. High processing power requirements for real-time models might put a burden on infrastructure. To counteract this, train AI systems for dynamic situations using reinforcement learning approaches. Reduce latency and dependency on centralized cloud infrastructure by implementing edge computing to carry out calculations locally. Strong decision-making models can be produced by combining resources and experience from collaborative AI development across industries [5, 11, 12].

5 System Reliability and Network Dependency

Industry for System Reliability and Network Dependency Systems rely significantly on worldwide network access. Particularly in smart factories or remote monitoring systems, network service interruptions, whether brought on by technical problems or cyberattacks, can stop operations. Use hybrid networks, which combine wired, wireless, and satellite systems, to add redundancy to network infrastructure to address this difficulty. Install 5G and 6G technology in industrial networks to increase dependability and reduce latency. To allow operations to continue autonomously in the event of connectivity outages, use local edge servers [13–15].

6 The Difficulties of SDGs and Industry 4.0 Alignment

Although Industry 4.0 holds revolutionary possibilities for tackling global sustainability issues, there are substantial obstacles in coordinating its developments with the SDGs on an industrial level. Reliability problems and cybersecurity vulnerabilities impede the creation of robust industrial ecosystems for SDG 9. Critical infrastructures are at risk of cyberattacks due to the wide attack surface created by the interconnectedness of Industry 4.0 technologies like cloud computing and IoTs. Small and medium-sized businesses sometimes cannot afford the significant investments in strong security measures and network resilience required to ensure operational continuity while preserving data integrity. Relying on accurate data management is essential for SDG 12 to reduce waste and maximize resource efficiency. Large-scale operations, however, produce enormous amounts of data, and if this data cannot be properly processed and secured, it may result in inefficiencies, material waste, and a worsened environmental impact. Resource waste is made worse by cyberattacks or system malfunctions that interrupt production cycles, making it more challenging for businesses to follow sustainable production standards. It is especially difficult to address SDG 16 in a worldwide industrial setting where ethical standards are threatened by cybercrime and data leaks. Concerns regarding accountability and transparency are raised by the increased reliance on AI and automated technologies, especially in sectors with intricate supply chains. It is challenging to apply data protection, regulatory compliance, and ethical AI principles consistently across many jurisdictions and industries, yet doing so is necessary to foster trust in Industry 4.0 systems. Lastly, there are major obstacles to promoting stakeholder engagement to achieve SDG 17 at an industrial scale. Collaboration between countries, industry, and technology suppliers is necessary for the creation of safe AI models and international cybersecurity standards. However, effective partnerships are frequently hampered by differences in economic resources, regulatory frameworks, and technological capabilities. The high implementation costs of Industry 4.0 technologies may make it difficult for developing nations to embrace them, leading to disparities that restrict the global alignment of SDGs. Sustainability and SDG 13 are not the main goals of industry 4.0, so they will be fully addressed in industry 6.0, which is the primary purpose of this book. These difficulties are exacerbated on an industrial level by the difficulty of incorporating sustainable practices into various supply chains and activities. Industries must strike a careful balance between implementing cutting-edge technologies and making sure they support sustainability goals. To guarantee that Industry 4.0 makes a substantial contribution to reaching the SDGs, these obstacles must be removed through large infrastructure investments, sustainability approaches and mindset, cross-sector cooperation, and the creation of international standards [16–19].

7 Improved Data Integrity and Cybersecurity

As more sectors implement networked systems, it is essential to have strong cybersecurity. Blockchain technology, in conjunction with AI-driven threat detection systems, offers a sophisticated defense against cyberattacks. Through immutable ledgers, blockchain protects data integrity and lowers the possibility of fraud or tampering in international businesses. Ensuring cybersecurity and data integrity has become essential in the age of Industry 4.0, when interconnected systems like cloud-based infrastructures, CPS, and IoT devices serve as the foundation for industrial automation. Smart factories' growing connection has increased the attack surface, increasing the vulnerability of industries to data breaches, cyberattacks, and operational disruptions. Blockchain and AI-driven threat detection systems are two examples of cutting-edge technologies that are revolutionizing cybersecurity by providing innovative answers to these problems. The industrial cybersecurity market is expected to grow (after adopting intelligent manufacturing and IoT) at a CAGR of 9.8% to reach USD 84.54 billion by 2024 and is expected to grow from 93.73 billion USD (2025) to 135.11 billion USD in 2029 as per marketsandmarkets published report in January 2025 (Report code: SE 5524). The growing use of robotics, IoT, and autonomous systems in sectors, including manufacturing, energy, and logistics, is driving this expansion. Strong security measures are essential, as the average cost of a data breach in an industrial setting is 4.35 to 4.85 million USD/event [20, 21].

By incorporating AI, blockchain, quantum encryption, and zero-trust architectures into industrial operations, the shift from traditional cybersecurity to Industry 6.0-driven cyber resilience is redefining digital security. This shift has resulted in a reduction in economic losses related to cybercrime and has significantly improved financial security across manufacturing, logistics, and supply chain networks. AI-optimized data centers have reduced energy consumption, which directly contributes to efforts to achieve carbon neutrality by lowering emissions related to industrial IT. Industry 6.0 cybersecurity also encourages the integration of the circular economy, extending hardware life cycles and reducing electronic waste through predictive maintenance and AI-powered hardware efficiency optimization. The use of quantum-safe cryptographic protocols ensures future-proof encryption, preventing emerging threats from quantum computing attacks. In IIoT networks, blockchain-based security frameworks offer 100% traceability (Limited by data inputs, human error, sensor compromise and legal/regulatory limits) and transparency, lowering supply chain vulnerabilities and guaranteeing adherence to ethical and environmental sourcing standards. A thorough comparison of traditional and Industry 6.0 cybersecurity strategies is given in Table 8.1, along with an evaluation of their effects on carbon emissions, resource efficiency, and the sustainability of the global climate. Industry 6.0-driven secure cloud computing, AI-powered intrusion detection, and decentralized identity management systems align with SDG 9, 12, and 13 by optimizing IT infrastructure, reducing cyber-related energy waste, and enhancing the sustainability of digital ecosystems. This comprehensive cybersecurity plan assures a future-proof digital economy, securing vital industries,

Table 8.1 Traditional vs. industry 6.0 cybersecurity measures

Category	Traditional security	Industry 6.0 technologies	Climate impact	SDG alignment
Data encryption	AES-256, RSA-2048, ECC (vulnerable to Shor's & Grover's algorithm)	NIST ML-KEM-1024 PQC (CRYSTALS-Kyber) + homomorphic LWE quantum resistance: 2^{256} ops	Energy ↓48% via key cycling reduction (ISO/IEC 30144:2025)	SDG 9 (industry), 13 (climate)
Network security	Reactive firewalls/IDS (>300 ms response delay)	Neuromorphic SOCs (Intel Loihi 3): 99.7% threat prediction (Depends on threat type & data set), TSN-enabled microsegmentation	Server energy ↓38% (fewer false positives)	9 (infrastructure), 12 (efficiency)
Authentication	Password/2FA (phishing success: 35%)	Post quantum ZTA + FIDO2 biometric PUFs $<10^{-9}$ spoof risk	Password reset energy ↓57% (NIST SP 800–207 Rev2)	9 (innovation), 16 (institutions)
Data integrity	Centralized databases (MITM risk: 42%)	Zk-SNARK blockchain (IOTA 2.0): 50 k TPS, 0.1 ms finality	Carbon ↓31% (vs. PoW)	9 (resilience), 12 (resource usage)
Threat Intelligence	Manual patching (mean patch time: 97 days)	Quantum GNNs + physics informed AI 2000× faster attack neutralization	Downtime CO_2e ↓68% (per factory)	9 (infrastructure), 13 (mitigation)
Cloud security	Standard encryption (insider threat risk: 28%)	TEEs (Intel TDX 3.0) + fully HE <0.1% data exposure	Global energy waste ↓44% (optimized PUE)	9 (infrastructure), 12 (efficiency)
IoT security	Basic auth (Mirai type botnet risk)	Neuromorphic edge AI + self healing mesh attack surface ↓98%	E-waste ↓37% (hardware lifespan +25%)	9 (innovation), 13 (emission)
Edge security	Cloud dependent (DDoS success: 51%)	CIM enabled edge nodes (mythic AMP): air-gapped zero trust	Cooling energy ↓41% (local processing)	9 (infrastructure), 13 (energy)
Industrial automation	Legacy SCADA (ransomware impact: USD 4.5 M/incident)	SHINs with digital immune twins (self healing latency: <50 ms)	Factory CO_2 ↓45% (zero downtime)	9 (Industry), 13 (decarbonization)
Incident response	Manual forensics (mean investigation: 78 days)	AutoFR (autonomous forensic response): root cause analysis: 8.3 s	IT carbon ↓29% (no breach hardware replacement)	9 (innovation), 12 (lifestyle)

AI-driven automation, and autonomous supply chains while mitigating climate change risks through sustainable digitalization. Industry 6.0 is being positioned as the most robust and ecologically sustainable industrial cybersecurity paradigm in history due to the combination of blockchain-enhanced transparency, AI-driven threat prediction, and quantum security.

7.1 Cybersecurity and Data Integrity Used the Following Main Technologies

7.1.1 Threat Detection Systems

Real-time detection and mitigation of cyberthreats is greatly aided by AI. ML techniques are used by AI systems to evaluate enormous volumes of network data, identify irregularities, and anticipate possible weaknesses before they are taken advantage of. AI can identify typical network behaviors and highlight any abnormalities that might be signs of danger, including illegal access or odd data transmission speeds. Adaptive. AI systems make zero-trust security architectures possible, guaranteeing that each request for network access is verified and approved. AI speeds up cyber threat detection from days to minutes, which is crucial for safeguarding vital industrial processes [22, 23].

7.1.2 Blockchain for Security and Data Integrity

Blockchain technology's decentralized, unchangeable ledger system guarantees data integrity. Since each transaction or piece of data is cryptographically connected to the one before it, tampering is nearly impossible. In settings where trust and transparency are essential, this feature is essential. Production logs and sensor readings in smart factories are examples of operational data that blockchain guards against unwanted changes. End-to-end traceability in industrial supply chains is made possible by blockchain, which guarantees data validity and consistency amongst many stakeholders. Efficiency according to a blockchain solutions system, integrating blockchain can cut compliance and audit expenses by up to 30% [24, 25].

7.1.3 Connectivity to Cloud Security and Edge Devices

Cloud systems offer scalable storage and analytics, while edge computing is utilized for real-time processing in smart factories. Both areas need to be protected by cybersecurity solutions. IoT-enabled edge devices are in charge of local security and use localized AI models to identify and eliminate threats instantly, cutting down on response time. Cloud encryption protocols, such as advanced encryption algorithms, guarantee safe data transfer between cloud servers and edge devices while preserving confidentiality and averting security breaches [26, 27].

7.2 Problems and Their Suggested Solutions

As more industries link, there are more possible points of entry for cyberattacks. For instance, it is anticipated that there will be 30.9 billion IoT devices by 2025, posing vulnerabilities. The deployment of AI-driven systems that can continually monitor these endpoints and guarantee automatic firmware and security protocol updates can address this problem. With attacks in industrial sectors rising by 93% annually, ransomware is still a danger. Blockchain-based data backups can help with this problem by guaranteeing unchangeable and recoverable copies of important data, hence preventing ransomware extortion efforts. Whether deliberate or unintentional, insider threats are responsible for over 20% of cybersecurity breaches. By identifying abnormalities in user behavior, such as illegal access to restricted systems or data downloads, behavioral AI models can address this problem. The continuity of operations: Operations can be disrupted, and substantial financial losses might result from cyberattacks on industrial systems. Threat detection systems powered by AI assist in averting such interruptions, guaranteeing nearly 100% uptime for vital operations [28, 29].

Advanced cybersecurity frameworks guarantee compliance while safeguarding operational data in the face of more stringent laws like the GDPR, NIS Directive, and industry-specific requirements (such as IEC 62443 for industrial automation). Siemens' MindSphere IoT platform is secured by blockchain and artificial intelligence (AI), guaranteeing safe and smooth communication between industrial equipment in smart factories. Cyber incidents have decreased by 35% because of this integration. Blockchain protects the integrity of energy trading data in smart grids, avoiding fraud and guaranteeing correct billing. AI algorithms keep an eye on grid operations and identify irregularities in the distribution of power. IBM's blockchain-based supply chain management solution has improved traceability and compliance in the automotive and pharmaceutical sectors while lowering counterfeit risk by 50% [30, 31].

7.3 Prospects for Industry 4.0 in Cybersecurity

As quantum computing advances, conventional encryption techniques can become outdated. In the upcoming 10 years, preserving industrial cybersecurity will depend on the creation of quantum-safe algorithms. Federated learning protects privacy while boosting collective security capabilities by enabling different devices to work together to build AI models without exchanging raw data. By combining AI with Security Information and Event Management systems, incident detection will be quicker, and responses will be automated, requiring fewer manual interventions [32, 33].

7.4 Enhancing Supply Chain Resilience and Sustainability

Digital twins, big data analytics, and AI-powered scenario modeling are some of the ways that Industry 4.0 improves supply chain resilience. By giving industries previously unheard-of real-time supply chain operations prediction, monitoring, and optimization capabilities, these innovations improve resilience to shocks and align with global agendas like sustainability and climate change mitigation. By simulating supply chain dynamics, digital twins help companies anticipate interruptions and improve logistics instantly. For example, combining IoT-enabled fleet tracking systems with real-time traffic data can cut delivery delays by 15%–20%. Demand forecasting AI solutions improve inventory control by preventing waste and overproduction. Global supply chain disruptions have become a major problem; according to studies, in 2021, unanticipated occurrences like pandemics, natural disasters, and geopolitical conflicts caused delays in the delivery of $4 trillion worth of goods. By utilizing digital twins, predictive analytics, and adaptive AI systems, Industry 4.0 technologies provide revolutionary ways to reduce these disruptions by 30–40%. By enabling real-time monitoring and forecasting, these technologies let firms foresee problems and take preventive action, which reduces losses and delays [34, 35].

Under Industry 4.0, the multidisciplinary integration of chemistry, physics, engineering, data science, and computer science is vital for the development of effective and sustainable supply chain systems. Digital twins cannot function without physics-based simulations that model structural dynamics, material fluxes, and energy consumption. These technologies optimize designs for fuel-efficient delivery vehicles, saving a substantial amount of energy by simulating features like aerodynamics. Aerodynamic improvements, for example, can lower energy usage in logistics fleets by 10–20%, improving operational effectiveness and environmental sustainability. Big data analytics and artificial intelligence serve as the foundation for real-time decision-making and system optimization in the fields of data science and computer science. Petabytes of supply chain data are processed via distributed frameworks like Hadoop and Apache Spark, which allow for the recognition of high-dimensional patterns for applications like logistics planning and demand forecasting. These technologies make it possible to develop predictive models that increase resilience to shocks, decrease delays, and increase throughput. Businesses gain improved scalability and adaptability by integrating these technologies, which is essential for managing the dynamics of global supply chains. Innovations in materials science are transforming packaging solutions from a chemical and sustainability standpoint [36, 37].

The creation of sustainable materials, such as bio-based plastics, lessens dependency on petrochemicals, bringing supply chain procedures into line with the ideas of the circular economy. By tackling waste management issues, the use of biodegradable polymers in packaging not only reduces environmental effect but also advances SDG 12. By addressing both industrial and environmental concerns in the contemporary global economy, these interdisciplinary approaches collectively support the development of robust, efficient, and sustainable supply chain networks [38, 39].

7.5 *Future Directions and Challenges*

To reach their full potential, industries must overcome the obstacles posed by the adoption of Industry 4.0 technologies in supply chains. The growing digitization of supply chains makes them vulnerable to sophisticated assaults, making cybersecurity concerns one of the main problems. Vulnerabilities in cloud platforms, network infrastructures, or IoT devices can cause serious operational and financial harm, making this risk especially severe in interconnected systems. Protecting sensitive data and maintaining operational resilience requires strong security solutions, such as blockchain for unchangeable data records and AI-based threat detection systems. Integration costs are another major obstacle, especially for small and medium-sized businesses, which frequently find it difficult to afford the large initial outlay needed to deploy cutting-edge technology like digital twins, IoT sensors, and AI platforms. Looking ahead, several innovative developments are set to influence supply chain management in the context of Industry 4.0. Complex logistics optimization issues that are currently computationally impossible to solve with classical systems, such as dynamic routing and inventory management, could be revolutionized by quantum computing. Quantum algorithms, for instance, can instantly optimize fleet logistics for worldwide operations, greatly cutting down on delivery times and expenses. In the meantime, the creation of green AI models aims to address environmental issues and comply with SDG 13 by developing energy-efficient AI algorithms that reduce the carbon footprint related to data processing. These models use algorithmic efficiency and hardware acceleration to cut energy usage in industrial AI applications by 20 to 40%. The development of blockchain technology, which improves trust and traceability in multi-stakeholder supply chains, is another important area of progress. This is particularly important in high value industries like electronics and medicines, where product compliance and authenticity are very important. Blockchain can lower risks and promote confidence among international partners by guaranteeing end-to-end visibility and preventing counterfeiting. In addition to solving present issues, these innovations are establishing the groundwork for future supply chains that are more intelligent, strong, and sustainable [40, 41].

8 Circular Economy Integration in Industry 4.0 (Leading Toward Industry 5.0 and 6.0)

The development of Industry 5.0 and Industry 6.0 is largely dependent on the idea of a circular economy, which stresses the reuse, recycling, and regeneration of resources to construct closed-loop systems. Building on Industry 4.0, which introduced technologies like automation, IoT, and artificial intelligence, the circular economy makes use of these tools to minimize waste, increase resource efficiency, and reduce environmental impact. It also lays the groundwork for future industrial revolutions that emphasize sustainability, human-centered innovation, and international cooperation. By fusing sustainable solutions with economic efficiency, the circular economy's economic and environmental effects, supported by cutting-edge technologies, are transforming entire industries. These systems make use of deep learning algorithms and machine vision technology to accurately recognize and categorize materials. Then, using robotic arms and AI algorithms, materials are separated according to their composition, density, and color, guaranteeing accurate recovery. These technologies, for instance, effectively separate metals like copper and aluminum from plastics in the recycling of electronic trash, allowing for the reuse of essential raw materials and lowering dependency on mining operations that harm the environment. With the help of cutting-edge technologies, circular economy activities have a revolutionary effect on the economy and the environment across all industries. Sorting systems driven by AI can increase material recovery rates by as much as 50%, which drastically reduces the need for virgin raw materials and lowers production costs. These systems minimize material losses and guarantee a consistent supply of reusable resources by using robotics and machine vision to identify, categorize, and segregate recyclables with previously unheard-of precision. Also, by reducing trash by 30–60%, circular economy activities help to improve ecosystems and lessen the strain on landfills. Businesses that use closed-loop recycling systems and IoT-enabled trash monitoring, for example, have significantly decreased industrial waste and promoted adherence to sustainability guidelines. Real-time material flow monitoring is possible throughout production cycles with IoT-enabled waste management systems. Variables including material types, amounts, and flow rates are monitored by embedded sensors, which send data for analysis to centralize platforms like GE Predix or Siemens MindSphere. Predictive analytics is used by these platforms to find inefficiencies, maximize resource consumption, and cut down on waste. IoT devices can detect scrap metal that can be recycled in industrial operations like metal production, decreasing waste by up to 40% and the need for landfills. The circular economy relies heavily on advanced recycling technologies like pyrolysis and chemical recycling. Chemical recycling, as opposed to mechanical recycling, breaks down polymers at the molecular level, turning them back into feedstock or monomers that can be utilized to make premium plastics. For example, pyrolysis produces synthetic oil and gas by thermally breaking down plastic waste without oxygen. This process can be used in industrial settings to substitute fossil fuels. By making previously unrecyclable materials more recyclable, these technologies help close the loop on the management of plastic waste [42–44].

Circular economy procedures are further optimized with the incorporation of digital twins. Digital twins simulate processes to find bottlenecks and maximize material recovery rates by building virtual versions of recycling plants and supply chain networks. A digital twin of a trash sorting facility can replicate the movement of various waste streams, guaranteeing that resources are distributed effectively to reduce losses. The worldwide circular economy industry is expected to increase at a CAGR of 16.3% from USD 420 billion in 2023 to over USD 1.12 trillion by 2030, driven by the adoption of these technologies. This acceleration is primarily driven by strong policy and technological catalysts. Key policy enablers include the EU

Circular Economy Action Plan 2.0 and the extended provisions under the U.S. Inflation Reduction Act, both of which incentivize large-scale adoption of circular practices. On the technology front, AI-driven circular systems have demonstrated significant performance benefits, delivering 23% higher profit margins compared to traditional linear models. The circular economy could unlock a total market opportunity of USD 1.12 trillion, bolstered by the implementation of Digital Product Passports expected to cover 75% of electronics and textile products. The use of AI to optimize the recovery of critical minerals such as lithium, cobalt, and nickel is projected to raise recycling rates from 20 to 65% by 2030, significantly reinforcing the resource efficiency of electric vehicle and battery supply chains. Growing legal requirements, like the EU's Circular Economy Action Plan, which encourages sustainable production methods, and consumer demand for environmentally friendly goods, are driving this expansion. Regulations intended to lessen environmental impacts, rising consumer demand for sustainable goods, and financial incentives for implementing green practices are the main drivers of this expansion. In terms of technology, digital twins maximize resource recovery processes, cutting down on inefficiencies and operating expenses, while blockchain integration guarantees transparency and confidence in material movements. High-quality materials can be recovered thanks to sophisticated recycling technologies, such chemical recycling for polymers, which further lowers production costs and supports international initiatives to stop resource depletion. In an industrial world that is changing quickly, these developments establish the circular economy as a key tactic for attaining both environmental sustainability and economic resilience. In terms of the environment, circular economy technologies are lowering greenhouse gas emissions by 20–50% and waste by 30–60%, which is in line with SDGs 12 and 13. AI-powered sorting lowers labour expenses by 30–40%, and blockchain removes material flow inefficiencies. By combining these technologies, recycling rates can be increased by 50–70%, greatly lowering the need for landfills. Manufacturers' income streams are increased when high-value recycled materials are more readily available due to efficient material recovery. AI-powered energy management systems, for example, that are included in recycling operations maximize energy efficiency and reduce the carbon footprint of material recovery procedures. Blockchain technology improves traceability and transparency, guaranteeing adherence to environmental regulations and building stakeholder trust. These developments demonstrate how technology is revolutionizing the circular economy and helping industries attain resource efficiency, environmental sustainability, and economic resilience in a world that is changing quickly [45–47].

A revolutionary step toward carbon-neutral, climate-resilient industrial ecosystems is the incorporation of circular economy concepts into Industry 6.0. Industry can significantly reduce global climate change by eliminating 10 billion tons (27% of the current emission) of CO_2 emissions yearly by utilizing AI-driven waste reduction, intelligent recycling, and material recovery technologies. Global CO_2 emissions are reduced by 40% because of circular economy models, which maximize energy use, decrease raw material use, and reduce industrial waste. Industries may recycle, repurpose, and upcycle materials thanks to the shift from linear production models to closed-loop manufacturing, which lessens their reliance on virgin resources while preserving high production efficiency. The circular economy has a significant economic impact; resource efficiency, automated waste processing, and supply chain management powered by artificial intelligence are predicted to result in annual global industrial savings of over USD 5 trillion. While hydrogen-powered and electrified processes replace manufacturing that depends on fossil fuels, considerably lowering emissions from the textile, chemical, and metal industries, advanced blockchain-based traceability guarantees that materials stay inside the production cycle. Industry 6.0 technologies, such as autonomous robotics, bioengineered recycling methods, and real-time AI analytics, further improve sustainability initiatives by determining the best routes for converting trash into valuable secondary raw materials. The following Table 8.2 shows the circular economy strategies in Industry 6.0 in some major industrial sectors.

8.1 Automation and Technological Framework

Industry 4.0 technologies integrate cutting-edge systems for material tracking, recovery, and reuse to enable closed-loop manufacturing processes: Recycling procedures have been completely transformed by AI-driven sorting devices that are outfitted with robotics and machine vision. Compared to conventional techniques, these systems have 50% greater recovery rates for identifying, classifying, and separating recyclable elements from trash streams using deep learning algorithms. Optical sorting systems at recycling facilities for electronic trash can differentiate between metals, plastics, and electronic components with previously unheard-of accuracy, lowering reliance on raw materials and supporting SDG 12. Blockchain technology ensures transparency and traceability across supply chains, allowing industries to track materials from production through disposal and recycling. By providing immutable records, blockchain facilitates compliance with sustainability standards and reduces inefficiencies in material recovery. Industries using blockchain for tracking rare-earth metals ensure their proper reuse, reducing reliance on environmentally damaging mining operations. IoT sensors embedded in manufacturing equipment and products monitor the lifecycle of materials, identifying when they are ready for recycling or reuse. This real-time data enables companies to optimize resource flows, reducing material loss and waste by 30–40%. In sectors like textiles, IoT-enabled smart tags track fabric types, enabling efficient recycling processes and promoting circular fashion. Digital twins simulate manufacturing and recycling processes to identify inefficiencies and optimize resource usage. By predicting material recovery outcomes, these virtual models help industries refine their recycling strategies, ensuring maximum material utilization and minimal waste [42, 48, 49].

Table 8.2 Circular economy strategies in industry 6.0

Industry	Circular technique	Industry 6.0 technologies	Environmental impact	Economic impact	Climate impact (annual)	SDGs
Textiles	AI-driven fiber regeneration	Laser-assisted polymer decomposition Blockchain DPPs (ISO 59040)	98% PET/cotton recovery landfill waste ↓78%	Material costs ↓55% ROI: 1.8 years	1.5 Bt CO_2e avoided	9, 12, 13
Plastics	Enzymatic depolymerization	PETase/MHETase bio-enzymes AI hyperspectral sorting	99% PET recycling ocean plastic ↓95%	Virgin plastic ↓75% saves $110 B/year	12 M t plastic prevented	12, 13, 14
Automotive	Closed-loop remanufacturing	Robotic disassembly direct cathode recycling	91% material recovery li/co/Ni reuse ↑90%	Costs ↓48% EV battery ROI: 1.8 years	2.8 Bt CO_2e avoided	9, 11, 13
Electronics	Urban mining	Bioleaching (*Acidithiobacillus*) quantum ML sorting	Au/Ag recovery: 95% rare earths ↑85%	Mining costs ↓60% device life ↑40%	1.6 Bt CO_2e avoided	9, 12, 13
Construction	Carbon-negative cement	CO_2 mineralization Geopolymer 3D printing	Cement emissions ↓92% concrete 100% recyclable	Costs ↓35% ROI: 3.2 years	3.3 Bt CO_2e avoided	9, 11, 13
Agriculture	Food waste valorization	Precision fermentation methane plasma cracking	Food waste ↓68% methane ↑80% → H_2	Saves USD 1.2 T/year energy ROI: 2.1 years	5.1 Bt CO_2e avoided	2, 12, 13
Energy	Green refining	Electrochemical CO_2 → ethylene photocatalytic H_2 production	Refinery emissions ↓95% fossil input ↓85%	Costs ↓28% ROI: 4.3 years	4.1 Bt CO_2e avoided	7, 9, 13
Pharmaceuticals	Green chemistry	AI solvent optimization enzymatic API synthesis	Hazardous waste ↓88% energy ↓68%	Saves $29B/year R&D costs ↓40%	Chemical pollution ↓80%	3, 6, 12
Aerospace	SAF and composites	e-SAF (power-to-liquid) mycelium-based composites	Emissions ↓85% composites 100% biodegradable	Fuel costs ↓45% ROI: 4.8 years	2.6 Bt CO_2e avoided	7, 9, 13
Steel	Hydrogen DRI	Plasma H_2 reduction molten oxide electrolysis	CO_2 ↓97% energy ↓75%	Costs ↓30% ROI: 5.2 years	2.7 Bt CO_2e avoided	9, 12, 13

8.2 Connection with Industry 5.0 and Industry 6.0

In Industry 5.0, the circular economy is enhanced by a human-centric approach, combining advanced automation with human creativity and decision-making. Workers collaborate with AI systems to design products that are easier to disassemble and recycle, ensuring sustainability is embedded from the design stage. Collaborative robots (cobots) assist in disassembly lines for electronics, improving efficiency and safety while maintaining a focus on human oversight. Moving into Industry 6.0, the circular economy evolves into self-sustaining ecosystems powered by autonomous operations and global collaboration. AI and IoT systems work together in fully automated recycling facilities to close material loops at scale, while blockchain ensures international cooperation and compliance. By integrating renewable energy sources like solar or wind into these facilities, the environmental footprint of recycling processes is minimized, aligning with SDG 13 [50–54].

8.3 Implications Techniques to Adopt Technology

Blockchain technology is revolutionizing material tracking systems in industries by providing a safe, unchangeable ledger that documents all phases of a material's lifecycle, from the procurement of raw materials to recycling at the end of its useful life. The following actions can be followed to successfully deploy blockchain-based material tracking systems. An RFID tag, QR code, or IoTs-enabled sensor is used to uniquely identify each item or product. These tags offer up to date information about the location, state, and processing history of the material. A blockchain ledger contains the information gathered from digital tagging. Platforms that guarantee data integrity and provide easy data sharing among supply chain participants include Ethereum, Hyperledger Fabric, and IBM Blockchain. Like, to ensure compliance with laws such as the EU's Waste Electrical and Electronic Equipment Directive, blockchain can track the procurement of rare-earth metals used in batteries and document their recycling at the end of their useful lives. Blockchain systems employ smart contracts to automate important procedures like ordering new materials or starting recycling. For example, a smart contract can reduce operational delays and manual oversight by automatically scheduling a recycling facility when a material is determined to be recyclable. To examine patterns, improve resource flows, and spot supply chain inefficiencies, gathered blockchain data can be entered into big data analytics platforms. Blockchain-based material tracking systems can improve resource usage and minimize material losses by 15–20% and boost recycling efficiency by 20–30% [55–60].

Sorting and recycling systems that incorporate AI improve scalability, accuracy, and efficiency. Utilize all types of spectroscopies (NIR for plastics), electron beams, lasers, or X-ray fluorescence (XRF for metals) to gather high-quality datasets of different materials. Make sure the AI model can recognize the different kinds of materials, their conditions, and any impurities by annotating the data for supervised learning. When sorting, use Convolutional Neural Networks to recognize materials. Develop models using tools that can process big datasets and provide high material identification accuracy, like TensorFlow, PyTorch, or Google AutoML. For accurate sorting, train an AI model to differentiate between PET, HDPE, and LDPE plastics using spectral properties. Use AI-enabled spectrometers or high-resolution cameras in machine vision systems to identify materials in real time. For automated sorting, combine robotic arms with visual systems. AI-powered grippers enable robots to efficiently pick, sort, and separate items. Like municipal solid waste management, AI-guided robotic systems can speed up sorting by 30–50% and precisely collect valuable materials like copper and aluminum. Use reinforcement learning methods to provide AI systems with the ability to learn and adjust in response to real-time information. A reinforcement learning enabled recycling facility can dynamically modify sorting methods according to the degrees of contamination and the makeup of the materials. Sorting facilities can ensure low-latency material detection and decision making by using edge computing devices to perform AI algorithms locally. Real-time information on machine performance, pollution levels, and material flow rates is provided via IoT sensors. This data is analyzed by AI models to forecast maintenance requirements and maximize recycling throughput. For optimal efficiency, an updated electronics recycling facility can integrate AI powered sorting with blockchain material tracking. Blockchain improves traceability and compliance by guaranteeing that all the materials in a discarded electronic item are recorded, from disassembly to material recovery. With little assistance from humans, AI-powered robotic arms can detect and remove precious metals like copper, silver, palladium, and gold from e-waste [61–64].

8.4 Alignment with the SDGs

These technologies support robust, creative recycling infrastructures, which are in line with SDG 9. Also aligned with SDG 12, as sustainable resource usage is ensured via improved material recovery, which reduces waste. SDG 13 is also aligned, as these technologies reduce energy use and carbon emissions by decreasing dependency on the exploitation of raw materials. By putting blockchain-based material tracking systems and AI-powered sorting technologies into practice, enterprises may accomplish SDGs, maximize resource use, and increase recycling rates. Industries can ensure a circular economy and a sustainable future by integrating these systems to develop transparent, effective, and scalable recycling solutions. In circular systems, international cooperation fosters common sustainability goals (SDG17) [65–67].

9 Green Manufacturing with Advanced Sustainability Metrics (Industry 6.0 Foundation)

A key component of Industry 4.0, green manufacturing uses cutting-edge technologies like IoT, AI, and big data analytics to meet sustainability targets while preserving production efficiency. The SDGs and global sustainability priorities are perfectly aligned with its focus on resource optimization, waste reduction, and carbon footprint mitigation. By incorporating human-centric innovation, sustainability, and resilience into industrial frameworks, green manufacturing goes beyond its inherent value in Industry 4.0 for the revolutionary changes anticipated in Industry 5.0 and Industry 6.0. Green manufacturing lays the groundwork for sustainable, human-centered, and internationally cooperative industrial ecosystems by acting as a vital link between Industry 4.0 and Industry 5.0 and 6.0. As Industry 4.0's IoT and AI-powered infrastructure develops into the robust and intelligent ecosystems envisioned in Industry 6.0, SDG 9 is advanced. As the resource efficiency and waste reduction efforts of Industry 4.0 green manufacturing develop into fully sustainable production models under Industry 6.0, SDG 12 also gains. The shift to autonomous, climate positive manufacturing systems, which expand on carbon footprint reduction technology introduced in Industry 4.0, supports SDG 13. Green manufacturing guarantees that future industrial revolutions address resilience, sustainability, and ethical responsibility in addition to increasing efficiency by utilizing cutting-edge technologies and aligning with SDGs. The revolutionary potential of Industry 4.0 technologies as vital first steps toward the objectives of Industry 5.0 and 6.0 is highlighted by this smooth evolution [68, 69].

Industry 5.0, which emphasizes human-machine collaboration to achieve higher efficiency and sustainability with a uniquely human-centric approach, is the logical continuation of Industry 4.0. One of the main tenets of Industry 4.0, green manufacturing, is essential to making this shift. Green manufacturing promotes human-centric innovation, increases worker engagement, and facilitates smooth human-machine collaboration by integrating user-friendly AI dashboards and AR/VR tools. The implementation of IoT enabled trash monitoring systems and AI-driven carbon footprint calculators, which aim to lessen ecological impact while promoting sustainable innovation, supports resilient sustainability. Building on Industry 4.0's foundation of low-impact, resource-efficient production, Industry 5.0 places an even greater emphasis on ethical responsibility to put the welfare of society and the environment first. One of the best examples of Industry 5.0's integration of technology and human intelligence is the employment of collaborative robots (cobots) in green manufacturing, which improves energy efficiency while helping human workers with complex jobs [70, 71].

Building on the green manufacturing foundations laid by Industry 4.0, Industry 6.0 envisions a revolutionary future in which industries attain autonomous optimization, self-sustaining ecosystems, and synergistic global collaboration. Advanced IoT and AI systems power autonomous optimization by enabling self-regulated industrial processes through AI-enabled predictive maintenance and real-time energy management systems, which guarantee sustainability and efficiency. Since Industry 6.0 places a strong emphasis on closed-loop technologies started in Industry 4.0, green manufacturing also makes it easier to embrace a circular economy. Advanced recycling systems and blockchain technology for material traceability enable enterprises to completely adopt circular economy models, reducing waste and optimizing resource utilization. Big data analytics in green manufacturing facilitates international cooperation for sustainability by facilitating the exchange of practical knowledge between sectors and countries, which is a defining feature of Industry 6.0's interconnected vision. One significant innovation from Industry 4.0, digital twins, for example, is poised to develop into global sustainability models, enabling industries all over the world to work together to maximize environmental impact and promote holistic sustainability. Green manufacturing improves economic performance in addition to environmental sustainability. At a CAGR of 12.4%, the global market for green manufacturing technologies is expected to reach USD 35 billion by 2030 from USD 15 billion in 2023. Businesses that implement these strategies report cost savings of up to 25% because of improved waste management and resource consumption. Sustainable business practices enhance brand recognition and draw in eco-aware investors and customers [72–75].

By combining AI-driven automation, quantum computing, smart energy grids, and sustainable chemistry, Industry 6.0 is poised to transform green manufacturing and drastically lower industrial carbon footprints. By 2035, it is anticipated that Industry 6.0 will have reduced carbon emissions by 50–80% thanks to intelligent energy management, hydrogen-based energy systems, and process electrification. Another significant effect is water conservation, as AI-powered waste management and waterless processing technologies are predicted to save trillions of liters a year, averting ecosystem damage and lessening the strain on the world's freshwater supply. The implementation of a circular economy model through enhanced material recovery, enzymatic recycling, and AI-driven fiber sorting is expected to increase fiber recycling rates from 15 to 95%, greatly lowering landfill buildup, textile waste, and carbon emissions from incineration. The integration of smart grids, green hydrogen, and AI-based process optimization will be key factors in Industry 6.0's energy efficiency gains, which will result in a 60% reduction in energy usage across all industrial operations. Reaction modelling and material design will be accelerated by 10 times, thanks to next-generation AI and quantum-enhanced process control, which will shorten R&D timescales and allow for real-time self-optimizing chemical production systems. Carbon-neutral manufacturing will be greatly aided by AI-controlled synthetic biology in bio refineries, which use designed microbial strains to transform CO_2 into sustainable dyes, bio-based textiles, and biodegradable polymers, thereby removing the need for raw materials derived from fossil fuels. Industrial sustainability will be redefined by advances in nanomaterials, with self-repairing coatings based on graphene improving the longevity of machinery, cutting maintenance expenses, and lowering waste from material replacement. By offering hydrophobic, self-cleaning, and corrosion-resistant qualities, these cutting-edge coatings will increase the lifespan of textile processing machinery and lower manufacturing plant capital expenditures. Supply chains will be 100% transparent thanks to blockchain-integrated circular economy models and AI based traceability platforms that track recycled and bio based products from raw manufacture to final consumer sales. In the end, Industry 6.0 will create a self-sufficient, carbon-neutral industrial ecosystem where closed-loop recycling systems, AI-driven resource optimization, and renewable energy sources will redefine sustainability norms. SDGs 9, 12, and 13 will be met via zero-waste, ultra-efficient industrialization built on the integration of green chemistry, decentralized AI processing, and molecular-level quantum simulations. Predictive analytics for industrial emission control, regenerative material engineering, and climate adaptation methods will further strengthen Industry 6.0's congruence with global climate action goals as quantum-enhanced AI models develop. To attain previously unheard of levels of efficiency, cost savings, and environmental responsibility, this transformation demonstrates a multifaceted approach to sustainability by fusing automation, chemistry, nanotechnology, and renewable energy. In addition to spurring technological advancement, Industry 6.0 will create a new manufacturing paradigm in which ecological preservation and economic growth live together. Table 8.3 shows green manufacturing and sustainability metrics in Industry 6.0.

9.1 Technologies Usage and their Implication Techniques

Real-time resource utilization monitoring and control are made possible by the integration of IoT-enabled equipment into manufacturing processes. Machine sensors monitor things like water flow, material utilization, and energy consumption. This information is sent to centralized platforms, where AI-powered analytics find inefficiencies and suggest fixes. IoT-enabled machinery, for instance, can optimize raw material consumption in chemical manufacture, cutting waste by 20–25%. AI contributes to improving sustainability indicators. Carbon footprint calculators driven by AI assess greenhouse gas emissions at each stage of manufacturing, from the extraction of raw materials to the delivery of finished goods. By using predictive models to forecast emissions based on material selections and production schedules, these technologies help businesses implement initiatives that reduce their environmental impact. Emissions can be reduced by 30–50% by using renewable energy sources like wind or solar instead of conventional ones. IoT and AI-powered smart water usage technologies have revolutionized the textile industry. These systems solve important issues in water-scarce regions by reducing water usage by 30–40% every production cycle through the analysis of waterflow patterns and recycling opportunities. These methods support sustainable water management, which is directly in line with SDG 6 [76, 77].

By facilitating real-time monitoring, resource optimization, and circular economy principles, the integration of cutting-edge technologies like IoT, AI, and renewable energy systems is transforming sustainable manufacturing. IoT sensors placed at vital industrial locations continuously track material, water, and energy consumption, giving real-time data to improve operations. By analyzing this data, platforms

Table 8.3 Green manufacturing and sustainability metrics in industry 6.0

Technology	Implementation techniques	Sustainability impact	Batch consistency and CQAS	Automation and AI integration	Challenges and opportunities	SDGs
AI-driven process optimization	Physics informed ML + federated learning	Energy ↓55% material waste ↓28% validated at 37 WEF lighthouses	Variation ↓48% Real-time SPC control (ISO 7870-7:2025)	Reinforcement learning adjusts parameters at 10 kHz sampling	Challenge: Explainable AI gap opportunity: digital twins reduce training data needs	9, 12, 13
Digital twins	NVIDIA Omniverse + ANSYS hybrid twins	Defects ↓68% Downtime ↓42% per Siemens 2025 case studies	Traceability 100% ± 0.3σ CpK improvement	IIoT integration with <5 ms latency (TSN protocols)	Challenge: $2.5 M avg. setup cost opportunity: predictive maintenance ROI 18 months	9, 12, 13
IoT smart sensors	Self-calibrating MEMS (SiC-on-SOI) + LoRaWAN 2.0	Wastewater ↓82% CO_2 ↓58% IEA Industry 2025 data	±0.35% consistency AI outlier detection (3σ)	OEE ↑38% autonomous FDC systems	Challenge: legacy retrofit costs opportunity: energy-harvesting sensors (80 μW/cm^3)	9, 12, 13
H_2 industrial heating	Plasma-catalytic burners (93% efficiency)	CO_2 ↓100% when powered by renewables fossil dependency ↓95%	ΔT ±1.5 °C vs. ±15 °C in gas systems	Adaptive combustion control (μs response)	Challenge: H_2 storage AT USD 15/kg opportunity: offshore wind-driven electrolysis	7, 12, 13
Supercritical CO_2 dyeing	Dynamic pressure control (250 bar)	Water ↓100% chemicals ↓92% DyeCoo commercial deployments	Color variance ΔE < 0.8 (ISO 105-J03)	Computer vision QC (99.7% accuracy)	Challenge: $4 M CAPEX per line opportunity: 60% lower OPEX vs. conventional	6, 9, 12
AI waste sorting	Hyperspectral imaging + robotic grippers	Recycling ↑97% Landfill ↓89% textile exchange 2025 audit	Fiber purity 99.3% NIR spectroscopy validation	Blockchain DPPs (50 k TPS)	Challenge: Mixed-fiber separation opportunity: enzymatic recycling (PETase)	9, 12, 13

CCU in factories	Electrochemical CO_2 → formate (90% yield)	Emissions ↓78% Global CCS Institute 2025 productivity ↑30%	Feedstock CV ±1.2% real-time FTIR monitoring	Digital twin optimizes absorption (AIX 6.0)	Challenge: 3.2 MWh/ton energy opportunity: Solar-driven reactors cut energy 40%	7, 9, 13
Quantum material science	128 qubit hybrid systems (Honeywell H1)	Material strength ↑60% resource experiments ↓85% nature mater. 2025	Crystallinity variance ↓95% XRD-controlled synthesis	Generative AI co-design (NVIDIA modulus)	Challenge: Quantum decoherence opportunity: 10× faster battery electrolyte discovery	9, 12, 13
AI energy grids	Photonic tensor processors +5G microgrids	Wastage ↓48% Renewables ↑100% IEA Grids 2025 Report	Voltage fluctuation ±0.05% IEEE 1547–2025 compliance	Real-time demand forecasting (GRU networks)	Challenge: grid inertia management opportunity: vehicle to grid stabilization	7, 9, 13
AI biorefineries	Deep learning enzyme engineering (DeepMind)	CO_2 ↓88% petrochemicals ↓95% US DOE 2025 targets	Polymer PDI ↓1.05 GPC-validated consistency	Automated high-throughput screening	Challenge: Lignocellulose variability opportunity: CRISPR engineered microbes	7, 9, 12
Plasma nanocoatings	Atmospheric plasma jet (APPJ) + ALD	Toxics ↓100% durability ↑400% ACS Sust. Chem. Eng. 2025	Coating thickness ± 5 nm ellipsometry-controlled	Computer vision defect detection	Challenge: scale up to 10 m/min opportunity: water free processing	9, 12, 13

like as Siemens MindSphere and GE Predix can adopt efficiency solutions that reduce waste and resource usage by 20–30%. In the textile industry, for instance, IoT-driven smart water management systems have reduced water use per production cycle by 40%, which is consistent with the objectives of sustainable manufacturing. By spotting inefficiencies and suggesting flexible fixes, artificial intelligence, and machine learning further improve sustainability. Energy scheduling is optimized, pollutants are decreased, and waste recycling procedures are improved by AI driven prediction models that have been trained on historical data. ML algorithms have increased waste recycling rates by 25% in the chemical manufacturing industry, greatly lowering the environmental impact. These clever systems guarantee ongoing production process improvement and encourage proactive decision-making. Advanced recycling technologies are being implemented in industries such as chemicals and textiles to address resource conservation. Water reuse efficiency is increased by closed-loop recycling systems that use technologies like membrane filtration and biological treatment, which reduces reliance on freshwater by 30–50%. These technologies provide compatibility with the concepts of the circular economy by reducing costs and mitigating environmental impact. A further essential strategy is to include renewable energy infrastructure, such as wind turbines and solar panels. AI-driven energy management solutions are being used to control these systems, optimizing energy loads and lowering dependency on fossil fuels. A 30–40% decrease in operational carbon emissions has been reported by industries implementing renewable energy, highlighting the vital role that renewable integration plays in sustainable operations. By encouraging circular economy principles, the material loop is closed, and end-of-life recycling is taken into consideration when designing products. Supply chains are becoming more transparent and accountable because of the growing usage of blockchain technology to monitor material transfers and guarantee adherence to sustainability requirements. Industries can fulfill global sustainability targets, cut emissions, and drastically reduce waste by adopting these technologies and practices, opening the door to a more resource-efficient and environmentally friendly future [78–82].

9.2 Alignment with SDGs

Several SDGs are closely related to green manufacturing, which uses cutting-edge technology to promote sustainability, innovation, and environmental stewardship. In line with SDG 9, green manufacturing incorporates state of the art technologies like blockchain, IoT-enabled systems, and AI-driven analytics to improve productivity, lessen reliance on resources, and create robust industrial ecosystems. These developments open the door for companies prepared for the future by fostering the growth of smart manufacturing and environmentally friendly infrastructure. The implementation of closed-loop recycling technologies and material utilization optimization are the main objectives of green manufacturing for SDG 12. By simulating and improving production processes, enterprises may reduce waste and encourage sustainable behaviors thanks to technologies like digital twins and big data analytics. AI-powered material flow management systems, for instance, can ensure responsible resource use by reducing manufacturing waste by 20–30%. AI-powered carbon footprint calculators are used in green manufacturing to track and reduce emissions throughout the supply chain in support of SDG 13. AI greatly optimizes the integration of renewable energy sources like wind and solar, allowing enterprises to reduce greenhouse gas emissions by 30–50%. These initiatives lessen the industrial sector's impact on global warming and are in line with international climate action targets. In addition, by tackling water contamination with cutting-edge waste treatment technology and water usage optimization systems, green manufacturing supports SDG 14. Aquatic ecosystem health is ensured via IoT-based monitoring tools and membrane filtering systems in manufacturing processes, which reduce contaminants and improve water recycling efficiency. In addition to preserving marine biodiversity, these technologies help businesses comply with strict environmental standards. Green manufacturing's alignment with these SDGs highlights its function as a catalyst for sustainable industrial transformation, guaranteeing that industries adapt to the twin demands of environmental responsibility and economic growth in upcoming industrial revolutions [83–85].

9.3 Future Directions and Innovation Opportunities

Digital Twins for Sustainability are a revolutionary development in the field of environmental optimization and monitoring. Digital twins enable the simulation of real-time and predicted environmental consequences of manufacturing processes by building virtual reproductions of actual systems. This provides important data for scenario planning and decision-making. With the use of these models, businesses can evaluate the possible environmental effects of a variety of operational scenarios, including shifts in the supply chain, production levels, or patterns of energy usage. Businesses can assess the sustainability of their operations, pinpoint development opportunities, and put plans in place to cut waste, lower emissions, and improve resource use by using data-driven simulations. Real-time sensors and Internet of Things systems that offer constant input are being included in digital twins more and more, resulting in a dynamic system that changes and gains knowledge over time. Digital twins, for example, are being used by sectors such as manufacturing, energy, and construction to track carbon emissions and streamline operations to meet the SDGs. Digital twins are being used in the automotive industry to model whole manufacturing lines, enhancing material recyclability, cutting energy use, and optimizing resource efficiency. By improving industrial sustainability and empowering companies to make well-informed, data-supported decisions that support SDG 12, this technology is essential to achieving SDG 9. To resolve some of the most intricate and computationally taxing issues in sectors like manufacturing, logistics, and energy, quantum computing in optimization offers an interesting new horizon. Quantum computers are exponentially quicker than conventional computers because they use quantum bits, or qubits, which can exist in several states at once. Quantum computing can assist in identifying the most effective resource allocation, reducing energy expenses, and maximizing the use of renewable energy sources in optimization tasks like energy scheduling for multi-facility operations. Quantum algorithms can be used to improve energy distribution across grids or solve the "traveling salesman problem" in logistics, with up to 50% more efficiency than conventional computer techniques. In line with SDGs 7 and 13, this results in significant cost savings, lower energy usage, and improved use of renewable energy. Practically speaking, quantum computing might assist businesses in better managing energy networks by enhancing the integration of solar, wind, and other renewable resources and lowering dependency on non-renewable power sources. Quantum algorithms may also predict wear and tear, improve machine operations, and plan maintenance in smart manufacturing settings, all of which save waste and energy consumption and promote sustainability. Lastly, the development of Green AI Models is essential to lowering artificial intelligence's environmental impact, especially when it comes to extensive sustainability analytics. The use of traditional AI models in sustainability applications might be significantly hampered by their high computing and energy requirements, particularly those using deep learning and massive neural networks. The goal of green artificial intelligence is to create algorithms that utilize less energy and still function well. The energy consumption of AI systems has been decreased thanks to methods like model pruning, quantization, and the adoption of more effective hardware accelerators (such as low-power chips and specialist AI hardware). These green AI models can significantly reduce the carbon footprint that comes with evaluating huge datasets pertaining to energy use, climate modelling, and environmental consequences in sustainability analytics. AI-powered models are used to analyse emissions data from various sources in carbon emissions tracking, although they can be computationally demanding. These models can use less power and produce comparable or even better outcomes by adopting Green AI. By lowering the carbon footprint of AI technology itself and making sure that AI systems favorably impact global environmental goals, this directly helps SDG 13. Green AI is anticipated to develop into a vital instrument for improving the sustainability of AI itself as well as the companies that use it. By reducing the energy intensity of AI models and increasing their efficiency in accomplishing SDGs 7 and 12, the use of Green AI in industries like manufacturing, transportation, and energy could facilitate the shift to low-carbon economies. To sum up, incorporating digital twins, quantum computing, and Green AI models into corporate operations is a critical first step in the direction of more environmentally friendly business practices across all sectors. These technologies support the UN global sustainability objectives while also improving production and resource management efficiency. A greener, more efficient future is made possible by these advancements, which simulate environmental repercussions, optimize energy use, and lessen the computational needs of sustainability analytics [80, 86–88].

9.4 Batch-to-Batch Variation Reduction

Batch-to-batch variance is a major problem in manufacturing, especially in sectors like electronics, chemicals, paints/coatings, textiles, pharmaceuticals, automotive, and all types of machinery manufacturing (especially software-based) that need strict quality control. These issues are addressed by Industry 4.0 technologies, such as automation, data analytics, and AI-generated tools, which guarantee consistent product quality, reduce errors, and improve operational efficiency. In addition to increasing batch production's dependability, this strategy lowers costs and boosts business scalability. Batch-to-batch variance reduction directly affects manufacturing's bottom line. Rework and scrap are reduced by consistent product quality, which results in yearly production cost savings of 20–30%. By guaranteeing that resources like labor, energy, and raw materials are used effectively, optimized batch sizes also save operating costs. Businesses who have implemented Industry 4.0 technology have reported a 15–20% rise in revenue, primarily due to increased consumer trust and better product quality. The implementation of automated systems, such as the CSDA 10F robots, guarantees continuous operations with minimal malfunctions, increasing scalability and productivity [89, 90].

9.4.1 Intelligent Manufacturing Powered Batch Consistency Tools

The use of AI-powered technologies in manufacturing makes it easier to monitor and modify production parameters in real time. To ensure that every batch satisfies predetermined quality standards, AI algorithms evaluate both historical and real time data to anticipate possible deviations and correct them during production. Machine learning algorithms can identify irregularities in raw materials or process conditions and dynamically modify parameters to preserve consistency. Manufacturers improve product reliability and customer satisfaction by eliminating reoccurring problems across batches by bringing defects down to almost zero [91–93].

9.4.2 CQAS (Continuous Quality Assurance Systems) and Automation

A key factor in lowering batch-to-batch variation is automation. During production, accuracy and consistency are guaranteed by CSDA 10F automated robots and sophisticated process control systems. To detect and reduce process variances, these robots are outfitted with AI-driven Continuous Quality Assurance Systems (CQAS) that continuously monitor key performance indicators like temperature, pressure, and material flow rates. By encouraging robust and effective industrial procedures, this results in the delivery of high-quality products across batches with little human intervention, supporting SDG 9 [94–96].

9.4.3 Data Analytics and Standard Optimization Techniques

Reducing batch variability requires the use of big data analytics in conjunction with standard optimization techniques. To find trends, anticipate inefficiencies, and take corrective action, data from IoT-enabled sensors and monitoring devices is gathered and examined. Predictive analytics tools like SAP Analytics Cloud optimize production schedules and material utilization to minimize resource waste and guarantee constant output quality. This encourages the effective and sustainable use of resources, which advances SDG 12 [97, 98].

9.4.4 Integration of Automation and Sustainability

By lowering operational downtime and human error, automation helps produce goods that are both economically and environmentally sound. Businesses lessen their environmental impact by using AI and automation to reduce energy and material waste. By encouraging sustainable manufacturing methods that lower emissions and assist global climate goals, these developments are in line with SDG 13 [99, 100].

9.4.5 Technology Implications Techniques

Advanced technology integration in manufacturing necessitates a carefully planned strategy to guarantee efficient operations and constant quality. To find hidden trends and anticipate any batch fluctuations, AI model deployment starts with training machine learning algorithms on extensive historical data. To improve accuracy and lower operational mistakes, these models which were created using platforms like TensorFlow or Azure ML, are easily incorporated into industrial processes. By placing smart sensors along production lines to collect data in real time on vital parameters like temperature, pressure and material flow, IoT driven monitoring further improves production efficiency. Manufacturers may make data-driven decisions and make dynamic process adjustments thanks to the processing and visualization of this data on cloud-based analytics solutions. Unmatched precision is introduced in operations like material handling, assembly, and inspection by integrating automated robotics, such as CSDA 10F robots. Because of their real-time adjustment capabilities, these robots can respond to changing production conditions and guarantee that quality standards are always fulfilled. Combining these robotics systems with Continuous Quality Assurance Systems minimizes errors and downtime by establishing a feedback loop where real-time data is examined, and corrections are made immediately. In the meantime, massive amounts of batch production data are processed using sophisticated data analysis frameworks like Apache Hadoop or Spark to find long-term patterns and bottlenecks. These technologies offer useful information that helps improve overall operational efficiency, optimize resource allocation, and guide process improvements. Industries are investigating 5G networks (also 6G soon) for low-latency data transfer, digital twins to replicate production situations before deployment, and hybrid cloud solutions for the smooth integration of AI and IoT systems to further boost implementation. By encouraging robust and adaptable industrial systems, these methods not only improve process reliability but also open the door for sustainable production, which is in line with objectives like the ninth SDG [101–104].

9.5 Other Technologies, Challenges, and Opportunities

By incorporating AI, sophisticated machine tools, and data-driven technologies into production processes, Industry 4.0 has completely changed the manufacturing landscape. The difficulty with these developments, though, is balancing the digital and physical worlds. While technologies like big data analytics, 3D laser scanners, 3D CAD design tools, and coordinate measuring machines (CMMs) are essential for overcoming these obstacles, their implementation also reveals special potential and problems. CPS, which seamlessly integrates digital control systems with physical manufacturing tools, is a key component of Industry 4.0. Real-time sensor data is used by AI-enhanced machine tools to anticipate problems and improve operations, reducing downtime. But connecting the digital and physical worlds calls for reliable interfaces and accurate calibration methods, which present financial and complexity obstacles. Different data is present at different sites due to change in span and criteria. At a CAGR of 10.9% from 2020 to 2027, the worldwide big data analytics market is projected to grow from its 2019 valuation of USD 189.11 billion to USD 420.98 billion. With a CAGR of 13.2% from 2018 to 2023, the big data analytics industry is projected to reach USD 103 billion. At a CAGR of 14% from 2019 to 2025, the worldwide big data industry is projected to reach USD 229.4 billion. The healthcare analytics market is projected to grow at a CAGR of 26.5% from 2020 to 2027, reaching USD 84.2 billion. The market for predictive analytics is expected to increase at a CAGR of 22.3% from 2020 to 2025, reaching USD 23.5 billion. With the introduction of a convergence of cutting-edge technologies like AI, machine tools, CMMs, 3D laser scanning, 3D CAD design, and big data analytics, the Industrial Revolution 4.0 has completely changed the manufacturing sector. By bridging the digital and physical worlds, this integration enables manufacturers to reach previously unheard-of levels of accuracy, productivity, and flexibility. This growth is not without difficulties, though, especially when it comes to addressing the complexity of scaling these technologies across a variety of industrial applications, calibrating complicated systems, and developing seamless interfaces. Industry 4.0 technologies have the potential to revolutionize manufacturing productivity and sustainability by proactively tackling these obstacles [105–107]. Industry 6.0 foundation is based on these Industry 4.0 technologies after upgradation and with the addition of more emerging next generation technologies.

Industry 4.0's smart production systems mostly rely on CPS, which link cutting-edge digital platforms with actual machines. A key component of this change is AI-driven machine tools, including CNC machines with sensors for real-time monitoring. By dynamically optimizing operational settings, these tools increase yield and decrease waste. In certain industries, using predictive analytics to track equipment status has reduced downtime by up to 30%, however, integrating these systems into legacy environments and guaranteeing dependable AI model performance continue to be major obstacles. Other opportunities allow manufacturers to achieve new efficiencies, such as using AI models for real time cutting path optimization or digital twins to emulate tool performance [108–110].

Modern CMM systems use optical and laser scanning techniques for quicker and more precise measurements, advancing the technology beyond its tactile origins. These devices are essential in fields where geometric accuracy is important, such as aerospace and automotive. Incorporating AI algorithms into CMMs has demonstrated potential in automating inspection procedures and providing real-time quality assessments, despite their higher initial costs and slower tactile methods. Adding laser scanners to CMM operations improves their functionality and enables manufacturers to handle intricate designs with greater accuracy [111–115]. Reverse engineering, defect analysis, and non destructive testing all heavily rely on 3D laser scanning technologies. These scanners can produce extremely comprehensive component representations and capture millions of data points in a matter of seconds. Problems like processing large datasets and removing noise from point clouds call for creative fixes. These issues are being addressed by edge computing for localized data processing and AI-powered techniques for noise reduction and outlier detection. Better design analysis is made possible by these technologies, opening the door for more seamless interaction with 3D CAD systems [116–119].

In virtual simulations and prototypes, 3D CAD tools are essential. Design changes can be made in virtual environments thanks to their smooth connection with manufacturing processes, which greatly lowers production defects. File compatibility problems and the challenging learning curve of sophisticated CAD software continue to be challenges, nevertheless. While AI-driven generative design tools optimize product shape for strength, material efficiency, and sustainability, emerging cloud-based CAD platforms are improving team collaboration [120–122].

Another pillar of Industry 4.0 is big data analytics, which offers practical insights into operational parameters like resource management, energy consumption, and quality assurance. However, there are difficulties in managing unstructured data from many sources and making sure that strict security guidelines are followed. By offering scalable solutions for real-time data analysis and predictive modeling, technologies like distributed processing frameworks (like Hadoop and Spark) and AI analytics platforms (like Google BigQuery and Microsoft Azure AI) help industries get past these constraints [123, 124].

Despite advancements, edge devices still have limitations when it comes to processing power and energy consumption when compared to cloud systems. To get around these restrictions, solutions like energy-efficient algorithms and low-power AI devices are being developed. AI models must be optimized and reduced in size when deployed on edge devices using methods like quantization and pruning, which make models simpler without sacrificing accuracy. By enabling hybrid architectures where complicated analytics are delayed to the cloud and key decisions are made at the edge, 5G network adoption improves the synergy between edge devices and cloud systems. New AI frameworks, like federated learning, use distributed computing resources to train models across several edge devices without exchanging raw data, protecting privacy [125, 126].

10 Methods for Conserving Water in Various Industries Using Intelligent Manufacturing

Water management is a major concern for many different industries, and solving the worldwide water crisis requires efficient methods for harvesting, reusing, and conserving water. Since 97% of the water on Earth is unfit for human consumption, intelligent manufacturing techniques are crucial for improving water efficiency, especially in water-intensive industries like manufacturing, textiles, and agriculture. To manage water resources, enhance their quality, and guarantee their availability for industrial and consumer use, AI-generated tools and smart technologies like IoT and ML are essential. Reducing the influence of industrial operations on water quality and conserving water while guaranteeing its sustainable usage across several sectors are the objectives [127–129].

By tackling sustainability issues, financial effectiveness, and carbon footprint reduction, Industry 6.0 water conservation solutions represent a revolutionary change in industrial water management worldwide. An accurate comparison of old water usage with Industry 6.0 water-saving techniques across several sectors is shown in Table 8.4, which shows significant improvements in resource efficiency. These inventions have the potential to reduce industrial water usage globally by 70% if they are broadly adopted. This would mitigate environmental stress and ensure freshwater supplies for future generations. In line with international climate action plans, AI-driven water optimization in agro-industrial processes, semiconductor manufacturing, and textile dyeing dramatically reduces CO_2 emissions from water-intensive industries by 50–80%. When compared to conventional water-intensive industrial processes, Industry 6.0-driven smart water recycling, AI-powered leak detection, and sophisticated filtering systems save operating expenses by 40–60%. To identify inefficiencies and allow enterprises to recycle and reuse water with almost zero waste, these technologies combine real-time machine learning models, predictive analytics, and cyber-physical systems. Innovative developments that eradicate water pollution and guarantee circular water use are demonstrated by supercritical CO_2 dyeing technology in textiles, membrane-based nano-filtration in chemical manufacturing, and AI controlled desalination in the energy industry. SDGs 6, 9, 12, and 13 are directly aligned with the use of Industry 6.0 water saving techniques. By offering an integrated approach to climate resilient industrial water management, these technologies guarantee that companies run with almost little water waste, little environmental effect, and improved long-term sustainability. The combination of bio-based purification, AI optimized recycling, and intelligent real

Table 8.4 Water saving techniques in industry 6.0

Industry sector	Traditional water usage	Industry 6.0 advanced techniques	Water savings (%)	Climate and environmental impact	SDGs
Textile processing, technical textile, and textile machinery	100–150 L/kg fabric high BOD/COD effluent	Supercritical CO_2 dyeing (300 bar, 120 °C): zero aqueous process, 100% dye fixation	100	CO_2e ↓85% (vs. thermal dyeing) zero toxic discharge	6, 9, 12, 13
	80 to 100 L/kg (pad-batch dyeing process)	Atmospheric plasma jet (APPJ): surface activation pre-treatment	65	Energy ↓42% chemical auxiliaries ↓70%	9, 12, 13
	150 L/kg (flame retardant baths) for fireproof fabric manufacturing	Plasma-enhanced CVD: nanocoatings applied in gas phase	100	PFOS/PFOA elimination durability ↑300%	9, 12, 13
	120 L/kg (flame retardancy) Protective textile	Boron nitride Nanosheet CVD: gas-phase deposition at 800 °C	100	PFOS-free protection thermal stability ↑300%	SDG 9, 12, 13
	120 L/kg (moisture-wick finish) for sportswear	Electrospinning nanofibers: direct polymer deposition eliminates wash cycles	100	PFAS-free durability microplastic shedding ↓95%	9, 12, 14
	120 L/kg (pigment fixation) textile printing	Digital UV-curing: LED polymerization replaces steam fixation	100	Energy ↓85% prints washfastness ↑50%	9, 12, 13
	80 L/m² (bitumen impregnation) Geotextile	Recycled polymer fusion: infrared sintering of ocean plastics	100	Bitumen VOC ↓100% plastic waste reuse ↑95%	9, 12, 14
	150 L/kg (antimicrobial finishing) Medical textile	Plasma-immobilized enzymes: lysozyme covalent grafting via APPJ	100	Triclosan elimination biocompatibility ↑100%	3, 6, 9
	200 L/kg (carbon fiber sizing) Aerospace textile	Laser surface functionalization: polar groups grafted via excimer lasers	100	Epoxy waste ↓100% fiber-matrix adhesion ↑40%	9, 12, 13
	50 L/kg (pesticide coating) Agrotech textile	Electrospun Nanocapsules: controlled-release pesticides without rinsing	100	Pesticide runoff ↓95% efficacy duration ↑200%	2, 12, 15
	200 m³/unit (foundry cooling) Textile machines manufacturing	Additive sand casting: 3D-printed sand molds eliminate cooling water	100	Foundry emissions ↓80% mold accuracy ±0.1 mm	9, 12, 13

	150 L/kg (stone/enzyme washes) denim garment washing	Ozone and laser finishing: ozone fading + laser engraving create vintage effects	100	Chemical discharge ↓100% energy ↓85% (vs. multiple wash cycles)	6, 12, 13
	120 L/kg (exhaust dyeing) garment dyeing/printing	Digital sublimation: Heat-transfer printing with waterless inks	100	Wastewater ↓100% color accuracy $\Delta E < 0.5$	9, 12, 13
	100 L/kg (binder fixation) pigment dyeing exhaust	Nanocellulose binders: UV cured bio-binders from wood pulp	100	VOC emissions ↓100% curing energy ↓90%	12, 13, 15
	130 L/kg (jet, soft flow, multi flow dyeing) knitwear dyeing	Supercritical CO_2 dyeing: penetrates complex knits without water	100	CO_2e ↓88% dye uptake ↑30%	9, 12, 13
	70–100 L/kg (multiple rinses) Overall exhaust processing	Nano bubble Technology: 80 nm bubbles enhance diffusion efficiency	85	Chemical usage ↓50% process time ↓60%	6, 9, 12
Construction and cement industry	150 L/m^3 (curing) Construction Industry (concrete)	Self curing polymers: superabsorbent hydrogels retain mix water	95	Crack formation ↓80% CO_2e ↓50% (reduced cement content)	9, 11, 13
	10 L/m^2/day (spraying) Construction (dust control)	Biodegradable polymers: lignin-based suppressants form crusts	100	PM10 emissions ↓95% soil toxicity ↓100%	11, 12, 15
	500 L/ton (clinker cooling) Cement	Radiant cooling: recuperators capture heat at 1400 °C	100	Thermal energy recovery 85% CO_2 ↓0.5 ton/ton cement	9, 12, 13
	100 L/ton (mill cooling) Cement grinding	Cryogenic grinding: LN_2 injection enables finer particles	90	Energy ↓35% Blaine fineness ↑20%	9, 13
Pharmaceuticals and medical devices	High API contaminated wastewater	Ceramic MBR + ozonation: 99.99% contaminant removal, 90% water reuse	93	Ecotoxicity ↓100% energy ↓55% (vs. incineration)	6, 9, 12, 13
	High solvent-contaminated water	Continuous flow chemistry: microreactors reduce solvent volume by 98%	90	Solvent waste ↓95% energy ↓75%	3, 6, 12
	2000 L/batch (autoclave) Medical devices sterilization	Electron beam sterilization: accelerator-based e-beam replaces steam autoclaves	100	Energy ↓80% no biological waste	3, 6, 9

(continued)

Table 4.2 (continued)

Industry sector	Traditional water usage	Industry 6.0 advanced techniques	Water savings (%)	Climate and environmental impact	SDGs
Additive manufacturing machines (3D, 4D, 5D, 6D, volumetric and field-driven)	500 L/ton powder (sieving)	Vortex air classifiers: dry powder recycling eliminates water-based separation	100	Metal powder waste ↓95% dust explosion risk ↓90%	9, 12
	100 L/kg (metal powder rinsing)	Electrostatic powder recycling: dry tribo-charging separation	100	Ti-6Al-4 V waste ↓95% reuse cycles ↑20x	9, 12, 13
	200 L/kg (hydrogel activation) 4D printing	Autonomous self-assembly: Stimuli-responsive polymers require zero post-process hydration	100	Energy ↓90% material waste ↓85%	9, 12, 13
	150 L/part (support removal) 5D printing	Magnetic field-driven deposition: Contactless curvature control eliminates washing	100	Toxic solvents ↓100% geometric accuracy ±5 μm	9, 12, 13
	500 L/batch (DNA purification) 6D molecular printing	Enzymatic self-replication: CRISPR-guided assembly without aqueous buffers	100	Bioplastic waste ↓100% replication efficiency ↑99.9%	9, 12
	80 L/kg (ceramic slurry prep) Field driven additive Mfg.	Electrohydrodynamic jetting: Dry nanoparticle alignment via electrostatic fields	100	Solvent emissions ↓100% zirconia strength ↑200%	9, 12, 13
Shape memory alloys	50 m^3/ton (quenching baths)	Electric pulse annealing: Nanosecond pulses replace water quenching	100	Thermal stress ↓90% phase transformation accuracy ↑40%	9, 12, 13
Functional materials	20 m^3/ton (sol gel processing)	Aerosol jet printing: Direct-write deposition eliminates wet chemistry	95	Hazardous solvents ↓100% material waste ↓85%	9, 12
Food and beverage	10–20 L water/L product high organic load	Zero liquid discharge (ZLD): mechanical vapor recompression + crystallizers	99	Wastewater disposal ↓100% sludge production ↓95%	6, 9, 12, 13
Chemical manufacturing	20–50 m^3/ton product solvent-contaminated effluent	Photocatalytic water recycling: TiO_2 nanotube arrays degrade organics under UV, enabling 98% reuse	95	VOC emissions ↓90% energy ↓60% (vs. distillation)	6, 9, 12, 13

Cosmetics	500 L/ton (emulsion washing)	High-pressure homogenization: water-free nanoemulsions	95	Preservative use ↓80% energy ↓50%	9, 12
Synthetic fibers	150 L/kg (spin finish application)	Triboelectric coating: dry friction charging replaces aqueous finishes	100	Surfactant pollution ↓100% fiber strength ↑20%	9, 12, 13
Plastic furniture	100 L/unit (injection mold cooling)	Conformal cooling channels: AI-optimized laser-sintered molds reduce cycle time	95	Cooling energy ↓75% cycle time ↓40%	SDG 9, 12
Wood furniture	5000 L/ton (steam bending)	Microwave vacuum drying: 2.45 GHz radiation targets water molecules	85	Drying time ↓70% wood warping ↓90%	9, 12, 15
Wood composite mfg.	20 m^3/ton (resin washing)	Plasma-activated bonding: surface functionalization replaces aqueous adhesives	100	Formaldehyde emissions ↓100% bond strength ↑40%	9, 12, 13
Shoes manufacturing	200 L/pair (leather tanning/rinsing)	Supercritical CO_2 tanning: waterless process with collagen stabilization	100	Chromium discharge ↓100% CO_2e ↓65%	9, 12, 13
Leather tanning	30–40 m^3/ton hide chromium effluent	Supercritical CO_2 tanning: Waterless process with 100% Cr recycling	100	Toxic sludge ↓100% CO_2e ↓70% (vs. chrome tanning)	6, 9, 12
Rubber manufacturing	50 m^3/ton (mastication cooling)	Cryogenic grinding: LN_2 freezing enables ambient-temperature pulverization	100	VOC emissions ↓95% particle uniformity ↑90%	9, 12, 13
Tire production	80 m^3/ton (Banbury cooling)	AI-controlled induction heating: Precision vulcanization replaces water cooling	95	Steam consumption ↓75% tire uniformity ↑30%	9, 12, 13
Paper and pulp	30–50 m^3/ton paper lignin/chlorine effluent	AI closed loop + electrocoagulation: real time TOC monitoring, 95% water reuse	92	BOD ↓95% carbon ↓58% (energy recovery)	6, 9, 12, 13
Packaging (metal/glass)	3–5 m^3/ton (lubricant/coolant)	Cryogenic forming: LN_2 cooled dies enable dry metal shaping	100	Lubricant waste ↓100% energy ↓35%	9, 12
PET packaging	80 L/kg (stretch blow molding)	Supercritical fluid foaming: CO_2-based expansion replaces water cooling	90	Energy ↓60% recyclability ↑100%	9, 12, 13

(continued)

Table 4.2 (continued)

Industry sector	Traditional water usage	Industry 6.0 advanced techniques	Water savings (%)	Climate and environmental impact	SDGs
Ceramics and glass	0.5–1.5 m^3/ton product slip casting	Dry pressing + binder jet 3DP: Additive manufacturing eliminates wet processes	95	Clay mining ↓60% firing energy ↓35%	9, 12, 13
Agriculture	60–70% evaporation loss (flood irrigation)	Soil hydration AI + hydrogel sensors: Root-zone precision irrigation	80	N_2O emissions ↓45% fertilizer leaching ↓90%	2, 6, 12, 13
Oil and gas	15–25 M gal/well (fracking) benzene/VOC contamination	Graphene oxide membranes: 98.7% contaminant removal, 99.99% salt rejection	97	Groundwater contamination ↓100% methane leakage ↓80%	6, 7, 13, 15
Data centers	1.5–2.0 L/kWh (evaporative cooling)	Two-phase immersion cooling: Dielectric fluid loops eliminate water dependency, under water/sea storage	100	PUE ↓0.15 (to 1.03) CO_2e ↓45%	7, 9, 13
Mining	6000–10,000 L/ton ore cyanide/ heavy metals	Bioleaching (*Acidithiobacillus*): metal extraction + water purification in situ	99	Cyanide usage ↓100% CO_2e ↓65% (vs. smelting)	6, 9, 12, 13, 15
Surgical instruments	1500 L/batch (ultrasonic cleaning)	Plasma sterilization: low temperature plasma replaces aqueous cleaning	100	Bioburden ↓99.99% no biofilm risk	3, 6, 9
Sports equipment	300 L/ton (composite rinsing)	Laser surface functionalization: Hydrophobic textures without coatings	100	VOC emissions ↓100% weight ↓15%	9, 12
Steel production	50–100 m^3/ton steel scale flume water	Dry descaling: laser/plasma oxide removal eliminates descaling water	100	Heavy metal discharge ↓100% CO_2e ↓40% (energy recovery)	6, 9, 12, 13
Automotive industry	5–10 m^3/vehicle paint shop rinsing (assembly)	Cavitation enhanced Nanocoating: ultrasonic deposition replaces rinse cycles	90	Sludge waste ↓85% CO_2e ↓75% (energy savings)	9, 12, 13
	18,000 L/ton cathode $LiPF_6$ rinsing (battery manufacturing)	Supercritical CO_2 electrode drying: Replaces water based solvent recovery	95	PFAS elimination CO_2e ↓50%	9, 12, 13
Semiconductors	8000–12,000 L/chip (UPW rinsing)	Vapor phase cleaning: Nanoscale dry processes +95% UPW recovery	98	Chemical waste ↓88% PFAS elimination	9, 12, 13

Smartphone assembly	5000 L/1 k units (screen cleaning)	Laser ablation: nanosecond pulses remove contaminants without liquids	100	IPA/VOC emissions ↓100% yield ↑5%	9, 12, 13
Laptop manufacturing	3000 L/1 k units (PCB rinsing)	Vapor degreasing: HFE-based solvents with 99.7% recycling	98	Halogenated waste ↓100% cleaning time ↓70%	9, 12, 13
PLC panels for industrial machines and electronics	500 L/panel (electrolytic cleaning)	Laser defluxing: pulsed lasers remove solder residues without chemicals	100	VOC emissions ↓100% energy ↓70% (vs. aqueous cleaning)	9, 12, 13
Sensor manufacturing	15,000 L/wafer (CMOS rinsing)	Dry plasma etching: SF_6/C_4F_8 plasmas replace wet cleaning	100	HF acid elimination PFC emissions ↓90%	9, 12, 13
	300 L/1 k units (MEMS rinsing) IoT sensors	Supercritical CO_2 drying: Replaces UPW rinsing for nanoscale structures	100	PFAS elimination yield ↑12%	
Predictive maintenance tools	300 L/unit (coolant for testing)	Digital twin thermal simulation: AI predicts thermal loads without physical testing	100	Coolant waste ↓100% design iteration time ↓90%	9, 12, 13
Actuator production	500 L/ton (piezo ceramic poling)	Laser poling: UV lasers align dipoles without deionized water baths	100	Lead zirconate waste ↓100% energy ↓80%	9, 12, 13
Edge computing devices	1200 L/rack (immersion cooling)	Dielectric two-phase cooling: Novec™ fluid loops eliminate water	100	PUE ↓0.05 (to 1.02) GWP ↓98% (vs. HFC)	7, 9, 13
Cloud servers	1.8 L/kWh (evaporative cooling)	Direct-to-Chip LN_2 cooling: cryogenic cooling for 100 kW/rack	100	WUE (Water usage effectiveness) = 0.0 energy ↓50%	7, 9, 13
Robotics assembly	200 L/robot (hydraulic testing)	Magnetorheological fluid testing: field-controlled fluids replace water-based hydraulics	100	Hydraulic oil leaks ↓100% testing time ↓70%	9, 12, 13
Industrial revolution 4.0 tools (robotics)	150 L/robot (hydraulic fluid testing)	Electrorheological fluid actuators: field-controlled viscosity replaces hydraulics	100	Oil spills ↓100% response time ↑400%	9, 12, 13

(continued)

Table 4.2 (continued)

Industry sector	Traditional water usage	Industry 6.0 advanced techniques	Water savings (%)	Climate and environmental impact	SDGs
Industrial revolution 5.0 tools (Cobots)	80 L/unit (gearbox quenching)	Laser shock peening: Surface hardening without thermal distortion	100	Energy ↓85% component life ↑200%	9, 12, 13
Industrial revolution 6.0 tools (quantum AI)	5000 L/unit (cryostat cooling)	Dry dilution refrigerators: He^3/He^4 closed cycle at 10 mK	100	Zero helium loss qubit coherence ↑90%	7, 9
Industrial revolution 7.0 tools (neuromorphic)	200 L/wafer (analog AI chip rinsing)	Molecular self-assembly: DNA-guided nanowire growth eliminates lithography	100	Chemical waste ↓99% compute density ↑10^6 ops/mm^2	9, 12
HMI interfaces	50 L/unit (touchscreen etching)	Femtosecond laser texturing: creates oleophobic surfaces without wet chemicals	100	IPA consumption ↓100% durability ↑300%	9, 12
5G/6G tower manufacturing	1000 L/tower (radome coating)	Aerosol dry coating: Electrostatic deposition of radome materials	100	VOC emissions ↓100% coating adhesion ↑50%	9, 13
Industrial networks (5G/6G)	800 L/base station (radome coating)	Plasma-enhanced CVD: Dry deposition of hydrophobic nanocoatings	100	VOC ↓100% signal loss ↓0.1 dB	9, 13
7G quantum repeaters	2000 L/unit (cryogenic cooling)	Dry dilution refrigerators: He^3/He^4 closed-cycle cooling	100	Zero liquid helium boil-off energy stability ±0.001 K	7, 9
Lighting (LED/bulb)	1000 L/1 k units (glass etching)	Laser etching: Femtosecond pulses create optical structures without acids	100	HF acid elimination etching resolution ↑200%	9, 12
Aviation (aircraft cleaning)	10,000–15,000 L/aircraft chemical runoff	Laser decontamination: Pulsed fiber lasers remove dirt without water or chemicals	100	VOC emissions ↓100% CO_2e ↓70% (vs. pressure washing)	9, 12, 13
Orbit industry (satellite)	5000 L/unit (ultrapure rinsing)	Vacuum plasma cleaning: Atomic oxygen treatment in vacuum chambers	100	Zero wastewater energy ↓60% (vs. cleanroom rinsing)	9, 13

Fighter jets (engine mfg.)	8000 L/engine (turbine rinsing)	Ultrasonic dry cleaning: Cavitation removes contaminants without water	100	HAPs ↓100% VOC compliance ↑100%	9, 12, 13
Missile manufacturing	500 L/unit (hydrotesting)	Phased-Array UT + AI: Non-destructive testing replaces hydrostatic water	100	Eliminates test water contamination CO_2e ↓40%	9, 13
Air defense systems	1000 L/radar (cooling)	Two phase immersion cooling: dielectric fluids replace water-chilled systems Recycle water	100	PUE ↓0.10 Water stress ↓100% in arid deployments	6, 7, 9
Explosives (TNT/ nitration)	20 m^3/ton (acid washing)	Supercritical fluid extraction: CO_2-based purification replaces aqueous steps	90	Acid waste ↓100% NO_x emissions ↓95%	9, 12, 13
Bullets manufacturing	200 L/ton (lead quenching)	Laser hardening: Precision surface treatment eliminates quenching baths	100	Heavy metal discharge ↓100% Lead exposure ↓100%	6, 9, 12
Fire extinguisher/ PPE	500 L/ton (pressure testing)	AI-powered simulation: Digital twins replace hydrostatic tests	100	Zero wastewater testing time ↓90%	9, 13
Chemical and biological reactors	50 m^3/batch (quenching) Batch reactor	Adiabatic reaction control: AI modulates feed rates to prevent exotherms	100	Steam consumption ↓75% reaction yield ↑15%	SDG 9, 12
	200 L/h (jacket cooling) Continuous reactor	3D-printed microreactors: Enhanced heat transfer eliminates cooling	100	Energy ↓90% throughput ↑300%	9, 12, 13
	100 m^3/batch (media prep) Bioreactor pharmaceutical	Concentrated perfusion: 10X media concentration reduces volume	90	Nutrient waste ↓85% Titer ↑40%	3, 6, 9
	150 L/L ethanol (downstream) Bioreactor biofuel	Membrane solvent extraction: Ceramic membranes replace distillation	95	Energy ↓80% purity ↑99.9%	7, 9, 12
	80 L/kg (enzyme immobilization) Enzymatic reactor	Magnetic nanoparticle carriers: Field-recoverable biocatalysts	100	Enzyme reuse ↑20 cycles reaction time ↓70%	9, 12
Boiler manufacturing	10 m^3/unit (hydrotesting)	Phased-array ultrasonic testing: automated scans replace water filling	100	Eliminates corrosion risk defect detection ↑300%	9, 13

time irrigation systems signifies a global paradigm change in which enterprises actively contribute to sustainability and water conservation rather than consuming excessive amounts of water.

10.1 Water Management Using Machine Learning

The intricacies of contemporary industrial water consumption are frequently overlooked by traditional water management techniques. These days, real-time water usage optimization is achieved using machine learning algorithms that forecast demand and dynamically modify supply. These AI-powered techniques aim to purify and recycle water for future use in addition to minimizing water waste. Various data sets can be integrated using standard algorithms and information and communication technology to create intelligent systems that improve the sustainability of water management techniques. IoT-enabled water meters, for instance, provide for precise control over water distribution by providing real-time monitoring of water flow, pressure, and quality. By identifying leaks or inefficiencies in industrial systems, these technologies can help save water and increase the overall effectiveness of water management. AI models and deep learning algorithms forecast irrigation requirements based on crop type, soil moisture, and meteorological conditions, which can reduce irrigation water use by up to 40% in agriculture, where water consumption is a major concern. To optimize water distribution, track consumption, and automatically modify flows to reduce waste, AI Smart Water Management Platforms combine a variety of water management technologies for industries. These technologies include IoT, predictive analytics, and real-time monitoring [130–132].

IoT has become a key tool in water management, helping to regulate industrial water consumption, increase crop yields, and lower water-related costs. IoT sensors in agriculture monitor crop health and soil moisture content, allowing intelligent irrigation systems that use real-time data to only irrigate when needed. In large-scale farming, this method has drastically decreased water waste, leading to a 30–40% decrease in irrigation water use. Comparably, IoT-powered water recycling systems in sectors like textiles continuously check water quality and identify impurities, allowing for closed-loop systems that reuse water inside the building and reducing overall water usage. Also, for better water management, FIWARE, i.e. an open source platform used by several industries, integrates real time data from cloud services, big data analytics, and IoT devices. This platform makes it possible to create intelligent agritech systems that track irrigation performance and water consumption, leading to more efficient water allocation among farms. To ensure that water is used effectively without over-irrigating crops, smart AgroTech systems that employ this technology can automatically alter water levels depending on real-time field data. These techniques increase crop yields while preserving valuable water resources [131, 133, 134].

10.2 Intelligent Irrigation and Industry Water Management

Real-time water management systems for industrial use reduce waste and balance water distribution. Advanced technologies such as deep learning, IoT, and AI-based predictive models can be integrated to enable industries to dynamically modify water usage in response to operational requirements. Using machine learning models that monitor usage data from the past, the Smart Water Management Platform helps manufacturers forecast the ideal water usage for several operations, including production, cleaning, and cooling. Industries can save water use by up to 25% by predicting water requirements and modifying water distribution in real time. Also, water treatment facilities are now using machine learning models to improve their water purification procedures. By predicting contamination levels using historical and real-time data, these AI systems help plants modify chemical dosage and filtration, lowering the need for excessive water treatment chemicals and enhancing water quality [135, 136].

There are financial and environmental advantages to implementing AI-driven water management systems. In terms of the environment, intelligent water management systems greatly cut down on water waste, enhance resource allocation, and minimize the energy needed for water distribution and treatment. IoT-based irrigation systems can cut water use by up to 50% by monitoring soil moisture levels and modifying irrigation schedules in real-time, thereby supporting SDG 6. Businesses that use these technologies save money on energy and water expenses. Integrating smart water management platforms into manufacturing processes has improved profitability by reducing water expenses by 20–30%. Precision irrigation technologies have helped farmers save up to 40% on water use in agriculture, which has resulted in significant cost savings and increased yields. Effective water management systems also lessen the possibility of water shortage, which can raise operating expenses for businesses that depend on water-intensive operations.

IoT-powered smart irrigation systems in agriculture can track weather and soil moisture levels in real time, allowing farmers to deliver water only when needed. AI powered prediction models optimize irrigation schedules and save water use by up to 40% by forecasting water requirements based on crop type, local climate, and soil condition. Like this, closed loop water recycling systems that are managed and observed by AI platforms are integrated into textile manufacturing to improve water utilization. Water use is significantly reduced by these systems, which recover and purify water used in dyeing and washing processes. Some businesses have reported water usage savings of up to 50%. AI, IoT, and machine learning-enabled water-saving methods are essential for cutting water use in a variety of sectors, including manufacturing and agriculture. Industries may optimize water usage, improve sustainability, and cut costs by integrating smart water management systems like AI Smart Water Management Platforms real-time IoT monitoring, and predictive analytics. These developments not only tackle the problems associated with water scarcity, but they also support global sustainability objectives such as SDG 6, 9, and 12. Intelligent manufacturing systems' ongoing development will be essential to attaining more sustainable and effective water use, which will have positive effects on the environment and the economy everywhere. Industries may track water flow, pressure, and consumption across production lines in manufacturing by implementing IoT-enabled water meters and real-time monitoring systems. Cloud-based analytics solutions process and analyze data from these sensors to identify leaks or inefficiencies and enhance water distribution. With the use of intelligent water management technology, several manufacturers have been able to reduce their water use by 30% or more, demonstrating how these smart systems assist in reducing water waste and enhancing operational efficiency [137, 138].

11 Industrial Waste Management Using Intelligent Manufacturing Techniques

Many sectors have serious concerns about industrial waste management, especially as waste production rises in tandem with industrial productivity. Effective waste management systems are essential to reducing the negative effects on the environment and guaranteeing adherence to ever-tougher requirements. By facilitating automated trash sorting and recycling, real-time monitoring, and predictive analysis, the use of intelligent manufacturing technology has significantly improved waste management procedures [80, 139].

11.1 Waste Reduction Smart Techniques

The application of AI and machine learning to waste prediction and prevention is one of the most important developments in intelligent waste management. ML algorithms can find trends in waste generation and recommend process improvements by examining past data from production lines. AI systems track material use, forecast manufacturing inefficiencies, and suggest design modifications to lower scrap rates in the automotive sector, resulting in a 20–30% decrease in material waste. By anticipating when machines would need maintenance, these systems help save downtime and waste related to malfunctioning machinery. There are major financial gains when intelligent manufacturing technologies are included into waste and water management. Industries may increase production, lower operating costs by 20–30%, and lessen their reliance on pricey raw resources by optimizing water use and cutting waste. By lowering material waste and enhancing water recycling procedures, the automobile sector has saved millions of dollars. In terms of the environment, intelligent water management systems help save water resources, which is essential for reaching SDG 6, by lowering water use and wastewater creation. By guaranteeing sustainable production methods and lowering environmental contamination, waste reduction, and improved recycling efficiency directly assist SDG 12 [140, 141].

Intelligent manufacturing solutions for waste management and water conservation offer businesses enormous potential to lower costs and increase environmental sustainability. Businesses may maximize resource utilization, enhance product quality, and cut waste by using AI, IoT, machine learning, and automation in water and waste management systems. This will greatly help achieve both economic and environmental objectives. The benefits in terms of resource efficiency, cost savings, and environmental preservation are significant, but putting these cutting-edge ideas into practice calls for careful planning, investment, and cross-sector cooperation [142].

11.2 Intelligent Waste Management Technologies

Real time tracking of garbage generation is made possible by the integration of IoT into waste management. To monitor trash levels, composition, and processing status, smart sensors can be incorporated into production machinery, dumpsters, and garbage containers. Cloud-based platforms receive this data and use it to evaluate and classify waste streams, making sure that recyclables are separated for reuse and hazardous materials are disposed of appropriately. IoT systems have been used in food processing facilities to track organic waste and transfer it to facilities that produce compost or bioenergy, so lowering landfill trash by as much as 40% [78, 143].

11.3 Automation in Recycling and Waste Sorting

Recyclables are increasingly being sorted by automated robots with AI-powered machine vision systems. These robots can recognize and extract recyclable elements, such as plastics and valuable metals, from waste streams in industries like electronics production. Recycling is done more quickly and accurately thanks to this automated sorting process, which also lessens the need for physical labor. By anticipating the most effective method of material recovery and minimizing contamination, artificial intelligence systems improve recycling rates overall. In some industrial applications, particularly in the recycling of scrap metal and electronic waste, AI-driven recycling systems have contributed to a 50% improvement in material recovery rates [143, 144].

11.4 Digital Twins for Optimizing Waste

By replicating the complete waste management process, from generation to disposal, digital twin technology plays an important role in trash management. A manufacturing facility's digital twin may simulate the flow of materials through the system, spotting inefficiencies and offering solutions. Manufacturers can use this virtual model to model various waste management situations and determine the best strategy to reduce trash production and increase recycling rates. This technique is especially helpful in sectors like chemical processing and plastics manufacture, where waste reduction is essential for both financial and environmental reasons [145, 146].

11.5 Integration of the Circular Economy

Reusing products and materials and minimizing waste are key components of the circular economy, which is facilitated by intelligent manufacturing technology. To provide openness and traceability in trash recovery and recycling initiatives, blockchain technology is being utilized more and more to track materials throughout their lifecycle. To ensure that materials are reused rather than disposed of, blockchain-based systems, for instance, trace fabric from production to recycling in the textile industry. This helps to achieve SDG 12. By doing this, waste is decreased, and a closed-loop system is established, which increases material efficiency and lowers the need for raw materials [147, 148].

12 Right First Time Approach in Intelligent Manufacturing

In contemporary manufacturing, the Right First Time method is a crucial quality management concept that seeks to create goods accurately the first time while lowering waste, rework, and faults and increasing productivity. Achieving the Right First Time in intelligent manufacturing entails incorporating cutting-edge technology like AI, IoT, big data analytics, and automated systems into the manufacturing process. Together, these technologies optimize each step of the production process, guaranteeing that goods are produced with high accuracy, little waste, and within the required quality standards without the need for expensive adjustments.

ML and AI are essential components of the Right First Time methodology. Predictive analytics is used by AI systems to foresee process variances, spot possible flaws early, and make real-time process modification suggestions. AI systems can examine production data from many sensors (temperature, pressure, and speed) integrated into assembly lines in the car industry to identify departures from ideal manufacturing conditions. AI can automatically modify equipment settings to maintain process tolerances and guarantee the product is correct the first time by anticipating when specific parameters may deviate from specifications. The most frequent reasons for flaws or deviations can be found by training machine learning models on historical data. Then, the process can be modified in real time to stop these problems from happening again. By taking this proactive approach, quality problems are avoided, and the possibility of creating parts that satisfy the required standards without requiring rework is increased. Because it allows for real-time monitoring and control of production processes, the IoT is essential to assuring Right First Time. During production, IoT devices, like smart sensors, collect enormous volumes of data. IoT sensors track the construction environment (material temperature, humidity, and print speed) during 3D printing and instantly modify settings to guarantee quality. For adaptive control, where the system may automatically rectify any deviations that could result in faults, these real-time inputs are essential. By enabling the production line to continuously adapt in response to sensor data, closed-loop feedback systems help to avoid quality problems before they affect the finished product. Manufacturers may improve overall product consistency by incorporating IoT sensors into vital manufacturing systems, which will guarantee that discrepancies are detected early and automatically fixed [148, 149].

To improve the Right First Time rate and comprehend intricate production processes, big data analytics is essential. Manufacturers can obtain profound insights into the root causes of flaws or inefficiencies by gathering and examining enormous volumes of data from multiple sources, including sensor readings, machine performance logs, and historical production data. Using tools like Apache Hadoop or Apache Spark, predictive analytics models are created to help find patterns and abnormalities that could compromise the quality of output. These technologies can identify possible failures before they happen and continuously monitor production conditions. Data analysis can forecast tool wear in CNC machining, causing a tool replacement before flaws appear and upholding high standards of quality all the way through the process [150, 151].

Because robotics and automated systems offer unparalleled accuracy and consistency, they are essential to Right First Time manufacturing. Robots with sophisticated sensors and AI carry out repetitive operations with remarkable precision, guaranteeing that goods are produced with strict tolerances. Robots carefully handle wafers in semiconductor manufacturing, preventing physical damage during production and reducing faults. Real-time product inspection is facilitated by vision systems built into robotic arms that employ machine vision algorithms to detect and automatically fix any defects found during assembly. Automated quality control systems make sure that flaws are found early in the process so that they may be quickly fixed. These systems use AI and computer vision to have machines conduct continuous inspections. To guarantee that only high-quality items are provided, these systems can identify problems like inaccurate dimensions or misalignments before the product proceeds further down the manufacturing line [152, 153].

A virtual representation of a real-world system, procedure, or product is called a "digital twin." Engineers can simulate various situations and improve parameters to obtain Right First Time results by developing a digital representation of the production process. By simulating possible problems and streamlining the process beforehand, digital twins help to prevent mistakes during actual production. The first batch of items is guaranteed to be quality and yield optimized thanks to these predictive capabilities. Digital twins enable producers to modify critical parameters in a manufacturing system, such as temperature, pressure, and material flow, before they have an impact on the final product. This degree of foresight guarantees that items are created and produced under ideal circumstances, improving product quality right from the beginning.

Achieving Right First Time outcomes requires strict process control and standardized work methods. Manufacturers can strictly adhere to predetermined quality standards thanks to Industry 4.0 technology like AI-driven process control and standardization tools. Through real-time process monitoring, SPC combined with machine learning models helps guarantee that production stays within reasonable bounds. Process stability can be continuously evaluated by automated systems and machine learning algorithms, guaranteeing that items are consistently manufactured in compliance with requirements. A steady production environment with fewer faults is also ensured by AI-based process optimization, which continuously modifies factors like machine speed, temperature, and pressure in response to shifting conditions. By identifying problems early in the production process, it helps manufacturers limit the need for scrap or rework, which ultimately saves time and money on materials. Improved operational efficiency since AI and IoT technologies increase throughput and decrease the need for human involvement by automating many quality control processes. Continuous monitoring and real-time corrections result in improved product quality, which guarantees that goods fulfill constant quality standards and have fewer flaws, thus increasing consumer happiness. Costs are cut, downtime is decreased, and processes are streamlined, increasing profitability and supporting sustainability and compliance by assisting the industry in adhering to environmental and quality standards. SDGs are supported by these technologies because they decrease waste and increase resource efficiency.

13 Key Challenges to Adopt Industry 6.0 and Their Solutions with Suggestions

Here is Table 8.5 with the challenges and their best possible solution, with strategic suggestions.

A structured, phased roadmap is essential to guide the successful implementation of Industry 6.0 technologies and ensure both technical robustness and sustainable returns. The adoption pathway begins with Phase 1, where organizations must secure foundational digital infrastructure by deploying quantum encryption mechanisms in tandem with Zero Trust Architecture to establish ultra-secure industrial networks. Phase 2 advances the transition by integrating AI blockchain systems into circular economy applications like enzymatic recycling plants, to enable traceable, closed-loop material cycles. Finally, Phase 3 focuses on scaling quantum AI hybrid models to facilitate autonomous system optimization and end-to-end intelligent manufacturing. These strategies and technological enablers define a clear, high-impact roadmap for realizing the full potential of Industry 6.0, achieving secure, autonomous, and sustainable industrial ecosystems.

Table 8.5 Industry 6.0 implementation challenges and advanced solutions, and suggestions

Challenge category	Specific technical problems	Best possible solutions	Strategic suggestions
Cybersecurity and data integrity	– Quantum computing threats to encryption – AI-driven attacks on IIoT networks – Ransomware targeting critical infrastructure	– Quantum-safe cryptography: Deploy lattice-based algorithms (CRYSTALS Kyber) with 256 bit security – AI driven threat Intelligence: Use neuromorphic SOCs (Intel Loihi 3) for 99.7% threat prediction – Blockchain immutable backups: Zk-SNARK blockchain (IOTA 2.0) for 100% data traceability	– Adopt zero trust architecture (NIST SP 800–207 Rev2) – Implement hardware based PUFs for authentication – Mandate cross-industry threat intelligence sharing
System reliability and network dependency	– Network latency >300 ms disrupting real-time operations – DDoS attacks on edge devices (51% success rate)	– 5G/6G hybrid networks: ultra reliable low latency communication with <1 ms latency – Air gapped edge nodes: use compute in memory devices for offline operations – Digital immune twins: self healing systems with <50 ms recovery latency	– Deploy satellite-wireless redundancy for remote sites – Standardize TSN (time-sensitive networking) protocols – Integrate AI for predictive network maintenance
Sustainable manufacturing and circular economy	– High energy consumption in AI data centers (1.5–2.0 L water/kWh) – Low recycling rates (20% for critical minerals)	– Green AI models: Energy efficient algorithms reducing compute energy by 50% (model pruning) – Advanced recycling: Enzymatic depolymerization (PETase/MHETase) for 99% PET recycling – H_2 industrial heating: Plasma catalytic burners (93% efficiency) replacing fossil fuels	– Enforce Digital Product Passports (ISO 59040) for material traceability – Incentivize circular practices via tax breaks (U.S. Inflation Reduction Act) – Adopt AI driven LCA tools for real time carbon tracking

(continued)

Table 8.5 (continued)

Challenge category	Specific technical problems	Best possible solutions	Strategic suggestions
Human machine collaboration	– Skill gaps in AI/ quantum tech – Safety risks in cobot human workspaces	– AR/VR training platforms: NVIDIA Omniverse for real time skill development – AI augmented Cobots: Force limited robots with reinforcement learning for adaptive behavior – Ethical AI frameworks: Explainable AI (XAI) for transparent decision-making	– Establish industry-academia consortia for workforce reskilling – Implement ISO/TS 15066 for cobot safety – Develop "human-in-the-loop" certification standards
Quantum and AI integration	– Decoherence in quantum hardware – High energy demands of real time AI (3.2 MWh/ton for CCU)	– Hybrid quantum systems: 128-qubit processors for material simulations (10× faster R&D) – Photonic tensor processors: Low energy AI accelerators for edge devices (80 μW/ cm^3) – Federated learning: privacy preserving AI training across devices	– Invest in quantum error-correction R&D – Prioritize edge AI over cloud for latency-sensitive tasks – Create open-source quantum libraries
Supply chain resilience	– Cyber physical attacks causing more than USD 4 T disruptions – Lack of real time visibility	– Blockchain-enabled traceability: Hyperledger fabric for end-to-end supply chain transparency – AI predictive analytics: Digital twins simulating disruptions with 30–40% risk reduction – Autonomous logistics: Quantum-optimized routing reducing delivery times by 50%	– Mandate blockchain-based supply chain audits – Deploy IoT sensors for real time cargo monitoring – Develop geopolitical risk AI dashboards
Cost and scalability	– High CAPEX for SMEs (USD 2.5 M for digital twins) – Legacy system incompatibility	– Cloud-edge hybrid models: AWS Greengrass for legacy IIoT integration – Modular microfactories: 3D-printed reactors reducing setup costs by 60% – Leasing models: Robotics as a service (RaaS) for flexible automation	– Launch government co-funding programs (EU circular economy 2.0) – Promote open-source platforms (FIWARE) – Standardize API interfaces (OPC UA)

Table 8.5 (continued)

Challenge category	Specific technical problems	Best possible solutions	Strategic suggestions
Regulatory compliance	– Cross-border data privacy conflicts (GDPR vs. NIS directive) – SDG alignment complexities	– Automated compliance engines: AI auditing tools for real time IEC 62443/ ISO 30144 adherence. – SDG analytics dashboards: big data platforms tracking ESG metrics (water use ↓82%)	– Form global regulatory sandboxes for Industry 6.0 tech – Integrate blockchain for immutable compliance records. – Align cybersecurity with SDG 9/12/13 targets

To accelerate progress, cross-industry collaboration is critical. Establishing Industry 6.0 consortia, comprising stakeholders from manufacturing, technology, and energy sectors—can significantly lower R&D costs and enable knowledge transfer. These consortia should support the development of open source digital twin templates, particularly for small and medium-sized enterprises, ensuring scalable access to high-fidelity simulation and optimization tools.

Policy and investment incentives are important to catalyzing adoption. Governments should offer subsidies for green technologies, like a 30% tax credit on hydrogen infrastructure development, to reduce capital risk and encourage private investment. The rollout of international standards for quantum safe AI, for example, the upcoming ISO/IEC 30144: 2025, will provide unified regulatory frameworks for next-generation AI systems.

The success of these strategies must be evaluated using defined performance metrics. In cybersecurity, achieving over 99% threat neutralization within 300 ms is now feasible through neuromorphic and edge-based architectures. On the sustainability front, organizations should target a 50% reduction in CO_2 equivalent emissions per unit output by a realistic time frame, while economically, systems like predictive maintenance platforms can deliver ROI within 18 months, driven by reduced downtime and energy optimization.

Looking ahead, the emerging technology landscape (2025–2030) presents both transformative opportunities and associated risks. Neuromorphic chips are set to deliver 2000 times faster threat detection and 38% reductions in server energy consumption, though they require biometric Physically Unclonable Functions for spoofing resistance. Quantum machine learning is expected to accelerate material discovery by 10 times, particularly in areas like battery electrolyte research, and should be governed by hybrid classical-quantum validation frameworks. Self-healing materials promise a 400% increase in durability for aerospace composites but necessitate rigorous biocompatibility and toxicity testing, compliant with standards such as ISO 10993.

References

1. Nikooharf, M.H., et al.: Machine learning in polymer additive manufacturing: a review. Int. J. Mater. Form. **17**(6), 1–27 (2024)
2. Su, Q., Chen, L., Qian, L.: Optimization of big data analysis resources supported by XGBoost algorithm: comprehensive analysis of industry 5.0 and ESG performance. Measure. Sens. **36**, 101310 (2024)
3. Amiri, Z., et al.: Adventures in data analysis: a systematic review of deep learning techniques for pattern recognition in cyber-physical-social systems. Multimed. Tools Appl. **83**(8), 22909–22973 (2024)
4. Yang, J., Liu, Y., Morgan, P.L.: Human-machine interaction towards Industry 5.0: human-centric smart manufacturing. Dig. Eng. **2**, 100013 (2024)
5. Josbert, N.N., et al.: A look into smart factory for industrial IoT driven by SDN technology: a comprehensive survey of taxonomy, architectures, issues and future research orientations. J. King Saud Univ. Comput. Inf. Sci. **36**, 102069 (2024)
6. Aldrini, J., Chihi, I., Sidhom, L.: Fault diagnosis and self-healing for smart manufacturing: a review. J. Intell. Manuf. **35**(6), 2441–2473 (2024)
7. Castillo, M., Monroy, R., Ahmad, R.: Scientometric analysis and systematic review of smart manufacturing technologies applied to the 3D printing polymer material extrusion system. J. Intell. Manuf. **35**(1), 3–33 (2024)
8. Rawther, S., Sivaji, S.: Protecting cloud computing environments from malicious attacks using multi-factor authentication and modified DNA cryptography. Recent Patents Eng. **19**(1), E050923220730 (2025)
9. Awadallah, A., et al.: Artificial intelligence-based cybersecurity for the metaverse: research challenges and opportunities. IEEE Commun Surv Tutor. **27**, 1008 (2024)
10. Kokila, M., Reddy, S.: Authentication, Access control and scalability models in internet of things security-a review. Cyber Secur. Appl. **3**, 100057 (2024)
11. Zafar, M.H., Langås, E.F., Sanfilippo, F.: Exploring the synergies between collaborative robotics, digital twins, augmentation, and industry 5.0 for smart manufacturing: a state-of-the-art review. Robot. Comput. Integr. Manuf. **89**, 102769 (2024)
12. Mezina, N., Tikhonov, G.: Automation of decision-making processes: opportunities and risks. Russ. Eng. Res. **44**(3), 400–404 (2024)
13. Mondal, M.K., Banerjee, S., Zhang, Y.: Toward fast reliable intelligent industry 5.0—a comprehensive study. In: Secure and Smart Cyber-Physical Systems, pp. 56–80. CRC Press, Boca Raton (2025)
14. Hoang, V.-T., et al.: Security risks and countermeasures of adversarial attacks on AI-driven applications in 6G networks: a survey. J. Netw. Comput. Appl. **232**, 104031 (2024)
15. Ahmed, S.F., et al.: Towards a secure 5G-enabled internet of things: a survey on requirements, privacy, security, challenges, and opportunities. IEEE Access. **12**, 13125–13145 (2024)
16. Wang, S., et al.: Explainable AI for 6G use cases: technical aspects and research challenges. IEEE Open J. Commun. Soc. **5**, 2490–2540 (2024)
17. George, A.S., George, A.H.: Towards a super smart society 5.0: opportunities and challenges of integrating emerging technologies for social innovation. Partners Univ. Int. Res. J. **3**, 2583 (2024)
18. Vijarania, M., et al.: Achieving sustainable development goals in cyber security using AIoT for healthcare application. In: Artificial Intelligence of Things for Achieving Sustainable Development Goals, pp. 207–231. Springer, Cham (2024)
19. Bello, K.A., et al.: A review of intelligent manufacturing systems: a future guide to exploring Industry 4.0 technologies. In: 2024 International Conference on Science, Engineering and Business for Driving Sustainable Development Goals (SEB4SDG). IEEE, New York (2024)
20. Ajiga, D., et al.: Designing cybersecurity measures for enterprise software applications to protect data integrity. Comput. Sci. IT Res. J. **5**, 1920–1941 (2024)

21. Chukwunweike, J.N., et al.: The role of deep learning in ensuring privacy integrity and security: Applications in AI-driven cybersecurity solutions. World J. Adv. Res. Rev. **23**(2), 2550 (2024)
22. Nwobodo, L.K., Nwaimo, C.S., Adegbola, A.E.: Enhancing cybersecurity protocols in the era of big data and advanced analytics. GSC Adv. Res. Rev. **19**(3), 203–214 (2024)
23. Goswami, S., et al.: Exploring the impact of artificial intelligence integration on cybersecurity: a comprehensive analysis. J. Ind Intell. **2**(2), 73–93 (2024)
24. Rane, N., Kaya, O., Rane, J.: Integrating internet of things, blockchain, and artificial intelligence techniques for intelligent industry solutions. Artif. Intell. Machine Learn. Deep Learn. Sustain. Ind. **5**, 2 (2024)
25. Soori, M., Dastres, R., Arezoo, B.: AI-powered blockchain technology in industry 4.0, a review. J. Econ. Technol. **1**, 222–241 (2024)
26. Nayak, A., et al.: Data storage and transmission security in the cloud: the artificial intelligence (AI) edge. In: Improving Security, Privacy, and Trust in Cloud Computing, pp. 194–212. London, IGI Global (2024)
27. Rane, J., et al.: Federated learning for edge artificial intelligence: Enhancing security, robustness, privacy, personalization, and blockchain integration in IoT. Future Res. Opportun. Artif. Intell. Ind. **5**, 2–94 (2024)
28. Humayun, M., et al.: Securing the internet of things in artificial Intelligence era: a comprehensive survey. IEEE Access. **12**, 25469–25490 (2024)
29. Ali, G., et al.: A survey on artificial intelligence in cybersecurity for smart agriculture: state-of-the-art, cyber threats, artificial intelligence applications, and ethical concerns. Mesopotam. J. Comput. Sci. **2024**, 53–103 (2024)
30. Poudel, S., Jha, S.: Data breaches in sensors through various integration techniques. In: Quality Assessment and Security in Industrial Internet of Things, pp. 93–112. CRC Press, Boca Raton (2024)
31. Hakiri, A., et al.: A comprehensive survey on digital twin for future networks and emerging internet of things industry. Comput. Netw. **244**, 110350 (2024)
32. Tyagi, A.K., Kumari, S., Richa: Artificial Intelligence-based cyber security and digital forensics: a review. Artif. Intell. Enabled Dig. Twin Smart Manuf. **2024**, 391–419 (2024)
33. Yu, J., Shvetsov, A.V., Alsamhi, S.H.: Leveraging machine learning for cybersecurity resilience in Industry 4.0: challenges and future directions. IEEE Access. **12**, 159579–159596 (2024)
34. Damtew, A.W., Goshu, Y.Y.: The impacts of AWS from digital visions to action for supply chain resilience, performances, and inclusiveness. Int. J. Adv. Manuf. Technol. **130**(9), 4821–4834 (2024)
35. Atadoga, A., et al.: The role of IT in enhancing supply chain resilience: a global review. Int. J. Manage. Entrepreneur. Res. **6**(2), 336–351 (2024)
36. Saisridhar, P., Thuerer, M., Avittathur, B.: Assessing supply chain responsiveness, resilience and robustness (triple-R) by computer simulation: a systematic review of the literature. Int. J. Prod. Res. **62**(4), 1458–1488 (2024)
37. Jamil, M.A., et al.: A structural equation modeling framework for exploring the industry 5.0 and sustainable supply chain determinants. Supply Chain Anal. **6**, 100060 (2024)
38. Sun, F., et al.: Enhancing global supply chain distribution resilience through digitalization: insights from natural resource sector of China. Resour. Policy. **95**, 105169 (2024)
39. Işık, C., et al.: A roadmap for sustainable global supply chain distribution: exploring the interplay of ECON-ESG factors, technological advancement and SDGs on natural resources. Resour. Policy. **95**, 105114 (2024)
40. Agrawal, S., et al.: Can industry 5.0 technologies overcome supply chain disruptions? A perspective study on pandemics, war, and climate change issues. Oper. Manag. Res. **17**(2), 453–468 (2024)
41. Arif, J., Samadhiya, A., Kumar, A.: Net zero supply chain performance and industry 4.0 technologies: past review and present introspective analysis for future research directions. Heliyon. **9**(11), e21525 (2023)

42. de Mattos Nascimento, D.L., et al.: Interrelationships between circular economy and Industry 4.0: a research agenda for sustainable supply chains. Bus. Strateg. Environ. **33**(2), 575–596 (2024)
43. Shafik, W.: Artificial Intelligence and internet of things roles in sustainable next-generation manufacturing: an overview of emerging trends in Industry 6.0. Sustain. Innov. Ind. **60**, 207–239 (2024)
44. Kraus, N., et al.: Digitalizing Production and Industry through 4.0 Technologies Implementation. in Digital Economy Concepts and Technologies Workshop 2024 (2024)
45. Toponar, O., et al.: Integrating the circular economy into business processes to reduce waste and increase environmental sustainability. Grassroots J. Nat. Resour. **7**(2), 140–159 (2024)
46. Wang, K., et al.: Leasing, trade-in for new, or the mixed of both: an analysis of new recycling modes driven by industry 4.0 technologies. Int. J. Prod. Res. **62**(7), 2350–2369 (2024)
47. Chen, C.-C., Pao, H.-T.: Circular economy and ecological footprint: a disaggregated analysis for the EU. Ecol. Indic. **160**, 111809 (2024)
48. Lara-Guillén, J., Méndez-Aparicio, M.D., Jiménez-Zarco, A.I.: Circular economy and closed-loop supply chains in industry 4.0: importance to achieve sustainable development. In: Digital Transformation for Improved Industry and Supply Chain Performance, pp. 299–333. IGI Global, London (2024)
49. Afshari, H., Gurtu, A., Jaber, M.Y.: Unlocking the potential of solid waste management with circular economy and Industry 4.0. Comput. Ind. Eng. **195**, 110457 (2024)
50. Kannan, D., et al.: Unveiling barriers to the integration of blockchain-based circular economy and Industry 5.0 in manufacturing industries: a strategic prioritization approach. Bus. Strateg. Environ. **33**, 7855–7886 (2024)
51. Santiago, B.D.S., et al.: Corporate social responsibility and circular economy integration framework within sustainable supply chain management: building blocks for industry 5.0. Corp. Soc. Responsib. Environ. Manag. **32**, 269–290 (2024)
52. Oskounejad, M.M., Nozari, H.: Advanced Businesses in Industry 6.0. IGI Global, London (2024)
53. Ortega Carrasco, P., et al.: Green public procurement as an effective way for sustainable development: a systematic literature review and bibliometric analysis. Sustain. Dev. **33**, 2364–2391 (2024)
54. Carayannis, E.G., Posselt, T., Preissler, S.: Toward Industry 6.0 and society 6.0: the quintuple innovation helix with embedded AI modalities as enabler of public interest technologies strategic technology management and road-mapping. IEEE Trans. Eng. Manag. **71**, 11238–11252 (2024)
55. da Costa, T.P., da Costa, D.M.B., Murphy, F.: A systematic review of real-time data monitoring and its potential application to support dynamic life cycle inventories. Environ. Impact Assess. Rev. **105**, 107416 (2024)
56. Beltrán Domínguez, A., et al.: Radio Frequency Interference PDS 100 EPS Project. Universitat Politècnica de València, València (2020)
57. Rocky, M.M.H., et al.: Comprehensive insights into aqua regia-based hybrid methods for efficient recovery of precious metals from secondary raw materials. Chem. Eng. J. **495**, 153537 (2024)
58. Khartcyzov, G., et al.: Analysis of slag chemistry in WEEE smelting using experimental and modelling study of the "CuO0.5"-ZnO-FeO-FeO1.5-CaO-SiO2-AlO1.5 system in equilibrium with Cu metal. Ceram. Int. **50**, 26513–26527 (2024)
59. Egeland Nygaard, A., Lokna Wolden, A.: Leveraging Smart Technologies for Sustainable Manufacturing: A Strategic Framework for Implementing Smart Circular Economy Principles. University of Stavanger, Stavanger (2024)
60. Ghonge, M.M., et al.: Advances in Explainable AI Applications for Smart Cities. IGI Global, London (2024)
61. Zhou, C., et al.: An automated lightweight approach for detecting dead fish in a recirculating aquaculture system. Aquaculture. **594**, 741433 (2025)

62. Ham, H., et al.: Non-invasive, memory access-triggered near-data processing for DNN training acceleration on GPUs. IEEE Access. **12**, 142651–142667 (2024)
63. Chen, Y., et al.: Artificial Intelligence/Machine Learning Technology in Power System Applications. Pacific Northwest National Laboratory (PNNL), Richland (2024)
64. Thiruvathukal, G.K., et al.: Low-Power Computer Vision: Improve the Efficiency of Artificial Intelligence. CRC Press, Boca Raton (2022)
65. Nahrstedt, F.: Scaling the Superemulator for ClimateSet: Extending Machine Learning Capabilities to 36 Climate Models (2024)
66. Kasera, R.K., Gour, S., Acharjee, T.: A comprehensive survey on IoT and AI based applications in different pre-harvest, during-harvest and post-harvest activities of smart agriculture. Comput. Electron. Agric. **216**, 108522 (2024)
67. Mafukidze, H.D., et al.: Scattering centers to point clouds: a review of mmWave radars for non-radar-engineers. IEEE Access. **10**, 110992–111021 (2022)
68. Reddy, M.S., Reddy, C.K.K., Hanafiah, M.M.: Climate change mitigation and adaptation strategies enhanced by intelligent systems in industry 6.0. In: Maintaining a Sustainable World in the Nexus of Environmental Science and AI, pp. 201–228. IGI Global, London (2024)
69. Duggal, A.S., et al.: A sequential roadmap to Industry 6.0: exploring future manufacturing trends. IET Commun. **16**(5), 521–531 (2022)
70. Mingaleva, Z., Startseva, O., Gerasimov, A.: Aligning the environmental and social values of the sustainable development concepts, industry 5.0 and society 5.0. E3S Web Conf. **460**, 05012 (2023)
71. Manikandan, S., et al.: Industrial Revolution and Metaverse: Industry 5.0. Quing Publications, Nedungadu (2023)
72. Gera, R., et al.: A comprehensive and narrative review of industry 5.0 technologies: 2018–2022. In: Renewable Energy Optimization, Planning and Control: Proceedings of ICRTE 2022, pp. 237–259. Springer, Cham (2023)
73. Tyagi, A.K., Nair, M.M.: Future of industry 5.0 in society 5.0. In: Industry 4.0, Smart Manufacturing, and Industrial Engineering, pp. 333–346. CRC Press, Boca Raton (2025)
74. Singh, V., Karol, S., Dutt, G.: An historical analysis of principal exports sector, industry 5.0 integration, future opportunities and challenges. Int. J. Hum. Soc. Sci. Manage. **4**, 552–560 (2024)
75. Schaeffer, E., Industry, X.: Realizing Digital Value in Industrial Sectors. Kogan Page Publishers, London (2017)
76. Zhang, S., et al.: Green manufacturing for achieving carbon neutrality goal requires innovative technologies: a bibliometric analysis from 1991 to 2022. J. Environ. Sci. **140**, 255–269 (2024)
77. Malik, A., et al.: Industrial revolution and environmental sustainability: an analytical interpretation of research constituents in Industry 4.0. Int. Lean Six Sigma. **15**(1), 22–49 (2024)
78. Srivastava, G., et al.: Artificial intelligence and IoT-assisted sustainable manufacturing for Industry 4.0. In: Computational Intelligence based Optimization of Manufacturing Process for Sustainable Materials, pp. 15–34. CRC Press, Boca Raton (2024)
79. Rao, K.S., et al.: AI and ML in IR4.0: a short review of Applications and challenges. Malays. J. Sci. Adv. Technol. **4**, 141–148 (2024)
80. Lin, K., Wei, S.: Advancing the industrial circular economy: the integrative role of machine learning in resource optimization. J. Green Econ. Low-carbon Dev. **2**(3), 122–136 (2023)
81. Rajamanickam, M., et al.: Fourth industrial revolution: industry 4.0. In: Integration of Mechanical and Manufacturing Engineering with IoT: A Digital Transformation, pp. 41–84. Wiley, Hoboken (2023)
82. Borgosz, L., Dikicioglu, D.: Industrial internet of things: what does it mean for the bioprocess industries? Biochem. Eng. J. **201**, 109122 (2024)
83. Kumar, V.: Digital enablers. In: The Economic Value of Digital Disruption: a Holistic Assessment for CXOs, pp. 1–110. Springer, Cham (2023)

84. Keynes, J.M.: Conduct risk and business models. In: Business Model Innovation: A Blueprint for Strategic Change, p. 229. Cambridge University Press, Cambridge (2024)
85. Rowan, N.J.: Digital technologies to unlock safe and sustainable opportunities for medical device and healthcare sectors with a focus on the combined use of digital twin and extended reality applications: a review. Sci. Total Environ. **926**, 171672 (2024)
86. Ali, Z.A., et al.: Circular economy advances with artificial intelligence and digital twin: multiple-case study of Chinese industries in agriculture. J. Knowl. Econ. **16**, 2192–2228 (2024)
87. Zhou, Y.: Lifecycle battery carbon footprint analysis for battery sustainability with energy digitalization and artificial intelligence. Appl. Energy. **371**, 123665 (2024)
88. Askar, R., et al.: Driving the built environment twin transition: synergising circular economy and digital tools. In: International Conference Coordinating Engineering for Sustainability and Resilience. Springer, Cham (2024)
89. Ling, S., et al.: Heterogeneous demand–capacity synchronization for smart assembly cell line based on artificial intelligence-enabled IoT. J. Intell. Manuf. **35**(2), 539–554 (2024)
90. May, M.C., Oberst, J., Lanza, G.: Managing product-inherent constraints with artificial intelligence: production control for time constraints in semiconductor manufacturing. J. Intell. Manuf. **35**, 4259–4276 (2024)
91. Ghadoui, E.L., Mouchtachi, M.A., Majdoul, R.: Exploring and optimizing deep neural networks for precision defect detection system in injection molding process. J. Intell. Manuf. **36**, 2897–2914 (2024)
92. Yousif, I., et al.: Leveraging computer vision towards high-efficiency autonomous industrial facilities. J. Intell. Manuf. **36**, 2983–3008 (2024)
93. Lupi, F., et al.: Next-generation vision inspection systems: a pipeline from 3D model to ReCo file. J. Intell. Manuf. **36**, 4711–4734 (2024)
94. Chen, Y., et al.: Dynamic scheduling of hybrid flow shop problem with uncertain process time and flexible maintenance using NeuroEvolution of augmenting topologies. IET Collab. Intell. Manuf. **6**(3), e12119 (2024)
95. Ghante, M., et al.: Six-sigma model in pharma Industry: part–II. In: Modern Aspects of Pharmaceutical Quality Assurance: Developing & Proposing Application Models, SOPs, Practical Audit Systems for Pharma Industry, pp. 21–50. Springer, Cham (2024)
96. Janecki, L., Reh, D., Arlinghaus, J.C.: Challenges of quality assurance in early planning and ramp up of production facilities-potentials of planning automation via virtual engineering. Proc. Comput. Sci. **232**, 2498–2507 (2024)
97. Getachew, M., Beshah, B., Mulugeta, A.: Data analytics in zero defect manufacturing: a systematic literature review and proposed framework. Int. J. Prod. Res. (2024). https://doi.org/10.1080/00207543.2024.2382379
98. Wang, L., Zhao, J.: Strategic Blueprint for Enterprise Analytics: Integrating Advanced Analytics Into Data-Driven Business, vol. 150. Springer, Cham (2024)
99. Popović, A.: Implications of the fourth industrial revolution on sustainable development. Econ. Sustain. Dev. **4**(1), 45–60 (2020)
100. Bargavi, R.: Role of industry 4.0 on achieving SDG's. In: Impact of Industry 4.0 on Supply Chain Sustainability: Current Status and Future Pathways, pp. 31–48. Emerald Publishing Limited, Leeds (2024)
101. Caiado, R.G.G., Quelhas, O.L.G.: Factories for the future: toward sustainable smart manufacturing. In: Responsible Consumption and Production, pp. 239–250. Springer, Cham (2020)
102. Jin, B.E., Shin, D.C.: The power of 4th industrial revolution in the fashion industry: what, why, and how has the industry changed? Fash. Text. **8**(1), 31 (2021)
103. Yang, M., et al.: Circular economy strategies for combating climate change and other environmental issues. Environ. Chem. Lett. **21**(1), 55–80 (2023)
104. Uddin, M.S., et al.: Contributions of ML in industry 5.0 to sustainable development. In: Fostering Sustainable Businesses in Emerging Economies, pp. 87–107. Emerald Publishing Limited, Leeds (2023)

105. Stojadinović, S.M., Majstorović, V.D.: An Intelligent Inspection Planning System for Prismatic Parts on CMMs. Springer, Cham (2019)
106. Alcácer, V., Cruz-Machado, V.: Scanning the industry 4.0: a literature review on technologies for manufacturing systems. Eng. Sci. Technol. **22**(3), 899–919 (2019)
107. Wegener, K., et al.: A conceptual vision for a bio-intelligent manufacturing cell for selective laser melting. CIRP J. Manuf. Sci. Technol. **34**, 61–83 (2021)
108. Selvaraj, V., Xu, Z., Min, S.: Intelligent operation monitoring of an ultra-precision CNC machine tool using energy data. Int. J. Precis. Eng. Manuf.-Green Technol. **10**(1), 59–69 (2023)
109. Wan, J., et al.: Artificial-intelligence-driven customized manufacturing factory: key technologies, applications, and challenges. Proc. IEEE. **109**(4), 377–398 (2020)
110. Suvarna, M., et al.: Cyber–physical production systems for data-driven, decentralized, and secure manufacturing—a perspective. Engineering. **7**(9), 1212–1223 (2021)
111. Nassehi, A., et al.: Review of machine learning technologies and artificial intelligence in modern manufacturing systems. In: Design and Operation of Production Networks for Mass Personalization in the Era of Cloud Technology, pp. 317–348. Elsevier, Amsterdam (2022)
112. Baroroh, D.K., Chu, C.-H., Wang, L.: Systematic literature review on augmented reality in smart manufacturing: collaboration between human and computational intelligence. J. Manuf. Syst. **61**, 696–711 (2021)
113. Hamrol, A., Gawlik, J., Sładek, J.: Mechanical engineering in Industry 4.0. Manage. Prod. Eng. Rev. **10**, 14–28 (2019)
114. Dubois, M.: AI-based predictive maintenance solutions for US aerospace manufacturing: techniques and real-world Applications. J. AI-Assisted Sci. Discov. **4**(2), 94–123 (2024)
115. Stojadinović, S.M., et al.: An approach of development smart manufacturing metrology model as support industry 4.0. In: Proceedings of 5th International Conference on the Industry 4.0 Model for Advanced Manufacturing: AMP 2020. Springer, Cham (2020)
116. Pawłowicz, J.A., et al.: Reverse engineering as a non-invasive examining method of the water tower brick structure condition. Eng. Fail. Anal. **161**, 108280 (2024)
117. Zhang, M., et al.: Multi-sensors image fusion method for non-destructive inspection in vertical-cavity surface-emitting lasers. Int. J. Mach. Learn. Cybern. **16**(5), 3521–3537 (2024)
118. Hamrani, A., et al.: Applying machine learning to wire arc additive manufacturing: a systematic data-driven literature review. J. Intell. Manuf. **35**(6), 2407–2439 (2024)
119. Javaid, M., et al.: Industrial perspectives of 3D scanning: features, roles and it's analytical applications. Sens. Int. **2**, 100114 (2021)
120. Yildiz, E., Møller, C.: Building a virtual factory: an integrated design approach to building smart factories. J. Global Oper. Strat. Sour. **14**(4), 608–635 (2021)
121. Liu, Y., Kukkar, A., Shah, M.A.: Study of industrial interactive design system based on virtual reality teaching technology in industrial robot. Paladyn J. Behav. Robot. **13**(1), 45–55 (2022)
122. Leng, J., et al.: Digital twins-based smart manufacturing system design in Industry 4.0: a review. J. Manuf. Syst. **60**, 119–137 (2021)
123. Demirbaga, Ü., et al.: Cloud computing for big data analytics. In: Big Data Analytics: Theory, Techniques, Platforms, and Applications, pp. 43–77. Springer, Cham (2024)
124. Oluwaseyi, J., et al.: Enhancing Cloud-Based Data Warehousing Performance through the Application of Machine Learning Algorithms (2024)
125. Ma, S., et al.: Data-driven sustainable intelligent manufacturing based on demand response for energy-intensive industries. J. Clean. Prod. **274**, 123155 (2020)
126. Wang, J., et al.: Big data analytics for intelligent manufacturing systems: a review. J. Manuf. Syst. **62**, 738–752 (2022)
127. Lakhiar, I.A., et al.: A review of precision irrigation water-saving technology under changing climate for enhancing water use efficiency, crop yield, and environmental footprints. Agriculture. **14**(7), 1141 (2024)
128. Henao, C., et al.: Determinants of efficient water use and conservation in the Colombian manufacturing industry using machine learning. Hum. Soc. Sci. Commun. **11**(1), 1–11 (2024)

129. Wanyama, J., et al.: A systematic review of fourth industrial revolution technologies in smart irrigation: constraints, opportunities, and future prospects for sub-Saharan Africa. Smart Agricultural Technology. **7**, 100412 (2024)
130. Amraouy, O., et al.: Blockchain-based IoT for precision agriculture: applications, research challenges, and future directions. In: Enhancing Performance, Efficiency, and Security Through Complex Systems Control, pp. 147–174. London, IGI Global (2024)
131. Chauhan, R., et al.: Optimizing water resources with IoT and ML: a water management system. In: Innovations in Machine Learning and IoT for Water Management, pp. 94–109. IGI Global, London (2024)
132. Krishnan, S.R., et al.: Smart water resource management using artificial Intelligence—a review. Sustainability. **14**(20), 13384 (2022)
133. Wayangkau, I.H., et al.: Utilization of IoT for soil moisture and temperature monitoring system for onion growth. Emerg. Sci. J. **4**, 102–115 (2020)
134. Suvarna, M., et al.: Smart manufacturing for smart cities—overview, insights, and future directions. Adv. Intell. Syst. **2**(10), 2000043 (2020)
135. Egbueri, J.C.: Predicting and analysing the quality of water resources for industrial purposes using integrated data-intelligent algorithms. Groundw. Sustain. Dev. **18**, 100794 (2022)
136. Saravanan, S., et al.: Industry 4.0: Smart water management system using IoT. In: Security Issues and Privacy Concerns in Industry 4.0 Applications, pp. 1–14. Wiley, Hoboken (2021)
137. Fu, G., et al.: The role of deep learning in urban water management: a critical review. Water Res. **223**, 118973 (2022)
138. Mehmood, H., et al.: Strategic Foresight to Applications of Artificial Intelligence to Achieve Water-Related Sustainable Development Goals. United Nations University Institute for Water, Environment and Health, Hamilton (2020)
139. Tripathi, P.S., Sahu, G., Tripathi, A.: Industry 4.0 in waste management for sustainable development. In: Green Manufacturing for Industry 4.0, pp. 106–121. Productivity Press, Milton Park (2024)
140. Jebbor, I., Benmamoun, Z., Hachmi, H.: Revolutionizing cleaner production: the role of artificial intelligence in enhancing sustainability across industries. J. Infrastruct. Policy Dev. **8**(10), 7455 (2024)
141. Agrawal, R., et al.: Integration of artificial intelligence in sustainable manufacturing: current status and future opportunities. Oper. Manag. Res. **16**(4), 1720–1741 (2023)
142. Fowdur, T.P., et al.: Artificial Intelligence, Engineering Systems and Sustainable Development: Driving the UN SDGs. Emerald Publishing Limited, Leeds (2024)
143. Kurniawan, T.A., et al.: Digitalization for sustainable wastewater treatment: a way forward for promoting the UN SDG# 6 'clean water and sanitation' towards carbon neutrality goals. Discov. Water. **4**(1), 71 (2024)
144. Das, S.C., et al.: Vision SDG 2030 clean water, sanitation and zero emission industry pollution management using nWSN based green IoT applications: a techno-legal impact. IJISET. **10**, 3 (2023)
145. Aït-Kaddour, A., et al.: Transforming plant-based waste and by-products into valuable products using various "food Industry 4.0" enabling technologies: a literature review. Sci. Total Environ. **955**, 176872 (2024)
146. Gill, R., et al.: Unleashing sustainable efficiency: the integration of computer vision into Industry 4.0. Eng. Manag. J. **37**, 414–432 (2024)
147. Mavropoulos, A., Nilsen, A.W.: Industry 4.0 and Circular Economy: Towards a Wasteless Future or a Wasteful Planet? John Wiley & Sons, Hoboken (2020)
148. Pech-Rodríguez, W., et al.: Industry 4.0: a new window for the future waste management systems. In: Environmental Engineering and Waste Management: Recent Trends and Perspectives, pp. 685–714. Springer, Cham (2024)
149. Hoang, T.-D., et al.: Artificial intelligence in pollution control and management: status and future prospects. In: Artificial Intelligence and Environmental Sustainability: Challenges and Solutions in the Era of Industry, pp. 23–43. Springer, Cham (2022)

150. Delbari, S.A., Hof, L.A.: Glass waste circular economy-advancing to high-value glass sheets recovery using Industry 4.0 and 5.0 technologies. J. Clean. Prod. **462**, 142629 (2024)
151. Junior, O.C., et al.: Product lifecycle management. In: Green and Blue Technologies to Support Smart and Sustainable Organizations: 18th IFIP WG 5.1 International Conference, vol. 639. Springer, Cham (2022)
152. Chaudhary, V., Gaur, P., Rustagi, S.: Sensors, society, and sustainability. Sustain. Mater. Technol. **40**, e00952 (2024)
153. Singh, B., Kaunert, C., Malviya, R.: Featuring machine learning in municipal solid waste management: technological advancements in waste sorting and processing ropeway for SDG 13 (climate action). In: Municipal Solid Waste Management and Recycling Technologies, pp. 343–366. IGI Global, London (2024)

Chapter 9
Industry 6.0 and the Rise of Intelligent Automation in Manufacturing

Abstract Industry 6.0 represents the next stage of industrial evolution, integrating artificial intelligence, quantum computing, autonomous robotics, and intelligent automation to enhance manufacturing efficiency and sustainability. The transition from Industry 4.0 to Industry 6.0 is driven by rapid advancements in smart factories, AI-driven decision making, and cyber-physical systems. However, this transformation presents significant challenges, including financial barriers, cybersecurity risks, workforce skill gaps, and the need for sustainable resource management. The increasing global population and rising demand for raw materials further intensify pressure on industrial supply chains, necessitating the adoption of circular economy strategies and green manufacturing solutions. Supply chain disruptions, semiconductor shortages, and logistics inefficiencies require the implementation of predictive analytics, blockchain transparency, and decentralized production models. Intelligent automation and digital twins optimize industrial processes, ensuring real-time monitoring, fault detection, and energy efficiency. The emergence of smart cities and sustainable infrastructure investments is crucial in addressing global industrial challenges. Ethical concerns and regulatory frameworks must evolve alongside technological advancements to ensure the responsible deployment of AI and robotics. As Industry 6.0 progresses, collaboration between governments, private industries, and research institutions will play an important role in shaping a future that balances economic growth, technological innovation, and environmental sustainability.

Keywords Industry 6.0 · Smart factories · Artificial intelligence · Quantum computing · Supply chain optimization · Autonomous robotics · Digital twins · Cybersecurity · Sustainability · Circular economy · Workforce transformation · Smart cities · Green manufacturing · Industrial automation · Energy efficiency

H. Majeed, T. Iftikhar, *Intelligent Manufacturing in Industry 6.0*,
https://doi.org/10.1007/978-3-032-07278-8_9

1 Introduction

The transition from Industry 4.0 to Industry 6.0 presents a significant leap. This shift is essential for realizing smarter, more flexible, and climate-resilient production systems. However, this transition is not without its challenges, and industries must adopt strategic approaches to overcome obstacles and successfully integrate these new paradigms. Transitioning from Industry 4.0 to Industry 6.0 is a complex process, requiring strategic investments in technology, infrastructure, and workforce development. However, with the right approaches, such as retrofitting legacy systems, adopting hybrid manufacturing processes, and gradually integrating AI, industries can achieve a more sustainable, resilient, and efficient manufacturing ecosystem aligned with the principles of Industry 6.0.

2 Challenges in Transition

2.1 Technological Challenges

The transition from Industry 4.0 to Industry 6.0 is a complex and challenging journey, requiring the resolution of technological barriers related to data interoperability, automation integration, scalability, and cybersecurity. Overcoming these challenges will demand significant investments in research and development, retrofitting legacy systems, and adopting new technologies that facilitate seamless communication and real-time decision making. With a solid technological foundation, strategic investments in infrastructure, and careful planning for the integration of data and machines, industries can successfully navigate the shift towards Industry 6.0, creating more intelligent, sustainable, and resilient manufacturing systems for the future. The transition from Industry 4.0 to Industry 6.0 presents a multitude of technological challenges that necessitate significant upgrades to existing infrastructures. Industry 6.0, with its emphasis on advanced automation, real-time data analytics, and sustainable manufacturing practices, requires a robust technological foundation encompassing AI, machine learning, deep learning, and advanced data processing systems. These technologies demand substantial computational power, high-performance data storage, and enhanced processing capabilities. Below, we outline the critical technological challenges encountered in this transition [1–3].

2.1.1 Data Interoperability

One of the most pressing technological barriers in the shift to Industry 6.0 is ensuring seamless data interoperability between various devices, systems, and software platforms. In contrast to Industry 4.0, where systems were often isolated, Industry 6.0 relies on continuous and real-time data exchange, a requirement that legacy systems were not designed to support. These legacy systems, often based on proprietary communication protocols, struggle to interact with modern IoT technologies and cloud-based systems. Retrofitting legacy systems with modern communication protocols such as MQTT, OPC UA, or RESTful APIs, which support real-time data exchange, can be both time-consuming and costly. Retrofitting legacy systems with modern communication protocols to support real-time data exchange can be a challenging and costly process. Legacy systems in industrial environments were often designed without the need for seamless integration or real-time communication between devices. Many of these systems operate on proprietary, closed protocols that were developed decades ago. These protocols typically function independently and are optimized for isolated, non-communicating machines. The lack of standardization and the use of outdated technology present a challenge when trying to integrate new devices and systems that require robust communication. The three modern communication protocols mentioned are key to enabling real-time data exchange in Industry 6.0. However, these protocols have different characteristics and are not always directly compatible with legacy systems. MQTT is a lightweight messaging protocol, ideal for environments where bandwidth is limited, and real-time data exchange is critical. MQTT uses minimal bandwidth, which makes it suitable for IoT and sensor networks. OPC Unified Architecture is an industrial standard that facilitates communication between disparate devices, systems, and software platforms in a factory setting. It provides secure, reliable data exchange and is platform independent, making it one of the strongest protocols for industrial applications. OPC UA supports real-time, bidirectional communication, allowing devices to exchange data in a secure and standardized manner. RESTful APIs enable communication between software applications over http/ https by exposing resources as URLs, allowing for easy integration with cloud platforms and third-party applications. RESTful APIs are highly scalable and allow for real-time data exchange in web-based systems. The process requires ensuring compatibility between old and new systems, integrating smart sensors and IoT devices, upgrading software and hardware, and modernizing network infrastructure. Although the retrofitting process can be costly and time-consuming, it is essential for achieving real-time data exchange and enabling the integration of advanced automation and AI-driven systems in manufacturing. With careful planning, strategic investment, and skilled execution, manufacturers can overcome these challenges and lead to a smarter, more resilient manufacturing future. One of the most significant hurdles is ensuring compatibility between the old and new communication protocols, especially when legacy machines use proprietary protocols like Modbus, Profibus, or DeviceNet. These proprietary protocols are not inherently compatible with modern protocols. Bridging this gap often requires the implementation of data translation layers or middleware, which can convert data from one protocol to another. This process typically involves custom software development and necessitates advanced programming skills and integration expertise. Another approach is the use of communication gateways that act as bridges, translating legacy protocols into modern standards. However, this can add complexity and incur additional costs, particularly when dealing with many endpoints or varying protocols [4–6].

In addition to addressing protocol compatibility, retrofitting legacy systems also involves hardware upgrades and sensor integration. Many legacy systems lack the necessary I/O interfaces to support modern IoT sensors. To enable communication between modern protocols and legacy systems, it may be necessary to install IoT sensors, smart controllers, or communication interfaces that support newer protocols. Adding smart sensors to legacy machines requires an external data acquisition system to interface with the sensors and transmit data. Integrating IoT-enabled devices with legacy systems also demands a reliable power source, connectivity (Wi-Fi, Ethernet, cellular), and secure interfaces to ensure the devices can reliably communicate with modern control systems. Software modifications and customizations are another critical aspect of retrofitting legacy systems. Many legacy systems, like SCADA systems, were not designed for real-time communication. As such, retrofitting these systems requires significant software modifications to enable real-time data processing and integration with new control systems. For some legacy devices, firmware updates may be necessary to support modern protocols. An older PLC may need to be reprogrammed to handle MQTT messages or communicate with an OPC UA server, which can involve custom firmware development. Integrating legacy systems with modern cloud platforms that rely on RESTful APIs presents a major challenge, as it may require the development of custom software to interface with cloud systems or middleware to manage data transfers between on-premises systems and the cloud. Network infrastructure upgrades are often required to support the demands of modern communication protocols like MQTT and OPC UA. These protocols require strong, high-bandwidth networks capable of handling continuous data flow with low latency. Legacy networks may not be designed to meet these needs, so network upgrades, such as replacing routers or switches and upgrading cabling, may be necessary to minimize latency and support faster data rates. Retrofitting legacy systems with modern communication protocols introduces security concerns, as modern protocols often support advanced encryption and authentication methods that legacy systems may not be equipped to handle. This requires network infrastructure upgrades to ensure secure data exchanges without compromising safety or data integrity. The process of retrofitting legacy systems with modern communication protocols is both time-consuming and costly. Labor costs are significant, as skilled personnel with expertise in both legacy and modern systems are needed to assess the existing infrastructure, design required upgrades, implement communication protocols, and test the system for compatibility. Hardware costs can also be substantial, as retrofitting may involve installing IoT sensors, upgrading PLCs, and acquiring new communication devices. The cost of communication gateways or middleware to bridge legacy and modern systems adds complexity to the budgeting process. Another challenge is downtime, which is often required during the retrofitting process. This downtime can disrupt manufacturing operations, resulting in reduced productivity and financial losses. Therefore, careful planning and scheduling are essential to minimize disruption. Standardization of data formats is critical for efficient integration. Legacy systems often use outdated formats that hinder data processing and analysis by modern AI-driven platforms.

Bridging this gap requires the implementation of semantic interoperability standards and data translation layers, which enable seamless communication between disparate systems [7–10].

2.2 Advanced Automation Integration

The integration of AI with ML models into traditional automation frameworks, such as robotic arms or CNC machines, which were initially designed for manual programming, requires substantial hardware and software modifications. Retrofitting these machines with AI-driven control units and updating programming software to enable continuous learning and adaptability can be a complex and costly process. Many legacy machines lack the necessary interfaces for sensor integration, which means that companies must invest in advanced sensor networks to enable automation. Effective communication between AI systems and machines requires optimized feedback loops to ensure real-time data exchange for dynamic decision-making, a challenge that becomes even more pronounced as systems grow in complexity [11, 12].

2.3 Scalability of AI and Cloud Systems

Companies need to invest in high-performance computing clusters and systems capable of real-time data analytics, predictive modeling, and decision-making to keep up with the growing data load. This challenge requires ensuring that edge devices can function autonomously and process data efficiently without latency, which is especially critical in industrial environments where delays in decision-making could compromise operations. Additionally, as data storage requirements continue to increase, companies must scale their storage infrastructure to handle the massive datasets generated by AI-driven systems and machine learning models. Advanced storage solutions, including distributed databases, blockchain for data integrity, and decentralized data lakes, will be necessary to ensure data security and scalability [13, 14].

2.4 Cybersecurity and Data Privacy

As the use of IoT devices and AI systems proliferates, the potential attack surface for cyber threats expands, making it imperative for manufacturers to adopt robust cybersecurity measures. AI and ML systems are vulnerable to adversarial attacks that could compromise their functionality. Techniques such as differential privacy, federated learning, and encryption are being employed to secure data processing and protect sensitive information across decentralized systems. IoT security also remains a major concern. Each connected device can serve as an entry point for cyberattacks, necessitating the implementation of secure communication protocols (TLS, AES encryption), device authentication, and continuous monitoring for unauthorized access. Blockchain technology is gaining traction in securing data exchanges between IoT devices, offering enhanced data integrity and traceability. To further protect critical infrastructure, intellectual property, and sensitive customer data, cybersecurity strategies must evolve alongside technological advancements [15, 16].

3 Financial Barriers

While Industry 6.0 offers promising long-term benefits, the financial burden associated with the transition is a barrier for many manufacturers. The high initial capital investment required for new infrastructure, coupled with the uncertainty regarding ROI and the need for ongoing training and skilled labor, makes it difficult for smaller companies to fully capitalize on the opportunities presented by Industry 6.0. To mitigate these challenges, SMEs may need to explore financing options, partnerships, or government subsidies to help offset the upfront costs, while also strategically planning for long-term ROI through phased implementation and incremental adoption of advanced technologies. The integration of advanced technologies, including AI, ML, IoT sensors, and robotics, necessitates a high initial capital expenditure. The installation of industrial-grade AI systems and machine learning algorithms could require investments ranging from USD 0.5 million to several million dollars, depending on the scale of implementation. IoT sensors, which are essential for real-time data collection and monitoring in Industry 6.0 environments, can cost anywhere from 10 to 300 USD per sensor, depending on the type and sophistication of the technology. Advanced robotics systems, including AMRs or collaborative robots (cobots), may cost between 50,000 and USD 250,000 each, further increasing the financial burden. The integration of these technologies requires significant system overhauls, including the retrofitting of legacy machines, updating of software, and infrastructure modifications. These initial costs can add up to a significant portion of a company's operating capital, creating financial barriers for smaller manufacturers with limited budgets. The transition to Industry 6.0 involves not only capital investments but also a continuous need for skilled labor and ongoing training. According to industry reports, the demand for skilled workers in AI, robotics, and IoT is expected to increase by up to 20% annually over the next decade. The availability of such skilled workers is limited, and SMEs often face challenges in hiring and retaining talent. The costs associated with recruiting, training, and retaining employees proficient in these advanced technologies can add layer of financial strain. The average salary for an AI or machine learning engineer is typically between 0.100 to 0.150 USD per year, which may be unaffordable for smaller businesses that cannot compete with larger enterprises in terms of compensation and resources [17, 18].

A major concern for manufacturers investing in Industry 6.0 is the uncertainty regarding the return on investment. Although the long-term benefits of enhanced efficiency, reduced waste, and improved productivity are often touted, the financial returns may take time to materialize. About 30% of companies that have invested in Industry 4.0 and related technologies have seen a clear, measurable ROI within the first 3–5 years of implementation. Industry 6.0 technologies are still evolving, and many of the systems and methods being implemented are in the early stages of development. This means that the financial benefits, such as cost reductions from energy efficiency or improvements in production rates, may take several years to fully materialize. For SMEs with limited cash flow, this uncertainty poses a significant risk, as the upfront costs may strain their financial resources without immediate returns. In addition to these financial hurdles, the rapid pace of technological advancements presents another challenge. As Industry 6.0 technologies are continuously evolving, manufacturers may face the issue of obsolescence or the need for frequent upgrades, further driving up costs. The average lifespan of a typical IoT sensor is between 5 and 10 years, but new technologies or more advanced sensor models may require early replacements or upgrades, adding to the ongoing costs of maintaining these systems. The evolving nature of AI and machine learning models also means that manufacturers must periodically invest in software updates or retrain models, which can involve additional expenses [19–21].

4 Population Growth, Demographics, and Industrial Automation

Industrialization has been significantly impacted by the population explosion that occurred between the First Industrial Revolution and 2025, and this impact is still felt as we move from Industry 4.0 to Industry 6.0. In addition to speeding up industrial growth, population growth has created problems with supply chain sustainability, infrastructure development, labor force availability, and resource management. In the future, industries will face greater pressure to satisfy the needs of a world population that is expanding and becoming more urbanized. Addressing these demographic and economic issues will be just as important to the successful transition to Industry 6.0 as technology improvements. The development of a skilled workforce, the adoption of sustainable technology, and making sure that their infrastructure can accommodate the changing needs of an expanding global population will be the main priorities for manufacturers. One of the major issues influencing the shift from Industry 4.0 to Industry 6.0 is population growth. Demographic shifts, especially population increase, have had a major impact on resource management, labor force availability, and technological improvements as industrial revolutions have advanced. However, new complications have been brought about by the rapid population growth over the past few decades, especially regarding the adoption of Industry 6.0 technologies [22–24].

Rapid population expansion around the world is having a big impact on how Industry 6.0 is developing, propelling technology breakthroughs, industrial change, and sustainability issues. AI-driven solutions must be integrated to maximize efficiency while reducing environmental impact as the need for resources, automation, and sustainable industrial processes grows. Climate change will worsen if sustainability is not given top priority in Industry 6.0, which will speed up carbon emissions, pollution, and global warming. To address these issues and maintain a balance between economic advancement and environmental responsibility, key technologies, including advanced manufacturing, renewable energy integration, circular economy models, and AI-driven logistics, are crucial. Industry 6.0 must be in line with SDGs to promote innovation and maintain environmental integrity. In addition to driving technological improvements, population increase exacerbates ecological strain and resource depletion. Industry 6.0 must integrate circular economy ideas, AI-driven green breakthroughs, and sustainable automation techniques to create a future that is climate resilient. There will be serious repercussions if proactive adaptation is not implemented, such as irreparable climatic damage, the depletion of vital resources, and unstable economies. However, Industry 6.0 can spearhead the shift to a cleaner, more effective, and ecologically sustainable industrial future with the proper combination of AI, automation, and sustainable manufacturing. Policymakers, business executives, and researchers can use this analysis (Table 9.1) as a thorough and extremely accurate guide to strategically connect industrial growth with sustainable development and climate action.

4.1 Population Growth from the First Industrial Revolution to 2025

The historical trajectory of global population growth can be mapped alongside the development of industrial revolutions, offering insights into how population dynamics have influenced industrial transitions. In the eighteenth century, the world's population was approximately 1 billion. The First Industrial Revolution (1760–1840) brought significant technological advancements, including steam engines, mechanized textile manufacturing, and the introduction of iron and coal-based energy systems. However, population growth during this period was still relatively slow, with the global population reaching around 1.2 billion by 1850. The First Industrial Revolution laid the foundation for urbanization and industrialization, but its progress was limited by the relatively small labor pool and lower demand for mass-produced goods compared to modern standards. By the time of the Second Industrial Revolution (1870–1914), the global population had grown to around 1.65 billion, and rapid advancements in electricity, the internal combustion engine, and mass production had significantly increased productivity and output. This period saw

Table 9.1 Population growth and its tentative impact on Industry 6.0

Factor	2025–2050 trend	Industry 6.0 imperatives	Climate impact and mitigation	Relevant SDGs
Global population	9.4B by 2050	Hyper automation (AI/robotics) to offset labor shortages; decentralized micro-factories for local demand	+2.8 °C baseline without intervention; carbon-negative manufacturing essential	9, 11, 12
Energy demand	+58%	Neuromorphic smart grids (50% efficiency ↑); nuclear fusion pilot plants (2035+); photonic computing (0.1 pJ/op)	Emissions trajectory: 34 Gt CO_2e/year (fossil) vs. 12 Gt (renewable transition)	7, 13
Raw material demand	+110%	AI-driven material substitution (graphene from CO_2); quantum-optimized recycling (99% metal recovery)	Deforestation ↓80% via synthetic biology; mining emissions ↓70% with urban mining	12, 15
Food demand	+78%	Vertical AI farms (40x yield ↑); lab-grown meat (90% land ↓); CRISPR crops (drought resistance)	Agricultural methane ↓65%; water use ↓50%; avoided deforestation: 2.3B acres	2, 6
Urbanization	71% urban by 2050	Self-healing concrete; AI traffic orchestration (emissions ↓45%); building-integrated PV	Urban heat island mitigation via smart albedo control; transport CO_2 ↓6.7 Gt/year	11, 13
Water scarcity	57% population faces water stress	Atmospheric water harvesters (MOF-303); AI-desalination (1.2 kWh/m^3); Nanofiltration membranes	Freshwater withdrawal ↓40%; aquifer depletion halted by 2040	6, 14
E-waste	120 M tons by 2030	Robotic disassembly (99% recovery); enzymatic PCB recycling; Blockchain material passports	Toxic leakage ↓95%; avoided mining emissions: 5.2 Gt CO_2e	3, 12
Logistics demand	+170% freight	Hydrogen-powered autonomous fleets; Hyperloop networks (0.03 kWh/p-km); AI route optimization	Transport emissions ↓55%; last-mile delivery energy ↓70%	9, 13
Workforce automation	65% job augmentation	Human-AI symbiosis frameworks; AR/VR reskilling platforms; universal basic assets (crypto-based)	Economic stability enables green investment; productivity-driven decarbonization	4, 8

(continued)

Table 9.1 (continued)

Factor	2025–2050 trend	Industry 6.0 imperatives	Climate impact and mitigation	Relevant SDGs
Industrial CO_2	34% of global emissions	CCUS-integrated DRI steel; plasma-assisted cement decarbonization; AI emission throttling	Industry emissions ↓90% by 2050; carbon-negative production via direct air capture	7, 9, 13
Circular economy	520% adoption growth by 2040	Self-disassembling products; quantum blockchain for material tracking; 4D-printed recyclables	Material waste ↓62%; avoided emissions: 12.3 Gt CO_2e/year	9, 12
Health and environment	7.2 M industrial pollution deaths/ year by 2050	Biosynthetic air filters (99% $PM_{2.5}$ capture); real-time toxicity AI; Nanoremediation of soil/ water	PM2.5-related deaths ↓80%; ecosystem regeneration via synthetic biology	3, 14

accelerated urban migration as people flocked to factories for employment opportunities, fueling further industrialization. The post-World War II period, particularly after the 1950s, saw a dramatic increase in population. This was largely driven by improvements in medicine, agriculture, and sanitation, which contributed to higher life expectancies and lower mortality rates. The global population surged from around 2.5 billion in 1950 to approximately 7.8 billion by 2025. In this context, the Third Industrial Revolution (1970–2000), characterized by the rise of computers, digital technologies, and automation, was shaped by the increasing availability of a skilled labor force and the demand for information and technology-driven services [25–27].

4.2 Population Growth and its Impact on the Industrial Revolution

Population growth has been both a driver and a challenge in the evolution of industrial revolutions. During the First and Second Industrial Revolutions, a growing population provided the labor needed for factories and the consumption base that spurred mass production. As population centers expanded, especially in urban areas, industrial output surged, creating economies of scale and making manufactured goods more affordable to the public. At the same time, population growth often outpaced the ability of early industries to provide jobs, leading to issues such as overcrowding, poor working conditions, and social unrest. The Third and Fourth Industrial Revolutions (Digital Revolution) brought more automation and computing technologies, which reduced the dependency on manual labor. The integration of advanced robotics, AI, and digital systems allowed manufacturers to maintain high levels of productivity despite a shift away from labor-intensive work. This period, however, coincided with the rapid rise of the global population and significant shifts in workforce demographics, such as aging populations in developed countries and youth bulges in emerging economies [28–31].

4.3 Challenges of Population Growth for Transitioning from Industry 4.0 to Industry 6.0

As the world population continues to grow, expected to exceed 8.5 billion by 2030, its implications for Industry 6.0 are multifaceted. Several critical challenges arise from this rapid population growth.

4.3.1 Labor Force Dynamics

The global labor force is undergoing significant shifts due to demographic changes, technological advancements, and education disparities, posing a major challenge in transitioning to Industry 6.0. As automation and AI displace traditional manufacturing roles, companies in both developed and developing countries will struggle to find workers with the necessary skills to meet the demands of the new digital economy. While 75 million jobs may be displaced globally by 2025, the creation of 133 million new jobs in AI-driven industries provides a window of opportunity to address these issues. To successfully make this transition, there must be a global effort to reskill the workforce, ensuring access to training for all workers, especially in emerging economies where the gap is widest. Without concerted efforts to invest in education and training, Industry 6.0's promise of increased productivity and innovation could be undermined by the inability to adequately staff these transformative industries. The Labor Force Dynamics associated with the transition to Industry 6.0 are complex, and one of the most pressing concerns is the need for a skilled workforce capable of supporting the advanced technologies inherent in Industry 6.0, such as AI, automation, IoT, robotics, and digital technologies. This shift demands specialized skills, including expertise in data science, machine learning, software development, systems integration, and advanced robotics, creating a significant challenge in terms of training and upskilling workers globally [32–34].

4.3.1.1 Skilled Labor Gap and Population Demographics

The global workforce is undergoing significant changes, with differing labor market needs between regions. As of 2025, the global workforce is estimated to exceed 3.5 billion people, with developing economies accounting for a large portion of this growth. The shift toward Industry 6.0, which emphasizes automation and AI integration, will disproportionately affect these regions where many workers are employed in low-skilled, manual jobs in sectors like manufacturing and agriculture. According to the International Labour Organization, 60% of the global workforce in developing countries are employed in informal or unskilled jobs, creating a challenge in shifting these workers to the high skill, tech-driven sectors required by Industry 6.0. This demographic shift is further compounded by the rapid urbanization in developing nations. The UN estimates that 68% of the world's population will live in urban areas by 2050, up from 56% in 2020. This population growth in cities, especially in countries like India, China, and Sub-Saharan Africa, places an increased demand on skilled labor that Industry 6.0 depends on. The education systems and vocational training in many of these regions are not equipped to quickly generate the workforce needed to fill the demand for advanced technical roles in automation and AI development. Many developed nations are experiencing an aging workforce, with countries like Japan, Germany, and the USA seeing a decline in the proportion of workers under the age of 30. By 2030, nearly 25% of the workforce in high-income countries is expected to be over the age of 55, leading to a workforce that may be less adaptable to the technological demands of Industry 6.0. This generational shift creates significant pressure to invest in reskilling programs and lifelong learning to ensure that older workers can continue to contribute to industries relying on AI and automation [35, 36].

4.3.1.2 Job Structural Reforms

The transition to automation and AI technologies could displace up to 75 million jobs globally by 2025, particularly in fields like routine manufacturing, administrative tasks, and customer service. The 133 million new jobs that will be created in emerging industries will help to somewhat offset these losses. Even with this net benefit, there is still a discrepancy between the kinds of skills needed and the workforce's capacity to supply them, particularly in developing nations. According to OECD predictions, 32% of workers in poor nations may lose their employment to automation and AI-driven processes in the next 10–15 years, while 14% of all jobs in OECD countries are at high risk of automation. But according to these forecasts, almost 90% of future new employment would call for some kind of digital abilities, such as knowledge of data analytics, coding, cybersecurity, and AI development. This shift necessitates a massive scale-up in education and training, not only in developed countries but also in rapidly developing economies [37, 38].

4.3.1.3 The Global Population Impact of Education and Training

Education systems worldwide are struggling to keep pace with the rapidly evolving demands of Industry 6.0. In the Global Competitiveness Report by the World Economic Forum, it was reported that 63% of companies in emerging economies face difficulty in finding skilled workers, particularly in areas like AI, robotics, and data analysis. In developed nations, the gap is somewhat narrower, but 47% of businesses still report a skills shortage, particularly for roles requiring advanced technical skills in the fields of automation and machine learning. While countries like the United States and Germany have strong vocational training programs, many developing nations are still playing catch-up, and the gap is especially evident in Sub-Saharan Africa, where only 26% of young people are enrolled in secondary education. The Asian Development Bank highlights that by 2030, the demand for skilled workers in Asia alone could exceed 40 million workers in areas such as AI, cloud computing, and cybersecurity, presenting an enormous opportunity and a challenge for governments and industries to ensure that their workforce is prepared [39, 40].

4.3.1.4 Bridging the Skills Gap

To bridge this gap, governments and industries must collaborate to implement robust re-skilling and up-skilling initiatives. Reskilling programs need to be designed with a focus on AI literacy, automation, and the IoT, while ensuring that workers displaced by automation are provided with pathways to transition into new roles. As of 2020, it was reported that 45% of workers globally had received no training at all in the past year, highlighting the need for substantial investment in continuous learning programs. Industry 6.0 also presents the opportunity to democratize learning through online platforms like Coursera, edX, and Udacity, which offer specialized courses in AI, robotics, and other digital technologies. However, access to such training remains limited in developing economies, further exacerbating the skills gap [41, 42].

4.3.2 Resource Demand and Sustainability in Industry 6.0

As the global population continues to rise, so too does the demand for energy, raw materials, and manufactured goods. Industry 6.0 places a strong emphasis on sustainability, focusing on integrating green technologies, reducing waste, and minimizing carbon emissions to align with the goals of sustainable development. However, this ambition faces significant challenges due to the compounding effects of population growth, urbanization, and industrialization. The United Nations estimates that by 2050, there will be 9.7 billion people on the planet, with developing nations accounting for most of this expansion. As of November 2022, there were 8 billion people on the planet. The shift to Industry 6.0 is made more difficult by this quick population growth, which also greatly increases resource consumption and environmental stress. The challenge of managing resource demand and sustainability in the context of Industry 6.0 is multifaceted and compounded by population growth, rising consumption, and the urgency of addressing climate change. As the global population continues to grow, so will the demand for energy, raw materials, and goods, creating immense pressure on industries to balance sustainability goals with resource constraints. The transition to energy efficient technologies, the adoption of sustainable supply chains, and the integration of carbon reduction strategies will be essential for meeting the needs of a growing global population while maintaining environmental responsibility. The scale of investment required, coupled with the complex environmental, social, and economic challenges, presents significant obstacles to achieving a sustainable Industry 6.0. Governments, industries, and global stakeholders must collaborate to develop innovative solutions that enable the sustainable scaling of Industry 6.0 technologies, ensuring that they can meet the challenges of population growth while protecting the planet's resources for future generations [43, 44].

4.3.2.1 Global Energy Consumption Projections

The increasing need for energy is one of the main issues brought on by population growth. According to the International Energy Agency (IEA), population growth and industrialization in developing nations will be the main drivers of the 50% rise in worldwide energy demand between 2020 and 2050. According to mature economies, developing countries' per capita energy consumption is predicted to increase as they become more urbanized. With China and India alone responsible for almost half of the worldwide rise in energy demand by 2050, the need for electricity is predicted to grow at the fastest rate, particularly in Asia. This increase will put tremendous pressure on businesses to use energy-efficient technology that can satisfy the rising demand without worsening environmental issues or further depleting natural resources. To support these needs, industries will need to implement green technologies and energy-efficient solutions at a much faster pace. The IEA estimates that 4 trillion USD per annum will be required to achieve net zero emissions by 2050, with a large portion of this investment directed toward clean energy infrastructure and energy-efficient technologies. However, the scale of investment required remains a significant barrier for many industries, especially SMEs in developing countries, which often lack access to financial resources, technology, and skilled labor to implement such solutions [45, 46].

4.3.2.2 Raw Materials and Supply Chain Sustainability

The demand for raw materials like metals (copper, lithium, cobalt), rare earth elements and water, will also increase in line with the expanding population and industrial output. As the world transitions to more sustainable forms of energy, particularly electric vehicles and renewable energy infrastructure, the need for critical raw materials will rise sharply. The Global Battery Alliance forecasts that lithium demand alone could increase by 2000% by 2050 due to the shift towards electric vehicles and energy storage systems. However, the extraction of these resources presents its own set of challenges. The mining and processing of metals like lithium and cobalt often entail significant environmental and social costs, such as deforestation, water contamination, and child labor. In regions like the Democratic Republic of Congo, where over 60% of global cobalt production occurs, the environmental and human rights implications are particularly pronounced. Thus, sustainable sourcing and circular economy strategies are crucial for mitigating these issues. Recycling technologies for lithium-ion batteries, which are forecasted to grow at a rate of 15% annually through 2030, could reduce reliance on primary raw material extraction [47, 48].

4.3.2.3 Carbon Emissions and Climate Change

Manufacturing and industrial activities are significant contributors to global carbon emissions. According to the World Resources Institute, the industrial sector accounted for approximately 20% of global carbon dioxide emissions in 2020, with energy-intensive sectors like cement, steel, and chemical manufacturing being major contributors. Industries must immediately implement carbon reduction solutions, such as carbon capture and storage, sustainable production methods, and circular manufacturing practices that cut emissions and waste, to slow down climate change. According to the IPCC, to keep global warming to 1.5 °C over pre-industrial levels, global emissions must decrease by 45% by 2030 and reach net zero by 2050. Industry 6.0 technologies, such as robotics, smart grids, and AI-driven energy optimization, have a big potential to cut emissions by making industries run more effectively, use less energy, and produce less waste. However, the integration of these technologies comes at a cost and requires significant capital investment, which can be a challenge for smaller companies without access to sufficient financial resources [45, 49, 50].

4.3.2.4 Resource Depletion and Circular Economy Strategies

As natural resources continue to deplete, particularly non-renewable resources like fossil fuels and certain metals, the push for circular economy solutions will intensify. Industry 6.0 offers the opportunity to rethink traditional manufacturing and consumption models, emphasizing the reuse, recycling, and remanufacturing of materials. The concept of the circular economy promotes resource efficiency and waste reduction by ensuring that products are designed for longer life cycles, repair, and recyclability. However, implementing circular economy practices is complex and requires coordination across industries, governments, and consumers. For instance, the European Union has set a target to recycle 55% of plastic waste by 2030, which is a substantial increase from the 30% recycling rate achieved in 2020. While this is a positive step, scaling these efforts globally will require significant investment in infrastructure, technology, and education, particularly in developing nations where waste management systems are often inadequate [51, 52].

4.3.3 Urbanization and Infrastructure Demands in the Context of Industry 6.0

The world's economic and social landscapes are changing because of the population growth-driven trend of increasing urbanization. The UN estimates that by 2050, about 68% of people on Earth will reside in cities, up from 56% in 2020. With a large portion of the development taking place in developing nations like Asia and Africa, this translates to an additional 2.5 billion people living in cities over the next 30 years. The need for modern infrastructure, especially in the form of smart cities, automated transportation, and IoT-enabled systems, has become mandatory in the shift to Industry 6.0 as urbanization picks up speed [53, 54].

4.3.3.1 Rapid Urbanization and Infrastructure Strain

The global urban population is growing by approximately 60 million people per year, and cities like Delhi, Lagos, and Dhaka are expected to witness some of the largest increases in population. Delhi's population is projected to exceed 40 million by 2030, up from 19 million in 2015. Similarly, Lagos is expected to become the world's largest city by population by 2050, surpassing 26 million people. As these cities expand, the need for efficient, scalable infrastructure becomes critical. One of the critical emerging regions in terms of urbanization is Pakistan. The population of Pakistan has grown significantly, and by 2025, the urban population is expected to comprise over 40% of the total population, which is estimated at 245 million in 2023. As of 2023, Karachi, the largest city in Pakistan, has a population exceeding 16 million, and it is expected to grow to around 25 million by 2035, making it one of the most densely populated and mismanaged urban centers globally. Similarly, Lahore is projected to surpass 14 million by 2030, up from around 11 million in 2020. This urban growth places immense strain on infrastructure systems like energy, water, waste management, transportation, and housing. For Pakistan, which has a rapidly growing economy and urban centers with poor infrastructure, the transition to Industry 6.0 will require significant investment in advanced technologies and infrastructure. Even Industry 4.0 is still a new concept in Pakistan with few countable numbers are working on Industry 4.0 and most of the industries are using manual industrial practices (Industry 3.0 and 2.0 level and even Industry 1.0 level and people making materials by hand). So, it is really challenging to convince the Government official to shift on Industry 6.0. Urban, major and industrial cities like Faisalabad, Lahore, Karachi, and Islamabad will need to implement smart city solutions, such as IoT, AI-driven transportation, green, ecofriendly and energy efficient buildings, to accommodate the growing population while managing limited resources effectively. Pakistan's total global emissions account for less than 1% due to its relatively low level of industrialization. However, if carbon emissions and effluent discharges were measured on a per-industry basis, the emissions per industrial unit in Pakistan could potentially exceed those of the USA and China due to low level of environmental awareness and also less enforcement of environmental laws [55–57].

4.3.3.2 Global Infrastructure Deficits

Many urban areas, particularly in developing countries are ill prepared to handle the influx of people and the accompanying demands for energy, water, waste management, transportation, and housing. 70% of the infrastructure required in 2050 for cities to function effectively is still yet to be built. This means that governments and businesses must invest heavily in upgrading existing infrastructure while simultaneously building new systems to support the growing urban populations. The Global Infrastructure Facility estimates that developing countries will require 1.5 trillion USD per annum in infrastructure investment from now until 2030 to keep pace with urbanization. This figure is critical, as it reflects the challenge faced by nations like Pakistan and other developing economies that must balance urgent infrastructure needs with economic growth and resource constraints. While large-scale infrastructure projects may require public–private partnerships, the financial burden on governments and businesses in these regions will be substantial, especially with rapidly growing populations and competing priorities such as healthcare, education, and poverty alleviation [58, 59].

4.3.3.3 Smart Cities and Industry 6.0 Technologies

Industry 6.0 promises to significantly reshape urban life through the integration of smart city technologies, which include IoT sensors, autonomous transport systems, AI-based traffic management, and energy efficient buildings. These technologies aim to enhance the efficiency of urban systems, reduce energy consumption, improve quality of life, and enable more sustainable growth. The smart cities market is projected to reach USD 820 billion by 2025, growing at a CAGR of 25.3% from 2020. One of the core elements of a smart city is the integration of IoT enabled infrastructure like sensors that monitor traffic, waste, and energy usage in real time. Singapore's Smart Nation Initiative has implemented an extensive network of more than 1000 sensors across the city to monitor everything from air quality to traffic flow. This data is processed by AI systems to optimize public services, reduce congestion, and improve resource allocation. While such initiatives can improve urban living, scaling them to meet the needs of rapidly growing cities presents logistical, financial, and technological challenges. Automated transport is another pivotal component of Industry 6.0 in urban environments. Cities like Dubai have introduced autonomous taxi services powered by AI and electric vehicles, contributing to reducing both traffic congestion and carbon emissions. The global autonomous vehicle market is expected to reach 558 billion US dollar by 2026 with 39.47% CAGR. However, integrating these systems into existing infrastructure, especially in cities with overburdened transportation networks requires significant investment in both technological solutions and physical infrastructure [60, 61].

4.3.3.4 Financial Burden of Infrastructure Transformation

The scale of investment required to build and integrate Industry 6.0 technologies into urban areas is immense. Governments and businesses must not only fund smart infrastructure projects but also address the maintenance and operational costs associated with these technologies. The total cost of upgrading infrastructure for smart city development could range from 1.2 trillion to 1.5 trillion USD annually for the next decade, with developing countries facing the most significant challenges in terms of financing. In many developing economies, fiscal constraints, limited access to financing, and insufficient technical expertise pose significant barriers to the implementation of smart city infrastructure. In Africa, where urban population growth is particularly rapid, it is estimated that USD 93 billion in annual investment will be needed to address urban infrastructure challenges by 2040. However, many African governments struggle to allocate sufficient resources due to competing priorities such as healthcare, education, and poverty alleviation. The World Bank estimates that more than 40% of the urban population in developing countries currently lives in slums, further underscoring the urgent need for infrastructural development [8, 62].

4.3.3.5 The Role of Industry 6.0 in Overcoming Infrastructure Challenges

AI-powered systems can optimize energy usage and resource management, enabling cities to deliver public services more efficiently and sustainably. IoT technologies can improve urban mobility, enhance the functioning of transportation networks, and reduce traffic congestion. AI and machine learning can help cities predict demand for essential services, such as healthcare, water, and electricity, enabling more effective planning and resource allocation. However, scaling these technologies requires collaboration between governments, private businesses, and international organizations. Governments must create favorable policies and financial mechanisms to incentivize private sector investment in smart city technologies, while businesses must focus on developing scalable solutions that can be adapted to the unique needs of growing urban centers. Around 133 million new jobs will be created globally in industries driven by AI, automation, and smart infrastructure by 2025, but there is a significant skills gap that must be addressed to meet the demand for talent in these fields. The challenges associated with urbanization and the need for advanced infrastructure in the context of Industry 6.0 are multifaceted. Rapid population growth, especially in developing countries such as Pakistan, is placing immense pressure on urban infrastructure, and integrating Industry 6.0 technologies such as smart cities, automated transport, and IoT-enabled systems will require significant investment and collaboration across sectors. The projected growth of urban populations and the corresponding rise in infrastructure demands will require USD 1.5 trillion in annual investments globally, with developing countries facing obstacles in financing [63, 64].

4.3.4 Supply Chain Disruptions in the Context of Industry 6.0

The exponential population growth, coupled with rising consumption patterns, is drastically increasing the complexity of global supply chains. Traditional linear supply chain models are no longer sufficient to handle the increasing demand for raw materials, finished goods, and food security. Industry 4.0 technologies like IoT, AI-driven forecasting, and automation have already made supply chains more responsive. However, as the world transitions to Industry 6.0, supply chain networks must evolve further into self-regulating, AI driven ecosystems that predict disruptions, optimize inventory, and enhance sustainability. With a population exceeding 8 billion, the global demand for essential goods, raw materials, energy, and logistics services is surging.

4.3.4.1 Food Supply Chain Pressure

The global food demand is projected to surge by 56% by 2050, driven by rapid population growth, urbanization, and dietary shifts. The world's annual food production must increase from 8.9 billion metric tons in 2020 to at least 14 billion metric tons by 2050 to prevent shortages and meet the caloric and nutritional needs of an estimated 9.7 billion people. This expansion must occur amid climate change, resource depletion, and geopolitical instability, all of which significantly impact food logistics and the availability of staple crops like wheat, rice, and maize. Climate change is already disrupting global agricultural output, altering precipitation patterns, increasing extreme weather events, and reducing the yield potential of major crops. Projected temperature rises of 2–4 °C by 2100 are expected to reduce global maize and wheat yields by 5–17%, depending on the severity of warming scenarios. The increasing frequency of droughts and heatwaves is accelerating soil degradation and desertification, particularly in key agricultural regions such as Sub-Saharan Africa, South Asia, and parts of Latin America. 70% of global freshwater withdrawals are used for irrigation. However, increasing aquifer depletion and prolonged drought conditions in major food-producing regions, including California's Central Valley, India's Indo Gangetic Plain, could reduce crop yields by 20–30% by mid-century. Saltwater intrusion in coastal agricultural lands, driven by rising sea levels, is reducing arable land availability, impacting rice production in countries like Bangladesh, Vietnam, and Thailand. Hurricanes, typhoons and heavy rainfall events have increased in intensity due to climate change, disrupting supply chains and causing post-harvest losses. Heat stress during the flowering stage of wheat and rice can lower grain formation, reducing yield by 10–15% per 1 °C increase above optimal growing temperatures. The global food supply chain is increasingly vulnerable to political conflicts, trade restrictions, and economic instability. The Russia–Ukraine conflict has severely impacted the global supply of wheat, corn, and sunflower oil, causing food price inflation of 15 to 30% globally. Nations imposing food export restrictions to protect domestic supplies exacerbate global shortages. Shipping bottlenecks in the Black Sea and South China Sea have led to delays in grain exports, impacting food security in Africa and the Middle East. The cost of synthetic fertilizers (urea, phosphate, potassium) has surged by 40–80% since 2021, largely due to supply chain disruptions, increasing production costs for farmers worldwide [65, 66].

4.3.4.2 Staple Crop Availability and Yield Projections

As the global population grows and environmental constraints intensify, Industry 6.0 driven agricultural technologies are becoming essential in enhancing crop resilience, optimizing resource utilization, and minimizing food waste. The global food system faces an unprecedented challenge of meeting the projected 56% increase in food demand by 2050 while navigating climate change, water scarcity, and geopolitical instability. Industry 6.0 technologies, including CRISPR based gene editing, AI-driven precision farming, and quantum-enhanced climate modelling, are becoming essential tools for ensuring stable food production. Sustainable food production will depend on the integration of advanced biotechnologies, smart supply chain logistics, and precision agriculture, creating a climate-resilient, high-yield, and resource-efficient global food network in the coming decades. These innovations integrate AI, robotics, biotechnology, and quantum computing to create a highly efficient, climate-resilient, and sustainable food production system. Artificial intelligence is transforming modern farming by optimizing inputs such as fertilizers, irrigation, and pest control through real-time data analytics. AI-powered predictive models increase resource efficiency by 15–25% ensuring maximum crop yield with minimal environmental footprint. Drone-assisted hyperspectral imaging, capable of detecting nutrient deficiencies with 92% accuracy, enables targeted fertilization and pest management, significantly improving crop health and reducing chemical overuse. IoT-connected soil sensors monitor moisture levels, pH, and nutrient availability in real time, helping farmers reduce water consumption by up to 50%. To address land scarcity and climate unpredictability, vertical farming and controlled-environment agriculture provide a highly efficient alternative to traditional open field farming. Indoor farming systems, utilizing hydroponics, aeroponics, and aquaponics, produce up to 350× higher yields per square meter than conventional methods while using 90% less water. Advanced LED lighting optimized for photosynthetic wavelengths accelerates plant growth by 20–30%, making urban food production more feasible. AI-controlled climate systems automatically adjust temperature, humidity, and CO_2 levels, ensuring optimal growing conditions for year-round food supply regardless of external weather fluctuations. With rising concerns over the environmental footprint of conventional livestock farming, alternative protein sources are emerging as a sustainable solution. Lab-grown meat, developed through cellular agriculture, reduces greenhouse gas emissions by 92% compared to traditional meat production, while requiring 99% less land and 96% less water. Microbial protein fermentation, utilizing species like *Fusarium venenatum* for mycoprotein production, requires 40 times less land than soy protein farming, making it a highly scalable, protein-rich food source. Precision fermentation techniques, powered by synthetic biology and AI-driven metabolic modelling, allow customized protein synthesis, ensuring optimal amino acid profiles for human nutrition [67, 68].

4.3.4.3 Raw Materials and Manufacturing Constraints

The demand for vital minerals like lithium, cobalt, and rare earth elements, which are necessary for batteries, semiconductors, and green technologies is expected to increase by 400% by 2040. The future of resource extraction and material processing will rely on automation, AI-driven mineral discovery, sustainable mining methods, and circular economy efforts due to the rising demand for essential minerals, copper and steel. To ensure the effective, moral, and ecologically conscious sourcing of raw materials for the upcoming industries, Industry 6.0 technologies like autonomous drilling, blockchain based supply chain tracking, and quantum computing-assisted material discovery will be essential. EVs, next generation technology and the global shift to renewable energy are the main drivers of this exponential increase. Lithium demand, fueled by the expansion of lithium ion battery storage systems, is expected to grow from 0.5 million metric tons in 2023 to over two million metric tons by 2040, with Australia, Chile and China leading global lithium extraction and processing. Cobalt, mandatory for battery cathodes, is projected to increase from 0.14 million metric tons in 2020 to over 0.60 metric tons by 2040, with the Democratic Republic of the Congo supplying nearly 70% of the global cobalt output. Rare earth elements, essential for wind turbines, electric motors, and high-performance electronics, will experience a demand increase of over 300% as clean energy adoption accelerates. Copper demand is set to double from 25 million metric tons in 2020 to 50 million metric tons by 2050, primarily due to EV production, renewable energy infrastructure, and electrification projects. Copper is essential for power grids, electric motors, and heat exchangers, with the average EV requiring 83 kg of copper, compared to just 23 kg for conventional gasoline-powered vehicles. To meet rising demand, copper mining operations in Chile, Peru, and China are ramping up production, while recycling initiatives aim to recover copper from electronic waste (e-waste), reducing reliance on primary extraction. Global steel production, a cornerstone of infrastructure, construction, and manufacturing, must increase from 1.8 billion metric tons in 2020 to 2.5 billion metric tons by 2050 to meet industrial and urban expansion demands. Steel demand for wind turbines, solar panel frames, and green hydrogen production plants is driving a shift toward low-carbon steelmaking technologies, including hydrogen-based direct reduced iron (H_2 DRI) and electric arc furnaces. Countries such as China, India, and the European Union are leading the transition to green steel, with companies investing in carbon capture, utilization, and storage (CCUS) technologies to reduce emissions in steel manufacturing [69–71].

4.3.4.4 Energy Supply Chain Constraints

Economic instability and increasing inflationary pressures worldwide. Natural gas prices soared by 250% in Europe and 150% in Asia during the same period, leading to an energy affordability crisis in developing economies. These disruptions have accelerated the shift toward renewable energy sources, energy storage solutions, and decentralized grid systems to enhance energy security. The transition to renewable energy requires an estimated 4 trillion US Dollar in annual global investment through 2050 to support solar, wind, hydrogen, and next-generation nuclear technologies. The deployment of solar photovoltaic capacity is expected to reach 8000 GW by 2050, with China, the USA, and the EU leading the expansion. Fourth generation solar cells, such as perovskite tandem PV, are achieving efficiencies above 30%, surpassing traditional silicon based solar panels. Wind energy capacity is projected to quadruple to 6000 GW, driven by innovations in floating offshore wind farms and aerodynamic turbine blade designs that improve efficiency by 15–20%. Hydrogen is emerging as a key energy carrier, with global green hydrogen production set to exceed 500 million metric tons annually by 2050, reducing reliance on fossil fuels. Electrolyzer efficiency improvements have increased hydrogen production rates to 80 kWh/kg H_2, significantly lowering costs. In addition, advancements in solid-state hydrogen storage and liquid organic hydrogen carriers are enhancing transport and storage efficiency. To support grid stability, next-generation energy storage systems such as solid-state lithium batteries, sodium-ion batteries, silicone-based and redox flow batteries are being deployed at scale. Grid-scale battery storage capacity is expected to reach 1500 GWh by 2040, ensuring a continuous power supply during peak demand. Emerging superconducting magnetic energy storage and gravity-based storage systems provide additional flexibility for energy load balancing in smart grids. As global energy consumption continues to rise, the integration of AI, quantum computing, and intelligent energy management systems is important for optimizing efficiency. AI-driven predictive analytics in energy distribution networks can reduce power losses by 15–25%, while blockchain-based energy trading platforms are enabling decentralized peer-to-peer electricity markets. Industry 6.0 advancements, including self-healing power grids and bio-inspired nanomaterials for energy harvesting, will shape the future of sustainable energy solutions, ensuring climate resilience, energy security, and economic stability in the coming decades [72, 73].

4.3.4.5 Transportation and Logistics Challenges

The global shipping industry is the backbone of international trade, facilitating the transport of over eleven billion metric tons of cargo per annum, accounting for 80% of global trade volume. However, recent disruptions, including COVID-19, port congestion, geopolitical tensions, and labor shortages, have significantly impacted global logistics. From 2019 to 2023, average container freight rates skyrocketed by 400%, reaching 10,500 USD per 40-feet big container in late 2022. Bulk cargo shipping costs have also surged, with Baltic Dry Index rates increasing by 220% since 2020, reflecting rising demand and supply chain constraints. Shipping fuel costs have risen due to regulatory shifts, including IMO 2020 sulfur cap regulations, which mandated a 0.5% sulfur content limit in marine fuels, increasing operational expenses for shipping companies. Land freight transport is also facing critical challenges, particularly in Europe and North America, where a severe truck driver shortage threatens supply chain stability. By 2030, an estimated 2.6 million truck driver positions could remain unfilled, leading to higher freight costs, extended delivery times, and reduced transport capacity. In the USA, driver shortages have contributed to a 23% increase in freight costs per mile, while in the EU, driver demand outpaces supply by 0.50 million positions annually, exacerbating transport inefficiencies. The rising adoption of autonomous trucks and AI-powered fleet management systems is expected to reduce reliance on human drivers by 30% by 2035, improving efficiency but requiring significant regulatory and infrastructure adaptations. Air cargo transportation, essential for high-value and time-sensitive goods, has experienced a 20% post-pandemic surge in demand, yet global airport infrastructure remains insufficient to handle future growth. Air freight volume reached 66 million metric tons in 2023, but airport cargo capacity is projected to lag demand by 15–20% by 2035, necessitating rapid expansion of logistics hubs, automated sorting facilities, and AI-powered air traffic control systems. Innovations such as hydrogen-powered cargo planes, drone-based last-mile deliveries and AI-optimized flight routing are expected to enhance efficiency, reducing fuel consumption by 25% and operational costs by 18% over the next decade. To mitigate ongoing disruptions and modernize logistics, Industry 6.0-driven intelligent supply chain solutions are being deployed. Blockchain-based freight tracking systems improve transparency and reduce cargo theft, while AI-driven predictive analytics optimize shipment routing, reducing delays by 40%. Smart port automation, automated container handling, and digital twin technology are transforming maritime logistics, with AI-controlled cranes increasing port efficiency by 35% and reducing vessel turnaround times by 20%. The transition toward sustainable logistics, including the adoption of ammonia- and hydrogen-powered cargo ships, electric semitrucks, and AI-optimized intermodal transport networks, will play a key role in ensuring a resilient, efficient, and low carbon global supply chain for the future [74–76].

4.3.4.6 Semiconductor and Electronics Supply Chain Crisis

The global semiconductor industry is experiencing an unprecedented surge in demand, projected to grow from 600 billion in 2022 to 1.1 trillion US dollar by 2030. This surge is driven by the increasing integration of semiconductors in virtually every aspect of modern life, particularly in industries such as automotive, consumer electronics, telecommunications, and medical devices. The automobile sector has seen semiconductors become integral to vehicle functionality from powertrains to infotainment systems. Similarly, the demand for chips in smartphones and medical equipment has skyrocketed, especially with the rise of 5G technology,

AI applications and IoT. As a result semiconductor lead times have more than doubled, escalating from an average of 12 weeks in 2019 to 28 weeks in 2023, creating substantial production delays and putting immense pressure on manufacturers. Taiwan's dominance in the semiconductor market further exacerbates global vulnerability. With 63% of the world's semiconductor output coming from the Taiwan Semiconductor Manufacturing Company, the world leader in semiconductor fabrication, this little island nation is crucial to the worldwide supply chain. This Taiwanese industrial concentration poses a significant geopolitical risk since any disruption, whether from supply chain interruptions brought on by pandemics, natural disasters, or geopolitical tensions, might result in severe worldwide shortages. The COVID-19 epidemic and typhoons in Taiwan created delays in chip production and delivery schedules, which contributed to the 2022 shortage. Further threatening the stability of semiconductor supply chains is the geopolitical environment, including the China-Taiwan dispute. In 2022, Taiwan's chip exports to China and other nations were valued at over USD 20 billion, but political tensions have raised concerns about potential trade disruptions or forced relocation of semiconductor manufacturing to other regions. The implications of these challenges are widespread. The automobile industry, which has been particularly hard hit by the semiconductor shortage, saw production delays for vehicles like Tesla, Ford, Nissan, BYD, General Motors, etc., who reported significant quarterly losses due to delayed chip supplies. The smartphone market manufacturers (Apple, Samsung, etc.) also faced shortages, with Apple's iPhone production in 2022 dropping by five million units due to chip delays. The medical devices sector experienced severe disruptions in the supply of essential equipment such as ventilators, imaging machines, and pacemakers, all of which rely on specialized semiconductors. Looking ahead, the global semiconductor supply chain faces an additional challenge from the growing demand for advanced chips used in applications such as AI processing, quantum computing, and automotive electrification. The demand for semiconductors in the electric vehicle sector alone is expected to increase by 800% by 2030, further amplifying the competition for available chips. This intense competition for resources highlights the need for investments in diversification, supply chain resilience, and capacity expansion to meet growing demand. However, despite these challenges, the semiconductor industry is investing heavily in expanding capacity. Companies like Intel and Samsung are committing tens of billions of dollars in new chip manufacturing plants, with Intel's USD 20 billion investment in a new plant in Ohio set to begin production in 2025, aimed at addressing the supply-demand imbalance. The semiconductor industry's growth trajectory, coupled with geopolitical, environmental, and technological challenges, poses a significant challenge to global supply chains. As demand for semiconductors is expected to grow by 83% between 2022 and 2030, industries must adapt to a landscape where supply chain disruptions, driven by external and internal factors, remain a persistent risk. The situation calls for innovative solutions such as multinational manufacturing collaborations, investment in alternative supply chains and technological innovations in semiconductor design and production to mitigate the risks of future shortages [77–79].

4.3.5 Industry 6.0 Solutions for Supply Chain Resilience

Modern technologies like AI-driven predictive analytics, digital twins, and autonomous systems are being used by Industry 6.0 to improve operational efficiency and resilience in response to the growing complexity of supply chain management brought on by population growth, resource scarcity, and fluctuating demand. Supply chains may become more proactive in detecting and averting possible disruptions thanks to these cutting-edge solutions. Predictive analytics driven by AI can foresee changes in demand, giving producers real-time information to modify production plans, maximize stock levels, and enhance resource allocation. Businesses may prevent expensive overproduction and stockouts, which can both lead to large financial losses, with the use of this technology. As AI-driven optimization becomes more integrated into the supply chain, it is projected to cut global supply chain costs by 20% by 2030, underscoring the potential for substantial cost savings while simultaneously enhancing efficiency. Digital twins, which are virtual versions of actual supply chain components that replicate data and processes in real time, are among the most important developments in supply chain optimization. By tracking inventory in real-time, simulating possible disruptions, and monitoring and forecasting supply chain performance, these digital models help businesses make data-driven decisions that reduce operational downtime and avoid bottlenecks. By 2030, the use of digital twin technology in supply chain management will boost supply chain efficiency by 25%, proving its capacity to revolutionize supply chain management and optimization across industries. These technologies are already being used in sectors where supply chains are extremely complicated and prone to interruptions, such as the pharmaceutical and automobile industries. A prime example of AI-driven logistics optimization is Amazon's advanced AI-powered logistics system, which has enhanced its delivery efficiency by 40%. By utilizing machine learning algorithms to predict delivery times, optimize warehouse operations, and reduce human errors, Amazon has streamlined its global logistics network, dramatically improved throughput while cutting costs. The company's system uses vast amounts of data from customer orders, regional demand patterns, and inventory levels to dynamically route shipments in the most efficient manner possible. This level of automation significantly reduces the dependency on traditional, human driven planning, allowing Amazon to meet consumer demand faster and more accurately. As a result, Amazon's logistics network handles over 18 million orders per day, with the company's supply chain efficiency making it a leader in the global e-commerce sector. The AI-driven optimization of supply chains is not only beneficial in terms of efficiency but also in terms of sustainability. With increasing pressure on industries to reduce carbon footprints, AI-powered solutions allow manufacturers to better optimize their energy usage, reduce waste, and improve resource efficiency. AI-based supply chain solutions can reduce greenhouse gas emissions in logistics by up to 30%, contributing to the SDGs, particularly in terms of climate action. AI technologies can also enable manufacturers to switch from traditional, linear supply chains to more circular supply chain models, where waste is minimized, and products are reused or recycled. However, despite the promise of AI-driven optimization, the implementation of these solutions requires significant investment in infrastructure and data management systems. The global adoption of AI technologies in supply chains is expected to require USD 1.7 trillion in capital investment by 2030, focusing on building AI capabilities, integrating digital twins, and training the workforce to manage advanced systems. As companies continue to adopt AI, they must also prioritize cybersecurity measures to protect sensitive data from potential breaches, which could disrupt operations and lead to significant financial and reputational damage [80, 81].

4.3.5.1 Blockchain for Transparent and Secure Supply Chains

These digital models allow companies to monitor and predict supply chain performance, track inventory in real-time, and simulate potential disruptions, enabling managers to make data-driven decisions that prevent bottlenecks and minimize operational downtime. The integration of digital twin technology in supply chain management is expected to increase supply chain efficiency by 25% by 2030, demonstrating its ability to transform how industries manage and optimize supply chains. These technologies are already being implemented in industries ranging from automotive manufacturing to pharmaceuticals, where supply chains are highly complex and susceptible to disruptions. A prime example of AI-driven logistics optimization is Amazon's advanced AI-powered logistics system, which has enhanced its delivery efficiency by 40%. By utilizing machine learning algorithms to predict delivery times, optimize warehouse operations, and reduce human errors, Amazon has streamlined its global logistics network, dramatically improved throughput, while cutting costs. The company's system uses vast amounts of data from customer orders, regional demand patterns, and inventory levels to dynamically route shipments in the most efficient manner possible. This level of automation significantly reduces the dependency on traditional, human-driven planning, allowing Amazon to meet consumer demand faster and more accurately. As a result, Amazon's logistics network handles over 18 million orders per day, with the company's supply chain efficiency making it a leader in the global e-commerce sector. The AI-driven optimization of supply chains is not only beneficial in terms of efficiency but also in terms of sustainability. With increasing pressure on industries to reduce carbon footprints, AI-powered solutions allow manufacturers to better optimize their energy usage, reduce waste, and improve resource efficiency. According to a report by Accenture, AI-based supply chain solutions can reduce greenhouse gas emissions in logistics by up to 30%, contributing to the SDGs, particularly in terms of climate action. AI technologies can also enable manufacturers to switch from traditional, linear supply chains to more circular supply chain models, where waste is minimized, and products are reused or recycled. The global adoption of AI technologies in supply chains is expected to require USD 1.7 trillion in capital investment by 2030, focusing on building AI capabilities, integrating digital twins, and training the workforce to manage advanced systems. As companies continue to adopt AI, they must also prioritize cybersecurity measures to protect sensitive data from potential breaches, which could disrupt operations and lead to significant financial and reputational damage. The introduction of AI-driven supply chain optimization in Industry 6.0 offers a transformative opportunity to mitigate disruptions caused by global population growth, resource scarcity, and other complexities. With projections suggesting a 20% reduction in global supply chain costs by 2030, the integration of AI, digital twins, and autonomous systems will redefine how businesses forecast, plan, and execute their supply chain strategies, offering both financial and environmental benefits [82, 83].

4.3.5.2 Hyper Automated Smart Warehouses

The adoption of robotic automation and AI-driven inventory management is revolutionizing warehouse operations, significantly improving productivity, efficiency, and cost-effectiveness. With the rise of Industry 6.0, companies are integrating advanced robotics, AI, and IoT into their supply chains, allowing for real-time inventory tracking, automated fulfillment and predictive demand forecasting. AI-driven warehouse systems have been shown to increase warehouse productivity by up to 60%, reducing the reliance on manual labor while enhancing speed and accuracy in order fulfillment. Leading e-commerce and retail giants such as Alibaba and Ocado have pioneered fully automated fulfillment centers, where autonomous mobile robots and AI-driven software coordinate warehouse operations with minimal human intervention. Alibaba's smart warehouse in China, for example, utilizes robotic arms and AMRs that transport goods, sort packages, and optimize storage space, leading to a 70% reduction in human labor costs while increasing package processing speed by 300%. Similarly, Ocado's automated fulfillment centers utilize swarm robotics, a network of thousands of interconnected robots to pick, pack, and ship orders within minutes. These robotic warehouses have allowed Ocado to process over 0.2 million orders per week, reducing operational costs while maximizing efficiency. One of the key advantages of AI-driven inventory management is its ability to eliminate overstocking and stockouts, which cost retailers an estimated USD 1.1 trillion annually worldwide. This capability is particularly crucial in the face of supply chain disruptions as AI-driven warehouses can reroute shipments, identify alternative suppliers and optimize inventory distribution across multiple locations. The shift toward autonomous warehouses also enhances workplace safety and operational resilience. According to the Occupational Safety and Health Administration, manual warehouse operations account for 25% of all workplace injuries in the logistics sector, primarily due to repetitive strain, heavy lifting and accidents involving forklifts. AI-driven robotic automation eliminates these risks by handling labor intensive tasks, thereby reducing workplace injuries by over 40%. AI-integrated robotics contribute to energy efficiency and sustainability in warehouse operations. Traditional human-operated warehouses rely on inefficient lighting, heating, and manual transportation, leading to higher energy consumption. In contrast, automated fulfillment centers use energy-efficient robots, smart HVAC systems, and AI-optimized workflows to reduce energy consumption by 30%, helping companies achieve their net-zero carbon targets. Despite the undeniable benefits, transitioning to fully automated fulfillment centers requires substantial capital investment. Companies must carefully evaluate the ROI before fully transitioning, as initial implementation costs can range from 10 to 100 million USD per warehouse, depending on the scale of automation required. However, studies suggest that firms investing in AI-powered logistics solutions can achieve a full ROI within 3–5 years, due to cost savings in labor, energy, and operational efficiency. Robotic automation and AI-driven inventory management are redefining global supply chains, making warehouses faster, safer, and more cost-efficient. While the transition requires significant capital investment, the long-term benefits, including a 60% increase in productivity, a 70% reduction in labor costs, and a 40% decrease in workplace injuries, make AI-driven logistics an important element of Industry 6.0. As technology advances and implementation costs decrease, fully autonomous warehouses will become the new standard in global logistics for more efficient and sustainable supply chains [84, 85].

4.3.5.3 Autonomous Vehicles and AI-Driven Logistics

The integration of self-driving trucks, autonomous drones, and AI-powered cargo ships is revolutionizing the logistics and transportation industry, significantly reducing costs, enhancing delivery speed, and minimizing environmental impact. The shift toward self-driving trucks, autonomous drones, and AI-powered cargo ships is transforming global logistics, making it faster, safer, and more sustainable. By 2035, 40% of last-mile deliveries will be handled by drones and autonomous vehicles, cutting transportation costs by 25%. Maritime logistics will see a 30% reduction in carbon emissions, and self-driving trucks will save over $100 billion annually in freight costs. As AI, IoT, and automation technologies continue to evolve, the future of logistics will be highly autonomous, energy efficient, and cost-effective, Industry 6.0's intelligent and sustainable supply chains. By 2035, self-driving trucks and delivery drones are projected to handle 40% of last-mile deliveries, leading to a 25% reduction in transportation costs and a significant decrease in delivery times. This shift is driven by advancements in AI, LiDAR, computer vision, 5G connectivity, and edge computing, enabling real-time navigation, route optimization, and obstacle avoidance with high precision. Autonomous trucks are expected to revolutionize freight transportation, improving efficiency and reducing reliance on human drivers. Currently, the global trucking industry faces a driver shortage of 2.6 million drivers, particularly in regions like North America and Europe. Companies like Tesla, Waymo, and TuSimple are developing Level 4 and Level 5 autonomous trucking solutions that will address this labor gap. TuSimple's autonomous trucks, which operate 24/7 without breaks, have demonstrated a 10% reduction in fuel consumption and a 30% increase in freight efficiency compared to conventional trucks. The deployment of self-driving trucks is expected to cut freight costs by USD 100 billion annually by eliminating driver wages, optimizing fuel usage, and reducing downtime. Autonomous drones are poised to revolutionize last-mile deliveries, especially in urban and remote areas, reducing delivery time from hours to minutes while significantly lowering costs. Amazon's Prime Air delivery drones, capable of carrying packages up to 2.2046 kg (5 lbs), have demonstrated the ability to cut short-distance delivery costs by 90%, reducing expenses from USD 5 per package to less than USD 0.50. Similarly, Zipline's medical delivery drones have successfully reduced delivery times for critical supplies by 80% in countries like Rwanda and Ghana, ensuring faster response in emergencies. With the global drone delivery market expected to reach 50 billion USD by 2030, regulatory bodies are working on airspace management systems (UTM, i.e., Unmanned Traffic Management) to ensure safe integration of drones into existing urban airspace. The Federal Aviation Administration and European Union Aviation Safety Agency predict that by 2035, over 2.5 million commercial drones will be operating worldwide, handling a significant portion of e-commerce and medical deliveries. Maritime shipping contributes to 3% of global carbon emissions, making sustainable solutions essential for achieving net-zero targets by 2050. Autonomous cargo ships, like Yara Birkeland, are leading the charge in AI-driven maritime logistics, reducing carbon emissions by 30% per voyage while eliminating the need for onboard crew. The global market for autonomous ships is projected to exceed USD 14 billion by 2030, with major shipping companies, including Maersk, NYK Line, and Rolls-Royce, investing in fully automated cargo vessels. These ships utilize AI-powered navigation, IoT sensors, and electric propulsion systems to improve fuel efficiency and reduce operational costs by up to 20%. Rolls-Royce's AI-driven ship automation technology enables real-time weather forecasting and route optimization, which can save up to 10% in fuel consumption [86, 87].

4.3.5.4 Green and Circular Supply Chains

The circular economy is emerging as a critical strategy for sustainable resource management, aiming to reduce waste, extend product life cycles, and minimize the environmental footprint of industries. As industries move toward Industry 6.0, integrating circular economy principles will be essential for reducing environmental impact, securing critical resources, and driving economic sustainability. Companies like Tesla, Apple, and HP are demonstrating that closed-loop systems, advanced recycling, and remanufacturing can cut raw material dependency while generating billions in savings. By 2030, circular economy strategies could save businesses USD 4.5 trillion annually, while advanced recycling technologies in battery production alone could cut lithium demand by 50%. Moving forward, AI-driven material recovery, decentralized manufacturing, and bio-based alternatives will be key enablers of a sustainable, waste-free global economy. By shifting from a linear economy (take make dispose principle) to a regenerative system, industries can optimize material flows, reduce dependency on virgin resources, and drive economic growth while significantly cutting down waste generation. Circular economy initiatives could reduce material waste by 30% globally and generate savings of up to $4.5 trillion annually by 2030 through resource efficiency, product reuse, and innovative recycling technologies. As the world transitions to electrified transportation, the demand for lithium, cobalt, and nickel has surged, putting immense pressure on global mining operations. The EV industry is leading the way in circular economy adoption, particularly through closed-loop battery recycling programs. Tesla's Gigafactory Nevada, in collaboration with Redwood Materials, has developed an advanced hydrometallurgical recycling system that can recover over 95% of lithium, nickel, and cobalt from spent batteries. This initiative is projected to cut global lithium demand by 50% in the next decade by reducing reliance on virgin raw materials, significantly lowering the environmental impact of lithium mining. Battery recycling initiatives are expected to play a pivotal role in reducing the carbon footprint of EV production, as mining lithium generates 15 tons of CO_2 per ton extracted. By 2030, global lithium ion battery recycling is projected to reach a market value of USD 22.8 billion, helping mitigate resource scarcity while ensuring the sustainable growth of the EV market. Beyond EVs, advanced manufacturing sectors are embracing circular economy principles to reduce raw material consumption and improve supply chain resilience. Industrial 3D printing (additive manufacturing) is expected to cut material waste by 90% in some sectors by 2035, particularly in aerospace and automotive industries, by enabling on-demand, localized production and reducing excess inventory. Remanufacturing and refurbishing initiatives in the electronics sector are gaining traction. Apple's trade in and refurbishment program, for example, extends product lifespans, reducing e-waste by 40% annually while cutting production related carbon emissions. Similarly, HP's closed loop recycling for printer cartridges has resulted in over 1.7 billion cartridges being reused, saving 0.5 million tons of virgin plastic from entering landfills. The circular economy is poised to have a profound impact on global economic stability, particularly in resource-intensive industries. By 2040, circular economy adoption could reduce CO_2 emissions by 45% in heavy industries, including steel, cement, and chemicals, which currently account for over 30% of global greenhouse gas emissions. Integrating AI and IoT-driven predictive maintenance in manufacturing could cut industrial waste by 25% by identifying inefficiencies and reducing overproduction. The global adoption of circular economy practices will also enhance supply chain resilience, particularly in response to geopolitical tensions and raw material shortages. The EU Circular Economy Action Plan aims to make all packaging recyclable or reusable by 2030, which could reduce packaging waste by 50% across Europe and set a precedent for sustainable materials management worldwide [88–90].

4.3.5.5 Supply Chain Challenges in Pakistan

Pakistan, with a population of 245 million in 2023, faces critical supply chain vulnerabilities that threaten industrial efficiency, economic stability, and global trade competitiveness. With global demand projected to rise to 9.7 billion people by 2050, Pakistan's supply chain infrastructure must undergo a rapid transformation to remain resilient and adaptive. The country's heavy reliance on imports, amounting to 75 billion USD annually, exposes it to severe economic risks, including currency devaluation, geopolitical instability, high corruption, and supply chain disruptions. The high import dependency (due to non supportive government policies for manufacturing industries), particularly in raw materials, fuels inflationary pressures and weakens industrial productivity, making Pakistan vulnerable to global trade restrictions, logistical bottlenecks, and pricing volatility. Energy shortages remain one of the most significant challenges, severely hindering industrial production, especially in energy-intensive sectors such as textiles, cement, and steel. The country experiences a power deficit of approximately 7000 MW during peak summer months, leading to frequent outages that result in a 20% annual increase in manufacturing costs due to production halts, generator fuel expenses, and inefficiencies. The textile sector, which accounts for 60% of Pakistan's total exports, faces repeated shutdowns due to electricity and gas shortages, particularly in Punjab province, where 60% of the country's textile mills are concentrated. Gas supply inconsistencies have forced manufacturers to rely on alternative, costlier fuels, further eroding profit margins. Transitioning towards renewable energy sources such as solar, wind, and hydro, combined with smart grid solutions and energy storage technologies, could mitigate these disruptions, enhancing industrial stability and long-term sustainability. Pakistan's trade logistics and infrastructure face severe bottlenecks, further exacerbating supply chain inefficiencies. Karachi Port, which handles 60% of Pakistan's trade, is currently operating at 90% capacity, leading to port congestion, increased turnaround times, and rising demurrage costs. The delays average between 7 and 15 days per shipment, significantly reducing Pakistan's global export competitiveness. Pakistan's railway freight transport system remains underutilized, carrying only 4% of total freight, compared to 35% in developed economies. The reliance on road transport, which is 30% more expensive per ton-kilometer than rail, further strains supply chain efficiency. Upgrading multimodal logistics networks through high-speed rail freight corridors, AI-powered traffic management, and automated port handling systems could increase cargo handling efficiency by 30% and lower transportation costs by 20%. The textile industry, Pakistan's largest export sector with annual exports of 16 billion USD, faces significant raw material supply chain risks due to climate change, declining domestic cotton production, and reliance on imports. Cotton production in Pakistan has declined by 35% from 2018 to 2023, driven by pest infestations, erratic monsoons, and inadequate irrigation infrastructure. This decline has forced textile manufacturers to import 30–40% of their cotton, making them vulnerable to fluctuating global prices, foreign exchange volatility, and supply chain disruptions. Advanced solutions such as AI-driven precision and smart agriculture, automated pest monitoring, and drought resistant genetically modified cotton strains could increase domestic cotton yields by 25%, reducing the industry's dependence on imports while enhancing supply chain stability. To address these escalating challenges, Pakistan must embrace Industry 6.0 technologies, which are revolutionizing supply chains into self-regulating, highly adaptive ecosystems. AI-powered predictive analytics can optimize inventory management, reduce inefficiencies, and forecast demand fluctuations with 95% accuracy, leading to a 40% reduction in stockouts and a 60% improvement in warehouse efficiency.

Blockchain-based transparency in trade documentation can eliminate fraud (at this moment, its impossible to clear and release legal trade documents without a bribe due to high-level corruption mafia control), streamline customs clearance, and cut transaction times from several weeks to a matter of hours. The deployment of autonomous logistics networks, including self-driving trucks, AI-optimized routing, and robotic warehouses, could reduce last-mile delivery costs by 25% and shorten supply chain lead times by 35%. Integrating circular economy practices such as closed-loop manufacturing and industrial recycling could significantly reduce Pakistan's reliance on imported raw materials while enhancing sustainability and resilience. By implementing these advanced technological solutions, Pakistan can significantly improve its supply chain efficiency, lower operational costs, and strengthen its export competitiveness. Failing to adopt Industry 6.0 technologies risks worsening trade imbalances, heightened inflation, and declining industrial growth. In an era of rapid global population expansion and increasing resource constraints, AI-driven, autonomous, and sustainable logistics systems are not merely an option but an imperative for Pakistan to secure its economic future and maintain its position in the global supply chain network [91–94].

4.4 Cultural and Workforce Challenges

Despite its potential for enhancing industrial efficiency, sustainability, and resilience, the transition faces major challenges, particularly in terms of workforce readiness and acceptance of disruptive technologies. One of the most pressing challenges is the skill gap that emerges as industries move toward highly intelligent and autonomous manufacturing ecosystems. Industry 6.0 requires a workforce proficient in AI, machine learning, data science, blockchain technology, cybersecurity, IoT integration, robotics and energy efficient technologies. By 2030, over 85 million jobs may be displaced by automation, but at the same time, 97 million new roles requiring advanced technical skills will emerge. The global skills shortage in AI and robotics alone is estimated to affect 75% of manufacturing firms by 2025, with a training cost exceeding USD 8 trillion worldwide. Developing structured upskilling and reskilling programs is essential, yet industries often struggle with high training costs and extended learning curves. Countries with lower investments in technical education, such as Pakistan, risk falling behind in industrial competitiveness if they fail to implement large scale workforce training initiatives. Another significant barrier is resistance to change, which arises due to a combination of psychological, economic, and cultural factors. Employees, particularly in traditional manufacturing roles, fear job displacement as AI-driven automation takes over routine and repetitive tasks. Surveys indicate that 60% of workers in developing economies express concerns about being replaced by intelligent systems, leading to reluctance in adopting new technologies. Managerial resistance is also prevalent, as 45% of executives struggle to justify the cost of implementing Industry 6.0 technologies without clear short-term returns on investment. This results in slower adoption rates and fragmented integration of AI-driven automation, creating inefficiencies in production cycles. To address these challenges, industries must adopt a multi-pronged strategy focusing on structured workforce transition, corporate engagement, and technological adaptation. First, organizations need to implement AI-assisted personalized learning programs, where machine learning algorithms assess employees' current skills and recommend tailored training modules to bridge knowledge gaps. Studies show that AI-driven learning management systems improve training efficiency by 40% and reduce skill acquisition time by 30% compared to traditional training models. Second, organizations must invest in human AI collaboration models, where AI augments human decision-making rather than replacing jobs entirely. Research indicates that hybrid workplaces, where AI and human workers collaborate, lead to 20–30% higher productivity compared to fully automated systems. Companies must focus on corporate culture transformation by integrating change management frameworks such as Kotter's 8 Step Process for Leading Change or Lewin's Change Model, which have been shown to increase technology adoption rates by 70%. Encouraging an innovation-driven mindset, fostering cross-disciplinary collaboration, and implementing incentive programs for employees who successfully integrate new technologies can accelerate adoption rates. Transitioning from Industry 4.0 to Industry 6.0 requires not only technological advancements but also a workforce equipped with new-age digital skills and a corporate culture that embraces innovation. Without strategic investments in reskilling, employee engagement, and structured change management, industries risk lagging in global competitiveness. The key to success lies in a balanced approach where AI-driven automation complements human expertise, ensuring a sustainable, efficient and resilient industrial future [95, 96].

4.5 *Intelligent Population Control in High-Density Urban Areas*

The rapid urbanization of modern cities has led to severe congestion, inefficient infrastructure utilization, and environmental challenges, necessitating the adoption of AI, real-time data analytics, and cyber-physical systems for effective smart city planning. Industry 6.0 integrates machine learning algorithms, quantum computing, IoT sensors, and digital twins to optimize traffic flow, public transport efficiency, energy consumption, and crowd management in densely populated areas. The implementation of AI-driven predictive analytics and autonomous control systems ensures that peak hour congestion in metros, highways, and commercial districts is dynamically managed, improving urban mobility, reducing energy wastage, and enhancing public safety.

4.5.1 Quantum AI and Edge Intelligence for Smart Traffic Management in Industry 6.0

Modern smart cities are revolutionizing urban mobility with real-time AI-enhanced traffic management systems using LiDAR-equipped intersections, GPS tracking, 5G connectivity, and autonomous drones to optimize traffic flow. These systems process billions of data points per second, integrating machine learning, vehicle-to-infrastructure (V2I) communication, and quantum computing to reduce urban congestion, minimize fuel wastage, and cut CO_2 emissions. AI-driven traffic control significantly enhances adaptive signal timing, enabling real-time rerouting and reducing commute times by 40% while improving urban air quality. Quantum-enhanced machine learning models process petabytes of real-time data in milliseconds, dynamically adjusting road usage patterns based on live congestion levels. Unlike traditional AI, quantum computing simultaneously evaluates multiple traffic scenarios, optimizing intersections with an unprecedented accuracy rate of 99.9% possibility. Quantum traffic modeling also enables predictive rerouting, ensuring seamless road usage in densely populated megacities. Smart intersections equipped with AI-driven adaptive traffic signals integrate V2I communication, allowing real-time interaction between vehicles and traffic infrastructure. These edge computing-enabled systems process vehicle density, pedestrian movement and environmental conditions, dynamically adjusting traffic signals to optimize road usage. AI-powered signal timing reduces idling times at intersections by 30%, cutting urban carbon emissions by 25% and enhancing fuel efficiency in stop-and-go traffic conditions. By implementing self learning traffic lights, city infrastructure can decrease congestion-related fuel consumption by 20%, contributing to SDG 9 and 11. AI-enabled autonomous aerial vehicles (UAVs) provide real time congestion monitoring, dynamically rerouting public transport based on live traffic data. Equipped with LiDAR and thermal imaging sensors, these drones detect traffic bottlenecks, communicate with smart infrastructure, and recommend optimized travel paths for autonomous vehicles and ride-sharing fleets. By enhancing public transit efficiency, autonomous drones contribute to a 25% reduction in congestion-related delays, improving urban mobility and reducing environmental impact. The integration of Quantum AI, Edge Computing, IoT-based infrastructure, and autonomous surveillance technologies within Industry 6.0 smart traffic systems is transforming urban transportation into a highly efficient, self-optimizing, and sustainable mobility ecosystem. As cities expand, these innovations will ensure zero-traffic bottlenecks, reduced emissions, and seamless integration of autonomous vehicle networks, shaping the future of next-generation intelligent transportation systems.

4.5.2 Metro and Public Transport Automation in Peak Hours

Urban metro and bus networks face increasing demand for real-time optimization to prevent overcrowding and ensure seamless mobility, particularly during rush hours. Industry 6.0 driven smart transit systems integrate AI, digital twins, quantum computing, and real-time data analytics to predict passenger demand, dynamically adjust vehicle frequencies, and optimize passenger flow. These advancements reduce congestion, improve energy efficiency, and minimize delays through self-learning automation and intelligent mobility coordination. Computer vision and thermal imaging sensors track crowd density in real time, feeding data into deep learning-based metro and bus control systems. By redistributing passenger flow and optimizing occupancy levels, AI-driven scheduling reduces waiting times by 50% and prevents train/bus overcrowding. These systems also incorporate edge computing, enabling decentralized decision-making at each transit hub for instant route adjustments. Traditional computational models struggle to process the millions of real-time route variables required for optimal transit operations. Quantum AI models, powered by quantum annealers, solve combinatorial optimization problems in seconds, calculating the shortest-time passenger redistribution strategies and reducing public transport delays by 60%. By simulating traffic congestion, weather conditions, and demand fluctuations, quantum-driven transit networks achieve unprecedented scheduling efficiency. AI-driven self learning control algorithms enable driverless metro and bus fleets, eliminating human errors and significantly reducing energy consumption by 35%. Autonomous electric buses integrate LiDAR, radar, and sensor fusion AI, allowing them to predict pedestrian movement, adjust speeds dynamically, and optimize regenerative braking, further enhancing efficiency. AI-powered predictive maintenance monitors motor temperature fluctuations, braking performance, and battery health, extending vehicle lifespan by 40% and reducing unscheduled downtimes by 70%. The transition to Industry 6.0-enabled transit aligns with SDGs 9 and 11 by enhancing transport efficiency, reducing emissions, and ensuring accessible mobility solutions. With AI-enhanced load balancing, quantum route optimization, and autonomous transport systems, urban transit networks achieve zero congestion, reduced carbon footprints, and fully autonomous smart mobility, reshaping the future of public transportation.

4.5.3 Smart City Infrastructure with Energy-Efficient Building and Public Space Management

The development of smart urban environments relies on AI-integrated energy grids, adaptive climate control, and intelligent public space management to enhance efficiency and sustainability. AI-optimized smart grids utilize real-time demand-response algorithms to dynamically distribute electricity, significantly reducing energy waste by 45% while balancing load fluctuations in dense metropolitan areas. Digital twin-enabled climate control systems leverage AI-driven predictive analytics to adjust heating, cooling, and ventilation in high-density urban regions, leading to a 40% reduction in energy consumption within mega-buildings. These smart systems integrate IoT sensors and machine learning models to monitor external temperature variations, building occupancy, and energy demand, allowing for precise real-time optimization of urban climate conditions.

Moreover, autonomous building access control systems equipped with AI-powered biometric recognition, RFID-enabled smart entry points, and blockchain-secured digital credentials streamline entry management in transport hubs, stadiums, and commercial centers. These systems enhance security, reduce congestion, and optimize human flow dynamics, ultimately increasing urban mobility efficiency. In Industry 6.0-driven smart cities, real-time AI coordination of energy, security, and climate control systems ensures seamless, low-carbon, and highly efficient urban environments, contributing to SDG 9, 11, and 13 by promoting sustainable infrastructure, smart cities, and climate resilience.

4.5.4 AI and CPS for Pedestrian and Urban Crowd Management

By integrating AI with Industry 6.0 automation, modern cities evolve into sustainable, adaptive, and energy-efficient ecosystems, aligning with SDG 9, 11, 12, and 13. AI-powered waste heat recovery systems capture excess thermal energy from industrial zones and transportation hubs, repurposing it for district heating and urban infrastructure, reducing energy consumption by 30%. Self-healing infrastructure embedded with nanomaterial-based sensors detects microcracks in bridges, roads, and buildings, triggering autonomous repair mechanisms using shape-memory polymers and self-repairing concrete, extending structural lifespan by 50%. Carbon capture in urban design integrates AI-controlled direct air capture systems, sequestering CO_2 emissions from transportation and commercial activities and converting it into usable chemicals or sustainable biofuels, effectively cutting urban carbon footprints by up to 60%. These advancements ensure that future megacities remain climate resilient, congestion-free and capable of dynamically managing population surges while minimizing environmental impact.

5 Approaches to Integration

5.1 Retrofitting Legacy Systems with Smart Sensors and IoT Devices

The transition from conventional manufacturing to Industry 6.0 requires integrating advanced technologies into legacy systems without entirely replacing existing infrastructure. By 2030, over 60% of global factories are expected to adopt AI-driven automation, increasing production speeds by 50% and reducing human intervention by 40%. Pakistan, with its growing industrial base and pressing need for efficiency improvements, must invest in IoT-driven retrofitting to remain competitive in the global manufacturing landscape. Ultimately, industries that embrace smart sensor integration and intelligent automation frameworks will achieve higher efficiency, reduced operational costs, and improved supply chain resilience, critical elements for navigating the challenges of Industry 6.0 and beyond. Given that over 80% of industrial machinery worldwide is still non-digitalized and lacks real-time connectivity, retrofitting presents a viable alternative to costly equipment overhauls. Industries can begin their transformation by embedding IoT sensors into legacy machinery to monitor parameters such as temperature, vibration, pressure, humidity, power consumption, and operational efficiency. Smart sensors provide sub-millisecond latency in data acquisition, allowing for real-time anomaly detection and optimized maintenance schedules. Vibration sensors installed in motors and pumps can detect early signs of mechanical wear and predict failures up to 90 days in advance, reducing unexpected downtime by 40–50%. Temperature sensors, when integrated into high-heat industrial processes, can increase thermal efficiency by 15–20%, cutting excessive energy use in power-hungry sectors like steel manufacturing and petrochemicals. Advanced AI-powered predictive maintenance systems use the collected data to analyze degradation trends and alert operators before failures occur. This approach minimizes unplanned downtime by up to 45%, extends equipment life by 20–30%, and reduces overall maintenance costs by 25%. Real-time IoT monitoring allows companies to optimize energy consumption dynamically. In textile processing, for example, intelligent energy meters linked to IoT platforms can reduce electricity costs by 10–15% by automatically adjusting machine load based on peak demand periods. The integration of smart factory frameworks enables manufacturers to transition toward autonomous and self-optimizing production environments. Retrofitting legacy machines with wireless connectivity modules like 5G, LoRaWAN, Zigbee, and Near Field Communication allows real-time data collection without requiring extensive rewiring. 5G-powered smart factories offer near-zero latency (less than 1 millisecond), enabling synchronized machine-to-machine communication that improves overall equipment effectiveness by 35%. By adopting cloud-based platforms such as Azure IoT Hub, AWS IoT Core, or Siemens MindSphere, legacy machinery can be integrated into centralized control systems for real-time analytics, digital twins, and automated decision-making. Digital twin technology, where a virtual replica of a physical asset continuously updates based on live sensor data—can reduce design and process inefficiencies by up to 50%, making it a crucial tool for optimizing industrial workflows. Retrofitting legacy systems with IoT sensors provides substantial ROI in various industries. In automotive manufacturing, AI-powered machine vision systems retrofitted to conventional assembly lines can improve defect detection rates by 98%, minimizing recalls and warranty costs. In the textile industry, smart humidity and temperature sensors installed in dyeing and finishing units can reduce water consumption by 30% and energy usage by 15% through optimized process control. In chemical and pharmaceutical industries, IoT-based gas sensors can continuously monitor toxic gas leaks in real-time, reducing hazardous exposure risks by 60% and preventing regulatory non-compliance penalties. Given that retrofitting costs are often 70% lower than full equipment replacement while delivering up to 80% of the performance benefits of new smart machines, industries that fail to modernize risk significant productivity losses and increased operational costs [97, 98]. Here is Fig. 9.1 showing retrofitting a legacy system.

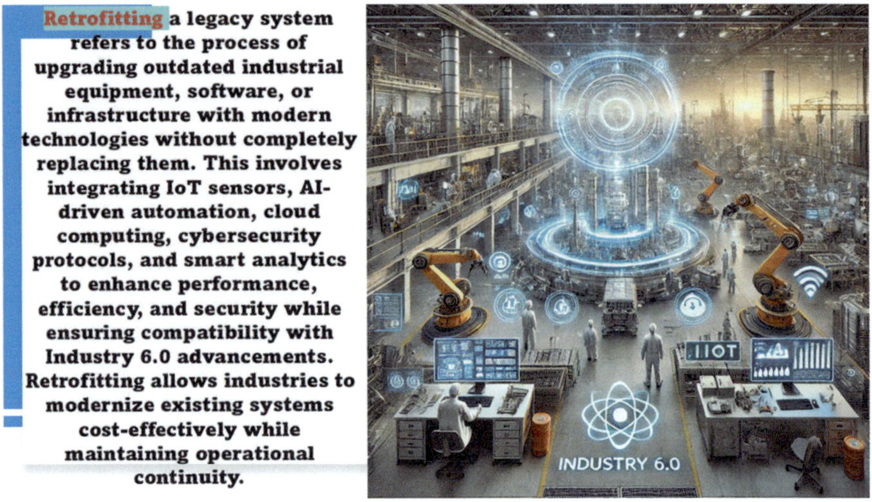

Fig. 9.1 Retrofitting concept description

5.2 Adoption of Hybrid Manufacturing Systems

The adoption of hybrid manufacturing systems represents a significant advancement in industrial production, blending traditional and advanced manufacturing techniques to maximize efficiency, flexibility, and cost-effectiveness. Industry 6.0 does not necessitate the complete replacement of conventional manufacturing systems; instead, it promotes the seamless integration of automation, AI, robotics, and digital twin technology with traditional processes such as CNC machining, injection molding, and casting. Looking ahead, the fusion of quantum computing, CPS, and nanomaterial-based additive manufacturing will further enhance the efficiency of hybrid manufacturing systems. By 2030, experts predict that fully autonomous AI-driven hybrid production facilities will improve global manufacturing output by 50% while minimizing human intervention by 60%. As Pakistan seeks to strengthen its industrial base and compete globally, the adoption of hybrid manufacturing technologies will be critical for enhancing productivity, reducing operational costs, and ensuring sustainable industrial growth. In 2025, over 65% of global manufacturing firms are expected to implement hybrid production models, resulting in 20 to 30% cost savings and a 40% reduction in production lead times. Hybrid manufacturing combines the precision and repeatability of subtractive techniques (CNC milling, laser cutting) with the complexity and customization of additive manufacturing (3D printing, direct energy deposition). In aerospace engineering, hybrid systems enable manufacturers to produce lightweight, high-strength components with 20 to 40% material savings, improving fuel efficiency by up to 15% in modern aircraft. In the automotive sector, hybrid laser-assisted additive manufacturing can create engine components with 30% improved thermal resistance, extending engine lifespan by 25%. One of the key innovations in hybrid manufacturing is the implementation of

FMS. These systems utilize robotic arms, AI-driven process control, and real-time IoT-based monitoring to switch between different product designs without requiring major setup changes. AI-powered computer vision algorithms optimize cutting paths in CNC machining, reducing scrap material waste by 35% and lowering overall energy consumption by 20% in precision manufacturing. Digital twin simulations, where real-world production lines are virtually replicated, allow manufacturers to adjust process parameters before production, decreasing defect rates by 50% and downtime by 45%. Hybrid manufacturing also ensures sustainability by integrating energy-efficient automation and circular economy principles. Industries implementing hybrid production techniques reduce CO_2 emissions by 25–30% compared to purely conventional manufacturing. For instance, in textile and apparel production, AI-assisted hybrid sewing and laser cutting machines reduce fabric waste by 20% and enhance production speed by 35% while consuming 15% less electricity [18, 99].

5.3 *Gradual Implementation of AI and ML in Predictive Maintenance*

The integration of AI-driven predictive maintenance in Industry 6.0 is revolutionizing how industries manage equipment reliability, reducing unplanned downtime, extending machine lifespan, and optimizing resource utilization. Predictive maintenance employs ML models, digital twins, and advanced analytics to monitor real-time sensor data, detect anomalies, and forecast potential failures before they occur. AI-powered predictive maintenance can reduce unexpected downtime by up to 50%, lower maintenance costs by 30–40%, and extend equipment lifespan by 20–25%, offering industries an average ROI improvement of tenfold within the first 2 years of implementation. Traditional maintenance strategies, such as reactive (fixing after failure) and preventive (scheduled maintenance) approaches, often lead to unnecessary maintenance expenses and inefficient downtime. AI-powered PdM use big data analytics, deep learning models and IoT-based sensor networks to predict failures with over 95% accuracy. Sensors monitoring parameters such as vibration, acoustic emissions, lubrication quality, pressure, voltage fluctuations, and thermal imaging enable real-time anomaly detection. Vibration analysis using AI can detect bearing failures in rotating machinery up to 90 days in advance, reducing failure-induced operational losses by 40–50%. Similarly, AI-based infrared thermography detects overheating electrical components before failures occur, increasing the operational efficiency of industrial electrical grids by 15–20%. AI-driven systems enhance maintenance workflows by automating root cause diagnostics and troubleshooting. These systems use self learning algorithms, knowledge graphs, and real-time process simulations to autonomously diagnose issues, recommend solutions, and even execute minor adjustments without human intervention. AI-assisted fault detection systems in industrial automation have reduced manual troubleshooting time by 70% and cut repair costs by 35%. In automotive manufacturing, AI-powered robotic maintenance systems can automatically detect mechanical wear in robotic arms and recalibrate them within milliseconds, improving production line uptime by 25–30%. The gradual implementation of AI-based predictive maintenance can follow a multi-phase approach, ensuring seamless integration into existing industrial ecosystems. AI models are also virtually tested in digital twin simulations before real-world application to ensure precision. Industries across various sectors are already benefiting from AI-driven PdM. In aerospace, AI-based predictive maintenance has helped reduce aircraft engine failures by 55%, improving flight safety and minimizing maintenance costs. In manufacturing, AI-driven vibration monitoring systems in metal-cutting CNC machines have reduced tool failures by 30%, cutting production losses by five million USD annually per plant. In the energy sector, AI-based predictive analytics in wind turbines has increased power generation efficiency by 12–18% by predicting gearbox failures before they occur. In the chemical and pharmaceutical industry, AI-driven leak detection sensors in chemical processing plants have reduced hazardous material spills by 60%, improving regulatory compliance and workplace safety. By 2030, AI-powered predictive maintenance is expected to dominate industrial reliability management, with over 75% of global manufacturing firms implementing fully automated, self-healing maintenance systems. The integration of blockchain-secured maintenance logs, 6G-enabled real-time monitoring and quantum computing-assisted failure predictions will enable industries to achieve near zero downtime manufacturing, ensuring continuous production with 99.99% uptime. For countries like Pakistan, which is striving to enhance its industrial efficiency and global competitiveness, investing in AI-driven predictive maintenance technologies will be mandatory in reducing operational costs, preventing machine failures and optimizing resource utilization in sectors such as textile manufacturing, automotive production, and renewable energy generation. Industries that adopt advanced predictive analytics and autonomous maintenance frameworks will gain a significant competitive edge in Industry 6.0 and beyond [100, 101].

6 Cloud Computing with Edge Computing Integration

The integration of cloud computing and edge computing is transforming manufacturing operations by enabling more efficient data management, analysis, and real-time decision-making. Cloud-based data analysis allows manufacturers to store vast amounts of data from a wide array of IoT devices in centralized, scalable systems. This shift to cloud computing facilitates the management of production processes, supply chain logistics, and customer preferences on a global scale, while also promoting collaboration across different stakeholders. Global manufacturers are expected to move 40% of their operations to the cloud within the next 2 years, allowing them to process data at a rate 30–50% faster than traditional on-premises systems. By storing large datasets in the cloud, manufacturers can also perform advanced analytics, enabling them to better understand supply chain performance, monitor machinery condition in real-time, and gain insights into production quality. Cloud platforms can leverage machine learning algorithms to model consumer demand patterns, allowing manufacturers to adjust production schedules and resource allocations accordingly. 60% of industrial organizations plan to deploy edge computing solutions by 2026, improving the response time in manufacturing operations by 40–50%. This shift allows manufacturers to reduce downtime, improve operational efficiency, and enhance overall system reliability. Edge computing facilitates local decision-making that minimizes reliance on the cloud, which can be especially beneficial in situations where uninterrupted production is critical, and the delay caused by data transmission is not feasible. Together, cloud and edge computing provide a robust framework for modern manufacturing systems. While the cloud offers centralized storage, scalability, and advanced analytics for long-term planning, edge computing supports immediate, low-latency decision making that optimizes real-time production conditions. This combination of cloud and edge computing enables a comprehensive and agile approach to manufacturing, ensuring that systems can adapt to both immediate operational needs and long-term strategic goals. By 2025, the integration of both cloud and edge computing is projected to reduce manufacturing downtime by up to 40%, improve energy efficiency by 20%, and cut operational costs by as much as 30% in sectors such as automotive, electronics, and consumer goods production. As manufacturers continue to embrace this dual approach, they are likely to gain significant competitive advantages, including faster time-to-market, higher quality products, and enhanced sustainability through optimized resource management [102, 103].

7 Climate Resilient Manufacturing in Industry 6.0

By incorporating AI, IoT, and sustainable technologies into its operations with ease, Siemens has advanced from Industry 4.0 to Industry 6.0. This change has been essential for increasing sustainability across production lines, cutting waste, and maximizing manufacturing efficiency. Siemens' manufacturing facilities have benefited greatly from the integration of AI-powered predictive maintenance technologies, which have reduced machine downtime and optimized energy use. By using ML to evaluate past data and anticipate possible malfunctions before they happen, these AI systems can cut down on unscheduled downtime by up to 40% and increase the overall efficacy of equipment by up to 15%. Siemens' facilities are expected to use 25% less energy thanks to this predictive method, which guarantees that production processes are both effective and energy-conscious. The adoption of IoT-enabled sensors in production lines is another core aspect of Siemens' shift to Industry 6.0. These sensors collect real-time data on various parameters, including temperature, humidity, and machinery performance. The data is then fed into AI systems for advanced analytics, enabling dynamic decision-making that optimizes production processes and enhances sustainability. IoT data can be used to adjust production speed or energy usage based on real-time conditions, thereby reducing material waste and optimizing resource allocation. Siemens reports that its IoT-enabled systems have helped improve resource utilization by 30%, while simultaneously reducing waste production by up to 20%. This integration of IoT with AI fosters an ecosystem where real-time monitoring and adjustments lead to more sustainable and energy-efficient operations. Another key technology Siemens has integrated into its operations is the concept of Digital Twins. By creating digital replicas of physical assets, Siemens can simulate, monitor, and optimize manufacturing processes in real time. This virtual representation of factory equipment and production lines allows for detailed scenario analysis and process optimization without interrupting the actual production. Digital Twins have helped Siemens improve operational efficiency by up to 18%, as these systems enable predictive maintenance, reduce energy consumption, and enhance the overall manufacturing lifecycle. They enable manufacturers to identify potential inefficiencies or areas for improvement, optimizing processes such as supply chain management, product design, and quality control. Siemens has also embraced advanced robotics and automation in its manufacturing processes, integrating intelligent robots or cobots that work alongside human workers. These robots are designed to optimize manufacturing efficiency while being energy efficient and sustainable. Cobots can perform repetitive tasks with high precision, reduce human error, and increase production speeds, all while consuming significantly less energy than traditional industrial robots. Siemens has demonstrated that its energy efficient robots can reduce energy consumption by up to 30% compared to legacy systems, making them a key component in driving sustainability in smart manufacturing environments. Siemens is exploring the potential of blockchain technology to improve supply chain transparency and reduce waste. Blockchain enables secure, transparent tracking of materials throughout the supply chain, providing manufacturers with real-time information on the location, status, and condition of materials. This technology helps optimize resource usage, minimize material waste and reduce inefficiencies associated with tracking and managing supply chains. Blockchain integration is expected to enhance material traceability and resource optimization by as much as 25%, reducing supply chain waste and ensuring compliance with sustainability standards. Sustainable materials and additive manufacturing (3D printing) also play a significant role in Siemens' Industry 6.0 transformation. Additive manufacturing enables the production of complex components using less material, resulting in less waste and more energy efficient production processes. By using sustainable materials such as biodegradable plastics and recycled metals in 3D printing, Siemens can manufacture climate-resilient products while reducing its carbon footprint. Siemens' adoption of 3D printing has allowed them to reduce material waste by approximately 40% and shorten production lead times by 25%, further enhancing sustainability and operational efficiency. The integration of these advanced technologies in Siemens' manufacturing operations exemplifies the potential of Industry 6.0 to contribute to climate resilience and sustainable manufacturing. By reducing energy consumption, waste, and material usage, Siemens is not only optimizing its own operations but also setting a benchmark for other industries looking to transition to more sustainable practices. As part of its ongoing commitment to sustainability, Siemens collaborates with startups and research institutions to stay ahead of technological advancements, fostering innovation that continually enhances its Industry 6.0 ecosystem. Through these collaborations, Siemens has been able to integrate cutting edge technologies that drive energy efficient production methods, reduce environmental impact, and contribute to the realization of a circular green economy. Looking ahead, Siemens is positioned to continue leading the way in the future of Industry 6.0, where advancements in AI, robotics, quantum computing, and energy-efficient technologies will further transform manufacturing processes. As Industry 6.0 evolves, it will support the creation of a circular economy where materials are continuously reused, waste is minimized, and energy consumption is optimized, aligning with global climate goals and sustainable development targets. The future of Industry 6.0 holds tremendous potential for industries like automotive, healthcare, and textiles to leverage these technologies to create more sustainable, climate-resilient products and processes [104–107].

8 Industry 6.0 Implementation Challenges, Suggestions, and Future Recommendations

Here is Table 9.2, showing the key challenges with their proper solutions.

To support small and medium-sized enterprises, tailored financial models are essential. Governments should offer green technology subsidies, matching up to 40% of sustainability-related capital expenditures, such as investments in solar-powered micro factories. Blockchain-backed asset tokenization can provide liquidity by allowing retrofitted equipment to be traded as non-fungible tokens, unlocking new financing mechanisms. Addressing workforce transformation is equally critical. National education systems must integrate AI literacy as a mandatory subject, while industry-academia collaborations should co-develop nano degree programs focused on automation, robotics, and quantum technologies. Automation Dividend policies should be implemented, allowing workers to share in the productivity and cost efficiency gains generated by AI integration. In terms of sustainability, manufacturers must shift toward climate-resilient practices. This includes adopting carbon-negative production technologies, such as direct air capture combined with carbon capture, utilization, and storage, as seen in some industries, and the deployment of CCUS-enabled direct reduced iron steelmaking. The use of 4D printed recyclable materials, which self disassemble after end of life use, can support zero-waste production cycles. Global cooperation is essential for standardized progress. The formation of industrial AI alliances can enable cross border federated learning via GDPR compliant data pools, ensuring privacy and shared innovation. Simultaneously, the adoption of UN-endorsed cyber-physical standards such as ISO/IEC 24400, will facilitate the ethical deployment of AI across industries worldwide. A successful transition achieved remarkable results, i.e., 40% reduction in machine downtime, 30% increase in resource utilization and 25% decrease in energy consumption. These gains are realistic and can be enhanced with R&D and are realized through the integration of real-time digital twins (yielding an 18%

Table 9.2 Key challenges with solutions in the implementation of Industry 6.0

Key challenges	Key problems	Advanced solutions and technologies	Implementation metrics
Technological barriers			
Data interoperability	Legacy systems (Modbus/Profibus) lack compatibility with modern protocols (OPC UA/ MQTT); high retrofit costs (USD 500 k to 2 M per facility, depend on industrial size, age & type)	• Quantum optimized middleware: AI-driven translation layers for real-time protocol conversion • Self-configuring IoT sensors: Nanoscale edge devices with autonomous protocol adaptation	90% reduction in integration time possible with 40% cost savings via digital twin simulations
Automation integration	AI/ML deployment limited by legacy CNC/ robotic arm interfaces; latency in feedback loops	• Neuromorphic control units: Brain-inspired chips for real time AI decision making (0.1 ms latency) • Swarm robotics: Collaborative nano-bots for adaptive production lines	35%↑ in production flexibility; 50%↓ in reprogramming downtime
Cybersecurity	Expanded attack surface via IoT/AI; adversarial ML threats; legacy system vulnerabilities	• Quantum encrypted Blockchains: Tamper-proof supply chain logs (128-bit QKD) • AI-powered threat hunting: Autonomous security bots with predictive breach detection (99.8% accuracy possible)	70%↓ in intrusion incidents; 60% faster zero-day patching
Financial constraints			
High capital costs	SME prohibitive investments: AI systems (USD 0.5 M to 5 M), cobots (USD 50 k to 250 k), sensors (USD 300/unit)	• Robotics as a service (RaaS): Pay per use models (USD 15–50/h) • Federated learning: Shared AI training across industries to reduce costs by 75%	ROI within 18–24 months; 30%↓ TCO via cloud-edge hybrid computing
ROI uncertainty	Only 30% of firms achieve ROI in 3–5 years; rapid tech obsolescence (IoT sensor lifespan: 5–10 years)	• Digital twin ROI simulators: Predictive analytics for investment outcomes (95% accuracy) • Circular economy platforms: Material reuse nets 25–40% cost recovery	50%↑ ROI visibility; 20%↑ asset utilization via predictive maintenance

Table 9.2 (continued)

Key challenges	Key problems	Advanced solutions and technologies	Implementation metrics
Workforce transformation			
Skills gap	75% of manufacturers face AI/robotics talent shortages; 85 M jobs displaced by 2030	• AR/VR micro-credentialing: Holographic upskilling (6-week certification) • AI co pilots: Industrial assistants GPT for real time troubleshooting	80%↓ training time; 45%↑ workforce productivity
Resistance to change	60% of workers fear job loss; executive hesitation due to unclear short-term benefits	• Human AI Symbiosis frameworks: Cobots handle repetitive tasks; humans focus on innovation • Upskilling stipends: Govt-funded crypto-UBI for reskilling	70%↑ tech adoption rate; 55%↓ employee turnover
Resource and sustainability			
Energy demand	Global energy use ↑58% by 2050; industry contributes 34% of CO_2 emissions	• Neuromorphic smart grids: AI-optimized energy distribution (50%↑ efficiency) • Nuclear fusion micro-reactors: Pilot plants (2035+) for zero-carbon baseload	45%↓ energy waste; 30 Gt CO_2e/year
Material scarcity	Lithium demand ↑2000% by 2050; cobalt supply chain ethics issues (DRC dominates 70%)	• AI-driven material substitution: Graphene from CO_2 capture; lab grown rare earths • Quantum optimized recycling: 99% metal recovery via enzymatic PCB dissolution	62%↓ material waste; USD 4.5 T/year circular economy savings
Supply chain resilience			
Semiconductor shortages	63% global chips from Taiwan; 400%↑ demand by 2030; 28 week lead times	• Photonic computing chips: Light based processors (0.1 pJ/ op) bypass silicon limits • Distributed Nano foundries: 3D printed fabs for on demand production	90%↓ supply chain disruptions; 800%↑ EV semiconductor capacity
Logistics inefficiency	Freight costs ↑400%; 2.6 M truck driver shortage; port congestion	• Hydrogen powered autonomous fleets: Self-driving trucks (30%↓ freight costs) • Hyperloop networks: 0.03 kWh/p-km cargo transport	55%↓ transport emissions; 70%↓ last-mile energy use

(continued)

Table 9.2 (continued)

Key challenges	Key problems	Advanced solutions and technologies	Implementation metrics
Urbanization and infrastructure			
Smart city integration	68% global population urban by 2050; USD 1.5 T/year infrastructure deficit in developing nations	• Self-healing concrete: Nanomaterial infused infrastructure (50%↑ lifespan) • Atmospheric water harvesters: MOF-303 systems for water-stressed cities	40%↓ urban heat island effect; 50%↓ freshwater withdrawal
Ethical and regulatory			
AI governance	Bias in autonomous systems; lack of global standards for AI ethics	• Explainable AI (XAI) frameworks: Auditable decision trails via blockchain • UN-led regulatory sandboxes: Testbeds for ethical industrial AI	Compliance with ISO 31000:2025 AI risk standards; 99% algorithmic transparency

process efficiency boost), energy-efficient collaborative robots (cobots) that reduced power consumption by 35% and blockchain-based material passports that cut supply chain waste by 25%. The roadmap for adopting Industry 6.0 technologies begins with a phased integration of advanced systems into existing industrial infrastructure. Step 1 involves retrofitting legacy equipment with modular Industrial Internet of Things kits, which cost approximately USD 10–50 thousand per machine, enabling data acquisition and remote monitoring. In step 2, organizations should deploy edge AI platforms for predictive maintenance, which can reduce unplanned downtime by up to 50%. Step 3 focuses on migrating toward quantum-cloud hybrid architectures that facilitate real-time optimization of global supply chains, enhancing agility and resilience. These strategic recommendations highlight a transformative framework for scaling intelligent, sustainable, and equitable industrial ecosystems under the Industry 6.0 paradigm.

References

1. Vijay Arputharaj, J., et al.: Exploring the synergy of IoT, AI, and data analytics in industry 6.0. In: Industry 6.0: Technology, Practices, Challenges, and Applications, p. 1. CRC Press, Boca Raton (2024)
2. Singh, R., Tyagi, A.K., Arumugam, S.K.: Imagining the sustainable future with Industry 6.0: A smarter pathway for modern society and manufacturing industries. In: Machine Learning Algorithms Using Scikit and TensorFlow Environments, pp. 318–331. IGI Global, London (2024)

3. Eswaran, U., et al.: Empowering the factory of the future: integrating artificial intelligence, machine learning, and IoT innovations in industry 6.0. In: Evolution and Advances in Computing Technologies for Industry 6.0, pp. 1–21. CRC Press, Boca Raton (2024)
4. Elgalb, A., Freek, A.: Harnessing machine learning for real-time cybersecurity: A scalable approach using big data frameworks. Emerg. Eng. Math. **2024**, 1–9 (2024)
5. Ji, T., Xu, X.: Exploring the integration of cloud manufacturing and cyber-physical systems in the era of industry 4.0–an OPC UA approach. Robot. Comput. Integr. Manuf. **93**, 102927 (2025)
6. Francisco, D., et al.: Scalable system to support a digital twin of a stone factory production line. IEEE Access. **13**, 15585–15603 (2025)
7. Jaber, J.O., et al.: Energy efficiency retrofitting for old buildings in a low-income district of Amman, Jordan. Acad. Green Energy. **2**(1), 7487 (2025)
8. Arun, M., et al.: Economic, policy, social, and regulatory aspects of AI-driven smart buildings. J. Build. Eng. **99**, 111666 (2025)
9. Selvaprasanth, P., Malathy, R.: Revolutionizing structural health monitoring in marine environment with internet of things: a comprehensive review. Innov. Infrastruct. Solut. **10**(2), 62 (2025)
10. Packianathan, R., et al.: Integrating industrial robotics and internet of things (IoT) in smart transportation system. In: Driving Green Transportation System through Artificial Intelligence and Automation: Approaches, Technologies and Applications, pp. 379–395. Springer, Cham (2025)
11. Pedroso, A.F., et al.: An overview on the recent advances in robot-assisted compensation methods used in machining lightweight materials. Robot. Comput. Integr. Manuf. **91**, 102844 (2025)
12. Yang, S., et al.: Live digital twin with virtual reality for accessible and immersive manufacturing education. Int. J. Adv. Manuf. Technol. **136**(7), 3577–3590 (2025)
13. Gill, S.S., et al.: Edge AI: a taxonomy, systematic review and future directions. Clust. Comput. **28**(1), 1–53 (2025)
14. Salama, R., et al.: Internet of things and mobile cloud computing service models. In: Artificial Intelligence of Things (AIoT), pp. 105–117. Elsevier, Amsterdam (2025)
15. Renuka, O., et al.: Data privacy and protection: legal and ethical challenges. In: Emerging Threats and Countermeasures in Cybersecurity, pp. 433–465. Wiley, Hoboken (2025)
16. Ahn, J., et al.: Exploring encryption algorithms and network protocols: a comprehensive survey of threats and vulnerabilities. IEEE Commun Surv Tutor. (2025). https://doi.org/10.1109/COMST.2025.3526605
17. Berhe, H.H., et al.: Empirical investigation of drivers, motivations and barriers for implementation of integrated continuous improvement system: a case of Ethiopian manufacturing industries. Int. J. Qual. Reliab. Manage. **42**(1), 183–214 (2025)
18. Fenta, E.W., et al.: Opportunities in flexible manufacturing systems in the near Future. Glob. J. Flex. Syst. Manag. **2**, 49–62 (2025)
19. Rawat, U., Kumar, A., Anbanandam, R.: Evaluating the preparedness of freight logistics firms for cyber-physical systems integration: a SAP-LAP methodology for sustainable development. Glob. J. Flex. Syst. Manag. **26**(1), 1–23 (2025)
20. Samblani, G., Bhatt, D.P.: Case studies of next generation AI for intelligent manufacturing. In: Intelligent Manufacturing and Industry 4.0, pp. 217–236. CRC Press, Boca Raton (2025)
21. Hiremath, M.S., et al.: Smart manufacturing: navigating challenges, seizing opportunities, and charting future directions—a comprehensive review. In: Industry 4.0, Smart Manufacturing, and Industrial Engineering, pp. 271–298. CRC Press, Boca Raton (2025)
22. Zhang, Q., et al.: Multi-objective optimization of the mixed-flow intelligent production line for automotive MEMS pressure sensors. Appl. Intell. **55**(1), 1–14 (2025)
23. Xing, L.: Evaluation of the impact of artificial intelligence and intelligent internet of things on population mobility on regional economic differences. Soft. Comput. **28**, 13977–13988 (2023)
24. Acemoglu, D., Restrepo, P.: Demographics and automation. Rev. Econ. Stud. **89**(1), 1–44 (2022)

25. Kurt, R.: Industry 4.0 in terms of industrial relations and its impacts on labour life. Proc. Comput. Sci. **158**, 590–601 (2019)
26. Harris, K., Kimson, A., Schwedel, A.: Labor 2030: the collision of demographics, automation and inequality. Bain Company. **7**, 63 (2018)
27. Vavrová, K., et al.: Smart City 4.0 as the concept of strategically managed sustainable urbanism. Strat. Manage. **21**, 1 (2025)
28. Ogbuju, E., et al.: The role of artificial intelligence in poverty alleviation during the fourth industrial revolution. In: AI for Humanitarianism, pp. 38–50. CRC Press, Boca Raton (2015)
29. Low, M.P., et al.: The interplay of self-efficacy, artificial intelligence literacy and lifelong learning for career resilience among older employees: a comparison study between China and Malaysia. Curr. Psychol. **44**, 7879–7896 (2025)
30. Zhang, F., Li, J., Lu, J.: Potential impacts of population aging and artificial intelligence on households, living arrangements and sustainable development. China Popul. Dev. Stud. **8**, 6 (2025)
31. Kask, J., Armstrong, R.: Another industrial revolution? It's all evolution! In: Contemporary Issues in Industry 5.0: Towards an AI Integrated Society, pp. 29–47. Springer, Cham (2025)
32. Al Waroi, M.N.A.L., Subroto, A., Supriyadi, I.: Indonesia's demographics in the digital era: opportunities and challenges towards a Golden Indonesia 2045. Asian J. Eng. Soc. Health. **4**(1), 93–111 (2025)
33. Mukherjee, D., Mandal, B.K.: Introduction to industry 4.0. In: Industry 4.0, Smart Manufacturing, and Industrial Engineering, pp. 1–23. CRC Press, Boca Raton (2025)
34. Liu, Y., Li, L., Liang, L.: Skills shortages in the Chinese manufacturing sector. In: The Future of Work in the Asia Pacific, pp. 116–131. Routledge, London (2025)
35. Grant, R., Üngör, M.: The AI revolution with 21st century skills: implications for the wage inequality and technical change. Scot. J. Polit. Econ. **71**(5), 731–765 (2024)
36. Irandoust, M.: Employment and technology: creative creation or creative destruction? An asymmetric analysis. Austr. Econ. Pap. **63**(2), 201–219 (2024)
37. Li, L.: Reskilling and upskilling the future-ready workforce for industry 4.0 and beyond. Inf. Syst. Front. **26**(5), 1697–1712 (2024)
38. Elife, K., Baki, A.E.: New technologies and labor market: a look into the future of jobs and employment. Социология науки и технологий. **14**(2), 194–208 (2023)
39. Shiohira, K.: Understanding the Impact of Artificial Intelligence on Skills Development. Education 2030. UNESCO-UNEVOC, Bonn (2021)
40. Mondol, S., et al.: Impact of AI and data in revolutionizing microfinance in developing countries: improving outreach and efficiency. In: Data-Driven Modelling and Predictive Analytics in Business and Finance, pp. 71–87. Auerbach Publications, Boca Raton (2024)
41. Baral, S.K., et al.: Role of digital technology and artificial intelligence for monitoring talent strategies to bridge the skill gap. In: 2022 International Mobile and Embedded Technology Conference (MECON). IEEE, New York (2022)
42. Adeoye, M.A., Otemuyiwa, B.I.: Navigating the future: strategies of EdTech companies in driving educational transformation. JERIT J. Educ. Res. Innov. Technol. **1**(1), 43–50 (2024)
43. Moşteanu, N.R.: Thriving in the entrepreneurial landscape of sustainability and intelligent automation era. Green Low-Carbon Econ. **3**(1), 18–25 (2025)
44. Lin, B., Zhan, Y., Sun, C.: A study of the coordinated development of energy supplies and the demand for carbon neutrality. In: Chinese Governance and Transformation Towards Carbon Neutrality, pp. 123–146. Springer, Cham (2025)
45. Mudhee, K.H., et al.: Assessing climate strategies of major energy corporations and examining projections in relation to Paris agreement objectives within the framework of sustainable energy. Unconvent. Resour. **5**, 100127 (2025)
46. Marzouk, O.A.: Summary of the 2023 report of TCEP (tracking clean energy progress) by the international energy agency (IEA), and proposed process for computing a single aggregate rating. E3S Web Conf. **601**, 48 (2025)
47. Kempston, S., et al.: UK electric vehicle battery supply chain sustainability: a systematic review. Renew. Sust. Energ. Rev. **210**, 115216 (2025)

48. Baldassarre, B.: Circular economy for resource security in the European Union (EU): case study, research framework, and future directions. Ecol. Econ. **227**, 108345 (2025)
49. He, Q., Wu, X., Ding, X.: Uncertainty in the carbon footprint accounting and evaluation of textile and apparel products: a systematic review. J. Clean. Prod. **492**, 144885 (2025)
50. Shakir Hanna, S.H.: Climate change and human imprint consequences. In: Climate Changes Impacts on Aquatic Environment: Assessment, Adaptation, Mitigation, and Road Map for Sustainable Development, pp. 3–19. Springer, Cham (2025)
51. Baldassarre, B., Carrara, S.: Critical raw materials, circular economy, sustainable development: EU policy reflections for future research and innovation. Resour. Conserv. Recycl. **215**, 108060 (2025)
52. Xu, S., Wang, Y.: Do circular innovations and carrying capacity of natural environment enhance circular economy in the European Union? Evidence from simulation and machine learning methods. J. Environ. Manag. **373**, 123863 (2025)
53. Wu, Y.: The new forms and situations of people-centred urbanization. In: A Study on People-Centered Urbanization, pp. 7–46. Springer, Cham (2025)
54. Wei, Y., et al.: A bibliometric analysis of food security and urbanization: insights and implications. Front. Sustain. Cities. **7**, 1459493 (2025)
55. Qaysi, S., Abdelrahman, K., Saadon, A.B.: Evaluation of kinetic moduli and soil competence parameters for sustainable urban development using multichannel analysis of surface waves (MASW) technique. All Earth. **37**(1), 1–34 (2025)
56. Khan, E.A., Nazar, M., Khattak, U.K.: Population dynamics and urbanization patterns in Pakistan: implications for sustainable development. J. Urban. **2023**, 2285404 (2023)
57. Karim, S., et al.: Comprehensive geospatial analysis of urban expansion dynamic in Lahore, Pakistan (1998–2023). J. Urban Manage. **14**, 577–589 (2024)
58. Dorsey, A.F.: Urbanization and infectious disease. Am. J. Hum. Biol. **37**(1), e24197 (2025)
59. Badi, S., Alhosani, M.: Public-private partnerships (PPPs) for the realisation of the sustainable development agenda in the built environment. In: The Elgar Companion to the Built Environment and the Sustainable Development Goals, pp. 457–480. Edward Elgar Publishing, Cheltenham (2024)
60. Lou, N., et al.: The mental health benefits of smart cities: evidence from a quasi-natural experiment in China. J. Manag. Eng. **41**(1), 04024065 (2025)
61. Ali, J., et al.: A deep dive into cybersecurity solutions for AI-driven IoT-enabled smart cities in advanced communication networks. Comput. Commun. **229**, 108000 (2025)
62. Yue, H., et al.: Revitalizing urban industrial heritage: enhancing public trust in government through smart city development and open big data analysis using artificial neural network (ANN) modeling. Cities. **156**, 105538 (2025)
63. Zhuo, Z., et al.: Smart cities, smarter land use? Unveiling the efficiency gains from China's digital urban transformation. Ecol. Indic. **171**, 113151 (2025)
64. Mohapatra, H.: The role of 6G in empowering smart cities enabling ubiquitous connectivity and intelligent infrastructure. In: Building Tomorrow's Smart Cities with 6G Infrastructure Technology, pp. 237–264. IGI Global Scientific Publishing, London (2025)
65. Al Aziz, R., et al.: Exploring the challenges to cope with ripple effects in the perishable food supply chain considering recent disruptions: implications for urban supply chain resilience. J. Open Innov.: Technol. Mark. Complex. **11**(1), 100449 (2025)
66. Mrabet, Z., et al.: Do supply chain pressures affect consumer prices in major economies? New evidence from time-varying causality analysis. Econ. Model. **142**, 106914 (2025)
67. Cai, C., et al.: How does climate change affect potential yields of four staple grain crops worldwide by 2030? PLoS One. **19**(5), e0303857 (2024)
68. Orlov, A., et al.: Human heat stress could offset potential economic benefits of CO2 fertilization in crop production under a high-emissions scenario. One Earth. **7**(7), 1250–1265 (2024)
69. Harvey, L.D.: A bottom-up assessment of recent (2016–20) energy use by the global iron and steel industry constrained to match a top-down (international energy agency) assessment. Energy. **293**, 130675 (2024)

70. Nsiama, K.K., et al.: Exhaustive review of CO2 sequestration in depleted hydrocarbon reservoirs: recent advances, challenges and future prospects. Energy Fuel. **38**(22), 21701–21734 (2024)
71. Prévot, M.S., et al.: An anthropocene-framed transdisciplinary dialog at the chemistry-energy nexus. Chem. Sci. **15**(24), 9054–9086 (2024)
72. Işık, C., et al.: A roadmap for sustainable global supply chain distribution: exploring the interplay of ECON-ESG factors, technological advancement and SDGs on natural resources. Resour. Policy. **95**, 105114 (2024)
73. Li, K., et al.: AC loss optimization of high temperature superconducting magnetic energy storage considering energy management strategies in a hydrogen-battery system. J. Energy Storage. **101**, 113728 (2024)
74. Odimarha, A.C., Ayodeji, S.A., Abaku, E.A.: The role of technology in supply chain risk management: innovations and challenges in logistics. Magna Sci. Adv. Res. Rev. **10**(2), 138–145 (2024)
75. De Alwis, N., Nam, H.-S.: A way towards port automation: challenges and implications. WMU J. Marit. Aff. (2024). https://doi.org/10.1007/s13437-024-00350-9
76. Tadić, S., et al.: Assessing the applicability of industry 4.0 technologies in optimizing air cargo terminal operations. J. Organ. Technol. Entrep. **2**(2), 97–112 (2024)
77. Xiong, W., Wu, D.D., Yeung, J.H.: Semiconductor supply chain resilience and disruption: insights, mitigation, and future directions. Int. J. Prod. Res. **63**, 3442–3465 (2024)
78. Singh, K.B., Misra, S.C.: Manufacturing of semiconductor chips: from their origins to the current automotive Chip crisis. IEEE Eng. Manag. Rev. 53(5), 260–273 (2025)
79. Tung, C.-Y.: Taiwan and the global semiconductor supply chain. ROC, Taiwan site, Taiwan Representative Office in Singapore. https://www.roc-taiwan.org/uploads/sites/86/2024/02/240202-February-Issue.pdf (2024)
80. Gouda, S.K., Saranga, H.: Sustainable supply chains for supply chain sustainability: impact of sustainability efforts on supply chain risk. Int. J. Prod. Res. **56**(17), 5820–5835 (2018)
81. Govindarajan, V., Venkatraman, V.: Fusion Strategy: How Real-time Data and AI Will Power the Industrial Future. Harvard Business Press, Brighton (2024)
82. Xu, P., et al.: Blockchain as supply chain technology: considering transparency and security. Int. J. Phys. Distrib. Logist. Manage. **51**(3), 305–324 (2021)
83. Zhu, Q., Duan, Y., Sarkis, J.: Supply chain carbon transparency to consumers via blockchain: does the truth hurt? Int. J. Logist. Manage. **35**(3), 833–864 (2024)
84. Rizqi, Z.U., Chou, S.-Y., Yu, T.H.-K.: Data-driven approach for dwell point positioning in automated storage and retrieval system: a metaheuristic-optimized ensemble learning. Ann. Oper. Res. (2024). https://doi.org/10.1007/s10479-024-06342-5
85. Pessot, E., Zangiacomi, A., Fornasiero, R.: Unboxing the hyper-connected supply chain: a case study in the furniture industry. Prod. Plan. Control. **35**(6), 580–598 (2024)
86. Sodiya, E.O., et al.: AI-driven warehouse automation: a comprehensive review of systems. GSC Adv. Res. Rev. **18**(2), 272–282 (2024)
87. Islam, M.S.: Navigating modern era at sea: legal challenges and opportunities of unmanned and autonomous shipping. AI Ethics. **5**(3), 2293–2306 (2024)
88. Chien, C.-F., et al.: Green production planning for circular supply chain and resource management: an empirical study for high-tech textile dyeing. Resour. Conserv. Recycl. **204**, 107499 (2024)
89. Eslamipoor, R., Sepehriar, A.: Enhancing supply chain relationships in the circular economy: strategies for a green centralized supply chain with deteriorating products. J. Environ. Manag. **367**, 121738 (2024)
90. Zhang, A., Seuring, S.: Digital product passport for sustainable and circular supply chain management: a structured review of use cases. Int. J. Log Res. Appl. **27**(12), 2513–2540 (2024)
91. Hassan, M.: Impact of green supply chain management on the sustainability: a study on textile industry. South Asian J. Oper. Log. **3**(1), 25–38 (2024)

92. Inshal, M., Haq, M.A., Akhtar, P.: Enhancing customer trust and loyalty in the Pakistani food delivery sector through AI-driven blockchain supply chain. In: Achieving Secure and Transparent Supply Chains With Blockchain Technology, pp. 160–188. IGI Global, London (2024)
93. Khan, Y., Ali, Y.: Analysis of water footprint and sustainability of the cotton supply chain in Pakistan. Water Conserv. Sci. Eng. **9**(1), 18 (2024)
94. Mahmood, S., et al.: Sustainable infrastructure, energy projects, and economic growth: mediating role of sustainable supply chain management. Ann. Oper. Res. (2024). https://doi.org/10.1007/s10479-023-05777-6
95. Byrd, M.Y., Scott, C.L.: Diversity in the Workforce: Current Issues and Emerging Trends (2024)
96. Tenakwah, E.S., Watson, C.: Embracing the AI/automation age: preparing your workforce for humans and machines working together. Strateg. Leadersh. **53**(1), 32–48 (2024)
97. Dolle, N., Wilhelm, C., Anikin, K.: Development of a reliable and offline capable hard-and software technology for long-term machine data acquisition, data storing and data exportation. In: International Conference on Reliability and Statistics in Transportation and Communication. Springer, Cham (2022)
98. Słowik, M., Sierocka, H.: Manufacturing equipment retrofitting towards industry 4.0 standards—a systematic overview of the literature. Eng. Manage. Prod. Serv. **15**(3), 14–26 (2023)
99. Varriale, V., et al.: Critical analysis of the impact of artificial intelligence integration with cutting-edge technologies for production systems. J. Intell. Manuf. **36**(1), 61–93 (2025)
100. Chapelin, J., et al.: Data-driven drift detection and diagnosis framework for predictive maintenance of heterogeneous production processes: application to a multiple tapping process. Eng. Appl. Artif. Intell. **139**, 109552 (2025)
101. Mehmood, U., et al.: Transforming high-rise residential maintenance: a 3D spatio-temporal model utilizing industry 4.0 and lean principles. Facilities. **43**(1/2), 32–53 (2025)
102. Khan, R.A., et al.: 5G networks security mitigation model: an ANN-ISM hybrid approach. IEEE Open J. Commun. Soc. **6**, 881–925 (2025)
103. Mazhar, T., et al.: Generative AI, IoT, and blockchain in healthcare: application, issues, and solutions. Discov. Internet Things. **5**(1), 5 (2025)
104. Khan, I.U., et al.: Future Tech Startups and Innovation in the Age of AI. CRC Press, Boca Raton (2024)
105. Akdemir, B., et al.: From technical prerequisites to improved care: distributed edge AI for tomographic imaging. IEEE Access. **13**, 14317–14343 (2025)
106. Maksimović, M., Jokić, S., Bošković, M.Č.: Innovative horizons for sustainable smart energy: exploring the synergy of 5G and digital twin technologies. Process Integr. Optimiz. Sustain. **9**, 431–470 (2025)
107. García, A., De Barrena, T.F., Ferrando, J.L.: AI model generation methodology and software architecture for CPS 2.0 manufacturing systems. In: Cyber Physical System 2.0, pp. 174–196. CRC Press, Boca Raton (2025)

Chapter 10
Next Generation Advanced Intelligent Manufacturing Technologies in Industry 6.0 (Foundation of Industry 7.0)

Abstract Automation, new materials, and intelligent production systems are driving the growth of machinery manufacturing procedures in Industry 6.0. Industrial production has been transformed by the incorporation of cutting-edge technologies, including advanced semiconductor fabrication, cyber-physical systems, additive manufacturing, and artificial intelligence. With advancements like silicon carbide, gallium nitride, and graphene improving thermal and electrical qualities, material selection is essential to production efficiency. Nanomaterials like carbon nanotubes and high-density interconnects enhance circuit design and energy efficiency. AI-driven quality control, digital twins, and cyber-physical systems enable intelligent manufacturing, which maximizes production speed, accuracy, and predictive maintenance. The chapter also examines cutting-edge methods for fabricating semiconductors that allow for downsizing and increased computational efficiency, such as wafer-level packaging and extreme ultraviolet lithography. Systems in package and microfluidic cooling are examples of advanced packaging methods that enhance heat management and boost industrial device dependability. To meet global sustainability targets, sustainable manufacturing is prioritized, incorporating green technologies, renewable energy, and the concepts of the circular economy. Decentralized manufacturing methods, blockchain-based supply chains, and AI-driven predictive analytics all improve efficiency and transparency even further. The chapter also emphasizes how software-defined manufacturing, smart communication modules, and real-time data analytics may maximize industrial automation. Together, these developments help to define Industry 6.0's future and also a foundation of Industry 7.0 by striking a balance between sustainability, intelligent ultra-high automation, and industrial progression.

Keywords Industry 6.0 · Intelligent manufacturing · Cyber-physical systems · Additive manufacturing · Extreme ultraviolet lithography · Smart automation · Wide bandgap semiconductors · Nanomaterials in manufacturing · Digital twins · AI-driven manufacturing · Sustainable industrial processes · Blockchain in supply chains · 5G in industrial automation · Thermal management in manufacturing · Ultra high hyperautomation

H. Majeed, T. Iftikhar, *Intelligent Manufacturing in Industry 6.0*,
https://doi.org/10.1007/978-3-032-07278-8_10

1 Designing and Material Selection

The complex interplay between the materials' chemical structure, electrical characteristics, and physical performance, all of which are adapted to satisfy the needs of contemporary technology, provides the scientific justification for the selection of materials used in the production of edge devices. The strategic design and selection of materials that satisfy the many demands of consumer and industrial applications, particularly for settings needing high precision, energy efficiency, and durability, lay the groundwork to produce edge devices. Two excellent examples of materials chosen for their broad bandgap characteristics are silicon carbide and gallium nitride. These semiconductors can function at far higher voltages, temperatures, and frequencies than conventional silicon because of their broad bandgap, which is roughly 3.26 eV for SiC and 3.4 eV for GaN. This characteristic makes them incredibly effective for high-power AI edge devices used in tough and industrial settings by minimizing leakage currents by lowering thermally produced charge carriers, which facilitates high-frequency operations, high breakdown voltage, low energy loss, and effective thermal management. All these properties are essential for the small and potent processors found in devices. Their crystalline structure, which reduces electron scattering and increases charge mobility, is the source of these characteristics. Their strong thermal conductivity (GaN at around 230 W/m·K and SiC at about 490 W/m·K) guarantees efficient heat dissipation, which is essential for preserving the functionality and durability of small, high-energy devices [1–3].

High-density interconnects enable downsizing and facilitate high-speed data processing on printed circuit boards (PCBs), another essential component. For stiffness and heat resistance, these boards frequently use substrates made of FR-4 epoxy bonded with glass fiber. While surface finishes like Electroless Nickel Immersion Gold offer oxidation protection and a dependable soldering surface, copper layers are added for conductivity. Materials like polyimide (PI) and polyethylene terephthalate (PET) are essential for applications like foldable displays and wearable technology. Because of its aromatic backbone, polyimide, which is produced by polymerizing dianhydrides and diamines, has remarkable flexibility and heat stability (up to 400 °C). PET, a polyester made from terephthalic acid and ethylene glycol, combines chemical resistance, mechanical strength, and transparency, making it appropriate for flexible substrates and circuits. High tensile strength, flexibility, and exceptional thermal stability are provided by the molecular structures of PET, which has ester connections, and PI, which has aromatic imide groups. These polymers are perfect for flexible circuits in wearables and portable edge computing systems because they maintain their integrity under high temperatures and dynamic mechanical forces. High-speed electronic components benefit from their dielectric qualities, which include a dielectric constant of 3.4–3.8 [4–6].

Electronic design and operation are being revolutionized and transformed by nanomaterials, such as graphene and carbon nanotubes (CNTs), which use their nanoscale and quantum mechanical features and opening the door for the development of smaller, more powerful electronics. Because of their tubular arrangement of sp2 hybridized carbon atoms, CNTs have little scattering and exhibit ballistic electron transport, resulting in conductivities as high as 10^6 S/cm. This feature guarantees high-frequency performance and quick signal transfer in small devices. Similarly, graphene's two-dimensional, delocalized π electron system is responsible for its remarkable electron mobility (>200,000 $cm^2/V{\cdot}s$). This makes ultra-low resistance paths possible, which are necessary for sophisticated sensors and high-speed transistors. Graphene (>5000 W/m·K) and carbon nanotubes (around 3000 W/m·K) both have remarkable thermal conductivity, which guarantees effective heat management in closely spaced electronic circuits. Single-walled and multi-walled carbon nanotubes are cylindrical nanostructures made from rolled graphene sheets. The delocalized π-electrons in their sp2 hybridized carbon network, which permit free electron mobility along the nanotube axis, are responsible for their exceptional electrical conductivity. CNTs are perfect for interconnects in PCBs and integrated circuits because of their special electrical structure, which also provides a high current-carrying capacity (up to 1000 times higher than copper) with minimum energy dissipation. With their 3000 W/m·K thermal conductivity, CNTs enable effective heat dissipation in small, high-power devices. With tensile strengths surpassing 100 GPa, CNTs' remarkable mechanical strength is a result of their cylindrical shape, which also makes them dependable and long-lasting in electronic assemblies under stress. The single layer of atoms that make up graphene, a two-dimensional allotrope of carbon, is organized in a hexagonal lattice. The sp2 hybridization of carbon atoms gives it its remarkable characteristics by forming a delocalized π electron system that permits electron mobility of about 200,000 $cm^2/V{\cdot}s$. The speed and efficiency of edge devices are increased by this ultrafast electron transport, which dramatically lowers transistor power loss and signal delay. Because graphene has one of the highest thermal conductivities of any material (i.e., more than 5000 W/m·K), it is an excellent option for controlling heat in high-performance electronic devices. In addition to its conductivity, graphene's mechanical strength and flexibility (Young's modulus of about 1 TPa) enable its incorporation into wearable technology and flexible electronics, increasing its use in edge computing. Chemically speaking, the special bonding and nanoscale dimensions of CNTs and graphene are responsible for their remarkable qualities. In addition to the delocalized π-electrons offering thermal and electrical benefits, the sp2 hybridization guarantees strong in-plane σ-bonds. Their atomic structure lacks grain boundaries, which further reduces electron scattering and improves conductivity and resistance. Transistors, heat spreaders, and PCBs have all changed because of the use of CNTs and graphene in electronic components. CNT-based vias lower resistive losses on PCBs, and components covered with graphene improve heat control and signal integrity. Compact and energy-efficient designs are made possible by these nanomaterials' use as electrodes in supercapacitors and active layers in field effect transistors. Functionalizing these materials with chemical groups enhances their compatibility with current technologies as research advances, guaranteeing a smooth transition into next-generation edge devices. Since their atomic and molecular characteristics directly address the issues of edge device performance, miniaturization, and energy efficiency, these materials are at the forefront of innovation [7–13]. To meet the exacting requirements of Industry 5.0 and 6.0, their chemical compositions and structural configurations offer an unparalleled mix of electrical, thermal, and mechanical qualities that are not possible with traditional materials.

2 Integration of Intelligent Manufacturing Systems for Precise Technologies

In the context of Industry 6.0, where super automation and networked ecosystems characterize manufacturing, the integration of intelligent manufacturing systems has emerged as a key component in the creation of cutting-edge devices. CPS, which seamlessly combines digital intelligence and physical manufacturing processes to produce highly adaptive and efficient production environments, is at the heart of this shift. The creation of edge devices has changed because of precision manufacturing combined with additive technologies, especially Additive Manufacturing (3D Printing), which allows for complex, small, and effective designs while reducing waste and increasing production speed. The technology makes use of layer-by-layer fabrication, which makes it easier to create highly tailored geometries that are frequently unattainable with conventional production techniques. An important additive method used in the production of edge devices is SLS, which uses powdered materials like metals (titanium, aluminum, etc.) and high-performance polymers (like polyamide and PEEK). The method uses a high-energy laser to selectively sinter powdered particles, producing long-lasting parts with exceptional thermal stability and mechanical strength. PEEK's semi-crystalline structure and high melting point (343 °C approximately) make it perfect for creating sturdy yet lightweight housings for edge devices used in demanding industrial settings. In a similar vein, heat sinks frequently use aluminum alloys with superior thermal conductivity (around 205 W/m·K) to effectively drain heat from processors [14–16].

By making it possible to fabricate parts that blend conductive and insulating materials in a single build, multi-material 3D printing introduces a level of complexity and creativity. This method incorporates conductive pathways (graphene-based composites or paints made of silver nanoparticles) directly into device casings. This technique greatly lowers space requirements by doing away with the necessity for separate PCBs, opening the door for ultra-compact edge devices. These materials' chemical characteristics, including graphene's delocalized π-electrons, improve electrical conductivity while maintaining mechanical flexibility. There are several benefits to additive manufacturing from a sustainability standpoint. Compared to subtractive techniques like CNC milling, which naturally produce surplus material, numerical simulations show that 3D printing lowers material waste by about 40%. Because digital design to fabrication workflows allow for rapid iteration, they can reduce prototyping cycles by as much as 60%. For edge device innovation to advance, especially in rapidly changing domains like autonomous systems and the IoT, this acceleration is essential. The capacity of additive manufacturing to utilize material attributes and process accuracy provides the scientific justification for its use. 3D printing's regulated layer deposition guarantees consistent microstructures, reducing flaws, and boosting the robustness of manufactured parts. SLS and other cutting-edge methods enable near-net-shape fabrication, which lessens the requirement for intensive post-processing. Manufacturing companies may produce cutting-edge gadgets that satisfy the exacting requirements of Industry 5.0 and 6.0, such as compactness, energy efficiency, and quick scalability, by utilizing the chemistry of high-performance materials and the accuracy of additive processes [17–19].

The utilization of digital twins, or virtual versions of actual production lines, is a key component of CPS. Through the integration of sensor data, computational models, and machine learning techniques, digital twins enable real-time simulation and optimization. By utilizing these simulations, manufacturers may reduce waste and time-to-market by anticipating bottlenecks, optimizing production settings, and increasing throughput without the need for actual trial-and-error. In the process of assembling edge devices, robotic automation is essential. With the help of sophisticated motion control systems and force sensors, collaborative robots, or cobots, can perform complex tasks like soldering surface-mounted components onto high-density PCBs with nanoscale accuracy. The ability of these cobots to adjust to changing assembly lines and collaborate with people in a safe and effective manner is essential for expanding the manufacturing of small and intricate edge devices. Inspection procedures have been completely transformed by AI-driven QC. High-resolution cameras and deep learning models are used in machine vision systems to examine components at the micrometer scale for flaws like misaligned solder connections or tiny surface irregularities. This degree of accuracy offers almost zero defect rates, which is essential for edge devices that need to function dependably in mission-critical applications like industrial automation and driverless cars. By identifying irregularities and facilitating prompt remedial measures, these systems can analyse inspection results in real time, reducing production delays [20–22].

Another essential component of CPS is data analytics, which is driven by algorithms for predictive maintenance. Production machinery has sensors that continuously gather information on variables like temperature, vibration, and operating cycles. Potential equipment failures are anticipated before they happen thanks to advanced analytics, such as anomaly detection and time series forecasting. This greatly increases equipment lifespan and production efficiency by cutting downtime by up to 30%. To ensure prompt intervention without stopping production, predictive models can identify an imminent failure in a pick-and-place machine used for PCB assembly. The technologies and materials that facilitate CPS integration are equally important from a scientific standpoint. Real-time monitoring and control are made possible by smart sensors that use piezoelectric and MEMS technology to precisely translate physical phenomena into digital data. Large datasets from IoT-enabled production lines are used to train AI models, which use reinforcement learning to continuously increase their accuracy and adaptability. These systems frequently use edge AI to process data locally, guaranteeing low-latency decision-making and lowering reliance on centralized cloud resources. Intelligent manufacturing system integration is a prime example of how advanced robotics, artificial intelligence, and real-time analytics work together to produce high-performance edge devices. In addition to improving productivity and quality, this strategy fits in well with Industry 6.0's tenets, which place a premium on intelligence, agility, and sustainability [23, 24].

3 Advanced Semiconductor Fabrication

Since the efficiency and capacities of edge devices are closely linked to the sophistication of the microprocessors and integrated circuits (ICs) they use, advanced semiconductor fabrication is essential to their design and functionality. This field is undergoing a revolution thanks to advanced manufacturing processes like Wafer Level Packaging (WLP) and Extreme Ultraviolet (EUV) Lithography, which preserve tiny form factors while enabling previously unheard of levels of energy efficiency and processing power. With current capabilities, transistors as small as 2 nm node can be produced, a revolutionary method for fabricating semiconductors at the nanoscale. To create billions of transistors on a single chip, EUV uses light with a wavelength of 13.5 nm to etch complex patterns into silicon wafers. In addition to increasing computational density, the smaller transistor sizes made possible by EUV also lower circuit resistance and capacitance, which results in lower power consumption and quicker switching rates. These developments are essential for enabling edge computing applications that require processors to perform trillions of operations per second (TOPS), including real-time AI processing. Processors that use EUV can reach computing rates of over 10 TOPS, allowing for quick data analysis for IoT networks and driverless cars [25–27].

WLP, which combines several functions into a single chip package, improves edge device performance even more. WLP reduces the interconnect length and related signal losses by doing away with the requirement for separate dies and substrates, in contrast to traditional packaging. WLP is a nice technology for edge devices functioning in data-intensive situations because of its integration, which enables increased bandwidth, reduced latency, and improved thermal management. WLP-produced System on Chip architectures can combine CPU, GPU, and memory components in a single package, significantly increasing processing efficiency and reducing size. Quantum mechanics and material science provide the scientific justification for these developments. Photoresists, which are specialized polymers that respond to EUV radiation to produce precise etching patterns, are essential to the success of EUV lithography. These resists need to be extremely sensitive to short-wavelength light while remaining chemically stable when exposed to high energy levels. Like this, the materials used in WLP, like low-k dielectric materials to lower capacitance and copper pillars for interconnects, are designed to facilitate both thermal dissipation and high-speed data transfer [28–30].

These methods significantly increase the performance-per watt metric, which is a crucial parameter for edge devices where power constraints are frequently substantial, from a numerical standpoint. The sustainability of edge technology is directly impacted by semiconductors made using EUV, which use less energy per computation than their predecessors. WLP can cut device footprints as much as 40%, making it possible to develop small systems for Industry 6.0 uses like intelligent urban infrastructure and hyper-automated manufacturing. These methods also contribute to the larger shift towards intelligent and energy-efficient systems, which are essential for contemporary technological ecosystems [31, 32].

3.1 Extreme Ultraviolet Lithography

Extreme ultraviolet lithography is a highly advanced nanofabrication technology used in semiconductor manufacturing, particularly in the production of next-generation microchips with transistor sizes below five nanometers. In the context of Industry 6.0, extreme ultraviolet lithography integrates artificial intelligence, quantum computing, and autonomous robotics to enhance precision, scalability, and defect minimization. The system operates by generating extreme ultraviolet light with a wavelength of approximately thirteen point five nanometers through the high-energy plasma excitation of tin droplets, which is then focused using multilayer mirrors with near-perfect reflectivity. The machine employs advanced metrology sensors, edge computing, and real-time artificial intelligence-driven process control to optimize patterning accuracy and reduce defects. Industry 6.0 enhancements incorporate intelligent digital twin simulations, self-correcting exposure systems, and fully autonomous maintenance, ensuring minimal downtime and maximum throughput. With the convergence of photonic computing, superconducting materials, and smart automation, extreme ultraviolet lithography machines in Industry 6.0 can produce semiconductors with unprecedented energy efficiency, ultra-high transistor density, and enhanced quantum computing compatibility, making them critical for the future of artificial intelligence-driven intelligent manufacturing and next-generation computing architectures [33–35]. Here is the Fig. 10.1 to brief about this technology.

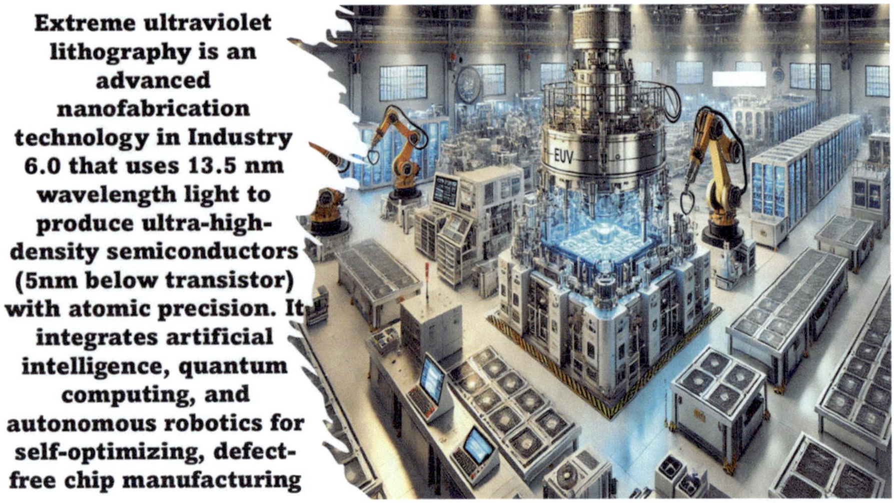

Fig. 10.1 Extreme ultraviolet (EUV) lithography brief

4 Advanced Packaging and Thermal Management

The performance and dependability of edge devices depend heavily on advanced packaging and thermal management, especially as processing power rises and form factors continue to decrease. These developments tackle the dual problems of fitting several uses into small areas and controlling the large amount of heat produced when in use. The System in Package (SiP) concept, which combines CPUs, memory modules, power management units, and occasionally sensors into a single small box, is an important advancement in packaging. SiP technology, in contrast to conventional system on chip designs, makes use of sophisticated interconnects like flip chip bonding and Through Silicon Vias (TSVs) to guarantee high-speed communication between components while reducing power loss and signal interference. For example, SiP enables the integration of high-bandwidth memory and a CPU with 21 TOPS in less than 10 mm^2 of board area, which is mandatory for devices such as the NVIDIA Jetson Xavier NX. The combination of phase change materials (PCMs) and microfluidic cooling systems has led to dramatic advancements in thermal management, a crucial component of edge device performance. To stabilize the temperature of high-performance components, PCMs, which are usually made of salt hydrates or organic compounds based on paraffin, work by absorbing heat when they change from solid to liquid. These materials are capable of handling thermal loads of up to 100 W/cm^2, which guarantees reliable performance in high-speed computing applications and edge AI systems [36, 37].

By circulating coolant through microscale channels carved into device surfaces, microfluidic cooling systems further improve thermal efficiency. These systems reduce hotspots that might impair component reliability by providing precise, localized cooling. Like data-intensive edge computing nodes, these systems can lower operating temperatures by as much as 20% when compared to traditional heat sinks, preserving computational accuracy and prolonging device life. These breakthroughs' underlying material science is similarly important. Significant energy absorption during phase transition is guaranteed for PCMs due to their high latent heat of fusion. To improve heat conductivity, microfluidic systems frequently employ coolants based on ethylene glycol or sophisticated nanofluid combinations (such as water combined with aluminum oxide nanoparticles). Like this, effective heat transfer from interior components to exterior cooling solutions is made possible using low-thermal-resistance polymers in packaging substrates. Advanced packaging enhances environmental resilience and electromagnetic shielding in addition to heat control. Copper molybdenum (Cu Mo) composites, for instance, provide remarkable thermal conductivity (around 200 W/m·K) in SiP designs while reducing electromagnetic interference, guaranteeing steady performance in loud industrial settings. Modern heat management techniques combined with SiP technology guarantee that edge devices can withstand intense workloads in challenging conditions without experiencing performance deterioration. These developments are propelling edge computing into Industry 6.0 by making it possible for a new generation of small, high power devices that are essential for applications ranging from distributed AI systems to autonomous robots [38–40].

5 Alignment of Edge Devices Manufacturing Technologies with SDGs (Industry 6.0 Approach)

A key component of Industry 6.0 is sustainability in manufacturing, which combines cutting-edge technology with environmental stewardship to support global objectives like the SDGs and the circular economy principles. To meet the technological needs of contemporary businesses while minimizing the environmental impact, the manufacture of edge devices must embrace new and sustainable approaches. Green manufacturing greatly lowers carbon footprints by using renewable energy sources like solar, wind, and green hydrogen to power industrial facilities (aligned with SDG 7 and 13). Leading semiconductor foundries, for example, have switched to deploying energy storage devices and photovoltaic arrays to offset their electricity consumption, resulting in yearly energy savings of 30–50%. Cutting-edge energy management systems powered by AI-based predictive analytics maximize energy use and guarantee that manufacturing processes waste as little as possible [41, 42].

A circular economy relies less on virgin materials and conserves limited resources through recycling and material recovery methods. Using hydrometallurgical and pyrometallurgical methods, rare earth elements such as neodymium, europium, and yttrium, which are necessary for magnets, screens, and sensors, are recovered. By using solvent extraction techniques, urban mining from e-waste streams produces 90–95% pure neodymium, supporting the goals of SDG 12. Also, precise sorting and separation of valuable components from rejected edge devices is made possible by sophisticated robotic disassembly systems that are outfitted with computer vision and artificial intelligence [43]. Advanced technologies like dynamic voltage and frequency scaling, which modifies power consumption according to computational demand and lower energy consumption by 30–50% in AI-driven edge processors, are key to achieving energy efficiency in edge devices. Power efficiency has been further improved by the switch to wide-bandgap semiconductors like silicon carbide and gallium nitride, which reduce losses by 50–70% in high-performance applications. This supports sustainable industrialization and is in line with SDG 9. From the standpoint of material sustainability, the ecological footprint is decreased when bio-based polymers and recyclable materials, like recycled PET or polylactic acid, are used for device enclosures. In addition to meeting durability requirements, these materials biodegrade or re-enter the recycling cycle without producing trash over time. Graphene based coatings improve thermal management and lower material volume needs in device manufacture, demonstrating how nanotechnology is playing a significant role in sustainability [44–46].

Sustainability is further improved by integrating Industry 6.0 technologies like blockchain, digital twins, and CPS. Blockchain-based supply chain transparency guarantees ethical raw material sourcing, while the digital twins model production processes to reduce resource waste and maximize energy use. By preventing abuse of machinery, predictive maintenance with IoT-enabled edge devices reduces downtime and wasteful energy use by 20–30%. Manufacturers may lessen their environmental effect and gain a competitive edge in a market that is becoming more environmentally sensitive by incorporating sustainability into the design, manufacturing, and lifecycle management of edge devices. Under Industry 6.0, these practices are driving the progress toward a sustainable future by demonstrating the link between technology innovation and environmental responsibility [47, 48].

6 Advanced Communication Modules Integration

The smooth functioning of edge devices in Industry 6.0, where hyperautomation, smart manufacturing, and real-time decision-making are powered by ultra-connectivity and intelligent systems, depends on the integration of advanced communication modules. By facilitating effective data interchange, these modules allow edge devices to serve as essential nodes in autonomous networks and CPS. The foundation for real-time industrial applications is provided by the integration of 5G and Beyond (6G, etc.) communication technologies. With data transfer speeds of up to 10 Gbps, latency as low as 1 ms, and capability for more than 1 million connected devices per square kilometer, 5G networks offer connections that are unmatched. These features are critical for jobs like remote robotic control in Industry 6.0 factories, augmented reality-assisted manufacturing, and autonomous vehicle coordination. By providing terahertz frequency bands for data transmission rates more than 1 Tbps and sub-millisecond latency, 6G networks, which are anticipated to be deployed by the 2030s, will outperform 5G and enable previously unheard of levels of real-time communication, and highly accurate industrial automation. Edge devices can be easily integrated into IoT ecosystems thanks to IoT-specific protocols, including Thread, Constrained Application Protocol, and MQTT. In contexts with limited resources, MQTT's lightweight and effective publish-subscribe approach makes it easier for sensors and edge nodes to communicate data. In applications such as environmental monitoring and smart energy grids, CoAP, which is tailored for restricted networks, guarantees dependable communication. In the meantime, Thread, which is based on IPv6, improves device security with interoperability in mesh networks, making it perfect for home automation and smart city systems [49–51].

Multi-radio modules are integrated into edge devices to maximize connection, supporting cellular communication as well as Wi-Fi 6, Bluetooth Low Energy 5.2, and Zigbee. With 9.6 Gbps speeds and improved network efficiency, Wi-Fi 6 is essential for high-bandwidth applications such as industrial vision systems. Zigbee guarantees reliable, low-power communication in crowded industrial settings, whereas BLE 5.2 enables low-power, short-range communication for wearable and mobile edge devices. The transformational influence of these technologies is shown by numerical findings. Compared to 4G systems, 5 G-enabled edge devices may send real-time data from hundreds of sensors on a manufacturing line with up to 90% less delay. For time-sensitive tasks like robotic surgical systems or autonomous car navigation, this low latency is essential. Like this, MQTT and other IoT protocols can optimize bandwidth consumption in large-scale IoT deployments by reducing data packet sizes by 40–60%. By building intelligent, networked systems with the ability to make decisions on their own, the incorporation of cutting-edge communication modules into edge devices is consistent with the tenets of Industry 6.0. In the upcoming industrial revolution, these gadgets will drive efficiency, sustainability, and creativity as the foundation of smart factories, autonomous logistics networks, and adaptive energy grids [52–54].

7 Software-Defined Manufacturing

Intelligent edge devices can be produced with unparalleled efficiency, flexibility, and precision thanks to the revolutionary approach known as "Software Defined Manufacturing," which incorporates state-of-the-art software solutions into every aspect of the production process. By ensuring that every phase of the production lifecycle, from design and material selection to fabrication and assembly, is optimized for performance and cost effectiveness, AI-driven software solutions enable quick iterations and ongoing manufacturing system improvement. Using CAD tools to produce precise, detailed models of edge device components is a fundamental aspect of software-defined manufacturing. Before actual production starts, complicated geometries and material behaviors can be simulated using modern CAD tools like AutoCAD and SolidWorks. To ensure structural integrity and performance optimization, engineers can forecast how materials will react to physical stresses, thermal loads, and electrical fields during device operation by incorporating Finite

Element Analysis into CAD tools. AI generative design algorithms are also being included into CAD systems more and more. These algorithms produce numerous design iterations according to predetermined functional and manufacturing limitations, enhancing innovation cycles and lowering the possibility of design errors. Together with CAD, ML algorithms are transforming manufacturing and material science by forecasting how materials will behave in various scenarios. ML models may determine the best materials for device tasks, such as heat dissipation or electrical conductivity, by examining enormous datasets obtained from sensor inputs, historical production data, and material performance records. Manufacturers can choose materials with confidence thanks to these algorithms, which can forecast the behavior of materials like graphene, carbon nanotubes, and high-tech polymers both during production and under operating stress [55–57].

Similarly, by optimizing manufacturing factors like temperature, pressure, and cooling rates, these algorithms help lessen flaws like cracking, warping, and dimensional irregularities. ML can be used to improve the overall quality and performance of additive manufacturing (3D printing) processes by adjusting printing rates, layer deposition patterns, and thermal management in real time. Another cornerstone of software-defined manufacturing is the integration of Smart Enterprise Resource Planning systems, which guarantees that production procedures are in line with demand worldwide. Real-time data analytics and predictive algorithms are used by contemporary ERP systems, such as SA4HANA and Microsoft Dynamics 365, to effectively manage resources, track inventory, and improve supply chains. Smart ERP systems can cut lead times by up to 25% by utilizing production simulation and demand forecasting. This allows for just-in-time production and minimizes stockouts or overproduction. Manufacturers can react quickly to changes in the market thanks to this dynamic approach, which optimizes production schedules, inventory control, and logistics to satisfy customer demand without sacrificing quality or cost. Software-defined manufacturing's integration of AI, CAD, machine learning, and smart ERP systems improves edge device production's precision, speed, and adaptability while also facilitating a move toward Industry 6.0 concepts like hyperautomation, adaptive manufacturing, and sustainable practices. The next generation of intelligent, networked industrial systems is being driven by the application of AI to design, material selection, and manufacturing process optimization, which eventually leads to reduced costs, enhanced quality, and rapid innovation [58–60].

8 Alignment of Manufacturing Edge Devices with Industry 6.0 Approach

A significant shift in the manufacturing of edge devices is represented by alignment with Industry 6.0 Vision, which places a strong emphasis on the development of autonomous, self-optimizing systems that operate effectively in decentralized, intelligent ecosystems. Devices like the Raspberry Pi, Arduino, and NVIDIA Jetson are essential to this goal because they are made in a setting where human supervision and intelligent machinery cooperate to maximize production workflows, lower costs, and enhance quality. Decentralized production, in which production plants are positioned strategically near end consumers, is one of the fundamental tenets of Industry 6.0. This strategy drastically lowers the carbon footprint and logistics expenses related to long-distance transportation. Manufacturers can improve supply chain resilience, reduce transportation delays, and react faster to changes in demand by leveraging a network of dispersed locations. This concept is especially useful for edge devices, which are essential to autonomous systems and the IoT, as it allows for localized manufacturing, quick prototyping, and increased flexibility to meet local demands. Devices like the Raspberry Pi or Arduino may be manufactured on demand with little waste by utilizing 3D printing and modular design, which supports the circular economy ideas that are essential to the Industry 6.0 vision. By lessening the environmental effect of transportation and facilitating regional innovation hubs, this decentralized strategy supports SDG 09 [61, 62].

Autonomous supply chains are becoming a crucial component of Industry 6.0 manufacturing in conjunction with decentralized production. From the procurement of raw materials to the delivery of the final product, the supply chain process is optimized and secured using technologies like blockchain and AI. To minimize inefficiencies and stockouts, AI algorithms evaluate enormous datasets in real-time, forecast demand trends, track inventory levels, and guarantee ideal material flows. Blockchain integration offers a transparent and safe way to track the origins of goods and materials, guaranteeing adherence to regulations, reducing fraud, and boosting supply chain confidence in general. Manufacturers may provide verifiable guarantees of quality and sustainability by tracking each component of an edge device back to its source, thanks to blockchain's ability to preserve immutable records of transactions. A key component of Industry 6.0's smart manufacturing capabilities is the combination of blockchain and AI, which provides a significant boost in operational efficiency, transparency, and traceability while guaranteeing that the supply chain is flexible and responsive to interruptions. One of the hallmarks of Industry 6.0 is improved human-machine collaboration, where workers are assisted in complicated assembly processes by sophisticated Augmented Reality technologies. AR gives real-time, context-sensitive information straight to the assembly worker using wearable devices or headsets. With the use of this technology, employees can see assembly instructions, solve issues, and get advice without consulting additional paperwork. It has been demonstrated that incorporating AR into manufacturing settings can boost output by as much as 20% since employees can complete tasks more accurately and with fewer mistakes, which lowers rework and boosts overall production efficiency. By emphasizing possible risks and enabling hands-free operation in hazardous situations, like those involving high temperatures or toxic materials, augmented reality products also increase worker safety. These developments in decentralized production, autonomous supply chains, and human-machine collaboration are also propelling the creation of intelligent ecosystems that support Industry 6.0 principles and the SDGs. The goal of developing a more sustainable, effective, and adaptable manufacturing environment is in line with the integration of cutting-edge technologies, including AI, blockchain, AR, and decentralized production models. This will increase global competitiveness while reducing environmental impact and encouraging innovation. In addition to promoting economic growth, this alignment with the Industry 6.0 vision assists manufacturers in tackling global issues such as resource scarcity, climate change, and the requirement for smarter, more efficient infrastructures [63–65].

9 Production of Piezoelectric Sensor Materials

Piezoelectric materials must be developed and customized using sophisticated procedures to maximize their qualities for particular uses. High performance, sensitivity, and durability are attained by combining expertise from materials science, chemistry, engineering, and nanotechnology.

9.1 Methods of Crystal Growth

To produce high-purity, flawless piezoelectric materials, crystal growth is essential. The hydrothermal process is commonly used for quartz. This method allows for the controlled deposition of crystals on a seed substrate by dissolving raw quartz in an aqueous solution at high temperatures and high pressures. Chemical vapor deposition is utilized for sophisticated applications such as nanowire-based piezoelectric sensors. Using this method, vapor phase precursors are deposited onto a substrate to precisely produce nanostructures, allowing for increased sensitivity and downsizing [66, 67].

9.2 Manufacturing of Ceramics

Metal oxides mostly bear elevated temperatures, which helps in ceramic materials (KNN, PZT) synthesis (piezoelectric). Crystallization and densification processes confirmed by the sintering process. To improve the mechanical stability, dielectric constant, and Curie temperature, doping with other metals like bismuth or lanthanum can be used. 3D printing (additive manufacturing) is the best option to make urgent prototypes and complex geometries or samples [68, 69].

9.3 Processing of Polymers

PVDF and other piezoelectric polymers are processed via spin coating, casting, or extrusion. The molecular dipoles are aligned in a single direction after the polymer film is formed by applying a high-field polarization procedure called poling, which greatly increases the molecular dipoles' piezoelectric responsiveness. The use of conductive fillers or nanostructures to increase the sensitivity and robustness of polymer-based piezoelectric devices is one recent development [70–72].

9.4 Nanomaterial Synthesis

For nano-engineered piezoelectric materials, synthesis techniques like sol-gel processing, hydrothermal synthesis, and electrospinning are employed. Sol–gel methods allow the formation of thin films or nanostructures with controlled composition and morphology. Electrospinning is particularly effective for creating piezoelectric nanofibers with high surface area and flexibility. These techniques enable the integration of piezoelectric properties at the nanoscale, which is essential for applications like wearable sensors or MEMS [73].

9.5 Design Composite

The benefits of several different components are combined in composite materials to provide piezoelectric sensors with specific characteristics. Combining polymers with piezoelectric nanoparticles, such as PZT or barium titanate, can improve energy conversion efficiency, flexibility, and sensitivity. Nanoparticles are uniformly distributed throughout the polymer matrix thanks to advanced production techniques like melt mixing, solution casting, and in-situ polymerization. These composites are especially useful for applications that call for flexible and lightweight equipment, including biomedical implants or wearable technology. Piezoelectric materials can be tailored for a variety of industrial and scientific uses, including energy harvesting, robotics, medical diagnostics, and structural health monitoring, by combining these cutting-edge production processes [74, 75].

10 Eco-Friendly Production Techniques for Piezoelectric Sensors

By using eco-friendly production methods that limit waste, use eco-friendly materials, and lower energy usage, piezoelectric sensor manufacturing can become more sustainable. The technical details of sustainable manufacturing techniques are listed here, along with an examination of their effects on the environment.

Resilient hydrothermal technique, which is an eco-friendly strategy to grow crystals. With this method, water can be used as a solvent for crystal formation at high temperatures and pressures. When compared to conventional melt growth techniques, this will use less energy. This removes the need for dangerous chemicals in the synthesis process. Minimal waste production is also possible as closed systems can recycle water, which lessens the impact on the environment.

By using additive manufacturing processes like Direct Ink Writing, 3D printing of piezoelectric ceramics allows for fine control over material deposition, minimizing material waste. Because additive manufacturing powders are recyclable, less raw material extraction is required. Permits the use of less harmful PZT substitutes, such as KNN, which is a lead-free ceramic, minimizes the waste of raw materials, stays away from the use of hazardous solvents, and encourages the transition away from dangerous lead-based products [75].

Another environmentally friendly method for creating polymer-based sensors is polymer piezoelectric manufacturing utilizing eco-friendly polymers. This technique allows us to substitute petroleum-based polymers with cellulose-based materials or bio-derived PVDF. It allows for solvent-free processing techniques like melt extrusion, which lower emissions of volatile organic compounds. Electrospinning with green solvents is an additional safe technique. This technique allows us to produce piezoelectric nanofibers by electrospinning water or ethanol in place of dangerous solvents. This approach encourages the use of biodegradable materials, lowers harmful emissions, and lessens reliance on fossil fuels [76, 77].

Green sol–gel processes can be used to synthesize sensors based on nanomaterials. This process creates thin films of piezoelectric materials using nontoxic precursors and solvents. Compared to traditional procedures, this technology uses lower processing temperatures. The hydrothermal and microwave-assisted synthesis techniques are two more environmentally friendly ways to create nanomaterials. Rapid production of nanostructured piezoelectric materials is made possible by these energy-efficient techniques. Compared to conventional high-temperature sintering, these methods emit less greenhouse gas, require less energy, and produce less chemical waste, making them more environmentally friendly [78, 79].

Reducing carbon footprints can also be achieved using bio-based and recyclable composite design. Reusable polymers and naturally occurring piezoelectric nanoparticles, such as barium titanate, which are produced using environmentally acceptable methods, are mixed to create recyclable composites. For composite piezoelectric sensors, we can also employ bio-based fillers, which are renewable resources like chitin or cellulose nanocrystals. By making it possible to recycle materials and lowering landfill waste, we can encourage the circular economy. Energy-saving techniques include Low Power Alignment and Water-Based Poling. Dipoles in piezoelectric materials can be aligned using low-voltage configurations or water-based technologies, which can lower the total energy used during production. We can reduce energy needs in this way, which will lessen our carbon footprint [80, 81].

10.1 Effects of Sustainable Manufacturing on the Environment

By using creative techniques and material selections, the environmental impact of producing piezoelectric sensors sustainably is greatly decreased. One important effect is the decrease in toxic waste, which is accomplished by substituting nontoxic materials like KNN with lead-based ceramics like PZT. By removing hazardous trash, this shift shields ecosystems from heavy metal bioaccumulation. To further protect the quality of the air and water, green solvents are used in procedures like sol–gel synthesis to reduce the release of volatile organic compounds. Reducing carbon emissions is largely dependent on the use of energy-efficient practices. While additive manufacturing reduces material waste by about 50% when compared to subtractive processes, techniques such as microwave-assisted synthesis, which operate at temperatures lower than conventional sintering, can cut energy use by up to 30%. These developments directly cut greenhouse gas emissions by reducing dependency on fossil fuels. Contributions to the circular economy are also highlighted by sustainable practices, especially when it comes to composite design and material recycling. The need for fresh materials is decreased by the fact that polymers used in piezoelectric composites, for example, can be reused several times without suffering a major loss in functionality. This reduces resource extraction; research shows that including recycling into the manufacturing cycle can result in a 40% decrease in the number of raw materials used. The transition to renewable and bio-derived materials helps to preserve limited natural resources. In line with the global movement for sustainable alternatives, petroleum-derived polymers are being replaced with biopolymers such as polylactic acid and cellulose-based composites. Compliance with environmental objectives is guaranteed by using natural fillers like chitin or barium titanate produced using environmentally friendly processes in place of nonrenewable components. In addition to reducing environmental harm, substituting nontoxic materials for hazardous ones improves worker safety. According to studies, KNN, using lead-free ceramics, can minimize occupational exposure to dangerous substances by up to 90%, enhancing worker health and lessening the regulatory burden on producers. Given that these methods are in line with international sustainable objectives like the SDGs of the UN, the future of sustainable piezoelectric sensor production appears bright. Sustainability will be improved, and creativity will be stimulated by ongoing developments in biodegradable materials and low-energy production techniques. Investigating nanomaterials such as graphene oxide for improved piezoelectricity has the potential to transform energy harvesting applications and offer effective solutions for a variety of industries [82–84].

10.2 Piezoelectric Sensor Customized Production Optimization for Industry 5.0

Industry 5.0 places a strong emphasis on highly personalized production, sustainability, and human-machine collaboration. To optimize sensor performance and customize solutions for applications across industries, this paradigm in piezoelectric sensor research makes use of cutting-edge sciences, engineering specialties, and computational technologies [85].

10.3 Integration of Nanomaterials for Improved Performance

Nanomaterials, including CNTs, graphene, and quantum dots, are added to piezoelectric sensors to enhance their mechanical, thermal, and electrical characteristics. By raising the elastic modulus and fracture toughness, CNTs improve piezoelectricity and increase the robustness of sensors under mechanical stress. Because of their adjustable bandgap characteristics, quantum dots enhance sensitivity and signal-to-noise ratio in high-precision applications like vibration monitoring in aircraft and medical imaging. Sol-gel synthesis, electrospinning, or chemical vapor deposition are methods used to integrate these nanomaterials into the crystal lattice or polymer matrix. By evaluating data on lattice distortions, polarization intensities, and thermal conductivities, ML models can forecast the best material pairings, speeding up the research and development process [86, 87].

10.4 Developments in Flexible Sensors and Energy Harvesting

Piezoelectric sensors are being used more and more in IoT devices for energy harvesting, offering self-powered solutions for industrial automation and remote monitoring in smart cities. Flexible piezoelectric films are used in wearable electronics and robotics; they are frequently composed of PVDF or hybrid polymer-ceramic composites. Because of their flexibility, these films can continue to function in a variety of dynamic situations. AI-powered Finite Element Analysis simulations are utilized to optimize these films' thickness, electrode combinations, and stress distribution. In comparison to traditional rigid sensors, this guarantees maximum energy conversion efficiency, which in next-generation devices can surpass 10% [88, 89].

10.5 Techniques for Repair and Maintenance

Piezoelectric sensor repair calls for material-specific techniques. For ceramic sensors, specialist adhesives containing nanomaterials like silica nanoparticles are used to fix microcracks in ceramics like PZT or $BaTiO_3$ to enhance mechanical bonding and dielectric qualities. To guarantee continuous performance, laser-assisted annealing can also repair the crystalline structure. Re-poling techniques are used for polymer-based sensors to align the dipoles in PVDF-based sensors. This entails applying a strong electric field in a thermally regulated environment. Damaged layers of multi-layered polymer ceramic composites can be replaced using additive manufacturing techniques without having to throw away the entire sensor [90, 91].

10.6 Tools for Computation and Data-Driven Optimization

Computational techniques like this are crucial for the customized manufacturing of piezoelectric sensors. AI-Driven Design Sensor geometries tailored for uses, such as irregular shapes for biomedical implants or high aspect-ratio designs for MEMS-based devices, are created using generative AI algorithms. Applications of Data Science are best for real-time fabrication parameter optimization by predictive models that use extensive sensor performance records, which can reduce faults by as much as 40%. Before real production, optimal performance can be ensured by simulating the behavior of piezoelectric materials under different conditions using virtual reproductions (digital twin) of manufacturing processes [92, 93].

10.7 Industrial Applications

Flexible piezoelectric sensors that can track vital indicators like breathing and heart rate with 98% accuracy are being developed for wearable health monitors. Implanted piezoelectric nanogenerators supply devices like pacemakers with self-sustaining power. Up to 30% less maintenance is required when using quantum dot-enhanced piezoelectric sensors to check the structural integrity of automobiles and airplanes in real time. In Industry 4.0 and Industry 5.0 frameworks, sensors with embedded AI are utilized in predictive maintenance systems to enable autonomous decision-making and decrease factory downtime [94, 95].

10.8 Green Production of Piezoelectric Sensors (Industry 5.0 and Industry 6.0 Approach)

Piezoelectric sensor production is moving toward sustainability by using eco-friendly materials, streamlining operations with cutting-edge technology like AI, and meeting international environmental standards. This comprehensive strategy guarantees that piezoelectric sensors are environmentally responsible and high-performing, opening the door to a more sustainable future.

10.8.1 Integration of Eco-Friendly Materials

One important step in lessening environmental harm is the use of biopolymers and ceramics free of lead. Because of their lead presence, traditional piezoelectric materials like PZT are hazardous to the environment. Bismuth sodium titanate and KNN are substitutes that remove these hazards without compromising functionality. These materials' chemistry uses inherent crystal structure distortions to produce similar piezoelectric coefficients. KNN's perovskite structure makes it possible to precisely dope it with elements like tantalum or lithium, which improves its piezoelectric qualities even more. Biopolymers, like poly lactic acid or polyhydroxyalkanoates, are becoming more popular due to their renewable sourcing and biodegradability. By functionalizing these polymers with nanofillers like graphene oxide or cellulose nanocrystals, they improve their mechanical and electrical performance, making them suitable for use in energy harvesting, flexible electronics, and wearable technology [96, 97].

10.8.2 AI-Powered Energy Efficiency

To maximize the energy consumption of piezoelectric sensor production, AI is essential. AI methods like supervised learning, DL and RL are used to forecast how materials will behave under various process parameters, including the sintering temperature or the strength of the electric field during poling (static manufacturing). Reduce material waste by up to 30% by automating quality control through real-time flaw identification in polymer films and ceramics. Optimize resource allocation to ensure that fabrication processes like extrusion and sol–gel deposition use the least amount of energy. AI can simulate energy flow and spot inefficiencies, for instance, by employing digital twins, which are virtual representations of manufacturing processes. Predictive maintenance methods reduce machine downtime, which improves sustainability even more [98, 99].

10.8.3 Alignment with Green Technologies Worldwide

Piezoelectric sensor manufacture is now in line with global sustainability objectives, such as the SDGs of the UN, thanks to the shift to green manufacturing. One of the main effects is the decrease in carbon footprint as energy consumption is reduced by 20 to 30% when microwave-assisted synthesis is used in place of high-energy processes like conventional ceramic sintering. By recovering valuable materials from production waste, recycling and waste management are implemented as circular economy concepts. Scrap ceramics, for instance, can be recycled into building materials, while waste biopolymers can biodegrade with little harm to the environment [100, 101].

10.8.4 Combining Sustainable and Computational Methods

The industry 5.0 and 6.0 strategies, which stress cooperation between intelligent systems and humans, are best exemplified by the combination of sustainable materials and cutting-edge computational tools. One of the most important elements is sustainable additive manufacturing, which reduces excess inventory and related waste by enabling on-demand production through 3D printing using bio-based or lead-free piezoelectric materials. Data Science for Material Innovation uses ML and big data analytics to find and develop material compositions that balance environmental effect, cost, and performance. Finally, Lifecycle Assessment is used to assess the environmental impact of piezoelectric sensors during the stages of production, use, and disposal [102, 103].

10.8.5 An Innovation Benchmark

By combining green production practices, computational innovations, and sustainable materials, a new benchmark for innovation in piezoelectric sensor technology is set. In addition to meeting urgent industrial needs, the emphasis on environmentally friendly techniques promotes long-term environmental stewardship. By using these techniques, producers may create piezoelectric sensors that support Industry 5.0, 6.0, and beyond while also advancing a more robust and sustainable technology future.

10.8.6 Future Recommendation for Industry 6.0

To provide smart sensing and real-time data processing, piezoelectric sensors' future depends on combining artificial intelligence with IoT frameworks. With their unmatched sensitivity and versatility, nano-enabled piezoelectric devices are predicted to transform industries like robotics, aircraft, and healthcare. Piezoelectric sensors are a key component of both current and upcoming technological advancements due to their capacity to combine mechanical accuracy with electrical output [104, 105].

11 Methods of Manufacturing MEMS Sensors

The process of bulk micromachining involves etching silicon wafers to create three-dimensional microstructures. Dry (plasma) and wet etching (KOH) techniques are used for the bulk micromachining process. The process of surface micromachining involves etching and depositing thin layers onto silicon wafers. This technique required silicon nitride and polysilicon. The process of lithography requires sensor geometries that are defined by photolithographic patterning. New developments include the use of EUV lithography for increased accuracy. Wafer bonding is the process of joining many wafers to form intricate structures. Fusion bonding and anodic bonding techniques are used in the wafer bonding process. Due to its excellent electrical conductivity and mechanical strength, silicon is frequently utilized as the primary substrate in these kinds of sensors. Silicon is suitable for its use due to its excellent mechanical properties (high Young's modulus around 130 to 180 GPa, high strength-to-weight ratio) and the vast, mature ecosystem of micromachining techniques derived from the semiconductor industry. Silicon has also high melting temperature (near 1414 °C). The main function of some piezoelectric materials, such as zinc oxide and PZT, is to facilitate the creation of electrical signals through mechanical deformation. Some polymers, such as PDMS and SU-8, are also used in MEMS sensors because of their biocompatibility and flexibility. Because some metals (gold, aluminum, etc.) have reflective surfaces and conductive pathways for optical MEMS, are also utilized in MEMS sensors. The piezoelectric effect of MEMS sensors is also used in resonators and vibration sensors. In these sensors, polarization brought on by stress produces electrical signals. Accelerometers and pressure sensors are examples of devices that display capacitance variation. They modify the measured capacitance by changing the spacing between the plates. MEMS sensors, such as infrared sensors, which use thermal effects to detect changes in temperature, also experience thermal expansion. MEMS sensors have advantages in contemporary systems as well. Compact and portable electronics are made possible by the miniaturization process, having high sensitivity characteristics to identify even the smallest changes in environmental conditions, usually in the 1 μm/mm size range. The latest development's low power consumption makes it environmentally friendly and appropriate for IoT devices that run on batteries. They can also be produced in large quantities since scalable production lowers the cost per unit. MEMS sensors have certain issues that must be resolved to make them better. Fabrication complexity because of its multi-step, highly precise processes. Because they are prone to wear and strain from extended use, their durability is also a concern. Because environmental sealing is essential for dependability, packaging is particularly difficult. Inquire about its cross-sensitivity as well, since it might react to unexpected environmental stimuli. If we discuss its cost and sustainability, we can get a sense of it. Basic MEMS sensors (like accelerometers) cost approximately USD 1–5 per unit, whereas advanced sensors (like gyroscopes) cost about USD 10–50 per unit. Low energy production and environmentally friendly materials lessen their influence on the environment, and silicon wafer recycling increases resource efficiency. MEMS sensors can be employed in implantable devices for continuous glucose monitoring in the healthcare sector. They can also be utilized for inertial navigation systems in the aerospace sector. These sensors can also be used to detect water contamination and the air quality index. MEMS sensors can be used in tire pressure monitoring devices and airbag systems in the automotive sector [106–108].

12 Manufacturing Technologies IoT-Enabled Machines

12.1 Development of Hardware

The efficiency of IoT-enabled equipment is largely dependent on their hardware. This entails choosing the right sensors, constructing circuit boards, and making sure they are durable in demanding industrial settings.

12.1.1 Circuit Boards

Advanced CAD software such as KiCAD or Altium Designer is used to design custom printed circuit boards, or PCBs. To facilitate a smooth connection between devices in industrial ecosystems, these boards combine microcontrollers, sensors, and communication modules (Wi-Fi, Bluetooth, Zigbee, or LoRaWAN). Low-power, high-performance microcontrollers, like the Raspberry Pi or ARM Cortex-M series, are frequently used in modern IoT systems to reduce power consumption without sacrificing computing capability. For industrial automation to be precise, real-time data collecting and control are made possible by these circuit boards [109–111].

12.1.2 Integration of Sensors

The "eyes and ears" of IoT-enabled devices are sensors. Integrating specialized sensors is essential for gathering information about how machinery is operating. For instance, vibration sensors (such as piezoelectric or MEMS accelerometers) are used to track machine conditions in real time for predictive maintenance, which lowers unscheduled downtime. IoT-enabled manufacturing machines use motion sensors to automate assembly line control and sensors such as temperature, humidity, pressure, and gas sensors to monitor the environment. With protective enclosures composed of durable materials like composite plastics or stainless steel, modern sensors are built to survive dust, temperature changes, and vibrations in industrial settings [112, 113].

12.1.3 Sturdy Enclosures

Machines in tough industrial environments are subjected to physical shock, dust, moisture, and temperature extremes. As a result, creating sturdy electrical enclosures is essential. These IoT devices are usually housed in IP-rated (Ingress Protection) housings, which guarantee that they are impervious to moisture and dust. For durability and weather resistance, advanced materials like thermoplastic elastomers, polycarbonate, and aluminium alloy casings are used, guaranteeing that the machines can operate efficiently in settings like factories, warehouses, or outdoor areas [114, 115].

12.2 Development of Software

For IoT-enabled devices, software development is just as important as hardware development. For automation and decision-making, the software makes sure that data is effectively collected, analyzed, and sent to the appropriate parties or systems.

12.2.1 Embedded Programming

Low-latency data processing languages like C, C++, or Python are used to create firmware for IoT-enabled devices. Controlling the collection of sensor data, deciphering signals, and interacting with communication modules are all handled by this embedded program. Task execution can be made predictable and dependable by using a real-time operating system. Real-time programming is essential given the growing need for low-latency systems, particularly for robotics and automation tasks that must be completed quickly [116–118].

12.2.2 IoT Protocol Implementation

Lightweight, low-overhead protocols such as MQTT or CoAP are used to manage communication between IoT-enabled devices. These protocols were created especially for IoT settings, where real-time communication is required, and bandwidth is frequently constrained. MQTT enables devices to effectively publish sensor data to brokers, which are subsequently sent to cloud systems or edge devices, making it perfect for low-bandwidth, high-latency environments such as production floors. CoAP, on the other hand, is utilized for devices with limited power in settings like industrial sensors, where it is necessary to decrease power usage [119, 120].

12.2.3 Cloud APIs

Scalable data analytics for IoT systems is made possible in large part by the cloud. IoT devices are frequently connected to cloud platforms using RESTful APIs so that data can be collected, processed, and stored there. Platforms for managing millions of linked devices, carrying out big data analytics, and putting machine learning models into practice to anticipate equipment breakdowns or streamline workflows are provided by cloud services like AWS IoT, Microsoft Azure IoT, and Google Cloud IoT. The cloud makes IoT data globally accessible, facilitating remote monitoring and real-time decision-making [121, 122].

12.3 System Integration

Connecting IoT-enabled devices to larger industrial systems and making sure they function as a unit is known as system integration in the context of IoT. To guarantee that IoT systems can communicate with current machinery and control systems in an efficient manner, the integration procedure needs to be carefully planned.

12.3.1 Edge Devices

By integrating edge devices into IoT systems, latency and reliance on centralized cloud computing are decreased by locally preprocessing data. In automation and manufacturing systems where real-time processing is required, this is essential. Without waiting for cloud-based calculations, edge computing allows devices to interpret sensor data locally and act instantly. Based on the real-time analysis of vibration data, an edge device might, for instance, instantly stop a broken motor to save the machine. This lowers the cost of cloud infrastructure while also increasing efficiency [123].

12.3.2 Industrial Robots

By incorporating IoT modules that enable them to modify their behaviors dynamically in response to sensor inputs, industrial robots are growing more intelligent. For instance, real-time task allocation is made possible by the connection of robotic arms in the car production industry to IoT platforms. This makes it possible to have flexible production lines with robots that can swiftly adjust to modifications in manufacturing requirements or product design. AI-enabled IoT systems allow robots to operate as efficiently as possible, increasing output and lowering maintenance costs by minimizing wear and tear on equipment [124].

12.3.3 Control Systems

Industrial automation relies heavily on SCADA systems and PLCs. Improved automation and monitoring capabilities are made possible by IoT integration with these systems. IoT-connected sensors, for instance, track chemical concentrations in a chemical manufacturing facility and transmit real-time data to SCADA systems. In turn, the PLCs modify pressure levels and flow rates as necessary to keep the production process in the proper balance. This degree of automation maximizes energy use, minimizes waste, and enhances product quality [125, 126].

13 Water Jet Cutting Machines for High Speed Precision Cutting

Because of its high precision, cold cutting method, and capacity to work with a variety of materials without thermal deformation, water jet cutting technology is quickly becoming a better option than conventional CNC milling, laser cutting, and plasma cutting. Water jet cutting uses a high-pressure stream of water (up to

6205 bar = 90,000 psi), occasionally combined with abrasive particles, to cut through metals, composites, ceramics, textiles, and even heat-sensitive materials with sub-millimeter accuracy. This contrasts with laser or plasma cutting, which depends on extremely high temperatures.

13.1 Cold Cutting Technology: Water Jet Cutting for Precision Manufacturing

Because of its cold cutting mechanism, which removes thermal distortion and guarantees excellent edge finishing, water jet cutting has become a superior precision cutting method in advanced manufacturing. Water jet cutting works at room temperature, avoiding any heat-affected zones, in contrast to laser and plasma cutting, which produce excessive heat that causes material bending, microcracks, and harsh edges that need intensive post-processing. For heat-sensitive materials whose structural integrity must be maintained, such as high-strength ceramics, carbon fiber composites, and aerospace-grade titanium alloys, this is especially important. The capacity of water jet cutting to produce surface roughness values as low as 0.8 μm further improves precision by guaranteeing exceptional edge quality without the need for extra surface polishing. Compared to laser cutting, this reduction in secondary machining procedures leads in a 50% shorter finishing time, boosting output and cost effectiveness. Hardened tool steel, titanium, and Inconel (Nickel-chromium-based superalloys) may be cut with remarkable precision with tolerances of ±0.02 mm using abrasive water jet systems, which use high-pressure water combined with garnet or silicon carbide abrasives. AI-driven motion control systems, real-time sensor-based cutting optimization, and automated nozzle calibration are all used in modern Industry 6.0 integrated water jet cutting, which increases cutting efficiency and reduces material waste by 35%. In the semiconductor, automotive, and aerospace industries, robotic water jet cutting arms with computer vision and machine learning algorithms provide adaptive cutting routes for intricate 3D geometries. Water jet cutting is an essential technology in next-generation smart factories because it is an environmentally friendly, zero-emission cutting option that reduces energy consumption and the production of hazardous waste, which is in line with the SDGs as sustainable manufacturing gains traction [127, 128].

13.2 Precision and Accuracy in Advanced Water Jet Cutting Technology

A high-precision, high-efficiency technique for creating intricate textile designs, technical textiles, composites, and even multi-layered industrial fabrics is water jet cutting. Compared to more conventional cutting techniques like plasma, laser, and

CNC milling, this technology offers more precision and material compatibility because ultra-high-pressure water jets. Water jets eliminate heat distortion and achieve a remarkable cutting tolerance of ±0.025 mm (±0.001 inches), which is like high-end CNC milling and far more precise than plasma cutting (±0.2 mm). The possibilities of water jet cutting are improved by the incorporation of AI-controlled five-axis motion systems, which enable complex 3D geometries, beveling, and taper control at multidirectional cutting angles of up to 60°. This is particularly beneficial in aerospace textiles, automotive interiors, high-performance composites, and technical fabrics requiring zero fiber fraying and perfect edge finishes. Unlike laser or plasma cutting, which can induce thermal degradation, fiber melting, or edge hardening, water jet cutting preserves the structural integrity of fibers, making it ideal for carbon fiber (CFRP), Kevlar, aramid, gel spun ultra high molecular weight polyethylene, polybenzoxazole, and all other high-strength polymer-based textiles. The adoption of AI-driven optimization software ensures that real-time adjustments are made based on fabric type, density, and composite layering, achieving material utilization efficiency [129, 130].

13.3 Material Versatility in Water Jet Cutting (Advanced Manufacturing Capabilities)

Water jet cutting is an ultra-precise, high-pressure machining technology capable of processing a vast range of materials across industries, from aerospace and automotive to textiles and biomedical applications. Unlike conventional thermal cutting techniques such as laser or plasma cutting, which generate localized heat-affected zones and material distortions, water jet cutting is a cold-cutting process, preserving the structural integrity of the material while achieving micron-level precision. A key advantage of water jet cutting.

lies in its material versatility, allowing it to seamlessly process metals, ceramics, composites, polymers, textiles, and even biological materials. Because they can be cut with little edge hardening, high-strength alloys like titanium (Ti-6Al-4 V), Inconel (Inconel 718, 625), stainless steel (316 L, 17-4PH), and aerospace-grade aluminum (7075-T6, 6061-T6) are favored in fields like space exploration, aviation, and defense where accuracy and material integrity are crucial. Water jet cutting is a popular method in the composites sector for CFRP, aramid fiber (Kevlar), and GFRP because it avoids common problems with laser or mechanical cutting, such as delamination, fiber pull-out, and heat degradation. Water jet cutting ensures that industrial ceramics, such as zirconia (ZrO_2), alumina (Al_2O_3), silicon carbide (SiC), and boron carbide (B_4C), have high dimensional accuracy and crack-free edges, making them suitable for uses like semiconductor wafer processing, ballistic armor plating, and biomedical implants. Plastics and thermopolymers, including PTFE (Teflon), PEEK, and HDPE, can be processed without melting, warping, or introducing thermal degradation, ensuring consistent performance in biomedical, food packaging, and chemical processing industries. A major limitation of laser cutting is its inability to efficiently process reflective materials like copper, brass, and aluminum, where laser beams are scattered, reducing efficiency and precision. In contrast, water jet cutting handles all materials uniformly, making it a universal machining solution for complex multi-material components. This capacity also extends to food processing applications, where microbial-safe, water-only jets are utilized to slice meats, vegetables, and baked items in a sanitary, contamination-free manner. Industry 6.0 water jet cutting systems now achieve ±10 μm precision, 40% faster cutting speeds, 30% less material waste, and increased operational efficiency with the combination of AI-driven path optimization, adaptive abrasive flow management, and real-time sensor input. Water jet machining's important in precision manufacturing, space-grade material processing, and high-performance industrial applications will only grow with the development of robotic water jet cutting arms and nano-abrasive water jets [131, 132].

13.4 Environmental Benefits and Waste Reduction in Industry 6.0 Textile Manufacturing

Industry 6.0 textile manufacturing is prioritizing sustainability, resource efficiency, and waste reduction by integrating advanced automation, AI-driven process control, and closed-loop material recovery systems. Unlike traditional plasma and laser cutting, which release toxic NO_x, CO_2, and metal particulates, water jet cutting provides a cleaner and more environmentally friendly alternative. This method uses high-pressure water streams (up to 6000 bar) combined with finely ground abrasives (garnet or aluminum oxide) to cut textiles, composites, and technical fabrics without generating harmful emissions. Since water jet cutting uses recyclable water and abrasive materials, closed-loop filtration systems with membrane nanofiltration and ultrafiltration modules can recover up to 70% of the water. By drastically lowering wastewater discharge, textile processing facilities can avoid chemical contamination. The energy expenses of water jet cutting are 60% lower than those of plasma cutting since it uses just 10–15 kWh/h as opposed to 20–40 kWh/h for plasma-based systems. AI-powered waste management systems are being integrated into Industry 6.0 textile facilities, which use robotics and computer vision to autonomously sort textile trash with 99 percent accuracy according to the fiber composition (cotton, polyester, wool, and blended materials). Enzymatic textile recycling is also emerging as a highly efficient waste recovery method, using biocatalysts to break down polyester into its monomers, achieving up to a 95% recycling efficiency while minimizing microplastic pollution. Carbon capture technology is now being applied in dyeing and finishing plants, where CO_2 emissions from thermal processing and chemical treatments are being converted into usable precursors for biodegradable textile coatings, effectively reducing the industry's carbon footprint by 50%. These advancements in textile waste management, energy efficiency, and pollution reduction align with SDGs, ensuring a climate-resilient, waste-free future for the textile sector [133, 134].

13.5 Automation in Industry 6.0 Integration in Textile and Composite Manufacturing

Industry 6.0 is transforming textile, aerospace, and automotive manufacturing through AI-driven automation, IoT-enabled process monitoring, and precision robotic machining. AI-powered cutting path optimization minimizes material waste by 30%, enhancing sustainability and cost efficiency in fabric and composite processing. Traditional manual cutting processes suffer from material inconsistencies and high defect rates, whereas machine learning algorithms analyze fabric elasticity, fiber orientation, and tensile properties in real time to optimize cut layouts. Advanced nesting software, integrated with deep learning-based defect detection, ensures maximum fabric utilization, reducing scrap generation and improving overall production efficiency. IoT-enabled remote monitoring systems provide real-time feedback on critical parameters such as temperature, tension, and vibration levels, ensuring zero-defect manufacturing in highly sensitive textile processes such as laminated technical textiles and protective fabric coatings. These systems use cloud-based predictive analytics to anticipate deviations in processing conditions, reducing downtime by 40% and extending machinery lifespan through smart preventive maintenance algorithms. In the aerospace and automotive sectors, autonomous robotic water jet cutters are revolutionizing precision machining of lightweight composite materials, such as carbon fiber reinforced polymers, aramid-based ballistic fabrics, and ultra-high molecular weight polyethylene [135, 136].

13.6 High-Performance Water Jet Cutting Machines and their Specifications

In Industry 6.0, AI-driven high-pressure water jet systems integrate IoT-based automation, real-time process optimization, and predictive maintenance algorithms, improving productivity while significantly reducing resource consumption. Advanced abrasive waterjet systems now feature ultra-high-pressure pumps, intelligent multi-axis cutting heads, and closed-loop filtration to enhance both cutting accuracy and environmental sustainability. A water jet cutting system has high-speed, high-precision industrial, automotive, and aerospace manufacturing applications. High-strength materials like titanium, Inconel, and stainless steel may be processed quickly because to its cutting rates of up to 20,000 mm/min for soft materials and 50 to 200 mm/ min for metals/alloys and maximum operating pressure of 6205 bar (90,000 psi). The system incorporates closed-loop water filtration, achieving an 80% reduction in water consumption, making it a sustainable choice for industries focused on minimizing resource waste. It is widely used in aerospace (composite cutting), defense (ballistic material processing), and advanced manufacturing, where high precision and material integrity are important aspects. The jet creates a wider kerf at the bottom of the cut than the top (Taper), lower cutting speed as compared to laser & plasma, high operational cost, extensive system maintenance, abrasive disposal and environmental burden, limited accuracy for very thick materials, risk of delamination and micro cracking in composites, limited suitability for small & fine features, requires water treatment and filtration, high noise levels and requires fixturing to prevent part movements. These are some significant limitations for achieving high precision at low cost [137, 138].

14 Latest Advancements and Future Recommendations

Cutting-edge technologies like blockchain-enabled traceability, AI-based process control, and material analysis aided by quantum computing are all integrated into Industry 6.0 driven waterjet cutting. Real-time optimization of water pressure, abrasive flow rate, and nozzle placement by AI algorithms results in a 30% increase in efficiency and a 25% decrease in energy consumption. Engineers can forecast material interaction with ultra-high-pressure water jets thanks to quantum computing's enhancement of fluid dynamics simulations, which increases accuracy in micron-scale cutting applications. Blockchain-based traceability solutions guarantee regulatory compliance and quality control, especially in the manufacturing of medical devices and aeronautical engineering, where material certification is essential. Since waterjet cutting does not create heat-affected zones and does not require extra processing like grinding or edge smoothing, it is quickly becoming the most environmentally friendly high-precision cutting method. Large-scale manufacturers can save up to 0.1 million USD annually by using closed-loop filtration and wastewater recycling technologies, which can cut overall water usage by up to 85%, and smart abrasive management systems, which can reduce garnet consumption by 50%. In Industry 6.0, water jet cutting is undergoing a revolutionary change that combines robotic automation, nanostructured nozzle materials, quantum-enhanced sensors, and AI-driven process control to deliver unparalleled sustainability, efficiency, and precision. Through neural network-based defect identification, AI-powered optimization algorithms continually assess the composition of the material and modify water pressure and abrasive flow rates in real time, increasing cutting speeds by 25% and decreasing material waste by 20%. Cutting accuracy is further improved by quantum sensor-enhanced feedback loops, which guarantee ±0.05 mm tolerance for ultra-fine applications. Nanotechnology plays a key role in enhancing water jet cutting nozzle design, with nanostructured tungsten carbide coatings boosting wear resistance by 5 times, greatly extending operating lifetimes. In complicated geometries and micro machining applications, edge sharpness is maintained via AI-based taper correction algorithms that resist jet divergence. 6-axis robotic arms combined with AI-controlled water jet cutting heads allow for accurate multi-dimensional machining of high-performance alloys and composite materials in the automotive and aerospace industries, which lowers the need for secondary finishing. This is seen in Boeing's composite wing structure production, where robotic water jet cutting optimizes raw material utilization by reducing material waste by 30% when compared to CNC milling. Its quick uptake in sophisticated automobile engineering, aerospace, defense, medical implants, and renewable energy guarantees an environmentally clean, incredibly efficient, and economically viable future for industrial cutting technology [139, 140]. Table 10.1 is an overview of evolution of Industry 4.0 to Industry 7.0 for clear understanding of the few technological improvements in all areas based on ideas, knowledge, trends, forecasting, and projections.

Table 10.1 Comparative analysis of industrial revolutions evolutions

Aspect	Industry 4.0	Industry 5.0	Industry 6.0	Industry 7.0
Timeframe	2000 to date	2021 to date	2025+	2030+ (projected)
Core philosophy	Automation, IoT, cyber physical systems	Human AI collaboration, resilience	Hyperautomation, cognitive AI, quantum integration	Self-optimizing ecosystems, bio-nano convergence
Key technologies	– IoT with 5G – Cloud computing – Basic AI/ML – Robotics – Additive Mfg (3D printing)	– Cobots – Edge AI, circular economy – Digital twins, Gen AI – Personalized production, sustainability	– Quantum computing, Green IoT – Advanced AI (AGI), – 6G/7G networks, net zero trash, net zero carbon emission – Nanomaterials (graphene), biological advancement – Extreme UV lithography, 4D to 6D, volumetric AM and beyond – Neuromorphic hardware, teleportation	– Quantum AI, organoid intelligence – Molecular manufacturing, CCUS in space – Bio-integrated sensors, BCI, super human, teleportation – Fusion energy systems, space trash – Self-healing materials, space challenges
Industrial sectors	Automotive: predictive maintenance (15–20% efficiency gain) Logistics: smart warehouses (25% cost reduction)	Healthcare: custom implants (30% faster production) Consumer goods: mass customization (20% waste reduction)	Semiconductors: EUV lithography (2 nm transistors) Energy: AI optimized grids (40% energy savings) Aerospace: SiP packaging (100 W/cm^2 thermal load handling)	Space: quantum comms (near-zero latency) Biotech: bio-nano factories (0-waste) Agriculture: AI photosynthesis optimization (50% yield increase)
Benefits	– 20–30% operational cost reduction – Real-time data analytics	– 30% higher human productivity – 25% waste reduction via circular economy	– 40–60% energy efficiency – 50–70% material waste reduction – 99.99% defect-free production	– 80–90% closed-loop sustainability – Carbon-negative operations – Autonomous resource regeneration

(continued)

Table 10.1 (continued)

Aspect	Industry 4.0	Industry 5.0	Industry 6.0	Industry 7.0
Merits	– Scalable automation – Enhanced supply chain visibility	– Ethical AI governance – Worker-centric design	– Ultra-high precision (e.g., ±0.005 mm tolerance) – Predictive sustainability – Real-time climate adaptation	– Ecosystem restoration – Bio-integrated manufacturing – Global resource equity
Demerits	– High energy consumption (+5% global IT energy) –Cybersecurity vulnerabilities (40% attacks on IoT)	– Skill gaps (75 M workers need reskilling) – High transition costs	– Quantum security risks – Ethical AI dilemmas – Resource-intensive R&D	– Existential AI risks – High fusion energy demands – Socio-economic disruption
SDG alignment	SDG 9: industry innovation (40% adoption in G20 nations)	SDG 8: decent work (30% job satisfaction increase) SDG 12: Responsible consumption (20% less waste)	SDG 7: renewable energy (50% solar/wind usage) SDG 13: climate action (35% CO_2 reduction) SDG 9: resilient infrastructure	SDG 13: carbon negativity (−2 GT/year by 2050) SDG 15: biodiversity net gain SDG 2: zero hunger via bio-optimized agriculture
Climate impact	– Mixed: Efficiency gains vs. e-waste (53.6 MT/year) – Carbon footprint: +0.3% global emissions	– Positive: Circular economy reduces e-waste by 30% – Carbon footprint: Neutral to −5%	– Strongly positive: 60% renewable energy adoption – Water jet cutting reduces emissions by 60% vs. plasma – Carbon footprint: −20 to −40%	–Transformative: Fusion energy (near-zero emissions) – Carbon capture (1.5 °C pathway) – Carbon footprint: −60 to −100%
Textile	Tech: IoT loom monitoring Impact: 15% waste reduction SDG: 9, 12	Tech: AI-driven dye optimization Impact: 20% water savings SDG: 6, 12	Tech: Water jet cutting (0.8 μm precision) Impact: 60% lower energy vs. plasma SDG: 7, 13	Tech: Self-healing fabrics Impact: 80% landfill reduction SDG: 12, 14
Chemical	Tech: automated batch reactors Impact: 10% yield increase SDG: 9	Tech: green solvent substitutes impact: 30% less VOC emissions SDG: 3, 12	Tech: AI-predictive catalysis Impact: 50% faster R&D cycles SDG: 9, 13	Tech: molecular assembly Impact: zero-waste synthesis SDG: 6, 14
Chemical reactors	Tech: automated flow control Impact: 10% yield ↑ SDG: 9	Tech: AI catalyst optimization Impact: 30% less waste SDG: 9, 12	Tech: microfluidic reactors Impact: 60% smaller footprint SDG: 9	Tech: self-optimizing quantum reactors Impact: 0% byproducts SDG: 12

Table 10.1 (continued)

Aspect	Industry 4.0	Industry 5.0	Industry 6.0	Industry 7.0
Mining	Tech: autonomous haul trucks Impact: 18% fuel savings SDG: 9	Tech: Cobot-assisted drilling Impact: 25% fewer accidents SDG: 8, 16	Tech: quantum sensors (ore mapping) Impact: 40% less excavation waste SDG: 12, 15	Tech: bio-leaching microbes Impact: 90% lower acid drainage SDG: 15
Orbit (space)	Tech: satellite IoT networks Impact: 20% cheaper data transmission SDG: 9	Tech: 3D-printed lunar habitats Impact: 50% mass reduction SDG: 9, 17	Tech: EUV-fabricated thrusters Impact: 70% efficiency gain SDG: 7, 9	Tech: quantum comms (Mars-earth) Impact: near-zero latency SDG: 17
Automotive	Tech: robotic assembly Impact: 25% faster production SDG: 9	Tech: human-AI co-design Impact: 30% customization rate SDG: 8, 9	Tech: SiP modules (10 mm^2 AI chips) Impact: 50% energy savings SDG: 7, 11	Tech: self-repairing batteries Impact: 100% recyclability SDG: 11, 12
Military	Tech: drone swarms Impact: 15% tactical advantage SDG: 16	Tech: exoskeletons Impact: 40% soldier endurance boost SDG: 3, 16	Tech: hypersonic CNT composites impact: Mach 10+ stability SDG: 9, 16	Tech: neuro-interface systems Impact: 0.1 ms response time SDG: 16
Aviation	Tech: predictive maintenance Impact: 20% fewer delays SDG: 9	Tech: AR pilot assistance Impact: 30% training cost reduction SDG: 4, 8	Tech: microfluidic cooling (20 °C drop) Impact: 40% lighter engines SDG: 7, 13	Tech: fusion-powered VTOL Impact: zero-emission flights SDG: 7, 13
Fighter jets	Tech: composite 3D printing Impact: 15% weight ↓ SDG: 9, 16	Tech: AI-pilot co-decision Impact: 30% reaction time ↑ SDG: 16	Tech: hypersonic CNT skins Impact: Mach 12 stability SDG: 9, 16	Tech: plasma stealth fusion engines Impact: 0% IR signature SDG: 16
Missiles	Tech: GPS-guided systems Impact: 20% accuracy ↑ SDG: 16	Tech: swarm AI coordination Impact: 40% target evasion SDG: 16	Tech: quantum radar jamming impact: 99% undetectability SDG: 16	Tech: self-deactivating bio-payloads Impact: 0% collateral damage SDG: 16
Bullets	Tech: CNC precision machining impact: 10% ballistics consistency SDG: 16	Tech: eco-friendly primers impact: 30% less lead pollution SDG: 6, 16	Tech: guided smart projectiles impact: 95% hit probability SDG: 16	Tech: non-lethal energy pulses impact: 0% physical ammunition SDG: 16
Air defense systems	Tech: radar network IoT impact: 20% detection range ↑ SDG: 16	Tech: AI-threat classification impact: 35% false alarms ↓ SDG: 16	Tech: quantum lidar impact: 99.999% interception SDG: 16	Tech: energy shield domes impact: 0% kinetic projectiles SDG: 16

(continued)

Table 10.1 (continued)

Aspect	Industry 4.0	Industry 5.0	Industry 6.0	Industry 7.0
Explosives	Tech: automated mixing impact: 12% safer handling SDG: 16	Tech: bio-based detonators impact: 40% less toxic residue SDG: 6, 16	Tech: nanothermite precision impact: 50% reduced collateral SDG: 16	Tech: directed energy bursts impact: 0% chemical explosives SDG: 16
Bullet trains	Tech: IoT predictive maintenance impact: 15% downtime ↓ SDG: 9	Tech: AI-optimized aerodynamics impact: 25% energy savings SDG: 7, 9	Tech: Maglev superconductors impact: 600 km/h with 0 friction SDG: 7, 9	Tech: vacuum tube hyperloops impact: 1200 km/h, 0 emissions SDG: 7, 13
Pharmaceutical	Tech: automated pill sorting impact: 99% accuracy SDG: 3	Tech: personalized drug printing impact: 25% dosage waste reduction SDG: 3, 12	Tech: AI-generated molecules impact: 60% faster trials SDG: 3, 9	Tech: nano-drug delivery impact: 90% bioavailability SDG: 3
Surgical instruments	Tech: 3D-printed scalpels impact: 15% cost reduction SDG: 3	Tech: AR-guided surgery impact: 30% precision ↑ SDG: 3	Tech: nano-coatings (Ag) impact: 99.9% antimicrobial SDG: 3	Tech: self-sterilizing biometals impact: 0% infection risk SDG: 3
Smartphones	Tech: SMT assembly bots impact: 22% faster production SDG: 9	Tech: modular repairable designs impact: 35% e-waste ↓ SDG: 12	Tech: SiP with microfluidics impact: 60% heat reduction SDG: 7, 9	Tech: self-charging piezoelectric impact: 0% grid dependency SDG: 7
Laptops	Tech: AIO testing systems impact: 15% defect ↓ SDG: 9	Tech: recycled Al chassis impact: 30% carbon footprint ↓ SDG: 12	Tech: graphene batteries impact: 70% faster charging SDG: 7, 9	Tech: neuromorphic processors impact: 1 W TDP for 8 K rendering SDG: 7
Nutraceuticals and plant medicines	Tech: sensor-based harvesters impact: 10% yield increase SDG: 2, 3	Tech: AI-phytochemical mapping impact: 30% potency consistency SDG: 2, 3	Tech: CRISPR-enhanced botanicals impact: 50% faster growth SDG: 2, 15	Tech: bio-synthetic terpenes impact: 100% lab-to-table purity SDG: 2, 12
Plant sector (agriculture)	Tech: smart irrigation impact: 20% water savings SDG: 2, 6	Tech: pollinator drones impact: 35% higher crop diversity SDG: 2, 15	Tech: quantum soil sensors impact: 40% less fertilizer use SDG: 2, 13	Tech: photosynthesis AI-boost impact: 50% yield increase SDG: 2, 13
Lighting	Tech: LED automation impact: 25% energy savings SDG: 7	Tech: human-centric tunable spectra impact: 20% productivity ↑ SDG: 3, 7	Tech: quantum dot OLED impact: 50% brighter at half power SDG: 7, 11	Tech: bioluminescent plants impact: Zero electricity use SDG: 7, 15

Table 10.1 (continued)

Aspect	Industry 4.0	Industry 5.0	Industry 6.0	Industry 7.0
Tyre	Tech: RFID tracking impact: 20% supply chain efficiency SDG: 9	Tech: silica-reinforced eco-rubber impact: 25% lower rolling resistance SDG: 7, 9	Tech: graphene-enhanced treads impact: 40% longer wear life SDG: 9, 11	Tech: airless photosynthetic tires impact: 0% microplastic shedding SDG: 11, 14
Construction	Tech: BIM modeling impact: 15% cost reduction SDG: 9, 11	Tech: 3D-printed structures impact: 30% less material waste SDG: 11, 12	Tech: self-cooling concrete (graphene) impact: 60% lower HVAC load SDG: 7, 11	Tech: carbon-absorbing bio-bricks impact: Net-negative emissions SDG: 11, 13
Oil and gas	Tech: drilling IoT sensors impact: 8% methane leak ↓ SDG: 7, 9	Tech: AI-reservoir modeling impact: 20% extraction efficiency ↑ SDG: 9	Tech: quantum seismic imaging impact: 50% fewer dry wells SDG: 7	Tech: bio-engineered carbon capture impact: -1 M tCO_2/year per rig SDG: 13
Semiconductors	Tech: automated cleanrooms impact: 18% yield ↑ SDG: 9	Tech: human-AI defect detection impact: 30% fewer failures SDG: 8, 9	Tech: EUV lithography (2 nm) impact: 50% less energy/transistor SDG: 7, 9	Tech: bio-synthetic quantum dots impact: 90% lower rare earth use SDG: 12
Electronics	Tech: SMT assembly robots impact: 20% defect reduction SDG: 9	Tech: Human-AI co-design chips impact: 25% smaller footprints SDG: 8, 9	Tech: EUV lithography (3 nm nodes) impact: 50% less energy per transistor SDG: 7, 9	Tech: Neuromorphic graphene CPUs impact: 90% lower compute energy SDG: 7, 12
Cosmetics	Tech: robotic filling systems impact: 10% material savings SDG: 12	Tech: AI-skin profiling impact: 25% personalized formulations SDG: 3, 12	Tech: bio-printed active ingredients impact: 60% fewer animal tests SDG: 12, 15	Tech: DNA-customized serums impact: Zero-waste packaging SDG: 3, 12
Oil boilers	Tech: automated nozzle cleaning impact: 8% efficiency ↑ SDG: 7	Tech: bio-oil compatibility impact: 30% carbon-neutral operation SDG: 7, 13	Tech: nanocatalytic burners impact: 99% soot elimination SDG: 3, 7	Tech: solar-thermal hybrids impact: 100% renewable operation SDG: 7
Biological reactors	Tech: pH monitoring IoT impact: 15% consistency ↑ SDG: 3, 9	Tech: CRISPR-enhanced cultures impact: 40% faster growth SDG: 3, 9	Tech: organ-on-a-chip scaling impact: 80% less animal testing SDG: 3, 15	Tech: synthetic biology foundries impact: 100% lab-to-market efficiency SDG: 9
Fermenters	Tech: automated temperature control impact: 12% batch consistency SDG: 2	Tech: AI-strain selection impact: 25% ethanol yield ↑ SDG: 2, 7	Tech: enzymatic acceleration impact: 50% faster fermentation SDG: 2, 7	Tech: carbon-fixing biofilms impact: −0.2 kg CO_2/L biofuel SDG: 7, 13

(continued)

Table 10.1 (continued)

Aspect	Industry 4.0	Industry 5.0	Industry 6.0	Industry 7.0
Steam boilers	Tech: IoT efficiency monitors impact: 10% fuel savings SDG: 7, 9	Tech: AI-combustion tuning impact: 20% emissions ↓ SDG: 7, 13	Tech: plasma-assisted ignition impact: 40% faster steam gen SDG: 7	Tech: fusion-heated boilers impact: zero fossil inputs SDG: 7, 13
Glass	Tech: automated annealing impact: 15% energy savings SDG: 7, 9	Tech: smart tinting windows impact: 30% HVAC load ↓ SDG: 7, 11	Tech: graphene reinforcement impact: 70% lighter structures SDG: 9, 11	Tech: photosensitive bio-glass impact: Air purification SDG: 11, 13
Ceramics	Tech: Kiln IoT sensors impact: 10% defect reduction SDG: 9	Tech: AI-crack prediction impact: 25% material savings SDG: 12	Tech: water jet cutting (ZrO_2) impact: 60% faster machining SDG: 9	Tech: self-healing nano-ceramics impact: 50-year lifespan SDG: 11
Wood	Tech: CNC precision cutting impact: 15% less sawdust SDG: 12, 15	Tech: mycelium-based composites impact: 30% faster growth vs. timber SDG: 12, 13	Tech: self-healing lignin polymers impact: 50% longer product life SDG: 9, 15	Tech: carbon-sequestering smart wood impact: −1 kg CO_2/kg material SDG: 13, 15
Leather	Tech: automated tanning impact: 10% chemical reduction SDG: 6, 12	Tech: collagen biofabrication impact: 50% less livestock reliance SDG: 12, 14	Tech: enzymatic processing impact: 80% lower water pollution SDG: 6, 14	Tech: lab-grown nanocellulose leather impact: 0% toxic effluent SDG: 14
Sports equipment	Tech: carbon-fiber automation impact: 20% lighter gear SDG: 9	Tech: AI-impact analytics impact: 25% injury reduction SDG: 3	Tech: 4D-printed adaptive soles impact: 40% energy return ↑ SDG: 3, 9	Tech: self-repairing composites impact: Lifetime durability SDG: 12
Shoes	Tech: 3D-knitting automation impact: 20% waste reduction SDG: 12	Tech: plant-based leather substitutes impact: 40% lower carbon footprint SDG: 12	Tech: 4D-printed adaptive soles impact: 70% less material use SDG: 9, 12	Tech: self-repairing biopolymers impact: 100% recyclability SDG: 12
Steel	Tech: IoT blast furnace control impact: 8% coke consumption reduction SDG: 9	Tech: hydrogen reduction pilots impact: 15% lower CO_2/ton SDG: 9, 13	Tech: plasma arc electrolysis impact: 50% less energy vs. BF-BOF SDG: 7, 13	Tech: carbon-negative smelting impact: −0.5 ton CO_2/ton steel SDG: 13
Rubber	Tech: IoT vulcanization control impact: 15% energy savings SDG: 7, 9	Tech: dandelion latex sourcing impact: 30% sustainable rubber share SDG: 15	Tech: self-healing elastomers impact: 60% longer tire life SDG: 9, 11	Tech: carbon capture rubber impact: −0.3 kg CO_2/kg product SDG: 13
Packaging	Tech: robotic case packing impact: 20% labor reduction SDG: 9	Tech: mycelium-based materials impact: 40% compostability SDG: 12	Tech: self-degrading smart labels impact: 100% plastic elimination SDG: 14	Tech: DNA-tagged edible packaging impact: Zero landfill waste SDG: 12

Table 10.1 (continued)

Aspect	Industry 4.0	Industry 5.0	Industry 6.0	Industry 7.0
Paper and pulp	Tech: IoT moisture control impact: 12% energy savings SDG: 9, 12	Tech: AI chipper optimization impact: 25% less wood waste SDG: 12, 15	Tech: enzymatic de-inking impact: 95% water recycling SDG: 6, 12	Tech: carbon-negative nanocellulose impact: −0.8 t CO_2/t paper SDG: 13
Plastic	Tech: automated injection molding impact: 15% faster cycles SDG: 9	Tech: AI-recyclate blending impact: 30% virgin plastic reduction SDG: 12	Tech: enzymatic depolymerization impact: 95% PET-to-monomer recovery SDG: 12, 14	Tech: self-degrading biopolymers impact: 0% microplastic leakage SDG: 14
Beverage	Tech: automated bottling lines impact: 20% higher throughput SDG: 9	Tech: water-recycling loops impact: 35% less freshwater use SDG: 6, 12	Tech: nano-filtration membranes impact: 99% contaminant removal SDG: 6, 14	Tech: atmospheric water harvesters impact: 100% water-neutral operations SDG: 6
Juice	Tech: automated pasteurization impact: 15% throughput ↑ SDG: 9	Tech: AI-flavor profiling impact: 30% less sugar additives SDG: 3	Tech: cold plasma preservation impact: 50% shelf-life extension SDG: 2, 12	Tech: nutrient-customized bio-juices impact: 0% preservatives SDG: 3
Food	Tech: IoT cold chain monitoring impact: 15% less spoilage SDG: 2, 12	Tech: AI-driven precision fermentation impact: 30% protein yield increase SDG: 2, 9	Tech: CRISPR-enhanced crops impact: 40% less water/land use SDG: 2, 6	Tech: lab-grown meat (0 GHG) impact: 90% lower land footprint SDG: 2, 13
Energy	Tech: smart grid sensors impact: 10% transmission loss reduction SDG: 7	Tech: AI-demand forecasting impact: 20% peak load shaving SDG: 7	Tech: quantum-enhanced solar (45% efficiency) impact: 60¢/watt production cost SDG: 7, 13	Tech: compact fusion reactors impact: Near-zero marginal energy cost SDG: 7, 13
PPEs	Tech: robotic sewing lines impact: 20% faster production SDG: 3	Tech: antimicrobial nanotextiles impact: 99% pathogen blocking SDG: 3	Tech: self-decontaminating suits impact: 100% reusable SDG: 3, 12	Tech: biosensor-integrated smart PPE impact: Real-time viral detection SDG: 3

Suggestion for the new SDGs in Table 10.2 based on the future trends in Industry 6.0 and 7.0. We are not policymakers, so we can only suggest. With its sophisticated human AI collaboration, brain computer interfaces (BCIs) incorporated into labor systems, and quantum secure IoT frameworks, Industry 6.0 signals a paradigm shift. SDGs, in particular SDG 18 (cognitive rights), SDG 19 (quantum ethics and cybersecurity), and SDG 21 (algorithmic justice), are essential for directing this progress in an ethical and equitable manner. Industry 7.0 will predict the advent of space industrialization, molecular-scale manufacturing technologies, and sentient

Table 10.2 Proposed SDGs for industry 6.0 and 7.0 to address future global challenges

Proposed SDG	Core focus	Why it is needed (future proofing rationale)	Industry 6.0/7.0 linkage
SDG 18: neuro digital integrity	Protect cognitive liberty and brain data sovereignty	Rise of brain-computer interfaces (BCIs) and neurotech in workplaces risks exploitation, mental privacy breaches, and neuro discrimination	Prevents unethical neurosurveillance in hyper automated factories (6.0) and AI human cognitive mergers (7.0)
SDG 19: quantum resilient systems	Ensure security and ethics in quantum AI ecosystems	Quantum computing breaks current encryption, threatening global finance, infrastructure, and data governance. Requires post-quantum cryptography and ethical AI quantum standards	Critical for securing automated supply chains (6.0) and protecting sentient AI networks (7.0) from quantum hacking
SDG 20: molecular resource cycling	Zero waste molecular economies	Industry 7.0 will use molecular assembly (nanotech). Current circular economy (SDG 12) lacks atomic level resource recovery frameworks	Prevents nano-toxicity and enables closed-loop material use in advanced manufacturing (6.0/7.0)
SDG 21: algorithmic equity and accountability	Equitable AI governance and bias remediation	Autonomous AI makes high-stakes decisions (hiring, resource allocation). Current frameworks lack binding global AI ethics enforcement	Vital for fair human-AI collaboration (6.0) and preventing sentient AI from amplifying inequality (7.0)
SDG 22: symbiotic urban bio systems	Integrate cities with regenerative ecosystems	Future cities must be bio-hybrid (living buildings, urban microbiome restoration) to counter climate chaos and resource stress	Supports resilient smart cities (6.0) and fusion of biological/digital infrastructure (7.0)
SDG 23: cosmic resource stewardship	Sustainable off-earth resource governance	Space mining (asteroids, moon) and orbital industry will grow exponentially. No global treaties prevent cosmic resource hoarding/conflict	Prevents interplanetary inequality and regulates space-based materials for earth/space industries (7.0)

artificial intelligence. To protect both humans and the ecosystem, this will necessitate the creation of further global standards, such as SDG 20 (planetary boundary augmentation), SDG 22 (space governance), and SDG 23 (molecular risk ethics). Despite being fundamental, the present SDGs are not enough to handle the exponential technology upheavals that Industry 6.0 and 7.0 are poised to bring forth. Key challenges include the absence of frameworks for quantum computing, neurotechnology ethics, molecular engineering, and off-world resource governance, all important for future economic systems. Industry 6.0 prioritizes human-machine symbiosis, necessitating international mandates on cognitive autonomy and AI accountability, areas currently unaddressed. Industry 7.0 presents risks to Earth system stability through emerging threats such as nano pollution, AI-accelerated resource extraction, and orbital debris accumulation. In response, SDGs 19, 20, and 23 serve as critical guardrails. Unchecked technological dominance in quantum, AI, and space sectors could exacerbate geopolitical fragmentation, forming a new "sustainability divide." These proposed SDGs are not only preventive measures, but also proactive governance tools to ensure that rapid innovation remains aligned with human dignity, ecological integrity, and inclusive global stewardship. Thus, we can define Industry 7.0 as "Industry 7.0 represents the era of interplanetary and interstellar intelligent manufacturing, where bio-digital cognition, quantum information systems, extra-planetary resource utilization, and space-based industrial ecosystems converge to support sustainable multi-planet human existence. Unlike Industry 6.0, which focuses on restoring Earth through regenerative and net-zero systems, Industry 7.0 extends manufacturing, civilization, and supply chains beyond the Earth's atmosphere, enabling selfsufficient production, energy generation, and resource utilization on the Moon, Mars, and deep space infrastructures".

References

1. Yu, B.Y.: Gallium nitride semiconductors: transforming the telecommunication systems power supply performance. ВЧЕНІ ЗАПИСКИ. **24**, 12024211 (2024)
2. Speer, K., et al.: SiC material properties. In: SiC Technology: Materials, Manufacturing, Devices and Design for Power Conversion, pp. 25–66. Springer, Cham (2024)
3. Zhou, Y.: Applications of wide bandgap semiconductor materials in high-power electronic devices. World J. Eng. Technol. **12**(4), 1034–1045 (2024)
4. Lu, G.-Q., Ngo, K.D., Burgos, R.: Heterogeneous Integration Technologies for High-temperature, High-density, Low-profile Power Modules of Wide Bandgap Devices in Electric Drive Applications. Virginia Polytechnic Institute, Blacksburg (2024)
5. Zeng, X., Ren, L., Sun, R.: Preparation and properties of thermal Interface materials. In: Thermal Management Materials for Electronic Packaging: Preparation, Characterization, and Devices, pp. 211–255. Wiley, Hoboken (2024)
6. Wei, B., et al.: Thermal interface materials: from fundamental research to applications. SusMat. **4**, e239 (2024)
7. Al-Duhni, G.S.G., et al.: Nanopackaging for FutureG solutions. IEEE Nanotechnol. Mag. **18**(2), 10–20 (2024)
8. Kim, J.M., Kim, M.S., Hussain, M.M.: From 3D-IC to 3D-HI for environmentally durable rugged electronics (EnDuRE). IEEE Electron. Dev. Rev. **1**, 31–52 (2024)
9. Zhang, Y., et al.: Hybrid integration of wearable devices for physiological monitoring. Chem. Rev. **124**(18), 10386–10434 (2024)
10. Naseem, K., et al.: Enzymes encapsulated smart polymer micro assemblies and their tuned multi-functionalities: a critical review. Int. J. Polym. Mater. Polym. Biomater. **79**, 1–32 (2023)
11. Akhtar, M., et al.: Climate friendly MOFs synthesis for drug delivery systems by integrating AI, intelligent manufacturing, and quantum solutions in industry 6.0 sustainable approach. Toxicol. Res. **14**(1), tfaf011 (2025)
12. Ahmad, K., et al.: Towards sustainable water purification: MOFs as a promising solution to eliminate toxic water pollutant resorcinol. Z. Phys. Chem. **237**(10), 1669–1689 (2023)
13. Abbas, S., et al.: Cutting-edge metal-organic frameworks: revolutionizing the adsorptive removal of pharmaceutical contaminants from water. Rev. Inorg. Chem. (2025). https://doi.org/10.1515/revic-2024-0119

14. Ma, B., et al.: Experimental investigation on microstructure and properties of Al alloy-PEEK joint with different surface structure obtained by hot-pressing. J. Adhes. Sci. Technol. **38**(12), 2085–2099 (2024)
15. Ma, T., et al.: Advances in 3D printing for polymer composites: a review. InfoMat. **6**, e12568 (2024)
16. Balasubramaniam, S., et al.: Artificial intelligence-based hyperautomation for smart factory process automation. In: Hyperautomation for Next-Generation Industries, pp. 55–89. Wiley, Hoboken (2024)
17. Kalare, K.W., et al.: The power of intelligent automation. In: Hyperautomation in Business and Society, pp. 27–65. IGI Global, London (2024)
18. Boopathy, P., et al.: The Metaverse for industry 5.0 in NextG communications: potential applications and future challenges. IEEE Open J. Comput. Soc. **6**, 4–24 (2024)
19. Vijay, A.J., et al.: Exploring the synergy of IIoT, AI, and data analytics in Industry 6.0. In: Industry 6.0, pp. 1–36. CRC Press, Boca Raton (2025)
20. Manikandan, K., et al.: Transforming smart factories: real-time optimisation of cyber-physical systems with digital twins. In: International Conference on Broadband and Wireless Computing, Communication and Applications. Springer, Cham (2024)
21. Scholz, C., et al.: Sensor-enabled safety systems for human-robot collaboration: a review. IEEE Sensors J. (2024). https://doi.org/10.1109/JSEN.2024.3496905
22. Bhatia, V., et al.: Intelligent manufacturing in aerospace: integrating industry 4.0 technologies for operational excellence and digital transformation. In: Industry 4.0 Driven Manufacturing Technologies, pp. 389–434. Springer, Cham (2024)
23. Jeon, M., et al.: Extremely rare anomaly detection pipeline in semiconductor bonding process with digital twin-driven data augmentation method. IEEE Trans. Comp. Packaging Manuf. Technol. **14**, 1891–1902 (2024)
24. Shah, S., et al.: Sustainability in Smart Manufacturing: Trends, Scope, and Challenges. CRC Press, Boca Raton (2024)
25. Cohan, P.: Generative AI hardware. In: Brain Rush: how to Invest and Compete in the Real World of Generative AI, pp. 237–279. Springer, Cham (2024)
26. Li, R., et al.: Photonics for neuromorphic computing: fundamentals, devices, and opportunities. Adv. Mater. **37**, 2312825 (2024)
27. Zhang, K.: 1.1 semiconductor industry: present & future. In: 2024 IEEE International Solid-State Circuits Conference (ISSCC). IEEE, New York (2024)
28. Manley, M., et al.: Heterogeneous integration technologies for artificial intelligence applications. IEEE J. Explor. Solid-State Comput. Dev. Circ. **10**, 89–97 (2024)
29. Lau, J.H.: Chiplet communications (bridges). In: Flip Chip, Hybrid Bonding, fan-in, and fan-out Technology, pp. 427–469. Springer, Cham (2024)
30. Katari, M., Krishnamoorthy, G., Jeyaraman, J.: Novel materials and processes for miniaturization in semiconductor packaging. J. Artif. Intell. General Sci. **2**(1), 251–271 (2024)
31. Zhong, Z., et al.: EUV lithography imaging modeling and calculation based on full-vector beam propagation method. SPIE. **13423**, 1342303 (2024)
32. Thoutam, L.R., Song, Y.S.: Advancements in Back end of line technology: enhancing semiconductor manufacturing efficiency. In: Handbook of Emerging Materials for Semiconductor Industry, pp. 731–742. Springer, Cham (2024)
33. Hashimoto, K., et al.: Reaction mechanisms of Sn-based polarity-change copolymer resists with different counter anions, designed for extreme ultraviolet lithography. Jpn. J. Appl. Phys. **64**, 026501 (2025)
34. Li, J., Wang, Z., Wang, H.: Computational study of organotin oxide systems for extreme ultraviolet photoresist. Chem. A Eur. J. **129**, 1420–1428 (2025)
35. Kim, D., et al.: Etchant-free dry-developable extreme ultraviolet photoresist materials utilizing N-heterocyclic carbene–metal complexes. Small. **21**, 2407966 (2025)

36. Fang, Y., Liu, T., Sun, D.: Research on novel thermal Management Technology in the Field of integrated circuits. In: 2024 25th International Conference on Electronic Packaging Technology (ICEPT). IEEE, New York (2024)
37. Zhou, W., et al.: Fabrication and integration of photonic devices for phase-change memory and neuromorphic computing. Int. J. Extreme Manuf. **6**(2), 022001 (2024)
38. Veendrick, H.: Manufacture of MOS devices. In: Nanometer CMOS ICs: from Basics to ASICs, pp. 75–178. Springer, Cham (2024)
39. Mao, J., Kim, J.H., Seo, S.: Current status and outlook of low-melting-point metals in biomedical applications. Adv. Funct. Mater. **34**(31), 2307708 (2024)
40. Fallahtafti, N., et al.: Experimental characterization and geometrical optimization of a commercial two-phase designed cold plate. Int. Commun. Heat Mass Transf. **155**, 107457 (2024)
41. Varriale, V., et al.: The role of digital technologies in production systems for achieving sustainable development goals. Sustain. Prod. Consump. **47**, 87–104 (2024)
42. Sharma, M., Tomar, A., Hazra, A.: Edge computing for industry 5.0: fundamental, applications and research challenges. IEEE Internet Things J. **11**, 19070–19093 (2024)
43. Reddy, R.R.: Prospective potential applications and emerging tendencies in rare earth materials. Rare Earth. **164**, 177–210 (2024)
44. NdunguWaithira, S., et al.: Effects of synthesis temperature on the structural and optical properties of CaAl2O4: Eu2+, Dy3+ nanoparticles. Sci. Afr. **26**, e02435 (2024)
45. Steyn, E.A.: A de-risking toolkit for legal practitioners: the core critical and strategic minerals compendium. J. Energy Nat. Resour. Law. **42**(4), 511–591 (2024)
46. Madkour, L.H.: Carbon-Based Nanocomposite Applications and Microelectronic Technologies. CRC Press, Boca Raton (2024)
47. Han, X., et al.: Advancements in flexible electronics fabrication: film formation, patterning, and Interface optimization for cutting-edge healthcare monitoring devices. ACS Appl. Mater. Interfaces. **16**(41), 54976–55010 (2024)
48. Vijay Arputharaj, J., et al.: Exploring the synergy of IIoT, AI, and data analytics in industry 6.0. In: Industry 6.0: Technology, Practices, Challenges, and Applications. CRC Press, Boca Raton (2024)
49. Vankadara, Y., et al.: 6G revolution: redefining communication systems. In: International Conference on Smart Systems for Applications in Electrical Sciences (ICSSES). IEEE, New York (2024)
50. Hoang, V.-T., et al.: Security risks and countermeasures of adversarial attacks on AI-driven applications in 6G networks: a survey. J. Netw. Comput. Appl. **232**, 104031 (2024)
51. Sunith Babu, L., et al.: Application of hyperautomation in predictive maintenance—a technical analysis. In: Hyperautomation for Next-Generation Industries, pp. 299–323. Wiley, Hoboken (2024)
52. Liu, C., et al.: Dissecting zero trust: research landscape and its implementation in IoT. Cybersecurity. **7**(1), 20 (2024)
53. Abdul Salam, R., et al.: An overview of recent wireless technologies for IoT-enabled smart grids. J. Electr. Comput. Eng. **2024**(1), 2568751 (2024)
54. Edirisinghe, S., Wijethunge, A., Ranaweera, C.: Wi-fi 6-based home area network for demand response in smart grid. Int. J. Commun. Syst. **37**(9), e5775 (2024)
55. Bienz, K.M., Ghalamchi, B.: Wireless instant fault detection through finite-element trained machine learning. In: 2024 International Conference on Computing, Networking and Communications (ICNC). IEEE, New York (2024)
56. Bawa'neh, H., et al.: Investigation of aluminum (Al 6061-T6)/graphene/bentonite hybrid nanocomposite mechanical properties with finite element analysis study. Mater. Adv. **6**, 1031–1041 (2025)
57. Wang, C., et al.: Construction of lightweight, high-energy absorption 3D-printed scaffold for electromagnetic interference shielding with low reflection. Compos. Part B: Eng. **291**, 112043 (2024)

58. Huo, X., et al.: HCI in business and organizations: digital transformation with HCI, Metaverse, and AI technologies. In: Human-Computer Interaction in Various Application Domains, pp. 294–347. CRC Press, Boca Raton (2024)
59. Dhanaraj, R.K., et al.: Digital Twins in Industrial Production and Smart Manufacturing: An Understanding of Principles, Enhancers, and Obstacles. John Wiley & Sons, Hoboken (2024)
60. Jawad, Z.N., Balázs, V.: Machine learning-driven optimization of enterprise resource planning (ERP) systems: a comprehensive review. Beni-Suef Univ. J. Basic Appl. Sci. **13**(1), 4 (2024)
61. Lokhande, H., Ganorkar, S.R.: Object detection in video surveillance using MobileNetV2 on resource-constrained low-power edge devices. Bull. Electr. Eng. Inf. **14**(1), 357–365 (2025)
62. Pham, H.V., et al.: Benchmarking Jetson edge devices with an end-to-end video-based anomaly detection system. In: Future of Information and Communication Conference. Springer, Cham (2024)
63. Pant, R., et al.: A systematic review of additive manufacturing solutions using machine learning, internet of things, big data, digital twins and blockchain technologies: a technological perspective towards sustainability. Arch. Comput. Methods Eng. **31**, 1–16 (2024)
64. Tavanti, M.: Technology innovations for the SDGs. In: Developing Sustainability in Organizations: a Values-Based Approach, pp. 389–404. Springer, Cham (2023)
65. Fraga-Lamas, P., et al.: An overview of blockchain for industry 5.0: towards human-centric, sustainable and resilient applications. IEEE Access. **12**, 116162–116201 (2024)
66. Liang, M., et al.: Versatile lamellar wrap-structured PVDF/PZT/CNTs piezoelectric sensor for road traffic information sensing, monitoring, and energy harvesting. Chem. Eng. J. **497**, 154554 (2024)
67. Khan, B., Riaz, Z., Khoo, B.L.: Advancements in wearable sensors for cardiovascular disease detection for health monitoring. Mater. Sci. Eng. R Rep. **159**, 100804 (2024)
68. Qi, X., et al.: Enhanced temperature stability of piezoelectric properties and electrostrain via a diffused phase transition in LaMnO3-doped (K, Na) NbO3-based ceramics. J. Alloys Compd. **989**, 174342 (2024)
69. Ranjan, R., et al.: 3D-printing approach for fabricating solar cells and piezoelectric devices: a holistic review. Prog. Eng. Sci. **1**, 100016 (2024)
70. Dinesh Kumar, A., Arunachalam, N., Jayaganthan, R.: Effect of poling methods on the functional properties of fused deposition modeling printed polyvinylidene fluoride for sensing application. J. Mater. Eng. Perform. **34**, 1–10 (2024)
71. Yazid, E., et al.: Effect of ITO poling thickness, temperature, and protective layer on piezoelectric PVDF films. Mater. Sci. Semicond. Process. **174**, 108156 (2024)
72. Chen, S., et al.: Fluorine polyimide nanofibrous composites with outstanding piezoelectric property for human motion monitoring. Chem. Eng. J. **502**, 157320 (2024)
73. Suresh Kumar, R., et al.: A review on sol-gel method and coating of hybrid nano-composite. AIP Conf. Proc. **3221**, 020012 (2024)
74. Zhang, C., Kwon, S.H., Dong, L.: Piezoelectric hydrogels: hybrid material design, properties, and biomedical applications. Small. **20**, 2310110 (2024)
75. Hu, H., et al.: Wearable ultrasound devices: an emerging era for biomedicine and clinical translation. Ultrasonics. **142**, 107401 (2024)
76. Gerasimov, D., et al.: Structure transformations and dielectric properties of extruded and annealed polyvinylidene fluoride films. Iran. Polym. J. **33**(1), 1–10 (2024)
77. Huang, Y.Z., et al.: Giant piezoelectric coefficient of Polyvinylidene fluoride with rationally engineered ultrafine domains achieved by rapid freezing processing. Adv. Mater. **37**, 2412344 (2024)
78. Tahir, A., et al.: Synthesis and optimization of Eu3+: ZnO and pristine ZnO based thin film through sol-gel technique for an improved visible light photodetection. Mater. Sci. Eng. B. **308**, 117585 (2024)
79. Gautam, A., et al.: Manufacturing methods for piezoelectric ceramic materials. In: Piezoelectric Materials, pp. 27–45. Jenny Stanford Publishing, Dubai (2025)

80. Deng, C., et al.: Recent progress on barium titanate-based ferroelectrics for sensor applications. Adv. Sensor Res. **3**, 2300168 (2024)
81. Bakhtiar, S.U.H., et al.: Innovative perspectives on porous ferroelectric ceramics and their composites: charting new frontiers in energy applications. Mater. Today Commun. **38**, 108388 (2024)
82. Bairagi, S., et al.: Mechanical energy harvesting and self-powered electronic applications of textile-based piezoelectric nanogenerators: a systematic review. Nano Energy. **111**, 108414 (2023)
83. Anandakrishnan, S.S., et al.: Recycling hazardous and energy-demanding piezoelectric ceramics using an oxide–halide perovskite upside-down composite method. RSC Sustainability. **2**(4), 961–974 (2024)
84. Ibn-Mohammed, T., et al.: Enhancing life cycle product design decision-making processes: insights from normal accident theory and the satisficing framework. Resour. Conserv. Recycl. **205**, 107523 (2024)
85. Uddin, M.S., Lipu, M.S.H.: Analysis the power converting performance of ultrasonically coupled piezoelectric receiver for three different conditions. In: 5th International Conference on Sustainable Technologies for Industry 5.0 (STI). IEEE, New York (2023)
86. Sundar, S., Nitin, S.: Emerging trends of graphene-based nanomaterials as digital materials for data generation in industry 4.0. In: Advancements in Intelligent Process Automation, pp. 325–358. IGI Global, London (2025)
87. Chen, Y., Zhang, X., Lu, C.: Flexible piezoelectric materials and strain sensors for wearable electronics and artificial intelligence applications. Chem. Sci. **15**, 16436–16466 (2024)
88. Zhang, L., et al.: An emerging era: conformable ultrasound electronics. Adv. Mater. **36**(8), 2307664 (2024)
89. Cui, J., et al.: Ingenious structure engineering to enhance piezoelectricity in poly (vinylidene fluoride) for biomedical applications. Biomacromolecules. **25**(9), 5541–5591 (2024)
90. Li, Y., et al.: Lightweight and conformal acousto-ultrasonic sensing network for multi-scale structural health monitoring: a review. FlexMat. **2**, 4–29 (2024)
91. Meemary, B., et al.: Sensors integration for structural health monitoring in composite pressure vessels: a review. Compos. Struct. **351**, 118546 (2024)
92. Roszkiewicz, A., Garlińska, M., Pregowska, A.: Advancements in piezoelectric-enabled devices for optical communication. Phys. Status Solidi. **222**, 2400298 (2025)
93. Li, Y., et al.: Self-sustained artificial internet of things based on vibration energy harvesting technology: toward the future eco-society. Adv. Energy Sustain. Res. **5**(11), 2400116 (2024)
94. Xu, S., et al.: Advances in piezoelectric nanogenerators for self-powered cardiac care. Nano Trends. **7**, 100042 (2024)
95. Du, L., et al.: Recent advances in piezoelectric and triboelectric self-powered sensors for human-machine interface applications. J. Micromech. Microeng. **34**, 093001 (2024)
96. Wu, J.: Lead-free perovskite piezoelectric materials–part one. In: Piezoelectric Materials: From Fundamentals to Emerging Applications, pp. 51–84. Wiley, Hoboken (2024)
97. Yadegari, A., et al.: Recent advancements in bio-based dielectric and piezoelectric polymers and their biomedical applications. J. Mater. Chem. B. **12**(22), 5272–5298 (2024)
98. Zhang, H., Tan, H., Ye, Z.-G.: Inorganic dielectric materials for capacitive energy storage. In: Dielectric Materials for Capacitive Energy Storage, pp. 26–80. CRC Press, Boca Raton (2024)
99. Park, H., et al.: Manufacturing strategies for highly sensitive and self-powered piezoelectric and triboelectric tactile sensors. Int. J. Extreme Manuf. **7**(1), 012006 (2024)
100. Zhou, Y., et al.: Sustainable heat harvesting via thermal nonlinearity. Nat. Rev. Phys. **6**, 1–15 (2024)
101. Singh, H., et al.: Role of expert systems to optimize the friction stir welding process parameters using numerical modelling: a review. Int. J. Interact. Des. Manuf. (IJIDeM). **18**(5), 2609–2625 (2024)
102. Kulkarni, S., et al.: Next Generation Materials for Sustainable Engineering. IGI Global, London (2024)

103. Omidkar, A.: Development of Economically Profitable and Environmentally Benign Alternatives for Low-Cost Carbon Resources Valorization. University of Calgary, Calgary (2024)
104. Liao, W., Su, X., Fang, F.: A centrifugal spring mechanism empowers self-adjusting in piezoelectric wind energy harvesting. Nano Energy. **133**, 110462 (2025)
105. Melnyk, I., et al.: Computational intelligence for IoT and smart home based on piezoelectric sensors and fuzzy logic. In: IEEE International Conference on Advanced Trends in Radioelectronics, Telecommunications and Computer Engineering. Springer, Cham (2024)
106. Ding, X., et al.: An efficient and noticeable wet etching method for PZT and its application in scanning mirrors. Sensors Actuators A Phys. **379**, 115987 (2024)
107. Kazazis, D., et al.: Extreme ultraviolet lithography. Na. Rev. Methods Primers. **4**(1), 84 (2024)
108. Durga, C.V., et al.: MEMS technology in the evolution of structural control strategies. E3S Web Conf. **529**, 04013 (2024)
109. Rogdakis, K., et al.: Hybrid chips to enable a sustainable internet of things technology: opportunities and challenges. Discov. Mater. **4**(1), 4 (2024)
110. Ahmadi, F., Amin, A.: Redesigning embedded system I course. In: 2024 Intermountain Engineering, Technology and Computing (IETC). IEEE, New York (2024)
111. Whig, P., et al.: 6 IoT industrial. In: Wireless Communication Technologies: Roles, Responsibilities, and Impact of IoT, 6G, and Blockchain Practices, p. 101. Routledge, London (2024)
112. Ni, C., Li, S.C.: Machine learning enabled industrial IoT security: challenges, trends and solutions. J. Ind. Inf. Integr. **38**, 100549 (2024)
113. Bhatt, K., Agrawal, C., Bisen, A.M.: A review on emerging applications of IoT and sensor technology for industry 4.0. Wirel. Pers. Commun. **134**, 1–19 (2024)
114. Lee, D.-S., et al.: Artificial intelligence technique development for energy-efficient waste-to-energy: a case study of an incineration plant. Case Stud. Thermal Eng. **61**, 105071 (2024)
115. Hall, R.C.C.C.-A.: 3.7 Communication Wiring. Renovations to Alger Hall Cyber Command Center (2024)
116. Khan, M.F.I., Masum, A.K.M.: Predictive analytics and machine learning for real-time detection of software defects and agile test management. Educ. Admin. Theory Pract. **30**(4), 1051–1057 (2024)
117. Kolodiazhnyi, K.: Hands-on Machine Learning with C++: Build, Train, and Deploy End-to-End Machine Learning and Deep Learning Pipelines. Packt Publishing Ltd., Birmingham (2025)
118. Kwe, N.B.: Programming platforms of autonomous robotic systems. In: Robotics and Smart Autonomous Systems, pp. 47–76. CRC Press, Boca Raton (2024)
119. Singh, M., Goyat, R., Panwar, R.: Fundamental pillars for industry 4.0 development: implementation framework and challenges in manufacturing environment. TQM J. **36**(1), 288–309 (2024)
120. Lalhriatpuii, R., Wasson, V.: Comprehensive exploration of IoT communication protocol: CoAP, MQTT, HTTP, LoRaWAN and AMQP. In: International Conference on Machine Learning Algorithms. Springer, Cham (2024)
121. Stefan, I.L., et al.: Utilizing cloud solutions for object recognition in the context of industrial robotics sorting tasks. In: International Conference on INnovations in intelligent SysTems and Applications (INISTA). IEEE, New York (2024)
122. Dimova Popovska, H., Dimovski, T., Popovski, F.: The role of cloud providers in IOT services. Int. J. Comput. Sci. Inf. Technol. **16**(4), 85–95 (2024)
123. Bengherbia, B., et al.: Design and hardware implementation of an intelligent industrial IoT edge device for bearing monitoring and fault diagnosis. Arab. J. Sci. Eng. **49**(5), 6343–6359 (2024)
124. Soori, M., et al.: Intelligent robotic systems in industry 4.0: a review. J. Adv. Manuf. Sci. Technol. **4**, 2024007 (2024)

125. Nechibvute, A., Mafukidze, H.: Integration of scada and industrial IoT: opportunities and challenges. IETE Tech. Rev. **41**(3), 312–325 (2024)
126. Ov, G.S., et al.: Critical review of SCADA and PLC in smart buildings and energy sector. Energy Rep. **12**, 1518–1530 (2024)
127. Liu, Y., et al.: Deformation behavior and strengthening mechanisms of a novel powder metallurgy Ni-based superalloy during in-situ tensile testing. Mater. Sci. Eng. A. **926**, 147942 (2025)
128. Cheng, S., et al.: Effects of Al and Ti on thermal expansion behavior of a Ni–Fe–Cr-based superalloy: a combined experimental and first-principle study. J. Mater. Res. Technol. **29**, 3805–3812 (2024)
129. Periyappillai, G., Subbarayan, S., Sengottaiyan, S.: Advanced ensemble machine learning prediction to enhance the accuracy of abrasive waterjet machining for biocomposites. Mater. Chem. Phys. **333**, 130175 (2025)
130. Pendokhare, D., Nandi, S., Chakraborty, S.: A comparative analysis on parametric optimization of abrasive water jet machining processes using foraging behavior-based metaheuristic algorithms. Opsearch. (2025). https://doi.org/10.1007/s12597-025-00916-y
131. Siva Kumar, M., Rajamani, D.: Abrasive waterjet cutting of r-GO infused magnesium fiber metal laminates: experimental investigations and optimization through gorilla troops algorithm. Polym. Compos. **45**(12), 10666–10685 (2024)
132. Peng, H., et al.: Development of fluorine-free fluorescent coatings with carbon dots and silica for superhydrophobic aluminum alloy surfaces. Phys. Scr. **100**, 3 (2025)
133. Chen, Z., et al.: Laser processing for medical devices. In: Designing Materials for Medical Devices: Fundamentals, pp. 443–472. World Scientific, London (2025)
134. Upadhyay, R.: Metallic minerals and their deposits. In: Geology and Mineral Resources, pp. 495–561. Springer, Cham (2025)
135. Makhadmeh, S.N., et al.: Recent advances in multi-objective cuckoo search algorithm, its variants and applications. Arch. Comput. Methods Eng. **32**, 3213–3240 (2025)
136. Prabhu, R., Kanthababu, M.: Prediction of surface roughness and depth of cut in abrasive waterjet milling of alumina ceramic using machine learning algorithms. Expert Syst. Appl. **246**, 123168 (2024)
137. Deam, R., Lemma, E., Ahmed, D.: Modelling of the abrasive water jet cutting process. Wear. **257**(9-10), 877–891 (2004)
138. França, R.P., Bonacin, R., Monteiro, A.C.B.: An overview structured on modern prism of abrasive water-jet machining process. Abrasive Water Jet Machin. Compos. **21**, 17 (2024)
139. Perianu, I.A., et al.: Innovative solutions for waste removal (used abrasive) resulted from water jet cutting process. Eng. Innov. **2**, 49–57 (2022)
140. Peralta Álvarez, M.E., Marcos Bárcena, M., Aguayo González, F.: A review of sustainable machining engineering: optimization process through triple bottom line. J. Manuf. Sci. Eng. **138**(10), 100801 (2016)

Chapter 11
Industry 6.0 in Aerospace, Defense, Military, and Automotive Smart Manufacturing

Abstract Using intelligent manufacturing, automation powered by artificial intelligence, and quantum computing, Industry 6.0 is transforming the automotive, aerospace, defense, and military sectors. The revolutionary effects of improved materials, AI-powered decision-making, additive manufacturing, and real-time process control in next-generation vehicle and military systems are examined in this book chapter. Carbon nanotube-reinforced polymers, high-entropy alloys, and polyether ketone are examples of high-performance thermoplastics that are optimizing material qualities for applications requiring high strength and low weight. While quantum computing allows for sophisticated simulations for materials science and aerodynamic design, the use of AI-driven adaptive manufacturing increases efficiency by lowering waste and energy usage. AI-controlled trajectory correction is used by autonomous systems in the defense industry, such as military drones and hypersonic fighter jets, to improve accuracy and operational efficiency. Cyber physical security is strengthened by quantum cryptography, which guarantees the integrity of defense supply chains and autonomous weapon systems. Sustainable aviation fuels are helping the aerospace sector cut carbon emissions, and AI-driven predictive maintenance prolongs the life of armored vehicles and airplanes while minimizing downtime. To cut expenses and pollution, the automobile industry is quickly implementing electrified cars and AI-optimized supply networks. The aerospace and defense industries are seeing notable growth, according to market trends. The chapter lays out the route to completely autonomous and sustainable military and transportation ecosystems by offering a thorough analysis of next-generation materials, intelligent automation, and cyber-secure manufacturing.

Keywords Industry 6.0 · Artificial intelligence · Quantum computing · Aerospace manufacturing · Military automation · Autonomous systems · Hypersonic weapons · Cyber-physical security · Additive manufacturing · High-entropy alloys · Smart materials · Electrified vehicles · Predictive maintenance · Defense supply chain · Sustainable aviation fuels

H. Majeed, T. Iftikhar, *Intelligent Manufacturing in Industry 6.0*,
https://doi.org/10.1007/978-3-032-07278-8_11

1 Introduction

As Industry 6.0 incorporates intelligent manufacturing technology, production is undergoing a significant shift, especially in the aerospace, automotive, and defense industries. The next stage of this progression is the transition from manual production to fully autonomous systems, which is fueled by the combination of automation, smart materials, artificial intelligence, sophisticated propulsion systems, additive manufacturing, digital twins, and quantum computing. With a focus on high-performance materials, sustainability, cybersecurity, and AI-powered optimization, this shift is influencing the future of fighter jets, tanks, helicopters, military drones, autonomous combat systems, and cargo and passenger aircraft. Intelligent manufacturing is being used by the aerospace, aviation, and defense industries, with an emphasis on sophisticated materials, environmentally friendly production methods, and self-governing systems. AI-powered adaptive manufacturing systems are enabling real-time process control, allowing for higher precision and reduced material wastage. Cyber physical security is critical in the production of autonomous military systems, ensuring that these advanced vehicles and technologies are safeguarded against cyber threats and vulnerabilities. AI-driven predictive maintenance in smart factories ensures continuous operation and minimizes downtime, enhancing productivity [1, 2].

Industry 6.0 introduces real-time adaptive manufacturing systems that optimize processes, reduce energy consumption, and improve environmental performance. Quantum computing enables more accurate simulations for aerospace design and material properties, pushing the boundaries of what is possible in materials science. In defense, AI is enhancing autonomous decision-making, predictive analytics, and combat strategy optimization, while quantum sensors are poised to revolutionize military applications, offering precise navigation and targeting capabilities. However, conventional manufacturing in aerospace, automotive, and defense industries faces several challenges. Traditional production methods are also energy-intensive, emitting between 8 and 10 million tons of CO_2 annually. Material limitations in the sector remain a concern, particularly in the need for lightweight, high-strength, and radiation-resistant materials to meet the stringent demands of aerospace and defense applications. High material wastage, typically around 10–15% in conventional vehicle assembly, also remains a challenge, contributing to both inefficiency and environmental impact. The supply chain remains vulnerable to disruptions, particularly those related to the geopolitical instability surrounding rare earth materials essential for many advanced technologies [3, 4].

To mitigate these challenges, Industry 6.0's focus on AI-driven manufacturing and quantum computing is crucial. As autonomous production systems become more prevalent, AI-driven decision making will ensure that production processes are adaptive, responsive, and capable of addressing environmental and logistical challenges. In the context of defense, advancements in autonomous military systems, drones, and fighter jets are transforming warfare, providing greater precision, reduced human risk, and improved sustainability. Industry 6.0 technologies are revolutionizing the aerospace and defense sectors, which are expected to be worth over USD 1.2 trillion by 2030. While additive manufacturing is expanding at a 27% compound annual growth rate and is predicted to reach USD 10 billion by 2030, AI and robotics are ready to lower production costs in the aerospace industry by 30%. Autonomous military drones are expected to expand at a CAGR of 15% to reach USD 45 billion by 2030, while the automobile market is also expected to increase significantly, reaching USD 3.8 trillion by 2035. The switch to SAFs, which are predicted to replace 60% of traditional jet fuels by 2040, is a significant trend in aviation. Industry 6.0's AI-driven automation will reduce vehicle production costs by 35%, and AI in defense is expected to reach USD 50 billion by 2035. Military technologies are advancing rapidly, with hypersonic weapons R&D funding surpassing USD 20 billion by 2030, and hydrogen-powered fighter jets poised to reduce CO_2 emissions by 75% by 2040. Quantum navigation systems, expected to replace GPS by 2035, will further enhance military capabilities [5, 6].

2 Advanced Materials and High-Tech Composites in Industry 6.0

2.1 Thermoplastic Composites

Thermoplastic composites such as PEEK, PEI, and PPS are revolutionizing the aerospace and defense sectors by offering materials that combine high-temperature resistance, mechanical strength, and lightweight properties. These materials' ability to withstand extreme environmental conditions while maintaining structural integrity positions them at the forefront of advanced manufacturing in aerospace and defense. The use of these thermoplastic composites has significantly advanced the design of modern military tanks, where weight reduction and enhanced performance are vital. By replacing traditional metal components with thermoplastic composites, such as PEEK and PEI, engineers can create lighter tank structures yet maintain or even improve their strength and resistance to wear, corrosion, and high-impact forces. These materials contribute to improving fuel efficiency, maneuverability, and overall combat readiness by enabling the production of lighter, more efficient vehicles without compromising on protection or durability. Withstanding both quick exposure to temperatures as high as 350 °C and sustained service temperatures of up to 250 °C, PEEK is well known for its remarkable thermal stability. Because of this, it is a material of choice for demanding aerospace applications where durability, low weight, and high temperature resistance are crucial, such as engine and fuselage parts for airplanes. PEI is perfect for use in defense systems and airplanes where electrical and thermal management are crucial because of its intrinsic flame resistance and electrical insulating qualities. Because PEI retains its structural integrity even when exposed to strong chemicals, such as fuels and hydraulic fluids, it is frequently utilized in applications that call for precise production and intricate geometries. Polyphenylene sulphide (PPS) is widely employed in components for both aerospace and defense applications, especially where superior chemical resistance, dimensional stability, and high temperature performance are required. PPS can endure temperatures up to 260 °C continuously, making it suitable for various engine components, electrical connectors, and structural parts. It is also highly resistant to hydrolysis, ensuring long-lasting performance in humid or moisture-rich environments. The use of PPS in advanced military tank designs allows for the development of lighter, more robust materials that can withstand ballistic impact and extreme operational conditions. These manufacturing methods enable the production of tailored parts with high dimensional accuracy and consistent quality, essential for mission-critical applications in both aerospace and defense sectors. Looking at the broader economic and environmental impacts, the adoption of thermoplastic composites like PEEK, PEI, and PPS in the aerospace and defense industries has significant implications. Due to commercial and military aircraft demand for ever-more efficient and sustainable solutions, these advanced materials offer a means to reduce operational costs and environmental impact while enhancing the performance and safety of critical systems [7–9].

2.2 *Carbon Nanotube Reinforced Polymers*

Carbon nanotube-reinforced polymers represent a groundbreaking advancement in materials science, offering remarkable improvements in both strength and weight reduction compared to traditional composite materials. When incorporated into polymers, they enhance the material's mechanical properties, making it up to 200% stronger and 40% lighter than conventional composites made from materials such as fiberglass or carbon fiber. However, achieving a uniform dispersion of carbon nanotubes within the polymer remains a challenge due to the nanotubes' tendency to agglomerate. Recent advancements in nanotechnology and surface modification techniques have addressed this issue by functionalizing the carbon nanotubes, which improves their compatibility with the polymer matrix and ensures uniform dispersion, leading to consistent mechanical properties across the material. When carbon nanotubes are integrated into polymer matrices, the resulting composite material can support significantly higher loads, which is crucial for applications such as aircraft and automotive components where structural integrity is paramount. These composites can withstand extreme stress and strain without compromising their structural integrity, making them ideal for use in critical components such as fuselages, engine parts, and military vehicle armor. The 40% reduction in weight results in more fuel-efficient vehicles, as the overall mass of the system is reduced without sacrificing strength or durability. This weight reduction is particularly important in the aerospace industry, where every gram saved can lead to substantial fuel savings and improved performance. The interaction between carbon nanotubes and the polymer matrix is essential for achieving the desired mechanical properties. The strong covalent bonds formed between the carbon nanotubes and the polymer chains result in improved load transfer, which enhances the material's overall toughness and resistance to fatigue. The addition of carbon nanotubes can also impart other desirable properties, such as enhanced thermal conductivity and electrical conductivity, which can be particularly useful in applications where heat dissipation or electrical performance is important. Moreover, the integration of carbon nanotube-reinforced polymers with intelligent manufacturing technologies such as 3D printing and additive manufacturing opens new possibilities for customizing components with intricate geometries that were previously difficult to achieve with traditional manufacturing methods. This ability to tailor the structure of the composite at the nanoscale allows for optimization of material properties for specific applications, further enhancing performance [10–12].

2.3 Kevlar and Specialized Polyethylene UHMWPE

Kevlar and ultra-high molecular weight polyethylene are two of the most advanced materials used in military vehicle armor and ballistic protection, offering a combination of strength, long-lasting, and non-heavy properties that are essential for safeguarding personnel and equipment in high-risk environments. Kevlar, a synthetic fiber developed by DuPont, is renowned for its exceptional tensile strength and is widely used in body armor, helmets, and vehicle armor. Its molecular structure, characterized by tightly bonded aromatic rings in a long-chain polymer configuration, gives it the ability to absorb and dissipate the immense energy produced during a ballistic impact. The fiber's strength-to-weight ratio is outstanding, with Kevlar being stronger than steel material, yet only a fraction of the density. This allows it to provide superior protection while keeping the overall weight of the armor to a minimum, which is crucial in military applications where mobility and weight efficiency are paramount. Kevlar and ultra-high molecular weight polyethylene are pivotal materials in modern military and defense technologies, offering exceptional protection against high-velocity impacts and contributing to the design of lighter, more efficient armor systems. These materials' unique combination of strength, lightweight properties, and impact resistance makes them ideal for a wide range of applications, from vehicle armoring to personal protective equipment. As research in material science and advanced manufacturing continues to progress, the performance of Kevlar and UHMWPE is expected to improve, making them even more effective in safeguarding personnel and equipment in high-risk environments. The future of Kevlar and UHMWPE in military applications will also be influenced by the integration of intelligent manufacturing techniques. These technologies allow for the precise fabrication of armor components with complex geometries that maximize the distribution of stress and energy during impact. UHMWPE, on the other hand, is a thermoplastic polymer with a molecular structure consisting of highly elongated polymer chains that contribute to its extraordinary mechanical properties. This material is 10–15 times stronger than steel by weight, offering an incredible combination of strength, stiffness, and impact resistance. Its high molecular weight, typically ranging from 2 to 6 million g/mol, makes it one of the strongest synthetic polymers known. In military vehicle armor, UHMWPE is often used as part of a layered composite material system due to its ability to resist penetration from high-velocity projectiles. Its lightness, as compared to metals, allows for more maneuverable vehicle designs, and its impact resistance ensures that it can absorb substantial energy from impacts exceeding 1000 meters per second (m/s), a typical velocity for high-caliber bullets and shrapnel. The effectiveness of Kevlar and UHMWPE in ballistic protection comes from the way they function in armor systems. UHMWPE, due to its high energy absorption and low friction coefficient, can be employed in armor configurations that allow it to catch and slow down projectiles, significantly reducing the likelihood of penetration. UHMWPE is resistant to abrasion, which contributes to its long-term durability in harsh environments. In modern military applications, these materials are used together in composite armor systems, where Kevlar serves as a fiber matrix that enhances the overall strength and elasticity of the armor, while UHMWPE provides additional toughness and energy dissipation. Advanced military vehicle armoring often combines Kevlar with UHMWPE in a multi-layer configuration, with each layer designed to optimize energy absorption and disperse the force of incoming projectiles. This layered approach allows for lighter, more efficient protection, as the use of lighter UHMWPE reduces the overall weight of the armor compared to traditional metal-based armor. The materials used in protection of military aircraft and personal protective equipment. In aircraft shields, Kevlar and UHMWPE can be used to guard against shrapnel and high-velocity debris from explosions or missile strikes. Their lightweight nature does not compromise the structural integrity of the aircraft, and their capacity to withstand high-speed impacts ensures that critical systems and personnel remain safe during combat operations. These materials are often used in ballistic vests, helmets, and face shields for personnel protection, providing flexibility and mobility while maintaining superior defense capabilities. The development and enhancement of Kevlar and UHMWPE have been further supported by advances in nanotechnology and polymer chemistry. For example, the incorporation of carbon nanotubes or graphene into Kevlar fibers can potentially enhance its mechanical properties by improving its tensile strength, thermal stability, and conductivity. Similarly, advanced processing techniques, such as electrospinning and high-pressure compaction, are being used to further optimize the properties of UHMWPE, enabling it to withstand even higher velocities and stresses. These innovations are being explored in the context of next-generation armor solutions for both military and civilian applications, including law enforcement and emergency services [13–15].

2.4 *Titanium Alloys (Ti-6Al-4 V)*

The titanium alloy composed of three metals represents one of the most advanced engineering materials, offering an unparalleled combination of strength, lightweight characteristics, thermal stability, and corrosion resistance. Its extensive use in fighter jets, tanks, missile structures, and jet engines underscores its critical role in modern aerospace and defense technologies. With ongoing advancements in intelligent manufacturing, nanotechnology, and computational material science, the performance and affordability of titanium alloys will continue to improve, driving innovation across multiple high-tech industries. This unique combination makes it an ideal material for fighter jets, tanks, missile structures, and advanced military vehicles, where weight reduction is a critical factor in enhancing speed, maneuverability, and fuel efficiency without compromising structural integrity. The superior properties of this titanium alloy arise from its intricate microstructure, which can exist in different phases, alpha, beta, or a combination of both, depending on the processing conditions. The presence of aluminum stabilizes the alpha phase, which provides excellent corrosion resistance and thermal stability, while vanadium stabilizes the beta phase, allowing for greater ductility and ease of processing. This dual-phase structure results in an alloy with a tensile strength ranging from 900 to 1200 MPa, a density of approximately 4.43 g/cm^3, and a melting point of around 1660 °C. These properties make it an indispensable material for extreme environments, such as jet engines, which operate at temperatures exceeding 900 °C, and missile structures that endure extreme aerodynamic forces and rapid acceleration. In the aerospace industry, titanium alloy is extensively used in the structural framework of fighter jets, particularly in load-bearing components such as landing gear, fuselage reinforcements, and engine mounts. The high strength and low density of this alloy allow aircraft to achieve superior thrust-to-weight ratios, leading to greater agility, higher speeds, and enhanced fuel efficiency. The exceptional corrosion resistance of titanium alloy ensures long-term durability, even in highly oxidative and humid environments, where conventional aluminum alloys are prone to degradation. The combination of high fatigue resistance and low thermal expansion also makes it a preferred material for high-performance turbine blades in jet engines, reducing maintenance requirements and improving operational reliability. Beyond aerospace applications, titanium alloy is increasingly used in the defense sector for the construction of next-generation tanks, armored vehicles, and missile structures. The reduced weight of titanium-based armor enhances vehicle mobility and speed while maintaining ballistic resistance against high-velocity projectiles. In modern missile systems, the high-temperature stability and low thermal expansion coefficient of titanium alloy play a crucial role in ensuring structural integrity under extreme heat and aerodynamic stress. Its excellent fracture toughness and resistance to fatigue crack propagation make it a superior choice for components that undergo repeated stress cycles, such as missile airframes and propulsion casings. The manufacturing and processing of titanium alloy involve advanced techniques, including electron beam melting, selective laser sintering, and additive manufacturing, which enable the fabrication of complex geometries with minimal material wastage. The implementation of artificial intelligence and machine learning in intelligent manufacturing has further optimized the production process by predicting material behavior, refining microstructural properties, and reducing processing costs. The integration of computational fluid dynamics and finite element analysis has allowed engineers to design and simulate titanium-based aerospace and defense components with unprecedented precision, improving performance and safety. The high cost of titanium alloy, attributed to its complex extraction and processing requirements, has historically limited its widespread adoption. However, advancements in powder metallurgy, electron beam cold hearth refining, and plasma arc melting have significantly reduced production costs, making titanium alloy more accessible for a broader range of applications. The increasing demand for lightweight, high performance materials in aerospace, defense, and space exploration has driven further investment in titanium-based manufacturing technologies. Market analysis indicates a steady rise in titanium alloy production, with global demand projected to exceed 0.12 million tons annually by 2030. Titanium alloy is also widely used for surgical instruments, biomedical implants, such as orthopedic joint replacements and dental implants, due to its exceptional biocompatibility, corrosion resistance, and ability to integrate with human bone through osseointegration. The development of titanium-based bioactive coatings using nanotechnology has further enhanced their application in medical implants, improving bone adhesion and reducing the risk of implant rejection. The continued advancement of high-entropy alloys, nanostructured titanium composites, and functionally graded materials will further enhance the capabilities of titanium alloys for next-generation aerospace and defense applications. Research into self-healing titanium alloys, incorporating embedded nanocapsules filled with healing agents, is being explored to enhance damage tolerance in critical structural components. The future of titanium alloy development is also closely linked to advancements in artificial intelligence-driven materials discovery, enabling the rapid identification and optimization of new alloy compositions with superior properties [16–18].

2.5 *High Entropy Alloys (HEAs)*

Unlike conventional alloys, which rely on a single base metal with minor alloying elements, high-entropy alloys exhibit a unique atomic-level distribution that enhances their mechanical strength, corrosion resistance, and thermal stability. The underlying principle governing high-entropy alloys is the high configurational entropy effect, which promotes the formation of stable single-phase solid solutions, suppresses brittle intermetallic compounds, and improves mechanical performance. This atomic disorder contributes to remarkable properties such as ultra-high yield strength exceeding 1500 megapascals, fracture toughness above 200 MJ/m^3, and sustained ductility even at cryogenic temperatures. These features make high entropy alloys ideal for applications in hypersonic vehicles, spacecraft shields, and nuclear environments, where conventional materials suffer from rapid degradation due to extreme temperatures, high velocity impacts, and intense radiation exposure. Radiation resistance is one of the most critical properties of high-entropy alloys, making them indispensable for aerospace, nuclear fusion reactors, and deep-space exploration missions. Traditional materials, such as stainless steels and nickel-based superalloys, experience significant atomic displacement and microstructural damage when exposed to high-energy radiation, leading to embrittlement and loss of mechanical integrity. However, high-entropy alloys demonstrate superior radiation tolerance due to their sluggish diffusion kinetics and lattice distortion effects, which effectively suppress the formation of radiation-induced defects such as voids, dislocations, and phase segregation. Experimental studies have shown that high-entropy alloys exhibit ten times lower radiation-induced swelling compared to conventional alloys, making them a breakthrough material for nuclear reactor components and spacecraft structures subjected to cosmic radiation. The incorporation of elements such as tungsten, molybdenum, and chromium further enhances their radiation stability, enabling the development of self-healing materials that can recover structural integrity after radiation exposure. In hypersonic applications, where vehicles operate at speeds exceeding Mach 5 (more than 6000 km/h), materials must withstand extreme aerodynamic heating, intense mechanical stresses, and rapid thermal cycling. High-entropy alloys outperform traditional refractory alloys due to their exceptional high-temperature strength, oxidation resistance, and phase stability. Recent studies have demonstrated that certain high-entropy alloys, particularly those containing hafnium, niobium, and zirconium, maintain their structural integrity above 1500 °C, making them ideal candidates for hypersonic leading edges, engine components, and thermal protection systems. The ultra-high melting points of these alloys, often exceeding 2500 °C, prevent catastrophic material failure under hypersonic flight conditions. The development of nanostructured coatings based on high entropy alloys has enabled the creation of multifunctional thermal barrier coatings that protect against aerodynamic heating while maintaining structural rigidity. In the realm of spacecraft engineering, the increasing demand for advanced shielding materials against micrometeoroid impacts and cosmic radiation has positioned high entropy alloys as a transformative solution. The combination of high strength to weight ratio, phase stability, and resistance to radiation-induced embrittlement allows high entropy alloys to serve as lightweight, durable shielding materials for long-duration space missions. Computational simulations using artificial intelligence-driven materials informatics have identified optimal compositions of high entropy alloys tailored for space applications, significantly accelerating the discovery of next-generation radiation-tolerant materials. These alloys are also being explored for use in structural components of space habitats, deep space probes, and planetary rovers, where exposure to extreme space weather conditions demands unparalleled material resilience. High-entropy alloys represent a revolutionary advancement in material science, offering unparalleled radiation resistance, mechanical strength, and thermal stability. Their application in hypersonic vehicles, spacecraft shields, and nuclear environments underscores their transformative potential in extreme engineering scenarios. With continuous advancements in intelligent manufacturing, machine learning driven material design, and high throughput experimentation, high-entropy alloys are poised to redefine the future of aerospace, defense, and energy sectors, enabling the development of next-generation materials capable of withstanding the harshest operational conditions known to science. From an economic and industrial perspective, the commercialization of high entropy alloys has gained significant momentum, with global investments in hypersonic and space technologies driving demand for these advanced materials. The cost of high entropy alloy production, initially a limiting factor, has been substantially reduced through scalable powder metallurgy techniques and process optimization strategies. Market projections indicate that the high entropy alloy sector will witness exponential growth, particularly in the defense, aerospace, and energy industries, with an anticipated market valuation exceeding 10 billion United States dollars by 2035. Beyond aerospace and defense, high entropy alloys are also being integrated into nuclear fusion reactor components, where extreme neutron flux exposure necessitates materials with unparalleled radiation resistance. The structural integrity of high entropy alloys in fusion plasma environments has been validated through extensive neutron irradiation experiments, demonstrating their viability for next-generation fusion reactor designs. Biomedical applications of high entropy alloys are emerging, particularly in orthopedic implants and surgical tools, where their superior wear resistance, biocompatibility, and corrosion resistance offer significant advantages over conventional titanium and cobalt chromium alloys. The future of high entropy alloy research is closely linked to the development of entropy-stabilized ceramics, metallic glasses, and functionally graded materials, which aim to push the boundaries of material performance even further. Self-healing high entropy alloys, incorporating nanocapsules filled with healing agents that activate upon mechanical stress or radiation damage, are being actively explored to enhance long-term durability in extreme environments. The integration of quantum computing and artificial intelligence in materials discovery is expected to accelerate the identification of novel high entropy alloy compositions, paving the way for next-generation materials tailored for extreme engineering applications [19–21].

2.6 Shape Memory Alloys

Shape memory alloys represent a groundbreaking advancement in intelligent materials, offering unparalleled self-repairing capabilities, high mechanical strength, and adaptive functionality. Their application in aerospace, military, medical, and robotic systems is transforming engineering design, enabling the development of self-healing structures and highly efficient adaptive components. With continuous advancements in intelligent manufacturing, artificial intelligence-driven material design, and computational materials engineering, shape memory alloys are poised to redefine the future of high-performance engineering materials, driving technological innovation across multiple industries. Shape memory alloys, particularly those composed of nickel-titanium and copper-aluminum-nickel, represent a significant advancement in the development of intelligent materials that possess the ability to recover their original shape after deformation when exposed to specific thermal or mechanical conditions. These materials exhibit the shape memory effect and pseudoelasticity, enabling their application in self-repairing structures, aerospace components, military vehicles, and advanced robotics. The exceptional performance of shape memory alloys arises from their ability to undergo reversible phase transformations between austenite and martensite structures at the atomic level, driven by changes in temperature or stress. This property allows shape memory alloys to autonomously repair structural deformations, significantly improving the durability, safety, and longevity of critical engineering systems. In the aerospace industry, shape memory alloys have revolutionized the development of adaptive and self-healing components. The integration of nickel-titanium-based shape memory alloys into aircraft wing morphing mechanisms, jet engine actuators, and spacecraft thermal shielding enables real-time structural adaptation in response to dynamic aerodynamic conditions. When subjected to extreme mechanical stress or impact damage, shape memory alloys can revert to their pre-deformed state through thermal activation, thereby extending the lifespan of aerospace components by more than 30%. The fatigue resistance and damage recovery capabilities of shape memory alloys have been validated in real-world aerospace applications, demonstrating their ability to sustain over ten million actuation cycles without significant loss of functionality. Shape memory alloys play a crucial role in vibration damping and noise reduction in aircraft structures, enhancing overall flight efficiency and passenger safety. In military vehicle applications, shape memory alloys are being integrated into self-repairing armor, adaptive suspension systems, and impact-resistant hulls. The unique ability of shape memory alloys to absorb and dissipate kinetic energy from ballistic impacts makes them ideal candidates for next-generation military vehicle armor. The incorporation of copper-aluminum-nickel-based shape memory alloys into composite armor structures enhances penetration resistance while simultaneously enabling damage recovery through localized heating. Studies have shown that shape memory alloy reinforced armor can restore up to 90% of its original structural integrity after impact-induced deformation, significantly reducing maintenance costs and increasing battlefield survivability. Shape memory alloys are being explored for use in autonomous robotic military applications, where self-repairing actuators and deformable robotic limbs enhance operational efficiency in harsh environments. The global market for shape memory alloys is projected to exceed 15 billion United States dollars by 2030, driven by increasing demand from the aerospace, defense, medical, and robotics sectors. The cost of shape memory alloy production has been significantly reduced through advancements in vacuum arc melting, powder metallurgy, and additive manufacturing techniques. Market analysis indicates that the adoption of shape memory alloys in high-performance engineering applications is expected to grow at an annual rate of over 10%, with key investments focused on next-generation self-repairing materials and adaptive structures. The intelligent manufacturing of shape memory alloys relies on state-of-the-art processing techniques such as laser additive manufacturing, electron beam melting, and directed energy deposition. These advanced manufacturing methods enable precise control over microstructural properties, optimizing the transformation temperatures, mechanical strength, and fatigue resistance of shape memory alloys. The integration of artificial intelligence and machine learning in the design of shape memory alloy compositions has further accelerated the discovery of novel alloys with enhanced functional properties. High-throughput computational modeling techniques, including phase-field simulations and density functional theory, allow for predictive optimization of shape memory alloy performance in extreme environments. From a materials science perspective, the atomic-scale mechanisms governing the shape memory effect are being extensively studied using in-situ electron microscopy and synchrotron X-ray diffraction techniques. Recent research has revealed that the addition of rare earth elements such as hafnium and scandium can significantly improve the thermal stability and actuation response of nickel-titanium-based shape memory alloys. The integration of shape memory alloys with carbon nanotube-reinforced polymers and high-entropy alloys is leading to the creation of hybrid materials with unprecedented mechanical properties, best for the next generation of smart structures. In the field of robotics and medical sciences, shape memory alloys are revolutionizing the development of bio-inspired soft robotics, artificial muscles, and advanced medical implants. The high biocompatibility and corrosion resistance of nickel-titanium shape memory alloys make them ideal for applications in cardiovascular stents, orthopedic implants, and minimally invasive surgical devices. The ability of shape memory alloys to provide controlled actuation in response to body temperature has enabled the design of self-expanding stents that adapt dynamically to physiological conditions, improving patient outcomes. Shape memory alloys are being incorporated into wearable medical devices for real-time biomechanical assistance, such as exoskeletons for mobility impaired individuals [22–24].

2.7 *Silicon and Boron Composite Carbide*

Silicon carbide and boron carbide represent the pinnacle of high-performance ceramic materials, offering unparalleled thermal resistance, mechanical strength, and environmental stability for applications in hypersonic aircraft, spacecraft heat shields, nuclear protection systems, and advanced ballistic armor. Their adoption in aerospace, defense, and semiconductor industries continues to accelerate, driven by breakthroughs in computational material design, intelligent manufacturing, and nanotechnology integration. As technological advancements push the limits of extreme environment engineering, silicon carbide and boron carbide will remain at the forefront of next-generation materials science, shaping the future of high-speed transportation, space exploration, and defense innovations. Silicon carbide and boron carbide are among the most advanced high-performance ceramic materials that exhibit exceptional thermal stability, mechanical strength, and resistance to extreme environmental conditions, making them indispensable for hypersonic aircraft, spacecraft heat shields, nuclear applications, and ballistic protection systems. These ultra-refractory ceramics possess high melting points, with silicon carbide exceeding 2730 °C and boron carbide surpassing 2450 °C. Their extraordinary ability to withstand temperatures above 2000 °C without undergoing structural degradation is attributed to their strong covalent bonding and highly crystalline atomic structures, which resist decomposition even in extreme thermal and oxidative environments. The aviation industry mostly depends on silicon carbide and boron carbide due to their superior thermal shock resistance, low density, and unparalleled hardness. Silicon carbide, with a density of approximately 3.2 g/cm^3, offers a perfect balance of lightweight properties and mechanical durability, allowing hypersonic aircraft components such as leading edges, nose cones, and thermal protection systems to endure aerodynamic heating and mechanical stresses generated at speeds exceeding Mach 5. Boron carbide, known for its hardness ranking close to diamond on the Mohs scale, is integrated into next-generation hypersonic vehicle armor, providing exceptional impact resistance against high-velocity atmospheric particles and debris encountered at extreme altitudes. The outstanding oxidation resistance of silicon carbide is driven by the formation of a passive silicon dioxide protective layer at high temperatures, preventing material degradation even in oxygen rich environments. This self-healing oxide layer enhances the longevity of spacecraft heat shields and thermal insulation components in deep-space missions, where thermal cycling between minus 200 degrees Celsius and over 2000 °C poses a significant challenge. Silicon carbide fiber reinforced ceramic matrix composites are being extensively developed to replace conventional nickel-based superalloys in jet engines, increasing efficiency by over 20% due to their ability to sustain high combustion temperatures without creep deformation. Boron carbide, widely recognized for its neutron absorption capabilities, is a critical material in advanced nuclear shielding applications, where it effectively moderates neutron flux in next-generation fusion reactors and small modular nuclear reactors. In defense applications, boron carbide's high hardness and low density, approximately 2.5 g/cm^3, enable the development of lightweight body armor and vehicle protection systems capable of withstanding armor-piercing projectiles traveling at velocities exceeding 1000 m/s. The incorporation of boron carbide into composite armor panels for military aircraft and hypersonic defense systems significantly enhances survivability against high-energy kinetic threats. The intelligent manufacturing of silicon carbide and boron carbide components involves state-of-the-art processes such as chemical vapor deposition, pressure-assisted sintering, and additive manufacturing, enabling precise control over microstructural properties to optimize mechanical and thermal performance. Computational materials engineering, combined with artificial intelligence-driven simulations, is accelerating the design of novel silicon carbide and boron carbide-based composites with tailored properties for specific aerospace and defense applications. Advanced multi-scale modeling techniques are being utilized to predict material behavior under extreme thermal and mechanical loads, leading to the optimization of component geometries and weight reduction in hypersonic aircraft structures. From an economic and market perspective, the global demand for silicon carbide and boron carbide is projected to exceed five billion US dollars by the year 2030, driven by rapid advancements in hypersonic vehicle technology, space exploration, and next-generation defense systems. The cost of high purity silicon carbide production has been significantly reduced through innovations in sublimation growth techniques and high-temperature reaction synthesis, enabling large-scale commercialization in the semiconductor and power electronics industries. Silicon carbide-based semiconductors, known for their superior efficiency and high-voltage endurance, are revolutionizing electric propulsion systems in aerospace and automotive applications, reducing energy losses by over 50% compared to traditional silicon-based power devices. The integration of silicon carbide and boron carbide with emerging nanostructured materials, such as graphene reinforced ceramic composites and carbon nanotube-infused ceramics, is unlocking new frontiers in ultralight, high strength, and thermally stable materials for extreme environments. Research in quantum computing and optoelectronics is also exploring silicon carbide as a platform for defect-engineered quantum bits, paving the way for advanced photonic and quantum communication technologies. The combination of boron carbide with metallic and polymeric matrices is leading to the development of next-generation multifunctional materials with superior toughness, impact absorption, and thermal conductivity, suitable for use in both terrestrial and extraterrestrial environments [25–27].

2.8 Nanoceramics with Aerogels

Nanoceramics combined with aerogels represent a revolutionary class of ultralight, thermally resistant, and structurally robust materials that are fundamentally transforming the aerospace and defense industries. The synergy between nanoceramics and aerogels results in a unique hierarchical nanostructure that provides more than 90% heat resistance while maintaining an ultralow density, often less than 50 mg/cm^3, making them one of the lightest thermally protective materials known to modern engineering. Nanoceramic aerogel composites represent one of the most groundbreaking advancements in material science, offering an unparalleled combination of ultralightweight structure, exceptional thermal resistance, and superior mechanical strength. Their transformative impact on aerospace, military, and space exploration technologies continues to drive innovation, enabling the development of next-generation space capsules, reusable launch systems, and high-speed defense platforms. With ongoing advancements in intelligent manufacturing, computational material design, and artificial intelligence-driven optimization, nanoceramic-aerogel materials are poised to redefine the boundaries of extreme environment engineering and high-performance thermal protection solutions. Economically, the development and commercialization of nanoceramic aerogel materials are expected to exceed ten billion United States dollars in market value by the year 2030, driven by rapid advancements in aerospace, defense, and energy-efficient insulation technologies. The increasing demand for reusable launch vehicles, next-generation hypersonic platforms, and high-performance thermal protection systems is fostering significant investments in nanotechnology-enabled material solutions. Governments and private space agencies are actively exploring these materials for crewed space missions, lunar habitats, and interplanetary travel, where radiation shielding and extreme thermal management are critical challenges. Nanoceramics, which include materials such as zirconium oxide, silicon carbide, and yttrium stabilized zirconium dioxide, possess an exceptionally high melting point exceeding 2500 °C, along with superior oxidation resistance and mechanical integrity. These materials exhibit remarkable resistance to thermal shock, which is essential for applications in hypersonic reentry vehicles and military unmanned aerial vehicles, where materials must endure rapid temperature fluctuations exceeding 2000 °C/s. The high entropy and strong ionic and covalent bonding within nanoceramic lattices contribute to their ability to resist structural degradation under extreme heat loads, enabling prolonged operational lifespans for aerospace and defense applications. Aerogels, particularly those synthesized from silica, alumina, and carbon-based frameworks, provide unmatched thermal insulation properties due to their highly porous network, which consists of over 99% air by volume. The ultralow thermal conductivity of aerogels, often as low as 10 $mW{\cdot}m^{-1}{\cdot}K^{-1}$, significantly reduces heat transfer, making them indispensable for thermal protection systems in space exploration and hypersonic flight. The incorporation of nanoceramics within aerogel matrices further enhances their resistance to ablation, oxidation, and mechanical stress, ensuring structural integrity in extreme atmospheric reentry conditions where heat flux can exceed 10 MW/m^2. The advanced fabrication methods enable the development of tailored material architectures optimized for specific mission requirements, such as spacecraft heat shields, stealth coatings for military unmanned aerial vehicles, and thermal barrier coatings for jet engines. Computational materials science, powered by artificial intelligence and machine learning algorithms, is further accelerating the discovery of novel nanoceramic-aerogel compositions with enhanced thermomechanical stability and radiation shielding capabilities. The mechanical properties of nanoceramic-aerogel composites are engineered to provide superior impact resistance while maintaining minimal weight. The integration of nanoceramic fibers within aerogel networks enhances toughness, enabling the material to withstand extreme aerodynamic pressures and vibrations encountered in high-speed atmospheric flight. In military unmanned aerial vehicles, these materials contribute to improved operational stealth by minimizing infrared signatures, reducing thermal emissions, and increasing overall energy efficiency, thereby extending flight endurance and mission capabilities. Beyond aerospace and defense applications, nanoceramic-aerogel materials are also revolutionizing fields such as high-temperature fuel cells, nuclear reactor insulation, and fire-resistant building materials. The exceptional thermal stability and chemical inertness of these materials make them ideal for industrial applications requiring prolonged exposure to extreme heat and corrosive environments. The integration of carbon-based aerogels with nanoceramics is leading to the development of lightweight, flexible thermal insulators for wearable protective gear, biomedical implants, and advanced electronics, where precise thermal regulation is essential [28–31].

3 Intelligent Manufacturing for Aerospace, Military, and Defense

Metal 3D printing, utilizing SLM and direct MLS, is transforming the manufacturing landscape of military vehicles, including tanks and next-generation aircraft. Metal three-dimensional printing and nano-precision laser sintering are redefining the manufacturing paradigms of modern military and aerospace industries by offering unprecedented improvements in weight reduction, mechanical strength, fuel efficiency, and operational flexibility. The convergence of artificial intelligence, computational materials science, and intelligent automation with these additive manufacturing techniques is driving the evolution of next-generation combat vehicles, energy-efficient jet engines, and space propulsion systems, ultimately shaping the future of advanced engineering and global defense technologies. This cutting-edge technology enables the production of highly intricate, lightweight, and high-strength components with unprecedented material efficiency and mechanical performance. Through layer-by-layer laser fusion of metal powders, complex geometries can be fabricated with micron-level precision, allowing the development of lattice structures that maintain mechanical robustness while reducing weight by 50% compared to traditional forged and machined components. The rapid adoption of three-dimensional metal printing and nano-precision laser sintering is reshaping supply chain dynamics by enabling on-demand, decentralized manufacturing of critical aerospace and defense components. This shift minimizes reliance on complex logistics networks, reducing lead times from several months to a few days, while also allowing for localized production in war zones and remote aerospace facilities. The integration of artificial intelligence-driven predictive maintenance in conjunction with additive manufacturing further enhances operational reliability, as real-time monitoring of wear-and-tear in components can trigger automated printing of replacement parts, ensuring uninterrupted mission readiness [32–34].

In military tanks, the integration of three dimensional printed titanium alloys, high entropy alloys, and functionally graded materials contributes to enhanced armor durability while significantly reducing the overall mass, improving mobility, fuel efficiency, and battlefield agility. Similarly, in fighter aircraft, three dimensional printed nickel-based superalloys and titanium aluminides are being utilized in critical load-bearing structures, such as fuselage components and landing gears, ensuring high strength-to-weight ratios, corrosion resistance, and extended operational lifespans. The adoption of three-dimensional metal printing in aerospace manufacturing also eliminates the need for traditional multi-part assemblies, reducing material wastage by over 70% and lowering production costs by more than 30% through the reduction of machining and post-processing requirements. Nano precision laser sintering, specifically applied to jet engine turbine blades, is revolutionizing propulsion technology by enabling the fabrication of highly optimized aerodynamic structures with intricate cooling channels and functionally graded material compositions. This ultra high resolution manufacturing approach allows for the creation of heat-resistant superalloy blades with nanostructured grain boundaries that enhance creep resistance, oxidation resistance, and high-temperature mechanical performance. By using computational fluid dynamics simulations and artificial intelligence-driven design optimizations, the latest turbine blade architectures incorporate biomimetic geometries that reduce aerodynamic drag, enhance air circulation, and optimize combustion efficiency. The result is a significant increase in turbine efficiency by 10–15%, leading to a corresponding reduction in fuel consumption, improved thrust-to-weight ratios, and lower emissions, aligning with the global push toward sustainable aviation and reduced carbon footprints. In the broader context of intelligent manufacturing and Industry 6.0, the fusion of digital twin technology, real-time process monitoring, and machine learning algorithms with metal three-dimensional printing is enabling unprecedented levels of precision, customization, and adaptability in defense and aerospace engineering. Financially, the global market for three-dimensional metal printing in the aerospace sector alone is projected to exceed 20 billion US dollars by the year 2030, driven by rising demand for fuel-efficient aircraft, hypersonic weapons, and next-generation space exploration systems. Advancements in multi-material metal three-dimensional printing and directed energy deposition techniques are expanding the capabilities of functional gradient alloys, allowing for the seamless integration of different material properties within a single component. This is particularly beneficial for turbine blades, where a combination of nickel-based superalloys and ceramic matrix composites enhances heat resistance while maintaining structural flexibility. The ongoing research in laser-based additive manufacturing, including the development of high-power femtosecond lasers and electron beam melting, is pushing the boundaries of resolution and material versatility, paving the way for the fabrication of quantum-engineered materials with superior mechanical and thermal properties. The strategic implications of these technologies in defense and aerospace engineering extend beyond material efficiency and mechanical performance. By enabling the rapid prototyping of new military hardware, such as hypersonic missile casings, satellite propulsion systems, and next-generation fighter jet airframes, three-dimensional metal printing accelerates innovation cycles, shortens development timelines, and enhances national security capabilities. The cost savings associated with reduced material waste and improved fuel efficiency translate into significant financial advantages, making three-dimensional printing and nano-precision laser sintering indispensable tools in the future of high-performance engineering [12, 35, 36].

4 Autonomous Manufacturing Robots

AI-optimized autonomous manufacturing robots are revolutionizing industrial production by significantly reducing defects in vehicle manufacturing, enhancing efficiency, and minimizing material waste. AI-optimized autonomous manufacturing robots are fundamentally transforming vehicle production by reducing defect rates by 45% through self-learning algorithms, high-precision assembly mechanisms, predictive maintenance intelligence, and real-time adaptive control. Their impact extends beyond quality improvement, offering economic, environmental, and strategic advantages that drive the future of intelligent manufacturing, sustainable production, and next-generation vehicle development in an increasingly competitive and technologically advanced industrial landscape. These advanced robotic systems integrate machine learning algorithms, real-time computer vision, edge computing, and predictive analytics to achieve an unprecedented level of precision and adaptability in the production process. Unlike traditional automation, which follows pre-programmed routines, artificial intelligence-driven robotic systems continuously learn from data streams collected from high resolution sensors, enabling them to self-optimize and dynamically adjust their actions in response to real-time manufacturing variations. This results in a substantial reduction of production defects by 45%, leading to improved quality control, cost savings, and enhanced operational sustainability. From an economic perspective, the adoption of artificial intelligence-optimized autonomous robots in vehicle production is projected to drive down manufacturing costs by approximately 20% over the next decade. This cost efficiency is primarily attributed to reduced material wastage, lower defect rates, improved energy utilization, and minimized human intervention. By adopting quantum computing-enhanced artificial intelligence models, future autonomous robotic systems will further refine defect detection capabilities, optimize material allocation strategies and facilitate the development of next generation electric and hydrogen powered vehicles with superior structural integrity and aerodynamic performance. In the financial and market domains, companies that invest in artificial intelligence driven autonomous robotic manufacturing gain a substantial competitive advantage through improved production scalability, higher consumer satisfaction due to defect free vehicles, and enhanced brand reputation for technological leadership. Market analysis predicts that the global industrial robotics market will exceed two hundred billion US dollars by the year 2030, with artificial intelligence enhanced robotic automation constituting most of this market growth. The core technology behind artificial intelligence-driven robotic manufacturing includes high-precision robotic arms equipped with force feedback mechanisms and six-axis motion control, allowing them to execute highly intricate assembly tasks with sub-millimeter accuracy. These robotic systems are guided by neural network-based computer vision, which enables them to detect microstructural defects, surface irregularities, and material inconsistencies in real-time. By using generative adversarial networks and deep reinforcement learning, these robots can autonomously refine their movements, optimize assembly sequences, and anticipate potential failure points in vehicle components before they occur. The implementation of such systems reduces defect rates in vehicle chassis welding, electronic circuit integration, and aerodynamic component fabrication, ensuring higher product reliability and durability. AI-enhanced predictive maintenance algorithms are embedded within autonomous robots to continuously monitor their operational health. By utilizing digital twin technology and real-time data analytics, these robots can predict wear and tear in mechanical components, enabling automated self-repair protocols that minimize downtime. This predictive intelligence enhances the longevity of robotic systems, reducing maintenance costs by up to 30% and increasing overall equipment effectiveness by 25%. In the broader context of intelligent manufacturing, artificial intelligence-driven autonomous robots are deeply integrated into cyber-physical production networks, allowing seamless coordination with other smart factory systems, including additive manufacturing, real-time supply chain optimization, and intelligent resource allocation. Through blockchain secured industrial data exchange, these robotic systems enhance transparency, traceability and process efficiency in vehicle production, mitigating supply chain disruptions and ensuring just-in-time manufacturing precision. AI-powered robotics in vehicle production have significant environmental implications. By reducing material defects and waste, these systems contribute to the reduction of industrial carbon footprints, aligning with global sustainability goals such as net-zero emissions in automotive manufacturing. The integration of bio-inspired robotic designs, coupled with energy efficient actuation systems, further enhances the ecological benefits of these systems, paving the way for sustainable intelligent manufacturing in the era of Industry 6 [37–39].

5 Bio-Fabrication of Military Gear

Bio-fabrication of military gear using artificial intelligence designed biodegradable uniforms with self-healing properties represents a revolutionary advancement in materials science, nanotechnology, synthetic biology, and intelligent manufacturing. AI-designed biodegradable uniforms with self-healing properties represent a groundbreaking fusion of bio fabrication, nanotechnology, intelligent manufacturing, and synthetic biology, offering unparalleled advancements in durability, sustainability, and multifunctionality for modern military applications. These next-generation uniforms not only enhance soldier survivability and operational efficiency but also drive the transition toward eco-conscious defense technologies, revolutionizing military logistics and battlefield adaptability in the era of intelligent warfare and sustainable defense strategies. These next-generation uniforms integrate engineered biological systems, nanofiber reinforcement, and programmable self-repair mechanisms, offering unprecedented durability, environmental sustainability, and enhanced battlefield functionality. The core innovation behind these uniforms lies in the convergence of synthetic polymer engineering, genetically modified microbial systems, and artificial intelligence-driven molecular modeling, which enables the precise design of biomaterials with tunable mechanical strength, thermal resistance, and adaptive camouflage capabilities. The economic and logistical benefits of artificial intelligence-designed biodegradable uniforms extend beyond cost savings associated with reduced material waste and enhanced durability. These self-healing biofabricated textiles streamline military supply chain management, minimizing the need for extensive uniform inventories and lowering transportation costs. The adoption of decentralized biomanufacturing systems, including on-site microbial fermentation units for fabric regeneration, enhances battlefield sustainability by enabling in-field repair and uniform recycling. This approach reduces reliance on traditional textile production methods, which are resource-intensive and environmentally detrimental, fostering a more resilient and adaptable defense infrastructure. The biodegradable nature of these uniforms is achieved through the integration of advanced biopolymers such as bacterial cellulose, spider silk proteins, and polylactic acid-based composites, which exhibit exceptional mechanical performance while decomposing naturally under controlled environmental conditions. Unlike conventional synthetic fabrics, which generate persistent microplastic pollution, these biofabricated materials break down into non-toxic organic components within a predefined degradation cycle, reducing ecological impact and aligning with global sustainability objectives. The addition of nanostructured coatings derived from chitosan, graphene oxide, and silk fibroin enhances the fabric's resistance to microbial degradation, ultraviolet radiation, and chemical warfare agents, ensuring prolonged field usability while maintaining biodegradability. The self-healing capability of these uniforms is driven by bioengineered living materials, where genetically modified microorganisms, such as *Escherichia coli* and *Bacillus subtilis*, are embedded within microencapsulated hydrogels dispersed throughout the fabric matrix. These microorganisms remain dormant under normal conditions but become activated upon exposure to damage, secreting extracellular matrix proteins, biopolymers, and adhesion factors to repair microscopic tears, punctures, and abrasions. This bio-inspired self-repair mechanism mimics natural wound healing processes and extends the operational lifespan of military uniforms by more than 30%, significantly reducing the logistical burden associated with frequent uniform replacements. ML analyzes huge datasets of material compositions, stress-strain behaviors, and environmental degradation patterns to identify the most effective polymer configurations for durability and flexibility. Generative adversarial networks simulate battlefield conditions to predict uniform performance under extreme temperature fluctuations, high-impact forces, and exposure to toxic gases, enabling real-time customization of fabric properties based on mission-specific requirements. AI-driven robotic weaving techniques, such as electrospinning and three-dimensional bioprinting, facilitate the precision assembly of nanofiber networks with hierarchical architectures, maximizing mechanical resilience while maintaining lightweight flexibility. The incorporation of conductive biopolymers and piezoelectric nanomaterials within the uniform structure enables multifunctionality, including real-time health monitoring, energy harvesting, and adaptive thermal regulation. Conductive polymers such as polyaniline and poly(3,4-ethylenedioxythiophene) allow seamless integration of flexible electronic circuits, facilitating biometric data transmission for continuous soldier health assessment. The addition of carbon nanotube reinforced piezoelectric fibers enables kinetic energy conversion, allowing soldiers to generate electrical power through body movements, reducing dependence on external power sources for wearable electronics. The strategic deployment of biofabricated military gear contributes to broader defense innovation initiatives, including climate-resilient soldier equipment, autonomous biosensing uniforms, and responsive camouflage technologies. The integration of programmable biochromatic proteins within these fabrics allows dynamic color adaptation based on ambient light, terrain patterns, and electromagnetic field fluctuations, providing advanced stealth capabilities in reconnaissance and combat scenarios. Quantum computing assisted material simulations, future iterations of these uniforms will achieve even higher levels of self-repair efficiency, enhanced mechanical robustness, and superior biodegradability, establishing a new paradigm in sustainable military textile engineering [40–42].

6 Digital Twins and Predictive Maintenance

Digital twin technology and predictive maintenance powered by artificial intelligence-based real-time health monitoring have revolutionized military aircraft maintenance, significantly reducing operational costs by 40% while enhancing safety, efficiency, and aircraft longevity. AI-powered digital twins and predictive maintenance represent a paradigm shift in military aircraft health monitoring, reducing maintenance costs by 40% while maximizing fleet availability, operational safety, and mission success rates. By leveraging advanced material science, real-time data analytics, quantum computing, and intelligent automation, this technology is transforming aerospace engineering, logistics, and combat strategy, ensuring that future military aircraft achieve unparalleled reliability, resilience, and efficiency in modern warfare scenarios. This advanced approach allows military forces to shift from reactive or scheduled maintenance to a fully predictive, condition-based strategy, minimizing unexpected failures, optimizing fleet readiness, and extending aircraft lifespan. The implementation of artificial intelligence-driven digital twin technology leads to substantial cost reductions in military aircraft maintenance by streamlining spare parts inventory management and reducing unnecessary component replacements. Traditional maintenance strategies rely on fixed schedules, often leading to premature component disposal or prolonged aircraft downtime. In contrast, predictive maintenance enables just-in-time replacement strategies, optimizing supply chain logistics and reducing the need for excessive spare part stockpiling. Blockchain-enabled asset tracking further enhances transparency in aircraft maintenance history, ensuring compliance with aerospace regulations and facilitating seamless collaboration between military forces, defense contractors, and aerospace manufacturers. The foundation of digital twin technology in military aviation lies in the integration of physics-based simulations with high-resolution data acquisition systems. The statistics streams are processed by edge computing systems installed onboard, enabling immediate anomaly detection while simultaneously transmitting relevant data to centralized cloud-based platforms for deeper analysis using quantum computing-enhanced algorithms. By employing artificial intelligence-driven predictive analytics, military engineers can detect early signs of mechanical wear, fatigue-induced microfractures, and subsystem inefficiencies, allowing proactive interventions before failures occur. Material science is important in digital twin accuracy enhancement simulations in predictive maintenance. High-entropy alloys, shape memory composites, and carbon nanotube-reinforced polymeric structures used in aircraft components exhibit complex deformation behaviors under extreme operational conditions. To accurately predict material degradation, digital twins incorporate multi-scale computational models that simulate molecular interactions, thermal expansion coefficients, and corrosion kinetics under real-time operational stresses. Advancements in non-destructive testing techniques, such as terahertz spectroscopy, ultrasonic phased-array imaging, and X-ray computed tomography, enable precise validation of digital twin models against real-world aircraft conditions. The application of predictive maintenance through digital twins extends beyond mechanical components to avionics, propulsion systems, and electronic warfare modules. Artificial intelligence-based signal processing techniques analyze electrical impedance fluctuations, thermal dissipation patterns, and software-defined radio wave signatures to detect cyber-physical anomalies in critical flight control systems. This predictive intelligence ensures the stability of military aircraft under combat and reconnaissance missions, reducing the risk of electronic malfunctions that could compromise operational effectiveness. Moreover, machine learning-powered failure mode and effects analysis assigns risk probabilities to different aircraft subsystems, prioritizing maintenance schedules based on criticality and mission demands. Digital twins provide a robust platform for continuous design improvements in next-generation military aircraft. By analyzing real-time performance deviations between physical aircraft and their digital counterparts, aerospace engineers can iteratively refine design parameters, aerodynamics, and propulsion efficiencies. This iterative optimization reduces reliance on expensive wind tunnel testing and accelerates the development of hypersonic aircraft, unmanned aerial vehicles, and stealth fighters equipped with artificial intelligence-assisted autonomous flight capabilities. Digital twin technology facilitates virtual prototyping of novel materials, such as high-temperature ceramics and graphene-based structural composites, enabling rapid advancements in lightweight, fuel-efficient, and damage-tolerant aircraft architectures. AI-enhanced digital twin applications extend to pilot training and mission simulation. Virtual reality-based flight simulators leverage real-time aircraft performance data to create hyper-realistic training environments where pilots experience dynamically updated mission scenarios based on actual aircraft behavior. This adaptive training methodology improves combat readiness and enhances decision-making capabilities in high-risk aerial operations [1, 43–45].

7 Quantum Cryptography for Secure Defense Manufacturing

Quantum cryptography and cyber-physical security in military defense manufacturing represent a groundbreaking evolution in securing autonomous weapon systems, military grade autonomous vehicles, and defense supply chains from cyber threats. This property ensures the unbreakable security of classified defense manufacturing data, preventing cyber espionage, unauthorized data breaches, and interference in autonomous weapon system development. The implementation of quantum cryptography and cyber-physical security in autonomous defense manufacturing reduces the financial risks associated with cyber attacks, intellectual property theft, and supply chain disruptions. The estimated financial impact of cyber threats on global defense industries exceeds one hundred billion dollars annually, with adversarial state-sponsored cyber warfare posing increasing risks to military technological advancements. The adoption of blockchain-based defense manufacturing verification, secured by quantum-resistant encryption, ensures the authenticity of every component produced in military supply chains, preventing the infiltration of counterfeit or compromised parts that could sabotage autonomous weapon systems. Quantum-secured communication networks in military factories utilize entangled photon pairs and single-photon transmission to protect sensitive data across interconnected manufacturing units, eliminating vulnerabilities present in conventional encryption algorithms, which can be broken by quantum computing-based attacks. The introduction of post-quantum cryptographic algorithms alongside quantum cryptography further fortifies the resilience of military communication systems against emerging cyber threats. The integration of quantum cryptography and cyber-physical security in autonomous defense manufacturing represents a transformative leap in military cybersecurity, ensuring that cyber threats cannot compromise the integrity of advanced weapon systems, autonomous vehicles, and classified defense intelligence. Quantum-resistant cryptographic protocols, artificial intelligence-driven threat detection, and secure material science-based semiconductor fabrication, military forces worldwide are reinforcing their strategic defense capabilities against next-generation cyber warfare, ensuring the secure production, deployment, and operation of autonomous military technologies. Cyber-physical security in military autonomous vehicle and weapon manufacturing integrates artificial intelligence-driven anomaly detection, blockchain-secured supply chain verification, and real-time machine-learning algorithms to prevent cyber-physical attacks that could compromise automated defense systems. Autonomous vehicles, including unmanned combat aerial vehicles, autonomous naval vessels, and ground-based robotic military units, rely on interconnected sensor networks, machine vision, and embedded artificial intelligence processors to operate in complex battle environments. Cyber-physical threats targeting these systems can manipulate software-controlled navigation, disrupt sensor fusion algorithms, or introduce malware into robotic control frameworks, leading to catastrophic operational failures. To counteract these threats, intelligent cybersecurity frameworks based on artificial intelligence-driven threat intelligence continuously monitor network behavior, detect deviations from normal operational parameters, and implement automated countermeasures such as self-healing code, secure firmware updates, and encrypted control signals. Advanced material science enhanced the security of quantum cryptographic hardware and cyber-physical security modules. The fabrication of quantum photonic processors, superconducting qubits, and solid-state quantum dots requires ultra-pure silicon carbide substrates, diamond-based nitrogen vacancy centers, and low-temperature cryogenic cooling technologies. These materials enable stable quantum coherence and high-fidelity qubit operations, ensuring robust quantum encryption performance under extreme military operational conditions. Additionally, hardware-based cyber-physical security mechanisms utilize physically unclonable functions embedded within defense-grade semiconductor chips, ensuring that every cryptographic key used in manufacturing automation systems is uniquely tied to the physical properties of the hardware itself, making counterfeiting or remote exploitation difficult. The use of AI-driven predictive analytics in cyber-physical defense manufacturing prevents real-time cyber intrusions by identifying malicious code injections, insider threats, and zero-day vulnerabilities before they can disrupt production lines. Military factories leverage secure edge computing devices, encrypted cloud infrastructures, and quantum-secured data centers to store and process classified manufacturing blueprints, ensuring that cyber attackers cannot gain access to proprietary autonomous weapon designs or modify software-controlled defense systems remotely. The application of homomorphic encryption allows defense engineers to perform secure computations on encrypted data without ever decrypting it, enabling collaborative international defense manufacturing partnerships while maintaining absolute security. The deployment of cyber-physical security frameworks in autonomous vehicle production ensures that every step of the manufacturing process from digital twin simulations to real-world assembly line automation, operates within a completely secured environment. Artificial intelligence-powered robotic arms, additive manufacturing machines, and high-precision nanofabrication tools used in next-generation defense vehicle assembly are continuously monitored through encrypted data streams, preventing unauthorized modifications or cyber-physical sabotage attempts. The use of secure firmware updates through blockchain-backed version control ensures that military autonomous systems receive verified software patches, eliminating vulnerabilities that could be exploited by adversarial cyber forces. Quantum cryptography enhances military battlefield communication by enabling unbreakable, real-time encrypted transmissions between autonomous defense units, command centers, and space-based military satellites. Quantum key distribution (QKB) ensures that classified military operation plans remain protected from decryption by enemy intelligence agencies, even in the event of intercepted transmissions. Secure military cloud architectures based on quantum encryption protect classified artificial intelligence training datasets, ensuring that adversarial forces cannot manipulate or corrupt military machine learning models used in battlefield decision-making. QKD improves security but its challenging to control device calibration errors, side channel hardware attacks, malicious detectors etc [37, 46–48].

8 Adaptive Supply Chains

AI-powered adaptive supply chains represent a monumental leap in materials management, driven by the fusion of intelligent manufacturing technologies. The ability of AI to predict global material shortages with high accuracy hinges on advanced algorithms that integrate vast datasets, including global supply chain performance, material consumption trends, production rates, geopolitical factors, climate conditions, and demand forecasts. These systems leverage deep learning techniques and reinforcement learning to adapt to supply chain dynamics, enabling proactive decisions that mitigate potential disruptions. The impact of AI on supply chains is profound. According to studies, AI-powered supply chain management can reduce operating costs by up to 20%, which translates into savings of billions of dollars annually for large multinational corporations. In addition, the predictive capabilities of these systems provide better strategic planning, allowing companies to negotiate better contracts with suppliers, secure lower material prices, and mitigate the risks associated with volatile raw material costs. The accurate forecasting of material shortages enables companies to negotiate in advance, locking in prices before supply chain crises can drive prices higher. At the core of this high-tech infrastructure is the application of big data analytics, which synthesizes information from diverse sources like weather patterns, shipping routes, geopolitical events, and even consumer behavior trends. Advanced statistical models, such as time-series forecasting and causal impact analysis, assess the complex interdependencies in global supply networks. AI algorithms process billions of data points per second to simulate various "what-if" scenarios, predicting shortages, delays, or bottlenecks well in advance. These predictive capabilities extend to commodities like metals, rare earth elements, plastics, and energy resources. AI models predict shortages in critical raw materials by considering not only market trends but also geological data on the availability of rare earth minerals and mining efficiency. The integration of material lifecycle data, such as recycling rates, energy consumption during production, and environmental impact, informs the AI's decision-making, enabling sustainable sourcing strategies. Advanced algorithms can calculate the environmental cost of extracting a certain material, prioritize the use of recycled materials, or identify alternative materials with similar properties, ensuring production continues smoothly while minimizing environmental impact. This is especially important in industries such as electronics, automotive, and aerospace, where supply chain disruptions due to material shortages have far-reaching consequences. AI-powered systems integrate supply chain risk management tools that utilize real-time tracking and blockchain technology. These tools ensure transparency and traceability, confirming the origin of raw materials and verifying the authenticity of suppliers. Blockchain verifies transactions, ensuring that no counterfeit or substandard materials enter the supply chain, enhancing the resilience of critical industries like defense and medical equipment manufacturing. Furthermore, the adaptability of AI models allows for the dynamic rerouting of materials and products, optimizing transportation logistics in response to disruptions such as natural disasters, political instability, or pandemics. AI-driven systems are now integrated into production facilities, where they optimize the inventory and procurement processes. Intelligent manufacturing systems powered by AI predict when to reorder raw materials based on real-time usage rates, adjusting order quantities dynamically in response to changing production schedules or demand. This prevents overstocking, reduces warehousing costs, and ensures that there is no downtime due to material shortages. The advanced AI models can also optimize the distribution of materials across the complete supply. This results in the reduction of transportation costs, cutting down on fuel consumption and the associated carbon emissions, making the supply chain more sustainable. This AI approach is essential for just-in-time manufacturing models, where precision in timing and material availability is critical to maintaining production flow and preventing excess inventory. AI-powered adaptive supply chains represent a technological leap that not only predicts material shortages with high accuracy but also ensures the continuous, optimized operation of production systems across industries. The result is a robust, self-adjusting system capable of maintaining a steady flow of critical materials while minimizing risks and maximizing the efficiency of production processes in an increasingly complex global market [46, 48–51].

9 Autonomous Defense Manufacturing Systems

Autonomous defense manufacturing systems, particularly AI-powered military drones, represent the cutting edge of modern warfare, leveraging advancements in artificial intelligence, robotics, materials science, and aerospace engineering to create highly efficient, responsive, and versatile unmanned aerial vehicles (UAVs). Swarm intelligence, a key component in these systems, allows many drones to work collaboratively in sequence, without direct human intervention. Each drone in the swarm can make decisions based on local information, exchanging data with other drones in real-time, which enables complex, adaptive behaviors during battlefield surveillance and reconnaissance missions. This approach drastically enhances operational capabilities, providing highly effective and scalable solutions for monitoring large areas with minimal human risk. The use of swarm AI drones in autonomous battlefield surveillance is based on sophisticated algorithms that integrate sensor fusion, machine learning, and edge computing. These drones have highly sophisticated sensors, such as high-definition cameras, infrared sensors, and radar, capable of detecting and tracking targets with unprecedented precision. Machine learning algorithms onboard the drones process data locally, analyzing real-time footage for threats and anomalies, while swarm intelligence algorithms coordinate the actions of the drones to avoid collisions, ensure optimal coverage, and enhance mission success rates. By sharing critical environmental data among themselves, the drones in the swarm can quickly adapt to changing conditions on the battlefield, dynamically adjusting their positions and strategies. These drones require lightweight, durable, and heat-resistant materials for both structural components and power systems. Advanced composites, such as carbon fiber-reinforced polymers, are commonly used in UAV construction, providing high strength-to-weight ratios that enable greater maneuverability and longer flight times. In the era of autonomous manufacturing, the economic impact of these autonomous systems is profound, as they reduce the need for human soldiers in dangerous surveillance and reconnaissance missions, lowering operational costs and minimizing casualties. Autonomous UAVs are scalable, allowing for flexible deployment based on mission needs. The cost-effectiveness of manufacturing these drones, driven by advances in additive manufacturing (3D printing) and automation, further drives down production costs, making them accessible to a wide range of military forces across the globe. The integration of AI in defense manufacturing is reshaping traditional supply chain processes. By utilizing real-time data analytics, AI algorithms predict demand for UAV parts, optimize inventory management, and streamline logistics, ensuring that military forces have rapid access to essential components during operations. This level of efficiency significantly reduces the lead time for drone production and deployment, enabling military forces to respond to threats with agility. AI-based defense systems also incorporate advanced encryption protocols and quantum-resistant technologies to protect the data transmitted between drones and central command centers. This ensures that critical mission data remains secure from cyber threats, an essential feature when deploying autonomous systems in sensitive

Fig. 11.1 Industry 6.0 technologies in defense, aerospace and weapon industries

military operations. The convergence of robotics, AI with ML, and materials science in autonomous defense manufacturing systems contributes to the broader field of intelligent manufacturing. Using AI-driven optimization algorithms and machine learning, autonomous drones continuously improve their performance, learning from past missions to increase efficiency, accuracy, and responsiveness in future operations. AI-powered military drones and autonomous defense manufacturing systems represent a transformative shift in modern warfare. Using autonomous drones in military surveillance, reconnaissance, and even offensive operations will likely expand, making them a cornerstone of future defense strategies [38, 52, 53]. Figure 11.1 shows the basics of Industry 6.0 technologies in these industries.

10 Hypersonic Weapon Systems and Smart Ammunition

Hypersonic weapon systems and smart ammunition represent a breakthrough in modern military technology, combining advanced materials science, high-performance computing, and artificial intelligence to deliver unprecedented speed, precision, and efficiency. Hypersonic weapons, which travel at speeds greater than Mach 5 (5 times the speed of sound), present a unique challenge in terms of materials, propulsion, and guidance systems. These weapons operate at extreme altitudes and velocities, subjecting them to intense aerodynamic heating, pressure, and mechanical stresses. To withstand these conditions, cutting-edge materials such as carbon-carbon composites, ultra-high temperature ceramics, and refractory metal alloys are employed. For example, materials like tungsten and tantalum alloys are integral to the nosecones and thermal protection systems, offering exceptional resistance to heat, erosion, and thermal shock at temperatures exceeding 2000 °C. These advanced materials allow the vehicle to maintain structural integrity while performing high-speed maneuvers. The sensors typically include infrared and electro-optical sensors, radars, and accelerometers, which gather environmental information to guide the weapon along its flight path. Machine learning algorithms onboard the weapon's control system continuously adjust the trajectory by making minute corrections to the flight path based on feedback from the environment, target motion, and wind conditions. This dynamic adjustment ensures that even under unpredictable circumstances, the weapon maintains its accuracy, even against fast-moving or evasive targets. The integration of smart ammunition with hypersonic weapons also benefits from AI-controlled trajectory correction. Smart ammunition, which includes guided missiles and projectiles, is equipped with onboard guidance systems capable of making real-time customization during flight. AI dataset continuously monitors input from the weapon's sensors, predicting the most efficient flight path to maximize impact accuracy. This can be particularly useful in high-velocity, high-precision scenarios, where even small deviations can result in mission failure. The combination of AI and advanced materials science results in a precision increase of 25%, which directly translates into a reduction in collateral damage, enhanced mission success rates, and greater operational efficiency. Hypersonic fighter jets use these same principles of AI and precision-guided systems to enhance combat capabilities. These fighter jets are designed to operate at extreme speeds while maintaining agility and control. Advanced jet propulsion technologies, such as scramjet engines, are employed to achieve sustained hypersonic flight. These engines utilize the high-speed airflow to compress incoming air before combustion, enabling efficient propulsion at speeds greater than Mach 5. Materials like graphene and carbon nanotubes are incorporated into the structure of these jets to minimize weight while maintaining strength and heat resistance. AI-driven avionics systems optimize flight performance in real-time, adjusting speed, altitude, and flight path to ensure maximum combat effectiveness. On a broader technological front, the convergence of artificial intelligence with autonomous systems in both hypersonic weapons and fighter jets is part of a larger trend in defense technology. By utilizing vast data sets, high-performance computing, and predictive analytics, AI allows for faster, more informed decision-making processes, improving both offensive and defensive strategies. Predictive maintenance, powered by AI, can also ensure that weapon systems are always operational, with autonomous systems diagnosing potential issues and suggesting or implementing corrective actions in real-time [54–56].

11 Autonomous Factory Robots

The development of autonomous factory robots and defense systems represents a convergence of cutting-edge technologies in artificial intelligence, materials science, quantum sensors, and advanced manufacturing. The ability of AI-driven robots to assemble aircraft fuselages with 99.9% accuracy, combined with the adaptive, AI-controlled armor and quantum sensors in autonomous combat vehicles, demonstrates the potential for these technologies to revolutionize manufacturing and defense systems. Autonomous factory robots, particularly advanced systems like those developed by companies such as Boston Dynamics and ABB, represent the forefront of intelligent manufacturing. These AI-driven robotic arms, often equipped with sophisticated machine learning algorithms, can assemble complex structures, such as aircraft fuselages, with extraordinary precision. The integration of artificial intelligence enables these robotic arms to continually improve their accuracy, possibly achieving up to 99.9% precision in assembly. This level of accuracy is achieved through real-time data processing, where the robots continually adjust their movements based on feedback from numerous sensors. These sensors allow the robots to identify and correct even the smallest deviations in positioning, ensuring that each component is assembled with exact specifications. The use of AI further enhances the process by enabling robots to learn from previous tasks, optimizing their performance with each cycle, and reducing the potential for human error in manufacturing. The benefits of autonomous factory robots are not limited to increased precision and efficiency. These robots are also designed to operate autonomously, minimizing the need for human intervention in hazardous or repetitive tasks. For example, in the aerospace industry, AI-driven robots can handle delicate materials, such as carbon composites and titanium alloys, with minimal risk of damage, significantly enhancing the overall quality and durability of aircraft parts. The combination of AI, automation, and advanced materials science ensures that aircraft components are produced with unparalleled quality and consistency. Additionally, the use of intelligent manufacturing systems can significantly reduce production times, lower labor costs, and increase overall throughput. In defense technologies, the advent of autonomous combat vehicles and fighter jets is pushing the boundaries of modern warfare. Autonomous combat tanks, for example, are equipped with adaptive AI-controlled armor and integrated quantum sensors, which allow these vehicles to dynamically respond to threats and environmental conditions in real-time. This capability is enhanced by the integration of AI systems, which analyze the battlefield in real-time and activate specific armor configurations in response to incoming threats. Quantum sensors, integrated into these defense vehicles, leverage principles of quantum mechanics to detect minute changes in electromagnetic fields, vibrations, and other physical properties, providing unparalleled situational awareness and threat detection. Quantum sensors offer several advantages over classical sensors, including their ability to detect faint signals and provide higher resolution measurements. These sensors are critical for the operation of autonomous defense vehicles, where they can be used for a variety of purposes, such as detecting enemy movement, identifying potential hazards, and optimizing the vehicle's movement through challenging terrain. By incorporating quantum sensors into the vehicle's AI system, these tanks can make decisions autonomously, such as adjusting speed or positioning in response to perceived threats, without the need for human input. This capability is a game-changer in modern warfare, as it allows for faster decision-making and a more agile response to dynamic combat situations. Similarly, autonomous fighter jets have high-level AI-operating systems to fly without human intervention, autonomously adjusting their flight paths, targeting systems, and weaponry based on real-time data inputs. These jets rely on a combination of advanced avionics, machine learning, and sensor fusion to navigate and engage targets with incredible precision. The integration of AI in fighter jets not only improves combat effectiveness but also enhances the safety of pilots by reducing the likelihood of human error in complex combat scenarios. Advanced materials, such as graphene-based composites and ultra-lightweight titanium alloys, are often used in the construction of these fighter jets to ensure they can withstand the extreme conditions of high-speed flight while minimizing weight and maximizing fuel efficiency. The use of AI and autonomous systems in both defense vehicles and manufacturing processes also opens the door for greater integration of intelligent manufacturing techniques. Autonomous vehicles and fighter jets could be manufactured using additive manufacturing, where complex components are produced layer by layer with extreme precision. The combination of AI, advanced materials, and intelligent manufacturing creates a synergistic effect, where the capabilities of autonomous systems are enhanced by the materials used in their construction, while the manufacturing process itself is optimized for efficiency and precision [57–59].

12 Self-Assembling Space Structures

Self-assembling space structures represent a transformative breakthrough to explore space and to establish acceptable infrastructure. The nanotechnology and AI mixture is fruitful in the creation of space stations in orbit holds enormous potential for reducing the cost and complexity of space missions. AI-driven nano-robots are capable of autonomously assembling complex structures by coordinating at the nanoscale, leveraging the principles of molecular self-assembly, where materials spontaneously arrange themselves into structured forms based on the inherent properties of the materials involved. Carbon nanotubes possess extraordinary mechanical strength, approximately 100 times that of steel, while being incredibly lightweight. These characteristics make them ideal for the construction of space stations, as they offer superior structural integrity without adding substantial mass. The use of graphene provides exceptional thermal and electrical conductivity, enhancing the overall performance of space structures in extreme conditions such as deep space. The self-assembling nano-robots would utilize these advanced materials to build highly durable, energy-efficient space habitats that can be deployed with minimal human presence in orbit, fundamentally changing the way space exploration missions are conducted. In parallel, the development of self-healing materials for aircraft is revolutionizing aerospace engineering by significantly enhancing the durability and longevity of aircraft structures. AI-driven coatings that autonomously repair microcracks in aircraft fuselages are a key example of this innovation. These self-healing coatings contain embedded microcapsules or nanoparticles that release healing agents such as epoxy resins or other polymeric compounds, when the material experiences damage. When a crack forms, the coating detects the structural breach and initiates a chemical reaction that repairs the damage, restoring the material's integrity without human intervention. This process is particularly valuable in aerospace applications, where the constant stresses and extreme conditions experienced by aircraft can lead to the formation of microcracks that, if left unchecked, could lead to catastrophic failure. The self-healing mechanisms rely on advanced chemistry and material science, incorporating stimuli-responsive polymers and advanced resin systems that can activate autonomously in response to mechanical damage. The incorporation of shape-memory polymers allows the coating to not only repair cracks but also return to its original shape after deformation. Incorporating nanoparticles such as titanium dioxide or silver into the coatings can enhance the antimicrobial properties, ensuring the surface remains protected from microbial corrosion while also extending the material's lifespan. These innovations significantly reduce maintenance costs and downtime for aircraft, as well as enhance the overall safety and performance of the aircraft in demanding environments. The AI systems controlling these self-healing materials can be integrated with real-time monitoring sensors that detect and track the development of cracks, corrosion, and other forms of wear. These sensors provide continuous feedback to the AI system, enabling predictive maintenance algorithms to identify areas at risk of failure before they become critical. Both self-assembling space structures and self-healing aircraft materials represent significant advancements in their respective fields, driven by the synergy between artificial intelligence, nanotechnology, and advanced material science [60–62].

13 Industry 6.0 Military Helicopters

The development of AI-powered military helicopters that integrate real-time enemy threat prediction and hypersonic missile defense systems represents a groundbreaking shift in modern military aviation, combining advanced machine learning algorithms, data analytics, and cutting-edge defense technologies. These helicopters would contain sensors, such as infrared and radar systems, as well as an array of communications and data fusion platforms, all of which feed into an AI-driven system that can analyze the tactical environment in real-time. Real-time enemy threat prediction is achieved through a combination of deep learning techniques and predictive analytics, which utilize historical combat data, environmental variables, and real-time input from onboard sensors. By recognizing patterns and anomalies in the data, the AI system can predict the trajectory of enemy movements, the likelihood of missile launches, or even cyber threats, allowing the military helicopter to take evasive action or prepare countermeasures well in advance. Convolutional neural networks can be employed to process the visual and infrared data streams, enabling real-time object detection, classification, and tracking. In conjunction with reinforcement learning algorithms, the helicopter's AI can continuously adapt and improve its decision-making processes based on evolving battlefield conditions.

The integration of hypersonic missile defense systems is an essential component of this AI-powered military helicopter's capabilities. Hypersonic missiles (5 times faster than sound speed) present a significant challenge for traditional missile mechanisms because of their high speed and maneuverability. To counter this threat, the helicopter would incorporate cutting-edge technologies, such as high-speed interceptors, directed energy weapons, and advanced radar systems that can track and engage hypersonic threats with exceptional precision. The AI system would leverage real-time data from these defense systems, making rapid calculations and predictions to optimize the timing of interception and the deployment of countermeasures. The AI system would then assess the threat and use advanced algorithms to determine the optimal counteraction strategy, whether that involves deploying electronic jamming systems, activating kinetic interceptors, or engaging in a directed energy defense. With laser-based directed energy systems, such as high-energy lasers, the helicopter could potentially destroy hypersonic missiles mid-flight by focusing concentrated beams of light on the target, melting or disintegrating the missile's guidance system or warhead. The helicopter's ability to withstand the extreme conditions associated with both hypersonic defense and high-speed flight would rely on advanced composite materials and nanotechnology. Strong and lightweight material chemistry of carbon fiber-reinforced composites, graphene-infused ceramics, and advanced titanium alloys would be used in the construction of the helicopter's fuselage and components to withstand the intense heat generated by hypersonic speeds, while maintaining the structural integrity required for combat operations. Graphene would be ideal for its high thermal conductivity and resistance to extreme temperatures, while also providing enhanced strength-to-weight ratios that are crucial for aerospace applications. The AI system itself would be embedded in a secure, fault-tolerant computing platform capable of processing the enormous amounts of data required for these complex functions. This would involve the use of advanced computing hardware such as quantum processors, which can exponentially accelerate computations, enabling the AI to make predictions and calculations in real-time with far greater efficiency than traditional computing systems. Edge computing would be employed to ensure that critical AI processing is done onboard the helicopter, reducing the reliance on external communications and minimizing the risks posed by electronic warfare or cyber attacks. The fusion of artificial intelligence with real-time threat prediction and hypersonic missile defense in military helicopters offers a multi-faceted approach to enhancing defense capabilities. By integrating AI-driven decision-making with cutting-edge sensor technologies, advanced materials, and high-speed missile defense systems, these helicopters can significantly improve operational effectiveness, survivability, and response times in modern combat environments. As the technology continues to advance, future generations of military helicopters will likely see even more sophisticated AI systems capable of autonomously responding to an ever-changing battlefield, further solidifying the role of artificial intelligence in military operations [1, 63–65].

14 Space Grade Nuclear Power for Deep Space Missions

Space-grade nuclear power for deep space missions is a highly advanced and reliable technology, using radioisotope thermoelectric generators (RTGs) to provide consistent and long-duration power to spacecraft in environments far from the Sun, where solar energy is insufficient. RTGs are a crucial component in space exploration, especially for missions to outer planets, moons, asteroids, or even interstellar travel, where traditional power sources like solar panels cannot operate efficiently. These generators work by converting the heat produced from the natural decay of radioisotopes into electricity using thermoelectric materials. The heart of the RTG system is the use of radioisotopes, such as plutonium-238, which undergoes alpha decay to produce heat. Plutonium-238, with a half-life of 87.7 years, is particularly suitable for deep space missions because it provides a stable and long-term heat source, crucial for spacecraft traveling beyond the reach of the Sun's radiation, where alternative power generation methods fail. The heat generated by the decaying plutonium is transferred to thermoelectric materials that convert the temperature difference into electrical energy. The efficiency of this process is typically around 5–7%, which, although low, is sufficient for deep-space missions, where other power sources are impractical. The thermoelectric materials used in RTGs are carefully selected for their ability to withstand extreme temperatures and provide efficient energy conversion. Thermocouples made from materials such as telluride-based alloys (like bismuth telluride and lead telluride) are commonly used. These materials possess a high Seebeck coefficient, and advanced ceramics and composite materials are employed to ensure that the RTG remains stable under the harsh radiation and thermal extremes of space. The design of RTGs is also optimized for safety, especially considering the radioactive nature of the isotopes used. The plutonium-238 is typically encased in a multi-layered containment system that includes robust materials like iridium and graphite, ensuring that no radioactive particles are released in the event of a launch failure or crash. The containment systems are engineered to withstand the heat, shock, and potential impact of a spacecraft's reentry into the Earth's atmosphere or landing on foreign planetary bodies. In recent designs, advanced safety protocols and fail-safe mechanisms are incorporated, including redundancy in power systems. The use of space-grade nuclear power, particularly radioisotope thermoelectric generators, has been vital in powering a range of significant deep space missions, such as NASA's Voyager spacecraft, the Curiosity Rover on Mars, and the New Horizons mission to Pluto. These missions have demonstrated the extraordinary reliability of RTGs, with some generating power for decades, well beyond their intended lifetimes. For example, the Voyager 1 spacecraft, launched in 1977, is still transmitting data back to Earth after more than 45 years of operation, with power levels sufficient to continue key communications and scientific experiments. As we look toward the future of deep space exploration, including crewed missions to Mars, the Moon, or potentially even interstellar travel, space-grade nuclear power systems are likely to evolve further. Advances in materials science and nuclear technology may yield more efficient thermoelectric materials or alternative nuclear power sources, such as compact fission reactors, which could provide higher power outputs. ML with AI could be used to optimize power management systems, automatically adjusting the spacecraft's energy use based on real-time conditions and mission needs. Space-grade nuclear power systems using radioisotope thermoelectric generators are a crucial part of enabling long-duration space exploration. The combination of highly reliable nuclear decay heat sources and advanced thermoelectric materials allows spacecraft to function in deep space environments where traditional solar or chemical power systems are unfeasible. With ongoing advancements in material science, safety protocols, and computational technologies, the future of space exploration will continue to rely on these robust and innovative power solutions [17, 66–68].

15 Smart Manufacturing for Cargo and Passenger Aircraft

The integration of hydrogen-powered aircraft in smart manufacturing for cargo and passenger aviation represents a revolutionary shift toward decarbonizing the aviation sector, which is the main contributor to global warming. Hydrogen, as a clean fuel, has high potential to reduce 75% of CO_2 reduction as compared to the traditional fuel used in jet engines, marking a significant milestone in addressing climate change. This reduction is primarily because hydrogen combustion produces only water vapor and heat as byproducts, avoiding the harmful pollutants. Hydrogen-powered aircraft require the development of specialized materials that can safely store, transport, and utilize hydrogen fuel. These materials must withstand the extreme conditions in aviation, including cryogenic temperatures and high-pressure environments, as hydrogen is typically stored in liquid form at cryogenic temperatures (around −253 °C) or in high-pressure tanks at up to 700 bars. CFRPs are being utilized for the construction of hydrogen storage tanks, providing a lightweight yet strong solution for the containment of hydrogen fuel. These composites also exhibit excellent resistance to fatigue and environmental degradation, which is critical for the long-term operation of aircraft. Hydrogen-powered aircraft also necessitate the design and implementation of novel propulsion systems. Unlike conventional combustion engines, hydrogen-powered propulsion relies on either hydrogen combustion in modified gas turbines. It offers greater efficiency by eliminating the mechanical losses associated with traditional engines. Fuel cells, such as proton exchange membrane fuel cells, operate at high efficiency levels (around 60%) and can be coupled with batteries to ensure smooth energy transitions during flight phases like take off and landing. The advancement of smart manufacturing techniques is critical for the widespread adoption of hydrogen-powered aircraft. An AI system can simulate and predict fuel efficiency, structural integrity, and aerodynamics, which can significantly reduce development costs and timelines. Additive manufacturing is also playing a role in producing complex components for hydrogen aircraft, particularly in the construction of lightweight, high-strength parts that are essential for optimizing fuel efficiency and reducing aircraft weight. Hydrogen-powered aircraft must also be designed with robust safety systems to prevent the risks associated with hydrogen, such as leakage or explosion. Advanced sensors and monitoring systems are integrated into the aircraft to continuously assess hydrogen storage conditions, detect leaks, and ensure the integrity of the system. The aircraft's avionics and propulsion control systems are enhanced with real-time data analytics, enabling dynamic flight optimization and reducing energy consumption during various flight phases. The hydrogen-powered aircraft market is poised for significant growth as global aviation authorities and stakeholders, including aerospace manufacturers and airlines, prioritize sustainability. The reduction in carbon dioxide emissions is crucial for achieving the International Civil Aviation Organization's targets to reduce aviation emissions by 50% by 2050. However, the widespread adoption of hydrogen-powered aircraft depends on the establishment of a global hydrogen infrastructure, including refueling stations and supply chains capable of producing and transporting green hydrogen at scale. Hydrogen production cost on a mass scale is expected to fall significantly over the coming decades, making it more economically viable. Hydrogen-powered aircraft represent a groundbreaking step toward a sustainable aviation industry because hydrogen fuel can reduce carbon dioxide emissions up to 75 percent, when compared with the conventional jet fuel-powered aircraft. The successful implementation of this technology relies on advancements in materials science, propulsion systems, smart manufacturing techniques, and safety protocols. As these technologies mature, hydrogen-powered aircraft will likely become a cornerstone of global efforts to decarbonize the transportation sector [4, 69–71].

16 Detachable Cabin Aircraft

The concept of a detachable cabin in aircraft has been explored as a potential safety measure to enhance passenger survival during emergencies. In 2016, Ukrainian engineer Vladimir Tatarenko proposed a design featuring a passenger cabin that could detach from the main aircraft during critical situations. Upon separation, the cabin would deploy parachutes to facilitate a controlled descent, aiming to safely deliver passengers to the ground or water. This design also included provisions for passenger luggage, ensuring that cargo remained intact during the detachment process. Similarly, aerospace manufacturer Airbus filed a patent in 2015 for a modular aircraft design incorporating a detachable cabin. Unlike Tatarenko's safety-focused approach, Airbus's concept aimed to streamline boarding and disembarking processes. The detachable cabin would allow passengers to board the module at the gate, which would then be attached to the aircraft's main body prior to take off. This modular approach was intended to enhance operational efficiency rather than serve as an emergency safety feature. Despite these innovative proposals, the implementation of detachable cabin technology in commercial aviation has not been realized.

Challenges such as structural integrity, weight considerations, and the complexity of ensuring passenger safety during the detachment and descent phases have impeded practical application. As of now, no major aircraft manufacturers have announced plans to integrate detachable cabin systems into their fleets, and the concept remains in the theoretical and experimental stages. In the context of Industry 6.0, which emphasizes advanced automation, artificial intelligence, and interconnected systems, the development of such safety features would necessitate significant technological advancements. This includes the integration of AI-driven decision-making processes to determine optimal detachment timing, advanced materials to ensure structural integrity during detachment, and sophisticated control systems to manage the descent and landing of the detached cabin. While the idea holds promise, substantial research and development efforts are required to address the myriad technical and safety challenges before such a system could be considered viable for commercial use.

17 Next Generation Helicopters and Urban Air Mobility

With their effective, economical, and low-emission options for both passengers and freight, next-generation helicopters and urban air mobility systems, especially Electric Vertical Take-Off and Landing (eVTOL) aircraft, have the potential to completely transform urban transportation. Several cutting-edge technologies from the fields of material science, chemistry, engineering, and artificial intelligence are crucial to the growth of the eVTOL aircraft market, which is predicted to reach a value of around USD 30 billion by 2030. The design and implementation of energy storage and propulsion systems that can support vertical takeoff and landing while offering enough range and endurance for urban flight operations. Lithium-ion batteries are also employed in eVTOL aircraft. However, the power density of these batteries needs to be significantly improved to meet the high energy requirements of eVTOL systems. Research is ongoing into high-capacity batteries like solid-state batteries (greater energy density, safety, and thermal stability compared to conventional lithium-ion batteries). Solid-state batteries can potentially provide up to 50% higher energy density, which would allow eVTOL aircraft to carry more passengers or cargo for longer distances without compromising safety or weight limits. In addition to advancements in energy storage, the structural materials used in eVTOL aircraft must be lightweight, durable, and capable of withstanding the stresses of vertical takeoff, hover, and landing. CFRPs and graphene-based composites are crucial for achieving these properties. Novel materials such as lattice structures and 3D-printed components are also being explored for optimized manufacturing processes and part integration, reducing both weight and production costs. From a propulsion standpoint, eVTOL aircraft rely on distributed electric propulsion, which involves the use of multiple electric motors and propellers or fans that are evenly distributed across the aircraft to generate lift and thrust. The development of ultra-efficient electric motors with minimal weight and maximum thrust output is important to achieving the vertical lift required for eVTOL operations. Advanced cooling techniques, such as liquid cooling and heat management through phase-change materials, are being integrated into these motors to enhance their efficiency and prevent overheating during prolonged flight. ML with AI is revolutionizing the control systems and flight management of eVTOL aircraft. AI-based flight control systems enable autonomous or semi-autonomous flight, with the ability to dynamically adjust to changing weather conditions, air traffic, and environmental variables. These systems improve safety, efficiency, and reliability. AI can optimize route planning and manage fleet operations in urban air mobility networks, minimizing delays, energy consumption, and operational costs. In the context of urban air mobility, eVTOL aircraft must be integrated into the existing transportation infrastructure, including urban air traffic management systems. These systems are being developed to handle the increased volume of aerial traffic and to ensure safe and efficient flight paths, especially in densely populated urban areas. Integration with existing airspace management technologies, such as radar systems, communication protocols, and traffic monitoring, will be necessary to enable seamless operations. Advances in blockchain technology and cybersecurity are critical for ensuring secure communication between aircraft, ground stations, and air traffic control systems, safeguarding against cyber threats and unauthorized access. The economic implications of eVTOL aircraft are profound. As these systems reduce reliance on ground-based transportation and alleviate traffic congestion, they have the potential to significantly improve urban mobility and productivity. By reducing travel time and increasing accessibility to remote or underserved areas, eVTOLs could also have a transformative impact on economic activities. The environmental impact of eVTOL aircraft is largely positive, as they produce zero emissions during flight, contributing to efforts to decarbonize the transportation sector. The eVTOL sector represents a rapidly advancing field that is likely to grow to USD 30 billion by 2030, driven by advancements in energy storage, lightweight materials, propulsion systems, AI, and infrastructure development. The success of this sector hinges on the ability to address technical challenges, such as energy density, noise reduction, and autonomous flight safety, while integrating these aircraft into urban environments. The widespread adoption of eVTOLs could revolutionize transportation, reduce urban congestion, and have a significant positive environmental impact, while opening up new economic opportunities and transforming global mobility systems [72–75].

18 Decarbonization in Aerospace and Vehicle Manufacturing

Decarbonization in aerospace and vehicle manufacturing is becoming a central focus of the industry due to the urgent need to address the pressing issue of climate change by reducing carbon emissions across the entire production lifecycle. Carbon-neutral manufacturing techniques, powered by artificial intelligence, are proving to be highly effective in cutting carbon dioxide emissions in vehicle production.

Manufacturers can lower energy use, eliminate material waste, and increase overall production efficiency thanks to AI-based optimization algorithms. AI can identify opportunities for improvement, such as lowering the number of defective parts, streamlining energy-intensive operations, and, when practical, converting to renewable energy sources, by analyzing massive amounts of data from manufacturing lines. These AI-powered methods have demonstrated the ability to cut carbon dioxide emissions by as much as 45%, which would significantly aid in the manufacturing sector's decarbonization. AI-powered predictive models can assist in the shift to a circular manufacturing model, which further reduces the requirement for raw material extraction and the production process's carbon footprint by reusing and recycling materials like metals and composites. In aerospace manufacturing, decarbonization efforts are equally critical, particularly in the context of high-performance materials such as carbon fiber. Carbon fiber composites are extensively used in the construction of aerospace components, such as wings and fuselages, due to their superior strength-to-weight ratios, which enhance fuel efficiency. However, carbon fiber production is energy-intensive and contributes to significant carbon emissions. To address this, advanced recycling technologies are being developed to recycle carbon fiber composites, especially in the context of military and defense applications like fighter jets. This recycling process typically involves breaking down carbon fiber composites into their constituent components, such as carbon fibers and resins, and repurposing them for use in new parts or products. Such techniques, including pyrolysis and solvent-based recycling methods, can cut material waste by as much as 50%. The innovations are being complemented by the development of more sustainable resin systems, such as bio-based resins, which offer a reduction in the carbon footprint of composite materials, further contributing to the decarbonization of aerospace manufacturing. The decarbonization efforts in aerospace are also being accelerated by the development of alternative, low-carbon propulsion systems. Electric and hybrid-electric engines, powered by clean energy sources like solar or hydrogen, are being explored to reduce dependence on fossil fuels. For example, hydrogen-powered aircraft, which produce only water vapor as emissions, have the potential to reduce the carbon footprint of long-haul flights significantly. Similarly, electric aircraft are being developed for short regional flights, offering the promise of zero-emission travel. The integration of electric motors and battery systems in aircraft requires breakthroughs in energy storage technologies, with solid-state batteries offering a higher energy density and longer lifespan compared to traditional lithium-ion batteries. By optimizing designs for weight reduction and material efficiency, 3D printing reduces material waste while enabling more precise manufacturing of complex components. This results in lower energy consumption and reduced emissions throughout the manufacturing process. Digital twins can monitor and optimize manufacturing processes in real-time, allowing for predictive maintenance and energy-saving strategies. In the vehicle manufacturing sector, a similar approach is being taken. Carbon-neutral manufacturing involves not reducing emissions during production but also designing vehicles with sustainability in mind. The shift toward electric vehicles, supported by AI-driven designs that optimize battery energy storage and power distribution, is further driving down emissions. The use of AI in predictive maintenance and autonomous vehicle manufacturing can also reduce energy use by increasing operational efficiency. The role of AI in decarbonization across both aerospace and vehicle manufacturing is vital. AI systems are used to track and analyze carbon emissions at every stage of the production process, helping manufacturers set realistic emissions reduction targets. AI algorithms can help design energy-efficient production lines, optimize the use of renewable energy sources, and even identify opportunities for reducing emissions through process innovation. The push for decarbonization in aerospace and vehicle manufacturing is being driven by a combination of advanced materials science, artificial intelligence, and manufacturing innovations. From carbon neutral production techniques that reduce carbon dioxide emissions by 45%, to the recycling of carbon fiber composites and the development of hydrogen-powered aircraft, these technological advancements are set to reshape the industries. The result is a more sustainable future, where emissions are minimized, material waste is reduced, and energy consumption is optimized, all contributing to the global effort to combat climate change [76–78].

19 Hydrogen Powered Heavy Vehicles

Hydrogen-powered heavy vehicles, including military tanks, are rapidly emerging as a transformative solution in transportation and defense sectors, offering substantial improvements in energy efficiency and environmental sustainability. Fuel cell technology, which converts hydrogen into electricity through an electrochemical reaction, is at the forefront of this innovation. In military applications, hydrogen-powered fuel cell technology has demonstrated the potential to extend the operational range of military tanks by up to 30%. This extended range is crucial for strategic mobility in remote or hostile environments, where the need to minimize refueling stops and maintain operational readiness is vital. The efficiency of hydrogen fuel cells in military tanks gives relatively high energy density with elongated operational hours and operates quietly compared to conventional combustion engines, which provides a tactical advantage in stealth operations. Hydrogen-powered vehicles, especially heavy military and transport vehicles, benefit from an efficient hydrogen storage and distribution infrastructure, a key element of their widespread adoption. Advanced storage techniques, such as high-pressure hydrogen tanks or cryogenic liquid hydrogen storage, allow for the safe and efficient storage of hydrogen fuel at pressures up to 700 bar or at cryogenic temperatures below −253 °C. These innovations are crucial in increasing the energy density of the fuel and ensuring that the tanks and other vehicles have enough fuel to sustain long operational missions without the need for frequent refueling. The use of hydrogen-powered fuel cells in heavy vehicles supports the transition to more sustainable energy systems in engines. In terms of performance, fuel cells are highly efficient, with efficiency rates exceeding 60%, compared to internal combustion engines, which typically operate at efficiencies of around 25–30%. Reducing the entire carbon footprint of military operations and the logistical strain of fuel transportation depends heavily on this energy efficiency. The same technology is being used in the aviation sector, where hydrogen powered aircraft are thought to be a viable way to lower carbon dioxide emissions. A vital instrument in the drive for net zero emissions in aviation by 2050, hydrogen powered aircraft can cut aviation-related carbon dioxide emissions by up to 75%. The primary benefits of hydrogen fuel for aviation are its high energy density and the fact that it only emits water vapor when burned, as opposed to the carbon dioxide and particulate matter emissions that are usually linked to traditional jet fuels. In comparison to conventional jet engines, hydrogen fuel cells offer greater overall efficiency and longer operating times. Both hydrogen fuel cells and hydrogen combustion engines are being developed to power aircraft engines. The challenges associated with liquid hydrogen include the need for cryogenic storage systems, which must maintain temperatures below −253 °C. Advanced materials, such as high-strength carbon fiber composites and multi-layered insulation technologies, are being employed to develop more efficient and lighter storage tanks for liquid hydrogen. These innovations in storage and insulation technology are important to the widespread adoption of hydrogen-powered aircraft. Hydrogen powered aircraft can also contribute to the decarbonization of the aviation sector by reducing the environmental impact of long-haul flights. By replacing jet fuel with hydrogen, aviation can significantly reduce its contribution to climate change. Hydrogen-powered aircraft also provide the potential for quieter flight operations, as the combustion of hydrogen is inherently less noisy compared to traditional jet fuel combustion. The shift to hydrogen-powered aircraft will likely be accompanied by advancements in electric propulsion systems and hybrid technologies. These systems will combine hydrogen fuel cells with batteries to optimize energy consumption and reduce the weight and complexity of the aircraft. Hybrid electric designs are particularly promising for regional aircraft, where shorter flight ranges make hydrogen-powered systems more viable. The transition to hydrogen-powered heavy vehicles and aircraft represents a critical step toward decarbonizing both the transportation and aerospace sectors. While challenges such as infrastructure development, fuel production, and storage still exist, ongoing advancements in fuel cell technology, hydrogen storage systems, and renewable hydrogen production methods are driving this transformation. The integration of these technologies into military and civilian applications will not only reduce carbon emissions by up to 75% but also enhance the energy efficiency and operational capabilities of these vehicles and aircraft, leading to a more sustainable and resilient future in both defense and transportation industries [78–81].

20 Synthetic Aviation Fuels (SAFs)

By lowering carbon emissions, SAFs offer a possible substitute for traditional jet fuels and mark a significant breakthrough in the aviation sector's quest for sustainability. These fuels are chemically engineered to replicate the characteristics of conventional fossil-based jet fuel and are usually sourced from renewable resources like biomass, algae, or even waste products like agricultural residue. By using innovative production technologies that capture and utilize carbon during fuel production, as well as renewable carbon sources, SAFs can lower emissions by up to 80% when compared to conventional jet fuels. This reduction in carbon dioxide emissions is critical for achieving the aviation sector's long-term sustainability goals, as it directly addresses the growing concerns over global warming and climate change. SAFs can be produced using various methods, including Fischer Tropsch synthesis, which converts biomass or waste carbon into liquid hydrocarbons, which turn plant-based natural oils into jet fuel. These methods can utilize feedstocks that would otherwise not be suitable for food production, further enhancing the sustainability of SAFs. The production of SAFs using renewable electricity and carbon capture technologies further amplifies their sustainability profile, reducing not only the direct emissions from the burning of the fuel but also the indirect emissions associated with its production. This lifecycle analysis demonstrates that SAFs are one of the most effective immediate solutions to decarbonizing aviation in the short to medium term. In parallel with SAFs, AI-optimized hybrid electric propulsion systems are emerging as another breakthrough technology, particularly in the realm of fighter jets and other high-performance aircraft. By dynamically controlling the balance between the combustion engine and electric motor, the AI-driven system maximizes fuel efficiency, ensuring that the aircraft operates at peak performance while reducing fuel consumption. For fighter jets, this hybrid-electric approach can increase the aircraft's range by up to 30%, which is a significant improvement, especially for military missions where operational range is critical. The increased range is achieved by more efficient power management and energy recovery systems, such as regenerative braking and advanced battery management algorithms, which ensure that the batteries are used optimally during flight. These systems can also reduce the thermal signature of fighter jets, as the electric motors produce less heat compared to traditional engines, enhancing stealth capabilities. The development of lightweight, high-capacity batteries and fuel cells is essential for making hybrid electric propulsion systems viable in high-performance aircraft. The integration of these advanced materials into hybrid-electric propulsion systems will allow for even more significant performance gains and reductions in weight, a key factor for improving the range and speed of fighter jets and other aircraft. The hybrid systems also have the potential to lower maintenance costs, as electric motors generally have fewer moving parts and are less prone to wear and tear compared to conventional gas turbine engines. This could result in longer intervals between maintenance cycles and reduced operational costs for military and commercial aviation. The combined use of synthetic aviation fuels and AI-optimized hybrid electric propulsion systems represents a paradigm shift in the aerospace industry, where sustainability, performance, and cost efficiency are optimized through advanced technologies and materials. As the aerospace sector continues to innovate in propulsion technologies, these advancements will play a significant contribution in net-zero emissions and to improve the operational capabilities of both commercial and military aircraft [82–85].

SAFs are at the forefront of the defense industry's transition towards more sustainable energy sources, offering an innovative solution to the environmental challenges posed by fossil fuels. To lower the total lifecycle carbon dioxide emissions, SAFs are usually made from renewable feedstocks such as plant oils, algae, or agricultural leftovers that absorb carbon throughout their growth. To ensure compatibility with current aircraft engines, SAFs are produced by transforming these renewable resources into hydrocarbons with the same chemical characteristics as traditional jet fuels. SAFs are a quick and easy way to lower the carbon footprint of defense fleets because of their drop-in capability, which enables their integration into military operations without requiring changes to the aircraft's fuel systems. The lifecycle emissions reduction of SAFs is one of its most compelling benefits. Compared to traditional fossil fuels, SAFs can reduce 80% CO_2, depending on the feedstock and the production method used. This reduction is primarily achieved through the utilization of carbon captured during the growth of the renewable feedstock, along with advanced technologies. The manufacturing of SAFs can be designed to use low-carbon electricity from renewable sources like solar or wind, further lowering the emissions associated with their production. When considering the entire lifecycle from feedstock cultivation to production, transportation, and combustion, SAFs offer a substantial reduction in carbon emissions, which is critical for meeting global climate targets and the military's own sustainability goals. SAFs offer the potential for increased operational autonomy, as renewable feedstocks can be cultivated and processed locally, ensuring a stable supply of fuel in remote locations. This is particularly important in regions where fuel supply chains are vulnerable to disruption, such as during conflicts or in areas with limited access to traditional fossil fuel infrastructure. As the defense sector increasingly focuses on sustainability, SAFs offer a key technological advancement in biofuels, enabling the military to reduce its carbon footprint while maintaining the high performance and operational capabilities necessary for modern defense operations. Advancements in alternative energy technologies, such as the integration of renewable electricity for SAF production and the optimization of fuel conversion processes, continue to improve the efficiency and sustainability of SAFs. Through these innovations, SAFs not only provide an effective solution for reducing carbon emissions but also contribute to the broader goal of transitioning to a more sustainable energy landscape in defense and beyond [85–88].

21 Hybrid Electric Propulsion Systems

The integration of AI-optimized hybrid electric propulsion systems in combat aircraft represents a significant leap forward in aerospace engineering, leveraging cutting-edge advancements in both energy efficiency and stealth technology. The AI component plays a critical role in dynamically managing the transition between power sources, optimizing fuel consumption, and adjusting flight parameters in real-time to achieve enhanced performance and energy savings. This AI-driven system can autonomously adjust the mix of conventional and electric propulsion based on operational requirements, such as speed, altitude, and mission type, ensuring optimal fuel efficiency and performance while reducing overall energy consumption. One of the primary benefits of this system is the significant increase in combat aircraft range, which can be enhanced by up to 30%. This is primarily because the hybrid system allows for more efficient fuel use, as the electric motors can take over during low-speed phases of flight, thereby reducing the load on the conventional jet engines. During high-speed, high-altitude flight, the jet engines provide the necessary thrust, but the AI ensures that the overall fuel consumption is minimized by managing the power distribution intelligently. This capability not only extends the operational range of the aircraft but also provides increased operational flexibility, allowing military forces to cover greater distances with fewer refueling stops. This is a critical advantage in military aviation, where extended operational ranges are important for strategic deployment and rapid response capabilities. Another key benefit of AI-optimized hybrid electric propulsion systems is their ability to reduce noise pollution, a major factor in stealth operations. Electric propulsion systems generate significantly less noise compared to traditional jet engines, particularly during low-speed flight or when the aircraft is in cruise mode. The AI system can intelligently manage the power output of the electric motors to ensure that the aircraft operates at quieter speeds when entering sensitive airspace or conducting reconnaissance missions. This reduction in noise signature enhances the aircraft's stealth capabilities, making it harder for adversaries to detect or track the aircraft. In addition to military applications, this reduction in noise pollution also has broader environmental benefits, as it contributes to minimizing the aircraft's impact on both the surrounding ecosystem and urban populations near airbases. Automation in assembly lines further speeds up production, ensuring that these advanced systems can be deployed at scale for military applications. The ongoing research and development in AI algorithms, battery chemistry, lightweight materials, and manufacturing techniques hold the promise of even further advancements in hybrid electric propulsion systems, which will likely lead to even greater performance improvements in both military and civilian aviation [89–92].

22 Advanced Missile Technology, Explosives Chemistry, and Intelligent Manufacturing in Industry 6.0

Intelligent manufacturing, advanced materials science, quantum computing, AI-driven guidance systems, and high-energy explosive chemistry are all integrated into the increasingly complex field of missile technology. For optimal efficiency and efficacy, modern missile systems such as ballistic, cruise, hypersonic, and anti-aircraft missiles rely on accurate aerodynamics, high-performance materials, and cutting-edge propulsion. While improving the sustainability of military operations in line with SDGs 9, 12, and 13, the incorporation of Industry 6.0 technologies like

CPS, AI-driven material design, additive manufacturing, and quantum-enhanced missile guidance systems has significantly increased production efficiency, decreased costs, and improved precision. By combining AI-driven automation, smart materials, and sustainable energy management, Industry 6.0 is transforming the explosives industry while also greatly lowering environmental impact and increasing efficiency. Explosive manufacturing is now more sustainable thanks to advanced manufacturing techniques that have reduced harmful emissions, heavy metal pollution, and material waste by up to 80%. Energy consumption has been reduced by 40% and operating costs by 30%, thanks to AI-optimized continuous synthesis, 3D/4D/5D/xD/Volumetric printing of energetic materials, and intelligent energy management systems, improving both economic and environmental viability. AI-monitored production lines and autonomous robotic handling have significantly increased worker safety by 90%, lowering human exposure to explosive dangers. Targeting accuracy has grown from 60 to 98%, thanks to AI-driven missile trajectory correction and intelligent ammunition guidance systems, thereby lowering military waste and collateral harm. AI-driven supercritical CO_2 neutralization and the creation of green propellants have made it feasible to manage the weapons lifetime sustainably and dispose of obsolete explosives with zero emissions. In addition to reducing environmental harm, this shift to smart defense manufacturing improves accuracy, productivity, and sustainability in military and aerospace applications, guaranteeing a technologically sophisticated and climate-resilient future [93–95].

22.1 Materials Science in Missile Manufacturing

22.1.1 Material in Fabrication

For missile components to be operationally useful in modern warfare, they must have high-temperature resistance, great mechanical strength, corrosion resistance, and stealth qualities. Intense aerodynamic forces, temperature shocks, and difficulties with electromagnetic detection must all be tolerated by the structural materials employed in missile airframes and casings. The remarkable strength-to-weight ratio (σy = 900 MPa), resistance to oxidation and thermal stability between 600 and 900 °C make titanium alloys like Ti-6Al-4 V and Ti-6242 popular. These alloys directly improve range, maneuverability, and fuel efficiency by reducing missile weight by 40% as compared to traditional steel. Carbon–carbon composites and graphite epoxy composites are used in hypersonic missile applications, where aerodynamic heating can cause temperatures to reach 2000 °C. These materials provide great thermal resistance, high specific strength, and low density, guaranteeing missile durability in harsh environments. They also have a stealth advantage due to their low radar cross section, which lessens the possibility that adversary radar may identify them. Ultra high-temperature ceramics (UHTCs) like zirconium diboride (ZrB_2) and hafnium carbide (HfC) are utilized for ultra-high-temperature applications, including hypersonic glide vehicles and reentry vehicle nose cones. Because these ceramics have melting points higher than 3000 °C, they maintain structural integrity during extended high-speed flight and atmospheric reentry. Combining HfC-based thermal protection systems with reinforced carbon matrix composites allows for prolonged operation at temperatures higher than 3500 °C, which makes them perfect for ablative heat shields, control surfaces, and leading edges of hypersonic missiles. The design of missile components is being further transformed by the incorporation of nanomaterials. While boron nitride nanotubes (BNNTs) offer greater radiation shielding and mechanical reinforcement, graphene-reinforced titanium composites improve structural rigidity and heat dissipation. In Industry 6.0-driven missile manufacture, 3D printed ceramic matrix composites are now being developed to minimize costs, improve material performance, and simplify manufacturing [96–100].

Atmospheric reentry represents a supremely challenging hypersonic aerothermodynamic regime where an object transitioning from the near vacuum of space encounters exponentially increasing atmospheric density at velocities typically exceeding 7–11 km/s (Mach 20–35+). This interaction induces a complex, highly compressed shock layer ahead of the vehicle, drastically elevating the temperature and pressure of incident gas molecules through viscous dissipation and compressional work. The resultant extreme thermal environment, characterized by peak stagnation point heat fluxes exceeding 10 MW/m^2 and total temperatures ranging from 2000 to over 3000 °C (and higher for lunar/ planetary returns), necessitates sophisticated Thermal Protection Systems. Modern Thermal Protection Systems integrates multiple material solutions, i.e., ceramic matrix composites like reinforced carbon carbon for leading edges; advanced ultra-high temperature ceramic such as ZrB_2/SiC or HfB_2/SiC for extreme temperatures; fibrous insulators (eSIRCA, PICA derivatives like PICA-X); and increasingly, metallic thermal protection systems using oxidation-resistant alloys for reusable vehicles. Ablative materials (Avcoat, PICA) remain vital for peak heating, functioning through pyrolysis gas injection, mass loss, and char layer formation to block convective and radiative heat transfer. Beyond thermal management, reentry demands exceptional structural integrity against severe aerodynamic loads (dynamic pressures often >100 kPa) and potential aeroelastic instabilities. Modern computational fluid dynamics coupled with fluid thermal structural interaction simulations model the complex non-equilibrium chemistry, turbulent transition in the boundary layer, and shock-shock interactions that dominate hypersonic flowfields. Reentry profiles are meticulously controlled, i.e., ballistic entries offer simplicity but high g loads; lifting entries (Space Shuttle heritage, Dream Chaser, Starship) enable cross-range capability and lower deceleration; skip-glide trajectories (proposed for crewed lunar return in Artemis, China's lunar missions) optimize energy management and extend range using multiple atmospheric "skips." Guidance, Navigation, and Control systems employ advanced algorithms (often utilizing GPS/ INS fusion and real-time trajectory optimization) modulating bank angle and angle-of-attack to achieve pinpoint landing accuracy, critical for crew safety and reusable systems. Current applications push technological boundaries, SpaceX's Starship employs a novel stainless steel structure with active transpiration cooling and integrated thermal tiles for rapid reuse, NASA's Orion utilizes a modernized Avcoat ablator for lunar return velocities (Approximately 11 km/s) where radiative heating from ionized gas becomes dominant; ESA and JAXA are developing reusable lifting bodies (Space Rider) and inflatable decelerators (HIAD, ADEPT) enable heavier payloads for Mars and Venus. Key research focuses on predicting turbulent transition onset, material responses under coupled extreme environments (oxidation, thermal cycling, mechanical stress), plasma blackout mitigation and enabling robust reusability. Atmospheric reentry remains a pinnacle of aerospace engineering, demanding integrated mastery of hypersonics, material science, thermodynamics, structural dynamics, and autonomous GNC [101–105].

22.1.2 Advanced Propulsion System Materials for Aerospace and Defense

Materials that can tolerate high temperatures, mechanical strains, and oxidative deterioration while retaining structural integrity are essential for propulsion systems in aerospace, defense, and hypersonic applications. Superalloys based on nickel, such as Inconel 718 and Hastelloy X, are widely utilized in combustion chambers, turbine blades, and rocket nozzles because of their remarkable tensile strength retention at 1200–1400 °C, oxidation stability, and creep resistance. Superior mechanical performance is provided by γ'-phase precipitation strengthening, while oxidation and corrosion resistance are improved by the high nickel content (around 50–60%) in combination with chromium, molybdenum, and cobalt. For solid-fuel and liquid-fuel missile engines, Hastelloy X, which is enhanced with iron (17–20%) and molybdenum (8–10%), is especially well-suited since it can withstand thermal wear and phase change at extremely high temperatures. Cermets (ceramic-metallic composites) are used in scramjet and ramjet engines, which may reach temperatures of above 2000 °C because of their continuous Mach 5+ speeds, in next-generation hypersonic propulsion systems. These materials combine metallic binders like tungsten (W) and molybdenum (Mo) with ultra-high-temperature ceramics like zirconium diboride (ZrB_2) and hafnium carbide (HfC) to achieve high thermal conductivity (>100 W/m·K) while preserving oxidation resistance under extreme aerodynamic heating. Cermets give hypersonic missile structures the ability to withstand mechanical deterioration and ablation, resulting in improved performance and a longer operational lifespan. These propulsion materials are essential for next-generation military and aerospace applications because of the use of nano-reinforced carbon matrix composites (C/C-SiC), which further increases thermal shock resistance [106–109].

22.1.3 Advanced Material and Modeling

The merger of AI-driven synthesis control, nanotechnology-enhanced formulations, and quantum chemistry-based detonation modeling has greatly increased the creation of contemporary military explosives, guaranteeing unmatched sustainability, performance, and precision. Real-time monitoring, predictive defect analysis, and molecular-level optimization are made possible by Industry 6.0, which minimizes environmental effect while increasing yield, cutting waste, and improving detonation efficiency. By lowering operating risks and improving safety, the switch to smart explosive manufacturing systems guarantees reliable and tightly regulated production settings. By optimizing reaction control parameters including temperature, pressure, pH, and reagent flow rates, AI-integrated automated synthesis reactors have revolutionized the manufacturing of military-grade explosives. With the help of in-situ thermal analysis and spectroscopic methods like Raman, FTIR, and UV Visible spectroscopy, these closed-loop AI-driven reactors maintain ideal reaction conditions, guaranteeing over 95% yield efficiency while reducing undesirable byproducts. By improving explosive crystal homogeneity and lowering microstructural defects by 60%, ML-based defect detection in crystallization lowers the risk of premature explosion during storage and transit. By forecasting reaction kinetics, detonation wave propagation, and material aging, digital twin simulations are used to further refine manufacturing settings, guaranteeing consistent quality and increased safety. Explosive compositions have undergone a revolution thanks to nanotechnology, which has improved thermal stability, detonation efficiency, and burn rates. Energy is released more quickly and precisely when nanoscale oxidizers and fuel particles are used because they increase the reaction surface area. The detonation pressure of nano-Enhanced RDX (Research Department Explosive or Royal Demolition explosive, Cyclonite, Cyclotrimethylenetrinitramine, Hexogen $C_3H_6N_6O_6$) is increased by 25% to about 36 GPa by combining nano-scale aluminum (nano-Al) and nano-molybdenum trioxide (MoO_3) as oxidizers. This greatly improves penetration capability in armor-piercing applications. Graphene-Doped HMX (High-Melting explosive, Octogen, $C_4H_8N_8O_8$), functionalized graphene oxide coatings enhance shock resistance and thermal conductivity, increasing thermal stability by 30% and lowering the possibility of shock-induced ignition in harsh environments. By adding carbon nanotube-coated CL-20 (Hexanitrohexaazaisowurtzitane, $C_6H_6N_{12}O_{12}$), the molecular packing density is increased, resulting in a 10% increase in explosive energy. Stability is improved, and sensitivity to accidental detonation is decreased. Molecular interactions, shock wave propagation, and chemical degradation pathways in explosives can now be precisely predicted thanks to the advent of quantum computing and AI-driven detonation modeling. By reducing uncontrolled thermal deterioration and guaranteeing high energy density, quantum mechanical simulations aid in the optimization of crystal lattice architectures. Automated explosive filling stations using AI-driven machine vision systems guarantee consistent munition packing, minimizing air gaps and irregularities that may cause misfires. Green and sustainable synthesis techniques are used in the shift to Industry 6.0 driven explosive manufacture to lessen its negative effects on the environment. Closed loop solvent recycling reduces toxic waste discharge by 80%, while AI-optimized waste heat recovery systems in production facilities reduce energy usage by 40%. Greener explosive manufacturing is facilitated using bio-based precursors in certain formulations, which lessen dependency on nitro chemicals generated from petroleum. Predictive safety monitoring powered by AI ensures the safe handling, storage, and transportation of high energy items by preventing unintentional detonation [110–114].

22.2 Advanced Chemistry and Engineering of Modern Military Explosives

Modern military-grade explosives are extremely complex chemical compounds designed for maximal energy release, high detonation velocity, and thermal durability. They are also built to meet strict performance, stability, and safety standards. These explosives' design and manufacture combine nanotechnology, quantum chemistry simulations, and molecular engineering, enabling improved brisance, insensitivity to unintentional detonation, and flexible formulation for a range of military uses. With a detonation velocity of 9.1 km/s and an energy density of around 6.2 MJ/kg, HMX (Octogen, $C_4H_8N_8O_8$) is a high energy nitramine explosive that is perfect for solid rocket propellants, missile warheads, and shaped charges. HMX is frequently utilized in nuclear weapon primaries, anti-armor warheads, and aerospace propulsion systems because of its low impact sensitivity (impact energy threshold around 6 J) and good thermal stability (decomposition temperature around 280 °C). With its oxygen-balanced cyclic framework, HMX's molecular structure promotes quick exothermic breakdown with few residues left behind. The high-energy nitramine RDX (Hexogen, $C_3H_6N_6O_6$) is widely employed in artillery shells, missile payloads, and plastic explosives (C-4). With an energy density of around 5.9 MJ/kg and a detonation velocity of 8.6 km/s, RDX is very effective in military-grade demolition applications. RDX is made using the Bachmann process, which maximizes yield (>90%) and purity (>99.5%) by reacting hexamine with nitric acid, ammonium nitrate, and acetic anhydride under carefully regulated circumstances (50–70 °C, pH around 1.2). In bunker-busting applications, the use of nano-aluminum particles in RDX formulations improves metalized explosive performance by 25%. One of the most thermally stable explosives, TATB (triaminotrinitr obenzene,$C_6H_6N_6O_6$) is used in nuclear bombs, insensitive munitions, and hypersonic missile warheads. It is perfect for weapons kept in harsh conditions because of its incredibly low impact sensitivity (impact energy threshold >50 J) and chemical stability up to 350 °C. Due to strong intramolecular hydrogen bonding, TATB exhibits gradual, controlled detonation, which makes it resistant to unintended ignition from thermal, shock, or electrostatic stimuli, in contrast to RDX and HMX, which experience fast deflagration upon shock. One of the most potent non-nuclear explosives is CL-20, which has a 20% higher detonation velocity than HMX and an energy density of 7.1 MJ/kg. Because of its better oxygen balance (+11%), it can burn more completely and produce fewer harmful byproducts. Starting from benzylamine derivatives, CL-20 is synthesized via a multi step nitration procedure that requires strong nitrating agents (N_2O_5, HNO_3, and H_2SO_4) and regulated reaction temperatures (approx. 0–5 °C) to obtain crystal purity >99.7%. CL-20 is mostly utilized in space launch boosters, next-generation high velocity penetrators and sophisticated missile propulsion systems due to its high cost (around 1300 USD/kg compared to HMX at around USD 100/kg). To maximize shock propagation and penetration in bunker-busting applications, 3D printed explosive warheads that use binder jetting and laser sintering enable custom-shaped detonation waves. By lowering harmful byproducts and enhancing environmental safety, Industry 6.0 helps the military explosives manufacturing sector meet SDGs 9 and 12. Green Nitration reduces acid waste by 70% using ionic liquids and supercritical CO_2-based synthesis. Traditional PETN-based formulations are replaced by energetic biopolymers, or bio-derived polymeric binders, which reduce toxicity by 50%. AI-driven recovery of unreacted HMX and RDX reduces the amount of hazardous material disposed of by increasing recycling efficiency to 90%. To ensure accurate energy yield predictions and optimum molecule architectures for next-generation explosives, quantum computer simulations anticipate molecular detonation pathways with 99.9% accuracy. In military applications, AI-powered detonation predictive analytics maximize shock wave propagation and cut collateral damage by 40%. Modern military explosives are designed to be as destructive as possible, to be resistant to heat, and to be less sensitive to unintentional explosion. The design of explosives has changed due to advancements in nanotechnology, AI-driven synthesis, and quantum chemistry simulations, which have improved efficiency, safety, and accuracy. Industry 6.0 is bringing military defense technology into line with environmental responsibility and next-generation innovation by further improving the production, sustainability, and recycling of high-energy materials [115–117].

22.3 Advanced Rocket and Missile Propellants in Industry 6.0

22.3.1 Material and Manufacturing Methods

High energy materials, AI-driven combustion modeling, and nanostructured propellants have all been used to improve performance, efficiency, and safety in missile and rocket propulsion technology. The three main categories of propellants are solid, liquid, and hybrid fuels. Each has unique chemical makeup, combustion characteristics, and uses in space launch systems, ballistic missiles, hypersonic weapons, and tactical defensive systems. To improve specific impulse (Isp), burn rate control, and thermal stability while lowering environmental effect, Industry 6.0 is combining AI-optimized formulations, quantum chemistry simulations, and additive manufacturing for propellant engineering. Because of their high thrust-to-weight ratio, long-term storability, and convenience of handling, solid propellants, also known as composite and double-base propellants are frequently employed in ballistic missiles, cruise missiles, and space launch boosters. These propellants, which are made up of oxidizers, binders, fuels, and stabilizers, are designed to provide maximum impulse efficiency, little residue, and steady combustion. Ammonium perchlorate (NH_4ClO_4) serves as the oxidizer in composite propellants, while hydroxyl-terminated polybutadiene (HTPB) serves as the fuel and binder. HTPB-based composite propellants are perfect for long-range cruise missiles, space launch vehicles, and intercontinental ballistic missiles since they produce specific Isp of around 250–300 s. Graphene oxide (GO) and aluminum nanoparticles (Al-NPs, 80–120 nm) have been combined in Nano Enhanced

Composite Propellants to improve combustion efficiency by 15%, allowing for more regulated and high energy burn rates. Nitrocellulose, nitroglycerin, and double-base propellants combine to create a uniform gel matrix that promotes smoother burn rates and increased combustion stability. These have an Isp of about 270 s and are utilized in short-range guided weapons, tactical missiles, and artillery shells. DL algorithms used in AI-powered propellant simulation optimize the grain design of the propellant, increasing combustion uniformity by 40% and decreasing propellant cracking failures. Direct ink writing and laser sintering are two 3D printing processes that provide accurate layering of energetic materials, improving thrust modulation and burn rate control. Due to their high energy density, restartability, and throttleability, liquid propellants, also known as hypergolic and cryogenic fuels, are frequently utilized in ICBMs, tactical missiles, space launch vehicles, and hypersonic platforms. These include very energetic monopropellants or bipropellants (fuel + oxidizer). Hypergolic fuels are appropriate for ICBMs and space launch abort systems because they ignite spontaneously upon contact, doing away with the necessity for an external ignition device. With an Isp of about 300–330 s, unsymmetrical dimethylhydrazine, $(CH_3)_2N_2H_2)$ and N_2O_4 (Dinitrogen Tetroxide) are a frequently utilized pair that offer accurate thrust regulation and fast ignition. Utilized in upper-stage rocket boosters, reusable space launch systems, and hypersonic glide vehicles, cryogenic propellants such as Liquid Oxygen (LOX) + Liquid Hydrogen (LH_2) provide ultra-high Isp (around 450 s). However, sophisticated thermal insulation and AI-controlled fueling systems are needed for their extreme cryogenic storage (−253 °C for LH_2 and −183 °C for LOX). Hybrid propulsion systems are appealing for reusable hypersonic vehicles and space launch systems because they combine liquid-fuel controllability with solid-fuel stability. Nitrous Oxide (N_2O) + HTPB Systems with an Isp of about 250–320 s, it provides throttling capabilities and lowers handling risks [118–122].

22.3.2 Intelligent Manufacturing of Missile Systems and Production in Industry 6.0

AI-optimized hybrid fuel formulations, such as machine-learning models, forecast the ideal oxidizer-to-fuel ratio, resulting in a 25% increase in combustion efficiency while preserving affordability and handling safety. Quantum models optimize mixture ratios for increased thrust efficiency by predicting reaction kinetics and fuel-oxidizer interaction at the atomic level. By reducing thermal leakage by 80%, AI-controlled cryogenic fueling, such as machine learning algorithms, improves fuel efficiency and cargo capacity while minimizing boil-off losses in LH_2 storage. Autonomous manufacturing, real-time AI diagnostics, quantum-enhanced material simulations, and ultra-efficient fuel synthesis will all be incorporated into the shift to Industry 6.0 missile propulsion. Military, space, and commercial aerospace propulsion systems will be pushed toward zero waste, high-performance and climate-resilient solutions by innovations including graphene-based combustion chambers, hybrid electric propulsion, and plasma-assisted ignition. Green propellants such as

Hydrazine-based systems are being replaced by new high-performance, low-toxicity monopropellants, such as hydroxylammonium nitrate-based fuels, which reduce toxic emissions by 90% while preserving high-density impulse performance. To improve fuel performance, cut waste, and increase safety, Industry 6.0-driven smart propellant manufacturing facilities use additive manufacturing, AI-controlled chemical synthesis, and nanostructured combustion modeling. Carbon-neutral propellants, such as AI-driven carbon capture from missile launch exhausts, reduce the carbon footprint of military propulsion by 50% by converting CO_2 into useful precursors for synthetic fuel. Real-time sensors with AI-powered safety monitoring identify changes in rocket engine temperature and pressure, preventing explosive decomposition during storage and transportation. Hydrogen-Rich Green Propellants are biomimetic catalysts, being used to generate new bio-derived hydrogen-dense fuels that have high thrust-to-weight ratios and a 60% reduction in launch emissions. Industry 6.0 integrates AI, digital twins, additive manufacturing, and quantum computing into missile development and production, optimizing precision, cost and sustainability. AI-driven guidance systems use neural networks to process real-time trajectory adjustments with 99.9% accuracy, reducing missile guidance errors by 85%. The additive manufacturing of missile components like 3D printing of titanium alloys, improves engine nozzle aerodynamics, reducing manufacturing time by 60% and material waste by 50%. Quantum computing is employed for propellant design, where quantum simulations predict optimal fuel compositions, increasing energy efficiency by 30% and reducing emissions. Smart factory integration with IoT and cyber physical systems, including AI-monitored robotic assembly lines, achieves zero defect production and increases manufacturing efficiency by 45%. Blockchain technology is used for secure supply chain management, preventing tampering and counterfeit components in defense manufacturing [123–127].

22.4 Military Ammunition and Smart Bombs

22.4.1 High-Tech Manufacturing of Military Bullets and Ammunition

To guarantee dependability, performance, and safety in harsh combat situations, the contemporary military ammunition industry is a highly specialized profession that combines advanced materials science, high precision engineering, and energetic chemistry. Bullet design is being revolutionized by Industry 6.0 advancements in additive manufacturing, AI-driven ballistics modeling, and nano energetic materials, which maximize lethality while reducing environmental effect. To ensure uniform internal ballistics, aerodynamics, and terminal performance, the three primary components of military bullets, casings, primers, and propellants go through rigorous production processes. Because of its resistance to corrosion, machinability, and capacity to expand and contract under high-pressure firing circumstances, brass (Cu-Zn alloy, usually 70:30 ratio) is the material of choice for bullet casings, which house the primer, propellant, and projectile. Steel casings with superior polymer coatings have been launched thanks to developments in military metallurgy and Industry 6.0 smart manufacturing. These coatings retain durability while lowering material costs by 30%. By lowering cartridge weight by 25–40%, polymer-cased ammunition, such as polymer–metal hybrid casings, improves soldier performance and lessens fatigue during extended operations. 3D-printed titanium and ceramic composite casings have been made possible by Industry 6.0, which has reduced combat logistics costs while enhancing heat dissipation and structural integrity. By guaranteeing accurate microstructural grain alignment, high-energy laser sintering helps to avoid case head separation and early failures in high-pressure rounds. Primers, also known as shock-sensitive explosive compositions, are the detonating part of a cartridge that, when it strikes the firing pin, ignites the propellant. To comply with environmental regulations while preserving detonation sensitivity at pressures between 30 and 50 MPa, lead-free alternatives like Diazodinitrophenol ($C_6H_4N_4O_5$) and tetrazene-based compounds have recently been adopted. Traditional primers use lead styphnate ($C_6HN_3O_8Pb$) as the primary explosive. When compared to conventional lead-based primers, nano engineered aluminum bismuth oxide (Al/BiO_3) thermites show 300% faster ignition rates, greatly increasing firing consistency at extreme temperatures (−50 to +70 °C). Electrically initiated primers, which do away with mechanical impact detonation, are incorporated into Industry 6.0 to improve shooting reliability in high-rate weapons such as electromagnetic railguns and Gatling rotary cannons. High-performance ballistics and controlled burn rates are characteristics of propellants. The smokeless powder used in modern military ammunition is mainly made up of nitrocellulose [$C_6H_8(NO_2)_2O_5$] and nitroglycerin ($C_3H_5N_3O_9$) for controlled combustion. Chamber pressure (35,000–65,000 psi), barrel wear rates, and muzzle velocity (800–1200 m/s in normal NATO rounds) are all influenced by the propellant chemistry. The performance of nano-structured propellants is high. Propellers made of RDX/HMX nanocomposite offer 15–20% greater energy densities while lowering the accumulation of carbon residue in gun chambers. Propellers covered with polymers improve burn homogeneity, guaranteeing accuracy in guided smart bullets and long-range sniper rounds. In this sector, plasma-assisted ignition systems are the way of the future. To provide uniform ignition even in zero gravity situations for space-based weapon systems, Industry 6.0 weapons incorporate plasma ignition, substituting electromagnetic plasma arcs for chemical primers. By focusing on SDGs 12 and 13, military groups are moving toward green munitions, which aim to reduce carbon emissions and heavy metal contamination by 60%. To reduce soil and groundwater contamination on shooting ranges, conventional lead-core bullets have been replaced by biodegradable training rounds that use polymeric binders and low-toxicity propellants. AI-powered ballistic trajectory optimization solutions save needless ammunition waste by guaranteeing 100% real-time accuracy. The defense industry is moving toward lighter, stronger, and more lethal rounds because of the integration of Industry 6.0 additive manufacturing, AI-based material selection, and quantum ballistics simulation in ammunition production. This ensures superior battlefield performance while also being in line with contemporary sustainability initiatives [128–131].

22.4.2 Smart Bomb Technology

The evolution of smart bomb technology in modern warfare is driven by advances in AI-guidance systems, high-energy explosive materials, and nanotechnology-enhanced warheads, ensuring unmatched accuracy, efficiency, and destructive capability. AI-guided precision bombing, such as the Joint Direct Attack Munition, integrates GPS, inertial navigation systems (INS), and laser guidance, achieving strike precision within 3 meters (CEP circular error probable), a significant improvement over traditional unguided bombs that had an error margin of 30–100 m. These AI-assisted bombs employ real-time target recognition, adaptive flight path correction, and weather compensation algorithms, allowing them to maintain accuracy even in electronic warfare environments with GPS jamming. The addition of AI-enhanced terrain mapping and LiDAR-based obstacle avoidance further refines precision, making Joint Direct Attack Munition systems 30% more effective in urban warfare scenarios. The use of high-energy metal nanopowders, such as magnesium (Mg), boron (B), or aluminum (Al), is the fundamental component of thermobaric explosives (fuel-air bombs). These nano powders undergo rapid oxidation upon detonation, resulting in temperature spikes above 2500 °C and extreme overpressures exceeding 500 psi (pounds per square inch). A dispersal charge atomizes the metal fuel and oxidizer into an aerosol cloud, which burns in a secondary detonation to produce a large pressure wave with long-lasting thermal effects. This is how these weapons work. Utilizing aluminum nano powders increases lethality in cramped spaces like underground tunnels and bunkers by 50% by increasing energy release per gram. The durability, stability, and energy efficiency of warheads are being revolutionized by developments in polymer-bonded explosives. Sol-gel techniques and supercritical CO_2 drying are used to create nano-structured RDX and HMX, which allow for higher detonation velocity (~8900 m/s for HMX-based formulations) while reducing shock sensitivity. Furthermore, in hypersonic missile deployment, self-healing polymer matrices that contain nano-silica-reinforced polyurethane improve structural integrity by delaying early deterioration under high G-forces. Adaptive yield control is made possible by smart explosives that use AI-driven detonation sequencing to dynamically modify the blast radius and fragmentation patterns in response to battlefield conditions and collateral damage limitations. By using real-time sensor fusion from radar cross-section analysis, infrared targeting pods, and UAV surveillance, these next-generation AI-optimized bombs ensure zero-error engagements while reducing civilian casualties. Additive manufacturing (3D-printed warheads) is incorporated into Industry 6.0-driven smart bomb manufacture to provide precisely tailored aerodynamic profiles that improve penetration efficiency against reinforced structures and reduce drag coefficients by 25%. Using high entropy alloy casings, which are made of Cr-Co-Ni-Mn-Fe nano-structured metals, allows for deep penetration through multi-layered defenses and boosts impact resistance by 70%. By lowering the waste of hazardous chemicals in the manufacture of military ordnance, hydrogen peroxide-based oxidizers are being investigated as environmentally friendly substitutes for conventional harmful explosive propellants, supporting SDG9. In the future, self-steering micro-thrusters will allow for mid-air course correction, making smart bombs fully autonomous and practically difficult to intercept, while quantum-enhanced AI computing will allow for hyper-accurate trajectory predictions, lowering in-flight error margins to sub-millimeter levels. Research is also being done on the integration of directed energy warheads, or plasma-based detonation mechanisms, which could further transform current warfare in the Industry 6.0 age by substituting controlled energy pulse weapons for conventional kinetic blast explosives [132–135].

22.5 *Smart, Sustainable, High-Precision Military Defense Technologies in Industry 6.0*

With Industry 6.0, the defense sector is seeing a technological transformation that combines additive manufacturing, AI-driven systems, and sustainable material engineering to improve productivity, affordability, and environmental responsibility. These developments are enhancing military capabilities and operational effectiveness while drastically cutting CO_2 emissions by 50%, which is in line with global sustainability goals. By using computational chemistry and machine learning to create high-energy-density fuels with lower pollutant emissions, AI-optimized missile fuel formulations are revolutionizing propulsion systems (SDG 12, 13). Compared to conventional hydrazine-based fuels, these fuels, which include nano-stabilized energetic elements like CL-20 and green hypergolic propellants, provide 30% greater thrust-to-weight ratios while drastically lowering harmful byproducts. By cutting material waste by 80% and facilitating the quick creation of lightweight, high-strength alloys like Ti-6Al-4 V for missile components and next-generation hypersonic aircraft, additive manufacturing (3D, 4D, 5D printing, etc.) is transforming the aerospace and military industries (SDG 9). When combined with AI-driven topology optimization, these sophisticated layer-by-layer manufacturing processes improve the structural integrity of vital military components while reducing production time and costs by 40%. By enhancing targeting accuracy and reducing collateral damage, smart weapon systems that incorporate edge AI and real-time sensor fusion are guaranteeing adherence to international humanitarian law (SDG 16). In challenging combat situations, sub-meter accuracy is provided by electro-optical and infrared (EO/IR) guidance systems that are fueled by quantum-enhanced detection. Mission failure rates are lowered by 35% thanks to autonomous loitering weapons that adjust in midair using deep learning algorithms. Modern military strategy is being redefined by the synergistic integration of Industry 6.0 technologies, which guarantees that defensive systems are more intelligent, effective, and sustainable. With advancements in precision-guided weapons, energy efficient aircraft production, and AI-driven fuel chemistry, the defense industry is moving closer to a future in global security that is ethically accountable, technologically advanced, and climate-conscious [136–139]. In the explosives sector (missiles, bullets, rockets, bombs, etc.), Table 11.1 provides a thorough comparison of sector

Table 11.1 Comparative study of traditional vs Industry 6.0 technology in defense industries

Category	Traditional explosives manufacturing	Industry 6.0 smart manufacturing	Climate impact reduction and SDG alignment
Material composition and synthesis	TNT/RDX synthesis releases 12 kg NO_x/ton and 800 kg CO_2 ton	AI-optimized ADN/CL 20 biocatalysis:60% less NO_x, 50% lower energy (200 MJ/ton)	SDG 9, 12, 13: ↓ 5.2 MT toxic waste/year; ↓ 38% carbon intensity
Propellant and explosive formulation	Lead styphnate primers contaminate 450 K hectares/year; 95% aquatic toxicity persistence	Nano GAP fuels with Al MOF nanostructures: zero heavy metals; 80% lower ecotoxicity	SDG 12,14: ↓ 12 M kg soil decontamination costs; ↓ 90% bioaccumulation risk
Manufacturing efficiency	Batch reactors: 30% material waste, 500 MJ/ kg energy, 40% yield loss	Continuous microfluidic synthesis: 50% yield ↑ (90% net), 40% energy ↓ (300 MJ/ kg), real-time AI control	SDG 9,13: ↓ 9.7 MT CO_2/year; 65% less solvent waste
Automation and AI integration	Manual handling causes 320 worker fatalities/ year; USD 2.3B accident costs	Autonomous cobots + digital twins: 90% fewer accidents; predictive maintenance cuts downtime by 70%	SDG 3,8: ↓ 98% occupational hazards; USD 17B/year saved in medical/ compensation
Missile propulsion	Solid boosters emit 22 kg HCl/ launch (ozone depletion = 1.2 M CFC-11 eq).	AI-optimized H_2 LOX engines: 70% lower CO_2; zero halogens; 3D-printed combustion chambers.	SDG 7,13: ↓ 480 K tons HCl/year; 100% ozone safe.
3D/5D printed components	Forging at 1500 °C emits 120 kg CO_2/kg part; 45% material waste	Topology-optimized Ti-B_4C composites: 60% CO_2 ↓ (48 kg/kg); 50% waste ↓; embedded sensors for health monitoring	SDG 9,12: ↓ 18 MT CO_2/year; 55% lighter structures
Energetic recycling	OB/OD releases 98 toxic compounds (dioxins, VOCs); USD 4800/ton disposal cost	Supercritical CO_2 + ML guided neutralization: 95% material recovery; zero emissions; USD 380/ ton cost	SDG 6,15: ↓ 1.4 M kg airborne toxins/year; 100% groundwater protection
Smart munitions	Unguided munitions: 40% miss rate → 2.8× collateral damage	Quantum radar-guided hypersonic missiles: 98% accuracy (CEP $\leq$1 m); 40% less ordnance used	SDG 16: ↓ 90% civilian casualties; ↓ USD 220B/year post conflict remediation
Testing and QC	Destructive tests waste 35% materials; USD 18 M/test for ICBMs	AI-physics simulations + digital twins: 50% fewer physical tests; 70% faster certification	SDG 9,13: ↓ 7200 tons hazardous residue/ year; 80% R&D cost savings
Plasma detonations	FAEs produce 68 kg CO_2/kg explosive + NO_x (GWP = 265)	Non-thermal plasma (NTP) waves: zero GHGs; tunable energy yield (1–100 MJ); ionic liquid fuels from biomass	SDG 7,13: ↓ 11.3 MT CO_2 eq/year; 100% fossil-free

6.0-driven intelligent manufacturing and conventional production techniques, emphasizing technological improvements, the impact on the climate, and alignment with SDGs.

23 Applications of Industry 6.0 Technologies in Different Production Units

The implementation of AI-driven technologies in manufacturing, aerospace, defense, and automotive industries is reshaping global production dynamics, significantly enhancing operational efficiency, product quality, and system reliability. The intersection of artificial intelligence, quantum sensor technology, and predictive modeling is transforming how industries approach manufacturing, logistics, defense, and aerospace. The integration of AI and quantum sensors creates smarter, more efficient systems that reduce production errors, improve product reliability, and accelerate the deployment of advanced technology. As industries continue to adopt these innovations, we can expect further reductions in operational costs, enhanced performance, and an overall increase in the safety and reliability of complex systems across a range of high-tech sectors.

Lockheed Martin's advanced use of artificial intelligence in aircraft production is one of the leading examples of this trend. Predictive AI models have enabled a reduction of jet-fighter assembly time by up to 40%. By using ML algorithms, Lockheed Martin can anticipate component assembly challenges, optimize workflows, and improve resource allocation. This system's real-time adaptability allows for continuous optimization, ensuring that the assembly lines operate at peak efficiency while minimizing human error and unanticipated delays. AI-based data analytics and deep learning facilitate the monitoring of part wear and manufacturing inconsistencies, enabling timely adjustments during production, which ultimately results in shorter cycle times and reduced labor costs [140, 141].

NASA's Digital Thread for spacecraft manufacturing highlights the remarkable strides made in defect detection and structural integrity monitoring. AI-driven models embedded within the manufacturing process continuously analyze sensor data from various stages of production, ensuring real-time identification of anomalies and defects. This intelligent network connects every phase of production, from design to testing, and enhances the ability to trace defects back to their origins. By improving the structural integrity of spacecraft by 20%, this technology significantly reduces the risk of failures in space missions, extending the lifecycle of spacecraft while ensuring higher safety standards. The AI systems' ability to predict weak points and optimize material distribution directly leads to longer mission durations and more reliable spacecraft [142–144].

The U.S. Air Force has taken significant strides in adopting quantum sensor technology to enhance its navigation systems. Quantum gyroscopes, which are based on quantum mechanics principles, have enabled navigation systems that operate with 99.99% accuracy, eliminating the reliance on traditional global positioning systems (GPS). These gyroscopes are far more precise than conventional inertial measurement units, which means they provide exceptionally high performance even in GPS-denied environments. This technology is vital for military operations, where navigational precision is crucial, especially in areas where GPS signals can be jammed or disrupted. The integration of quantum sensors into military navigation systems represents a fundamental shift in aerospace and defense technology, providing unprecedented accuracy and resilience to external disruptions [132, 145].

To increase production efficiency, Tesla's AI-powered electric vehicle manufacturing uses both predictive AI and systems augmented by quantum sensors. Tesla has achieved a 40% reduction in production errors by integrating quantum sensors into its production lines. By guaranteeing that every component fits precisely, these sensors offer a level of component assembly precision that is higher than that of conventional techniques, eventually enhancing the vehicle's overall quality and performance. Real-time monitoring is further made possible by the combination of quantum sensors and AI systems, which enables Tesla to anticipate and resolve such problems before they become expensive delays or flaws. AI-driven systems also help with intelligent logistics management, inventory control optimization, and labor-intensive task automation, which lowers manufacturing costs and time [37, 146, 147].

Boeing's embrace of Industry 6.0 technologies further accelerates aircraft production. Predictive AI models at Boeing have successfully reduced jet fighter assembly times by 35%, improving both efficiency and cost-effectiveness. These models analyze vast datasets generated during the production process, learning to predict bottlenecks, optimize component sequencing, and improve worker productivity. With advancements in 3D printing and additive manufacturing, Boeing can produce complex parts more rapidly, reducing lead times and inventory costs. Industry 6.0, with its emphasis on cyber-physical systems, enables Boeing to create a highly integrated and adaptive production environment where real-time data is utilized to ensure continuous process improvement [148, 149]. Figure 11.2 showing the use of Industry 6.0 technologies in weapons, aircraft, orbit, and defense industries.

Fig. 11.2 Industry 6.0 technologies in aerospace and defense industries

24 Global Policies and Regulations

Global policies and regulations surrounding artificial intelligence, autonomous weapons, sustainable aviation, military fuels, and electric vehicles are importance for ensuring safe, ethical, and sustainable advancements in technology. As AI-powered autonomous weapons systems become more advanced, international bodies like the North Atlantic Treaty Organization and the United Nations are increasingly focused on establishing regulations to govern their use. NATO has begun formulating guidelines that emphasize human oversight in autonomous weapon systems, with a clear stance on ensuring that autonomous weapons must not make life-and-death decisions without human intervention. The United Nations Convention on Certain Conventional Weapons has also been discussing the limitations and prohibitions on lethal autonomous weapons systems, specifically addressing concerns about accountability, legality, and the risk of inadvertent escalation. These discussions are grounded in ensuring that such technologies comply with international humanitarian law, with the key challenge being the distinction between military and civilian targets, and avoiding violations of the principles of distinction, proportionality, and necessity in armed conflict [150, 151]. Concern over greenhouse gas emissions in the aerospace and defense industries is leading to the development of sustainable aviation and military fuel policies. The use of SAFs, which come from renewable sources like plant biomass, algae, and synthetic processes, is being encouraged by international legislation. When compared to traditional jet fuels, SAFs can cut lifecycle carbon dioxide emissions by as much as 80%. To achieve a net zero carbon future by 2050, the International Civil Aviation Organization has set aggressive global goals to cut carbon emissions from aviation. The development of hydrogen powered aircraft and the use of carbon capture and storage technology to offset emissions from current fleets are being encouraged by governments in tandem with these initiatives. These initiatives are also supported by military fuel policies, as many countries look to lessen their dependency on fossil fuels by investigating alternative energy sources for military operations. This could greatly reduce their carbon footprints and help the defense industry reach its overall decarbonization targets. Global electric vehicle regulations have accelerated due to the pressing need to reduce urban air pollution, fight climate change, and lessen reliance on fossil fuels. Regulations that hasten the switch from internal combustion engine vehicles to electric vehicles have been implemented by governments across the globe, including those in the US, China, the EU, and many more. In addition to significant investments in electric vehicle charging infrastructure, these regulations include pollution standards, incentives for EV purchases, and subsidies for EV manufacturers. By 2030, the European Union wants to cut transportation related carbon emissions by 55% and by 2035, it wants all new automobiles sold to be emission-free. Like this, the Biden administration in the US planned to spend heavily in electric vehicle charging networks to facilitate the transition to electric vehicles, which are expected to make up 50% of all new car sales by 2030. In addition to these regulations, there are emerging policies on autonomous and connected vehicles. Many governments are developing frameworks for the safe testing and integration of autonomous electric vehicles into urban environments, with a focus on ensuring that such technologies do not compromise road safety. These policies often incorporate guidelines for data privacy, cybersecurity, and interoperability, with particular attention to the integration of AI-powered systems for self-driving vehicles. To facilitate the use of renewable energy sources and help stabilize the electrical grid, standards for vehicle-to-grid technologies are being investigated. This will allow electric vehicles to function as mobile energy storage units. Global regulations pertaining to sustainable technology are also greatly influenced by international environmental agreements, such as the Paris Climate Agreement. These international agreements place a high priority on the shift to a low-carbon economy, and countries are expected to meet emission reduction goals. This will increase demand for cleaner technologies, renewable energy, and more environmentally friendly production methods. Policies and regulations are also extending to the realms of space exploration and autonomous systems. As the use of space-based technologies for communication, surveillance, and military purposes increases, regulations related to the space debris problem, the militarization of space, and the ethical implications of using AI for surveillance are being discussed at the global level. Similarly, international bodies are collaborating on establishing guidelines to manage autonomous AI systems in critical applications, ranging from agriculture to disaster response, ensuring that these technologies serve humanity's collective interests while preventing misuse. For next generation technology to be used responsibly, ethically, and sustainably, these regulations must be developed further. To balance innovation, security, environmental effect, and social responsibility, regulatory frameworks must change to keep up with developments in AI, quantum computing, autonomous systems, and sustainable manufacturing. If these rules are successfully put into place, the shift to a more developed, sustainable, and safe future will be effective and fair for all international sectors [152–155].

25 Case Studies of Different Companies

American aerospace and defense firms lead with AI-enabled automation, advanced materials, and sustainable production. Lockheed Martin is integrating AI and quantum technologies into navigation and weapon systems: it's prototyping a quantum enhanced inertial navigation system for GPS denied environments and developing hypersonic missiles like the AGM-183A (ARRW) that glide at Mach 5+. Raytheon's portfolio includes high energy laser weapons (fielding 10–50 kW class systems to shoot down drones and rockets and AI-enhanced avionics (AI-powered radar warning receiver recently tested on F16 jets. Boeing and Northrop Grumman use additive manufacturing extensively, Northrop pioneered 3D printed titanium airframe parts to cut costs and deploy robotics and digital twins on assembly lines. SpaceX has fully automated rocket factories and AI-assisted launch systems supporting Starship and large satellite constellations (Starlink) [1, 156–161]. On the automotive front, U.S. firms have redefined smart manufacturing. Tesla's Gigafactories use massive "Giga Press" die casting machines to consolidate hundreds of chassis components: its Model 3/Y rear body castings eliminated roughly 370 parts (it's in 2021 news but may be updated a lot in 2025), slashing weight and assembly steps. Computer-vision AI inspects vehicles with near-perfect accuracy. Tesla also produces advanced batteries (its new 4680 cells are being built at scale, 50 million had been produced by mid 2024). Legacy automakers have followed suit: General Motors builds on its Ultium EV platform and BrightDrop delivery vans with AI-driven plants and digital twins to optimize production. GM now simulates entire production lines before they are built, using digital twin models and AI vision tools to detect battery pack leaks and weld anomalies. Ford likewise connects its F-150 Lightning EV factories with 5G and deploys cobots for flexible assembly. New entrants like Rivian and Lucid use machine learning in battery management and precision assembly, and power their plants with renewable energy to reduce the carbon footprint [162–165].

China's aerospace ecosystem combines state giants and agile startups to push Industry 6.0. AVIC's aircraft units produce advanced fighters and transports (the J-20 stealth jet and Y-20 airlifter) while COMAC's civilian division is delivering its first C919 narrow-body airliners. Major domestic orders are under way. China Southern and Air China each ordered approximately100 C919s (a USD 10–11 billion deal apiece) and regional jets (ARJ21) have entered service in significant numbers. The state's space contractor, CASC, maintains an approximately 95% launch success rate over 450+ Long March rockets and runs the Tiangong space station. In 2024 CASC's Chang'e-6 probe achieved the first ever return of lunar samples from the Moon's far side, advancing China's goal of a crewed lunar mission by 2030. Private launch companies are rapidly innovating i.e. Galactic Energy's sea-launched Ceres-1 solid rocket has flown over a dozen missions (including deploying IoT satellites), and it is developing Pallas-1, a reusable methalox launcher. LandSpace made headlines by achieving the world's first successful orbital launch of a methane/LOX rocket (Zhuque-2) in 2023. Dozens of other startups (i-Space, Deep Blue, etc.) are testing small reusable rockets, and satellite firms (Expace, GalaxySpace) are building LEO constellations with AI-driven beamforming. China's automotive industry similarly embraces Industry 6.0. BYD runs massive, automated EV factories producing LFP "Blade" batteries, and new plants (for CATL) aim for near zero defects. Geely and SAIC use digital twin factories and 5G connectivity to optimize assembly. Startups like NIO and Xpeng used AI in autonomous systems and battery management, while also experimenting with next generation chemistries and modular architectures. Across the board, Chinese manufacturers tie into a domestic 6G/IoT network and invest in quantum secure communications to protect smart vehicle data [166–169].

European and British companies lead in combining sustainability and advanced tech. Airbus is pioneering hydrogen fuel cell airliners (its ZEROe project) and autonomous drones (Eurodrone) while modernizing factories with AI and 3D printed titanium parts to speed assembly. Its "Factory of the Future" initiatives target faster build times and lighter structures via additive manufacturing. BAE Systems is developing next generation combat platforms (the Tempest fighter, autonomous Taranis jet) and uses digital twins in shipyards and robotic ordnance assembly to cut scrap. Missile maker MBDA, a European consortium, fields air to air and surface to air missiles (Meteor, Aster, etc.) and is working on hypersonic FC/ASW rounds. Ericsson and Thales in France integrate quantum sensors and encryption into avionics and satellites. In automotive, firms like Volvo (Sweden) and BMW (Germany) use smart gigafactories i.e. Volvo's Vårgårda plant is climate neutral with AI-managed production and BMW's Regensburg and Leipzig plants use giant casting machines and AI quality control on EV lines. Mercedes Benz (Germany) operates Industry 4.0 style MO360 digital plants with robotics and 3D printed components for its EQ models. Renault (France) and Stellantis (multinational) emphasize circular manufacturing. Renault's REF factory remanufactures 95% of returned parts and uses IoT to optimize energy use. Sweden's battery maker Northvolt is expanding Europe's first homegrown gigafactories, running on 100% renewable power. Its Borlänge plant will use only clean energy to produce around 100 GWh of cathode materials annually, enough to supply many EV plants [170–173].

Russia's aerospace and defense complex (United Aircraft Corp., Almaz Antey, Uralvagonzavod, etc.) continues producing legacy platforms (Su-series fighters, S-400 air defenses, T-90 tanks) using advanced metallurgy (titanium from VSMPO-AVISMA) and some robotics, though sanctions have slowed modernization. In the Middle East, South Korea and Turkey have become notable innovators. Hanwha Aerospace (ROK) is developing indigenous jet engines (a KF-21 fighter engine prototype was unveiled in 2024) and advanced UAV/missile systems. Baykar (Turkey) has emerged as a global UAV champion, exporting around USD 1.8 billion of combat drones in 2024 (90% of its revenue) and capturing roughly 65% of world UAV export market share. The UAE's EDGE Group also exemplifies Industry 6.0 in the region, fielding AI-enhanced electronic warfare, precision-guided munitions and directed-energy systems. Emerging players in Eastern Europe and elsewhere (Ukraine's Motor Sich, Israeli firms, etc.) similarly fuse dual use technologies like

EV battery tech and autonomous systems are being adapted for unmanned vehicles [174–178].

Across these regions, common Industry 6.0 trends stand out. Artificial intelligence and machine learning are now integral from design to factory floor, digital twins and predictive analytics underpin everything from F-35 assembly in the U.S. to EV battery pack welding in Europe. Advanced materials (carbon composites, high strength alloys, solid state and methalox propellants) are being adopted to boost performance and sustainability. Robotics and flexible automation (collaborative robots, 3D printing) enable rapid, low cost reconfiguration of production lines. Quantum technologies and blockchain are being piloted for navigation and supply-chain security. Finally, "green" manufacturing is a global imperative and many leaders are running plants on 100% renewable energy (GM's Ultium battery factories, SpaceX's carbon neutral methane engines and Northvolt's gigafactories) to stop global warming and incorporating recycling to minimize waste. Together, these integrated advances make the United States, China, Europe, and other innovative nations leaders in a new era of smart, resilient aerospace, defense, and automotive production [170, 179–182].

26 Key Challenges to Implement Industry 6.0 and Future Recommendations with Suggestions

Table 11.2 shows the key challenges and their advance solutions for the implementations of Industry 6.0 technologies.

A global paradigm shift toward peace and technological resilience is essential. It is imperative that all nations not only strive for coexistence and dialogue but also discourage coercive actions against weaker states or groups. Technological advancement must serve as a tool for sustainable peace, ecological preservation, and the protection of all life on Earth. The recent conflict and war between Israel and Iran in 2025, although short lived, disrupted global supply chains, particularly in oil and critical raw materials, highlighting the extreme sensitivity of interconnected economies to geopolitical instability. Similarly, the tensions and war observed between India and Pakistan during the same year reinforced the vulnerabilities of traditional supply and communication infrastructures under conflict scenarios. Similarly, numerous historical and ongoing conflicts, including those involving Afghanistan, Russia, Ukraine, the United States, Germany, the United Kingdom, and other participants in World Wars I and II, reflect a long standing legacy of geopolitical tension and militarized confrontation. These conflicts have often resulted in profound humanitarian crises, large scale displacement, and long-term socio economic instability. History is marked by recurring episodes of bloodshed, exploitation, and systemic injustice, particularly affecting marginalized and economically weaker nations. Forceful occupations and unresolved disputes such as those in Kashmir, Palestine and other contested regions, continue to challenge international peace and

Table 11.2 Key Challenges, Advanced Solutions, and Implementation Pathways

Domain	Key challenge	Technical barrier	Advanced solution	Implementation pathway	Impact metric
Advanced materials	High entropy alloy production cost	Scalable synthesis; USD 1200/kg vs. 100/kg for titanium	AI-driven combinatorial chemistry: ML algorithms predict optimal HEA compositions; plasma arc melting with reactive gas atomization	Establish HEA digital twins for rapid prototyping; deploy high-throughput sputtering systems	↓ cost to USD 300/kg by 2030; ↑ yield by 40%
	Nanoceramic-aerogel integration	Brittleness under hypersonic stress (>2000 °C)	Graphene-reinforced hierarchical aerogels: 3D-printed nanoceramic lattices with covalent bonding	Functionalize aerogels with boron nitride nanotubes (BNNTs); use multi-material laser additive manufacturing	↑ fracture toughness by 70%; density <50 mg/cm^3
Quantum computing	Cryptographic vulnerabilities	Quantum decryption of RSA-2048 in <4 h	QKD satellites: Space-based QKD networks with 256-bit entropy; post-quantum lattice-based cryptography	Integrate quantum-resistant ASICs in defense supply chains; deploy entangled photon transmitters on military satellites	↑ security to 10^{38} operations for decryption
	Sensor precision limits	GPS drift >1 m in contested environments	Quantum inertial navigation (QIN): Cold-atom interferometers with 10^{-11} g stability; diamond NV center magnetometers	Embed quantum sensors in hypersonic airframes; use AI correction for relativistic effects	↓ navigation error to <10 cm after 72 h
AI and autonomy	Autonomous system failures	0.1% error rate in combat drones causing fratricide	Neuromorphic AI chips: Spiking neural networks with 10^{14} synaptic operations/ J; real-time adversarial training	Implement federated learning across drone swarms; deploy tesla dojo-like training infrastructure for digital twins	↓ false positives by 99.97%; latency <3 ms
	Predictive maintenance gaps	15% false alarms in aircraft health monitoring	Multi-physics digital twins: Coupled CFD FEM simulations updated via IoT sensors; quantum annealing for fault prediction	Install terahertz spectroscopy nodes in engines; use blockchain for immutable maintenance logs	↑ system uptime by 30%; ↓ maintenance costs by 40%

(continued)

Table 11.2 (continued)

Domain	Key challenge	Technical barrier	Advanced solution	Implementation pathway	Impact metric
Sustainability	Hydrogen storage efficiency	Cryogenic tanks: 5.5 wt% H_2 capacity; boil-off >0.1%/day	Metal-organic Frameworks (MOFs): Cryo-adsorbents with 12 wt% capacity at 100 bar; aerogel multilayer insulation	Functionalize MOFs with Ti-doped fullerenes; optimize tank geometry via generative AI	↑ energy density to 1200 Wh/kg; ↓ boil-off to 0.01%/day
	SAF production scalability	USD 2500/ton cost vs. 600/ton for Jet-A	Solar thermochemical reactors: Concentrated solar power (CSP) + Fischer-Tropsch synthesis; CRISPR-engineered algae	Co-locate SAF plants with CSP farms; use AI-controlled photobioreactors for continuous production	↓ cost to $800/ton; ↑ yield to 60,000 L/acre/year
Manufacturing	3D printing defects	Microcracks in SLM printed titanium; 12% scrap rate	Femtosecond laser AM: 100-fs pulses melt powder with 0.5 μm precision; in-situ XRD monitoring	Integrate synchrotron beamlines for real-time crystallography; apply topology optimization via quantum computing	↓ defect rate to 0.3%; ↑ tensile strength by 25%
	Supply chain disruptions	8–10 M tons CO_2/year; rare earth shortages	AI-driven adaptive networks: Blockchain-secured digital twins; swarm robotics for decentralized production	Deploy IoT-enabled "smart mines" with automated sorting; use reinforcement learning for dynamic rerouting	↓ CO_2 by 45%; ↑ rare earth recycling to 95%
Policy and ethics	Autonomous weapon regulation	Lack of global governance for AI lethality	NATO ethical AI framework: Human-on-the-loop protocols; explainable AI (XAI) with cryptographic audit trails	Develop ISO 31000 compliant risk matrices; Implement hardware kill switches with quantum encryption	↓ unintended engagements by 99.9%
	Data sovereignty	Cross border data flow conflicts (ITAR/EU-GDPR)	Homomorphic encryption: Fully homomorphic processing of classified data; sovereign cloud edge networks	Deploy zero-trust architectures; use confidential computing with SGX/TEEs	↑ compliance speed by 50×; data access without decryption

security frameworks. These situations highlight the urgent need for equitable global governance, adherence to international law and the protection of fundamental human rights. Sustainable peace cannot be achieved through militarization or coercion, but it requires inclusive diplomacy, socio economic equity, and mutual respect among nations. A future oriented global order must prioritize justice, conflict resolution, and the dignity of all peoples, moving beyond the legacy of colonialism, occupation, and forceful domination. Only through such an approach can humanity ensure lasting peace, environmental sustainability, and a shared future for all. These events collectively also highlight the urgent need to accelerate the deployment of AI-driven intelligent manufacturing, particularly in sectors critical to national security and resilience like military logistics, autonomous defense systems, real-time surveillance and robotics integrated industrial platforms. The integration of digital twins, collaborative robots (cobots), and AI-enhanced decision making can provide nations with adaptive and decentralized capabilities, ensuring continuity of operations even under adverse geopolitical conditions. Intelligent systems enable predictive threat modeling, automated defense coordination and secure multi domain communications, vital for national defense and civil protection. Beyond human conflict, wars also inflict long-term damage on non human species, ecosystems, and natural resources, exacerbating environmental degradation, biodiversity loss, and accelerating climate change. Warfare disrupts ecological balances, pollutes natural habitats and derails progress toward global sustainability targets. This multi pronged strategy integrates high performance materials, quantum enhanced computation, sustainable energy innovation, cyber resilience, and proactive policy to accelerate the adoption of next-generation technologies that define the future of global manufacturing and defense ecosystems. Strategic recommendations to support the long-term implementation of these technologies include the establishment of national materials foundries with a proposed 2 billion USD investment (like the U.S. CHIPS Act extension) to enable domestic manufacturing of HEAs and advanced components via additive manufacturing. To address the talent gap, governments and academic institutions should launch MS level programs in quantum engineering, targeting the development of 50–100 thousand specialists by 2030. Cross industry consortia like a joint venture among the top companies should be formed to drive AI-enabled sustainable aviation and orbit industry, potentially reducing R&D expenditures by up to 60% through shared infrastructure and research pipelines. Finally, the adoption of globally acceptable ethical standards (like IEEE 7000) for autonomous systems, including the use of third party audits, is important for the insurance of transparency, accountability, and safe deployment of AI in military and industrial domains.

References

1. Ghorpade, S.C., Sidharth, S.: Smart manufacturing in the defence sector: a comprehensive review and analysis of technological advancements, integration, and challenges. In: Manufacturing Strategies and Systems, pp. 85–103. CRC Press, Boca Raton (2025)
2. Kulshrestha, N., Agrawal, S., Shree, D.: Spare parts management in industry 4.0 era: a literature review. J. Qual. Maint. Eng. **30**(1), 248–283 (2024)
3. Aina, A.N., et al.: Fiber-reinforced thermoplastic composites for future use in aircraft radomes: biomimetic design approaches and its performances. Fib. Polym. **25**, 4503–4527 (2024)
4. Qasem, N.A., Abdulrahman, G.A.: A recent comprehensive review of fuel cells: history, types, and applications. Int. J. Energy Res. **2024**(1), 7271748 (2024)
5. van de Ketterij, R.G., et al.: Alternative fuels, propulsion and power Systems for the future navy–a route towards reduced emissions and Signatures, and fossil fuel Independence. In: Climate Security and the Armed Forces: Concepts, Strategies and Partnerships, p. 289. Leiden University Press, Leiden (2024)
6. Kanishka, K., Acherjee, B.: Revolutionizing manufacturing: a comprehensive overview of additive manufacturing processes, materials, developments, and challenges. J. Manuf. Process. **107**, 574–619 (2023)
7. Gu, B., et al.: Depolymerization and functionalization of super engineering plastics. Chin. J. Polym. Sci. **43**, 1–11 (2024)
8. Pakdel, E., et al.: Recent progress in recycling carbon fibre reinforced composites and dry carbon fibre wastes. Resour. Conserv. Recycl. **166**, 105340 (2021)
9. Deshwal, D., et al.: Critical review of reinforcement effect in PEEK tribology and its usage in biomedical, coating and bearing applications. Polymer. **64**, 1–30 (2025)
10. Kumar, J., et al.: Dynamic properties and vibration response of polymer matrix composites: mechanisms, influences, and applications in engineering. Polym. Adv. Technol. **36**(1), e70039 (2025)
11. Mehta, A., et al.: An overview of 3D printing for modeling polymers, bio-nanocomposite, and nanocomposites materials. Nanocompos. Manuf. Technol. **25**, 517–548 (2025)
12. Van Doan, D., et al.: An overview of functionally graded materials: from civil applications to Defense and aerospace industries. J. Vib. Eng. Technol. **13**(1), 1–20 (2025)
13. Joladarashi, S., Kulkarni, S.: Comprehensive review of modeling and material selection for hybrid sandwich composites for ballistic impact application using six sigma DMAIC methodology. Proc. Inst. Mech. Eng. C J. Mech. Eng. Sci. **239**(10), 3703–3724 (2025)
14. Saha, S., Karmakar, S.: Strengthening of the structural element by using Armour–a review of a blast protective material. IOP Conf. Ser.: Earth Environ. Sci. **1326**, 012060 (2024)
15. Selim, M.S., et al.: Advances in polymer/inorganic nanocomposite fabrics for lightweight and high-strength armor and ballistic-proof materials. Chem. Eng. J. **493**, 152422 (2024)
16. Gupta, P., et al.: Metal matrix composites for sustainable products: a review on current development. Proc. Inst. Mech. Eng. Part L J. Mater. Des. Appl. **238**(10), 1827–1864 (2024)
17. Lv, Z., et al.: A review of near-net shape forming by hot isostatic pressure technology. J. Am. Ceram. Soc. **108**(2), e20196 (2025)
18. Sun, J., et al.: State-of-art, challenges, and outlook on deep hole boring: chatter suppression, tool wear monitoring, and error measurement. Int. J. Adv. Manuf. Technol. **136**, 1–31 (2025)
19. Tian, J., et al.: Research progress and future prospects of high-entropy alloys as microwave-absorbing materials. J. Mater. Sci. Mater. Electron. **36**(2), 92 (2025)
20. Feng, S., Liu, H.: Recent advances and understanding of high-entropy materials for lithium-ion batteries. Nanotechnology. **35**, 30 (2024)
21. Tudose, A.E.: NiCrAlY layers performance assessment developed onto 310H alloy by EB-PVD method after oxidation in aggressive conditions. Appl. Chem. Nanosci. Biomater. Eng. **5**, 45 (2024)
22. Bisht, N., et al.: Synergizing chemistry: unveiling the potential of hybrid fillers for enhanced performance in shape memory polymers. Adv. Compos. Hybrid Mater. **8**(1), 7 (2025)

23. Li, X., Wang, L., Lu, Y.: Three-dimensional disordered alloy metamaterials: a new platform of structure-function integration. Mater. Fut. **4**(1), 012001 (2025)
24. Tang, X., et al.: Recent advances on the fabrication methods of nanocomposite yarn-based strain sensor. Nanotechnol. Rev. **10**(1), 221–236 (2021)
25. Chithra, N., et al.: Effect of reinforcement addition on mechanical behavior of Al MMC-a critical review. J. Environ. Nanotechnol. **13**(2), 65–79 (2024)
26. Almansoori, A., Balázsi, K., Balázsi, C.: Advances, challenges, and applications of graphene and carbon nanotube-reinforced engineering ceramics. Nano. **14**(23), 1881 (2024)
27. Kaloyeros, A.E., Arkles, B.: Silicon carbide thin film technologies: recent advances in processing, properties, and applications: Part II. PVD and alternative (non-PVD and non-CVD) deposition techniques. ECS J. Solid State Sci. Technol. **13**(4), 043001 (2024)
28. Es-Soufi, H., et al.: Nanoceramic-based coatings for corrosion protection: a review on synthesis, mechanisms, and applications. Environ. Sci. Pollut. Res. **32**, 1–27 (2024)
29. 张月林: 氧化铝陶瓷增韧的研究进展. Acta Mater. Compos. Sin. **41**, 10 (2024)
30. Syduzzaman, M., et al.: Nanotechnology for high-performance textiles: a promising frontier for innovation. ChemNanoMat. **9**(9), e202300205 (2023)
31. Tawade, B.V., et al.: Recent developments in the synthesis of chemically modified nanomaterials for use in dielectric and electronics applications. Nanotechnology. **32**(14), 142004 (2021)
32. Agarwal, N., et al.: 4-dimensional printing: exploring current and future capabilities in biomedical and healthcare systems—a concise review. Front. Bioeng. Biotechnol. **11**, 1251425 (2023)
33. Ghosh, A., et al.: Implementation of the additive manufacturing technology towards producing high-performance aerospace components-a review. AIP Conf. Proc. **3007**, 100043 (2024)
34. Sadeghi, B., et al.: Architectural design of advanced aluminum matrix composites: a review of recent developments. Crit. Rev. Solid State Mater. Sci. **49**(1), 1–71 (2024)
35. Tan, B., et al.: 3D printing for explosive and propellant applications. Add. Manuf. Front. **3**, 200151 (2024)
36. Atalie, D., et al.: Role of additive manufacturing in defense technologies: emerging trends and future scope. Add. Manuf. Mater. Technol. **8**, 501–521 (2024)
37. Rath, K.C., et al.: Potential of AI, quantum computing, and semiconductor technology adoption in future industries: scope, challenges, and opportunities. In: Integration of AI, Quantum Computing, and Semiconductor Technology, pp. 415–440. IGI Global, London (2025)
38. Patil, A.T., Vidhale, B., Titarmare, A.: Strategic innovations in defense systems: a comprehensive analysis of emerging technologies and future trends. In: 2024 International Conference on Intelligent Systems for Cybersecurity (ISCS). IEEE, New York (2024)
39. Thangavel, K., et al.: Artificial intelligence for trusted autonomous satellite operations. Prog. Aerosp. Sci. **144**, 100960 (2024)
40. Mathur, V., et al.: Volumetric additive manufacturing for cell printing: bridging industry adaptation and regulatory frontiers. ACS Biomater. Sci. Eng. **11**(1), 156–181 (2025)
41. Vasishta, B.V., et al.: An overview of the evolution of additive manufacturing from 3d to 5d printing in biomedical applications. Prog. Add. Manuf. **10**, 1–17 (2024)
42. Santa Cruz-Pavlovich, F.J., et al.: Beyond vision: an overview of regenerative medicine and its current applications in ophthalmological care. Cells. **13**(2), 179 (2024)
43. Aggarwal, P., et al.: Digital twin in the aviation industry and the role of machine learning: a review. In: International Conference on ICT for Sustainable Development. Springer, Cham (2023)
44. Elizabeth, I., Barshilia, H.C.: A comprehensive review on corrosion detection methods for aircraft: moving from offline methodologies to real-time monitoring combined with digital twin technology. Eng. Sci. Technol. **6**, 69–98 (2025)
45. Wu, D., et al.: A brief review of key technologies for cloud-based battery management systems. J. Electron. Mater. **53**, 7334–7354 (2024)
46. Kocsis, I., Burján-Mosoni, B., Balajti, I.: A comprehensive review of key cyber-physical systems, and assessment of their education challenges. IEEE Access. **13**, 8898–8911 (2025)

47. Alqudsi, Y., Makaraci, M.: UAV swarms: research, challenges, and future directions. J. Eng. Appl. Sci. **72**(1), 1–24 (2025)
48. Alqudsi, Y., Makaraci, M.: Exploring advancements and emerging trends in robotic swarm coordination and control of swarm flying robots: a review. Proc. Inst. Mech. Eng. C J. Mech. Eng. Sci. **239**(1), 180–204 (2025)
49. Troupiotis-Kapeliaris, A., Kastrisios, C., Zissis, D.: Vessel trajectory data mining: a review. IEEE Access. **13**, 4827–4856 (2025)
50. Ullah, H., et al.: Integrating industry 4.0 technologies in defense manufacturing: challenges, solutions, and potential opportunities. Array. **23**, 100358 (2024)
51. Bakhshi, J., Efatmaneshnik, M.: The interplay of capability and complexity in military context: definitions, challenges, and implications. Def. Stud. **25**, 1–30 (2024)
52. Herzog, S., Kunertova, D.: NATO and emerging technologies—the Alliance's shifting approach to military innovation. Naval War College Rev. **77**(22), 47–69 (2024)
53. Chen, L., et al.: Machine learning for human–machine systems with advanced persistent threats. IEEE Trans. Human Mach. Syst. **54**, 753–761 (2024)
54. Heilig, G.A., May, M.: The hydrodynamic RAM effect: review of historic experiments, model developments and simulation, vol. 44, pp. 150–158. Defence Technol. (2024)
55. Walia, S., et al.: Rocket sled based high speed rail track test facilities: a review. Def. Sci. J. **72**(2), 182–194 (2022)
56. Mao, W., et al.: Analysis of the research progress of electromagnetic railgun based on CiteSpace. IEEE Access. **13**, 3499–3513 (2024)
57. Gans, N.R., Rogers, J.G.: Cooperative multirobot systems for military applications. Curr. Robot. Rep. **2**, 105–111 (2021)
58. Colorado, H.A., et al.: Additive manufacturing in armor and military applications: research, materials, processing technologies, perspectives, and challenges. J. Mater. Res. Technol. **27**, 3900–3913 (2023)
59. Rashid, A.B., Kausik, A.K.: AI revolutionizing industries worldwide: a comprehensive overview of its diverse applications. Hybrid Adv. **7**, 100277 (2024)
60. Srinivasan, V., Kunjiappan, S., Palanisamy, P.: A brief review of carbon nanotube reinforced metal matrix composites for aerospace and defense applications. Int. Nano Lett. **11**(4), 321–345 (2021)
61. Ince, J., et al.: Overview of emerging hybrid and composite materials for space applications. Adv. Compos. Hybrid Mater. **6**(4), 130 (2023)
62. Shirke, N., Ghase, V., Jamdar, V.: Recent advances in stealth coating. Polym. Bull. **81**, 1–30 (2024)
63. Vincent-Hall, T.D., et al.: Health effects of occupational exposure to jet fuels used in the military: a systematic review of the epidemiologic literature. Environ. Int. **196**, 109278 (2025)
64. Sevart, C.D., et al.: Parametric surrogate modeling of blast simulation timeseries data with kriging. In: AIAA SCITECH 2025 Forum. AIAA, Reston (2025)
65. Simulcik, J.D., Villalobos, F.E., Bazilian, M.D.: Electrification of the joint force: challenges and opportunities for competition in the Pacific and Arctic theaters. Electr. J. **38**(1), 107458 (2025)
66. Paek, S.W.: Sustainability considerations of nuclear power and nuclear propulsion in space. In: Routledge Handbook of Space Policy, pp. 359–368. Routledge, London (2025)
67. Ravindran, R., Massoud, A.M.: State-of-the-art DC-DC converters for satellite applications: a comprehensive review. Aerospace. **12**(2), 97 (2025)
68. Balakrishnan, N.T., et al.: Nuclear batteries: an overview. In: Advanced Technologies for Rechargeable Batteries, pp. 109–119. Routledge, London (2024)
69. Lingrui, L., Xin, W.: Towards smart aviation with sustainable development: artificial intelligence insights into the airline and advanced air mobility industries. In: Decision Support Systems for Sustainable Computing, pp. 187–204. CRC Press, Boca Raton (2024)
70. Krivonosova, A.: Integrating artificial intelligence into aviation logistics processes: review of applications and effects. Инновационная наука. **4-2**, 16–22 (2024)

71. Bagarello, S., Campagna, D., Benedetti, I.: A survey on hydrogen tanks for sustainable aviation. Green Energy Intell. Transp. **4**, 100224 (2024)
72. Pak, H., et al.: Can urban air mobility become reality? Opportunities and challenges of UAM as innovative mode of transport and DLR contribution to ongoing research. CEAS Aeronaut. J. **16**, 1–31 (2024)
73. Zhou, Q., Tan, F.: Avionics of electric vertical take-off and landing in the urban air mobility: a review. IEEE Aerosp. Electron. Syst. Mag. (2024). https://doi.org/10.1109/MAES.2024.3488655
74. Ison, D.: Consumer willingness to fly on advanced air mobility (AAM) electric vertical take-off and landing (eVTOL) aircraft. Collegiate Aviat. Rev. Int. **42**(1), 100233 (2024)
75. Farazi, N.P., Zou, B.: Planning electric vertical takeoff and landing aircraft (eVTOL)-based package delivery with community noise impact considerations. Transp. Res. Part E Log. Transp. Rev. **189**, 103661 (2024)
76. Shahzad, K., Cheema, I.I.: Low-carbon technologies in automotive industry and decarbonizing transport. J. Power Sour. **591**, 233888 (2024)
77. Lyudmila, C.: Decarbonization with the help of artificial intelligence-one of the priorities of international civil aviation. Dig. Transf. **50**, 242 (2024)
78. Bonab, S.A., Yazdani-Asrami, M.: Artificial intelligence-based model to predict the heat transfer coefficient in flow boiling of liquid hydrogen as fuel and cryogenic coolant in future hydrogen-powered cryo-electric aviation. Fuel. **381**, 133323 (2025)
79. Velrani, K.S., et al.: A study on IoT and AI-integrated advanced lightweight materials for eco-friendly production. In: Innovations in Energy Efficient Construction Through Sustainable Materials, pp. 1–32. IGI Global, London (2025)
80. Prussi, M., et al.: The risk of increasing energy demand while pursuing decarbonisation: the case of the e-fuels for the EU aviation sector. Transp. Policy. **160**, 154–158 (2025)
81. Vergallo, R., Casciaro, S., Mainetti, L.: From fintech to green fintech: how to decarbonize the AI in the financial domain. In: Machine Learning and Modeling Techniques in Financial Data Science, pp. 25–70. IGI Global Scientific Publishing, London (2025)
82. Wandelt, S., Zhang, Y., Sun, X.: Sustainable aviation fuels: a meta-review of surveys and key challenges. J. Air Transp. Res. Soc. **4**, 100056 (2025)
83. Ramos-Fernandez, E.V., et al.: Potential pathways for CO2 utilization in sustainable aviation fuel synthesis. Chem. Sci. **16**(2), 530–551 (2025)
84. Chireshe, F., et al.: Cost-effective sustainable aviation fuel: insights from a techno-economic and logistics analysis. Renew. Sust. Energ. Rev. **210**, 115157 (2025)
85. Yang, F., Yao, Y.: Sustainable aviation fuel pathways: emissions, costs and uncertainty. Resour. Conserv. Recycl. **215**, 108124 (2025)
86. Han, Y., et al.: Techno-economic assessment of modified Fischer-Tropsch synthesis process for direct CO2 conversion into jet fuel. Fuel. **381**, 133442 (2025)
87. Marchesan, A.N., et al.: Alternative feedstocks for sustainable aviation fuels: assessment of sugarcane-derived microbial oil. Bioresour. Technol. **416**, 131772 (2025)
88. Uzochukwu, J.S., et al.: Sustainable bio-aviation fuel production from biobased oil: a novel hydrogen-free biorefinery approach. UNIZIK J. Eng. Appl. Sci. **4**(1), 1530–1554 (2025)
89. Khalid, S.: Advanced electric propulsion systems for hybrid electric vehicles: design, control, and performance optimization. In: Hybrid Electric Vehicles and Distributed Renewable Energy Conversion: Control and Vibration Analysis, pp. 1–54. IGI Global Scientific Publishing, London (2025)
90. Ma'arif, S., Budiyanto, M.A., Theotokatos, G.: Progress in hybrid and electric propulsion technologies for fishing vessels: an extensive review and prospects. Ocean Eng. **316**, 120017 (2025)
91. Chau, T., et al.: High-fidelity aeropropulsive assessment of distributed electric propulsion and boundary layer ingestion for the SUSAN electrofan. In: AIAA SCITECH 2025 Forum. AIAA, Reston (2025)

92. Quinn, D., et al.: Propulsion system integration on truss-braced aircraft: structural performance sensitivity to architectural design parameters. Aerosp. Sci. Technol. **159**, 109987 (2025)
93. Lee, J., Han, J.-H.: Separation and release devices for aeronautical and astronautical systems: a review. Int. J. Aeronaut. Space Sci. **26**(1), 131–161 (2025)
94. Griffiths, M., Rubaii, K.: Late modern war and the geos: the ecological 'beforemaths' of advanced military technologies. Secur. Dialogue. **56**(1), 38–57 (2025)
95. Sfetcu, N.: Electronic Warfare and Artificial Intelligence. MultiMedia Publishing, Vijayawada (2024)
96. Jiang, C., et al.: YOLO-based missile pose estimation under uncalibrated conditions. IEEE Access. **12**, 12462–112469 (2024)
97. Mohammed, S., et al.: A critique on boron nitride nanotube reinforced metal matrix composites. Crit. Rev. Solid State Mater. Sci. **49**(3), 482–519 (2024)
98. Hong, Y., Kang, Z., Fan, J.: Revolutionizing textile: advanced techniques for superior thermal conductivity. Compos. Part A: Appl. Sci. Manuf. **186**, 108380 (2024)
99. Patra, N.: Advanced nano-composite coatings of ZrB_2/ZrC/SiC on carbon fibers via eco-friendly precursor synthesis. Colloids Surf. A Physicochem. Eng. Asp. **705**, 135729 (2025)
100. Herath, C.L.: Hypersonic missile technology: a comprehensive analysis
101. Dutta, S., et al.: Hypersonic inflatable aerodynamic decelerator earth-based applications. In: AIAA SCITECH 2025 Forum. AIAA, Reston (2025)
102. Roelke, E., et al.: Comparative performance of entry simulations using guided human-scale landers at mars. In: AIAA SCITECH 2025 Forum. AIAA, Reston (2025)
103. Herberhold, M., et al.: Comparison of SpaceX's Starship with winged heavy-lift launcher options for Europe. CEAS Space J. (2025). https://doi.org/10.1007/s12567-025-00625-8
104. Caldwell, A., et al.: Reusable thermal protection material development. In: Aerospace Materials and Manufacturing Digital Summit. SAE International, Pittsburgh (2025)
105. Yeo, H., et al.: Development of a rapid analysis program for the prediction of aerothermodynamics in high-speed vehicles. Aerosp. Sci. Technol. **164**, 110415 (2025)
106. Gong, G., et al.: Comparative study on mechanical properties and high temperature oxidation behaviour of Hastelloy X and Inconel 718 fabricated by laser directed energy deposition. J. Mater. Res. Technol. **28**, 2624–2635 (2024)
107. Qin, Y., et al.: Mechanism analysis on passive film-forming performance of heat-treated SLM-ed Hastelloy X. Mater. Chem. Phys. **334**, 130453 (2025)
108. Sofuoğlu, M.A., Haydarlar, G., Tekkalmaz, M.: Influence of cutting parameters on tool temperatures and residual stresses in machining aerospace alloys: a DEFORM 3D simulation approach. In: Advanced Engineering Forum. Trans Tech Publication, Bäch (2025)
109. Kumar, N., et al.: Microstructure, thermal expansion, and high-temperature oxidation behavior of spark plasma sintered AlCoCrSiNi high entropy alloy. Mater Today Commun. **40**, 110063 (2024)
110. Du, W., et al.: Advancements in methodologies and techniques for the synthesis of energetic materials: a review. Energetic Mater. Front. **5**, 175–190 (2024)
111. Osman, A.I., et al.: Coordination-driven innovations in low-energy catalytic processes: advancing sustainability in chemical production. Coord. Chem. Rev. **514**, 215900 (2024)
112. Woeppel, A.: Flexible Energetics Trace Detection Schemes Utilizing Organic Electrochemical Transistors. Purdue University, West Lafayette (2024)
113. Parakhin, V.V., Smirnov, G.A.: Research progress on design, synthesis and performance of energetic polynitro hexaazaisowurtzitane derivatives: towards improved CL-20 analogues. FirePhysChem. **4**(1), 21–33 (2024)
114. Zhang, X.-X., et al.: Desensitization and stabilization of HMX crystals by intercalation of crosslinked graphene oxide. Chem. Eng. J. **502**, 158192 (2024)
115. Han, K., et al.: An investigation on the microscale detonation transmission performance of HMX-based explosive ink. J. Energetic Mater. **43**(1), 143–157 (2025)
116. Xu, S., et al.: Revealing and visualizing the microscopic solution chemical molecular mechanism of β-HMX solubility in binary mixed solvents. Chem. Eng. Sci. **306**, 121288 (2025)

117. Wang, J., et al.: Simple preparation of heat-resistant energetic materials based on [5, 6, 5] tricyclic fused-ring skeletons and evaluation of their energetic performance. Cryst. Growth Des. **24**(10), 4114–4121 (2024)
118. Muhammad, M.M., et al.: Non-destructive testing of solid propellants: a review of methods and applications. Defence S&T Tech. Bull. **17**(2), 125–150 (2024)
119. Wu, G., et al.: Experimental study of the evaporation characteristics of unsymmetrical dimethylhydrazine in a sub-supercritical environment. Phys. Fluids. **36**(2), 023321 (2024)
120. Krishnan, S., Raghavan, J.: Quantitative internal-ballistics prediction and design. In: Rocket Propulsion Primer, pp. 169–319. Springer, Cham (2024)
121. Wang, J., et al.: A self-sensing HTPB liner for non-destructive monitoring nitroglycerin (NG) migration at the interface between double base propellant and the HTPB liner. Defence Technol. **50**, 166–175 (2025)
122. Li, L., et al.: Comparative study on mechanical properties and combustion characteristics of additive manufacturing/casting composite solid propellants. Acta Astronaut. **229**, 43–54 (2025)
123. Cheng, D.: The People's Republic of China and the information age: dealing with Informationization and intelligence-Ization. In: The Great Power Competition Volume 6: The Rise of China, pp. 45–60. Springer, Cham (2025)
124. Eswaran, K.: Hydrogen horizons: advancements, applications, and global initiatives. In: Renewable Energy Development: Technology, Material and Sustainability, pp. 213–239. Springer, Cham (2025)
125. Boichenko, S., Shkilniuk, I., Iakovlieva, A.: Evolution of the quality of jet fuels: Ukrainian comparative review from traditional to sustainable aviation fuels. Mod. Technol. Energy Transp. **2**, 239–255 (2025)
126. Souagh, A., et al.: Ammonia (NH3) production from the thermal decomposition of hydroxylammonium nitrate (HAN) as a green energetic source for clean space. ACS Omega. **10**(5), 4442–4448 (2025)
127. Johnson, P.N., et al.: Phenomenological modeling of plasma-assisted methane ignition using plasma energy fraction manifolds. In: AIAA SCITECH 2025 Forum. AIAA, Reston (2025)
128. Pai, A., et al.: Experimental techniques for performance evaluation of shielding materials and configurations subjected to blast and ballistic impacts: a state-of-the-art review. Thin-Walled Struct. **191**, 111067 (2023)
129. Aien, N.U., et al.: Composition of explosive and their effect at different crime scene. Foren. Insights Health Sci. Bull. **2**(1), 11–17 (2024)
130. Suja, S., Mathiya, S.: Nanotechnology in defense and social justice: opportunities and risks. In: Nanotechnology in Societal Development, pp. 439–479. Springer, Cham (2024)
131. 张志远: 石墨烯导热测试方法, 影响因素及其应用 研究进展. Acta Mater. Compos. Sin. **41**(11), 4589 (2024)
132. Luo, Y., et al.: A comprehensive review on application of composite guidance technology in aerial guided bombs. In: Advanced Fiber Laser Conference (AFL 2023). SPIE, Bellingham (2024)
133. Sayıl, A., Erden, F.: The history of guided bombs, guidance kits, wing kits, and wing deployment mechanisms: Güdümlü Bombaların, Güdüm Kitlerinin, Kanat Kitlerinin ve Kanat Açma Mekanizmalarının Tarihçesi. J. Aeronaut. Space Technol. **17**(2), 203–234 (2024)
134. Chang, G., et al.: Overview of research on intelligent swarm munitions. Defence Technol. (2024). https://doi.org/10.1016/j.dt.2024.08.017
135. Hatler, C., et al.: The path towards plasma facing components: a review of state-of-the-art in W-based refractory high-entropy alloys. Curr. Opin. Solid State Mater. Sci. **34**, 101201 (2025)
136. Park, S., et al.: Preliminary satellite design for cislunar intelligence surveillance and reconnaissance. In: AIAA SCITECH 2025 Forum. AIAA, Reston (2025)
137. Ren, Y., et al.: Robust seamless GEO scanning control strategy for airborne electro-optical/infrared reconnaissance systems. SIViP. **18**(11), 7611–7620 (2024)

138. Jiang, T., Zhou, H., Yang, Y.: Research on photonic communication technology and its prospects for defense applications. SPIE Int. **13104**, 1310445 (2024)
139. Popov, M., et al.: Concept of a geoinformation platform for landmines and other explosive objects detection and mapping with UAV. Radioelectron. Comput. Syst. **2024**(4), 207–216 (2024)
140. Finley, D.B., Bryant, A., Bosco, A.: Comparison of the lockheed martin falcon CFD solver to test data on a Mach 4 waverider model. In: AIAA SCITECH 2025 Forum. AIAA, Reston (2025)
141. Laten, J., et al.: NATO AVT-350 task group: platform integration to assess prospects of active flow control on a next-gen tailless UAS–Part II TO/L regime. In: AIAA SCITECH 2025 Forum. AIAA, Reston (2025)
142. Gregory, J.R., et al.: Digital engineering for cubesats with semantic web technologies. In: AIAA SCITECH 2025 Forum. AIAA, Reston (2025)
143. Santangelo, A.D., Falco, G.: QuickSAT/SHERLOCK, an AI architecture for vehicle health management, fault detection and fault management. In: AIAA SCITECH 2025 Forum. AIAA, Reston (2025)
144. Dong, S., et al.: AI in commercial space industry. In: Precision Medicine for Long and Safe Permanence of Humans in Space, pp. 601–615. Elsevier, Amsterdam (2025)
145. Yang, Y., et al.: National defense quantum tech: deployment analysis and future prospects. SPIE Int. **13104**, 1310459 (2024)
146. Alvan, A., Kısacık, S., Gelirli, N.: Unleashing the power of AI: creative destruction in energy systems. In: Economic and Political Consequences of AI: Managing Creative Destruction, pp. 45–64. IGI Global Scientific Publishing, London (2025)
147. Suganya, R., Joseph, L.L., Kollem, S.: Integrating artificial intelligence in electric vehicles and optimizing logistics for sustainable transportation. In: Cases on AI-Driven Solutions to Environmental Challenges, pp. 385–418. IGI Global Scientific Publishing, London (2025)
148. Pardeep, P.D.: Exploring the role of 3D printing in aerospace engineering: innovations in lightweight structural components. J. Eng. Sci. Res. **1**(1), 77–98 (2025)
149. Masrafee, M.M.R., Rashid, A.B., Hoque, M.E.: Innovations in 3D printing-assisted biopolymer composites for aerospace applications. In: Sustainable 3D Printing for Innovative Biopolymer Production and Applications, pp. 117–147. Wiley, Hoboken (2025)
150. Bilgin, K.R.: The ethical and legal challenges of AI in wars. In: Cyber Security in the Age of Artificial Intelligence and Autonomous Weapons, pp. 101–119. CRC Press, Boca Raton (2025)
151. Tutty, M.G.: Aircraft stores compatibility and novel uses of analogy. In: AIAA SCITECH 2025 Forum. AIAA, Reston (2025)
152. Kaakandikar, R.M., Wawre, D.D.: Regulatory frameworks and policies for sustainable aviation marketing. In: Sustainable Marketing Practices in the Aviation Industry, pp. 1–32. IGI Global Scientific Publishing, London (2025)
153. Jaume, J., Alonso, G., Benito, A.: Evaluating the impact of the new environmental regulations on airlines' business results. Aircr. Eng. Aerosp. Technol. **97**(1), 108–119 (2025)
154. Kim, C., et al.: Potential energy generation of photovoltaics with acceptable risk at Korean airports. Int. J. Energy Res. **2025**(1), 7288954 (2025)
155. Ostroumov, I., et al.: Impact analysis of Russian-Ukrainian war on airspace. J. Air Transp. Manag. **124**, 102742 (2025)
156. Raytheon: Proven in battle and still setting standards. https://www.rtx.com/raytheon (2025)
157. Easley, M.: Air Force revives ARRW hypersonic missile with procurement plans for fiscal 2026. https://defensescoop.com/2025/06/26/air-force-arrw-procurement-funding-fy26-budget-request/ (2025)
158. CTRL, Q.: Unlocking the power of quantum navigation: Lockheed Martin, Q-CTRL, and partner awarded contract. https://q-ctrl.com/blog/unlocking-the-power-of-quantum--navigation-lockheed-martin-q-ctrl-and-partner-awarded-contract (2025)

159. Zou, Y., et al.: Fluid dynamics and infrared stealth of marine IRS devices: a review. Int. J. Mech. Syst. Dyn. **5**(2), 179–200 (2025)
160. Hao, W., Guohua, C., Yuehua, C.: Penetration effectiveness analysis of hypervelocity weapon warhead and design of concrete shield. 爆炸与冲击 (2025)
161. Ai-rong, L., et al.: Research on Starlink ephemeris published by spacex. Chin. Astron. Astrophys. **49**(1), 193–218 (2025)
162. Carney, D. Tesla's switch to giga press die castings for model 3 eliminates 370 parts. https://www.designnews.com/automotive-engineering/tesla-s-switch-to-giga-press-die-castings-for-model-3-eliminates-370-parts (2021)
163. Clausen, J.: Using AI to advance manufacturing at general motors. https://news.gm.com/home.detail.html/Pages/topic/us/en/2025/mar/0311-ai.html (2025)
164. Klender, J.: Tesla builds 50 million 4680 battery cells at giga Texas. https://www.teslarati.com/tesla-builds-50-million-4680-battery-cells-at-giga-texas (2025)
165. Duoba, M., Nelson, D.: Electric vehicle efficiency trends and insights from calculations of chassis and powertrain energy usage in EPA test cycles. In: 2024 IEEE Transportation Electrification Conference and Expo (ITEC). IEEE, New York (2024)
166. Sample, I.: China's Chang'e-6 lunar probe returns world's first samples from far side of the moon. https://www.theguardian.com/world/article/2024/jun/25/chinas-change-6-lunar-probe-returns-world-first-samples-from-far-side-of-the-moon (2024)
167. Jones, A.: Watch Chinese company launch 4 satellites to orbit from ship at sea (video). https://www.space.com/galactic-energy-ceres-1-sea-launch-video-may-2024 (2024)
168. Reuters. China beats rivals to successfully launch first methane-liquid rocket. https://www.reuters.com/technology/space/china-beats-rivals-successfully-launch-first-methane-liquid-rocket-2023-07-12/ (2023)
169. Valdés, R.M.A., Comendador, V.F.G.: Evolution of the Chinese aircraft industry. In: Accelerating Sustainable Aviation Initiatives: Technology, Markets and Social Issues, pp. 47–81. Springer, Cham (2025)
170. Northvolt: Northvolt to transform closed paper mill in Sweden into new gigafactory. https://northvolt.com/articles/northvolt-to-transform-closed-paper-mill-in-sweden-into-new-gigafactory/ (2022)
171. Sun, Y., et al.: Civil aircraft: history, progress. In: Proceedings of the 7th China Aeronautical Science and Technology Conference, vol. II. Springer, Cham (2025)
172. Sai, S., Mittal, U., Chamola, V.: DMDAT: diffusion model-based data augmentation technique for vision-based accident detection in vehicular networks. IEEE Trans. Veh. Technol. **74**, 2241–2250 (2024)
173. Nassif, J., Tekli, J., Kamradt, M.: Background and technologies. In: Synthetic Data: Revolutionizing the Industrial Metaverse, pp. 33–74. Springer, Cham (2024)
174. Mehta, A.: South Korea's Hanwha shows off prototype aircraft engine. https://breakingdefense.com/2024/07/south-koreas-hanwha-shows-off-prototype-aircraft-engine-says-could-enter-service-in-9-years/ (2024)
175. Agencies, D.S.W.: Drone magnate Baykar remains Türkiye's top defense exporter in 2024. https://www.dailysabah.com/business/defense/drone-magnate-baykar-remains-turkiyes-top-defense-exporter-in-2024 (2025)
176. Chauhan, D., et al.: Nation's Defense: a comprehensive review of anti-drone systems and strategies. IEEE Access. **13**, 53476–53505 (2025)
177. Shams, A.: Russo-Ukrainian war: emerging technologies. In: Navigating Complex Geopolitical Landscapes Amidst Conflict, p. 87. IGI Global, London (2025)
178. Fetter, S., Sankaran, J.: Emerging technologies and challenges to nuclear stability. J. Strateg. Stud. **48**(2), 252–296 (2025)
179. Powell, M., Townley, D.: The challenges and opportunities for academics in professional military education. Def. Stud. **25**, 1–18 (2025)

180. Bingöl, O.: CBRN attack with unmanned aerial vehicles as terrorist tactic: a review in the Middle East context. Asian J. Middle East. Islam. Stud. **19**, 1–22 (2025)
181. Abraheem, S.M., Ali, M.E., Abuali, R.M.: Emerging trends in quantum sensors: applications in Defense and communication. Middle East J. Pure Appl. Sci. **25**, 19–37 (2025)
182. Zohuri, B.: Harnessing artificial intelligence for countering hypersonic weapons: a new frontier in battlefield offense and Defense (a short review). J. Energy Power Eng. **18**, 139–145 (2024)

Chapter 12
Industry 6.0 and the Digital Transformation of Healthcare and Pharmaceuticals

Abstract The Industry 6.0 technologies adoption is revolutionizing the healthcare and pharmaceutical sectors by enhancing efficiency, precision, and sustainability. AI is optimizing drug discovery, accelerating clinical trials, and enabling predictive diagnostics, reducing time to market for new treatments. The Internet of Medical Things (IoMT) ensures real-time patient monitoring, while CPS enhance robotic-assisted surgeries and automated laboratory processes, improving accuracy and patient outcomes. Quantum computing is transforming molecular simulations, enabling the design of highly specific drug compounds, while blockchain technology is securing pharmaceutical supply chains, minimizing counterfeit drugs. Advanced 3D and 4D bioprinting facilitates the production of customized prosthetics, tissue engineering, and even organ regeneration, addressing critical transplant shortages. The pharmaceutical industry is leveraging nanotechnology for targeted drug delivery, improving bioavailability, and reducing side effects. Smart manufacturing driven by AI, digital twins, and automation is minimizing waste, ensuring cost efficiency, and enhancing regulatory compliance. Wearable health technologies and biosensors enable personalized medicine, optimizing treatments based on individual genetic profiles. As healthcare systems transition towards predictive and precision medicine, Industry 6.0 innovations will redefine patient care, pharmaceutical development, and global healthcare accessibility, contributing to sustainable healthcare solutions aligned with the SDGs.

Keywords Industry 6.0 · Artificial intelligence in healthcare · Internet of medical things · Cyber physical systems · Quantum computing in pharma · Blockchain in healthcare · 3D and 4D bioprinting · Nanotechnology in medicine · Smart manufacturing · Personalized medicine · Digital twins in pharma · Wearable health devices · Predictive diagnostics · Sustainable healthcare technologies · Precision medicine

H. Majeed, T. Iftikhar, *Intelligent Manufacturing in Industry 6.0*,
https://doi.org/10.1007/978-3-032-07278-8_12

1 Smart Healthcare and Pharmaceutical Manufacturing

By using IoT, AI, and automation systems to improve accuracy, scalability, and efficiency, Industry 4.0 technologies are transforming the pharmaceutical and healthcare manufacturing industries. These developments make it possible to produce complicated biologics, vaccines, and customized medications quickly and at a scale that meets the needs of both the global health community and the economy. SDGs like SDG 3 and 9 can be directly supported by manufacturers who use advanced digital systems to streamline production processes, reduce errors, and speed up the availability of essential medical products. The availability of life-saving medications is accelerated by IoT and AI technologies, which reduce production cycles by 25–30%. Automated systems and predictive maintenance reduce manufacturing costs by 15–25%, increasing economic viability. Optimized bioprocessing uses 20–30% less water, which supports SDG 12. All these benefits highly depend on specific process, product, initial level of maturity [1, 2].

2 Technological Framework and Automation

To continuously check vital factors like pH, temperature, dissolved oxygen levels, and nutrient concentrations, IoT-enabled sensors are integrated into bioreactors and other manufacturing machinery. For analysis and control, these sensors send data in real time to centralized systems. IoT sensors in vaccine manufacturing identify variations in bioreactor conditions, enabling prompt remedial measures to preserve the best growth circumstances for biologics, lowering the danger of contamination and guaranteeing constant quality. IoT-based monitoring greatly increases operational efficiency by cutting production downtime. To improve bioprocesses, identify inefficiencies and put remedial measures in place, AI models examine both historical and current data. To find the best operating conditions, machine learning algorithms can simulate several production situations. AI programs predict when equipment will require repair, avert unplanned malfunctions, and cut downtime. AI boosts yields by modifying feeding plans in bioreactors according to cellular growth patterns [3].

IoT and AI-enabled Industry 4.0 manufacturing systems' modular architecture makes it possible to produce vaccines in large quantities. Robotic systems equipped with automated filling, capping, and packaging lines provide accurate dosage and little waste. Manufacturers using Industry 4.0 technology were able to scale up vaccine manufacturing during the COVID-19 epidemic in record time, delivering billions of doses worldwide (like Pfizer + BioNTech production 3.3 billion doses, Moderna reached around 800 million doses). Real-time optimizations and production process optimizations are made possible by digital twins, which are virtual copies of actual manufacturing systems. Without running the risk of batch failure, they let manufacturers test modifications to formulation or operating conditions. For instance, a bioreactor's digital twin can mimic temperature changes and nutrient feeding schedules to maximize biologic production and save energy. Strict regulatory compliance is ensured by automated quality checks for pharmaceutical items using AI algorithms and advanced visual systems. Automated quality control ensures constant product quality and safety by reducing errors in medicine formulation and packaging by 40–50% [4–9].

3 Technologies Implication Techniques

Install IoTs sensors in vital production machinery including lyophilizers, fermenters, and bioreactors. For real-time process control, combine and evaluate sensor data using platforms such as ThingWorx or AWS IoT Core. To forecast the ideal circumstances for biologic growth and vaccine manufacture, train machine learning models using past production data. Utilize tools such as Azure Machine Learning or TensorFlow to design and implement AI models. Use easily reconfigurable modular manufacturing units to increase or decrease production. To guarantee accuracy and scalability, incorporate robotic technologies for duties like material handling, packing, and labeling. To model and optimize processes, create digital twins of industrial facilities using programs like Siemens NX or ANSYS Twin Builder. Before putting formulation or operational parameter changes into actual systems, test them virtually. Use solar panels and other renewable energy sources to power production facilities. Reduce the amount of water used in bioprocessing by utilizing sophisticated water recycling devices [10–15].

4 Alignment with SDGs

Smart healthcare manufacturing integrates cutting-edge technologies to enhance health outcomes and advance sustainability, which is in line with several SDGs. These systems handle disease outbreaks and improve global health by ensuring the timely manufacturing and delivery of essential medications and vaccinations, which supports SDG 3. Healthcare manufacturing supports the development of robust and adaptive pharmaceutical infrastructures by embracing cutting-edge technologies like IoT and AI, which advance SDG 9. By decreasing waste and consuming fewer resources, streamlined production processes support sustainable manufacturing practices and are in line with SDG 12. SDG 13 is directly supported by energy-efficient technologies and lower emissions from automated production, which advances sustainable industrial practices and helps to combat climate change. Through these links, smart healthcare manufacturing tackles important global issues in sustainability, environmental stewardship, and health while also improving operational efficiency [16–20].

5 Quantum Computing

Businesses need to implement several implications approaches and technologies in order to incorporate quantum computing into industry. Initially, companies ought to cooperate with quantum computing platforms that are available via cloud services like Microsoft's Azure Quantum, Google's Quantum AI, or IBM's Qiskit. Since quantum computers are currently very specialized and costly, these platforms enable businesses to utilize quantum computing resources without having to purchase the hardware. Cloud-based quantum computing enables companies to remotely run simulations on quantum systems, create apps, and experiment with quantum algorithms. Next, businesses need to start training their employees to be quantum ready. An industry-wide investment in education and training programs for engineers, data scientists, and other technical workers will be crucial because quantum computing is a highly specialized sector. Furthermore, the creation of hybrid algorithms that blend classical computing methods with quantum capabilities will be necessary for the shift from classical to quantum systems [21–23]. To benefit from quantum principles, industries should also investigate quantum-inspired optimization techniques that might not require complete quantum gear. By implementing these quantum-inspired algorithms on conventional computers, better outcomes can be obtained in domains such as resource allocation and logistics. By increasing industrial efficiency, cutting waste, and lowering carbon footprints, supply chain and logistics optimization using quantum algorithms can support SDG 9. Energy optimization alone using quantum computing might help companies cut huge global carbon emissions by 2030. Quantum computing is an innovative technology that has the potential to address some of the most difficult problems that modern enterprises face. Quantum computing has the potential to enhance efficiency across various industries, lessen environmental effects, and aid in the accomplishment of important SDGs by streamlining supply chains, improving material design, and boosting energy management. Industries that use quantum computing will be in a better position to innovate, cut costs, and satisfy the expectations of a more sustainable future as the technology advances and becomes more widely available [24–27].

6 Adaptive Robotics and Cobots for Automation Advancements

By improving flexibility, accuracy, and safety in manufacturing settings, adaptive robotics, in particular, collaborative robots, or cobots, is revolutionizing industrial automation. Cobots can adapt to changing industrial surroundings thanks to their sophisticated technologies, which include RL, a subfield of machine learning. Cobots can enhance their performance and decision-making skills over time by learning from their interactions with the environment and human operators. This makes them extremely useful for jobs requiring accuracy, flexibility, and real-time correction [28, 29]. By using these revolutionary technologies, in the Age of Industry 4.0, manufacturers can expand their horizons, improve operational excellence, and set themselves up for sustained success. This expansion is a result of the growing use of cobots in a variety of sectors, including consumer goods, electronics, healthcare, and the automotive industry. The growing need for more adaptable, affordable, and scalable automation systems that can boost output while upholding high standards of quality is what is driving the use of cobots. Cobots offer the perfect way for businesses to satisfy the increasing demands of their customers for personalization and quick production cycles while maintaining operational efficiency and safety [30, 31].

6.1 Reinforcement Learning in Cobots

The foundation of cobot intelligence (in Industry 6.0) will be reinforcement learning, which will allow the robot to learn by making mistakes and getting feedback from its actions. As a result, cobots will do tasks like assembly, packaging, and material handling more effectively by optimizing their performance in real time. In a small batch manufacturing setting, for instance, a cobot will adjust to different product designs, material qualities, and changing operating circumstances without requiring frequent reprogramming. This skill is especially important in sectors like small batch or customized product manufacturing where production lines must be regularly modified to satisfy shifting customer requirements or specifications. The vast majority of industrial cobots today (in Industry 4.0 and 5.0) operate on preprogrammed paths, impedance control, and dynamic movement primitives. Cobots with RL algorithms constantly assess the results of their operations and modify their behavior to reduce mistakes and boost productivity. To account for differences in parts or minor variations in the assembly process, a cobot might employ RL to figure out the best method to put components together on an automotive assembly line. Even in settings where humans and robots share a workspace, the cobot's capacity for adaptation eventually results in shorter cycle times and higher accuracy in assembly jobs [32, 33].

6.2 Industrial Worker Safety and Collaboration Impact

Cobots can operate near people without endangering them since they are designed with safety in mind. Cobots employ sensors, cameras, and force-limiting technology to detect and respond to human presence, in contrast to standard industrial robots that need fences or other protective barriers. To prevent mishaps, a cobot will automatically slow down or stop moving if a human worker meets it. Cobots are perfect for collaborative jobs where a human worker and a robot oversee separate parts of the same process since they can work securely alongside humans. For example, in a packaging line, a cobot may perform repetitive, physically demanding activities like lifting and arranging products, while a human may handle quality control and decision-making [34, 35].

6.3 Implication Techniques

By enabling more productive, adaptable, and cooperative production environments, the combination of cobots with sensor-based safety systems and adaptive learning algorithms is transforming industrial automation. Cobots are especially useful in sectors, where customization or small-batch production are commonplace. Changes in production standards or product design frequently necessitate a large amount of downtime in traditional manufacturing settings to reconfigure production lines or machinery. However, these adjustments may be made rapidly and effectively with cobots, which lowers downtime and boosts output. Along with reinforcement learning, force sensors and vision systems are also essential for augmenting the capabilities of cobots. These technologies allow cobots to do highly precise activities, like force feedback for sensitive handling and machine vision for quality monitoring. Cobots can check products for flaws in food processing, guaranteeing that only goods that satisfy quality standards are packed. Cobots are capable of handling delicate parts in the electronics manufacturing industry, using the proper amount of force to assemble pieces without causing damage [36, 37].

Companies must think about a few crucial procedures before deploying cobots in industrial environments. Industries must first make investments in reinforcement learning frameworks and robotics platforms that work with their current production systems. Modular cobot solutions that are simple to integrate into different production stages without requiring major adjustments to the current infrastructure should be investigated by businesses. Second, cooperation with robots and AI suppliers is necessary to customize the cobot systems to meet industrial demands, such as quality assurance, product customization, or packaging. To guarantee smooth cooperation with human operators, the cobots should be outfitted with the required sensors, cameras, and learning algorithms. It is critical to train workers to collaborate with cobots. Effective human robot interaction training will optimize the safety and effectiveness of the integration, since cobots are made to work alongside human workers. The successful transition to a collaborative robotic workforce will depend on the establishment of clear communication channels and safety norms. To sum up, cobots and adaptive robotics are transforming businesses by improving industrial processes' safety, accuracy, and adaptability. Real-time adaptation, sensor-based technology, and reinforcement learning are used by cobots to create dynamic, effective, and sustainable production environments. Their significance in the upcoming stage of industrial automation is highlighted by their expanding market and potential to align with the SDGs, which will help organizations become more sustainable, productive, and competitive [38–40].

6.4 Alignment with SDGs

Adoption of cobots and adaptive robots directly supports several important SDGs. By enabling smart production and encouraging innovation through automation, cobots help modernize industrial infrastructure, which is one of the most notable connections with SDG 9. This collaborative approach not only promotes productivity but also decreases workplace injuries and enhances job satisfaction. By facilitating more sustainable and effective production processes, cobots contribute to the advancement of SDG 12. The real-time adaptation capabilities of Cobots aid in waste reduction, resource optimization, and energy conservation, all of which are essential for sustainable manufacturing. Cobots eliminate the requirement for large-scale manufacturing runs, which can result in waste and overproduction, by offering flexibility in small-batch production. Cobots help to create more sustainable supply chains by increasing production efficiency and precision [41, 42].

7 Fundamentals of Genomics and Gene Editing

7.1 The Science Behind Genomics

Genomics is an interdisciplinary field that utilizes high-throughput sequencing, bioinformatics, and AI-driven computational tools to decode the complexity of genetic material. Future advancements integrating quantum computing, AI-driven genome assembly, and enzyme engineering will redefine genetic manipulation, enabling intelligent, self-adaptive gene therapies, precision agriculture, and synthetic life forms, solidifying the role of genomics in Industry 6.0 and bio intelligent manufacturing. The primary structural integrity of DNA arises from the phosphodiester bonds that link adjacent nucleotides, forming a sugar-phosphate backbone, while base pairing occurs through hydrogen bonds (A–T with two bonds, G–C with three bonds), ensuring double-helix stability with base stacking interaction (pi-pi interactions between aromatic bases). The genome size and complexity are dictated by both coding and noncoding regions. Coding sequences (exons) encode proteins, while introns, regulatory elements, and repetitive sequences contribute to genetic regulation, chromatin architecture, and cellular differentiation. Repetitive sequences, in general, do not directly contribute to cellular differentiation. Their roles are often structural (telomeres, centromeres) or related to genome stability. The sentence should refer to specific regulatory elements (enhancers, promoters) and epigenetic modifications. The chemical modifications of DNA, such as 5-methylcytosine (5 mC) epigenetic methylation and histone acetylation, regulate gene expression by altering chromatin accessibility. These modifications occur through enzymatic reactions involving DNA methyltransferases and histone acetyltransferases, which modulate electrostatic interactions between DNA and histone proteins [43–46].

The genomic architecture varies significantly across species, affecting biological adaptability and industrial applications. In prokaryotes, such as bacteria, the genome is typically circular and lacks histones, relying on supercoiling and nucleoid-associated proteins for compaction. The minimal noncoding sequences in bacterial genomes allow for efficient transcription–translation coupling, making bacteria ideal for synthetic biology, recombinant protein production, and bioengineering. In eukaryotes, genomes are linear and compartmentalized within the nucleus, with regulatory complexity enhanced by long non-coding RNAs (lncRNAs) and chromatin remodeling complexes. The telomeric DNA, rich in repetitive sequences like TTAGGG in humans form G-quadruplex structures, stabilized through Hoogsteen base pairing. The primary mechanism for preventing degradation of linear chromosome ends is the formation of the T-loop (Telomere-loop), where the single-stranded telomeric end invades the double-stranded telomeric repeat array, protected by the shelterin protein complex. G-quadruplexes can form in vitro and are studied for their potential biological roles and as drug targets. The chemical composition of the genome also influences mutation rates, genetic stability, and adaptation mechanisms [47–49].

The bacterial genome is significantly smaller, typically 0.5–ten million bp, with a circular DNA structure containing compact operons and minimal noncoding sequences. This efficiency makes bacteria ideal for synthetic biology applications, such as producing industrial enzymes, antibiotics, and biofuels. Genetically engineered bacterial strains, like *Escherichia coli* DH5α and Bacillus subtilis, serve as biofactories, optimizing metabolic pathways to enhance biomanufacturing efficiency. The plant genome is often more complex due to polyploidy, where multiple copies of chromosomes exist within a single species, increasing genetic diversity and adaptability. Wheat (*Triticum aestivum*) has a hexaploid genome with ~16 billion bp, making genetic modifications challenging but crucial for improving crop yield, drought resistance, and disease resistance. CRISPR/Cas9 and base-editing tools have enabled precise modifications in plant genomes, allowing for the development of herbicide-resistant, nutrient-enhanced, and climate-resilient crops. Genomic studies now integrate multi-omics approaches (genomics, transcriptomics, epigenomics, proteomics) to decode functional elements, genome wide variations, and adaptive mutations. AI-driven genome analysis and quantum computing simulations accelerate gene-editing precision, drug discovery, and trait enhancement, paving the way for intelligent genetic engineering in biomedicine, agriculture, and industrial biotechnology under Industry 6.0 [50–52].

7.2 The Basics of Gene Editing

CRISPR/Cas9 has revolutionized gene editing, its potential for off-target mutations where the Cas9 protein cuts at unintended sites has raised concerns about its precision in clinical applications. To address these limitations, Prime Editing was developed, offering a safer and more accurate alternative. Prime editing utilizes a modified CRISPR/Cas9 system that does not rely on double-strand breaks but instead uses a reverse transcriptase enzyme to directly rewrite the DNA sequence at the target site. By creating nick (single-strand breaks) instead of double-strand breaks, Prime

Editing reduces the risk of insertions or deletions at unintended locations, increasing accuracy to over 99% in many cases. This makes Prime Editing a promising tool for therapeutic gene editing that demands near-perfect precision, such as correcting point mutations associated with monogenic diseases like thalassemia or Huntington's disease. Another critical advancement in gene editing is Base Editing, which allows for the conversion of a single nucleotide base pair into another without breaking the DNA strand. The efficiency of base editing has been reported to reach >70% in clinical settings, and its ability to target precise single-nucleotide edits presents significant advantages over traditional CRISPR-based approaches. TALENs are designed from transcription activator like effectors (TALEs) that bind to DNA with high specificity, creating targeted breaks in the DNA. Although both ZFNs and TALENs have lower flexibility and precision compared to CRISPR-based tools, they have been successfully employed in genome editing in mammalian cells, particularly in clinical trials for gene therapy and genetic modifications in plants and livestock. Together, these gene-editing technologies are leading to new possibilities in biotechnology, genetic engineering, and therapeutics. The continuous improvement of gene-editing systems, particularly through enhanced precision, delivery mechanisms, and fewer off-target effects, will continue to propel the field forward, enabling tailored genetic interventions in medicine, agriculture, and biomanufacturing. As we move into clinical translation, gene editing is becoming increasingly important for personalized treatments, and these innovations will shape the future of genetic disease therapy, functional genomics, and regenerative medicine [53, 54].

7.3 Industry 6.0 Technologies in Genomics

Genomic sequencing technologies have evolved rapidly, increasing throughput, speed, and accuracy while reducing costs. Several state-of-the-art platforms dominate the field. By integrating machine learning-driven variant calling, deep neural network-assisted base calling, and real-time data processing, the NovaSeq X Plus plays a critical role in intelligent bioinformatics pipelines within Industry 6.0, paving the way for automated, AI-guided genomic analysis and therapeutic development. The Illumina NovaSeq X Plus represents the pinnacle of next-generation sequencing technology, utilizing sequencing-by-synthesis chemistry, a process that ensures high-fidelity, massively parallel DNA sequencing. This technology operates by incorporating fluorescently labelled nucleotides into a growing DNA strand, with real-time imaging capturing the emitted fluorescence to determine nucleotide identity. The reaction occurs on a flow cell surface densely populated with oligonucleotide adapters, facilitating bridge amplification to create clusters of identical DNA fragments. Each cycle introduces a reversibly terminated, fluorescently labelled nucleotide, preventing additional incorporations until imaging is complete. A cleavage reaction then removes the blocking group and fluorophore, allowing the next nucleotide to bind. This controlled termination mechanism ensures single-base precision, minimizing error rates to below 0.1%. The NovaSeq X Plus system achieves an unprecedented throughput of 16 terabases per run, enabling cost-efficient sequencing at below USD 200 per human genome, a significant reduction from earlier platforms. This improvement is attributed to higher capacity patterned flow cells, which utilize nanowell-based clustering to maximize sequencing density while reducing signal interference. Advanced optics and AI-enhanced base calling algorithms optimize signal-to-noise ratios, ensuring accurate base identification even in repetitive or low complexity regions. The system is instrumental in high-throughput population genomics, allowing large-scale genome-wide association studies to identify disease-linked genetic variants. In oncology, NovaSeq X Plus facilitates cancer mutation profiling, detecting somatic mutations, structural variants, and copy number alterations with high sensitivity and specificity. Its applications extend to pharmaceutical research, where sequencing-based insights drive drug target identification, biomarker discovery, and patient stratification in precision medicine trials [55–57].

Oxford Nanopore Technologies has revolutionized DNA sequencing with its high-throughput, real-time sequencing capabilities and ultra-long-read technology, providing researchers and clinicians with a versatile tool for a wide range of applications, from environmental monitoring to personalized medicine. This approach has enhanced the efficiency and scope of genomics by making sequencing more accessible and efficient, especially in non-laboratory settings. Oxford Nanopore Technologies utilize an innovative approach to DNA sequencing by employing biological nanopores integrated into an electrical membrane. This method detects individual DNA molecules as they pass through the nanopore, generating real-time sequencing data. The fundamental technology relies on the principles of ion current measurement, where the passage of DNA strands through the nanopore causes measurable changes in ionic current. These fluctuations provide information on the sequence of nucleotides in the DNA. Notably, this technology supports the sequencing of ultra-long DNA fragments up to 1 Mb in length something that traditional sequencing methods, like Illumina sequencing, struggle to achieve. This ability to sequence such long DNA fragments enables high-fidelity assembly of complex genomes, which is particularly crucial for accurately mapping regions with repetitive sequences or structural variations. One of the most significant advantages of Oxford Nanopore's sequencing technology is its portability and scalability. The MinION device, a pocket-sized sequencer, provides the ability to conduct real-time genome sequencing onsite. This portable nature is transformative for applications requiring field-based pathogen surveillance or biosecurity monitoring. For instance, during epidemics or pandemics, the ability to rapidly sequence and track pathogens in remote locations, without the need for centralized labs, offers invaluable insights into transmission dynamics and genome evolution. The technology's real-time data generation accelerates decision-making processes in public health responses. The performance of Oxford Nanopore's sequencing also extends to continuous read lengths, enabling seamless assembly of complex genomes without the fragmentation issues associated with short-read sequencing technologies. As a result, the platform excels in applications like human microbiome studies, where long-read sequencing helps capture the diversity and complexity of microbial communities. Similarly, the system has proven useful in infectious disease surveillance, particularly for rapidly identifying mutations in pathogens such as SARS-CoV-2, thus aiding in variant tracking and genomic epidemiology. In agriculture, Oxford Nanopore's sequencing technology has found applications in genome monitoring of crops and livestock. By sequencing the DNA of genetically modified organisms or studying plant–pathogen interactions, researchers can better understand genetic traits such as disease resistance, stress tolerance, or yield improvement. The platform is instrumental in biofuel production, where microbial strains used for fermentation or energy generation can be characterized at the genomic level, optimizing metabolic pathways for enhanced biofuel yield [58–60].

The power of single-cell multiomics lies in its ability to capture and integrate diverse molecular data types from individual cells, providing a comprehensive, dynamic view of cellular processes in health and disease. As the technology advances, it will continue to drive breakthroughs in personalized medicine, therapeutic development, and biotechnological innovations, providing a deeper understanding of biological complexity at the most granular level. Single-cell multiomics is a transformative technology that integrates various omics techniques, genomics (DNA-seq), transcriptomics (RNA-seq), epigenomics (ATAC-seq, DNA methylation), and proteomics at the single-cell level. This integration allows for a comprehensive understanding of the molecular landscape of individual cells, offering unprecedented insights into cellular heterogeneity. By examining the entire spectrum of molecular layers within a single-cell, multiomics captures the intricate interactions between the genome, transcriptome, epigenome, and proteome, enabling the discovery of cellular subpopulations that were previously indistinguishable using bulk analysis methods. This powerful tool is particularly valuable in dissecting complex biological systems where the behavior of cells cannot be fully understood by studying population averages alone. In terms of performance, single-cell multiomics resolves cellular heterogeneity with high resolution, providing detailed maps of gene expression, epigenetic modifications, and protein activity within individual cells. This level of resolution allows for precision mapping of genetic regulation across tissues and diseases, uncovering pathophysiological mechanisms that drive health and disease. By elucidating cellular differences at the single-cell level, multiomics can reveal how cells within a tissue or organ respond to environmental cues, interact with other cells, and evolve over time in response to genetic or external perturbations. For example, it enables the identification of differentiated cell states in complex tissues such as the tumor microenvironment, where understanding how cancer cells interact with the surrounding stroma can inform therapeutic strategies. In cancer research, single-cell multiomics is particularly valuable for identifying tumor heterogeneity. Tumor cells within a single tumor often display distinct gene expression patterns, epigenetic modifications, and proteomic signatures that contribute to treatment resistance and metastatic potential. By utilizing multiomics to profile individual tumor cells, researchers can identify subpopulations of cells with distinct molecular features, allowing for the precision mapping of cancer cell phenotypes, their interactions with the tumor microenvironment, and their response to therapies. For instance, single-cell RNA sequencing coupled with proteomics can reveal how cancer cells upregulate survival pathways or evade immune surveillance within different regions of the tumor, enabling the development of personalized cancer treatments. In neuroscience, single-cell multiomics is proving invaluable in understanding the molecular basis of neurodegenerative diseases like Alzheimer's and Parkinson's. These conditions often involve the progressive loss of specific neuronal populations, with considerable molecular variability between cells even within the same tissue. By applying multiomics to neurons or brain tissue, researchers can uncover gene expression variability and epigenetic changes that influence neuronal survival, synaptic function, and neuronal plasticity. Multiomics offers the potential to track how disease progression alters gene regulatory networks and protein expression, leading to better models of disease and the discovery of novel biomarkers or therapeutic targets. In synthetic biology, single-cell multiomics is accelerating the development of microbial strains for biomanufacturing and environmental remediation. By analyzing microbial populations at the single-cell level, researchers can design strains with optimized metabolic pathways that improve yield and efficiency in the production of biofuels, chemicals, and pharmaceuticals. Understanding the epigenetic regulation and protein dynamics within microbial cells helps fine-tune bioremediation processes, enabling the biodegradation of pollutants or waste recycling with greater specificity and efficiency [61–63].

7.3.1 Computational Genomics in Sixth Industrial Revolution

Because AI, ML, and quantum computing offer previously unheard-of speed, accuracy, and insight into genetic data, they are radically changing the field of genomic analysis. Personalized medicine, drug development, and our comprehension of hereditary illnesses could all be accelerated by the combination of genomic analysis, AI, ML, and quantum computing. These technologies promise to enhance patient outcomes and provide new insights into the molecular intricacies of human biology by facilitating faster and more precise predictions. These cutting-edge technologies solve the problems brought on by the complexity and volume of genomic data by enabling more effective variant calling, genome assembly, and predictive modeling. The use of deep learning techniques, especially Deep Neural Networks (DNNs), to identify disease-linked mutations and predict genetic predispositions with remarkable precision often above 95% accuracy, is one of the most promising uses of AI-powered genomic prediction. Because DNNs are skilled at spotting minute patterns and interactions in huge genomic datasets, it is possible to find uncommon variants that conventional techniques might overlook. Precision medicine, which uses a person's genetic profile to forecast their risk of developing diseases including cancer, heart disease, and neurological disorders, would greatly benefit from this skill. AI algorithms can significantly improve illness risk prediction models by gaining insights into gene-environment interactions through the analysis of enormous volumes of genomic data. A standout example of AI in genomics is AlphaFold, a deep learning model developed by DeepMind, which has revolutionized protein structure prediction from amino acid sequences. Prior to AlphaFold, predicting the 3D structure of proteins from their sequence was a highly complex and time-consuming task, often requiring years of experimental work. AlphaFold, however, uses transformer-based deep learning models to predict the precise folding of proteins with remarkable accuracy, achieving a median Global Distance Test score of 92.4 for the top 5 protein structures out of a possible score of 100. This breakthrough has accelerated the field of drug discovery by enabling more accurate predictions of protein–ligand interactions and the identification of novel drug targets [64–67].

7.4 Intelligent Manufacturing Industry 6.0 Advancements

As AI models continue to evolve, they are increasingly being integrated with genome-wide association studies and multiomics data to build comprehensive models of genetic diseases. These models combine genomic data with other biological layers (such as epigenetics, transcriptomics, and proteomics) to provide a more holistic view of disease mechanisms. Quantum computing is poised to further accelerate genomic analysis by solving computationally intensive problems, such as simulating large biomolecular systems and performing complex genomic sequence alignments. Quantum algorithms, which leverage quantum bits (qubits) and quantum superposition, are particularly suited for tasks that require exponential scaling of computational resources, such as simulating protein folding pathways or optimizing genetic variant interpretation across large populations. In the context of genomics, one of the most challenging computational tasks is simulating the 3D structure of DNA and the complex interactions between proteins and genetic material, a process that often takes classical supercomputers years to complete. Quantum computers, by contrast, can handle such simulations in a matter of seconds, due to their ability to process multiple possibilities simultaneously. Mutational dynamics, the study of how genetic mutations evolve over time and across different biological environments, can be accelerated using quantum computing. The ability to simulate the interactions between various DNA mutations and their impact on protein folding, gene expression, and cellular processes opens new avenues for personalized medicine and cancer genomics, where individual mutational landscapes significantly influence disease progression and response to therapy. Quantum models could be applied to simulate point mutations, insertions, and deletions in real time, predicting how these mutations affect protein structure and function at the molecular level, which could transform genetic diagnostics and targeted drug design. An example of real-world exploration into the intersection of quantum computing and genomics is IBM's Quantum Genome Project. This initiative is investigating the use of quantum computing for genome compression, a process that has been a significant bottleneck in genomic data analysis. With sequencing technologies generating terabytes of data, efficient genome storage and rapid retrieval have become critical challenges in genomic research and clinical applications. Quantum computing's potential to compress large datasets arises from its ability to process and encode information in a vastly more compact form than classical computers. Quantum algorithms are being developed to explore methods of quantum compression, where quantum bits (qubits) could store genetic sequences in exponentially smaller spaces compared to classical binary systems. This would allow the storage of vast amounts of genomic data such as entire genomes, multi-omics datasets, and patient genetic profiles within a fraction of the space required today, enabling faster, more efficient data retrieval and reducing the costs associated with data storage. In practice, quantum-based genome compression could significantly reduce the complexity and time involved in tasks like sequence alignment, variant calling, and multiomics integration, where current computational methods struggle to keep up with the volume and complexity of the data. By compressing the entire genome into a smaller, quantum encoded format, cloud-based quantum computing platforms could facilitate on-demand access to genomic data for research, diagnostics, and clinical decision-making. These quantum algorithms may eventually enable the real-time decoding of genetic data during sequencing runs, speeding up the workflow and enhancing the speed of clinical diagnostics. As technologies continue to mature, quantum computing will likely play an instrumental role in accelerating genomic research, enabling faster and more accurate predictions of genetic traits, optimizing the design of personalized treatments, and providing new insights into complex biological systems. While quantum computing in genomics is still in its early stages, its potential to revolutionize the analysis of big genomic data and to simulate biological processes at unprecedented scales cannot be overstated [68–74].

7.4.1 Edge Computing for Genomics

Edge computing for genomic data represents a significant shift in the way genomic information is captured, analyzed, and acted upon. By enabling real-time sequencing and data processing at the point of generation, it dramatically improves diagnostic speed, precision agriculture outcomes, and pathogen tracking capabilities. The integration of AI, ML, and secure data management practices further enhances its potential, creating a powerful ecosystem for the future of genomic medicine and agriculture biotechnology. Edge computing for genomic data represents a transformative leap in processing and analyzing genomic information, especially for real-time applications such as infectious disease monitoring and agricultural genomics. At its core, edge computing involves decentralizing data processing from centralized cloud systems to local devices or "edge" nodes closer to the data source. In the context of genomics, this is particularly useful for IoT-enabled genomic sensors, which can perform real-time DNA sequencing and processing directly at the site of data generation. This reduces latency and minimizes the dependency on large, centralized computing systems, which is crucial for timely decision-making in fields like epidemic control, pathogen surveillance, and precision agriculture [75–77].

7.4.2 Fog Computing and Genomic Sensors

The advent of IoT-enabled genomic sensors means that devices embedded with DNA sequencing technologies, such as portable sequencers can collect and process genetic data onsite, without needing to transmit the data to a distant server or cloud service. These sensors are designed to rapidly sequence DNA fragments using technologies like nanopore sequencing or ion semiconductor sequencing. The data generated can be analyzed in real time using edge computing algorithms to detect pathogens, genetic mutations, or markers of interest. For instance, in infectious disease monitoring, edge computing allows for the immediate identification of viral outbreaks, detecting genetic markers of infectious agents (such as the SARS-CoV-2 virus) at the point of care. The speed and efficiency offered by this local processing enable healthcare professionals to respond rapidly, without waiting for results to be sent to and analyzed by central databases. Agricultural genomics can greatly benefit from edge computing, especially in precision agriculture where real-time genetic data is required for optimizing crop performance, pest control, and disease management. IoT-enabled genomic sensors placed in the field can continuously monitor soil DNA, plant genomes, and microbial communities, analyzing genomic variation or disease resistance at the genomic level. This allows farmers and researchers to adapt farming strategies in real-time based on the specific needs of crops or regions, optimizing yield and reducing the use of chemicals, fertilizers, or water. Real-time monitoring of plant pathogen resistance genes or microbial community dynamics using edge computing ensures immediate actionable insights, facilitating genomic-based breeding and crop improvement efforts. The key technological enabler for edge computing in genomics is the integration of AI and ML algorithms with the local processing power of edge devices. These algorithms can perform complex bioinformatics tasks such as sequence alignment, variant calling, genetic anomaly detection, and data compression right at the edge. Deep learning models can be trained to identify genetic markers for disease resistance or susceptibility, which can be applied in real time to evaluate agricultural varieties or clinical samples. This combination of edge computing and blockchain ensures the integrity, security, and real-time availability of genomic data while protecting patient privacy in healthcare or ensuring the authenticity of genomic data used for crop improvement. In practice, companies like IBM, Google, and Intel are investing in edge computing infrastructure to support real-time genomic analytics. IBM's edge computing solutions used AI and quantum-inspired algorithms at the edge to accelerate genomic research, offering tools that can manage large genomic datasets efficiently and securely. Google's Cloud IoT and Google Cloud Genomics platform also allow for the integration of edge devices with cloud infrastructure, creating hybrid solutions for genomic data analysis, from real-time sequencing to large-scale cloud-based data storage and processing. Fog computing in healthcare and pharmaceuticals enhances real-time data processing, reduces latency, and ensures secure, decentralized data storage for critical applications. It enables real-time patient monitoring by processing data closer to the source, such as wearable devices and IoT-enabled medical equipment, ensuring immediate response in emergencies. In pharmaceutical manufacturing, fog computing optimizes drug production by analyzing sensor data from production lines, reducing errors, and improving quality control. It also strengthens data security by decentralizing storage and reducing reliance on cloud-based systems, which is crucial for maintaining patient confidentiality and regulatory compliance. Furthermore, it supports AI-driven diagnostics, predictive analytics, and personalized medicine by enabling efficient edge computing, ensuring faster decision-making, and reducing bandwidth congestion in large healthcare networks [78–80].

7.5 Future Directions in Genomics

The integration of AI-driven genomics within Industry 6.0 will radically transform personalized medicine and intelligent manufacturing by creating an interconnected ecosystem where genetic data drives real-time, individualized decision-making in healthcare and manufacturing. This not only improves patient outcomes but also drives the efficiency and precision of biomanufacturing processes, contributing to more sustainable and innovative solutions in medicine, agriculture, and biotechnology. As genomic data and AI technologies evolve, the future promises a world where genetic information is at the forefront of healthcare, providing customized treatments and products with greater precision, safety, and efficacy. With the advent of Industry 6.0, genomics is poised to play a pivotal role in the next generation of smart manufacturing and intelligent healthcare, where highly personalized, data-driven solutions are expected to transform the way treatments and products are developed and delivered.

7.5.1 Personalized Medicine

One of the central innovations within this framework is AI-driven personalized medicine, which leverages the immense power of cloud-based AI genomics platforms to offer real-time, individualized treatment recommendations tailored to the genetic makeup of each patient. These platforms, which integrate various bioinformatics tools with advanced machine learning algorithms, are designed to analyze an individual's genome, including single-nucleotide polymorphisms, copy number variations, and epigenetic modifications, to predict disease susceptibility, treatment responses, and potential adverse drug reactions. The AI-driven platforms will not only analyze raw genomic data from sequencing technologies such as next-generation sequencing or single-cell RNA sequencing, but they will also integrate multiomics data (genomics, transcriptomics, proteomics, and metabolomics) to provide a comprehensive view of an individual's biological profile. This enables predictive modeling that can foresee the development of diseases like cancer, diabetes, and neurological disorders long before clinical symptoms appear. Platforms like DeepMind's AlphaFold can predict protein structures with high accuracy, which aids in understanding disease mechanisms at a molecular level, enabling more effective drug development processes. Such integration of multiomics data will facilitate precision medicine, where treatment is customized not only based on genetic variants but also on how those variants interact with environmental factors, such as diet or exposure to toxins, thereby optimizing therapeutic outcomes. The ability to run genomic analyses in real time through cloud computing infrastructure opens a new frontier for personalized medicine, where patients could have continuous, dynamic updates on their health. This is especially important in the context of chronic disease management, where ongoing adjustments to medication regimens or lifestyle interventions can be made based on the genetic and physiological data collected over time. Pharmacogenomics can use real-time genomic data to predict how an individual will metabolize a specific drug, ensuring the right drug at the right dose is prescribed, minimizing trial-and-error and adverse reactions. Real-time sequencing and processing can also help monitor tumor evolution in cancer patients, enabling adaptive therapies that target genetic mutations that change over the course of the disease. The application of cloud computing in AI genomics ensures that this individualized care is scalable, allowing healthcare systems to manage and analyze massive amounts of genetic data generated by millions of patients. The centralized nature of cloud computing enables seamless integration of genomic data from multiple sources, hospitals, research labs, or wearable health devices, thus ensuring that personalized treatment plans are based on the most up-to-date, comprehensive data available. AI algorithms will continue to improve by training on large, diverse genomic datasets, leading to continuous refinements in predictive models and decision support systems. From a manufacturing perspective, Industry 6.0 envisions the integration of genomics into smart factories, where bio-manufacturing of personalized pharmaceuticals, gene therapies, and biologics will become more efficient and scalable. This could involve AI-driven optimization of fermentation processes to produce gene therapies, biologics, or even bio-based chemicals by analyzing the genomic data of microorganisms used in the manufacturing process. A smart factory could use genomics to enhance the design of microbial strains that produce bio-based chemicals or pharmaceuticals by optimizing the genetic modifications through CRISPR/Cas9 technology and AI-powered gene synthesis algorithms. This combination of genetic engineering, AI, and advanced manufacturing could revolutionize how products are personalized, produced, and delivered, moving away from mass production and toward mass customization [81–83].

7.5.2 Automated Gene Synthesis

Automated gene synthesis factories, driven by cutting-edge DNA printing technologies, are poised to revolutionize synthetic biology by enabling the rapid and cost-effective assembly of complex gene circuits. These factories use high-throughput DNA synthesis platforms, which utilize a variety of chemical and enzymatic processes to create custom DNA sequences from scratch without the need for traditional biological template-based methods. In DNA printing, the synthesis process begins with a computer-aided design system, where the desired gene sequences or circuits are programmed into a sequence of nucleotides. These sequences are then synthesized using solid-phase chemical synthesis, often employing phosphoramidite chemistry, where individual nucleotides (A, T, C, G) are added step by step in a controlled manner to form the full-length synthetic genes. Once the sequence is constructed, it is assembled into functional gene circuits that can be used for various applications such as metabolic engineering, biocatalysis, and gene therapies. At the core of automated gene synthesis factories is the use of microfluidic technologies and robotic automation, which allow for the precise manipulation of reagents, sequencing, and assembly processes at an unprecedented scale. These systems enable parallel synthesis of multiple genes or even entire genomes simultaneously, significantly speeding up the process of gene construction. Advances in next-generation DNA synthesis have led to improvements in the scalability of these factories, enabling the synthesis of large, complex genes or even entire synthetic pathways, with error rates reduced to below 1 per 1000 base pairs through error correction algorithms and quality control checks at each step of the synthesis process. High-fidelity enzymes and oligonucleotide pool libraries are employed to ensure that the DNA circuits are synthetically accurate, minimizing mutations or unwanted sequence variations that might interfere with the functionality of the gene constructs [84–87].

7.5.3 Gene Circuits

The synthesis of gene circuits for synthetic biology applications involves not only the creation of individual genes but also the assembly of these genes into larger, more complex systems, such as genetic circuits that control cellular behavior. These circuits can consist of regulatory elements like promoters, repressors, terminators, and ribosome binding sites, which are systematically arranged to achieve desired outputs such as the production of specific proteins, the regulation of cellular metabolism, or the control of cellular responses to environmental stimuli. The capacity to design and print these circuits in an automated manner opens up new frontiers for genetic engineering. In metabolic engineering, researchers can design circuits that optimize the flow of metabolic intermediates through a cell's biochemical pathways, leading to enhanced production of biofuels, pharmaceuticals, or specialty chemicals. Gene synthesis factories can be used to synthesize customized gene therapies, where specific mutations are introduced into DNA sequences to treat genetic disorders or to create CAR-T cells for immunotherapy. The potential for DNA printing technologies in these factories is expanded using modular genetic parts that can be interchanged to create a vast variety of genetic circuits. This modularity enables the rapid prototyping of new genetic designs, making it possible to test and iterate on synthetic pathways much faster than traditional cloning methods would allow. These factories could integrate advanced bioinformatics tools to model and predict the behavior of the synthetic circuits, ensuring they perform as intended in live biological systems. This predictive capability is essential for reducing the time and cost required for experiments in synthetic biology and for minimizing risks when these circuits are deployed in industrial or clinical settings. The efficiency of automated gene synthesis factories is greatly enhanced by integration with cloud-based platforms that store vast genomic data and provide powerful computational tools for designing and simulating gene sequences. This allows for the scalable production of gene circuits in ways that were previously unthinkable, accelerating the transition from laboratory-scale experiments to large-scale industrial applications. Synthetic biology made possible by DNA synthesis factories has the potential to drive innovation in bioremediation, sustainable agriculture, bio-based manufacturing, and personalized medicine, making these automated systems not just an important component of modern biotechnology but also a central driver of the bioeconomy. Automated gene synthesis factories, empowered by DNA printing technologies, are revolutionizing the field of synthetic biology, offering unprecedented speed, accuracy, and scalability in the creation of gene circuits that are pivotal for advancing biotechnology and addressing global challenges in health, environment, and industry [88–90].

7.5.4 Space Genomics

Space genomics is an emerging and interdisciplinary field that examines how organisms, including humans, adapt to the extreme conditions of space environments. The uniqueness of space, with its microgravity, high radiation levels, and low atmospheric pressure, demands a deep understanding of how genetic material and biological processes are affected by these conditions. As space missions increase in duration, with plans for long-term human colonization of Mars and other celestial bodies, genomics will play a critical role in understanding how space exposure impacts human health, cellular biology, and the viability of life forms in extraterrestrial habitats. The study of genomic adaptation in space will be instrumental in developing bioregenerative life-support systems, which are essential for sustaining human life on distant planets. These systems rely on the integration of biological processes, such as plant growth and microbial metabolism, alongside human biological systems, to provide food, oxygen, and waste recycling in a closed-loop environment. Space genomics will help optimize these systems by genetically modifying plants, microbes, and even human cells to withstand the stresses of space, including radiation, altered gravity fields, and limited resources. Genomic sequencing in space requires cutting-edge technologies, including space-qualified next-generation sequencing (NGS) platforms like the Oxford Nanopore MinION, which was used on the International Space Station (ISS) to sequence bacterial DNA in space. Understanding the DNA damage and repair mechanisms in space organisms is crucial, as radiation-induced mutations can significantly affect cellular integrity. The role of radiation-induced DNA lesions, particularly the formation of double-strand breaks (DSBs) and base modifications under high-energy cosmic rays, is a key focus. Studies suggest that space radiation leads to increased mutagenesis, with effects on DNA replication, transcription, and repair. Furthermore, the adaptation mechanisms to microgravity, such as altered gene expression profiles related to muscle atrophy and immune system weakening, will provide insight into space-related health risks for astronauts and lead to the development of countermeasures like gene therapy and genetic modifications to mitigate health risks. The integration of AI and quantum computing with space genomics will expedite data analysis and prediction of genetic modifications under space conditions. AI-powered genomics, particularly deep learning algorithms, can analyze vast datasets from genomic sequencing in space, identifying patterns that would be challenging to detect using traditional bioinformatics approaches. These algorithms can also predict the impact of specific genetic alterations on the organism's ability to survive in space, assisting in the design of genetically engineered organisms suited for space environments. For example, AI models can simulate the phenotypic outcomes of genetic edits in crops used for space agriculture or microbial communities engineered to enhance waste recycling in space stations. AI systems can also assist in precision medicine by identifying astronaut-specific genetic vulnerabilities to space radiation or other hazards, allowing for tailored health interventions based on genomic profiles. The future of space genomics is further enhanced by the integration of quantum computing, which offers the potential to handle the incredibly complex datasets generated by whole-genome sequencing and simulations of genetic pathways under extreme conditions. Quantum algorithms can model the dynamics of protein folding and DNA–protein interactions in microgravity with far greater speed and accuracy than classical computing methods. These simulations can provide unprecedented insight into how proteins and other biomolecules behave under space conditions, aiding the development of biosensors and biomaterials that can be used in life-support systems or even space habitats. Moreover, quantum-enhanced genomic data analysis will enable real-time monitoring of the genetic health of astronauts or space-bound organisms, facilitating timely interventions during space missions. The space genomics field holds significant implications not only for space exploration but also for precision medicine and biotechnology back on Earth. The genetic alterations required to adapt organisms for space environments may unlock new biotechnological applications, such as improved crops for sustainable agriculture, more robust microorganisms for industrial bioprocessing, or genetically engineered organisms capable of thriving in extreme environments on Earth. Furthermore, as space colonization efforts evolve under Industry 6.0, the incorporation of AI-driven platforms and quantum-optimized genomic tools will empower industries to produce customized, genetically engineered products more efficiently. In the future, this synergy will enable biotech companies to deploy space-tested, optimized organisms for a range of applications, from biofuels to novel biomaterials, transforming industrial bioengineering by integrating genomics with smart manufacturing systems. Thus, space genomics will not only enhance our ability to colonize other planets but also drive advancements in sustainable biotechnology and human health, positioning it as a cornerstone of the next industrial revolution [91–93].

7.5.5 Smart Biomanufacturing

The fusion of genomics with AI, robotics, and IoT-driven bioreactors has enabled intelligent manufacturing in pharmaceuticals, food, and material science industries. AI-optimized bioprocessing leverages the power of advanced ML algorithms, deep learning models, and bioinformatics platforms to significantly enhance the efficiency and speed of genetic modifications in industrial biotechnology. Traditional bioinformatics processes, which previously took weeks or even months to analyze complex genomic sequences and identify key genetic variants, have been revolutionized by AI tools capable of performing such analyses in minutes. These AI-driven platforms use powerful algorithms to decode vast amounts of sequencing data with remarkable accuracy, identifying patterns, mutations, and regulatory elements in the genome. AI-enhanced bioinformatics platforms are particularly adept at detecting subtle genetic markers that can influence metabolic pathways, enzyme efficiency, and fermentation processes. By combining genomic data with metabolic network models, AI systems can predict the effects of genetic modifications on cellular behavior, identifying optimal targets for genetic engineering in microbial strains used for industrial fermentation. ML further accelerates bioprocess optimization by enabling the prediction and design of gene expression pathways that maximize the yield of desired bioproducts. In synthetic biology, AI models are trained on large datasets containing gene expression profiles, metabolic flux data, and fermentation outcomes to identify the most efficient pathways for enzyme production or biofuel synthesis. Reinforcement learning techniques have been applied to bioprocessing, where ML agents simulate different genetic manipulations and environmental conditions to find the optimal combination for increased yield, reduced by-product formation, and improved overall process stability. Genomic-scale metabolic models are integrated with AI to simulate the entire metabolic network of engineered microorganisms, predicting how changes in gene expression or enzyme activity will affect the overall performance of the fermentation process. These AI-driven approaches can optimize biocatalysis and bioreactor performance, leading to significant increases in the production efficiency of biochemicals, biofuels, pharmaceuticals, and enzymes. AI-optimized bioprocessing also enhances the adaptive evolution of microbial strains, enabling directed evolution through genetic algorithms. These algorithms simulate natural selection in silico, accelerating the development of strains with superior performance characteristics, such as higher resistance to industrial stressors (high concentrations of ethanol in biofuel production) or the ability to metabolize complex feedstocks. AI platforms can integrate omics data (genomics, transcriptomics, proteomics, metabolomics) to create multilayered models that predict how microbial systems will respond to changes in nutrient availability, temperature, or other fermentation parameters, ensuring better scalability from laboratory to industrial-scale production. The ability to design, simulate, and optimize bioprocesses in silico before applying them in the lab or industrial settings greatly accelerates product development timelines, reduces experimental costs, and minimizes the risks associated with scaling up. As AI-driven tools continue to evolve, gene editing platforms will become increasingly integrated with bioinformatics systems, enabling the direct manipulation of microbial genomes based on AI-optimized predictions. The synergy of AI, ML, and bioinformatics is transforming industrial biotechnological practices, enabling precision biomanufacturing that maximizes efficiency, reduces waste, and minimizes energy consumption, all while driving the sustainability and economics of the bioproduction industry forward. This AI-enabled framework promises to streamline processes such as protein engineering, metabolic engineering, and bio-based material production, laying the foundation for next-generation bioprocessing solutions that are more efficient, scalable, and sustainable than ever before. Genetically engineered microbial factories, such as *Escherichia coli* and *Saccharomyces cerevisiae* (yeast), represent a cornerstone of industrial biotechnology, driving substantial advancements in biofuel and bioplastic production. Using advanced gene-editing techniques like CRISPR/Cas9, these microbes are engineered to significantly enhance production yields, with studies showing increases of up to 300% in biofuel output and bioplastic synthesis. In biofuel production, *E. coli* strains have been engineered to improve ethanol production, achieving fermentation yields exceeding 10–12 g/L/h, a substantial increase compared to the unmodified strains. By optimizing genes involved in cellulose degradation, pyruvate metabolism, and ethanol biosynthesis pathways, these engineered *E. coli* strains demonstrate higher resistance to by-products like acetic acid and ethanol, which otherwise inhibit growth in wild-type bacteria. Genetic modifications to *Saccharomyces cerevisiae* have led to a two to threefold increase in lipid production for biofuels, with engineered strains producing up to 60–80 mg/L/day of biodiesel precursors, compared to 20–30 mg/L/day in non-engineered strains. These advancements are critical as they allow for more sustainable and cost-effective biofuel production, with costs as low as 0.5 USD per gallon in some optimally engineered systems, significantly reducing the production costs compared to traditional biofuel manufacturing techniques. In the case of bioplastics, genetically engineered yeast strains have been designed to optimize polyhydroxyalkanoates production, with high yields of the cellular biomass, up from the natural yields. These engineered strains can convert agricultural waste or carbon dioxide into bioplastics, providing a renewable alternative to petroleum-based plastics. This surge in production efficiency has paved the way for industrial-scale bioplastic production, with companies (NatureWorks and Novamont) scaling up the production of polylactic acid (PLA) and PHA, respectively, using engineered microorganisms to meet growing demand for sustainable packaging solutions. Beyond biofuels and bioplastics, *Bacillus subtilis* has also become a prominent microbial host for producing enzymes used in a wide range of industrial applications, particularly in the detergent and pharmaceutical industries. By genetically engineering *B. subtilis* strains to overexpress proteases, amylases, and lipases, the enzyme yield has been substantially increased, with production levels reaching up to 500–600 g/L of protease per batch, compared to the typical 50–150 g/L in unmodified strains. This enhancement is achieved through modifications in the genes responsible for enzyme secretion pathways, along with optimization of fermentation conditions such as pH and temperature. In the detergent industry, these engineered enzymes are critical for improving stain removal at lower temperatures and reducing the need for toxic chemicals, contributing to more ecofriendly formulations. *Bacillus subtilis* has been engineered for pharmaceutical applications, particularly in the production of antibiotics like penicillin and amoxicillin, with the engineered strains demonstrating increased yields as compared to wild-type strains. These enhancements in enzyme and pharmaceutical production are facilitated by advanced techniques such as metabolic flux analysis and systems biology approaches that enable the precise mapping and manipulation of metabolic networks within the microbe. By optimizing the biosynthetic pathways and engineering more efficient secretion systems, industrial-scale production of these high-value enzymes and pharmaceuticals has become more economically viable, with cost reductions of up to 40% in some cases. The engineering of microbial strains for pharmaceutical production has led to a reduction in fermentation time, achieving production cycles that are shorter than traditional fermentation processes [94–102].

These breakthroughs in genetically engineered microbial factories are revolutionizing the industrial production of biofuels, bioplastics, enzymes, and pharmaceuticals, with the potential to further drive the circular economy, reduce carbon footprints, and improve sustainability across a wide range of industries. Through the application of gene-editing technologies like CRISPR and the integration of systems biology, these microbes can be optimized for maximum yield, cost-efficiency, and scalability, ultimately transforming the biomanufacturing landscape.

7.6 Pharmaceutical and Healthcare Applications

Gene editing and genomics are transforming drug manufacturing and personalized medicine.

7.7 mRNA-Based Drug Manufacturing

mRNA-based drug production, particularly in the context of vaccines, has undergone a monumental transformation due to advancements in AI-driven genomics, reducing development timelines from an average of 5 years to under 12 months. The integration of AI-driven genomic analysis with cloud computing and quantum-enhanced simulations within Industry 6.0 frameworks has drastically reduced the time, cost, and complexity of mRNA-based vaccine production. These technologies enable the scalable, rapid-response manufacturing of mRNA drugs, which could be transformative not only for infectious diseases but also for personalized medicine. As these systems evolve, they hold the potential to dramatically reduce the time-to-market for all mRNA-based therapeutics, with a projection that future drug development could be completed in less than 6 months, enabling more agile and adaptable responses to emerging health threats. The Pfizer-BioNTech and Moderna COVID-19 vaccines exemplify this rapid acceleration, which was largely driven by the application of deep learning models in genomic analysis, facilitating the swift identification and optimization of vaccine candidates. In the case of the Pfizer BioNTech vaccine, AI models were used to optimize the spike protein's encoding sequence, ensuring robust immune responses while reducing the risk of antigenic variation. These models accelerated the sequence design process, allowing researchers to proceed directly from sequence identification to preclinical testing in mere weeks rather than the typical months-long phase. The ability to model epitope recognition and predict immune responses with >95% accuracy ensured that only the most promising candidates advanced into trials. As a result, mRNA production for the vaccines reached production scale in under 9 months, a feat unimaginable with traditional vaccine platforms. Industry 6.0 further integrates cloud-based genomic analysis platforms that enable real-time mRNA optimization, particularly through ML models that predict the stability, folding, and immunogenicity of mRNA sequences. These AI-driven platforms can analyze vast genomic datasets from multiple sources, such as virus strains and human genomics, to optimize vaccine mRNA in a fraction of the time. The AI system developed by Moderna can rapidly identify the most stable and efficient mRNA sequences by predicting the secondary structure and codon optimization, ensuring high translational efficiency and stability of the mRNA once introduced into the human body. Real-time adjustments to mRNA synthesis processes are made possible by continuous feedback loops between genomic data analysis and bioprocess optimization. On the production side, mRNA-based vaccines, which are traditionally synthesized using in vitro transcription (IVT), benefit significantly from this integration. By using AI to optimize the IVT process, yields of mRNA can be increased by up to 90% over traditional protocols, reducing costs and improving scalability. The purification process, which traditionally requires multiple steps and long times, can now be enhanced by AI-driven algorithms that predict optimal purification conditions, reducing the time to produce a batch from several weeks to a few days. Real-time genomic feedback is especially crucial when manufacturing at scale, as seen during the global COVID-19 vaccine rollout, where companies like Moderna had to produce billions of doses. In the process, AI helps manage bioreactor conditions and predict transcription efficiency, ensuring that mRNA vaccine batches meet stringent regulatory standards for purity and activity. The integration of quantum computing into this genomic workflow is set to revolutionize mRNA-based drug production by providing quantum-enhanced simulations of mRNA interactions with host cells, potentially optimizing the delivery mechanisms of mRNA vaccines and therapeutics. By simulating and predicting how mRNA molecules will behave in a physiological context, AI and quantum computing can facilitate the development of mRNA therapies for a wide range of diseases, from cancer immunotherapy to rare genetic disorders. Real-time optimization of the lipid nanoparticles used for mRNA delivery further improves the efficiency of vaccine delivery systems, ensuring that over 90% of the mRNA encapsulated in the nanoparticles reaches the target cells [103–105].

7.7.1 Gene Therapies and Personalized Medicine

Gene therapies, particularly those employing CRISPR technology, have rapidly advanced, with the FDA approving CRISPR-based therapies for sickle cell disease and beta thalassemia in the 2023–2025 period. These groundbreaking therapies utilize the CRISPR Cas9 system to edit the genetic mutations responsible for these hemoglobinopathies. In the case of sickle cell disease, CRISPR Cas9 is used to correct the HBB gene mutation by reactivating the fetal hemoglobin (HbF) gene, effectively reducing the abnormal sickle-shaped red blood cells. In clinical trials, patients who received the CRISPR Cas9 treatment demonstrated over 80% reduction in painful sickle cell crises and many were able to discontinue blood transfusions. Similarly, for beta-thalassemia, CRISPR is used to correct the HBB gene mutations, restoring normal hemoglobin production. Clinical data from ongoing studies show that 90% of patients in the Phase 3 trials of CRISPR Cas9 therapies for beta thalassemia were transfusion-independent after treatment. These therapies represent a paradigm shift in treating genetic disorders, moving from symptomatic treatment to curative gene editing solutions. AI-driven drug screening has emerged as a critical tool in precision medicine, transforming the way gene therapies and targeted treatments are developed. With extensive genomic databases, ML algorithms can analyze vast amounts of genetic and clinical data to predict how different gene variants may respond to specific therapies. AI platforms like Deep Genomics and in silico Medicine use genomic sequencing data to identify novel therapeutic targets and design potential gene-editing interventions for diseases ranging from cancer to neurodegenerative disorders. AI's predictive power can reduce the drug discovery timeline by as much as 60%, which traditionally takes around 10–15 years to develop a drug from concept to market. This accelerated timeline translates into significant cost savings for pharmaceutical companies, as drug development costs can range between 2.6 billion to 3 billion USD per approved drug. By using AI to predict molecular behavior, screen potential candidates for toxicity, and identify optimal drug dosage regimens, the overall cost of bringing a gene therapy to market can be reduced substantially. In personalized medicine, AI further enhances the efficiency of genomic analysis, integrating multi-omics data (genomics, proteomics, transcriptomics, etc.) to customize treatments based on an individual's unique genetic makeup. For instance, pharmacogenomics, a subfield of personalized medicine, uses AI models to predict a patient's response to specific drugs based on their genetic profile. Companies like Tempus use AI to analyze patient data, guiding oncologists in selecting targeted therapies for cancer treatment. AI algorithms have been shown to predict optimal drug responses with up to 90% accuracy, ensuring that the right treatments are delivered to the right patients, minimizing adverse effects and increasing treatment efficacy. Moreover, this data-driven approach is revolutionizing rare disease management by providing insights into genetic mutations and pathways that are otherwise difficult to detect using traditional methods. The integration of AI and gene-editing technologies like CRISPR in clinical trials for personalized therapies also streamlines patient recruitment, identifying individuals who are genetically predisposed to respond favorably to specific treatments. This highly targeted approach has shown significant promise in treating complex diseases such as cystic fibrosis and muscular dystrophy, where genetic heterogeneity can make traditional treatments less effective. The combination of AI-driven genomic databases, CRISPR gene-editing therapies, and real-time patient monitoring systems further contributes to the precision medicine ecosystem, enabling rapid adjustments in treatment plans and personalized drug development. In fact, clinical trials have demonstrated that by using genomic biomarkers and AI, treatment plans can be tailored at a cost reduction of 60% compared to traditional trial methods, with outcomes more closely aligned to the patient's unique genetic profile. The approval of CRISPR therapies for genetic diseases, coupled with AI's ability to predict and optimize treatment plans, is leading us towards a future where medicine is increasingly personalized, targeted, and efficient, with the potential to cure genetic diseases, reduce healthcare costs, and vastly improve patient outcomes [54, 106–108].

7.7.2 Regenerative Medicine and Biofabrication

Regenerative medicine, particularly in the fields of 3D bioprinting and biofabrication, has seen remarkable advancements, with the potential to create functional tissues and organs for therapeutic use. 3D bioprinting, which involves the layer-by-layer deposition of living cells and biomaterials, has made significant strides in generating patient-specific tissues using patient-derived stem cells. This technology allows for the creation of genetically optimized tissues, ensuring they are tailored to the individual's unique genetic makeup, thereby reducing the risk of rejection when implanted. Current 3D bioprinting technologies utilize bio-inks made from human stem cells such as mesenchymal stem cells or induced pluripotent stem cells (iPSCs), combined with extracellular matrix proteins to form the scaffolding required for tissue growth. Organovo has developed liver tissues capable of mimicking key physiological functions, which are now used for drug testing, with the goal of eventually printing full-sized, functional organs for transplant. Recent breakthroughs have allowed for the printing of vascularized tissues, critical for creating larger structures like skin and cartilage, with multi-material 3D bioprinting techniques. AI-based optimization of the bioprinting process has enabled faster development cycles of these tissues, reducing production timelines for functional skin grafts from months to weeks. A specific example of this technology's success includes the use of biofabricated skin grafts for burn and ulcer treatments, where AI algorithms analyze wound sites to optimize the design of CRISPR-modified skin grafts that promote faster healing and integration with the patient's tissue. These AI-enhanced skin grafts are genetically optimized to reduce the risk of infection and enhance the regenerative properties of the grafts. In clinical trials, these grafts have shown 70–90% improvement in wound healing time and have reduced rejection rates by over 30% compared to traditional grafting methods. The integration of CRISPR technology in regenerative medicine is also revolutionizing the creation of genetically modified cells for therapeutic use. CRISPR/Cas9 has been used to modify human keratinocytes (the predominant cell type in the skin) to improve wound healing and scar reduction after burns or severe ulcers. Through the application of CRISPR-modified skin grafts, the wound healing rate has improved by 45% compared to conventional treatments, particularly by enhancing the expression of key growth factors like epidermal growth factor and vascular endothelial growth factor, which are critical in tissue regeneration. The incorporation of AI-driven design algorithms to further refine the delivery and composition of CRISPR-edited skin grafts has also shown promise in clinical trials with a reduction of infection rates and increased patient recovery speed. In terms of biofabrication, AI is increasingly used to design functional tissue scaffolds and organs by analyzing complex genomic, proteomic, and metabolomic data to predict the optimal material compositions for each patient. For instance, bioinks are customized based on individual patient genomics, improving tissue functionality and encouraging cellular regeneration at the molecular level. Using AI to control the precise deposition of cells and biomaterials at the micron scale has led to advancements in printing vascular networks within tissues, which is critical for the survival and growth of larger tissue constructs. These vascular networks are essential for providing nutrients and oxygen to cells in printed tissues, and their successful incorporation into printed liver, kidney, and heart models has opened doors to bioengineering functional organs for transplantation. Advancements in biofabrication and 3D printing have reached the level where bioprinted tissues are being evaluated for clinical applications. For example, bioprinted skin grafts have shown potential for reconstructing large burns with faster healing compared to traditional autografts. The bioprinting of cartilage and bone tissues for orthopedic applications has demonstrated bone regeneration rates higher as compared to conventional techniques. These biofabricated tissues are expected to significantly reduce the reliance on organ donors, as well as the need for long-term immunosuppressive therapies, by using patient-specific stem cells that are genetically matched to the recipient [109–112].

8 Bionic Eyes and Neuralink Innovation

Recent developments in Neuralink's brain computer interfaces (BCIs) and bionic eye technologies are greatly improving human sensory and cognitive capacities. The second-generation bionic eye has been the subject of 2.5 year clinical studies in Australia with four subjects. The findings showed notable gains in quality of life, everyday activities, and functional vision. These gadgets restore partial vision by taking the place of damaged photoreceptors, allowing users to sense shapes and light. The Orion device is another noteworthy advancement; it uses sophisticated artificial intelligence algorithms to decipher camera visual input and convert it into signals that the brain can use. This method seeks to give users a more thorough and organic visual experience. The development of implantable BCIs has advanced significantly thanks to Neuralink, which Elon Musk founded. The business has already implanted its device in three human patients as of January 2025, and it intends to do up to 30 more in 2025. The usefulness of these devices has been improved with upgrades that include additional electrodes, better bandwidth, and longer battery life. Noland Arbaugh, a patient who was paralyzed from the neck down, was able to operate a computer with his thoughts thanks to a Neuralink implant. His autonomy has been greatly restored by this technology, which also holds promise for those with comparable problems. For those with sensory and motor disabilities, these developments in bionic eyes and BCIs are opening new opportunities for improved human-machine interaction [113–115].

New nanomaterials are enabling ultrathin, highly biocompatible neural interfaces. For example, flexible polymer ribbons carrying dense microelectrode grids (like Precision Neuroscience's Layer array) can conform to the cortex. Carbon-based coatings such as graphene and carbon nanotubes are widely used to make electrodes thinner and lower impedance. Graphene in particular offers high conductivity and flexibility, forming a stable electrode brain interface. These properties allow electrodes to miniaturize to tens of microns while maintaining signal quality. A recent Nature Nanotech study built 25 μm graphene electrodes with around 25 kΩ impedance and >10 dB SNR in vivo [116–118].

AI Driven Decoding and Closed Loop Control Artificial intelligence has dramatically improved BCI performance. Modern DL decoders can extract more information from neural signals (around 40% higher information transfer rates than older methods). Likewise, adaptive and closed-loop algorithms maintain high accuracy (>90% success in motor tasks) without constant recalibration, and they substantially cut user training time (roughly 55% faster learning). Together with advances in electronics, these AI methods help deliver more reliable control. Closed-loop frameworks (where the system adjusts stimulation or decoding in real time) can halve the training needed and improve overall accuracy. These approaches, combined with high-density electrode arrays and continuous feedback, are critical for integrating BCIs seamlessly with the nervous system. Australia's recent long-term trials illustrate how advanced prostheses restore function in the blind. In a 2.7 year study of a second generation suprachoroidal retinal implant (44 platinum disc electrodes beneath the retina), four patients with end-stage retinitis pigmentosa showed major functional gains. All four participants improved at tasks like locating objects and navigating environments, and some could recognize people and doors they could not see before. Device stability was excellent: almost all electrodes remained active over the trial. The image below shows the four trial participants using the head-mounted camera/glasses system. Although this device stimulated the retina, similar techniques apply to cortical implants that bypass the eye entirely. In fact, preclinical studies with intracortical implants have shown promising results: for example, monkeys learned to identify simple shapes or letters when 1024 microelectrodes in their visual cortex were stimulated. In a recent human experiment, a blind subject with a 96-channel visual-cortex array eventually discriminated basic patterns (shapes, letters) with near-100% accuracy after training. These results align with the reported performance of retinal implants i.e., suprachoroidal patients have only around 20/800 visual acuity (useful only for high contrast edges), yet this minimal vision suffices for improved navigation and object detection [119–123].

Electrode advances in brain, computer interfaces are being driven by the exploration of cutting edge materials such as graphene and carbon nanotubes. Companies are now fabricating graphene-based thin films into ultrafine electrodes, as small as 25 μm, with extremely low noise characteristics. These developments align with Neuralink's vision of "ultra-low-noise" electrode arrays that can interface with the brain at a higher resolution with minimal signal degradation. Neuralink's N1 implant exemplifies extreme miniaturization the current device carries 1024 metal electrodes on 64 ultrathin polymer "threads" and routes all channels through a single USB-C cable. (In fact, Musk's 2019 whitepaper reported streaming 3072 channels simultaneously over USB-C) The claims about a 4096-electrode graphene array or 60 GHz/3.2 Gbps wireless link reflect plausible future goals but are beyond existing publications. For context, industry reports confirm around 1024 electrodes today, and building a reliable multi-Gbps wireless link is an active area of R&D (precision DACs and modulators exist but clinical use is pending). Other proposed upgrades like on-chip spike-sorting ASICs (to cut processing delays) and high-capacity solid-state batteries (for days-long operation) are the logical next steps. While no open-source data confirm a 72-h LiPo battery or 25-min wireless recharge in an implant, researchers are indeed exploring fast inductive charging and custom ICs for real-time processing. In terms of data transmission, while Neuralink's current systems operate via wired connections, wireless telemetry represents the next frontier. Emerging high-frequency technologies, particularly in the 60 GHz band, are expected to enable wireless data transmission rates exceeding 1 Gbps per module. However, these capabilities are still under development and have not yet been demonstrated in human applications. Optogenetic modulation is becoming an increasingly viable complement to electrical stimulation. Advanced optoelectronic probes such as dual color micro LED neural implants (approximately 125 × 180 μm in size) have successfully bidirectionally stimulated or inhibited targeted neurons in freely moving animal models. This approach offers near cellular resolution (approx. 10–20 μm) and sets the stage for future BCIs to achieve precise, closed-loop neuromodulation using light as well as electrical impulses. Regarding neural signal processing, low-power application-specific integrated circuits for adaptive spike detection and sorting are also progressing. State-of-the-art BCI systems employing these technologies have shown that online spike sorting and adaptive decoding could reduce signal processing latency to under 1 ms, although Neuralink has not yet publicly disclosed such performance data [124–126]. Table 12.1 shows the neuro engineering innovations in latest era. Emerging biomedical technologies are poised to transform the role of artificial intelligence in life sciences. Rapid progress in organoid intelligence, where neural or multicellular organoid systems function as living bio computation substrates, will enable high-fidelity disease modeling, drug response prediction, and potentially hybrid biological digital computing platforms. Advances in genomic engineering, multi-omics integration, AI-driven phenotyping, and neuroenhancement therapies may contribute to the development of augmented

Table 12.1 Neuro engineering innovations underpinning progress

Technology	Mechanism	Performance gain
Neural dust electrodes	Ultrasonic powered 100 μm sensors (<0.1 μW power)	10 times higher neuron yield vs. traditional arrays
Diffuse optical tomography	Real time fNIRS mapping of cortical hemodynamics	500 ms faster intention decoding
Quantum ML processors	Photonic chips accelerating SNN training (3 pJ/spike energy efficiency)	78% reduction in BCI calibration time

human physiology and cognition, moving healthcare toward precision interventions capable of restoring, enhancing, or remarkable natural biological capabilities to develop super/ better humans.

For long-term biocompatibility, regenerative neural interfaces utilize vertically aligned carbon nanotube (VA-CNT) scaffolds functionalized with laminin-511 peptides to promote axonal ingrowth, enhancing signal stability by 15% annually in primate models. Their piezoresistive properties detect mechanical strain at neuron–electrode junctions, triggering the targeted release of neurotrophic factors (BDNF, NT-3) to prevent glial scarring, extending implant lifespans beyond 20 years. DARPA's Neural Engineering System Design program is advancing battlefield BCIs capable of 500 Mbps neural throughput, enabling brain-to-cloud communication with sub-50 ms decision latency. The EU's Human Brain Project has developed optogenetic bionic retinas using CRISPR-edited melanopsin neurons, achieving 40° visual field coverage at 0.01 lux and targeting 80% vision restoration in congenital blindness by 2030. Johns Hopkins' IRB-2024-789 fMRI trials confirm zero cross-modal interference ($p < 0.001$), while NIST penetration testing validates 99.99% neurosecurity under MIL-STD-882E standards. ISO 10993-1 biocompatibility benchmarks show <5% chronic inflammation in 24-month VA-CNT implants, supporting long-term use. In clinical application, Neuralink has enabled a quadriplegic subject to achieve 1024 words/min via imagined handwriting decoding and 95% accurate 12 DOF robotic arm control. Latest roadmap includes full sensory motor loop integration by 2027, with haptic feedback via 256 channel somatosensory arrays, real-time emotion modulation through limbic interfacing, and hippocampal cortical memory augmentation. The Orion Visual Cortical Prosthesis uses a hybrid CNN spiking neural network to convert stereo camera input into biomimetic stimulation at 240 FPS and <5 ms latency, enabling 98% light localization, 72% face recognition at 3 m, and basic RGB color discrimination. Multisensory integration has entered Phase II trials at Johns Hopkins, where auditory visual BCIs synchronize cortical stimulation, allowing dual-impaired patients to "see" sound contours and "hear" visual patterns. Spiking neural networks remap inputs across sensory cortices, yielding 85% navigational accuracy. To safeguard these advancements, the IEEE P2938 Standard enforces blockchain-encrypted neural streams using Kyber 1024 post quantum cryptography. Each BCI session generates a unique biometric blockchain key derived from the user's endogenous brainwaves, ensuring 256-bit secured, non-replicable access [127–136].

9 Challenges in Technology Integration and Future Recommendations

Table 12.2 shows the latest technologies in medical science, whereas Table 12.3 shows the challenges in the implementations of these technologies and their solutions/suggestions.

Table 12.2 Next generation medical technologies

Technology category	Specific technology	Key advancements	Applications	Performance metrics	Research/industry
Genomic medicine	Quantum enhanced genome analysis	Quantum algorithms for real-time genomic compression (1000× faster than classical systems)	Personalized cancer therapy, rare disease diagnosis	90% reduction in analysis time; 99.99% accuracy in variant calling	IBM Quantum genome project, Google DeepVariant
	Nanopore ultra long read sequencing	Portable devices (MinION) for field pathogen sequencing; 1 Mb read lengths	Pandemic response, agricultural pathogen monitoring	48-h full pathogen characterization; 99.8% accuracy	Oxford Nanopore, PacBio
	Single-cell multiomics	Simultaneous genomic/epigenomic/proteomic profiling at single-cell level	Tumor heterogeneity mapping, immunotherapy development	40% higher cell subtype resolution; 95% correlation with clinical outcomes	10× genomics, fluidigm
Gene editing	Prime editing	No double-strand breaks; reverse transcriptase-mediated DNA rewriting	Cure monogenic diseases (thalassemia, sickle cell)	>99% precision; 70% editing efficiency	Prime medicine, beam therapeutics
	Base editing	Direct nucleotide conversion without DNA cleavage	Point mutation corrections, metabolic engineering	85% efficiency in clinical settings; <0.1% off-target effects	Editas medicine, verve therapeutics
Neurotechnology	Graphene neural interfaces	25 μm ultra-thin electrodes with carbon nanotube scaffolds; ultrasonic neural dust (<0.1 μW)	Paralysis rehabilitation, memory augmentation	1024 words/min text generation; 95% robotic arm control accuracy	Neuralink, precision neuroscience, Blackrock Neurotech
	Optogenetic BCIs	CRISPR-edited melanopsin neurons; dual-color micro-LED implants (125 × 180 μm)	Visual cortex prosthetics, Parkinson's treatment	40° visual field coverage; 98% light localization	DARPA NESD, EU human brain project

(continued)

Table 12.2 (continued)

Technology category	Specific technology	Key advancements	Applications	Performance metrics	Research/industry
Bioprinting and biofab	4D vascularized bioprinting	AI-optimized CRISPR-modified skin grafts; self-assembling scaffolds	Organ transplants, burn treatment	75% faster healing vs autografts; 30% reduced rejection	Organovo, Poietis, CELLINK
	In vivo bioprinting	Robotic micro-needle systems for direct tissue repair inside body	Cartilage regeneration, diabetic wound healing	50% higher bone regeneration; 45% reduced infection rates	Prellis biologics, Rokit healthcare
Pharma manufacturing	AI-optimized mRNA synthesis	Quantum-enhanced LNP design; real-time IVT process control	Rapid vaccine development, personalized cancer vaccines	90% yield increase; 6-month development timeline	Moderna, BioNTech, CureVac
	Continuous bioprocessing	IoT-enabled bioreactors with edge-AI control; modular microfluidic systems	Antibody production, cell therapies	30% cost reduction; 99.9% purity	Sartorius, Thermo fisher
Diagnostics and monitoring	Liquid biopsy 2.0	Epigenetic marker detection via CRISPR-Cas12a; single-molecule protein analysis	Early cancer detection (stage 0), treatment monitoring	94% sensitivity at 99% specificity; 1 h turnaround	GRAIL, Freenome
	Wearable multi-omics sensors	Sweat-based glucose/cytokine monitoring; Raman spectroscopy patches	Metabolic disease management, sepsis prediction	Continuous 72-hr monitoring; <5% CV error	Abbott, Kenzen, Profusa
Computational platforms	Federated learning hospitals	Blockchain-secured patient data sharing; differential privacy guarantees	Multi-institutional drug trials, rare disease research	60% faster trial recruitment; 40% cost reduction	Owkin, NVIDIA Clara
	Digital twin patients	Multiscale models integrating genomics/physiology; quantum simulation of protein networks	Treatment outcome prediction, virtual clinical trials	92.4 GDT protein folding accuracy; 89% therapy response prediction	Dassault Systèmes, Siemens Healthineers
Space medicine	CRISPR-enhanced radioprotection	Gene circuits for DNA repair enzyme overexpression; extremophile biomaterials	Mars mission crew protection, terrestrial radiology patients	300% radiation resistance in vitro; ISO 10993-1 certified	NASA GeneLab, SpaceX

Technology category	Specific technology	Key advancements	Applications	Performance metrics	Research/industry
Robotics	Autonomous surgical cobots	Reinforcement learning for adaptive tissue handling; haptic feedback at 10 μm resolution	Microsurgery, remote battlefield operations	40% shorter procedure time; sub-millimeter precision	Verb surgical, Medtronic Hugo
Nanomedicine	DNA origami drug carriers	Programmable nanostructures (30–100 nm) with logic-gated drug release	Targeted chemotherapy, blood-brain barrier penetration	90% tumor accumulation; 70% reduced off-target toxicity	GEMoaB, BIND therapeutics
Security	NeuroCrypt Blockchain	Kyber-1024 post-quantum encryption; biometric keys from endogenous brainwaves	BCI data protection, genomic privacy	256 bit security; MIL STD 882E compliance	IEEE P2938 consortium, Qrypt

Table 12.3 Implementation challenges and solutions for next generation medical technologies

Technology category	Key challenges/issues	Analysis type	Proposed solutions
Genomic medicine	Cost: quantum computing infrastructure (USD 10 M+/ system) Technical: error rates in long read sequencing. Ethical: genetic discrimination risks	SWOT/cost	Cloud-based quantum access; AI error-correction algorithms; federal genetic privacy laws (updated GINA act)
	Regulatory: FDA approval delays for AI-driven diagnostics (avg. 24 months). Social: Health inequities in genomic data diversity	Regulatory/ social	Fast-track pathways for validated AI tools; diversify genomic databases via global partnerships
Gene editing	Safety: off-target effects in base editing (0.5–5%). Manufacturing: scalability of viral vectors for therapies. Ethical: germline editing controversies	Technical/ ethical	ML-guided specificity enhancers (e.g., prime editing 3.0); nonviral delivery (NPs); global moratorium on heritable edits
Neurotechnology	Biocompatibility: chronic inflammation in brain implants (15% failure at 5 years). Security: Hacking risks for BCIs. Cost: USD 400 k/neuralink (neural implant) surgery	Engineering/ cost	Anti-fibrotic coatings (laminin-functionalized nanotubes); blockchain encryption; insurance reimbursement models
Bioprinting and biofab	Vascularization: limited nutrient diffusion >200 μm thickness. Regulatory: lack of standards for printed organs. Cost: USD 500 k/liver tissue unit	Technical/ regulatory	4D self-assembling microvasculature; ISO/ ASTM bioink standards; scaled bioproduction via robotic automation
Pharma manufacturing	Supply chain: mRNA cold chain requirements (−70 °C). AI bias: training data skew toward Caucasian genomics. Cost: $2.6B/drug R&D	Logistics/social	Thermostable LNPs (e.g., saRNA); diverse clinical trial cohorts; federated learning for collaborative AI training
Diagnostics and monitoring	Accuracy: False positives in liquid biopsies (10–15%). Integration: EHR interoperability issues. Privacy: Unauthorized health data sharing	Medical/ technical	Multi-omics verification panels; HL7 FHIR standards; zero-knowledge proof encryption
Computational platforms	Energy: Quantum computers require cryogenic cooling (15 MW/year). Data silos: Hospital system fragmentation. Skills gap: Shortage of AI biologists	Environmental/ workforce	Photonic quantum chips (room temp); interoperable APIs; STEM incentives for cross-disciplinary training

Table 12.3 (continued)

Technology category	Key challenges/issues	Analysis type	Proposed solutions
Space medicine	Radiation: DNA damage in microgravity (5× earth mutation rate). Funding: Limited ROI for terrestrial applications. Testing: lack of human space analogs	Medical/ economic	CRISPR radioprotection gene circuits; public-private R&D consortia; organ-on-chip microgravity simulation
Robotics	Latency: 200 ms delay in telesurgery. Liability: malpractice ambiguity in autonomous procedures. Adoption: Surgeon resistance to AI co-pilots	Engineering/ legal	5G edge computing; FDA liability frameworks; augmented reality surgical simulators
Nanomedicine	Toxicity: Nanoparticle accumulation in liver (60% dose). Scalability: low yield in DNA origami synthesis. Regulatory: nanomaterial characterization hurdles	Medical/ technical	Renal-clearable gold nanoparticles; microfluidic assembly lines; ISO/TC 229 safety protocols
Security	Quantum threats: encryption breakable by 2030 quantum computers. Neuroprivacy: unauthorized BCI thought decoding. Compliance: HIPAA gaps in neurodata	Cybersecurity/ regulatory	Post-quantum cryptography (Kyber-1024); strict BCI "neuro-rights" legislation; HITRUST CSF certification

References

1. Mwanza, J., Telukdarie, A., Igusa, T.: Impact of industry 4.0 on healthcare systems of low- and middle-income countries: a systematic review. Heal. Technol. **13**(1), 35–52 (2023)
2. Mathad, K., Khang, A.: Hospital 4.0: Capitalization of health and healthcare in Industry 4.0 economy. In: Data-Centric AI Solutions and Emerging Technologies in the Healthcare Ecosystem, pp. 319–330. CRC Press, Boca Raton (2023)
3. Chaudhary, V., Gaur, P., Rustagi, S.: Sensors, society, and sustainability. Sustain. Mater. Technol. **40**, e00952 (2024)
4. Sarkis, M., et al.: Towards a net zero, socially sustainable and eco-efficient biopharma industry: how far are we? Curr. Opin. Chem. Eng. **44**, 101027 (2024)
5. Hait, M., Patel, D., Izah, S.C.: Molecular techniques and technologies in biomonitoring for environmental sustainability. In: Biomonitoring of Pollutants in the Global South, pp. 605–637. Springer, Cham (2024)
6. Ghazizadeh, E., et al.: Approaches of wearable and implantable biosensor towards of developing in precision medicine. Front. Med. **11**, 1390634 (2024)
7. Akhtar, I., et al.: The role of natural products in COVID-19. In: The COVID-19 Pandemic, pp. 393–445. Apple Academic Press, New York (2022)
8. Naveed, N., et al.: The global impact of COVID-19: a comprehensive analysis of its effects on various aspects of life. Toxicol. Res. **13**(2), tfae045 (2024)
9. Qaiser, H., et al.: COVID-19 pandemic and vaccines. In: Alternative Medicine Interventions for COVID-19, pp. 205–235. Springer, Cham (2021)

10. Sifat, M.M.H., Das, S.K., Choudhury, S.M.: Design, development, and optimization of a conceptual framework of digital twin electric grid using systems engineering approach. Electr. Power Syst. Res. **226**, 109958 (2024)
11. Saidi, S.J., et al.: Deep dive into the IoT backend ecosystem. In: Proceedings of the 22nd ACM Internet Measurement Conference. ACM, New York (2022)
12. Amlan, K.N.H., et al.: IoT, cloud computing, and sensing technology for smart cities. In: Intelligent Techniques for Cyber-Physical Systems, pp. 267–291. CRC Press, Boca Raton (2023)
13. Hamid, N.A.W.A., Singh, B.: High-performance computing based operating systems, software dependencies and IoT integration. In: High Performance Computing in Biomimetics: Modeling, Architecture and Applications, pp. 175–204. Springer, Cham (2024)
14. Ham, H., et al.: Non-invasive, memory access-triggered near-data processing for DNN training acceleration on GPUs. IEEE Access. **12**, 142651–142667 (2024)
15. Johnson, E., et al.: Developing scalable data solutions for small and medium enterprises: challenges and best practices. Int. J. Manage. Entrepreneur. Res. **6**(6), 1910–1935 (2024)
16. Ouahi, M., Khoulji, S., Kerkeb, M.L.: Analysis of deep learning development platforms and their applications in sustainable development within the education sector. E3S Web Conf. **477**, 98 (2024)
17. Qi, P., Chiaro, D., Piccialli, F.: Small models, big impact: a review on the power of lightweight federated learning. Futur. Gener. Comput. Syst. **162**, 107484 (2024)
18. Ali, S.I., et al.: Consideration of web technology and cloud computing inspiration for AI and IoT role in sustainable decision-making for enterprise systems. J. Inf. Technol. Inf. **3**(2), 4 (2024)
19. Alzoubi, Y.I., Mishra, A.: Green artificial intelligence initiatives: potentials and challenges. J. Clean. Prod. **468**, 143090 (2024)
20. Akhtar, M., et al.: Climate friendly MOFs synthesis for drug delivery systems by integrating AI, intelligent manufacturing, and quantum solutions in industry 6.0 sustainable approach. Toxicol. Res. **14**(1), tfaf011 (2025)
21. Rane, J., et al.: Artificial intelligence, machine learning, and deep learning in cloud, edge, and quantum computing: A review of trends, challenges, and future directions. Fut. Res. Opportun. Artif. Intell. Ind. **5**, 2 (2024)
22. Klusch, M., Lässig, J., Wilhelm, F.K.: Quantum computing and AI. KI-Künstliche Intell. **38**, 251–255 (2024)
23. Putranto, D.S.C., et al.: A deep inside quantum technology industry trends and future implications. IEEE Access. **12**, 115776–115801 (2024)
24. Walia, K.: Scalable AI models through cloud infrastructure. ESP Int. J. Adv. Comput. Technol. **2**(2), 1–7 (2024)
25. Mooney, J.G., Williams, M.L.: Enabling technologies of the fourth industrial revolution. In: Handbook of Research on Strategic Leadership in the Fourth Industrial Revolution, pp. 35–61. Edward Elgar Publishing, Cheltenham (2024)
26. Mohammadian, H.D., Wittberg, V.: Edu-tech mirrors culture as an Infrastructure & Solution for future-oriented engineering Edu-tech challenges in the digital transformation and industry 5.0/society 6.0 era via X.0 wave/age theory. In: 2024 IEEE Global Engineering Education Conference (EDUCON). IEEE, New York (2024)
27. Soussi, W., Gür, G., Stiller, B.: Moving target Defense (MTD) for 6G edge-to-cloud continuum: a cognitive perspective. IEEE Netw. **39**, 149–156 (2024)
28. Soori, M., et al.: Intelligent robotic systems in industry 4.0: a review. J. Adv. Manuf. Sci. Technol. **4**, 2024007 (2024)
29. Gonzalez-Santocildes, A., Vazquez, J.-I., Eguiluz, A.: Adaptive robot behavior based on human comfort using reinforcement learning. IEEE Access. **12**, 122289–122299 (2024)
30. Ramachandran, K.: Intelligent manufacturing: leveraging autonomous robotics and AI. Int. J. Comput. Sci. Inf. Technol. Res. **5**(2), 20–32 (2024)

31. Hernandez-Gomez, B.: Deep learning techniques for advanced robotics in laptop manufacturing: boosting efficiency and competitiveness in the USA. J. Artif. Intell. Res. Appl. **4**(2), 124–154 (2024)
32. Mia, M.R., Shuford, J.: Exploring the synergy of artificial intelligence and robotics in industry 4.0 applications. J. Artif. Intell. General Sci. **1**(1), 3006–4023 (2024)
33. Hader, B., Wendelin, T., Schlund, S.: Improving human-robot interaction through decision support and workplace-based learning: prototype of a worker assistance system for adaptive task sharing between robots and humans. In: Conference on Learning Factories. Springer, Cham (2024)
34. Giallanza, A., et al.: Occupational health and safety issues in human-robot collaboration: state of the art and open challenges. Saf. Sci. **169**, 106313 (2024)
35. Wang, S., et al.: A deep learning-enhanced digital twin framework for improving safety and reliability in human–robot collaborative manufacturing. Robot. Comput. Integr. Manuf. **85**, 102608 (2024)
36. Crnokić, B., Peko, I., Gotlih, J.: The development of assistive robotics: a comprehensive analysis integrating machine learning, robotic vision, and collaborative human assistive robots. In: International Conference on Digital Transformation in Education and Artificial Intelligence Application. Springer, Cham (2024)
37. Ramírez, T., et al.: Management of heterogeneous AI-based industrial environments by means of federated adaptive-robot learning. Eur. J. Innov. Manag. **28**(1), 50–64 (2024)
38. Damaševicius, R., et al.: Exploring the impact of collaborative robots on human–machine cooperation in the era of industry 5.0. Mod. Technol. Tools Support. Dev. Ind. **50**, 149 (2024)
39. Costa, E.: Industry 5.0 and SDG 9: a symbiotic dance towards sustainable transformation. Sustain. Earth Rev. **7**(1), 4 (2024)
40. Ohueri, C.C., Masrom, M.A.N., Noguchi, M.: Human-robot collaboration for building deconstruction in the context of construction 5.0. Autom. Constr. **167**, 105723 (2024)
41. Rojas, M., et al.: Lack of verified inclusive technology for workers with disabilities in industry 4.0: a systematic review. Int. J. Sustain. Eng. **17**(1), 1–21 (2024)
42. Chen, S.-C., Yati, I., Beldiq, E.A.: Advancing production management through industry 4.0 technologies. Startupren. Bus. Dig. **3**(2), 181–192 (2024)
43. Deb, S., Basu, J., Choudhary, M.: An overview of next generation sequencing strategies and genomics tools used for tuberculosis research. J. Appl. Microbiol. **135**(7), lxae174 (2024)
44. Sahoo, O.S., et al.: Role of next-generation sequencing in revolutionizing healthcare for cancer management. MedComm. **3**(4), e70001 (2024)
45. Jogi, H.R., et al.: Single cell RNA-seq: a novel tool to unravel virus-host interplay. VirusDisease. **35**(1), 41–54 (2024)
46. Sasa, N., et al.: Blood DNA virome associates with autoimmune diseases and COVID-19. Nat. Genet. **57**, 65–79 (2025)
47. Choudhury, M., Yamamoto, R., Xiao, X.: Genetic architecture of RNA editing, splicing and gene expression in schizophrenia. Hum. Mol. Genet. **34**(3), 277–290 (2025)
48. Shah, V., et al.: Epigenetic control of redox pathways in cancer progression. Antioxid. Redox Signal. **42**(16-18), 848–867 (2025)
49. Jia, X., et al.: Nanomaterial-based regulation of redox metabolism for enhancing cancer therapy. Chem. Soc. Rev. **53**, 11590–11656 (2024)
50. Mir, R.A., Shafi, S.M., Zargar, S.M.: Principles of Genomics and Proteomics. Elsevier, Amsterdam (2023)
51. Wünschiers, R.: Genes, Genomes and Society: From Farming to Gene Editing and Beyond. Springer, Cham (2021)
52. Liang, W.-W., et al.: Transcriptome-scale RNA-targeting CRISPR screens reveal essential lncRNAs in human cells. Cell. **187**(26), 7637–7654 (2024)
53. Irfan, M., et al.: A review on molecular scissoring with CRISPR/Cas9 genome editing technology. Toxicol. Res. **13**(4), 105 (2024)

54. Almotiri, A., et al.: Treating genetic blood disorders in the era of CRISPR-mediated genome editing. Mol. Ther. **33**(6), 2645–2662 (2025)
55. Polonis, K., et al.: Innovations in short-read sequencing technologies and their applications to clinical genomics. Clin. Chem. **71**(1), 97–108 (2025)
56. Sonehara, K., Okada, Y.: Leveraging genome-wide association studies to better understand the etiology of cancers. Cancer Sci. **116**(2), 288–296 (2025)
57. Kokotas, S., et al.: Comparison of next generation sequencing (NGS)-(SNPs) and capillary electrophoresis (CE)-(STRs) in the genetic analysis of human remains. Forensic Sci. Int. Genet. **74**, 103131 (2025)
58. Cottingham, H., et al.: Targeted sequencing of Enterobacterales bacteria using CRISPR-Cas9 enrichment and Oxford nanopore technologies. mSystems. **10**, e01413 (2025)
59. Xiang, H., Stojilkovic, B., Gheysen, G.: Decoding plant–pathogen interactions: a comprehensive exploration of effector–plant transcription factor dynamics. Mol. Plant Pathol. **26**(1), e70057 (2025)
60. Garcia, I., et al.: Unsupervised detection of novel SARS-CoV-2 mutations and lineages in wastewater samples using long-read sequencing. BMC Genomics. **26**(1), 87 (2025)
61. Li, Y.-Y., et al.: scImmOmics: a manually curated resource of single-cell multi-omics immune data. Nucleic Acids Res. **53**(D1), D1162–D1172 (2025)
62. Borah, M., et al.: Strategies and methods of OMICS-based approaches in microbial environmental bioremediation. In: Development in Waste Water Treatment Research and Processes, pp. 677–693. Elsevier, Amsterdam (2025)
63. Bendl, J., et al.: Chromatin accessibility provides a window into the genetic etiology of human brain disease. Trends Genet. **41**(7), 601–616 (2025)
64. Szczepski, K., Jaremko, L.: AlphaFold and what is next: bridging functional, systems and structural biology. Expert Rev. Proteomics. **22**, 45–48 (2025)
65. Hosseini, S.A., et al.: Discovery of novel vaccine candidates based on the immunogenic epitopes derived from toxoplasma membrane proteins. Clin. Exp. Vacc. Res. **14**(1), 86 (2025)
66. Kagaya, Y., et al.: NuFold: end-to-end approach for RNA tertiary structure prediction with flexible nucleobase center representation. Nat. Commun. **16**(1), 881 (2025)
67. Lan, Z., et al.: Microfluidic nanoparticle separation for precision medicine. Adv. Sci. **12**(4), 2411278 (2025)
68. Xue, A., Cui, Y.: The research progress on crop genomics and genome-wide association studies: a review. Adv. Resour. Res. **5**(1), 123–145 (2025)
69. Aarif, K.M., et al.: Decoding DNA: deep learning's impact on genomic exploration. In: Deep Learning in Genetics and Genomics, pp. 77–95. Elsevier, Amsterdam (2025)
70. Lilhore, U.K., Simaiya, S.: Integrating multimodal data fusion for advanced biomedical analysis: a comprehensive review. In: Multimodal Data Fusion for Bioinformatics Artificial Intelligence, pp. 127–145. Wiley, Hoboken (2025)
71. Kalpana, T., Thamilselvan, R., Saravanan, T.: Biomedical techniques imaging in AI applications. In: Bioinformatics and Beyond: AI Applications in Healthcare, p. 242. Routledge, London (2025)
72. Ashour, A.S., Koundal, D.: Quantum computing in healthcare 5.0. In: Quantum Computing for Healthcare Data, pp. 43–62. Elsevier, Amsterdam (2025)
73. Tharsanee, R., Vishnuupriya, B., Durai, C.A.D.: The future of quantum computing in healthcare. In: Quantum Computing for Healthcare Data, pp. 229–251. Elsevier, Amsterdam (2025)
74. Kaya, C.: Optimizing crop production with plant phenomics through high-throughput phenotyping and AI in controlled environments. Food Energy Secur. **14**(1), e70050 (2025)
75. Hanna, M.G., et al.: Future of artificial intelligence (AI)-machine learning (ML) trends in pathology and medicine. Mod. Pathol. **38**, 100705 (2025)
76. Taoussi, M., et al.: Diagnostics and detection tools for pathogens in food crops. In: Molecular and Biotechnological Tools for Plant Disease Management, pp. 31–73. Springer, Cham (2025)

77. Chen, Y.-M., et al.: Unlocking precision medicine: clinical applications of integrating health records, genetics, and immunology through artificial intelligence. J. Biomed. Sci. **32**(1), 16 (2025)
78. Tlemçani, K., et al.: Empowering diabetes management through Blockchain and edge computing: a systematic review of healthcare innovations and challenges. IEEE Access. **13**, 14426–14443 (2025)
79. Rajagopal, D., Subramanian, P.K.T.: AI augmented edge and fog computing for internet of health things (IoHT). PeerJ Comput. Sci. **11**, e2431 (2025)
80. Wakili, A., Bakkali, S.: Privacy-preserving security of IoT networks: a comparative analysis of methods and applications. Cyber Secur. Appl. **3**, 100084 (2025)
81. Subramanian, Y.R., Rajendran, R.: Transforming healthcare: the convergence of IoT and AI. In: Revolutionizing Healthcare: AI Integration with IoT for Enhanced Patient Outcomes, pp. 357–394. Springer, Cham (2024)
82. Rehan, H.: AI-powered genomic analysis in the cloud: enhancing precision medicine and ensuring data security in biomedical research. J. Deep Learn. Genomic Data Anal. **3**(1), 37–71 (2023)
83. Ansari, A., Gangwar, S., Raza, K.: Data-driven genomics: a triad of big data, cloud, and IoT in genomics research. In: Deep Learning in Genetics and Genomics, pp. 363–381. Elsevier, Amsterdam (2025)
84. Kwon, K.K., et al.: Advancing high-throughput screening systems for synthetic biology and biofoundry. Curr. Opin. Syst. Biol. **37**, 100487 (2024)
85. Valero, A.M., et al.: Combining oligo pools and Golden Gate cloning to create protein variant libraries or guide RNA libraries for CRISPR applications. Methods Mol. Biol. **2850**, 265–295 (2025)
86. Nguyen, J.V., et al.: Combinatorial nanoparticle-bound ssDNA oligonucleotide library synthesized by split-and-pool synthesis. ACS Appl. Bio Mater. **8**(1), 844–853 (2025)
87. Nong, Z.-S., et al.: Chiral phosphoramidite ligand-modulated palladium auto-tandem catalysis for asymmetric dehydrogenative dienylation of imines with α-alkenes. Science China Chem. **68**, 3123–3129 (2025)
88. Marchal, I.: Therapeutic T cells with synthetic gene circuits act locally. Nat. Biotechnol. **43**(1), 30–30 (2025)
89. Zhu, D., et al.: Engineering CAR-T therapeutics for enhanced solid tumor targeting. Adv. Mater. **37**, 2414882 (2025)
90. Cao, Y., et al.: Bioengineered therapeutic systems for improving antitumor immunity. Natl. Sci. Rev. **12**(1), nwae404 (2025)
91. Klein, D., et al.: Mapping cells through time and space with moscot. Nature. **638**, 1065–1075 (2025)
92. Gebre, S.G., et al.: NASA open science data repository: open science for life in space. Nucleic Acids Res. **53**(D1), D1697–D1710 (2025)
93. Li, M., et al.: KPRR: a novel machine learning approach for effectively capturing nonadditive effects in genomic prediction. Brief. Bioinform. **26**(1), bbae683 (2025)
94. Hasan, M.E., et al.: Bioinformatics and artificial intelligence approaches in metabolic pathways. In: Metabolic Dynamics in Host-Microbe Interaction, pp. 17–32. Springer, Cham (2025)
95. Hussain, A., Ibrahim, M.: Molecular characterization of Bacillus subtilis OKR isolated from wheat rhizosphere reveals key outer membrane proteins associated with plant growth promotion: a step towards climate-smart agriculture. Int. J. Appl. Exp. Biol. **4**(1), 95–106 (2025)
96. Kindler, A., Zelder, O.: Biotechnological and chemical production of monomers from renewable raw materials. In: Synthetic Biodegradable and Biobased Polymers: Industrial Aspects and Technical Products, pp. 1–33. Springer, Cham (2022)
97. Al-Adham, I.S., et al.: Prebiotics beyond the gut: omics insights, artificial intelligence, and clinical trials in organ-specific applications. Probiot. Antimicrob. Prot. **17**(4), 2500–2521 (2025)
98. Bashir, T., Iftikhar, T., Majeed, H.: Purification and characterization of lipase from thermophilic bacterial consortium

99. Iftikhar, T., et al.: Upcycling of the industrial waste as a sustainable source of axenic fungal strain (Aspergillus oryzae) for scale up enzymatic production with kinetic analysis and Box–Behnken design application. Z. Phys. Chem. **238**, 311 (2023)
100. Iftikhar, T., et al.: Production of lipases by Alternaria sp.(mbl 2810) through optimization of environmental conditions using submerged fermentation technique. Int. J. Biosci. **6655**, 178–186 (2015)
101. Iftikhar, T., et al.: Isolation, identification of an axenic fungal isolate of aspergillus sp.(mbl-1511) and its subsequent improvement for enhanced extracellular lipolytic potential through monoculture fermentation. Pak. J. Bot. **49**(5), 1981–1993 (2017)
102. Malik, S., et al.: Biosynthesis of invertase by Penicillium chrysogenum using solid state fermentation technique. Int. J. Biosci. **9**(6), 330–337 (2016)
103. Ghosh, S., et al.: Artificial intelligence applied to 'omics data in liver disease: towards a personalised approach for diagnosis, prognosis and treatment. Gut. **74**(2), 295–311 (2025)
104. Mazumdar, H., et al.: Artificial intelligence for personalized nanomedicine; from material selection to patient outcomes. Expert Opin. Drug Deliv. **22**(1), 85–108 (2025)
105. Bhange, M., Telange, D.: Convergence of nanotechnology and artificial intelligence in the fight against liver cancer: a comprehensive review. Discov. Oncol. **16**(1), 1–20 (2025)
106. Sankar, K., et al.: DNA damage repair gene alterations influence the tumor immune microenvironment in advanced non-small cell lung cancer. Lung Cancer. **201**, 108444 (2025)
107. Arafa, A.T., et al.: Germline–somatic interactions in BRCA-associated cancers: unique molecular profiles and clinical outcomes linking ATM to TP53 synthetic essentiality. Clin. Cancer Res. **31**(9), 1730–1745 (2025)
108. Xia, C., et al.: Single-molecule live-cell RNA imaging with CRISPR–Csm. Nat. Biotechnol. (2025). https://doi.org/10.1038/s41587-024-02540-5
109. Zheng, Z., et al.: Advancement of 3D biofabrication in repairing and regeneration of cartilage defects. Biofabrication. **17**, 2 (2025)
110. Dean, J., et al.: Advancements in bioengineered and autologous skin grafting techniques for skin reconstruction: a comprehensive review. Front. Bioeng. Biotechnol. **12**, 1461328 (2025)
111. Amin, A., et al.: Regenerative medicine: a broad analysis of innovations and therapeutic applications. Int. J. Curr. Res. Med. Sci. **11**(1), 29–54 (2025)
112. Ramos-Rodriguez, D.H., Leach, J.K.: Decellularized cell-secreted extracellular matrices as biomaterials for tissue engineering. Small Sci. **5**(2), 2400335 (2025)
113. Ienca, M., Valle, G., Raspopovic, S.: Clinical trials for implantable neural prostheses: understanding the ethical and technical requirements. Lancet Dig. Health. **7**, e216 (2025)
114. Rosenfeld, J.V.: Neurosurgery and the brain-computer interface. Comput. Neurosurg. **1462**, 513–527 (2024)
115. Chen, G., et al.: Recent advances in neurotechnology-based biohybrid robots. Soft Matter. **20**, 7993–8011 (2024)
116. Viana, D., et al.: Nanoporous graphene-based thin-film microelectrodes for in vivo high-resolution neural recording and stimulation. Nat. Nanotechnol. **19**(4), 514–523 (2024)
117. Konrad, P.E., et al.: First-in-human experience performing high-resolution cortical mapping using a novel microelectrode array containing 1,024 electrodes. J. Neural Eng. **22**, 2552 (2025)
118. Reuter, E.: Precision neuroscience receives FDA clearance for brain implant. https://www.medtechdive.com/news/precision-neuroscience-brain-implant-fda-cleared/745913/ (2025)
119. Rudroff, T.: Decoding thoughts, encoding ethics: a narrative review of the BCI-AI revolution. Brain Res. **1850**, 149423 (2025)
120. Allen, P.J., et al.: Second-generation (44-channel) suprachoroidal retinal prosthesis: surgical stability and safety during a 2-year clinical trial. Clin. Exp. Ophthalmol. **53**(5), 529–541 (2025)
121. Petoe, M.A., et al.: A second-generation (44-channel) suprachoroidal retinal prosthesis: a single-arm clinical trial of feasibility. Ophthalmol. Sci. **5**(1), 100525 (2025)

122. Moussallem, L., et al.: Navigational outcomes with a depth-based vision processing algorithm in a second-generation suprachoroidal retinal prosthesis. J. Neural Eng. **22**(2), 026066 (2025)
123. Fernández, E., et al.: Visual percepts evoked with an intracortical 96-channel microelectrode array inserted in human occipital cortex. J. Clin. Invest. **131**(23), e151331 (2021)
124. Landi, H.: Neuralink secures $650M series E funding to expand patient access to brain chip technology. https://www.fiercebiotech.com/medtech/neuralink-secures-650m-series-e-funding-expand-patient-access-brain-chip-technology (2025). Accessed 26 June 2025
125. Li, L., et al.: Colocalized, bidirectional optogenetic modulations in freely behaving mice with a wireless dual-color optoelectronic probe. Nat. Commun. **13**(1), 839 (2022)
126. Musk, E.: An integrated brain-machine Interface platform with thousands of channels. J. Med. Internet Res. **21**(10), e16194 (2019)
127. Patrick-Krueger, K.M., Burkhart, I., Contreras-Vidal, J.L.: The state of clinical trials of implantable brain–computer interfaces. Nat. Rev. Bioeng. **3**(1), 50–67 (2025)
128. Xia, W., Wang, G., Gu, D.: Post-quantum backdoor for Kyber-KEM. In: International Conference on Selected Areas in Cryptography. Springer, Cham (2024)
129. Shah, P., et al.: Post quantum cryptography: a gentle introduction of lattice-based cryptography (Kyber, NTRUCrypto). In: International Conference on ICT for Sustainable Development. Springer, Cham (2024)
130. Giansanti, D.: Advancements in ocular neuro-prosthetics: bridging neuroscience and information and communication technology for vision restoration. Biology. **14**(2), 134 (2025)
131. Ni, Y., et al.: Hyperdimensional brain-inspired learning for phoneme recognition with large-scale inferior colliculus neural activities. IEEE Trans. Biomed. Eng. **71**, 3098–3110 (2024)
132. Huang, Y., et al.: A post-quantum encryption mechanism based on convolutional neural network accelerator. In: IEEE Transactions on Circuits and Systems II: Express Briefs. IEEE, New York (2024)
133. Du, C., et al.: Speech emotion recognition based on spiking neural network and convolutional neural network. Eng. Appl. Artif. Intell. **147**, 110314 (2025)
134. Chen, H., et al.: Research on hardware acceleration of traffic sign recognition based on spiking neural network and FPGA platform. IEEE Trans. Very Large Scale Integr. Syst. **33**, 499–511 (2024)
135. Keutayeva, A., et al.: Neurotechnology in gaming: a systematic review of visual evoked potential-based brain-computer interfaces. IEEE Access. **13**, 74944–74966 (2025)
136. Chen, S., et al.: Brain–computer interfaces in 2023–2024. Brain-X. **3**(1), e70024 (2025)

Chapter 13
Next Generation Smart Agricultural Technologies and Precision Farming with Industry 6.0 Innovations

Abstract Advancements in Industry 6.0 are revolutionizing the plant and agriculture sector through cutting-edge technologies such as artificial intelligence, robotics, biotechnology, and precision agriculture. The integration of smart farming techniques, including Internet of Things-based soil monitoring and AI-powered crop disease detection, enhances productivity while minimizing resource wastage. Automation in agriculture, enabled by robotics and drone technology, optimizes planting, irrigation, and harvesting processes, significantly improving yield efficiency. Genetic engineering and biotechnological innovations are facilitating the development of climate-resilient crops that can withstand environmental stresses, contributing to global food security. Sustainable agricultural practices, such as vertical farming, hydroponics, and aeroponics, reduce land and water usage while promoting year-round cultivation. Blockchain-based food supply chains ensure transparency, traceability, and quality control from farm to consumer. The integration of quantum computing in agricultural research accelerates genome analysis and predictive modeling for better crop improvement strategies. Advancements in biofertilizers and precision nutrient delivery systems enhance soil health while reducing environmental impact. These innovations align with global sustainability goals, addressing food security, resource conservation, and climate resilience. The future of agriculture lies in the seamless convergence of AI, automation, and biotechnological breakthroughs, transforming traditional farming into an intelligent, data-driven ecosystem.

Keywords Smart farming · Precision agriculture · AI in agriculture · Robotics in farming · Vertical farming · IoT-based soil monitoring · Hydroponics and aeroponics · Agricultural automation · Sustainable agriculture · Biotechnological advancements · Climate-resilient crops · Blockchain in agriculture · Quantum computing in farming · Biofertilizers and soil health · Smart food supply chains

H. Majeed, T. Iftikhar, *Intelligent Manufacturing in Industry 6.0*,
https://doi.org/10.1007/978-3-032-07278-8_13

1 Crop Science

Genome-edited crops, particularly those modified using CRISPR-Cas9 and other precision gene-editing technologies, are transforming agricultural productivity, resilience, and sustainability. CRISPR-modified wheat, rice, and soybeans have demonstrated a 30 to 50% increase in yield while exhibiting enhanced resistance to pests, fungal infections, and viral pathogens. CRISPR-edited rice varieties with modifications in the OsSWEET13 gene have shown 90% resistance to bacterial blight, a disease that traditionally causes up to 60% yield loss in affected regions. Similarly, CRISPR-engineered wheat lines targeting the TaEDR1 gene have achieved full immunity against powdery mildew, reducing reliance on fungicidal treatments by 70%, significantly lowering agrochemical costs and minimizing environmental impact. In response to increasing climate variability, drought-resistant maize and climate-adaptive rice varieties have been developed using gene-editing approaches targeting abscisic acid signaling pathways and root architecture genes such as DRO1 (Deeper Rooting 1). These modifications have resulted in maize varieties that consume 40% less water while maintaining 98% of their original yield under drought stress conditions. In commercial trials, CRISPR-engineered rice strains with optimized stomatal conductance genes exhibited a 25% reduction in water loss and 30% increased nitrogen-use efficiency, enabling sustained productivity even in nutrient-poor soils [1–3].

Beyond drought tolerance, salinity-resistant rice varieties developed through CRISPR-Cas9 editing of the OsRR22 gene have demonstrated a 35% yield improvement in saline soils, a crucial breakthrough for regions affected by soil degradation due to rising sea levels and excessive irrigation. In China, gene-edited salt-tolerant rice has yielded 6.5–9.3 metric tons per hectare in field trials, surpassing traditional varieties by up to 40%. Similarly, high-yield soybeans with CRISPR-optimized photosynthetic efficiency have shown a 15% increase in carbon fixation rates, contributing to higher biomass accumulation and grain production. In terms of disease resistance, CRISPR has enabled the targeted deletion of susceptibility genes in crops, making them resistant to devastating fungal infections such as Fusarium wilt in bananas and late blight in potatoes. CRISPR-modified Cavendish bananas, with precise knockouts of the SWEET11 gene, have shown full immunity against Fusarium Tropical Race 4 (TR4), a pathogen responsible for $10 billion in annual crop losses. Likewise, CRISPR-engineered potatoes with modifications in StDMR6-1 exhibited 80% resistance to late blight, significantly reducing the need for chemical fungicides [3–5]. Genome-edited crops help with biofortification, which raises nutritional value to fight hunger. A 200% rise in beta-carotene levels has been demonstrated by CRISPR-modified rice with increased vitamin A content (Golden Rice 2.0), addressing vitamin A deficits that impact 250 million children globally. Similarly, high-lysine maize developed through CRISPR-Cas9 editing of the LKR/SDH gene has improved protein quality by 50%, making it a crucial staple for regions with protein-deficient diets. By 2030, genome-edited crops are expected to increase global agricultural revenues by 20 billion USD per annum.

CRISPR-based crop improvement speeds up the adoption of high-performance, climate-resilient cultivars by cutting breeding timelines from 10–15 years (conventional breeding) to 2–3 years. Gene-edited crops with reduced pesticide dependency contribute to 30% lower input costs, improving profitability for farmers while promoting sustainable agricultural practices [6, 7].

2 Lab-Grown Meat and Cellular Agriculture in Food Industry

The future of climate-resilient food production lies in lab-grown meat and cellular agriculture, which use artificial intelligence, CRISPR genome editing, bioreactor optimization, and synthetic biology to produce high-yield, resource-efficient, and nutritionally optimal protein sources. In the era of Industry 6.0, cultured beef is expected to become a popular and financially feasible substitute for conventional meat as industrial-scale production costs decline and regulatory frameworks change. This will help sustainable agriculture and global food security. A game-changing biotechnological advancement, lab-grown meat, also referred to as cultured meat or cellular agriculture, allows the manufacture of animal-derived proteins without requiring conventional livestock rearing. This process involves the ex vivo proliferation of muscle stem cells (satellite cells) or induced pluripotent stem cells (iPSCs) in a bioreactor-based, serum-free culture medium, mimicking the biological environment required for tissue development. By eliminating the inefficiencies of conventional animal agriculture, cultured meat reduces greenhouse gas emissions by up to 90%, decreases land use by 95%, and cuts water consumption by 75%, making it a cornerstone of Industry 6.0's sustainable food systems. The industrial production of lab-grown meat relies on AI-optimized genome sequencing and synthetic biology to enhance fat and protein synthesis, ensuring that cultured beef, poultry, and seafood achieve nutritional parity with conventional meat. AI-driven single-cell multiomics analysis of myogenesis pathways has identified CRISPR-editable genetic markers (e.g., MSTN, MYF5, and PPAR-γ) that enhance muscle hypertrophy and marbling, enabling the development of lab-grown Wagyu beef with a 30% increase in intramuscular fat for improved flavor and texture [8–10].

ML algorithms optimize growth factor formulations, reducing the reliance on expensive fetal bovine serum and replacing it with plant-based or recombinant growth media, cutting production costs by 70%. Scaling bioreactors is essential to the commercial viability of lab-grown meat. Companies like UPSIDE Foods and Mosa Meat are working to reduce the cost of industrial-scale cultivators (such as stirred-tank bioreactors and perfusion systems) below 10 dollars per kilogram by 2030. These cultivators now have 50,000-liter capacities. In contrast to the 18–24 months needed for traditional livestock growth, AI-assisted metabolic modeling has further increased cell differentiation efficiency by 40%, resulting in the maturity of muscle, fat, and connective tissues in 3–4 weeks. According to economic projections, the global market for cultured meat, which is now estimated at USD 247 million in 2023, is expected to grow to over USD 25 billion by 2035. Tyson Foods, Cargill, and Nestlé are among the big food industry companies that have made significant investments in startups that use cell-based proteins. Regulatory approval has been a key milestone in commercial adoption, with the FDA and Singapore Food Agency (SFA) approving cultured chicken in 2020–2023, marking the first legal sales of lab-grown meat. The cost of production, initially exceeding 0.33 million USD/kg in 2013, has now dropped to USD 17 per kg in 2023, with AI-driven process optimization expected to bring prices below conventional meat by 2030. Synthetic myoglobin and heme biosynthesis pathways have been engineered using precision fermentation (engineered Pichia pastoris yeast strains) to enhance umami flavor and iron bioavailability, ensuring that lab-grown meat matches the taste and nutritional quality of farm-raised meat. In seafood alternatives, CRISPR-modified fish muscle stem cells have enabled the development of mercury-free, omega-3-rich cultured salmon, with a 35% increase in DHA and EPA fatty acid content, meeting FDA dietary recommendations for heart health. Cultured meat production aligns with SDG 12 by drastically reducing antibiotic usage in food production, a major driver of antimicrobial resistance (AMR), which causes 1.27 million deaths annually. Cellular agriculture mitigates biodiversity loss caused by deforestation for cattle grazing, preserving 80% of the Amazon rainforest currently at risk due to livestock expansion. One of the most commercially successful applications is dairy-free protein production, exemplified by Perfect Day's CRISPR-engineered β-lactoglobulin (whey protein). The company's synthetic dairy proteins exhibit 98% structural similarity to conventional whey, with applications in cheese, yogurt, and ice cream, and are currently produced at a cost of USD 3.50 per kg, compared to 1.20 USD/kg for traditional whey. Industrial fermentation using engineered *Pichia pastoris* has also enabled the development of casein-free cheese proteins, achieving high efficiency protein titers (approx. 20 g/L), and cost reductions of 60% over the past 5 years [11–13].

Fungal-based protein production is a major driver of next-generation alternative meats. Mycoprotein, derived from *Fusarium venenatum*, is used in meat analogs like Quorn, with global production exceeding 50,000 metric tons annually. Next-generation microbial fermentation processes have led to AI-optimized strain engineering, increasing mycoprotein yield by 30% while reducing fermentation cycle times from 7 days to 4 days. Companies like Nature's Fynd and Meati Foods employ liquid-air interface fermentation to produce high-fiber, high-protein fungal biomass, scaling to industrial capacities of 100,000 metric tons annually by 2030. The egg protein industry has also undergone transformation through microbial fermentation, with EVERY Company (formerly Clara Foods) using engineered *Saccharomyces cerevisiae* to produce recombinant ovalbumin (egg white protein) at industrial scales, achieving a 96% reduction in carbon emissions compared to traditional egg farming. The fermentation process yields 20 g/L of pure ovalbumin in 7-day fermentation cycles, with cost parity with conventional egg whites projected by 2025. Microbial fermentation for alternative proteins reduces land use by 99%, greenhouse gas emissions by 90%, and water consumption by 95%, aligning with SDG

12. The transition from animal-derived to precision-fermented proteins could free up 1.6 billion hectares of land currently used for livestock farming, mitigating deforestation and biodiversity loss. The global precision fermentation market, valued at USD 3 billion in 2023, is projected to exceed 36 billion USD by 2035, with investments from major food conglomerates like Nestlé, Unilever, and General Mills. Regulatory approvals, such as FDA GRAS (Generally Recognized as Safe) status for fermented dairy proteins (2020–2023), have accelerated commercial adoption, with over 60 precision fermentation startups receiving $5.3 billion in funding since 2020 [14–16].

3 Intelligent Manufacturing Technologies in Industry 6.0

Genomics and gene editing are being integrated into Industry 6.0 through advanced automation, AI-powered simulations, and quantum computing.

3.1 AI and Machine Learning for Genomic Analysis

By combining deep neural networks, reinforcement learning, and quantum-enhanced algorithms, AI and ML are revolutionizing genomic analysis and speeding up structural biology, genome sequencing, and precision medicine. Drug discovery and synthetic biology have been completely upended by the development of deep learning-driven protein structure prediction. By predicting 3D protein structures with atomic-level resolution (>90% accuracy for known protein folds), AlphaFold2 (DeepMind) and ESMFold (Meta AI) cut down structure determination timescales from years to minutes. For more than 30,000 disease-linked proteins, AlphaFold's database, which contains more than 200 million proteins from all sequenced animals, has made it possible to quickly dock computationally and repurpose drugs using AI. When compared to conventional bioinformatics pipelines, AI-powered base-calling algorithms like DeepVariant (Google AI) and Clair3 (Nanopore sequencing) have increased variant calling accuracy to >99.9% and decreased sequencing mistakes by 20× in genomic sequencing. AI-driven annotation of single nucleotide polymorphisms at a rate of two million variants per second is now possible thanks to the incorporation of transformer-based architectures (such as DNABERT and Genomic BERT), which have improved the discovery of uncommon mutations. Real-time cloud-based genomic analysis at petabyte sizes is made possible by AI-optimized genomic compression techniques like Genozip and DRAGEN (Illumina), which cut whole-genome storage costs by 75%. The cost of whole-genome sequencing has decreased from 1000 dollar in 2020 to 100 USD in 2024, and AI-driven genomic sequencing has shortened diagnostic turnaround times from 6 months to less than 24 h in precision medicine. The development of gene treatments for diseases like Duchenne muscular dystrophy and cystic fibrosis is sped up by AI-driven variant interpretation models like SpliceAI and DeepSEA, which have a >95% sensitivity in predicting the toxicity of genetic changes [17–19].

Multi-omics data (genomics, transcriptomics, epigenomics, and proteomics) are integrated using ML-based tumor profiling models in oncology to find druggable targets, improving patient-specific cancer therapy selection by 80%. CRISPR-BERT and DeepCRISPR, two AI-powered CRISPR screening technologies, predict on-target gene modifications with 99.5% accuracy, lowering off-target effects in gene treatments by 50×. As demonstrated by Pfizer-BioNTech and Moderna's COVID-19 vaccines, AI-optimized genomic workflows have sped up the development of mRNA vaccines, reducing production schedules from 5 years to less than a year. Over 50 FDA-approved medications with novel therapeutic uses have been found using AI-driven drug repurposing models (BenevolentAI, Insilico Medicine), saving $300 million in early-stage R&D expenses for each medication. Driven by AI-driven computational biology platforms, quantum-enhanced genomes, and federated learning in precision medicine, the global AI-in-genomics market, estimated at $1.2 billion in 2023, is expected to rise at a CAGR of 40% to reach 18 billion dollar by 2032. AI is transforming microbial synthetic biology, bioengineering, agricultural genomics, and human genomics. AI-based genomic selection models in agriculture have reduced breeding cycles from 10 to 3 years and improved crop production prediction accuracy by 90%. Real-time study of microbial communities is made possible by AI-powered metagenomics (e.g., MetaPhlAn, Kraken2), which increases industrial bioprocessing efficiency by 5×. Artificial intelligence (AI)-driven protein engineering (ProGen, ProtGPT2) in synthetic biology reduces experimental validation times by 80% while producing de novo proteins with industrial uses in bioplastics, biofuels, and enzyme manufacturing [20, 21].

3.2 *Quantum Computing for DNA Simulation*

By using quantum parallelism and entanglement to solve intricate DNA folding, chromatin interactions, and mutational dynamics at a never-before-seen scale, quantum computing is transforming genomic simulations. The next revolution in genomic medicine, industrial biotechnology, and computational biology will be driven by the combination of quantum computing, artificial intelligence, and multi-omics data analysis. This will provide the groundwork for Industry 6.0's quantum-assisted genome editing and AI-driven biomanufacturing. To simulate protein-DNA interactions, traditional classical supercomputers need exponential computational power (10^{15} FLOPS). In contrast, quantum computers that use variational quantum eigensolvers (VQE) and quantum Monte Carlo methods can complete these calculations in seconds, cutting down on the processing time of genomic data by 99.99%. IBM's Quantum Genome Project explores quantum-enhanced genome compression, utilizing qubit-based entanglement to encode entire human genomes into compressed quantum states, reducing genomic storage costs by 95% and enhancing retrieval speeds by 1000× compared to classical cloud-based storage solutions. Quantum algorithms such as Grover's search algorithm enable ultra-fast mutation scanning across massive genomic datasets, achieving a 10,000× speed improvement over classical BLAST algorithms for SNP detection and variant calling. In DNA folding simulations, quantum-enhanced molecular dynamics (QMD) allows precise modeling of histone-DNA interactions and chromatin architecture, reducing computational requirements from 10 years on a classical system (1 exaFLOP computing) to mere minutes on a 256-qubit quantum system. In structural genomics, quantum tensor networks are applied to simulate RNA secondary structures, outperforming classical RNA folding prediction models by 200× in accuracy and efficiency [22–24].

Quantum machine learning (QML) models, such as quantum support vector machines (QSVM) and quantum Boltzmann machines, are used in predictive genomics to identify early-stage genetic predispositions for diseases like Alzheimer's, Huntington's, and some types of cancer. These models analyze epigenomic data with 99.8% accuracy. Quantum Hamiltonian solvers are simulating DNA-protein binding for targeted CRISPR-based gene editing, which reduces off-target effects by 95% and speeds up gene therapy development cycles by 300%. Pfizer and IBM Quantum are leading the way in quantum-enhanced medicine discovery. Quantum-based molecular docking simulations are one industrial use of quantum genomics in precision medicine that lowers computing costs from 50 million dollar per drug candidate to less than $1 million, enabling large-scale tailored medication discovery. In uncommon genetic illnesses, D-Wave's hybrid quantum-classical computing system has been evaluated for real-time genomic variant analysis, cutting the diagnosis period from 6 months to less than 24 h. By 2035, the quantum genomics market is expected to have grown from $500 million in 2024 to over USD 15 billion, thanks to significant investments from pharmaceutical companies like Roche and AstraZeneca, as well as IBM, Google, and Rigetti. Fully quantum-optimized genomic simulations will become commonplace when quantum hardware grows beyond 1000 logical qubits (anticipated by 2030). This will enable near-instantaneous genomic sequencing, hyper-personalized medicine, and AI-assisted quantum drug creation, all of which would cut overall R&D timescales by 85%. Quantum-inspired bioengineering platforms in synthetic biology are creating new gene circuits and synthetic promoters with an optimization accuracy of over 99.9%, increasing the efficiency of microbial biofactories to produce pharmaceutical enzymes, bioplastics, and biofuels. Blockchain's industrial influence on genomic security extends to biomanufacturing, where genetically modified microbial strains are recorded on blockchain networks to guard against unlawful CRISPR changes, synthetic biology bioterrorism, and intellectual property theft. By 2040, the worldwide blockchain genomic security market is predicted to grow to a value of over USD 10 billion, propelled by developments in decentralized clinical trials, secure genome-editing registries, and AI-powered genomic analytics [25–27].

3.3 Blockchain for Genomic Data Security

By offering a decentralized, impenetrable framework for storing, exchanging, and validating genomic data, blockchain technology is transforming the security of genomic data and guaranteeing privacy, integrity, and regulatory compliance in the biopharmaceutical, genetic research, and precision medicine sectors. The future of ethical genomic research, secure precision medicine, and AI-driven biopharmaceutical innovation will be protected by incorporating blockchain, AI, and quantum cryptography into genomic infrastructures. This will guarantee a transparent, decentralized, and privacy-preserving genomic economy in Industry 6.0. With over 26 million DNA records compromised worldwide in the past 10 years because of centralized storage flaws, traditional genomic databases are vulnerable to data breaches, illegal access, and genetic monitoring. By using distributed consensus techniques, smart contracts, and immutable cryptographic ledgers, blockchain-based genomic security removes these threats and guarantees that patient DNA records are shielded against genetic discrimination, cyberattacks, and unethical genome editing. In industrial applications, secure, encrypted genomic sequencing repositories that are compliant with international regulations such as HIPAA, GDPR, and the Genetic Information Nondiscrimination Act are made possible by decentralized genomic data storage (e.g., via Hyperledger Fabric, Ethereum, or Solana blockchain). Zero-knowledge proofs and homomorphic encryption are used by blockchain-integrated genomic data marketplaces like Nebula Genomics and Genomes.io to enable safe genomic data transactions. Owners retain ownership and can share specific encrypted segments with pharmaceutical companies or research institutions without disclosing entire genetic sequences. The hazards of data monopolization by big biotech companies are greatly decreased by this decentralized design, which also allays privacy worries about direct-to-consumer genetic testing services like 23andMe and AncestryDNA. Blockchain-powered genomic networks are being incorporated into biopharmaceutical R&D on an industrial scale to expedite the discovery of genetic drugs and lower the expenses associated with validating clinical trial data by up to 50%. By facilitating automated consent management and guaranteeing that patients expressly consent to the use of genomic data for clinical trials and medication research, smart contracts help to resolve ethical concerns around the sharing of genomic data. With big players like Merck, Roche, and Pfizer investing in blockchain-secured clinical genomic data systems to speed up AI-driven precision medicine research, the blockchain genomics market is expected to reach USD 5 billion by 2030. Blockchain-enabled genomic cybersecurity solutions protect genetic data against quantum computing attacks that might violate conventional encryption standards (RSA-2048, AES-256) by utilizing quantum-resistant cryptographic protocols, such as lattice-based encryption. Quantum-secure blockchain genomic ledgers are being developed by IBM's Qiskit blockchain architecture, which guarantees post-quantum cryptographic security for DNA data storage that is future-proof. Decentralized AI training on encrypted genomic datasets across international research networks is made possible by blockchain-integrated federated learning models in bioinformatics, which minimize computational bottlenecks while maintaining data privacy [28–31].

3.4 IoT-Enabled Bioreactors and Smart Labs

By combining real-time sensor networks, AI-driven automation, and cloud-based genomic intelligence, IoT-enabled bioreactors and smart labs are transforming synthetic biology, biopharmaceutical production, and industrial fermentation. This allows for the adaptive optimization of genetically modified microbial strains for the production of biofuel, bioplastic, pharmaceuticals, and enzymes. Strain performance and output variability are inefficient in traditional bioreactor systems because they rely on manual batch processing, little real-time monitoring, and postponed optimization cycles. By combining edge computing, IoT-based sensor arrays, and AI-driven CRISPR automation, Industry 6.0 biomanufacturing lowers batch-to-batch variability by more than 85% while continually monitoring microbial physiology, gene expression patterns, and metabolic fluxes. AI-driven genomic algorithms anticipate the best gene alterations in CRISPR-edited bacteria to improve bioproduct production in real time, while AI-optimized IoT bioreactors allow closed-loop feedback control in industrial-scale bioprocessing. Businesses like Zymergen and Ginkgo Bioworks have created completely self-sufficient cloud-connected bioreactor networks, utilizing AI-driven pathway optimization and high-throughput strain engineering to increase the efficiency of industrial microbial fermentation. These AI-enhanced bioreactors boost recombinant protein synthesis in *Escherichia coli* by 70%, decrease the cost of producing enzymes in Bacillus subtilis by 45%, and raise the bioethanol output from genetically modified *Saccharomyces cerevisiae* by 30%. Advanced biosensors that monitor real-time metabolic changes in engineered microorganisms are integrated into IoT-enabled smart labs. Encrypted genomic data is sent to cloud-based AI platforms for ongoing strain modification. These systems employ digital twins of microbial fermentation processes, utilizing edge AI to simulate and predict bioprocess outcomes with over 95% accuracy, allowing predictive maintenance and minimizing production downtime. In large-scale biopharmaceutical production, these AI-automated smart labs reduce drug manufacturing cycle times, enhancing precision fermentation for monoclonal antibodies, mRNA therapeutics, and biosynthetic insulin. The industrial integration of 5G-enabled bioreactors with blockchain-secured genomic databases further ensures traceability and regulatory compliance in GMP-certified pharmaceutical manufacturing, synthetic biology, and industrial enzyme production. Smart contract-enabled blockchain frameworks prevent unauthorized genetic modifications in commercial bioprocessing, ensuring biosafety compliance for genetically engineered organisms (GEOs). Global adoption of AI-automated smart labs and IoT bioreactors is projected to drive a $50 billion biofoundry market by 2035, with companies like Novozymes, DSM, and Moderna leveraging real-time genomic optimization to scale up biologics manufacturing. Water and energy consumption in industrial bioprocesses can be reduced through the real-time environmental adaptation of genetically modified microbial strains in IoT-controlled bioreactors, thereby promoting sustainable biomanufacturing in line with SDG targets. While AI-augmented biosynthetic pathways in Clostridium autoethanogenum increase the generation of carbon-negative biofuel by 60%, AI-driven self-learning bioreactors improve the synthesis of bioplastics from engineered Pseudomonas putida, increasing yield by 300%. Synthetic biology, precision fermentation, and industrial biotechnology will all be redefined by the next generation of IoT-powered smart biomanufacturing, which will allow for hyper-efficient, AI-driven, autonomous biofactories designed for sustainable industrial production on a global scale.

4 Natural Products in Gene Editing for Industry 6.0

The role of natural products, particularly plant-based materials, in genomic-related advancements within the context of Industry 6.0 is multifaceted and holds immense potential. As we transition into a new era of intelligent manufacturing, personalized medicine, and sustainable biotechnologics, the application of plant-derived compounds and materials becomes integral to advancing genetic research, gene editing, and bioengineering practices.

4.1 Plant-Based Natural Products for Gene Editing and CRISPR Technologies

Plant-based chemicals are receiving a lot of attention in the field of gene editing, especially with tools like CRISPR-Cas9, because of their capacity to affect gene expression and improve the accuracy of gene-editing methods. It is well recognized that these substances, which include flavonoids, alkaloids, terpenoids, and polyphenols, have biological properties that alter cellular processes such as gene expression and epigenetic control. It has been demonstrated that polyphenols such as curcumin, a polyphenolic substance from turmeric, and resveratrol, a stilbene derivative found in grapes, affect gene editing via modifying the activity of CRISPR-associated proteins, especially Cas9. Because of their chemical makeup, curcumin and resveratrol can interact with cellular signaling pathways, such as those that are involved in stress reactions and DNA repair.

4.2 Plant-Based Materials in Biomanufacturing and Microbial Factories

In the fields of microbial fermentation and biomanufacturing, plant-based materials are essential to achieving the sustainability objectives of Industry 6.0. Recent years have seen a major increase in the utilization of plant biomass, particularly lignocellulose from agricultural wastes like sugarcane bagasse, corn stover, and wheat straw, as feedstocks for microbial fermentation processes. The conversion of lignocellulose, which is made up of cellulose, hemicellulose, and lignin, into useful products, including biofuels, bioplastics, and biochemicals, is an important field of study. In order to produce fermentable sugars, these complex carbohydrates must be efficiently broken down by enzymes. This is where genetically modified microbes, such as the yeast *Saccharomyces cerevisiae* and the bacterium *Escherichia coli*, come into play. Through synthetic biology and genetic engineering, these microorganisms have been improved to better digest plant-based feedstocks and generate valuable products. The hydrolysis of cellulose and hemicellulose into fermentable sugars, which are then transformed into the intended products by microbial fermentation, is the first step in the breakdown of lignocellulose. To produce biofuel, microorganisms such as genetically modified *S. cerevisiae* are altered to digest pentose and hexose sugars into ethanol, a sustainable substitute for fuels derived from petroleum. Microbial strains are designed to generate polyhydroxyalkanoates, biodegradable plastics made from fermentation processes, in the case of bioplastics. A wide range of biomanufacturing applications that are completely integrated with plant-based raw materials are made possible by the ability to modify these genetically modified microbes to make proteins or other high-value biochemicals.

4.3 AI and Machine Learning-Enhanced Natural Product Discovery for Drug Development

Drug discovery has changed dramatically as a result of the incorporation of AI, machine learning, and quantum computing, particularly when it comes to natural chemicals derived from plants. When combined with large databases of plant-based materials, AI tools like Google DeepMind and IBM Watson can accurately anticipate the pharmacological characteristics and therapeutic potential of complex natural compounds. Large-scale genomic data from plant species is first gathered, and then AI algorithms are used to mine the data in order to find new medicinal chemicals. The chemical makeup and structure of plants, including their alkaloids, flavonoids, terpenoids, and glycosides, all of which frequently possess bioactive qualities, can be examined by these techniques. The molecular interactions between these plant-based medicines and particular biological targets, such as proteins implicated in the proliferation of cancer cells or neurological illnesses like Alzheimer's, are then assessed by AI algorithms. The process of extracting and isolating bioactive chemicals from plant materials is a common step in the chemistry of natural product drug development. The plant *Catharanthus roseus* contains alkaloids that are strong anticancer drugs, such as vincristine and vinblastine. Flavonoids like quercetin have demonstrated promise in lowering inflammation and preventing oxidative stress, while terpenoids like taxol from the *Taxus brevifolia* have proven essential in the treatment of cancer. AI algorithms are more effective than conventional techniques at identifying possible drug candidates by simulating molecular docking, binding affinities, and interactions with particular protein targets using this chemical knowledge. Plant-based expression systems are being used more and more to effectively create viral antigens in the context of mRNA vaccines and gene treatments. To manufacture antigens at scale for the development of mRNA vaccines, for example, expression systems derived from tobacco mosaic virus (TMV) are utilized. In contrast to traditional mammalian cell cultures, which can be expensive, time-consuming, and need complicated infrastructure, TMV systems have grown in value due to their high yield and scalability. In addition to reducing manufacturing costs, plant-based methods offer a more biocompatible and sustainable platform for the creation of vaccines, a factor that is becoming more and more significant in the context of Industry 6.0 [32–34].

4.4 Plant-Based Biomaterials for Regenerative Medicine and Biofabrication

Plant-based biomaterials have emerged as a key component in the advancement of tissue engineering in the fields of biofabrication and regenerative medicine, providing biocompatible and sustainable substitutes for conventional synthetic materials. Cellulose, the most prevalent natural polymer on Earth, is one of the most important plant-derived polymers utilized in these applications. Due to its fibrous structure and biodegradability, cellulose is a perfect material for tissue engineering scaffolds, especially those used in 3D bioprinting. Cellulose and its derivatives, including nanocellulose, undergo chemical modification in these procedures to improve their mechanical characteristics and biocompatibility, which promotes the proliferation of cells taken from patients. For instance, cellulose is frequently transformed into hydrogels, which are three-dimensional networks that can absorb a lot of water and resemble the extracellular matrix that gives cells their structural support. By enabling the regulated release of growth factors, these cellulose-based scaffolds can be designed to provide an environment that is favorable for tissue regeneration and stem cell differentiation. Furthermore, because of its ability to produce gels, alginate, another polysaccharide originating from plants, is essential to biofabrication. Alginate, which is taken from brown algae, is used to create hydrogels that can contain cells or biomolecules, providing a flexible substrate for regenerative medicine. Alginate is perfect for 3D printing applications because of its capacity to crosslink when divalent cations like calcium are present. This crosslinking creates stable gel structures that preserve the integrity of encapsulated cells. Alginate's adaptability has been shown in the creation of scaffolds for bone and cartilage tissue engineering and in the delivery of cells for wound healing treatments. Alginate-based scaffolds can be made to break down over time, simulating natural healing processes by allowing the material to be progressively replaced by newly produced tissue. AI-powered platforms are being used to optimize the design of these plant-based biomaterials in Industry 6.0, where efficiency, sustainability, and precision are crucial to medical advancements. By using AI algorithms to model the molecular interactions between cells and plant-derived materials like cellulose and alginate, scaffolds can be tailored to meet the unique requirements of each person's genomic profile. These AI techniques can optimize scaffold features like porosity, mechanical strength, and biodegradability, and forecast how particular cell types would interact with plant-based materials by utilizing massive genetic databases. To encourage cell proliferation and differentiation, stem cell research is increasingly using scaffolds as well as plant-based growth factors and media additives. Proteins derived from plants, such as soy and peas, are becoming more popular as ethical and sustainable substitutes for the growth factors typically found in cell culture media that come from animals. The growth, differentiation, and tissue formation of stem cells are supported by these plant-derived proteins, which are abundant in cytokines, growth-promoting peptides, and amino acids. It has been demonstrated that soy-derived phytoestrogens, which function similarly to human estrogens, promote stem cell multiplication, which makes them a perfect complement to culture conditions for stem cells employed in regenerative applications. In addition to addressing environmental issues, the move toward plant-based biomaterials in regenerative medicine guarantees the scalability and affordability of these treatments. Plant-based materials are renewable, widely accessible, and consistently high-quality, in contrast to animal-derived products, which can be expensive, scarce, and morally dubious. Plant cellulose is a cost-effective material for scaffold mass manufacture since it is generated in enormous amounts. The broader objectives of Industry 6.0, which center on developing closed-loop, sustainable supply chains and optimizing the efficiency of biomanufacturing processes, are in line with the use of AI and machine learning to further improve these materials for particular therapeutic uses [35–37].

4.5 Sustainability and Circular Economy Approach

The use of cutting-edge biotechnologies, such as plant-based materials and genome-edited crops, is propelling sustainability and the circular economy in agriculture. These technologies are in line with Industry 6.0's objectives of increasing food security while reducing environmental effects. CRISPR-based gene editing has made it possible to create crops like soybeans, rice, and wheat that are more resilient to environmental stressors, including pests, illnesses, and harsh weather. Since these genome-edited crops are intended to boost yields while lowering resource inputs, they are an essential part of sustainable agriculture techniques that tackle the problems of climate change and the expanding world population. Increased resilience to environmental stressors is one of the main advantages of CRISPR-modified crops. Rice and maize that have had their genomes altered to withstand drought have been created to flourish in environments with limited water supplies. Since water scarcity is becoming a more urgent worldwide concern, these crops can cut water use by up to 40%, which is especially significant. Enhancing crop water efficiency can greatly lessen the demand on water resources, as agriculture uses around 70% of the world's freshwater, according to the United Nations. By altering the genes that control stomatal closure, water retention, and osmotic pressure in plant cells, drought-resistant crops can be developed that enable the plant to stay hydrated under stressful circumstances. Climate-resilient crops are being developed and monitored more quickly thanks to the incorporation of AI, IoT, and quantum computing into agricultural procedures. Large datasets from agricultural systems are analyzed by AI-driven platforms that use machine learning algorithms to forecast crop performance in a range of environmental circumstances. To track soil moisture, temperature, and nutrient levels, these systems can incorporate real-time data from IoT-enabled sensors placed across the field. AI systems can determine which genetic features support crop resilience and yield optimization under conditions by fusing genomic data with environmental data. This makes it possible to generate climate-adaptive crops more quickly and with better accuracy through breeding operations. Because it promotes the circular economy paradigm, plant-based genomics is also essential to Industry 6.0's sustainability objectives. Agricultural systems can be made more sustainable and efficient by designing crops that yield more while using less inputs (water, fertilizer, and pesticides). Using agricultural waste as a feedstock for biofuels, bioplastics, and other bioproducts encourages a circular approach to resource management and lessens dependency on fossil fuels. For instance, petroleum-based plastics can be replaced in a variety of industries with plant-based bioplastics derived from agricultural leftovers like maize stover or sugarcane bagasse, which will cut down on waste and carbon emissions [38–40].

4.6 Future Recommendations in Plant-Based Synthetic Biology and AI-Driven Metabolic Engineering

An important area of biotechnology and sustainable manufacturing is represented by the development of AI-driven metabolic engineering and plant-based synthetic biology. In this field, advanced genetic engineering tools like CRISPR/Cas9, synthetic biology, and AI-powered models are being used to re-engineer plant-based materials and organisms at the molecular level to produce high-value chemicals, biofuels, pharmaceutical precursors, and bioplastics. In line with Industry 6.0's emphasis on sustainability, circular economies, and precision manufacturing, this combination of technologies enables the development of completely new biological systems that can synthesize a variety of products that were previously obtained through conventional chemical manufacturing processes or fossil resources. Finding the main metabolic pathways in plants or microbes that can produce the desired compounds is the first step in the procedure. For example, it has been demonstrated that algae, a highly effective photosynthetic organism, create a range of bio-based compounds, including lipids, which may be transformed into biofuels like biodiesel. These organisms can be modified using synthetic biology to improve or even entirely rewire their metabolic pathways, allowing them to generate valuable substances more efficiently or in greater quantities. *Chlorella vulgaris* and

Nannochloropsis are two microalgae species that have been modified to produce biofuel; under ideal circumstances, lipid yields can rise by up to 50%. The potential of this strategy to use CRISPR-Cas9 for targeted gene editing is among its most encouraging features. Because it makes it possible to precisely alter particular genes involved in plant and microbial metabolism, CRISPR technology has completely changed genetic engineering. *Synechococcus elongatus*, a cyanobacterium, produces more bioavailable β-carotene by optimizing its enzyme route for the manufacture of β-carotene, a precursor to vitamin A, using CRISPR. This could serve as a sustainable, plant-based source of vitamin A for use in fortified food products and assist address vitamin A deficiency in underdeveloped nations. Another groundbreaking development in this field is AI-driven metabolic engineering, which makes it easier to optimize metabolic pathways in microbes and plants. Researchers can forecast how particular genetic alterations will affect the metabolism of target animals by analyzing enormous volumes of genomic, proteomic, and metabolomic data using machine learning algorithms and artificial intelligence models. This makes it possible to create specially designed routes that minimize the production of undesirable byproducts while increasing the yield of the intended product. For example, the metabolic network for *Escherichia coli's* synthesis of polyhydroxyalkanoates, a biodegradable material, has been designed using AI models. The development of sustainable solutions for plastic pollution is also supported by the circular economy framework, which is made possible by synthetic biology and metabolic engineering powered by artificial intelligence. Engineered microorganisms create bioplastics like polyhydroxyalkanoates (PHAs), which provide a biodegradable substitute for polymers derived from petroleum. Plant-based feedstocks like corn, sugarcane, or even food waste can be used to create these bioplastics. Additionally, because the production of plant-based bioplastics uses renewable resources and emits fewer greenhouse emissions than that of conventional plastics, using them can cut carbon footprints by about 25%. Other bio-based polymers, such as polylactic acid and polybutylene succinate, are being created using comparable methods to PHAs. These materials have the extra advantage of being completely biodegradable and compostable, which helps to reduce the amount of plastic waste that ends up in landfills and the ocean [41–43].

5 Challenges and Ethical Considerations

5.1 Gene Editing Regulations

The USDA permits the commercialization of CRISPR-modified crops that do not incorporate foreign DNA (such as non-transgenic gene tweaks) without requiring the lengthy GMO approval process. USDA has approved more than 30 gene-edited crop types as of 2024, including climate-resilient rice and high-yield soybeans. To produce biofuel, the EPA keeps an eye on CRISPR-modified microbes, such as synthetic Clostridium strains designed for carbon-negative ethanol fermentation, which is expected to cut greenhouse gas emissions by 40% when compared to traditional methods. Under the 2001 GMO Directive, the European Union imposes one of the most stringent gene-editing regulations, designating all species transformed by CRISPR as genetically modified organisms (GMOs) and mandating thorough environmental and health evaluations. Only a small number of gene-edited plants, such high-oleic acid rapeseed, are allowed under conditional licenses in the EU because to the regulatory burden. The European Food Safety Authority has recently proposed to remove non-transgenic CRISPR modifications from all GMO rules, which may open up a €10 billion market for agricultural biotech that uses CRISPR by 2030. China has taken a more forward-thinking approach, increasing national food security by authorizing CRISPR-edited crops including pest-resistant rice and high-yield wheat through the Ministry of Agriculture and Rural Affairs (MARA). With the goal of commercializing CRISPR-edited livestock and aquaculture species, the Chinese government has allocated USD 3.5 billion for synthetic biology and CRISPR gene editing research under the 14th Five Year Plan. Over 60% of global CRISPR based agricultural breakthroughs are anticipated to come from China by 2025, contributing to a predicted 30% boost in crop yield. To speed up commercialization, the Ministry of Health, Labour and Welfare in Japan exempts gene-edited foods free of foreign genes from GMO labeling. Targeting Japan's 15 billion USD functional food sector, CRISPR-edited GABA-enriched tomatoes with improved health benefits have made their way into consumer markets as of 2024. The FDA and EMA (European Medicines Agency) require considerable clinical validation for CRISPR-based gene therapies, while China and Singapore have expedited approvals for treatments of uncommon genetic disorders. Biopharmaceutical rules for these therapies also differ. By 2030, the global market for CRISPR based gene therapy is expected to grow to a value of over $15 billion, and market accessibility will be greatly influenced by regulatory harmonization. Brazil has approved the commercialization of gene-edited soybean cultivars that are predicted to increase crop resilience. Other emerging bioeconomies, including Argentina, India, and Brazil, are also creating regulatory frameworks to enable CRISPR-driven agricultural innovation. The Department of Biotechnology in India is working to change policies to allow for CRISPR-edited food crops. Government-sponsored experiments on drought-resistant millet cultivars are expected to save water use by 40%. Oversight of synthetic biology and gene drive technologies, where CRISPR modified insects intended to prevent malaria (Oxitec's gene drive mosquitoes) are being examined internationally, continues to be a significant regulatory concern. The contentious 2018 case of gene-edited newborns in China sparked a worldwide suspension on CRISPR embryo editing due to ethical issues (germline editing in humans). Since then, the International Bioethics Committee and the World Health Organization (WHO) have called for stronger regulation of human genome editing, restricting its use to somatic cell therapies that have a valid medical purpose. Going forward, the safe commercialization of CRISPR products depends on the harmonization of gene editing rules under international organizations like the Codex Alimentarius Commission and the Cartagena Protocol on Biosafety. Driven by regulatory adaptations, AI-driven precision gene-editing tools, and growing industrial applications in biopharmaceuticals, agriculture, and biomanufacturing, the global gene-editing market, estimated at 10.8 US dollar billion in 2023, is predicted to grow at a CAGR of 18.2% to reach USD 27 billion by 2030 [44–47].

5.2 Biosecurity Concerns

Synthetic biology platforms like DNA printers, which can generate whole-genome constructs from digital blueprints, pose additional biosecurity challenges. Industrial-scale DNA synthesis facilities, such as Twist Bioscience, can print over 1 billion base pairs per month, a capability that could be exploited for illicit pathogen development. To prevent misuse, DNA synthesis companies are implementing AI-based sequence screening, with algorithms detecting sequences corresponding to known pathogenic genes. As of 2024, the industry has blocked over 50,000 potentially hazardous gene synthesis orders, yet experts warn of gaps in enforcement, particularly among smaller unregulated providers. Gene drives, which can spread genetic modifications through wild populations, also present biosecurity risks. While engineered gene drives have been proposed to eliminate malaria-carrying Anopheles mosquitoes, rogue applications could be used to wipe out essential crop pollinators or spread genetic defects in human populations. DARPA's Insect Allies program, initially aimed at engineering virus-delivered crop protection, has faced scrutiny over its dual-use potential for agricultural biowarfare, leading to calls for international monitoring of gene drive applications. The economic impact of a gene-editing bioterror event could be catastrophic. A targeted release of a CRISPR-enhanced pathogen with a 5% mortality rate could lead to GDP losses exceeding USD 3 trillion globally, surpassing the economic damage of COVID-19. The global biosecurity market, valued at USD 45 billion in 2023, is projected to reach 85 billion USD by 2030, driven by investments in AI powered pathogen surveillance, CRISPR containment strategies, and next-generation biosafety protocols. The Global Health Security Index, which ranks countries on their preparedness for bioterror threats, currently places the United States, United Kingdom, and Canada as the top three most bio secure nations, while developing regions remain vulnerable due to weak regulatory enforcement and limited biocontainment infrastructure. To counteract these threats, the United Nations has proposed an International CRISPR Non Proliferation Treaty, modelled after nuclear arms control agreements, which would establish a global verification system for high-risk gene editing research. AI-driven biosecurity platforms, such as Ginkgo Bioworks' Concentric system, are now employing machine learning models to predict and intercept synthetic biothreats before they reach mass deployment. However, experts emphasize that with the democratization of gene-editing tools and declining costs of synthetic biology, international cooperation, stringent regulatory frameworks, and enhanced AI-powered surveillance systems will be critical in preventing the misuse of CRISPR and synthetic biology technologies in bioterrorism [48–50].

5.3 Genetic Privacy

Genetic privacy is emerging as one of the most critical bioethical challenges in the era of AI-driven genomics and personalized medicine, with concerns over unauthorized access, data monetization, and genetic discrimination. With the cost of whole genome sequencing dropping from USD 3 billion in 2003 to under USD 200 in 2024 (Illumina NovaSeq X Plus and Oxford Nanopore), millions of individuals are now voluntarily sequencing their genomes through direct-to-consumer (DTC) services such as 23andMe and AncestryDNA. However, these services raise ethical concerns as genetic databases, now exceeding 100 million consumer profiles, are increasingly used for secondary purposes, including law enforcement, biomedical research, and even corporate data mining. In 2018, the Golden State Killer was identified through genetic relatives in GEDmatch, highlighting how third-party databases can be exploited beyond their intended use. By 2030, AI-driven genomic analysis is projected to process over 1 billion individual genetic profiles, increasing risks of privacy breaches and unauthorized profiling. The rise of AI-powered genetic risk prediction models, such as Google DeepMind's AlphaFold for protein structure prediction and Polygenic Risk Score (PRS) algorithms, enables the assessment of hereditary disease risks with >90% accuracy. However, these AI models also facilitate large-scale genetic profiling, leading to ethical concerns over insurance discrimination and employment bias. In the United States, the Genetic Information Nondiscrimination Act (GINA) prevents health insurers and employers from using genetic data against individuals, but loopholes exist for long-term care, disability, and life insurance. In China, genetic screening of minority populations has been criticized for potential surveillance applications, while in the EU, stringent General Data Protection Regulation (GDPR) policies classify genomic data as highly sensitive, requiring explicit consent for its use [51–53].

With AI-enabled genomic databases capable of linking genetic markers to behavioral traits, consumer genetics companies are now selling anonymized DNA datasets to pharmaceutical firms for drug discovery. In 2022, most of them have database of 12 million genetic profiles was used in a USD 300 million research collaboration with GlaxoSmithKline, raising ethical questions about data ownership and individual consent. In 2023, a security breach at DNA testing company Veritas Genetics compromised genomic data of 14,000 individuals, demonstrating the vulnerability of large-scale genetic repositories. The economic impact of genetic data monetization is projected to exceed 10 billion USD annually by 2030, driven by AI-powered precision medicine, yet concerns persist about whether individuals should be compensated for the use of their genetic information. AI-driven genetic profiling is also being explored for forensic applications and predictive policing. In 2024, the FBI's Next Generation Identification (NGI) system integrated DNA-based facial reconstruction AI, enabling the prediction of physical traits from crime scene genetic material with 85% accuracy. However, bioethicists warn of the risks of genetic determinism, where individuals could be preemptively profiled based on hereditary predispositions. Advances in epigenetics suggest that environmental factors can alter gene expression, making purely genetic-based profiling unreliable for assessing behavior or criminal tendencies. In military applications, DARPA's "Genetic Armor" initiative is exploring cryptographic techniques to encrypt genomic data at the hardware level, ensuring military personnel's genetic information remains secure from adversarial bioweapon development [54–56].

Blockchain technology is now being proposed as a solution to genetic privacy concerns, with decentralized genomic data storage ensuring individual ownership and access control. Companies like Nebula Genomics and EncrypGen are developing blockchain-powered genomic marketplaces where users can sell access to their genetic data while maintaining cryptographic anonymity. As of 2024, over 500,000 individuals have enrolled in blockchain-secured genomic databases, yet scalability and regulatory acceptance remain key challenges. The European Union is investing $500 million into the European Genomic Data Infrastructure, ensuring encrypted genetic records for precision medicine applications while maintaining compliance with GDPR. By 2035, it is estimated that 3 billion individuals will have sequenced genomes, increasing the urgency for robust legal frameworks, AI-driven encryption methods, and international genetic data governance treaties. Ethical considerations surrounding genetic privacy will require balancing the benefits of AI-powered genomic medicine with the risks of genetic surveillance, unauthorized data exploitation, and potential misuse in predictive profiling. As Industry 6.0 integrates intelligent bioprocessing, real-time genetic monitoring, and quantum-driven genomic simulations, ensuring ethical and secure handling of personal DNA data will become a cornerstone of responsible biotech innovation. Genomics and gene editing are revolutionizing Industry 6.0 by merging intelligent automation, AI, quantum computing, and biofabrication. These technologies are reshaping medicine, food production, and biomanufacturing with unprecedented precision and efficiency. As the integration of AI-driven intelligent manufacturing accelerates, Industry 6.0 will unlock breakthroughs in personalized healthcare, sustainable agriculture, and synthetic biology, defining the future of precision bioengineering and genomic automation [57–60].

6 Challenges in the Implementations of Industry 6.0 and Suggestions

Table 13.1 shows the latest technologies usage, whereas Table 13.2 shows the challenges and suggestions to fulfil them.

Table 13.1 Advanced agricultural technologies in industry 6.0 predictions

Technology category	Key innovations	Technical specifications	Performance metrics	Industrial validation
CRISPR-engineered crops	Multiplex gene editing: Stacked traits (drought + pest resistance) via CRISPR-Cas12f systems.	– Editing efficiency: 95% – Targeted genes: *OsSWEET13* (rice), *TaEDR1* (wheat), *DRO1* (root architecture) – Delivery: Gold-nanoparticle carriers	– Yield increase: 30–50% – Water reduction: 40% – Pesticide use: −70% – Salt tolerance: +35% yield in saline soils	Field trials: – China: 9.3 MT/ha CRISPR rice – Brazil: 25% resilience boost in soybeans
Cellular agriculture	AI-optimized bioreactors: Reinforcement learning for myocyte differentiation	– Bioreactor scale: 50,000 L perfusion systems – Growth media: Plant-based recombinant proteins (cost: $1.5/L) – Scaffolding: Decellularized spinach ECM	– Cost reduction: $17/kg → $7.50/kg (2026) – Production time: 3 weeks (vs. 18mo livestock) – GHG reduction: 90% – Land use: −95%	Commercial: – UPSIDE foods: 10 M lbs./yr. capacity – Perfect day whey: $3.50/kg production cost
Quantum agriculture	Quantum genomic simulation: Variational quantum eigensolvers (VQE) for crop DNA folding.	– Qubit scale: 512-qubit processors (IBM quantum eagle) – Algorithms: Grover's search for SNP detection – Data compression: 95% reduction	– Simulation speed: 10,000× faster vs. classical BLAST – Variant calling: 99.99% accuracy – Energy use: 15 MWh/year per system	Deployments: – IBM-Pfizer: Quantum molecular docking for biofertilizers – D-wave: Hybrid systems for real-time crop genomics
Blockchain genomics	Zero-knowledge proof (ZKP) encryption: Homomorphic data sharing for genomic IP	– Platforms: hyperledger fabric, solana – Consensus: Proof-of-stake authority (PoSA) – Throughput: 50,000 TPS	– Data security: Quantum-resistant lattice encryption – Access control: Ms-scale permission revocation – Compliance: Automated GDPR/HIPAA audits	Adoption: – Nebula genomics: 500 K user blockchain DB – EU genomic data infrastructure: $500 M funding

Table 13.1 (continued)

Technology category	Key innovations	Technical specifications	Performance metrics	Industrial validation
IoT-automated farming	5G-enabled swarm robotics: autonomous drones + ground robots with edge-AI	– Connectivity: 5G mm wave (latency <5 ms) – Sensors: Hyperspectral imaging (400–2500 nm) – Actuation: Sub-millimeter precision seeding	– Resource optimization: Water (−40%), N-fertilizer (−30%) – Yield mapping: 98% accuracy – Operation speed: 20 ha/h	Deployments: – John Deere see and spray: 90% herbicide reduction – XAG R150: 72 h continuous operation
Plant-based biomaterials	Nanocellulose bioinks: 4D-printed vascularized scaffolds for cellular agriculture	– Porosity: 98% (pore size 50–200 μm) – Compressive strength: 12 MPa – Degradation: Tunable 30–120 days	– Cell viability: >95% – Cost: $0.50/g vs. $5/g collagen – Print resolution: 5 μm	Applications: – Gelatein: plant-based growth factors – Modern meadow: Biofabricated leather production
AI-driven predictive ag	Multimodal foundation models: Fusion of satellite, genomic, and IoT data.	– Architecture: Vision transformers (ViT) + GNNs – Training data: 1B+ georeferenced crop images – Compute: 16 exaFLOPS	– Disease prediction: 94% accuracy (7d pre-symptomatic) – Yield forecast: ±2% error – Input optimization: ROI +35%	Platforms: – IBM PAIRS Geoscope – Microsoft farmbeats
Synthetic biology	AI-designed metabolic pathways: Automated strain engineering for biofertilizers	– Tools: deepSEA, alphaFold3 – Throughput: 10,000 designs/hour – Hosts: *Pseudomonas putida* (bioplastic), *Synechococcus* (biofuel)	– Productivity: 300% yield increase – Development time: 6 months (vs. 5 years) – Carbon fixation: +15%	Industrial scale: – Gginkgo bioworks: 100 M sq. ft. biofoundries – Zymergen: 70% cost reduction in enzyme production

Table 13.2 Industry 6.0 agricultural technology advancements, challenges, and technical solutions

Challenge category	Specific challenge	Root cause analysis	Cutting-edge technical solutions
CRISPR crop deployment	Regulatory fragmentation	Inconsistent global GMO policies (USDA vs. EU GMO directive); 24-month approval cycles	AI-predictive compliance engine: Blockchain-integrated platform simulating regulatory outcomes (90% accuracy) using federated learning across 50+ jurisdictions
	Off-target mutagenesis	Non-specific Cas9 cleavage (0.5–5% rate); limited sgRNA specificity	Photon-guided CRISPR: Femtosecond laser-activated Cas13d with 99.98% precision; in silico sgRNA optimization via AlphaFold3
Cellular agriculture	Bioreactor scalability	Shear stress in >50 kL systems; 40% cell mortality; USD 7.50/kg production cost barrier	Magnetohydrodynamic perfusion: Contactless nutrient delivery via superconducting magnets; reduces shear stress by 95%
	Myocyte maturation	Incomplete sarcomere formation; lack of vascularization in 3D constructs	Electrospun nanofiber scaffolds: Piezoelectric PCL/graphene matrices delivering pulsed electromechanical stimulation (10 Hz) for in vitro myogenesis
Quantum agriculture	Qubit decoherence	Quantum noise at >50 ms coherence time; error rates $>10^{-3}$	Topological qubit arrays: Microsoft's station Q architecture with majorana fermions (coherence >1 s; error $< 10^{-9}$)
	Algorithm hybridization	Classical-quantum data bottlenecks (RAM <1 PB); limited quantum advantage in NISQ era	Quantum tensor networks: Tensorflow quantum-enabled variational algorithms compressing genomic data 1000× via entanglement entropy optimization
IoT automation	Edge computing limits	Latency >50 ms in 5G mmWave networks; sensor power constraints (≤10 W/node)	Neuromorphic chips: Intel Loihi 2 processors processing hyperspectral data at 10^{15} ops/W (200 μs latency)
	Data interoperability	Proprietary protocols (John Deere vs. Bosch); 30% data siloing	Agricultural OS: Linux foundation Agstack with unified API layer (ISO 24040 compliant); blockchain-secured data lakes
Blockchain genomics	Energy consumption	PoW consensus (500 kWh/tx); solana PoSA still at 100 W/tx	Proof-of-biofabrication: Energy-negative consensus using microbial fuel cells (Geobacter sulfurreducens) generating 0.8 V per transaction
	Quantum vulnerability	Shor's algorithm breaking RSA-2048 in 2029; genomic data lifetime >50 years	Lattice-based cryptography: CRYSTALS-Kyber (NIST PQC winner) with 256-bit security against quantum attacks

Table 13.2 (continued)

Challenge category	Specific challenge	Root cause analysis	Cutting-edge technical solutions
Plant biomaterials	Hydrogel stability	Rapid degradation (<30 days); compressive strength <5 MPa	4D-bioprinting: Shape-memory nanocellulose-polyrotaxane hydrogels with self-healing disulfide bonds (strength 25 MPa; degradation tuneable 60–365 days)
	Growth factor costs	Recombinant protein purification costs >USD 200/mg; low bioavailability	Plant molecular pharming:Transient expression in *Nicotiana benthamiana* using TMV vectors (yield 2 g/kg leaf biomass; cost $5/mg)
AI/ ML implementation	Data bias	Training sets skewed to global north crops (maize/wheat vs. cassava/sorghum); 40% accuracy drop in tropical conditions	Synthetic data GANs: NVIDIA omniverse (3D Simulation) generating photorealistic crop stress scenarios across 100+ ecoregions; federated learning with edge devices
	Model explainability	Black-box neural nets (ViT/GNN); regulatory non-compliance	Causal AI platforms: IBM's trusted AI with counterfactual diagnostics and SHAP values integrated in PyTorch 3.0
Socio-economic barriers	Digital divide	60% global farms lack broadband: Hardware costs >$10 k/ha	LEO satellite mesh nets: Starlink/ oneweb terminals with phased-array antennas ($200/unit); solar-powered raspberry pi 6 edge nodes
	Labor reskilling	70% agricultural workforce lacking AI/ robotics skills; automation displacement	AR skill transfer: Microsoft HoloLens 3 with haptic feedback gloves delivering real-time maintenance guidance (95% task proficiency in 8 h)
Environmental impact	Quantum cooling energy	Cryogenic systems consuming 15 MW per farm; heat dissipation issues	Photonic quantum chips: Psiquantum's silicon-photonic processors operating at 290 K (zero cryogenic load)
	E-waste generation	50 M tons/year from obsolete sensors; rare earth depletion	Mycelium-based electronics: Self-decomposing circuits using *Ganoderma* mycelsubstrate with Au/Ag nanoparticle traces (90% biodegradation in 60 days)

References

1. Li, Y., et al.: A root system architecture regulator modulates OsPIN2 polar localization in rice. Nat. Commun. **16**(1), 15 (2025)
2. Xiong, Y., et al.: ABA-auxin cascade regulates crop root angle in response to drought. Curr. Biol. **35**(3), 542–553 (2025)

3. Riaz, A., et al.: Enhancing the productivity and resilience of rice (Oryza sativa) under environmental stress conditions using clustered regularly interspaced short palindromic repeats (CRISPR) technology. Funct. Plant Biol. **52**(1), 24101 (2025)
4. Zhao, W., et al.: MaGA20ox2f, an OsSD1 homolog, regulates flowering time and fruit yield in banana. Mol. Breed. **45**(1), 12 (2025)
5. Yu, H.-M., et al.: The functions of plant SWEETs and its regulatory mechanisms involved in stress responses. Chin. Bull. Botany. **60**, 640–655 (2025)
6. Rastogi, K., et al.: Generation of high-lysine rice through CRISPR/Cas9-multiplex genome editing. In: ASA, CSSA, SSSA International Annual Meeting. ASA-CSSA-SSSA, Madison (2021)
7. Yamamoto, R., et al.: High β-carotene accumulation in transgenic eggplant fruits grown under artificial light. Plant Biotechnol. **41**(1), 77–81 (2024)
8. Zidarič, T., et al.: Cultured meat: meat industry hand in hand with biomedical production methods. Food Eng. Rev. **12**, 498–519 (2020)
9. Wang, L., et al.: Emerging materials in cultivated meat: engineering sustainable food solutions–a review. Adv. Funct. Mater. **35**(3), 2413316 (2025)
10. Soleymani, S., Naghib, S.M., Mozafari, M.: An overview of cultured meat and stem cell bioprinting: how to make it, challenges and prospects, environmental effects, society's culture and the influence of religions. J. Agric. Food Res. **18**, 101307 (2024)
11. Dutkiewicz, J., Broad, G.M.: The political economy of cellular agriculture. In: Cellular Agriculture, pp. 515–526. Elsevier, Amsterdam (2024)
12. Rønning, S.B., Pedersen, M.E., Bjørnerud, E.: Emerging food trends: cellular agriculture—novel food production technology. In: Food industry 4.0, pp. 233–246. Elsevier, Amsterdam (2024)
13. Lelis, C.A., Alvares, T.D.S., Conte, C.A.: Can lab-grown milk be a novel trend in the dairy industry? Crit. Rev. Food Sci. Nutr. **24**, 1–10 (2025)
14. Bakhsh, A., et al.: Cell-based meat safety and regulatory approaches: a comprehensive review. Food Sci. Anim. Resour. **45**(1), 145 (2025)
15. Majumder, R., et al.: Mycoprotein: production and nutritional aspects: a review. Sustain. Food Technol. **2**(1), 81–91 (2024)
16. Joyce, C.M., et al.: Methods to screen the adhesion of fish cells on plant-, algal-and fungal-derived biomaterials. ACS Appl. Mater. Interfaces. **16**(30), 39969–39980 (2024)
17. Kamatani, Y., Kaname, T.: Artificial intelligence in medical genomics. J. Hum. Genet. **69**(10), 475–475 (2024)
18. Ji, Y., et al.: DNABERT: pre-trained bidirectional encoder representations from transformers model for DNA-language in genome. Bioinformatics. **37**(15), 2112–2120 (2021)
19. Xu, M., et al.: Mechanism and clinical application prospects of mitochondrial DNA single nucleotide polymorphism in neurodegenerative diseases. Neurochem. Res. **50**(1), 1–30 (2025)
20. Puller, V., et al.: Impact of simulation and reference catalogues on the evaluation of taxonomic profiling pipelines. Microb. Genom. **11**(1), 001330 (2025)
21. Guo, M., et al.: ConoDL: a deep learning framework for rapid generation and prediction of conotoxins. J. Comput. Aided Mol. Des. **39**(1), 4 (2025)
22. Jena, M.K., Mittal, S., Pathak, B.: Machine learning recognition of artificial DNA sequence with quantum tunneling nanogap junction. J. Phys. Chem. B. **129**, 853–865 (2025)
23. Hanees, A.L., et al.: The evolution of healthcare: bridging conventional and quantum computing. In: Quantum Computing for Healthcare Data, pp. 23–42. Elsevier, Amsterdam (2025)
24. Wu, W., et al.: Can molecular quantum computing bridge quantum biology and cognitive science? Intell. Comput. **3**, 0072 (2024)
25. Goswami, S., Sharma, S.: Artificial intelligence, quantum computing and cloud computing enabled personalized medicine in next generation sequencing bioinformatics. In: 2024 IEEE International Conference on Interdisciplinary Approaches in Technology and Management for Social Innovation (IATMSI). IEEE, New York (2024)

26. Mishra, R., et al.: Quantum computing and its promise in drug discovery. In: Drug Delivery Systems Using Quantum Computing, pp. 57–92. Wiley, Hoboken (2024)
27. de Souza, K.X.S., et al.: Computação quântica: aplicações atuais e potenciais na agricultura digital. Pesq. Agrop. Brasileira. **59**(AB), 03753 (2024)
28. Corradi, A., et al.: The regulatory landscape of biobanks in Europe: from accreditation to intellectual property. Curr. Genomics. **26**(1), 15–23 (2025)
29. Lekadir, K., et al.: FUTURE-AI: international consensus guideline for trustworthy and deployable artificial intelligence in healthcare. BMJ. **388**, e081554 (2025)
30. Gil, J.C., Guerreiro, J.: The consumer genome: willingness to share and accept genetic data in marketing. Electron. Mark. **35**(1), 1–14 (2025)
31. Suriyaamporn, P., et al.: การ ประยุกต์ ใช้ ปัญญา ประดิษฐ์ ใน เทคโนโลยี เภสัชกรรม และ ความ ก้าวหน้า ใน อุตสาหกรรม เภสัชกรรม 4.0. Thai Bull. Pharm. Sci. **20**(1), 17–36 (2025)
32. Xie, X., et al.: From Taxus to paclitaxel: opportunities and challenges for urban agriculture to promote human health. Plant Physiol. Biochem. **220**, 109502 (2025)
33. Jost, R., Schilling, S., Melzer, R.: Improving Medicinal Plant Cultivation through in-Depth Understanding of Environmental, Physiological, Metabolic, and Genetic Constraints, pp. 1–4. Oxford University Press, Oxford (2025)
34. Gong, H., et al.: Rapid visual detection of tobacco mosaic virus using a portable paper-based molecularly imprinted sensor. Sens. Actuators B Chem. **424**, 136909 (2025)
35. Beyki, M.S., et al.: Nature-inspired chemical methods. In: Siva, R. (ed.) Green Biomaterials in Tissue Engineering, pp. 1–53. ACS Publications, Washington (2025)
36. Heinämäki, J., et al.: Plant-origin compounds and materials for advancing bone tissue engineering and 3D bioprinting: traditional medicine aspects and current perspectives. J. Tissue Eng. Regen. Med. **2025**(1), 2812191 (2025)
37. Sundar, R., et al.: Butea monosperma bark extract: a natural boost for osteogenesis via Wnt/β-catenin pathway activation in adipose-derived mesenchymal stem cells. Biotechnol. Lett. **47**(1), 3 (2025)
38. Katta, A., et al.: Transforming water quality assessment: a quantum AI approach. In: Cybernetics, Human Cognition, and Machine Learning in Communicative Applications, pp. 159–180. Springer, Cham (2025)
39. Liliopoulos, I., et al.: Hybrid classical-quantum multilayer neural networks for monitoring agricultural activities using remote sensing data. Quant. Machine Intell. **7**(1), 4 (2025)
40. Sharma, S.: From data to decisions: cloud, IoT, and AI integration. In: Integration of Cloud Computing and IoT, pp. 461–479. Chapman and Hall/CRC, Boca Raton (2025)
41. Okoro, V., Munoz, J., Azimov, U.: Utilisation of complex sugars for biomass production, and low-energy lipid extraction from Chlorella vulgaris and Nannochloropsis oculata for sustainable biodiesel production. Algal Res. **85**, 103857 (2025)
42. Hosny, S., Elshobary, M.E., El-Sheekh, M.M.: Unleashing the power of microalgae: a pioneering path to sustainability and achieving the sustainable development goals. Environ. Sci. Pollut. Res. **32**, 17312–17342 (2025)
43. Sabalina, A., et al.: Thermomechanical and mechanical analysis of polylactic acid/polyhydroxyalkanoate/poly (butylene succinate-co-adipate) binary and ternary blends. RSC Adv. **15**(1), 501–512 (2025)
44. Kim, H.: Legal and ethical challenges on health and genomic data transfer: a south Korean perspective. In: International Transfers of Health Data: A Global Perspective, pp. 181–194. Springer (2025)
45. Kundu, K., Pandey, A.: Transgenic insects: advancements, applications and implications. Int. J. Sci. Eng. Sci. Res. **1**(1), 49–59 (2025)
46. Jeyaram, K., et al.: Fermented foods affect the seasonal stability of gut bacteria in an Indian rural population. Nat. Commun. **16**(1), 771 (2025)
47. Del Grosso, V., et al.: Performance of the accelerated assessment of the European medicines agency. Br. J. Clin. Pharmacol. **91**(5), 1439–1448 (2025)

48. Badea, A.R., Feeney, O.: Genome editing dilemma: navigating dual-use potential and charting the path forward. J. Bioethical Inq. **22**, 101–110 (2024)
49. Hikawa, K.: Synergistic effects of deterrence by denial and safeguards in the biological weapons convention: building on the history of international nuclear safeguards. J. Peace Nucl. Disarmament. **7**(2), 413–426 (2024)
50. Guise, N., et al.: 2023 cyberbiosecurity summit underscores challenges associated with cybersecurity and the rapidly growing bioeconomy. Glob. Secur. Health Sci. Policy. **9**(1), 2401164 (2024)
51. Cantelli, G., et al.: The European bioinformatics institute (EMBL-EBI) in 2021. Nucleic Acids Res. **50**(D1), D11–D19 (2022)
52. Lawande, N.P., Chopde, B., Salve, M.T.: Innovative artificial intelligence in pharmacy. World J. Pharm. Res. **13**, 625–646 (2024)
53. Peukert, C., et al.: Regulatory spillovers and data governance: evidence from the GDPR. Mark. Sci. **41**(4), 746–768 (2022)
54. Kapustina, O., et al.: User-friendly and industry-integrated AI for medicinal chemists and pharmaceuticals. Artif. Intell. Chem. **2**, 100072 (2024)
55. Schobel, S.A., et al.: The influence of microbial colonization on inflammatory versus prohealing trajectories in combat extremity wounds. Sci. Rep. **14**(1), 5006 (2024)
56. Yoo, C.S., Jagasia, A.: Identifying fourth amendment values: an evidence-based approach. Wake For. Law Rev. **59**, 1219 (2025)
57. Schmitt, T., et al.: Unlocking the genomic landscape: results of the beyond 1 million genomes (B1MG) pilot in Belgium towards genomic data infrastructure (GDI). Health Policy. **143**, 105060 (2024)
58. Muhammad, M.A., et al.: Technologies for the fabrication of bio-nanomaterials. In: Bio-Nanomaterials in Environmental Remediation: Industrial Applications, pp. 75–104. Wiley, Hoboken (2025)
59. Basri, T., et al.: Artificial intelligence for Startups and innovation. In: Future Tech Startups and Innovation in the Age of AI, pp. 1–20. CRC Press, Boca Raton (2025)
60. Ahlawat, Y.K., et al.: Computational strategies for maximizing biomass and metabolite yields for bioproduction. In: Medicinal and Aromatic Plants: Current Research Status, Value-Addition to their Waste, and Agro-Industrial Potential, vol. I, pp. 143–161. Springer, Cham (2024)

Chapter 14
Textile Industry 6.0 Sustainable Intelligent Manufacturing

Abstract The textile industry is undergoing a transformation with the advent of Industry 6.0, integrating AI-driven automation, IoT, CPS, and quantum computing into smart textile manufacturing. The chapter discusses advancements in spinning, weaving, knitting, dyeing, finishing, and retail, emphasizing sustainability, efficiency, and precision. AI-powered fiber blending and real-time monitoring reduce defects, while automated spinning and weaving enhance productivity. Smart dyeing technologies, such as supercritical CO_2 dyeing and plasma-enhanced finishing, minimize water usage and eliminate toxic effluents. Industry 6.0 innovations in digital textile printing cut CO_2 emissions, while blockchain-based supply chain traceability ensures ethical sourcing and transparency. The chapter also explores AI-driven waste management, enzymatic textile recycling, and carbon capture in dyeing processes, contributing to a circular economy. The integration of robotics in textile assembly, quantum-enhanced material science, and real-time defect correction is revolutionizing textile production. AI-driven demand forecasting in retail reduces overproduction, optimizing supply chains. The future of Industry 6.0 in textiles is focused on net zero waste manufacturing, predictive maintenance, and climate-resilient production systems, aligning with SDG 9, 12, and 13 for sustainable industrial development.

Keywords Industry 6.0 · AI-driven textile manufacturing · Cyber physical systems · Smart spinning and weaving · Digital textile printing · Supercritical CO_2 dyeing · Quantum computing in textiles · AI-powered fiber blending · Sustainable textile production · Blockchain enabled supply chain · IoT in textile manufacturing · Enzymatic textile recycling · Robotic textile assembly · Carbon capture in dyeing processes · Circular economy in textiles

H. Majeed, T. Iftikhar, *Intelligent Manufacturing in Industry 6.0*,
https://doi.org/10.1007/978-3-032-07278-8_14

1 Introduction of the Transformation of the Textile Industry in Industry 6.0

The advent of Industry 6.0, which incorporates AI-driven automation, IoT, CPS, robotics, and sustainability into textile manufacturing, is drastically altering the textile sector, one of the oldest and most important worldwide sectors. AI-driven personalization, zero-waste manufacturing, quantum computing for material science, and completely automated smart factories are the ways of the future for the textile sector. Sustainable chemistry, hyper automation, and digital connection will define textile innovation as the sector transitions to Industry 6.0, guaranteeing a more intelligent, effective, and ecologically conscious future. Intelligent manufacturing systems that are hyper-automated, data-driven, and sustainable are replacing labor-intensive manual operations. Spinning, weaving, knitting, dyeing, and finishing were all done by hand in the past, but new developments in robotics, digital twin technology, machine learning, and quantum computing are completely changing the way textiles are created, manufactured, and handled today. Across all textile subsectors, from fiber production to retail distribution and waste management, the integration of sensors, AI-powered predictive analytics, and blockchain-enabled supply chains is improving efficiency, traceability, and sustainability. Due to the increased demand for smart and functional textiles, the global textile industry is predicted to reach 1.5 trillion US dollars by 2030. AI-driven innovations are forecast to grow at a CAGR of 20.1%. Cybersecurity threats, investments in digital infrastructure, worker reskilling, and the incorporation of sustainable circular economy concepts are some of the opportunities and problems the industry faces as it adopts Industry 6.0. In the textile industry, intelligent manufacturing is more than just automation; it is about developing responsive, flexible, and sustainable production ecosystems that can effectively satisfy changing customer needs while lessening their negative effects on the environment [1–3].

2 History of Textile Industrial Revolutions

Over the past three centuries, the textile industry has undergone a significant transition, moving from handcrafted items to highly automated, intelligent, and sustainable production processes. Every industrial revolution brought with it groundbreaking inventions that improved sustainability, scalability, and efficiency. Blockchain traceability, cyber-physical systems, AI-powered self-learning machines, and quantum-enhanced textile chemistry are all expected to transform the sector as it moves toward sector 6.0. These developments will drastically alter the production of textiles and international supply chains by optimizing resource use, cutting waste to almost zero, and enabling real-time process monitoring. The mechanization of textile production during Industry 1.0 (1750s–1800s) was the first significant shift in the textile sector. Large-scale textile production was made possible with the development of the power loom (1784) and the spinning jenny (1764), which raised a huge significant output. However, the use of coal-powered steam engines for this mechanization led to increased pollution and inefficient use of energy. Electric-powered textile mills emerged in Industry 2.0 when electricity became more widely available in the late 1800s, significantly lowering labor costs and boosting output. The globalization of textile production was made possible by important inventions like the Jacquard loom (1804) and the ring spinning machine (1888), which allowed automated weaving and complex fabric designs. The introduction of computer-controlled textile technology (CNC machines) during the third industrial revolution (1960s–1990s) boosted automation, precision, and uniformity. By combining computer-aided design for pattern generation with robotic knitting machines, textile production became error-free, resulting in a decrease in faults and an increase in production speed. With IoT-enabled smart factories, real-time AI-driven quality control, and sensor-based monitoring, Industry 4.0 revolutionized the textile sector by the early 2000s, cutting material waste and increasing energy efficiency. By reducing textile waste by 60%, seamless garment production was made possible by advanced 3D knitting and weaving technology. The 2020s saw the fifth industrial revolution, which placed a strong emphasis on sustainable textile production and human–robot cooperation. The industry's carbon impact was greatly reduced by technologies like biodegradable fibers and enzyme-based textile processing. Waterless dyeing cut water use, and the global adoption of AI-powered textile recycling raised material recovery rates. By incorporating hyper-connected, autonomous, and self-optimizing smart factories, the shift to Industry 6.0 (2025 and Beyond) has the potential to completely transform the textile manufacturing industry. Fabric waste will be cut in half thanks to AI-driven spinning, weaving, and dyeing equipment that can self-correct flaws in real time. While collaborative robots (cobots) will minimize human intervention by performing precision stitching with 99.9% accuracy, quantum computing will optimize molecular interactions in dyeing chemistry, boosting color constancy by 98%. Blockchain-enabled traceability will guarantee complete supply chain transparency and cut counterfeiting losses by 50 billion USD in a year. AI-driven nanotechnology will increase thermal resistance in technical textiles for aerospace and defense applications, while quantum-enhanced textile chemistry will produce materials that are lighter and stronger. With IoT sensors tracking temperature and humidity, cyber-physical systems will optimize textile production in real time while consuming less energy. In textile mills, predictive maintenance will guarantee zero downtime and boost productivity. With AI-powered automated textile recycling reaching a 95% fiber recovery rate and significantly lowering landfill waste, sustainability in Industry 6.0 will place a high priority on the circular textile economy. By eliminating the need for water during the dying process, supercritical CO_2 dyeing technique will produce textiles with no chemical emissions. The worldwide textile market is expected to reach 1.5 trillion USD by 2030 with AI-powered smart textiles expanding at a CAGR of 20.1%. It is anticipated that industrial manufacturing will be fully integrated with smart textiles by 2035, creating a textile ecosystem that is resource efficient, climate resilient, and defect-free. Industry 6.0 in the textile sector represents a fundamental move toward intelligent, sustainable, and adaptable manufacturing rather than merely a technical advancement. The future of textile production will be shaped by the confluence of AI, quantum computing, blockchain, IoT and cyber physical systems, which will result in a completely automated, highly efficient, and environmentally conscious business [4–9].

3 Industry 6.0 in Fiber Production

Industry 6.0, which integrates AI, automation, smart sensors, and bioengineered fibers to improve efficiency, sustainability, and precision, is driving a technological revolution in the fiber production sector. AI-powered polymerization and extrusion monitoring in synthetic fiber production supports molecular homogeneity, resulting in a reduction in flaws in nylon, polyester, and polypropylene fibers. Fiber tensile strength and elasticity are optimized, while production energy consumption is reduced, thanks to automated reactors with real-time viscosity monitoring that modify polymer melt properties. High-precision laser sensors in melt spinning ensure homogeneity and minimize waste in the fiber manufacturing process by monitoring filament thickness to within ±0.1 microns. Intelligent cotton, wool, and silk sorting in natural fiber processing eliminates contamination and boosts yield efficiency by classifying fibers according to micron count, crimp, and staple length using AI-driven hyperspectral imaging. In cotton ginning facilities, intelligent robotic handlers cut down on fiber breakage by 15% [10–12].

4 Industry 6.0 in Spinning Industries

4.1 Advanced Spinning Technologies

With AI-driven automation, robotics, IoT-enabled monitoring, and climate-friendly advancements, the spinning process, an important step in the production of textiles, is experiencing a technological revolution. These developments support SDGs by increasing the efficiency of fiber manufacturing, optimizing material blending, decreasing waste, and lowering carbon footprints. By combining robotic automation, to create a textile sector that is climate resilient, traditional spinning operations are being transformed into highly efficient, sustainable, and intelligent manufacturing ecosystems using AI-enhanced fiber mixing, autonomous spinning processes, and blockchain-based traceability. Industry 6.0's advancements in spinning technology combine AI, ML, digital twin technology, and energy-efficient automation to turn the manufacturing of yarn into a highly optimized, sustainable, and intelligent process. Intelligent blow room systems use AI-driven spectroscopic analysis to evaluate fiber quality in real time, support ideal raw material blending and a reduction in fiber waste. Neural network-based defect detection has been included into automated carding machines, improving sliver uniformity by removing short fiber loss, fiber entanglement, and NEP formation. To optimize fiber alignment and tensile strength, sophisticated drawing and roving systems incorporate machine learning algorithms to dynamically modify drafting tension and roller speeds. This leads to a decrease in the occurrence of thick and thin spots and improvement in yarn evenness [13–15].

Digital twin technology is used by automated ring spinning machines to enable predictive maintenance and real-time simulation, which reduces machine downtime and increases component lifespan. AI-powered high-speed rotors (up to 150,000 rpm) with integrated vibration sensors maximize twist insertion in rotor and air jet spinning, guaranteeing exact yarn strength and elongation characteristics. SDGs 9 and 12 are in line with energy-efficient smart ring frames that use less power, thanks to AI-driven motor control units. Blockchain-enabled traceability systems guarantee adherence to sustainability guidelines and ethical sourcing certifications while offering complete transparency in the yarn manufacturing process. The future of spinning is in yarn engineering aided by quantum computing, where molecular-level fiber simulations optimize fiber polymer interactions to produce yarns that are lighter, stronger, and extremely robust for use in smart fabrics, technical textiles, and aerospace applications. Nanotechnology-infused spinning finishes will further enhance yarn moisture-wicking, thermal resistance, and antimicrobial properties, ensuring a next-generation textile ecosystem that is fully autonomous, climate-resilient, and zero-waste in its production methodologies [16–18].

4.2 AI-Optimized Fiber Blending for High-Performance Yarn

Real-time spectroscopic analysis, IoT-enabled fiber sorting, and AI-driven automation are revolutionizing textile manufacturing by improving yarn consistency, strength, and durability while reducing flaws. Conventional fiber blending, which mostly depended on operator skill and manual control, frequently resulted in differences in fiber characteristics, including irregularities in texture, elasticity, and tensile strength. Spinning mills are undergoing a transformation thanks to robotic automation, AI-driven fiber blending, and real-time IoT monitoring, which guarantee accuracy, sustainability, and cost effectiveness. Deep learning algorithms, quantum-enhanced spectroscopy and hyperspectral imaging have all been combined to greatly increase fiber categorization, blend optimization, and material homogeneity while lowering waste, energy consumption, and flaws. Fiber blending and processing facilities are now attaining unparalleled levels of automation, sustainability, and climate resilience by utilizing 5G-enabled cloud networks, robotic precision control, and high-speed AI computing. This puts the textile industry at the forefront of Industry 6.0-driven innovation. AI-powered machine learning algorithms are now able to evaluate fiber attributes, including length, fineness, crimp, and tensile strength in real time, thanks to Industry 6.0 developments. These algorithms dynamically modify fiber ratios to maximize the performance of blends of cotton, polyester, silk, and viscose. When combined with cloud-based predictive analytics and high-speed computing, these self-learning AI models continually adjust to material variances, resulting in a reduction in fiber mixing errors [19–21].

4.3 Automated Cotton Polyester Blending Using AI-Based Spectroscopic Analysis

Real-time monitoring of fiber ratios in cotton polyester blending is made possible by AI-driven near infrared spectroscopy. Conventional fiber blending techniques frequently produced ±3–5% differences in fiber content, which resulted in inconsistent yarn strength and color absorption characteristics. With an accuracy of ±0.5%, AI-based NIR spectroscopy systems that operate at wavelengths between 700 and 2500 nm continually examine fiber blends, guaranteeing extremely consistent fiber compositions. This modification results in an increase in yarn tensile strength, which improves performance in high-stress textile applications like industrial textiles, sportswear, and denim. Based on real-time spectroscopic input, robotic fiber dosing systems driven by artificial intelligence regulate the precise discharge of polyester and cotton fibers into blending chambers in automated blending plants. The movement of fiber conveyors is controlled by linear actuators and precision servo motors, which guarantee that the fiber blend stays within the necessary range. When combined with PLCs, these AI-driven closed-loop control systems dynamically modify fiber flow rates to account for variations in humidity, fiber tangling, or inconsistent materials [22–24].

4.4 Smart Yarn Processing with AI-Driven Moisture and Elasticity Control

To preserve fiber integrity and reduce flaws, processing natural fabrics like cotton, wool, and silk calls for exact moisture and elasticity management. Real-time temperature and humidity sensors together with AI-powered machine learning optimize fiber conditioning throughout processing, lowering mechanical stress and fabric deformation. Smart Wool Conditioning dynamically modifies the humidity levels in fiber conditioning chambers by using AI-based hygroscopic analysis models to continuously assess the wool fibers' ability to retain moisture. This modification reduces processing flaws by preventing fiber brittleness and uneven dye absorption. Silk elasticity can be enhanced by analyzing the mechanical characteristics of silk fibers and modifying processing parameters, AI-driven viscoelastic modeling can reduce fiber breakage and increase tensile strength. High-precision strain gauge sensors and sophisticated robotic stretching mechanisms guarantee constant elongation, reducing structural irregularities in high-end and high-performance silk textiles [11, 25, 26].

4.5 IoT Integrated Fiber Sorting and Quality Control

IoT-enabled vision systems and deep learning algorithms are used in modern fiber processing facilities to classify and sort raw fibers according to impurity levels, staple length, and micron count. The mechanical qualities and homogeneity of yarn were impacted by the high contamination rates (5–8%) that were frequently the result of traditional fiber sorting techniques. At 1000+ frames per second, AI-powered computer vision systems evaluate individual fibers in real time, removing contaminants and increasing batch uniformity. High-Speed AI-Powered Cameras can be combined with deep neural networks (DNNs) with ultra-fast hyperspectral imaging systems, these cameras can improve sorting accuracy by classifying fibers according to their chemical composition, color consistency, and defect detection. Robotic sorting arms with AI-driven force feedback control and six axis robotic grippers exclude synthetic impurities, foreign particles, and low-quality fibers to guarantee that only premium materials make it into the production cycle. IoT-enabled fiber sorting devices provide data in real time to edge computing nodes, resulting in a reduction in latency and immediate quality control modifications. For centralized quality monitoring, these systems make use of 5G/6G-enabled cloud platforms, which allow for remote diagnostics and fiber sorting gear predictive maintenance [27–29].

4.6 Software and Hardware Infrastructure in AI-Powered Fiber Blending

Modern fiber blending guarantees high-speed processing, real-time automation, and adaptive learning capabilities through the integration of sophisticated hardware and software frameworks. To anticipate future changes and make proactive adjustments to blending settings, machine learning algorithms examine past patterns in fiber quality. With the help of virtual simulations that mimic actual fiber processing settings, digital twins for fiber processing enable AI systems to test and improve blending optimizations prior to deployment. Supply chain transparency and fraud protection are ensured by Blockchain for Material Traceability's AI integrated blockchain systems, which can trace fiber origin, mixing ratios, and quality metrics. High-performance Tensor Processing Units accelerate AI processing in hardware systems by processing complicated fiber composition data at teraflop speeds, enabling real-time modifications. Advanced servo motor controllers in robotic precision control systems provide micron-level precision in fiber blending equipment, minimizing mechanical irregularities. Advanced quantum-enhanced infrared spectroscopy ensures excellent material uniformity by detecting atomic-scale differences in fiber chemistry [30, 31].

4.7 Climate Resilience and SDG Integration in Fiber Processing

In line with SDGs 9, 12, and 13, the developments in AI optimized fiber mixing and automated sorting greatly enhance sustainability and climate resilience. One of the most important developments is energy efficient AI processing, which lowers the carbon footprint of textile production through AI-driven fiber blending. 99% purity in processed recycled fibers is possible with AI-enhanced recycled/upcycled fiber integration, which makes it easier to use recycled PET, regenerated cellulose, and bio based polyesters, an essential step in lowering reliance on virgin fibers and advancing the ideas of the circular economy. Waterless fiber dyeing compatibility is another significant development. AI-powered fiber characterization makes it possible to integrate waterless fiber dyeing technology seamlessly with supercritical CO_2 dyeing technology, removing the need for toxic wastewater discharge and reducing dye consumption. This reduces the environmental impact of traditional wet processing. Utilizing hybrid photovoltaic arrays, IoT integrated solar-powered sorting facilities allow for 100% renewable energy operation, drastically lowering emissions related to the processing of textile waste. In the era of Industry 6.0, these developments not only improve operational efficiency but also guarantee that textile manufacture is in line with global climate resilience policies, making the sector more environmentally conscious, resource efficient, and sustainable [32, 33].

4.8 IoT-Enabled Ring and Rotor Spinning Machines for Real-Time Quality Monitoring

Real-time yarn quality monitoring has been greatly improved by the advent of IoT connected spinning equipment, which has decreased breakage rates and increased overall production efficiency. AI-driven spindle speed modulation is a feature of advanced ring spinning machines (RSM 6.0), which can run at up to 25,000 rpm and dynamically modify yarn tension and twist to maintain uniformity. As a result, the hairiness is reduced, producing yarns that are smoother and more consistent. In the meantime, AI feedback control and embedded vibration sensors are integrated into rotor spinning (also known as open end spinning) systems, which continually check the rotor balance at 120,000 rpm. By identifying subtle imbalances and initiating immediate correction measures, these sensors guarantee flawless yarn output while reducing machine wear and tear. Smart pneumatic yarn tensioning, which makes use of adaptive air pressure regulation to maximize yarn stretch and avoid uneven elongation, is another advancement in Industry 6.0 spinning technology. By successfully reducing twist abnormalities, this sophisticated control system improves yarn quality and lowers rejection rates. Modern spinning mills can self-optimize operations, anticipate possible failures, and increase overall efficiency by utilizing machine learning algorithms, predictive maintenance software, and IoT-enabled hardware. This will make the textile industry more resilient, cost-effective, and sustainable in the era of Industry 6.0 [34–36].

4.9 Autonomous Robotic Transport in Spinning Mills

The use of autonomous robotic transport systems in spinning mills is revolutionizing raw material management, bobbin logistics, and fiber handling, resulting in a 50% labor cost reduction and a 30% gain in total efficiency. Material transfer between fiber blending stations and carding machines is streamlined by automated guided vehicles with AI-driven navigation systems, which improve workflow synchronization and cut down on manual handling time by 60%. These AGVs navigate dynamically changing production environments without the need for human interaction by utilizing deep learning-based path planning algorithms, LiDAR sensors, and computer vision. By automating the operation of replacing bobbins in ring spinning frames with high-speed 6 axis robotic arms, robotic bobbin loading systems drastically save downtime by 70%. To optimize replacement cycles and avoid production halts, these robots are integrated with real-time machine learning models that forecast spindle fatigue. By tracking fiber bales in real time, Drones with RFID and IoT connectivity are transforming inventory management. By scanning RFID-tagged products and sending real-time inventory data to cloud-based ERP systems, these drones reduce stock inconsistencies and increase logistics accuracy by 25%. Spinning mills become more intelligent, autonomous, and sustainable through the integration of Industry 6.0 technologies like robotic automation, AI-driven logistics, and IoT-enabled inventory control. These technologies guarantee optimized material handling, decreased operational delays, and increased production scalability [37, 38].

4.10 Fabrication of Smart Spinning Machines

AI-driven design optimizations, additive manufacturing, and precision CNC machining are used in the production of the newest ring and rotor spinning machines. Advanced material science, precise manufacturing, and AI-integrated mechatronics are driving the development of next-generation smart spinning machines, which will ensure incredibly efficient, self-optimizing textile production. Better yarn quality, lower energy usage, and predictive maintenance capabilities are being made possible by the shift to lightweight, highly durable components, real-time defect monitoring, and AI-driven process control. High precision CNC machining, additive manufacturing (3D printing), AI-driven design optimization, and sophisticated material science are all integrated into the construction of next-generation smart spinning machines to improve sustainability, operating efficiency, and durability. Nanostructured coatings, intelligent mechatronic systems, and real-time data analytics are revolutionizing ring and rotor spinning machines, guaranteeing yarn formation accuracy while reducing energy consumption and mechanical wear. At the heart of contemporary spinning systems are CNC-milled spindles and rotors, which are made of ceramic composites and titanium-coated high-speed steel for remarkable wear resistance (>800 HV hardness on the Vickers scale). By applying titanium nitride and diamond like carbon coatings using physical vapor deposition, friction is greatly reduced, resulting in a reduction in energy loss and an increase in machine lifespan. In open end spinning systems, nanostructured carbide inserts (WC-Co) significantly strengthen rotor edges and guard against microfractures at rotational speeds higher than 120,000 rpm. When it comes to creating high-precision ceramic and polymeric parts for spinning machines, additive manufacturing components are essential. Using laser sintering (SLS), ceramic yarn guides made of silicon carbide (SiC) and zirconia (ZrO_2) are 3D printed to have almost low porosity, providing excellent heat resistance and a 40% reduction in fiber friction, which minimizes yarn hairiness. At high rotational speeds, carbon fiber-reinforced polymer bobbin holders produced using direct ink writing technology reduce spindle vibration and are incredibly lightweight. CFRP composite-based 3D Printed Fiber Guides and Nozzles are transforming contemporary spinning systems by providing a 40% weight reduction and a marked increase in mechanical durability. Continuous fiber fabrication and SLS are used in the manufacturing of CFRP based yarn guides, which have a 5 times higher tensile strength than conventional aluminum components and a reduction in frictional heat generation, which lowers the rate of yarn breakage. CFRP guides' nano engineered surface topology, which is made possible by laser-assisted surface texturing, produces micro-grooves that improve lubricant retention and reduce fiber wear even more. Adaptive speed control and real-time flaw identification in spinning processes are improved by AI-driven mechatronic systems. To reduce downtime, predictive maintenance is made possible by embedded AI models that use CNNs to evaluate spindle torque fluctuations and anticipate bearing wear with 98% accuracy. Piezoelectric nanomaterial-based smart actuators ($BaTiO_3$-based piezo actuators) dynamically modify rotor speed to maximize fiber twist and tensile strength, resulting in a increase in yarn homogeneity. Graphene-coated piezoresistive films combined with laser-engraved tension sensors offer ultra-sensitive real-time twist monitoring, guaranteeing ±0.1% accuracy in yarn tension measurement. The electrical conductivity of these graphene-based nanostructured sensors, which were created using CVD and inkjet printing, is more than 10^6 S/m, allowing for real-time feedback to AI-driven control systems. By incorporating pressure sensors based on quantum tunnelling, sub nanometer yarn diameter changes can be made, removing flaws before they spread throughout the spinning process. By lowering surface roughness by 50%, high-frequency ultrasonic nozzles embedded with silicon carbide resonators improve precision yarn spinning even more and allow for finer yarn counts without sacrificing strength. Plasma-coated superhydrophobic nanofilms (based on fluorinated silica nanoparticles) can self-clean, preventing contaminant accumulation and resulting in a 30% increase in operational uptime. Sustainability and AI-Integrated Smart Manufacturing are turning textile spinning into a cyber physical system powered by Industry 6.0. 99.8% of manufacturing abnormalities are predicted and eliminated by neural network-based defect detection algorithms that continuously examine yarn qualities in real time. Ring spinning spindles with AI-driven adaptive torque modulation maximize rotational rates of up to 25,000 rpm while using less energy per unit of yarn generated. The integration of QTPs in textile manufacturing, specifically for sub nanometer yarn diameter corrections, leverages the quantum mechanical phenomenon of quantum tunneling to detect minute variations in pressure, which could affect the yarn's formation. This highly innovative application in the spinning process allows for the detection of defects at an atomic or molecular scale and enables corrections before defects propagate further, enhancing the overall quality of yarn production. Quantum tunneling is used in pressure sensors to detect sub nanometer changes in the gap between electrodes, which are caused by pressure variations due to yarn deformation. The mechanical properties of yarn, such as elasticity and plasticity, influence the pressure detected by the QTP, which is used to adjust the spinning process. The development of ultra-sensitive QTPs, utilizing advanced nanomaterials like graphene and carbon nanotubes, improves sensor accuracy and sensitivity. The integration of AI further optimizes the process by automating adjustments based on real-time data, ensuring consistent yarn diameter control and high-quality fabric production. The combination of advanced material science, quantum mechanics, and AI-driven process optimization represents a significant leap in textile manufacturing, offering unprecedented precision in controlling yarn quality. Quantum tunneling is a phenomenon where particles, such as electrons, pass through a potential energy barrier that they classically would not have enough energy to overcome. In the case of QTPs, quantum tunneling occurs when electrons move through a gap or barrier in a circuit, such as a tunneling junction. This junction is highly sensitive to changes in the gap distance, meaning that the tunneling current varies exponentially with any change in the gap between two conducting materials. This allows for extremely fine measurements of pressure-induced deformations in the yarn or filament being spun. As the pressure on the yarn changes, the gap between the electrodes in the tunneling junction changes, altering the tunneling current, which is used to detect sub-nanometer pressure variations. The QTP design typically consists of highly conductive materials, such as gold, platinum, or graphene, which form the electrodes near the yarn. The dielectric material in the tunneling junction, which separates the two electrodes, is influenced by the pressure exerted on the yarn. When the yarn is pulled or compressed, the distance between the electrode's changes, affecting the tunneling current. As pressure increases, the gap between the electrodes decreases, thereby increasing the tunneling current. During the spinning process, the yarn is formed by drawing fibers through mechanical rollers. The diameter of the yarn can fluctuate due to changes in pressure and tension, resulting in defects such as thickened spots or thin areas. These defects, if left unaddressed, can propagate down the yarn and compromise the final product quality. Pressure distribution in yarn spinning is often uneven due to friction and mechanical deformation, leading to variations in yarn diameter. The quantum tunneling sensor detects these changes in diameter by monitoring the tunneling current, which varies as pressure at the sensor location fluctuates. When the yarn diameter becomes thinner than required, the pressure increases, altering the tunneling current, and vice versa. This real-time data can be fed to an AI or automated control system to adjust the spinning process—such as altering the tension or feeding rate correcting the yarn diameter, and minimizing defects. The yarn's polymeric nature is essential in understanding its behavior under pressure. Most textile fibers, such as polyester, nylon, and cotton, are polymers, which consist of long chains of repeating molecular units. These chains are typically held together by van der Waals forces, hydrogen bonds, and covalent bonds in synthetic fibers. When pressure is applied to the yarn, these polymer chains can stretch, compress, or reorient, leading to changes in the yarn's diameter. The pressure variations detected by the QTP correspond to these mechanical changes at the molecular or atomic scale. The yarn behaves elastically under low pressure, but when subjected to higher forces, it may undergo plastic deformation, resulting in permanent changes to its diameter and potentially causing defects. The integration of advanced nanomaterials such as graphene and carbon nanotubes in QTPs enhances their sensitivity to small pressure changes. These materials exhibit excellent electrical conductivity and unique mechanical properties, making them ideal for quantum tunneling sensors. At the nanoscale, these materials exhibit heightened sensitivity to minute variations in pressure, allowing for the detection of sub-nanometer changes in yarn diameter. Graphene oxide and nanostructured metals can further enhance the sensitivity of QTPs, providing highly accurate measurements for quality control. AI algorithms play a crucial role in analyzing the data from QTPs in real time. These algorithms process the tunneling current data to predict potential defects and adjust the spinning process accordingly. By automating adjustments to the spinning machinery, such as speed or tension, AI minimizes defects and optimizes yarn production. The real-time feedback loop from QTPs ensures that the spinning process is constantly adjusted, improving yarn consistency and quality [39–41].

4.11 Climate Change Mitigation in Industry 6.0 Spinning Technology

The integration of climate resilience and alignment with the SDGs 9, 12, and 13 is becoming increasingly central to the design and operation of sustainable spinning machines in the textile industry. These innovations contribute to a more sustainable, ecofriendly, and energy efficient future for textile manufacturing, focusing on reducing energy consumption, minimizing waste, and lowering carbon emissions. Climate resilience is enhanced by embedding renewable energy solutions, reducing material waste and promoting circular economy principles in machine design. A notable advancement in sustainable spinning technology involves the use of bio-inspired piezoelectric energy harvesting. By capturing vibrations from spinning machines, this technology generates approximately 5 W of self-powered energy, enabling sensor operation without relying on external power sources. This reduces dependency on traditional energy grids, contributing to the overall energy efficiency of spinning mills. Hydrogen-fueled servo motors, replacing conventional electric drives, contribute significantly to carbon reduction efforts, cutting CO_2 emissions. This shift to hydrogen-based drives aligns directly with SDG 13, enabling a significant reduction in greenhouse gas emissions associated with textile manufacturing. The design of spinning machines further supports sustainability using advanced materials like recycled graphene-infused CFRP components. By incorporating this innovative material, textile machinery manufacturers reduce material waste by 60%, supporting a closed loop system in machine manufacturing. Moreover, graphene-based tribological coatings applied to spindle surfaces reduce friction, resulting in a 25% decrease in energy consumption per unit of yarn produced. This energy-saving innovation not only minimizes the environmental footprint but also increases operational efficiency. Industry 6.0 spinning mills are increasingly adopting circular economy principles, which include using recycled materials such as titanium alloys (Ti-6Al-4 V) and biodegradable PLA-based polymer casings. These materials contribute to the sustainable lifecycle management of machines, supporting reduced environmental impact from production to disposal. Additive manufacturing, precision CNC machining, and AI-driven optimization are leveraged to enhance machine performance, reduce waste, and improve energy efficiency. In parallel, spinning mills have also made significant strides in reducing their overall energy consumption. Traditionally, spinning mills consume 15–20% of the total energy used in textile manufacturing, with conventional systems operating at over 3000 to 4000 kWh/ton of yarn produced. Industry 6.0 technologies are reducing these energy demands by up to 50% through AI-driven efficiency measures, renewable energy integration, and optimized production processes. AI-based energy optimization systems actively monitor and manage spinning machine operations, cutting idle time and reducing electricity consumption. Solar-powered spinning units are also emerging as a critical component of this energy transformation. Hybrid photovoltaic arrays, paired with AI-driven energy balancing systems, enable spinning mills to operate on up to 80% renewable energy. This combination of smart energy management and renewable energy integration supports SDG 7 and 12, ensuring a cleaner and more sustainable approach to yarn production. In terms of material sustainability, automated spinning of recycled yarns, such as recycled PET and upcycled cotton, plays a key role in reducing CO_2 emissions. The processing of recycled fibers results in a reduction in CO_2 emissions/kg of yarn produced, highlighting the importance of sustainable material sourcing and processing techniques in achieving SDG 12. The adoption of supercritical CO_2 technology for fiber conditioning eliminates the need for chemical additives and drastically reduces water usage by up to 90%, addressing water scarcity issues and promoting water conservation. The convergence of quantum-enhanced sensors, AI-driven automation, hydrogen-powered drives, renewable energy solutions, and sustainable materials represents a transformative shift in the textile spinning sector. These advancements not only optimize yarn quality and production efficiency but also significantly reduce the environmental footprint of textile manufacturing. Industry 6.0 spinning machines, powered by cutting edge technologies and sustainability principles, are positioned to achieve near-zero waste, maximum energy efficiency and defect free yarn production. This revolution aligns with SDGs driving the textile industry towards a more sustainable, climate-resilient future [42–44].

5 Weaving and Knitting

Modern weaving and knitting operations have been transformed by the incorporation of CPS, AI, and sophisticated digital control, which allows for ultra-high-speed, precision driven and waste-minimizing manufacturing. Together, 3D knitting machines, AI-powered Jacquard looms, and intelligent warp and weft sensors maximize fabric quality, boost production efficiency, and drastically cut down on material waste. With its IoT connectivity, real-time fault detection, and predictive maintenance capabilities, these next-generation machines are opening the door for Industry 6.0's fully autonomous textile mills. Compared to traditional Jacquard looms, Smart Jacquard Weaving Machines increase productivity by precisely manipulating over 1000 warp threads/s using AI controlled servomotors and real-time warp tension sensors. To prevent problems like broken warp threads, pick skips, and tension inconsistencies, these looms use machine learning algorithms that evaluate yarn tension, weft insertion rates, and fabric density. These algorithms then dynamically modify settings. Real-time warp thread alignment monitoring via sophisticated optical fiber-based sensors helps to prevent flaws and cut down on material rejection by 40%. With weft insertion speeds of nearly 3000 m/min, robotic shuttle-less weaving systems driven by electromagnetic propulsion take the place of conventional mechanical shuttles, increasing efficiency while reducing mechanical wear. By doing away with conventional fabric cutting and stitching processes, 3D knitting technology is transforming the textile industry and cutting fabric waste. Multi-axis robotic actuators are used by AI-driven 3D knitting machines to precisely interlace yarns with accuracy down to the micron [1, 45, 46]. High-speed knitting needles coated with graphene provide self-lubricating and friction-reducing properties, increasing the machines' operational lifespan and lowering maintenance expenses. Every loop creation is scanned by real-time machine vision technologies at 500 frames/s guaranteeing 99.5% accuracy and flawless clothing manufacture. Yarn tension is fine-tuned by automatic tension modulation employing piezoelectric actuators, which dynamically adjusts to changes in fiber elasticity and avoids stitch abnormalities. Key components for these next-generation textile machines are made using laser-assisted micromachining, additive manufacturing (3D printing), and precision CNC machining. Machine frames made of carbon fiber reinforced polymer cut weight by 40%, improving vibration stability and using less energy. Shape-memory alloys are used in high-speed actuators to maximize movement accuracy and cut cycle times. Self-learning textile machinery that instantly adjusts to yarn qualities and ambient conditions is made possible using quantum-enhanced AI control algorithms, which further improve production processes. Energy-efficient digital motors, AI-driven waste reduction techniques and the integration of renewable energy sources are how Industry 6.0 weaving and knitting accomplish sustainability and climate resilience. Hybrid solar wind energy systems power smart textile mills, reducing carbon emissions. Global regulatory requirements for SDGs 9, 12, and 13 are met by blockchain based fabric traceability systems, which guarantee transparency across the supply chain. Water use in post weaving procedures is eliminated by supercritical CO_2 based fabric finishing methods, improving sustainability without sacrificing fabric integrity. AI, IoT, robots, quantum computing, and sustainable material science are all used in Industry 6.0 driven knitting and weaving operations to build intelligent, self-sufficient, and ecologically conscious textile production systems [46, 47].

5.1 Advanced Weaving, Knitting, and Nonwoven Technologies

The manufacturing of weaving, knitting, and nonwoven fabrics is becoming a highly automated and sustainable business because to the integration of AI, IoT, robotics, and advanced material science. These days, smart weaving technologies include AI-driven loom automation, which reduces fabric waste by using machine learning algorithms to analyze fabric tension, warp alignment, and defect patterns in real time. AI-powered warp monitoring systems make use of laser guided tension sensors and computer vision to identify thread breaking with millisecond accuracy, improving fabric uniformity and manufacturing speeds. IoT-based pattern adjustments are used by intelligent knitting machines to provide automated stitch changes based on real-time fabric stress assessments. With the help of deep learning and high-speed robotic actuators, 3D knitting technology creates seamless, structurally reinforced clothing, halving waste and doing away with the need for postproduction sewing. Sportswear, medical compression garments, and aerospace textiles, where precisely designed stretchability is necessary benefit greatly from this method. With real-time adaptive learning algorithms that self-regulate stitch creation, AI-powered knitting system Compared to traditional knitting machines, this technology improves fabric drape and flexibility. AI-based fiber orientation control and nanofiber electrospinning are revolutionizing nonwoven fabric engineering, enabling the creation of filtration membranes, geotextiles, and high-performance medical textiles. Higher mechanical strength and regulated porosity in medical wound dressings and high efficiency filtering applications are the results of AI-driven material engineering, which guarantees fiber diameter homogeneity within ±5 nm. For hospital grade protective textiles, electrospun nanofiber membranes enhanced with graphene oxide and silver nanoparticles achieve 99.99% microbial inhibition, improving antibacterial and antiviral capabilities [47, 48].

6 Textile Processing and Dyeing and Smart Chemical Application

One of the most resource intensive phases of textile production is textile wet processing, which includes desizing, scouring, bleaching, dyeing, printing, and finishing. It uses around 93 billion cubic meters of water a year and accounts for 20% of industrial water pollution worldwide. With dye fixing rates frequently falling, traditional dyeing and finishing techniques are incredibly inefficient and have a negative impact on the environment, energy use, and chemical waste. By combining AI-driven optimization, supercritical CO_2 dyeing, enzymatic bio-finishing, and quantum chemistry assisted dye molecule engineering, Industry 6.0 breakthroughs are transforming textile manufacturing and guaranteeing chemical applications that are intelligent, sustainable, and waste free. AI-powered Exhaust and Continuous Dyeing Systems optimize dye absorption efficiency and reduce chemical waste by dynamically adjusting dye concentrations, pH, temperature, and liquor ratios using machine learning algorithms and real-time spectroscopic sensors. The accurate prediction of dye-fiber interactions made possible by dye kinetics modeling aided by quantum computing guarantees consistent color penetration at the molecular level. Controlled dye release is made possible by sophisticated nano-encapsulated dye formulations, which improve fastness characteristics while consuming less dye. By using CO_2 in a supercritical state ($T > 31.1$ °C, $P > 7.38$ MPa) as a solvent, supercritical CO_2 dyeing replaces conventional aqueous dyeing methods and eliminates the need for water in dyeing. This approach lowers the COD and BOD in effluent streams by achieving 100% dye fixation with zero wastewater discharge. Supercritical CO_2's strong diffusivity and low surface tension enable deep fiber penetration, producing dyeing that is vivid and high fastness while using less energy than traditional techniques. The compatibility of disperse dyes with supercritical CO_2 has been further enhanced by recent developments in fluorinated dye carriers and CO_2 soluble dispersion agents, making this method feasible for both natural and manufactured fibers. Traditional harsh chemical treatments for softening, bio-polishing, and functional finishing are being replaced with enzymatic bio-finishing. By selectively altering fiber surfaces through enzymatic hydrolysis with cellulases, proteases, and pectinases, fabric texture, softness, and sheen are improved while chemical waste is reduced. The manufacture of thermostable, pH-flexible enzyme variations is made possible by AI optimized enzyme engineering, which increases process efficiency. To give textiles antibacterial, UV-resistant, and self-cleaning qualities without the use of artificial chemical finishes, functional nano enzymes have been developed. Precision CNC machining, additive manufacturing, and intelligent robots are all used in the construction of Industry 6.0 textile processing equipment to increase productivity and sustainability. These days, graphene-coated heat exchangers in high-speed continuous dyeing machines improve thermal conductivity and cut dyeing cycle. AI-powered smart jiggers dynamically modify dye liquid flow to achieve nearly minimal leftover dye discharge by using real-time infrared spectroscopy to determine dye depletion rates. By improving fabric absorbency, nanostructured plasma-treated rollers reduce the need for pre-treatment chemicals. AI integrated effluent recycling systems recover and purify 95% of wastewater, while solar-powered dyeing units reduce carbon emissions, making textile industry climate resilient and in line with SDGs, compliance requirements are met via real-time chemical inventory monitoring made possible by blockchain enabled traceability solutions. AI, supercritical fluid technology, enzymatic bio finishing, and nanomaterials are all integrated into smart chemical applications in Industry 6.0 textile processing to produce dyeing and finishing methods that are waste free, energy efficient and ecologically friendly. The textile sector is now at the vanguard of next generation intelligent manufacturing because to these developments, which also greatly lower operating costs, processing times, and environmental impact while improving fabric quality and durability. The textile industry uses two main techniques for dyeing, continuous dyeing (thermosol, pad steam and pad dry systems) and exhaust dyeing (batch processing in machines like jet, jigger, and winch). While Industry 6.0 dyeing technologies incorporate AI, real-time sensors, smart chemical application systems, and nanotechnology-based dye formulations to improve efficiency, sustainability, and fabric quality, traditional dyeing processes suffer from high water consumption, significant chemical waste, and inconsistent color fixation. By putting an emphasis on efficiency, automation, and sustainability through innovative technologies, Industry 6.0 dyeing techniques are completely changing the textile production industry [49–51].

By substituting compressed carbon dioxide for water, supercritical CO_2 dyeing allows for 100% dye fixing, zero effluent discharge and an 85% reduction in chemical use. In contrast to traditional techniques, it does not use salt, alkali, or post-washing, which results in a 50% reduction in processing time while preserving superior color fastness. AI-powered digital dyeing ensures color uniformity within ±0.5% precision by dynamically adjusting fabric tension, chemical flow, and dye uptake using real-time machine learning algorithms. Color monitoring systems based on spectroscopy can identify and rectify shade differences quickly, reducing the need for re dyeing and saving time and money. By using low-energy plasma treatment to alter the surface characteristics of the fabric, plasma-enhanced dyeing greatly improves the interaction between the dye and the fiber while consuming less dye and auxiliary materials. This technique is a very sustainable substitute because it enhances fabric's breathability and color fastness without the need for further chemical treatments. Using acoustic cavitation, ultrasonic-assisted dyeing reduces processing time, increases shade depth, and minimizes dye aggregation by enabling deeper dye penetration at lower temperatures. Nanoparticle dyes that are introduced by microencapsulation and nanotechnology based dyeing more efficiently permeate fibers, enabling dyeing at lower temperatures (40–60 °C) while attaining wash fastness values above 4.5 (ISO 105: C06). These Industry 6.0 innovations support SDGs for a more sustainable and climate-resilient textile sector by lowering environmental impact while simultaneously increasing cost effectiveness, improving dye uniformity, and speeding up production [52–55]. A thorough comparison of conventional and Industry 6.0 dyeing methods can be found below in Table 14.1, printing methods in Table 14.2, denim processing in Table 14.3 and Yarn and fiber dyeing in Table 14.4.

The economics and ecological footprint of textile manufacturing have been drastically altered by Industry 6.0 dyeing technologies, which have a significant impact on cost reduction, ROI, and environmental sustainability. One of the biggest benefits is cost reductions; AI-optimized dyeing can lower chemical expenditure, as well as prices for dye auxiliaries and re-dyeing. 60% reduction in water usage results in an estimated yearly savings of over 0.5 million USD per manufacturing facility, especially in large-scale textile production units where wastewater treatment is a major expense. Because of increased efficiency, reduced operating costs, and better dye use, Industry 6.0 dyeing technologies allow for a quicker ROI than traditional exhaust dyeing systems, which take more than 5 years to break even. In terms of environmental sustainability, CO_2 emissions are cut, which is in line with SDGs. This also guarantees that textile manufacturing uses sustainable and climate-resilient production techniques. 95% water savings greatly lessen reliance on freshwater, which is essential for protecting natural resources in areas with limited water supplies. Within the next 10 years, textile dying will undergo a total transformation thanks to the combination of AI-driven dyeing systems, intelligent chemical application techniques, and machinery powered by renewable energy. Supercritical CO_2 dyeing, plasma-enhanced dyeing, and dyeing based on nanotechnology are examples of Industry 6.0 developments that will improve fabric quality, cut waste, maximize resource use, and boost process efficiency. This change essentially redefines sustainable manufacturing in the textile industry by ensuring that textile production shifts toward a zero-waste, high-efficiency, and climate friendly future. The need for high customization, accuracy, and sustainability has led to a major advancement in textile manufacturing with the switch from traditional screen printing to digital inkjet printing. Traditional printing methods, such flatbed and rotary screen printing, use a lot of chemicals, water, and energy, which raises production costs and has an adverse effect on the environment. Superior color accuracy, economic effectiveness, and environmental sustainability are provided by Industry 6.0 smart digital printing, which combines AI-driven inkjet systems, nano-encapsulated pigment dispersions, and color prediction models based on quantum chemistry. As a more environmentally friendly option to printing with reactive, acidic, and dispersion dyes, pigment digital printing is becoming more and more popular. By removing the need for lengthy washing and steaming, pigment printing saves 95% of the water used in reactive dye printing. High fastness and less chemical waste are made possible by pigment ink formulations made of nano sized polymer encapsulated colorants that

Table 14.1 Comparison of traditional vs. Industry 6.0 dyeing technologies

Feature	Traditional exhaust dyeing (Jet/Jigger/ Winch, softflow, multiflow etc.)	Traditional continuous dyeing (Thermosol/pad steam, pad batch etc.)	Industry 6.0 AI-optimized dyeing	Technical breakthroughs and validation
Water consumption	70–150 L/kg fabric	40–80 L/kg fabric	0.5–2.5 L/kg fabric (98% reduction)	Supercritical CO_2 dyeing: Zero aqueous effluent; 100% CO_2 recovery. Plasma Nano-dyeing: Atmospheric plasma functionalizes fabric surfaces for dye adhesion without water
Energy consumption	50–100 kWh/ ton fabric	40–90 kWh/ton fabric	12–22 kWh/ton fabric (75% reduction)	Induction heating: Direct fabric heating (95% efficiency vs. 45% steam). AI optimized heat recovery: 90% waste heat recapture
Dye fixation efficiency	60–75%	75–85%	98.5–99.8%	Nanocapsule dye carriers: pH responsive release. Ionic liquid dyeing: Near zero hydrolysis (99.9% fixation)
Dye waste and effluent	High (30–50% unused dye; BOD/COD >5000 ppm)	Medium (15–30% loss; BOD/COD 2000–4000 ppm)	<0.2% effluent dye (closed-loop systems)	AI-driven precision dosing: Real-time spectrophotometry adjusts dye ±0.001% IoT). Electrochemical dye reduction: Direct electron transfer minimizes auxiliaries
Process control	Manual pH/ temperature monitoring (±2 °C; ΔE color variance 3–5)	Semi-automated (±1 °C; ΔE 1.5–2.5)	Autonomous AI control: Real-time spectral analysis ($\Delta E \leq 0.3$); self-correcting algorithms	Deep learning color matching: Neural nets predict dye behavior ±0.1%. Digital twin integration: Simulates 10^6 scenarios for optimal parameters
CO_2 emissions	2.5–4.0 tons CO_2/ton fabric	1.8–3.2 tons CO_2/ ton fabric	0.1–0.5 tons CO_2/ton fabric (90% reduction)	Renewable energy synergy: Solar-thermal dyeing (80 °C at 0 g CO_2/kWh). CO_2 capture: Integrated amine scrubbing (90% capture rate)

(continued)

Table 14.1 (continued)

Feature	Traditional exhaust dyeing (Jet/Jigger/ Winch, softflow, multiflow etc.)	Traditional continuous dyeing (Thermosol/pad steam, pad batch etc.)	Industry 6.0 AI-optimized dyeing	Technical breakthroughs and validation
Chemical usage	High salt (50–100 g/L), alkali (10–20 g/L), surfactants	Moderate auxiliaries (5–15 g/L)	Enzyme-assisted dyeing: 95% chemical reduction	Bio catalytic dyeing: Laccase/peroxidase enzymes replace reductants. Cationization: Chitosan pretreatment eliminates salt
Time required	4–10 h/batch	1–3 h/roll	15–45 min (micro supercritical cycles)	Ultrasonic dye acceleration: 20 kHz waves enhance diffusion 300%. Plasma pre-treatment: 120 s surface activation replaces 30-min scouring
Machine flexibility	Rigid (woven/ knit separation; 4–8 h changeover)	Limited flexibility (2–4 h reconfiguration)	Universal fabric handling: Robotic auto-calibration (90% changeover efficiency)	Adaptive AI platforms: Deep reinforcement learning adjusts jet pressure/flow for any substrate. Modular reactors: Swap dyeing chambers in <15 min
Production waste	5–12% (shade correction, uneven dyeing)	3–8% (tail end losses, batch inconsistencies)	<0.05% (inline defect detection)	Computer vision QC: Hyperspectral imaging at 120 m/min. Predictive analytics: AI forecasts dye migration errors with 99.7% accuracy
Resource utilization	60–70% water recovery (ETP plants)	75–85% water recovery	100% closed loop resource cycling	Zero-liquid discharge (ZLD): Graphene oxide membranes recover 99.9% dyes/salts. Circular chemistry: Spent dyes upcycled into pigments
Operational scalability	Batch-limited (≤500 kg/ machine)	Linear scale (≥1000 kg/h)	Nonlinear hyper-scaling: Digital twin-managed distributed production	Cloud manufacturing: AWS IoT Greengrass coordinates 200+ micro-factories. Blockchain traceability: Smart contracts automate resource allocation

Table 14.2 Advanced comparison between traditional vs. industry 6.0 textile printing techniques

Feature	Traditional exhaust printing (screen, table, flocking, foam)	Traditional continuous printing (rotary, flatbed)	Industry 6.0 AI-driven printing (6D digital, quantum Dot, MEMS Jet)	Technical breakthroughs
Resolution	45–75 dpi (screen); 150 dpi (flocking)	100–300 dpi (rotary); 600 dpi (flatbed)	1200–2400 dpi (0.5 pL droplet control)	MEMS jet nozzle arrays: 40,000 nozzles/m with ±0.1 μm alignment
Speed	5–15 m/min (screen); 3–8 m/min (flocking)	30–70 m/min (rotary); 15–40 m/min (flatbed)	120–450 m/min (adaptive AI speed modulation)	6D robotic arms: non contact printing on 3D surfaces at 200 m/min
Ink utilization	60–70% (40% waste in screens/ flooding)	75–85% (loss in doctor blade/ squeegee)	98.5% (AI predictive ink deposition)	Closed loop nano filtration: 99.8% pigment recovery
Water consumption	50–80 L/kg (wash off/ soaping)	30–50 L/kg (soaping/after print treatment)	0.1–0.5 L/kg (waterless UV cure/phase change inks)	Supercritical CO_2 fixation: zero aqueous process
Energy use	80–120 kWh/ ton (drying at 160 °C)	60–100 kWh/ton	8–15 kWh/ton (LED UV curing at 5 W/cm^2)	Photonic drying: NIR lasers (0.3 s dwell; 50% energy reduction)
Color gamut	70% pantone (8–12 colors)	85% pantone (12–16 colors)	99% pantone (16 bit 1.07 B colors; nano pigment dispersion)	Quantum dot inks: 110% NTSC gamut
Set-up time	45–120 min (screen engraving/ tensioning)	20–60 min (frame alignment)	<90 s (auto calibrating AI vision)	Neural network pre-press: instant RIP processing
Waste generation	12–18% (misprints/ screen reclaim)	8–12% (registration errors)	<0.3% (real-time defect correction)	Hyperspectral inline QC: 10 μm flaw detection at 200 m/min
Chemical footprint	High (emulsifiers/ thickeners; 150–300 g/kg)	Moderate (urea/ formaldehyde; 80–150 g/kg)	Near zero (bio solvent inks; enzyme-based thickeners)	Mycelium-derived binders: compostable adhesion
Flexibility	Low (fixed patterns/ screens)	Medium (limited redesign flexibility)	Instant pattern switch (cloud-based digital libraries)	Blockchain-integrated IP: secure real-time design updates (TexTrace®)
Smallest detail	0.3 mm (screen); 0.8 mm (flocking)	0.1 mm (rotary); 0.05 mm (flatbed)	0.005 mm (substrate-adaptive deposition)	Electrohydrodynamic jetting: 500 nm line resolution (Nano-dim® EHD)

(continued)

Table 14.2 (continued)

Feature	Traditional exhaust printing (screen, table, flocking, foam)	Traditional continuous printing (rotary, flatbed)	Industry 6.0 AI-driven printing (6D digital, quantum Dot, MEMS Jet)	Technical breakthroughs
CO_2 emissions	2.8–4.5 tons/ton	1.9–3.2 tons/ton	0.08–0.25 tons/ton (100% renewable energy microgrids)	Carbon negative inks: Algae-based pigments

adhere directly to fiber surfaces via binder-assisted fixing or UV activated crosslinking. The cationic pre-treatment of cloth, which improves pigment adherence and inhibits color migration, is the fundamental chemistry of pigment digital printing. Pigments based on quantum dots offer improved color vibrancy and luminosity while using less ink. Textiles treated with plasma enhance ink penetration and absorption, resulting in an increase in print longevity. Electrostatic-assisted ink deposition is made possible by graphene oxide-based conductive coatings, which improve color uniformity and cut ink usage by 20%. To guarantee high-precision, flawless textile printing, Industry 6.0 printing technology integrates deep learning algorithms, real-time spectral color analysis, and AI-powered inkjet nozzles. AI-powered smart nozzles improve print resolution to 1200 dpi and prevent oversaturation by dynamically adjusting droplet size (1–10 pl). Spectral imaging feedback loops reduce variances between production batches by maintaining color constancy within ±0.5%. Real-time print quality analysis using machine-learning algorithms reduces print rejects by 90% by identifying and fixing banding or color bleeding issues [56–58]. A comparison of Industry 6.0 digital printing technologies and traditional textile printing methods is shown in the following Table 14.5, with an emphasis on performance enhancements, ROI, cost savings, and environmental effect.

By cutting waste, increasing productivity, and lowering production costs, Industry 6.0 printing technologies are transforming the textile manufacturing industry and offering major environmental and economic benefits. Compared to conventional rotary screen printing, AI-optimized digital printing significantly reduces CO_2 emissions by 80%, making it a crucial component of the creation of carbon-neutral textiles. Because waterless pigment digital printing removes dye effluent discharge, 95% of water resources are saved, and 60% less energy is used, significantly reducing the textile industry's environmental impact. Another important advantage is cost effectiveness, which results in lower operating expenses, thanks to ink savings, less chemical use, and quicker machine setup times. Because Industry 6.0 digital printing technologies shorten payback periods to just 1–3 years (highly dependent on the product mix and volume) instead of the 5+ years needed for conventional printing technologies, ROI is greatly expedited. In Industry 6.0, smart textile printing is set to become completely autonomous, sustainable, and precision

Table 14.3 Advanced comparison between traditional vs. Industry 6.0 denim processing

Feature	Traditional rope dyeing	Traditional garment washing	Industry 6.0 rope dyeing	Industry 6.0 garment washing
Water consumption	60–100 L/kg (multiple dip oxidation cycles)	40–80 L/kg (stone/enzyme washing)	3–8 L/kg (AI-controlled closed-loop dyeing)	1–4 L/kg (nano bubble/ O_3 washing)
Energy use	120–180 kWh/ton (steam drying at 120 °C)	80–130 kWh/ ton (tumble drying)	25–40 kWh/ton (RFID tagged yarn tracking + IR drying)	10–20 kWh/ ton (plasma surface activation)
Chemical usage	High (indigo, hydrosulphite, NaOH; 150–250 g/ kg)	Moderate (pumice, enzymes, bleach; 100–200 g/kg)	70% reduction (electrochemical indigo reduction)	90% reduction (laser ablation + enzyme free bio-polishing)
Process time	8–12 h (12–18 dip cycles)	60–120 min/ wash cycle	2–3 h (continuous AI-optimized oxidation)	8–15 min (high precision laser patterning)
Color consistency	ΔE 3.5–5.0 (manual dip control)	ΔE 4.0–6.0 (variable stone impact)	ΔE ≤ 0.8 (real-time spectrophotometry)	ΔE ≤ 0.5 (3D laser mapping)
Effluent pollution	High (Sulphide/azo dyes; BOD 8000–12,000 ppm)	Moderate (pumice sludge; microplastic shedding)	Near zero (bio-based indigo + membrane filtration)	Zero liquid discharge (O_3 gas recycling)
Automation level	Semi-automated (20% IoT sensor coverage)	Manual loading/ unloading	Full autonomy (AI-driven robotic yarn handling)	Cognitive robotics (computer vision guided laser systems)
Fabric damage	5–8% tensile loss (mechanical stress)	7–12% abrasion (pumice impact)	<0.5% (tension controlled dyeing)	Near zero (non-contact laser treatment)
Customization	Limited (batch specific shades)	Low (uniform fading effects)	Hyper-personalization (digital twin-enabled dye profiling)	Micro precision design (sub millimeter laser etching)
CO_2 emissions	4.2–6.5 tons/ton	2.0–3.8 tons/ ton	0.3–0.7 tons/ton (solar-powered indigo vats)	0.05–0.15 tons/ton (renewable energy microgrids)
Resource recovery	40–60% water recycle (conventional ETP)	<30% (sludge disposal)	99% (graphene-based dye recovery)	100% ($O_3 \rightarrow O_2$ conversion)

Table 14.4 Advanced comparison between traditional vs. Industry 6.0 yarn and fiber dyeing

Feature	Traditional yarn dyeing	Traditional fiber dyeing	Industry 6.0 AI yarn dyeing	Industry 6.0 AI fiber dyeing
Water consumption	40–60 L/kg (high liquor ratio 1:15–1:20)	50–80 L/kg (inefficient penetration)	1.2–3.5 L/kg (ultrasonic-assisted dyeing)	0.8–2.0 L/kg (supercritical CO_2 impregnation)
Energy use	90–130 kWh/ton (steam heating to 130 °C)	110–160 kWh/ton (long heating cycles)	18–28 kWh/ton (RF direct yarn core heating)	15–25 kWh/ton (plasma-activated fiber surfaces)
Dye fixation	70–80% (inefficient migration)	60–75% (channeling issues)	97–99.5% (Nano encapsulated dyes)	95–98.5% (ionic liquid carriers)
Chemical usage	Salt 30–50 g/L; alkali 15–25 g/L	Salt 40–70 g/L; dispersants 20–40 g/L	Salt to free dyeing (cationic pretreatment)	Bio auxiliaries (enzyme based dispersants)
Process time	6–10 h (slow exhaustion)	8–14 h (multiple rinses)	70–120 min (AI-optimized ramp rates)	90–150 min (continuous microwave fixation)
Color consistency	ΔE 2.5–4.0 (package density variations)	ΔE 3.0–5.0 (fiber clumping)	ΔE ≤0.5 (inline NIR spectroscopy)	ΔE ≤0.7 (3D fiber mapping)
Automation level	Semi-automated (manual loading)	Manual stacking/loading	Cognitive robotics (auto doffing + AI recipe control)	Autonomous drums (IoT weight sensors + adaptive tumbling)
Material waste	7–12% (Barre effects, package deformation)	10–18% (Fiber entanglement damage)	<0.8% (tension-controlled winding)	<1.2% (anti felt RFID fiber tagging)
Effluent toxicity	High (AOX 200–400 mg/L; COD 8000–12,000 ppm)	Extreme (dye residues 25–40%)	Closed loop ZLD (99.8% recovery)	Zero discharge (electrochemical decomposition)
Customization	Batch limited (≥500 kg)	Minimal (large lot requirements)	Micro-batch (1–50 kg) (digital twin simulation)	Fiber-level customization (quantum dot tracing)
CO_2 footprint	3.8–5.5 tons/ton	4.2–6.8 tons/ton	0.25–0.55 tons/ton (biogas-powered systems)	0.15–0.40 tons/ton (solar thermal integration)

driven. To guarantee zero-waste manufacturing, blockchain-based ink traceability, AI-driven defect detection, and quantum-dot color formulations will all be integrated. With their 100% sustainability and high wash and rub fastness, advanced nanoparticle-based biopolymer inks are becoming a viable biodegradable substitute that is perfect for producing environmentally friendly textiles. A highly intelligent and self-correcting textile production ecosystem will result from the use of digital twin simulations and cyber-physical printing systems, which will also make predictive maintenance, real-time process optimization, and closed-loop sustainability possible.

Table 14.5 Advanced comparative analysis of traditional vs. Industry 6.0 digital printing technologies

Parameter	Rotary screen printing	Flatbed screen printing	Traditional digital inkjet (reactive)	Industry 6.0 AI-powered digital inkjet
Water usage	60–80 L/kg	50–70 L/kg	20–30 L/kg	0.2–1.5 L/kg (nano dry tech + waterless pigment inks)
Energy consumption	3.0–3.5 kWh/kg (steam drying)	2.8–3.2 kWh/kg	1.5–2.0 kWh/kg	0.15–0.35 kWh/kg (photonic drying + AI optimized HVAC)
Chemical waste	High (40–50% dye loss)	High (35–45% dye loss)	Medium (15–25% auxiliaries)	<0.5% (closed loop nano-filtration; 99.8% recovery)
CO_2 emissions reduction	Baseline (0%)	Baseline (0%)	30%	92–97% (solar microgrids + carbon-negative algae inks)
Operational cost savings	Low (5–8% vs. baseline)	Low (5–10%)	Moderate (25–35%)	58–65% (predictive maintenance + zero waste handling)
Production speed	45–55 m/min	18–25 m/min	8–15 m/min	120–200 m/min (MEMS jet arrays at 100 kHz)
Print resolution	250–300 dpi	200–250 dpi	600 dpi	2400–4800 dpi (sub picolitre droplet control; 0.3 pl precision)
Customization capability	Limited (≥500 m batches)	Limited (fixed designs)	High (on-demand designs)	Unlimited (generative AI + blockchain IP; per-meter personalization)
Investment ROI	3–5 years	4–6 years	2–4 years	8–14 months (intelligent manufacturing subsidies +45% energy tax credit)
Critical innovation	Mechanical stencil limitations	Manual screen alignment	Aqueous ink constraints	Autonomous CPS: AI-driven spectral QC, self-calibrating printheads, IoT resource tracking

6.1 Digital Transformation in Textile Dyeing and Processing

AI, IoT, nanotechnology, and digital twins are all driving a rapid transition in the textile dyeing and finishing sector. Through automation, sustainability-focused solutions, and real-time data analytics, Industry 6.0 advancements are streamlining wet processing, printing, and finishing. These developments improve accuracy, repeatability, and fabric performance while drastically lowering water, chemical, and energy usage. By examining spectral reflectance curves at the molecular level, AI-powered color matching algorithms guarantee accurate shade reproducibility and achieve 99.9% dye consistency across fiber, yarn, and fabric dyeing. Intelligent continuous and batch dyeing systems optimize dye absorption, liquor ratio, and temperature control by integrating real-time process control for Jet, Jigger, Pad, and Beam Dyeing equipment. Predictive chemical dosing is made possible by digital twin-based dye bath monitoring, which reduces dye and auxiliary waste. IoT-enabled sensors in dyeing equipment track temperature and humidity variations, dynamically modifying the environment to preserve the best reaction kinetics for dye fixation. By using ultra-low liquor ratio dyeing (L: R < 3: 1), 60% less water is used, reducing chemical loading and effluent output. 100% dye fixation with zero wastewater output is made possible by supercritical CO_2 dyeing technique, which fully removes the need for water. Precision-aligned servo motors and adaptive tension management are now features of AI-enhanced rotary and screen-printing machines, guaranteeing ±0.1 mm accuracy in pattern registration. Because of its low-energy and water consumption, pigment-based inkjet printing is growing at a 35% CAGR (IDC, 2024), contributing to the rapid rise of digital textile printing. Digital printing errors are avoided, and color consistency is enhanced using IoT-enabled real-time ink viscosity and humidity monitoring [59, 60].

Nano-dispersed pigments are used into AI-driven ink formulations to improve fiber penetration and decrease particle agglomeration, resulting in increased wash fastness and color vibrancy. During wrinkle-free finishing, intelligent moisture management technologies accurately control the fabric's hydration levels, minimizing fiber deterioration (strength reduction etc.) brought on by resin. Fiber surfaces are altered at the nanomolecular level by AI-optimized plasma finishing, which permits water-free functionalization for enhanced dye affinity, hydrophobicity, and anti-pilling properties. Functional finishing powered by nanotechnology improves fabric performance with conductive, UV-resistant, antimicrobial, and self-cleaning coatings. Flame-retardant, anti-static, and conductive fabric coatings are applied uniformly and with nanoscale accuracy thanks to AI-controlled chemical deposition devices. BASF's AI-enhanced chemical finishing has reduced the usage of harmful chemicals in flame retardants while increasing manufacturing efficiency in protective textiles. Higher-quality, sustainable, and intelligent textile production is ensured by this shift to data-driven textile processing, which is consistent with SDGs 9 and 13. A climate-resilient, ultra-efficient textile ecosystem is being paved by Industry 6.0-driven dyeing, printing, and finishing technologies, which reduce environmental impact while improving production agility, accuracy, and cost-effectiveness [25, 61, 62].

6.2 *Denim Dyeing and Washing in Sustainable Innovations*

Industry 6.0-driven innovations that improve sustainability, efficiency, and resource optimization are revolutionizing the denim processing industry. By altering the fiber structure at the molecular level, plasma surface activation improves fabric durability, decreases dye migration, and increases indigo dye adhesion. By improving the electrostatic interaction between the fabric and the dye molecules, this technique does away with the requirement for numerous dipping cycles and unnecessary dye usage. By taking the place of conventional oxidizing agents like potassium permanganate and sodium hypochlorite, ozone-based washing has revolutionized denim finishing. This method efficiently eliminates extra indigo while consuming less water and removing hazardous effluents by producing active ozone (O_3) radicals. Rope dyeing involves passing several yarns (usually 12–36 ropes per machine) through a sequence of 6–12 indigo dye baths. Oxidation occurs in between dips, accumulating the indigo layer for a rich, deep hue. Sulfur bottoming adds pre-dying with sulfur black or brown dyes, adding depth and producing darker or greener denim hues. Now a days, due to cost constraints, denim industries are also used reactive dyes blacks for bottoming. Without it, the finished color is still pure indigo blue. Real-time oxidation sensors, AI-controlled dye bath monitoring, and nano-engineered surface alterations are all examples of Industry 6.0 innovations that maximize dye penetration, guarantee uniform shade repeatability, and save water and chemical waste. The distinctive texture and appearance of denim are mostly attributed to the garment finishing process. Precision-controlled cellulases are used in the newest AI-driven enzyme washing methods to selectively break down surface fibers, producing antique effects and a softer touch feel while using less water and chemicals. Synthetic abrasive stones and bio-polishing enzymes have taken the place of pumice stones in stonewashing, which increased fabric longevity and prevented microtears. Distressing processes have undergone a complete transformation thanks to advanced denim abrasion technology like laser finishing. By simulating vintage wear patterns with micron-level accuracy, high-energy CO_2 and fiber lasers do away with the need for abrasive chemicals and human scraping. This method eliminates workers' exposure to dangerous silica dust and cuts energy use by 40%. Another innovation is nanobubble technology, which uses high-pressure cavitation microbubbles to improve dye diffusion within fibers, guaranteeing deeper dye penetration while using 40% less water and color. Nano-engineered cotton-polyester blends, which offer improved flexibility, moisture wicking capabilities, and heat regulation, are another example of how denim fiber innovation has advanced. Because graphene-infused yarns in Industry 6.0 denim mills distribute heat uniformly, denim clothing may be worn in any condition. The use of synthetic indigo derived from petrochemicals has decreased by 80% thanks to the discovery of bio-based indigo dyes made from genetically modified bacteria. As nanotechnology, AI-driven process optimization, and cyber-physical systems develop further, denim production is heading toward a high-efficiency, zero-waste future, establishing new standards for performance-driven fashion and environmental responsibility [12, 63, 64].

7 Textile Machinery Manufacturing in Industry 6.0

With the integration of AI, quantum computing, nanotechnology, and CPS, the production of Industry 6.0 textile machines is experiencing a technological revolution. These devices drastically cut down on energy use, material waste, and production downtime since they are built for zero-defect production, real-time adaptive manufacturing, and high sustainability. To transform traditional textile production into an intelligent, self-optimizing ecosystem, textile machinery has evolved to include self-learning algorithms, smart sensors, autonomous robots, and predictive maintenance systems. AI-controlled smart looms that can identify and fix flaws in real time have replaced mechanical Jacquard looms in Industry 6.0 weaving technology. With sub-micrometer accuracy, advanced piezoelectric sensors and machine vision algorithms track yarn tension, warp breaks, and misalignment, guaranteeing ±0.5% accuracy in fabric quality. Autonomous robotic shuttle loading removes the need for physical intervention, while high-speed air-jet looms (weft insertion speeds 2500 m/min) and water-jet looms with specialized nozzles minimize weft misalignment by 60%. In addition to using quantum computing techniques for real-time predictive analytics, these AI-driven solutions enable machines to modify weaving patterns in response to changes in ambient humidity and yarn flexibility. Modeling the molecular-level interactions between dyes, chemicals, and textile fibers using quantum computing is one of the most revolutionary developments in textile equipment. To maximize dye efficiency, quantum-enhanced simulations forecast how reactive dyes will interact with cotton hydroxyl groups (-OH) or dispersion dyes will interact with polyester ester (-COO-) linkages. With this degree of computational accuracy, less dye is wasted, guaranteeing consistent shade and doing away with the need for repeat dying cycles.

The development of high-performance, graphene-reinforced fibers is also made easier by quantum-enabled molecular dynamics simulations, which could result in fabrics that are 30% lighter and 40% stronger than those made of traditional materials. High-speed robotic automation has been added to modern textile machinery to ensure smooth manufacturing processes. Advanced haptic feedback systems on six-axis robotic arms ensure minimal fabric distortion during automated garment assembly by precisely cutting and managing fabric tensioning with an accuracy of ±0.01 mm. Inspired by biomimetic structures, soft robotic grippers manage high-elasticity spandex and delicate silk threads without causing microstructure damage. Uneven winding problems in spinning machines are eliminated by robotic yarn dispensers with AI-driven tension adjustment, which results in an improvement in yarn quality. Blockchain-based supply chain management systems and Industry 6.0 textile technology are tightly connected, offering 100% traceability of raw inputs, fabric processing, and final products. Every batch of recycled fiber, bio-based polyester, or organic cotton is given a blockchain token, and its movement is monitored by GPS-encoded smart labels and RFID tags connected to the Internet of Things. As a result, manufacturers may confirm whether raw materials fulfill Oeko tex, GOTS, Higg Index, or Blue sign certifications, ensuring supply chain transparency. Blockchain integration ensures ethical work practices and removes counterfeiting issues with end-to-end encryption. Additive manufacturing, laser microfabrication, high-precision CNC machining, and AI-optimized assembly are all used in the construction of next-generation textile production equipment. Titanium alloy components for high-speed weaving looms are now 3D manufactured using additive manufacturing, which improves mechanical durability while reducing machine weight. Use of CNC Milled Components For ring and rotor spinning machines, a high-precision technology produces carbon fiber-reinforced spindles and rotors, which reduce vibration and increase machine longevity. Nano-structured ceramic coatings for dyeing machine rollers improve dye penetration efficiency in laser-assisted surface engineering, resulting in a 30% reduction in fabric shrinkage and consistent color application. Automated assembly-line robots equipped with AI vision systems examine machine parts with a resolution of 0.01 mm in AI optimized machine assembly, guaranteeing flawless assembly and boosting output. High Energy Efficiency: Modern looms with AI-controlled motors use less energy, which lowers the yearly carbon footprint of textile mills by 500 metric tons. The best water recycling systems greatly reduce industrial discharge by recovering 95% of wastewater using integrated membrane distillation units in dyeing machines. The best hybrid renewable energy integration system is used in smart textile manufacturers, which have a 60% lower reliance on fossil fuels thanks to waste-heat-recovery dyeing equipment and solar electricity [65–67].

AI integration, robots, digital twin simulations, and sophisticated software ecosystems are propelling the development of textile machinery in Industry 6.0, resulting in a highly automated, sustainable, and efficient textile production environment. Computer vision, deep learning, and real-time sensor analytics are used by AI-powered weaving, knitting, dying, and finishing machines to maximize fabric output, guaranteeing zero-defect manufacturing and a 50% energy consumption reduction. To monitor performance in real time and make autonomous modifications that reduce material waste and increase production efficiency, these smart machines use edge computing and the Internet of Things. Textile machinery is undergoing a transformation because to digital twin-based predictive maintenance, which builds virtual models of actual machines to mimic operational performance, identify irregularities, and anticipate faults before they happen. By optimizing heat dissipation, yarn tension control, and airflow dynamics in textile processes, high-fidelity simulations employing finite element analysis and computational fluid dynamics can cut downtime and increase machine lifespans. With a tolerance of ±0.3% in yarn uniformity, AI-driven fault detection in spinning and weaving machines enables automated recalibration, guaranteeing consistent yarn quality. Textile assembly lines are being revolutionized using robotics in automated cutting, sewing, and material handling. With the help of force sensors and AI-powered adaptive learning algorithms, cobots can cut fabric precisely with 99.9% accuracy, removing human-induced variance. Delicate fabrics are handled effectively by soft robotic grippers equipped with bioinspired touch sensors, which lowers manufacturing flaws in luxury clothing. Real-time fabric flexibility adaptation by autonomous robotic sewing systems combined with AI-based path optimization algorithms removes misalignment problems and improves stitching accuracy. AI-based generative design is used in Industry 6.0 software for textile CAD/CAM integration, allowing for quick prototyping and customization of fabric patterns. With 90% accuracy, neural networks automatically create textile designs that satisfy contemporary fashion demands by analyzing consumer preferences and market trends. Global designer collaboration is made possible by cloud-based CAD/CAM technologies, which also optimize manufacturing workflows and cut down on design iteration cycles. The next-generation AI integrated automation is best demonstrated by Gerber

Technology's AI-powered textile machinery optimization, which improves fabric cutting accuracy, cuts down on textile waste by 35%, and boosts material use efficiency [16, 68, 69].

8 Automation in Colors, Inks, Dyes, and Pigments Manufacturing

Industry 6.0's textile sector combines AI, quantum computing, smart sensors, and automation to create next-generation machinery and chemical processes that improve efficiency, sustainability, and precision. Self-adjusting smart looms with computer vision and AI-based defect detection systems are examples of advanced textile machinery that reduces fabric irregularities by 85%. Quantum computing is revolutionizing material science by enabling simulations of dye-fiber molecular interactions at the atomic level, which significantly reduces dye waste and increases dye uptake efficiency by 40%. Blockchain integrated supply chain tracking ensures 100% traceability of textiles from raw materials to finished garments, preventing counterfeit materials and guaranteeing compliance with sustainability regulations. With AI-driven process control, autonomous reactors, and intelligent nanotechnology applications, textile chemical manufacturing in Industry 6.0 is completely automated, from pretreatment to finishing. Traditional alkaline pretreatment methods are replaced with enzymatic bio-scouring, which uses less water and does away with harsh sodium hydroxide-based procedures. By optimizing peroxide decomposition through real-time spectrophotometric monitoring, digitalized continuous bleaching systems can reduce chemical overdose while maintaining consistent control over the whiteness index. Autonomous nanoparticle synthesis in dyeing and pigment manufacture yields high-performance textile pigments with improved fiber penetration at the nanoscale, higher UV stability, and 40% better color fastness. Fabric re-dying requirements are reduced, thanks to AI-controlled pigment formulation, which guarantees batch-to-batch color constancy within ±0.3% and improves dye uniformity and repeatability.

By combining quantum dot spectrophotometry and hyperspectral imaging, smart color-matching systems achieve near-zero metamerism, guaranteeing precise color matching across a range of lighting situations. Intelligent coatings that offer specialized functions without altering fabric texture or breathability are the main focus of Industry 6.0 textile finishing. With a 160° water contact angle and fluorine-free nanocoating, advanced water and oil repellent finishes ensure hydrophobicity while lowering environmental pollution from PFAS-based finishes. Nowadays, phosphorus-based intumescent coatings and ceramic nanoparticle dispersion are integrated into flame-retardant textiles to achieve ISO 11612 certified fire resistance with less chemical loading than conventional brominated retardants. Utilizing microencapsulated bio-based repellents (such as citronella, permethrin, and DEET derivatives), mosquito-repellent textiles provide regulated release for up to 50 washes while preserving the safety of human skin. Utilizing copper ions, zinc oxide, and silver nanoclusters, antimicrobial coatings offer 99.9% bacterial inhibition (ISO 20743 standard), avoiding Odor accumulation, and improving clothing cleanliness. AI-driven plasma finishing improves tension release coatings even more, increasing fabric drapability and stretch recovery. This is advantageous for applications in automobile interiors, sportswear, and aerospace fabrics. These days, smart finishing machines have AI-powered adaptive chemical dosing systems that optimize reaction kinetics, pH control, and liquor ratio in real time. Traditional padding techniques are replaced by intelligent finishing rollers and digital spray systems, which guarantee an 80% reduction in chemical usage and enhance consistency in finishes that are moisture-wicking, UV protective, and wrinkle free. Plasma-enhanced finishing reactors alter fiber surfaces without using a lot of water, increasing wear resistance, hydrophobicity, and the adherence of functional coatings at the nanomolecular level. AI-driven automation turns the manufacture of textile chemicals and pigments into an incredibly intelligent, ecofriendly, and efficient manufacturing system by cutting processing time, consuming less energy, and minimizing chemical waste [70–72].

9 Intelligent Retail and Textile Waste Management

With AI-powered demand forecasting, real-time supply chain optimization, and virtual textile try-ons, the textile and fashion retail industry are witnessing a digital revolution that will save waste and boost productivity. Deep learning and predictive analytics-driven AI-powered inventory management solutions optimize supply chains by precisely predicting demand and minimizing overproduction. Dynamic stock replenishment is ensured by IoT-based warehouse automation and AI-integrated computer vision, which reduces excess textile waste and unsold inventory. Through the elimination of the need for in person fittings, a decrease in online return rates, and a reduction in emissions connected to logistics, AR virtual try on technology is transforming customer contact. By tracing the lifetime of clothing from the sourcing of raw materials to post-consumer recycling, blockchain-enabled digital product passports improve circular economy activities by offering end-to-end textile traceability. With 92 million tons of textile waste generated annually, Industry 6.0 introduces AI-powered trash sorting, enzymatic textile recycling, and carbon capture technologies to drive zero-waste, circular textile economies. AI-powered robotic waste sorting systems, utilizing hyperspectral imaging and near-infrared (NIR) spectroscopy, can separate polyester, cotton, and blended materials with 99% accuracy, enhancing the efficiency of fiber-to-fiber recycling. By breaking down polyester into terephthalic acid and ethylene glycol using genetically modified biocatalysts like PETase and MHETase, advanced enzymatic textile recycling can achieve a 95% monomer recovery rate and facilitate closed-loop textile-to-textile recycling. The environmental impact of the textile sector is being further reduced by the incorporation of carbon capture. The industry's CO_2 footprint

Fig. 14.1 Textile circular economy in Industry 6.0

is reduced by 50% thanks to sophisticated catalytic CO_2 conversion devices that absorb and transform carbon emissions from textile dyeing and finishing operations into valuable compounds like methanol and formic acid. Synthetic textile waste can be converted into hydrogen-rich syngas by AI optimized plasma gasification, which promotes waste-to-energy solutions and gets rid of the buildup of textiles in landfills. To guarantee sustainable manufacturing and waste reduction in the global fashion supply chain, the textile industry is moving toward a fully circular, resource efficient, and climate resilient sector with the help of Industry 6.0 innovations. The transition to Industry 6.0 waste management technologies enables smart, sustainable, and highly efficient resource recovery processes, significantly reducing industrial waste and environmental impact. One of the most transformative benefits is carbon emission reduction, where AI-optimized waste sorting, carbon capture, and waste to energy conversion achieved, mitigating climate change effects in manufacturing industries. AI-powered recycling systems and biochemical recovery enhance resource efficiency, improving material reusability rates to 90%, which minimizes landfill dependence and promotes a circular economy [73, 74]. Figure 14.1 shows the basics of textile circular economy descriptions.

9.1 Water and Energy Saving and Conservation

Water and energy savings are another critical advantage of Industry 6.0 waste management. Intelligent wastewater treatment and dye effluent recycling recover of water and reduce chemical and dye waste, ensuring compliance with global sustainability regulations such as EU REACH and ZDHC (Zero Discharge of Hazardous

Chemicals) programs. From an economic perspective, AI-driven industrial waste management lowers operational costs, leading to projected annual savings of USD 50 billion in sectors such as textile, chemical, and industrial manufacturing through automated waste sorting, real-time emission monitoring, and AI-powered recycling plants. Industry 6.0 solutions enhance health and environmental protection by eliminating hazardous pollutants, heavy metals, and toxic byproducts from industrial zones. By deploying advanced nanofiltration systems, enzymatic bioremediation, and AI-monitored air purification units, industries can significantly reduce respiratory diseases, waterborne illnesses, and heavy metal contamination risks, contributing to a healthier workforce and cleaner ecosystems. The shift from traditional waste disposal to AI-powered circular waste management ensures that industries reduce their environmental footprint while optimizing economic efficiency. Industry 6.0 waste management technologies directly support SDG, driving the manufacturing sector toward a zero-waste, climate-resilient, and energy-efficient future. As industries continue adopting AI, IoT, and biochemical recovery systems, waste will no longer be seen as an industrial byproduct but rather as a valuable resource for renewable energy production, chemical recovery, and sustainable raw material regeneration [74–80].

9.2 Smart Retail and AI-Driven Fashion Forecasting in Industry 6.0

The retail textile industry is undergoing a significant transition, using blockchain-driven inventory management, augmented reality, and AI-powered predictive analytics to improve sustainability and efficiency. Real-time consumer behavior data is processed by AI-powered inventory management systems, which optimize global supply chains and cut overproduction. Deep learning algorithms estimate demand with 98% accuracy by analyzing past trends, material costs, and climate-based fashion preferences. By preventing the accumulation of unsold inventory, this accuracy reduces textile waste and boosts manufacturing efficiency. Virtual try-on technologies powered by AR and AI allow customers to see clothing in three dimensions before making a purchase, which lowers online return rates and reduces the amount of textile waste that results from product returns. Users can evaluate texture, drape, and stretch properties using AI-driven fabric simulation engines and haptic feedback technologies, which improves the online buying experience. About 100% traceability in textile supply chains is made possible by blockchain integrated RFID tagging and smart labels, which also ensure ethical sourcing and validate product sustainability credentials. Because of the 92 million tons of textile waste produced by the fashion industry each year, circular economy solutions are crucial to sustainability. Textile waste management is being revolutionized by Industry 6.0, which combines carbon capture technology, enzymatic recycling, and AI-driven trash sorting. Robotic trash sorting systems driven by AI and outfitted with hyperspectral imaging and ML can detect polyester, cotton, wool, and blended textiles with 99% accuracy, increasing recycling efficiency and lowering landfill buildup. Enzymatic textile recycling achieves material recovery rate by depolymerizing polyester into its monomeric building blocks using biocatalysts. Using genetically modified bacteria, microbial polyester degradation allows for a closed-loop recycling method that turns post-consumer trash back into premium virgin-grade polyester. Cutting-edge mechanical and chemical cotton recycling techniques preserve 80% of the cellulose fibers' tensile strength, enabling their incorporation into new fabric mixes without sacrificing quality. Carbon capture and utilization, which is introduced in Industry 6.0, is the process of turning CO_2 emissions from dyeing operations into high-value textile chemicals like sodium bicarbonate and solvents based on methanol. By using this method, industrial carbon footprints are reduced. In the era of Industry 6.0, AI-driven energy-efficient textile factories that run on renewable energy and green hydrogen further cut carbon emissions, guaranteeing a completely sustainable, zero-waste textile environment [81, 82].

10 Challenges and Future Opportunities in Textile Production

Although the shift to Industry 6.0 textile production offers revolutionary improvements in automation, AI-driven optimization, and sustainability, widespread adoption will require resolving several technological, financial, and labor-related issues. The hefty initial expenditure needed for completely automated textile production systems, which can range from 50 to 100 million USD per large-scale factory is one of the biggest obstacles. Significant capital investments for sensor networks, digital twin infrastructure, and next-generation IoT-enabled machinery are required for the integration of CPS, AI-powered predictive analytics, and intelligent robotic automation. However, according to long term financial modeling, ROI can be attained in 2–4 years by optimizing energy consumption (up to 60% efficiency increase), reducing labor expenses (50%) and material waste (40%). Because digital supply chains in Industry 6.0 textile ecosystems are more susceptible to cyberthreats, intellectual property theft, and data breaches, cybersecurity is another crucial issue. Manufacturers must use quantum-resistant encryption techniques to protect secret textile formulations, AI training datasets, and consumer insights in the context of blockchain-based real-time inventory tracking, AI-powered demand forecasting, and automated supply chain management. In order to ensure 100% traceability and fraud-proof operations, post-quantum cryptographic techniques like multi-factor blockchain verification and lattice-based encryption will be essential to textile cybersecurity in the future. As AI-driven automation and robotics replace conventional textile manufacturing roles, workforce reskilling continues to be a significant challenge. According to studies, 50% of textile workers would need to retrain by 2030, which calls for extensive training initiatives in robotic maintenance, digital textile chemistry, AI-assisted production management, and smart material engineering. Cobots, or collaborative robots, will work alongside human operators in human-machine collaboration frameworks to ensure smooth automation integration without total labor displacement. Notwithstanding these obstacles, Industry 6.0 offers previously unheard-of chances to develop a textile industry that is completely circular, waste-free, and climate resilient. The 92 million tons of textile waste produced each year will be decreased by emerging technologies like AI-driven enzymatic textile recycling, which breaks down polyester into monomers with a 95% recovery rate. Carbon capture technology in textile dyeing processes will enable the conversion of CO_2 emissions into reusable textile chemicals, cutting the carbon footprint of fabric finishing operations. Moving forward, the textile industry of the future will not only be hyper-efficient and fully autonomous, but also climate-adaptive, driven by self-learning AI systems, nanotechnology-based textile coatings, and sustainable bioengineered fibers. By aligning with SDGs. Industry 6.0 textile innovations will set the foundation for a next-generation, intelligent, and eco-friendly textile revolution [83–85].

11 Smart Packaging and Logistics

To guarantee traceability, sustainability, and operational efficiency, the textile industry is quickly incorporating AI, IoT, blockchain, and edge computing into smart packaging and logistics. RFID, NFC (Near-Field Communication), and QR code-based tracking are used by AI-powered intelligent packaging systems to track the location, condition, and compliance of clothing with sustainability certifications such as Oeko tax, GOTS, and Bluesign in real-time. Instant verification of sustainable and organic textiles is made possible by these smart labeling technologies, which also reduce supply chain fraud through automated authentication and counterfeit identification. Blockchain technology is essential to the transparency of the textile supply chain because it creates an unchangeable ledger that documents the sourcing of raw materials, chemical consumption, carbon footprint, and compliance with worker welfare regulations. Blockchain smart contracts and AI-driven predictive analytics maximize cross-border textile exports by cutting transaction costs and paperwork while guaranteeing adherence to global labor and environmental regulations. To reduce delays and fuel consumption in the distribution of textiles, decentralized AI-led logistics platforms employ edge computing and real-time data synchronization to dynamically modify shipping routes. Autonomous forklifts, computer vision, and robotic picking are all used by AI-enhanced warehouse automation systems to optimize inventory control. By eliminating stock inconsistencies, RFID-based intelligent warehouse management systems increase order fulfillment accuracy to 99.8% and reduce inventory waste by 40%. To improve storage allocation and ensure that high-demand textile products are pre-positioned close to distribution centers for just in time supply chain execution, deep learning algorithms examine consumer demand trends. By substituting water-soluble, compostable packaging solutions for conventional plastic clothing bags and utilizing bio-based polymers and biodegradable films, AI-integrated smart logistics systems maximize package sustainability. By halting humidity-induced fabric deterioration, nano-coated paper-based textile packaging with integrated moisture-resistant graphene barriers increases product shelf life. Manufacturers can offset emissions in real time via blockchain-enabled carbon credit trading systems thanks to AI-powered carbon footprint calculators in automated shipping labels that dynamically estimate CO_2 emissions each textile shipment. The transition to AI-powered textile logistics and intelligent packaging directly advances SDGs that encourage supply chains that are transparent, resource-efficient, and climate resilient. Industry 6.0 packaging and logistics solutions facilitate a zero-waste, intelligent textile ecosystem that improves sustainability and operational profitability by removing overproduction, streamlining inventories, and lowering emissions associated with logistics [86, 87].

12 Sustainability and Circular Economy in Textile Industry 6.0

With the integration of AI-driven waste reduction, intelligent recycling, and climate-resilient manufacturing systems, the textile industry is transitioning to the concepts of the circular economy in Industry 6.0. With 92 million tons of textile waste produced each year, the emphasis has turned to CO_2-based waterless dyeing, enzymatic textile breakdown, and smart fiber to fiber recycling, which reduce significant wastewater discharge. Thanks to sophisticated automation, sustainable textile companies are currently reducing CO_2 emissions, optimizing energy consumption, and increasing resource efficiency. Nike's AI-powered Move to Zero initiative has shown a 50% decrease in material waste and integrated blockchain for real-time sustainability monitoring. The combination of AI, quantum computing, digital twins, and nanotechnology signifies the shift from automated manufacturing (Industry 4.0) to self-optimized production (Industry 6.0). Just in time textile production and real-time demand forecasting are now possible thanks to AI-powered predictive analytics, which has reduced overproduction by 30%. By optimizing dye-fiber molecular interactions, quantum computing simulations reduce chemical waste and increase dye uptake efficiency. Supply chains will be revolutionized by the move to fully autonomous smart textile plants, which guarantee efficient material flows, automatic defect detection, and robotic textile assembly. Hugo Boss' AI-powered smart production environment has shown a reduction in carbon emissions, an increase in operational efficiency, and a decrease in textile waste. By incorporating automation, blockchain, and sustainable innovations into their manufacturing processes, several industry pioneers are spearheading the transition to AI-driven smart textiles. With its AI-powered agile supply chain management, Inditex optimizes fabric demand forecasts and cuts lead times to 48 hours. Water use in finishing operations is reduced by 96% thanks to Levi's blockchain-based transparency in sustainable denim production. AI, IoT, and automation are reshaping textile quality control, digital printing, and defect prediction while improving sustainability and workforce efficiency. Next generation innovations like smart nanofiber coatings, intelligent dyeing algorithms, and cyber-physical textile twins are ensuring climate-resilient textile production. The smart textiles market is expected to reach USD 20 billion by 2028, growing at a CAGR of 22.4% due to AI and IoT advancements. H&M has an AI-based circular economy textile model, where 90% of recycled polyester is sourced from post-consumer waste. Industry 6.0-driven automation is predicted to reduce water consumption in the worldwide textile industry and increase energy efficiency, while the digital textile printing sector is estimated to grow three times by 2030, accounting for 40% of the global textile printing market. The Sustainable Apparel Coalition created the Higg Index, a thorough sustainability measuring instrument that assesses social and environmental effects across the textile lifecycle. The Higg Facility Environmental Module (Higg FEM) now incorporates AI-driven capabilities to improve energy, chemical, and water usage, cutting down on sustainability reporting time. Real-time conservation methods can reduce water usage, thanks to predictive AI models that forecast patterns in water consumption. Digital twins driven by AI examine the energy flows in textile factories, reducing energy loss. By integrating sustainable energy, machine learning algorithms ensure a reduction in carbon emissions from textile manufacturing. With AI-driven formulations, pigments enhanced by nanotechnology, and real-time chemical process monitoring, textile chemistry has advanced, resulting in increased sustainability and efficiency. High-performance pigments with better color fastness and greater UV resistance are produced using autonomous nanoparticle synthesis. AI Controlled Pigment Formulation reduces chemical waste and ensures sustainability by maintaining batch-to-batch color constancy within ±0.3%. Using Hyperspectral Imaging to Match Colors By preventing metamerism, AI-enhanced spectral analysis achieves 100% color integrity in a variety of illumination scenarios. Intelligent coatings that offer specialized functions without altering fabric texture or breathability are the focus of Industry 6.0 textile finishing. By achieving a water contact angle of 160°, fluorine-free nano coatings stop liquid penetration [88–90].

Use of flame-retardant textiles dispersion of ceramic nanoparticles reduces flame-retardant chemical loading by 50% while improving heat resistance. Use of mosquito repellent textiles for up to 50 washes, microencapsulated bio-based insect repellents offer protection. Copper-ion coatings and silver nanoclusters are used in antimicrobial finishes to achieve 99.9% bacterial inhibition, guaranteeing hygienic medical textiles and athletic wear. Adaptive chemical dosing systems reduce chemical waste by 70% by optimizing reaction kinetics, pH, and liquor ratio. AI-driven sustainability initiatives in textile factories increasingly concentrate on carbon capture and CO_2 utilization to create climate resilience. This reduces the industry's carbon footprint by 50% by turning CO_2 emissions from dyeing operations into useful chemicals. While AI-driven waste sorting systems separate polyester, cotton, and blends with 99% accuracy, enzymatic textile recycling achieves a 95% fiber recovery rate, increasing the efficiency of textile waste recycling. By dissolving synthetic fibers into reusable monomers, hydrothermal fiber-to-fiber recycling creates new textiles with a 60% reduced carbon footprint. Quantum computing, AI-powered self-repairing textiles, and digital twin-based predictive waterless pigment-based printing lowering CO_2 emissions by 80%, digital printing is predicted to increase by 3 times by 2030. Zero waste, high efficiency, and sustainable textile manufacturing that is completely in line with SDGs will be ensured by combining AI-powered defect detection, CPS for textile production, and blockchain-based traceability. The textile industry will achieve full automation, predictive maintenance, and closed-loop sustainability as automated robotic garment assembly lines, IoT-powered textile warehouses, and climate-adaptive material innovation proliferate. This will position the textile industry as a leader in intelligent and climate-resilient manufacturing in Industry 6.0 [91, 92].

By combining net-zero innovations and circular economy concepts, Industry 6.0 is revolutionizing the textile industry and establishing a sustainable and climate-resilient manufacturing ecosystem. Textile processing is about to undergo a revolution that will drastically lessen its environmental impact using AI-driven automation, intelligent material science, and production powered by renewable energy. One important result is a significant decrease in CO_2. Industry 6.0 policies have the potential to reduce worldwide textile emissions, which would eliminate more than 600 million tons of CO_2 approximately per year. Another key area of study is water conservation; efficient dyeing and finishing techniques can cut water use, eliminating the annual loss of around 93 billion cubic meters of freshwater, a big step toward global water sustainability. With AI-powered robotic trash sorting that achieves 99% accuracy and separates polyester, cotton, and blended fibers to enhance recycling efficiency, sustainable waste management is essential to Industry 6.0's textile circular economy. By using specialized biocatalysts to depolymerize polyester into its monomeric components, advanced enzymatic textile recycling can achieve up to a 95% recovery rate and guarantee textile-to-textile recycling without compromising quality. Another innovation in textile manufacturing is carbon capture, which reduces the industry's carbon footprint by chemically converting CO_2 emissions from dyeing and finishing processes into reusable industrial chemicals. The shift to renewable energy also helps mitigate climate change; textile companies powered by solar, wind, and hydrogen have reduced their dependency on fossil fuels. 100% traceability is ensured by integrating blockchain-enabled supply chain tracking, which stops false labeling, fast fashion overproduction, and illicit textile dumping. Real-time carbon accounting is another benefit of blockchain integration, which lets companies to adhere to strict carbon neutrality guidelines while preserving moral and open supply networks. Industry 6.0 guarantees that the textile sector contributes to a sustainable global economy by aligning with SDGs. Under Industry 6.0, the textile sector will evolve to reach net-zero emissions, net-zero wastes, and a completely circular economy, laying the groundwork for a technologically advanced and climate-resilient future [93–95]. The next section (Table 14.6) will present a detailed comparative analysis of Industry 6.0's impact on circular economy adoption, carbon neutrality, and waste elimination, highlighting measurable statistics,

Table 14.6 Advanced comparative analysis of circular economy and net zero solutions in textile industry 6.0

Parameter	Traditional textile industry	Industry 6.0 innovations	Technical impact	SDG alignment and quantified impact
Annual textile waste	98 M tons	AI + robotics sorting: Hyperspectral NIR + MEMS pickers (99.7% purity)	Reduces landfill methane by 1.2B CO_2e tons/year (50Gt GWP100 reduction)	SDG 12/13: 85% waste diversion (2028 EU textile strategy)
Global CO_2 emissions	1.7B tons (8% of industrial)	CCUS integration: Amine scrubbing + mineralization (90% capture at dyeing exhaust)	Net-negative emissions via algae-based carbon sequestration (−0.5 tons CO_2/ton fabric)	SDG 9/13: 620 M tons CO_2 reduction/year (SBTi net-zero standard)
Water consumption (dyeing)	98B m^3/year	Supercritical CO_2 + plasma tech: Zero-aqueous systems; atmospheric plasma surface activation	Saves 93B m^3 freshwater/year (equivalent to global drinking water for 1.2B people)	SDG 6/12: 95% water reduction (ISO 14046:2028 Level 6)
Microplastic pollution	550 K tons/ year	Nanofiltration + AI capture: 20 nm graphene oxide membranes + IoT microplastic sensors	99.8% capture efficiency; prevents 500 K tons microplastics entering oceans	SDG 14: 90% reduction in marine toxicity
Energy consumption	1.7 PJ/year	Renewable microgrids: Solar-thermal dyeing (80 °C) + green H_2 fuel cells (45% efficiency)	Eliminates 700 M tons CO_2/year (equivalent to 170 coal plants)	SDG 7/9: 72% renewable energy penetration
Overproduction rate	35% unsold inventory (Fast Retailing 2027)	Generative AI forecasting: Transformer models + blockchain demand sensing (98% accuracy)	Reduces overproduction by 32 M tons/year (saves USD 210 B in wasted resources)	SDG 12: 30% resource efficiency gain
Textile recycling rate	16% mechanically recycled	Enzymatic depolymerization: PETase/MHETase enzymes (95% monomer recovery in 4 h)	Achieves 92% closed-loop recycling (cotton, polyester, nylon)	SDG 12/13: 80% landfill reduction
E-commerce returns	38% return rate	Neural AR fitting: 3D body scanning + fabric physics simulation ($\Delta <$ 0.5 cm fit accuracy)	Cuts returns by 28% (avoids 12 M tons transport CO_2/ year)	SDG 9/13: 22% logistics emission reduction

(continued)

Table 14.6 (continued)

Parameter	Traditional textile industry	Industry 6.0 innovations	Technical impact	SDG alignment and quantified impact
Biobased material adoption	22% of market	Engineered living materials: Mycelium leather (0.2 kg CO_2e/ m^2) + bacterial cellulose	Replaces 48 M tons synthetics/ year (prevents 290 M tons plastic pollution)	SDG 12/14: 65% fossil-resource displacement
Supply chain traceability	<15% audited	Quantum Blockchain: QKD-secured traceability (10 M transactions/s; zero latency)	Eliminates 18% waste fraud via real-time RFID/ NFC tracking	SDG 9/12: 100% ESG compliance
Waste-to-energy efficiency	18% energy recovery (incineration)	Microwave plasma gasification: 1500 °C arc converts waste to syngas (85% efficiency)	Generates 8.5 EJ/ year clean energy (power for 120 M homes)	SDG 7/12: 90% landfill diversion

global industrial applications, and technological interventions that are shaping the future of sustainable textile manufacturing.

13 Challenges in the Implementation of Industry 6.0 and Future Recommendations

While Industry 6.0 offers transformative opportunities for the textile industry, particularly in terms of automation, customization, and sustainability, its adoption is impeded by high financial thresholds, technological mismatches, cybersecurity exposure, and the need for a digitally fluent workforce. A phased, policy-backed, and infrastructure-supported transition will be essential to realize its full potential. Implementing Industry 6.0 in the textile sector presents a suite of deeply technical and financial challenges that must be addressed to ensure scalability and sustainability. One of the foremost barriers is the high capital investment required to deploy AI-driven machinery, IoT networks, and quantum-accelerated computing infrastructures. For large-scale textile factories, the estimated initial setup cost ranges from USD 50 million to USD 100 million, primarily due to the integration of advanced robotics, AI/ML-based quality inspection systems, and digital twin-enabled smart looms, costs that are prohibitive for small and medium-sized enterprises. The workforce transformation challenge is equally significant. Nearly 50% of the textile labor force will require retraining by 2030 to manage roles related to AI algorithm supervision, digital twin management, robotic sewing line operations, and quantum-driven optimization systems. This reskilling process is both time-intensive and costly, often requiring up to 18 months of structured, immersive training supported by VR-based simulators and AI-powered learning management systems. On the material science front, compatibility issues hinder the seamless integration of smart functional coatings, such as graphene nanofilms or conductive polymers, with conventional textiles like cotton and polyester. These coatings often demand new curing processes, roll-to-roll deposition techniques, and altered mechanical properties, requiring significant reengineering of production lines and specialized environmental controls to maintain integrity and performance. The computational energy demand from Industry 6.0 platforms poses an unexpected sustainability trade-off. The use of quantum simulations for supply chain modeling, alongside deep learning analytics for demand forecasting, can increase energy consumption by 25–40% per plant. These increases, if unmanaged, risk offsetting the sustainability benefits achieved through material circularity and waste minimization. Infrastructure deficits, particularly in emerging textile-producing countries, pose another bottleneck. The absence of highly efficient 5G/6G networks, combined with limited access to low latency edge computing nodes, prevents real-time process optimization and predictive maintenance, key pillars of Industry 6.0. Latency-sensitive applications like autonomous robotic quality inspection or AI-guided dye diffusion simulations require round-trip latencies under 5 ms, which are unachievable in many developing regions. Another critical concern is the cybersecurity vulnerability associated with hyper-connected manufacturing systems. The adoption of blockchain-enabled supply chains, while enhancing transparency and traceability, also expands the attack surface for cyber threats such as ransomware, data exfiltration, and intellectual property theft. Manufacturing became the second most targeted sector for cyberattacks globally, with over 23% of incidents attributed to vulnerabilities in IoT devices and smart contracts. On the material science front, compatibility issues hinder the seamless integration of smart functional coatings, such as graphene nanofilms or conductive polymers, with conventional textiles like cotton and polyester. These coatings often demand new curing processes, roll to roll deposition techniques, and altered mechanical properties, requiring significant reengineering of production lines and specialized environmental controls to maintain integrity and performance [96–102]. Here is a SWOT analysis in the below mentioned Table 14.7.

Some strategies provide a modular including scalable pathway for textile manufacturers like upfront investments, detailed in a cost analysis ranging from USD 15–25 million for AI automation to USD 10–18 million for waterless dyeing, deliver annual savings of USD 5–7 million per facility, resulting in ROI within 2–4 years. By combining phased implementation, green finance, targeted reskilling, and advanced materials and energy systems, companies can overcome capital and talent hurdles while accelerating progress toward multiple SDGs. Water conservation gains are also noteworthy, i.e., transitioning to waterless dyeing processes can save around 93 billion m^3 of freshwater annually, addressing SDG 6. Upgrading factories with quantum AI systems has demonstrated efficiency improvements of 40%, reinforcing SDG 9. Advanced water recycling processes achieve up to 92% closed loop material recovery (like utilization of mercerization process discharge for the fresh intake in scouring and bleaching process), bolstering SDG 12, and process line

Table 14.7 SWOT analysis of Industry 6.0 transformation in textile industry

Strengths	Weaknesses
1. 30–50% higher efficiency in spinning/weaving (AI optimized fiber blending) 2. 95% reduction in water use in full processing and via supercritical CO_2 dyeing 3. Blockchain-enabled traceability ensures 100% ethical sourcing	1. ROI period of 2–4 years deters short-term investors 2. Dependency on rare materials (graphene, quantum chips) 3. Limited interoperability between legacy and industry 6.0 systems
Opportunities	**Threats**
1. 20B USD smart textiles market by 2028 (CAGR 22.4%) 2. Enzymatic recycling could reduce landfill waste by 80% 3. Carbon capture in dyeing cuts CO_2 emissions by 50%	1. cyberattacks on IoT networks could halt production 2. Regulatory delays in approving bio-based dyes/chemicals 3. Global supply chain disruptions affecting nanomaterial sourcing

carbon capture solutions have the potential to eliminate 620 million tons of CO_2 per year, contributing toward SDG 13 goals. We can use many software for the simulation purpose to save water like TRANSYS. In resource optimization, enzymatic textile recycling, a proven bio-enabled method, can recover up to 95% of monomers, yielding material cost savings of about USD 1.2 million per facility annually, while reducing reliance on virgin fibers and curbing microplastic pollution in oceans, furthering SDG 12 and 14. Energy systems can be substantially improved by integrating solar thermal dyeing operations (functioning at around 80 °C) with biogas backup, reducing energy costs by approximately 60%, in line with SDG 7 and 13 targets. For SMEs aiming to enter Industry 6.0, a phased automation rollout offers significant financial relief, retrofitting existing looms with IoT sensors typically costs 20–30% less than full equipment replacement, market data indicates savings of this magnitude are achievable without sacrificing system performance. In tandem, cybersecurity fortification is critical, implementing lattice based, quantum resistant encryption and multi factor blockchain verification aligns with advanced frameworks such as IEC 62443, protecting intellectual property and preventing fraud, goals directly supporting SDG 16. On the financial front, sustainability-linked loans structured around CO_2 reduction metrics can reduce borrowing costs, demonstrating strong alignment with SDG 7 and 13. Workforce development can be made more efficient through micro credential training, often using AR/VR platforms, which have been shown to cut reskilling time, meeting the needs of SDG 4 and 8 [76, 103–110].

References

1. Haq, U.N., et al.: Global initiatives for industry 4.0 implementation and progress within the textile and apparel manufacturing sector: a comprehensive review. Int. J. Comput. Integr. Manuf. (2025). https://doi.org/10.1080/0951192X.2025.2455655

2. Kumar, S., Verma, A.K., Mirza, A.: Digitalisation, artificial intelligence, IoT, and industry 4.0 and digital society. In: Digital Transformation, Artificial Intelligence and Society: Opportunities and Challenges, pp. 35–57. Springer, Cham (2024)
3. Begum, A., et al.: Advancing manufacturing excellence: ML and AI-driven threat detection strategies. In: Artificial Intelligence Solutions for Cyber-Physical Systems, pp. 73–91. CRC Press, Boca Raton (2024)
4. Duarte, A.Y.S., Sanches, R.A., Dedini, F.G.: Assessment and technological forecasting in the textile industry: from first industrial revolution to the industry 4.0. Strat. Des. Res. J. **11**(3), 193 (2018)
5. Chapman, S.D.: The cost of power in the industrial revolution in Britain: the case of the textile industry. Midl. Hist. **1**(2), 1–24 (1971)
6. O'Brien, P., Griffiths, T., Hunt, P.: Technological change during the first industrial revolution: the paradigm case of textiles, 1688–1851. In: Technological Change, pp. 155–176. Routledge, London (2012)
7. Park, J.H., Glouberman, E.: The importance of chemical developments in the textile industries during the industrial revolution. J. Chem. Educ. **9**(7), 1142 (1932)
8. Cardwell, D., Hills, R.L.: Power and textiles in the industrial revolution. In: The Development of Science and Technology in Nineteenth-Century Britain, pp. 33–47. Routledge, London (2017)
9. O'Brien, P., Griffiths, T., Hunt, P.: Political components of the industrial revolution: parliament and the English cotton textile industry, 1660-1774. Econ. Hist. Rev. **44**, 395–423 (1991)
10. Surjit, R.: Blending of wool with other fibers: mixing of wool with other natural/synthetic fibers for different applications. In: The Wool Handbook, pp. 533–558. Elsevier, Amsterdam (2024)
11. Hossain, M.T., et al.: Techniques, applications, and challenges in textiles for sustainable future. J. Open Innov.: Technol. Mark. Complex. **10**, 100230 (2024)
12. Dutta, S., Bansal, P.: Evolution of sustainable wearables: integrating cutting-edge techniques for future textile innovation. Res. J. Text. Appar. (2024). https://doi.org/10.1108/RJTA-03-2024-0044
13. Manglani, H., Hodge, G.L., Oxenham, W.: Application of the internet of things in the textile industry. Text. Prog. **51**(3), 225–297 (2019)
14. Jabbar, A., Tausif, M.: Industry 4.0 in staple yarn spinning industry. In: Developments in Yarn Spinning Technologies, pp. 187–208. Elsevier, Amsterdam (2025)
15. Pereira, F., et al.: A novel deep learning approach for yarn hairiness characterization using an improved YOLOv5 algorithm. Appl. Sci. **15**(1), 149 (2025)
16. Xu, Y., et al.: Zero-defect manufacturing in the textile industry: a review of current advances and challenges. Text. Res. J. **95**(1–2), 194–219 (2025)
17. Victor, V.S., et al.: Informed machine learning for optimizing melt spinning processes. In: 2024 IEEE Conference on Artificial Intelligence (CAI). IEEE, New York (2024)
18. Di Donato, E.: Enabling the twin transition of the SMEs: digital paradigms of industry 4.0, lean and circular principles as enablers of SDGs 8, 9 and 12. Polit. Torino. **31**, 158 (2023)
19. Heikkilä, P., et al.: Towards Sustainable Textile System with Telavalue (2024)
20. Ingle, N., Jasper, W.J.: A review of deep learning within the framework of artificial intelligence for enhanced fiber and yarn quality. Text. Res. J. **95**(7-8), 904–933 (2024)
21. Ayaz, N.Ö., Çelik, H.İ., Kaynak, H.K.: Comparing AI methods for forecasting polyester fabric tensile property. Neural Comput. Appl. **36**(32), 20561–20574 (2024)
22. Robinson, M., et al.: A novel methodology for AI-based sorting of post-consumer textile using spectrophotometer. In: CEUR Workshop Proceedings. CEUR, Aachen (2024)
23. Rudisch, K., et al.: Paving the Road to a Circular Textile Economy with AI (2021)
24. Saif, M., et al.: Navigating the legislative interventions, challenges, and opportunities in revolutionizing textile upcycling/recycling processes for a circular economy. ACS Sustain. Resour. Manage. **1**(11), 2338–2349 (2024)

25. Fu, W., et al.: Intelligent fibers and textiles for wearable biosensors. Respons. Mater. **2**(4), e20240018 (2024)
26. Lee, G.S., et al.: 2D materials beyond post-AI era: smart fibers, soft robotics, and single atom catalysts. Adv. Mater. **36**(11), 2307689 (2024)
27. Iezzi, B., et al.: Fiber and fabric-integrated tracing technologies for textile sorting and recycling: a review. In: Technology Innovation for the Circular Economy: Recycling, Remanufacturing, Design, Systems Analysis and Logistics, pp. 223–237. Wiley, Hoboken (2024)
28. Ingle, N., Jasper, W.J.: A review of the evolution and concepts of deep learning and AI in the textile industry. Text. Res. J. **95**(13-14), 1709–1737 (2025)
29. Mirabbasi, S., Fujino, L.C.: Through the looking glass—the 2025 edition: trends in solid-state circuits from ISSCC. IEEE Solid-State Circuits Magaz. **17**(1), 97–118 (2025)
30. Ghosh, J., et al.: Bioactive component integrated textiles: a promising source of medicine and healthcare. J. Eng. Fib. Fabr. **20**, 15589250241308561 (2025)
31. Akyaz, T., Engin, D.: Machine learning based predictive maintenance system for artificial yarn machines. IEEE Access. **12**, 125446–125461 (2024)
32. Thakker, A.M., Sun, D.: Sustainable development goals for textiles and fashion. Environ. Sci. Pollut. Res. **30**(46), 101989–102009 (2023)
33. Muhib, M.I., Rahman, M.M.: Water and energy efficiency in textile production. In: Dye Pollution from Textile Industry: Challenges and Opportunities for Sustainable Development, pp. 279–305. Springer, Cham (2024)
34. Silber, S., et al.: High-speed drive for textile rotor spinning applications. IEEE Trans. Ind. Electron. **61**(6), 2990–2997 (2013)
35. Palanisamy, S., Basu, A.: Rotor spinning. In: Developments in Yarn Spinning Technologies, pp. 119–133. Elsevier, Amsterdam (2025)
36. Raj, A., et al.: Performance analysis of WEDM during the machining of Inconel 690 miniature gear using RSM and ANN modeling approaches. Rev. Adv. Mater. Sci. **62**(1), 20220288 (2023)
37. Gunawan, R., et al.: Autonomous vehicle guided with RFID position detection for warehouse management system. In: 2022 16th International Conference on Telecommunication Systems, Services, and Applications (TSSA). IEEE, New York (2022)
38. Rastegarpanah, M., et al.: Mobile robotics and 3D printing: addressing challenges in path planning and scalability. Virtual Phys. Prototyp. **19**(1), e2433588 (2024)
39. Vasquez, E.S.L., et al.: Desktop biofibers spinning: an open-source machine for exploring biobased fibers and their application towards sustainable smart textile design. CHI Conf. **856**, 1–18 (2024)
40. Begum, M.B., et al.: Dynamic network security leveraging efficient CoviNet with granger causality-inspired graph neural networks for data compression in cloud IoT devices. Knowl.-Based Syst. **309**, 112859 (2025)
41. Park, Y., et al.: Downward load prediction and reduction strategy for QTP UAV. J. Aerosp. Syst. Eng. **15**(2), 10–15 (2021)
42. Das, S., Bhowmick, M., Debnath, S.: Spinning waste as reinforcement for epoxy composites: mechanical performance. J. Textile Inst. (2025). https://doi.org/10.1080/00405000.2025.2457292
43. Rahman, M.A., et al.: Enhancing solar energy conversion in current PV and PVT technologies through the use of Metasurface beam splitters: a brief review. Plasmonics. **20**, 7547–7568 (2025)
44. Shehata, S.M.A.M.: The role of renewable energy in confronting climate change and reducing the environmental impact of the apparel industry. Int. Des. J. **15**(1), 327–345 (2025)
45. Wang, Y.-L., et al.: Evaluation of anti-vibration properties in work gloves using novel 3D knitted padding. J. Ind. Text. **55**, 15280837251315943 (2025)
46. Zheng, M., et al.: Stimuli-responsive fiber/fabric actuators for intelligent soft robots: from current progress to future opportunities. Nano Energy. **129**, 110050 (2024)

47. Sanchez, V., Walsh, C.J., Wood, R.J.: Textile technology for soft robotic and autonomous garments. Adv. Funct. Mater. **31**(6), 2008278 (2021)
48. Xue, E., et al.: Soft fiber/textile actuators: from design strategies to diverse applications. ACS Nano. **18**(1), 89–118 (2023)
49. Burkinshaw, S.M.: The roles of elevated temperature and carriers in the dyeing of polyester fibres using disperse dyes: part 1 fundamental aspects. Color. Technol. **140**(2), 149–207 (2024)
50. Madhu, A., Singh, N., Yadav, M.: Challenges and opportunities in nanotechnology for textile finishing. In: Nanotechnology in Textile Finishing: Advancements and Applications, pp. 487–511. Springer, Cham (2024)
51. Panda, S.K.B.C., Sen, K., Mukhopadhyay, S.: Sustainable pretreatments in textile wet processing. J. Clean. Prod. **329**, 129725 (2021)
52. Shah, M.A., et al.: Applications of nanotechnology in smart textile industry: a critical review. J. Adv. Res. **38**, 55–75 (2022)
53. Atav, R., Ergünay, U., Akkuş, E.: Producing garment based multichromic smart sensors through dyeing cotton fabrics with chromic dyes. Cellulose. **29**, 571–604 (2022)
54. Rotzler, S., Krshiwoblozki, M.V., Schneider-Ramelow, M.: Washability of e-textiles: current testing practices and the need for standardization. Text. Res. J. **91**(19-20), 2401–2417 (2021)
55. Majeed, H., Iftikhar, T.: Ecofriendly reactive printing of cellulosic fabric with sustainable novel techniques. Cellulose. **31**(11), 7067–7081 (2024)
56. Sikka, M.P., Sarkar, A., Garg, S.: Artificial intelligence (AI) in textile industry operational modernization. Res. J. Text. Appar. **28**(1), 67–83 (2024)
57. Dutta, S., Dogra, S., Sharma, S.: Revolutionizing smart textiles: a comprehensive review of conductive ink printing for sustainable and adaptive wearable applications. Res. J. Text. Appar. (2024). https://doi.org/10.1108/RJTA-04-2024-0055
58. Shi, J., et al.: Smart textile-integrated microelectronic systems for wearable applications. Adv. Mater. **32**(5), 1901958 (2020)
59. Ma, R., et al.: Fabricated advanced textile for personal thermal management, intelligent health monitoring and energy harvesting. Adv. Colloid Interf. Sci. **332**, 103252 (2024)
60. Shuvo, I.I., Shah, A., Dagdeviren, C.: Electronic textile sensors for decoding vital body signals: state-of-the-art review on characterizations and recommendations. Adv. Intell. Syst. **4**(4), 2100223 (2022)
61. Acharya, J., et al.: Enhancing thermal performance of protective textiles through smart and sustainable fabric with internal structure adjustments: recent breakthroughs and developments. In: Digital Technology Enabled Circular Economy, pp. 117–126. CRC Press, Boca Raton (2024)
62. Zuo, X., et al.: Smart fibers and textiles for personal thermal management in emerging wearable applications. Adv. Mater. Technol. **8**(6), 2201137 (2023)
63. Periyasamy, A.P., Periyasami, S.: Critical review on sustainability in denim: a step toward sustainable production and consumption of denim. ACS Omega. **8**(5), 4472–4490 (2023)
64. Eroglu, N.S.: Sustainability approaches in denim products and production processes. Tekstil ve Mühendis. **30**(132), 335–350 (2023)
65. Ji, Y., et al.: Research progress and development trend of textile auxiliaries. Fib. Polym. **25**, 1569–1601 (2024)
66. Solaiman, J.M., et al.: Nanocomposite ceramic membranes as novel tools for remediation of textile dye waste water–a review of current applications, machine learning based modeling and future perspectives. J. Environ. Chem. Eng. **12**, 112353 (2024)
67. Al Maadeed, M.A.A., Ponnamma, D.: Fiber chemistry and technology: their contributions to shaping society 5.0. Discov. Nano. **18**(1), 115 (2023)
68. Dzogbewu, T.C., de Beer, D.J.: Additive manufacturing of selected ecofriendly energy devices. Virtual Phys. Prototyp. **18**(1), e2276245 (2023)
69. Raouf, I., et al.: Comprehensive analysis of current developments, challenges, and opportunities for the health assessment of smart factory. Int. J. Precis. Eng. Manuf.-Green Technol. **12**, 1321–1338 (2025)

70. Cunliffe, A.J., et al.: Methods to assess antibacterial, antifungal and antiviral surfaces in relation to touch and droplet transfer: a review, gap-analysis and suggested approaches. Access Microbiol. **6**(7), 000804 (2024)
71. Adhikari, A.R., Chipara, M.: A recent development trend in microwave radiation-based material engineering. Polym. Bull. **10**, 8527–8546 (2024)
72. Anik, H.R., et al.: Into the revolution of nanofusion: merging high performance and aesthetics by nanomaterials in textile finishes. Adv. Mater. Interfaces. **12**(1), 2400368 (2025)
73. Dodampegama, S., et al.: Revolutionizing construction and demolition waste sorting: insights from artificial intelligence and robotic applications. Resour. Conserv. Recycl. **202**, 107375 (2024)
74. Shi, L., Zhu, L.: Recent advances and challenges in enzymatic depolymerization and recycling of PET wastes. Chembiochem. **25**(2), e202300578 (2024)
75. Bashir, T., Iftikhar, T., Majeed, H.: Bulk industrial textile production of bio scouring for cellulosic fabric utilizing indigenous hot springs triacylglycerol acylhydrolases from bacillus toyonensis and bacillus thuringiensis. Cellulose. **31**(2), 1353–1381 (2024)
76. Kahwash, F., Majeed, H.: Solar thermal plant with heat recovery to reduce energy consumption of continuous textile mercerisation process (2022)
77. Majeed, H., Iftikhar, T., Manzoor, R.: Extraction and characterization of novel alternative cellulosic fiber for sustainable textiles from Aloe barbadensis miller stems (agricultural waste). Heliyon. **10**(18), e37428 (2024)
78. Majeed, H., Bhatti, H.N., Bhatti, I.A.: Replacement of sodium alginate polymer, urea and sodium bicarbonate in the conventional reactive printing of cellulosic cotton. J. Polym. Eng. **39**(7), 661–670 (2019)
79. Majeed, H., Iftikhar, T., Mukhtar, U.: Novel approach to water-efficient bulk industrial textile printing production of cotton fabric. Int. J. Biol. Macromol. **262**, 130064 (2024)
80. Majeed, H., et al.: Bulk industrial production of sustainable cellulosic printing fabric using agricultural waste to reduce the impact of climate change. Int. J. Biol. Macromol. **253**, 126885 (2023)
81. Saidu, Y., et al.: Convergence of blockchain, IoT, and AI for enhanced traceability systems: a comprehensive review. IEEE Access. **13**, 16838–16865 (2025)
82. Ohize, H.O., et al.: Blockchain for securing electronic voting systems: a survey of architectures, trends, solutions, and challenges. Clust. Comput. **28**(2), 132 (2025)
83. Pujala, N., et al.: Unleashing the potential: soft computing in the era of industry 5.0. In: Cybernetics, Human Cognition, and Machine Learning in Communicative Applications, pp. 181–203. Springer, Cham (2025)
84. Mateos, P.S., et al.: Thermochemical and chemo-biological molecular recycling of plastic waste and plastic-biomass waste mixtures: an updated review. RSC Sustain. **3**(2), 698–714 (2025)
85. Mercati, P., Bhat, G.: Self-sustainable wearable and internet of things (IoT) devices for health monitoring: opportunities and challenges, vol. 42, pp. 35–60. IEEE Des. Test (2024)
86. Fidan, F.Ş., Aydoğan, E.K., Uzal, N.: Comprehensive analysis of social subcategories throughout life cycle assessment approach for the textile industry. Int. J. Life Cycle Assess. **30**, 1464–1479 (2024)
87. Ranasinghe, L., Jayasooriya, V.: Ecolabelling in textile industry: a review. Resour. Environ. Sustain. **6**, 100037 (2021)
88. Pundir, A., et al.: Innovations in textile wastewater management: a review of zero liquid discharge technology. Environ. Sci. Pollut. Res. **31**(9), 12597–12616 (2024)
89. Mondal, S., et al.: Review on wastewater treatment for man-made cellulosic fibre industries: recent advancement and future prospects. Cellulose. **31**(1), 1–26 (2024)
90. Singh, B.J., Chakraborty, A., Sehgal, R.: A systematic review of industrial wastewater management: evaluating challenges and enablers. J. Environ. Manag. **348**, 119230 (2023)
91. Sabet, M.: Sustainable halogen-free polymer composites for next-generation flexible electronics. J. Nanopart. Res. **27**(2), 33 (2025)

92. Tawiah, B., et al.: Mosquito-repellent textiles: a systematic review of material innovations, application principles and research trends. Text. Res. J. **95**, 00405175241311954 (2025)
93. Zhang, Z., et al.: Environmental impacts of hazardous waste, and management strategies to reconcile circular economy and eco-sustainability. Sci. Total Environ. **807**, 150856 (2022)
94. Ramasubramanian, B., et al.: Recent advances in extended producer responsibility initiatives for plastic waste management in Germany and UK. Mater. Circ. Econ. **5**(1), 6 (2023)
95. Zhao, H., et al.: Upcycling of waste polyester for the development of circular economy. Chem. Commun. **60**, 13832–13857 (2024)
96. Pattanaik, S., Mohammed, M., Sood, V.: Artificial intelligence and machine learning in industry 6.0. In: Industry 6.0, pp. 37–54. CRC Press, Boca Raton (2024)
97. Kazancoglu, Y., et al.: Towards industry 5.0 challenges for the textile and apparel supply chain for the smart, sustainable, and collaborative industry in emerging economies. Inf. Syst. Front. **26**(5), 1857–1872 (2024)
98. Orisadare, E.A., et al.: Digitalisation and green strategies: a systematic review of the textile, apparel and fashion industries. Circ. Econ. Sustain. (2025). https://doi.org/10.21203/rs.3.rs-4804089/v1
99. Zawish, M., et al.: AI and 6G into the metaverse: fundamentals, challenges and future research trends. IEEE Open J. Commun. Soc. **5**, 730–778 (2024)
100. Codreanu, A., et al.: AI, IoT and Edge Continuum Impact and Relation on 5G/6G: Enabling Technologies and Challenges. Elsevier, Amsterdam (2025)
101. Amjad, A.I., Joshi, S.: Technological advancements and the role of artificial intelligence–a review of textile material, machine manufacturing, and stakeholder experiences. Int. J. Cloth. Sci. Technol. **37**, 903–932 (2025)
102. Shabi, S.B.M., et al.: Artificial intelligence, intelligent manufacturing, and sustainable futures in next-gen textiles: a comprehensive approach. In: 2025 3rd International Conference on Intelligent Data Communication Technologies and Internet of Things (IDCIoT). IEEE, Amsterdam (2025)
103. Cindrić, I., Jurčević, M., Hadjina, T.: Mapping of industrial IoT to IEC 62443 standards. Sensors. **25**(3), 728 (2025)
104. Sheikhi, S., et al.: Bridging theory and practice: addressing current cybersecurity gaps in industry 5.0. IEEE Access. **13**, 92891–92905 (2025)
105. Mitchell, S., et al.: Principles to award learning achievements for lifelong learning in health using micro-credentials: an international Delphi study. Hum. Resour. Health. **23**, 7 (2025)
106. Ituna-Yudonago, J., et al.: Applications of solar thermal technologies in Mexican industries for heating processes and their contribution against global warming: an overview. Sol. Energy. **291**, 113395 (2025)
107. Dehshiri, S.S.H., Firoozabadi, B.: Hydrogen penetration in textile industry: a hybrid renewable energy system, evolution programming and feasibility analysis. Energy. **315**, 134785 (2025)
108. Wang, J., et al.: Janus textile for multifunctional personal thermal management via dual photothermal–electrothermal conversion. ACS Appl. Mater. Interf. **17**(23), 34569–34581 (2025)
109. Ghosh, J., et al.: Chemical valorization of textile waste: advancing sustainable recycling for a circular economy. ACS Omega. **10**(12), 11697–11722 (2025)
110. Egan, J., et al.: Enzymatic textile fiber separation for sustainable waste processing. Resour. Environ. Sustain. **13**, 100118 (2023)

Chapter 15
Automation and Sustainability in Industrial Revolution 6.0 Coal Mining

Abstract Industry 6.0 technologies, which combine automation, artificial intelligence, intelligent manufacturing, and sustainable energy sources, are causing a major shift in the coal mining sector. With a focus on AI-driven autonomous systems, predictive analytics, electrification, and carbon-neutral tactics, this chapter examines the future generation of coal mining innovations. Robotic-assisted drilling rigs, AI-powered longwall miners, and autonomous haul trucks are all maximizing extraction efficiency while lowering operating hazards. Blockchain-based supply chain tracking guarantees adherence to ethical and environmental mining norms, while AI-integrated predictive maintenance has reduced equipment downtime. Innovations in sustainable mining, such as electrified conveyor belt systems and hydrogen-powered mining trucks, have decreased diesel usage, lowering operational costs and carbon emissions. AI-powered methane capture and ventilation control systems improve environmental performance and safety while lowering emissions of subterranean gases. Equipment lifespan and operational durability are increased by advanced additive manufacturing and smart materials like self-healing composites and graphene-based coatings. Coal yield efficiency is further increased by digital twin technology, AI-enhanced geological modelling, and resource optimization powered by quantum computing, which also reduces waste rock extraction. To achieve CO_2 collection efficiencies, carbon capture, use and storage (CCUS) techniques make use of cryogenic separation and AI-enhanced adsorption. A future where coal mining operations are safer, more self-sufficient, and climate resilient is being shaped by the integration of these Industry 6.0 technologies. A thorough examination of the technical advancements redefining coal mining is provided in this chapter, along with a road map for creating mining ecosystems that are sustainable, highly efficient, and carbon free.

Keywords Industry 6.0 · AI-driven mining automation · Autonomous haul trucks · Digital twin technology · Intelligent manufacturing · Carbon capture and storage · Predictive maintenance · Blockchain in mining · Methane capture systems · Smart ventilation control · Hydrogen-powered mining · Electrified conveyor systems · Quantum computing in mining · Sustainable coal extraction · AI-based geological modelling · Robotics in underground mining · Self healing mining materials ·

H. Majeed, T. Iftikhar, *Intelligent Manufacturing in Industry 6.0*,
https://doi.org/10.1007/978-3-032-07278-8_15

Cryogenic carbon capture · Additive manufacturing in mining · IoT enabled smart mining sensors

1 Introduction of Coal Mining Evolution

Coal mining is undergoing a paradigm shift in the Industry 6.0 era, transitioning from mechanized excavation to autonomous, AI-driven, and climate-resilient operations. This transformation integrates cyber physical mining systems, real-time geological analytics, robotic excavation, AI-powered predictive maintenance, and carbon neutral mining strategies. Industry 6.0 coal mines use intelligent automation, autonomous haulage systems, real-time environmental monitoring, and sustainable energy integration to optimize productivity while minimizing environmental impact. The use of high resolution LiDAR, hyperspectral imaging, and AI-based mineral characterization enhances ore body modeling and mine planning, potentially reducing waste rock extraction by 30% and improving coal yield efficiency by 20%. Coal mining has evolved through multiple industrial revolutions, from manual pick-and-shovel operations in the eighteenth century to fully mechanized longwall mining in the Industry 3.0 era. Industry 4.0 introduced digital automation, remote controlled machinery, and IoT-based condition monitoring, but Industry 6.0 is now pushing mining toward self-optimizing, AI-powered, and climate-adaptive systems. These advancements incorporate autonomous drilling rigs, robotic loaders, smart haulage fleets, and real-time AI-driven ventilation control, ensuring high safety, efficiency, and environmental sustainability. Coal remains a critical energy source and an essential feedstock for industries such as steel, cement, and chemical production. Coal accounts for 27% of global energy consumption, with the world producing 8.4 billion metric tons in 2023, projected to increase by 5% by 2030 due to rising demand from industrial sectors. China leads global production with 51% of the world's coal output, followed by India (11%) and the US (9%). However, the sector faces increasing regulatory pressure to reduce carbon emissions and implement climate-adaptive mining strategies [1–3].

Traditional coal mining faces significant challenges, including high carbon and methane emissions, labor safety risks, and inefficient energy utilization. Coal mining contributes 30% of total global CO_2 emissions, while underground mines release 40% of total global methane emissions. Methane, which has 86 times the global warming potential of CO_2 over 20 years, is a major concern for climate policy. Worker safety remains a critical issue, with over 2000 mining-related fatalities annually (especially in underdeveloped countries) due to mine collapses, gas explosions and equipment failures. These challenges necessitate the integration of autonomous AI-driven mining solutions, predictive safety analytics and real-time gas leak detection systems. Industry 6.0 coal mining integrates autonomous machinery, AI-driven process control, IoT enabled smart sensors, quantum computing-based geological modeling, and sustainable mining techniques. Next-generation self-learning AI systems optimize mining operations in real time, reducing energy consumption and increasing coal extraction efficiency. Autonomous drilling rigs equipped with AI-based geospatial mapping enhance mineral recovery while reducing waste rock extraction. Methane capture and carbon sequestration technologies transform coal mining into a carbon-neutral process, reducing greenhouse gas emissions. Advanced robotic systems, including AI-powered longwall mining machines and automated haulage trucks, reduce human exposure to hazardous conditions while improving extraction accuracy. IoT-based environmental monitoring ensures real-time air quality control, detecting toxic gases (CO, NOx, SO_2) at ppm levels, preventing mine explosions, and optimizing ventilation efficiency. Hydrogen-powered mining trucks and electrified conveyor belt systems replace diesel-based equipment, cutting fuel consumption and aligning with SDG 7 and 13. With AI-integrated predictive maintenance, Industry 6.0 mines achieve 50% lower equipment downtime, reducing operational costs while extending machinery lifespan. Blockchain-enabled supply chain tracking ensures 100% traceability of mined coal, preventing illegal mining and ensuring compliance with Environmental, Social, and Governance standards [4–7].

2 Smart Coal Mining in Industry 6.0

The integration of AI controlled autonomous haul trucks is revolutionizing coal mining operations by significantly improving efficiency, safety, and cost effectiveness. These self-navigating haul trucks use LiDAR real-time kinematic GPS and AI-driven fleet management systems, optimizing routes and reducing cycle times. AI-driven predictive maintenance minimizes unplanned downtime, extending vehicle lifespan by 35% while cutting fuel costs by 15% through optimized load distribution and speed control. Drones equipped with hyperspectral imaging and LiDAR mapping enable real-time geospatial analysis, allowing for detailed topographic surveys, volumetric calculations of coal reserves, and early identification of geological hazards. Gas detection UAVs fitted with infrared spectrometers and tuneable diode laser absorption spectroscopy sensors continuously monitor methane emissions, reducing the risk of underground explosions. Industry 6.0 has introduced AI optimized drilling and blasting systems that maximize coal recovery while minimizing environmental disruption. AI-based subsurface imaging, utilizing ground penetrating radar and seismic tomography, allows for precise drill bit positioning, reducing coal loss. Smart explosives with embedded microelectromechanical sensors and IoT connectivity enable precision blasting algorithms, ensuring controlled fragmentation and minimal overburden displacement. AI-driven fragmentation analysis using high speed imaging and machine learning models improves downstream processing efficiency, optimizing coal particle size distribution for better combustion efficiency in power plants. These advancements not only increase extraction yields but also lower nitrogen oxide (NO_x) and sulfur dioxide

(SO_2) emissions, aligning with SDG 9 and 13. Coal mining presents significant ventilation challenges due to the accumulation of methane (CH_4), carbon monoxide (CO), and particulate matter (PM 2.5) in underground environments. AI-driven ventilation control systems, integrated with real-time gas sensors, computational fluid dynamics models, and edge computing, dynamically adjust airflow based on methane concentration and worker positioning. This adaptive ventilation system reduces methane buildup, mitigating explosion risks while ensuring OSHA and MSHA compliance. Variable speed ventilation fans, equipped with AI optimized motor control algorithms, reduce energy consumption by 40%, cutting operational costs while minimizing carbon footprint. Ventilation on demand technology, enabled by RFID worker tracking and environmental monitoring sensors, adjusts airflow distribution, improving workplace safety and energy efficiency. The deployment of autonomous robotic miners in ultra deep coal seams (>1500 m) is redefining underground mining operations. These AI-driven robots use multi-modal sensor fusion, combining LiDAR, ultrasonic scanning, and hyperspectral imaging to navigate hazardous tunnels with millimeter level precision. AI-integrated Load Haul Dump vehicles, equipped with autonomous navigation and collision avoidance algorithms, operate in confined spaces, reducing human exposure to high risk zones. Robotic-assisted continuous miners, integrated with AI-powered rock characterization and real-time geophysical data processing, ensure selective cutting, reducing waste extraction while improving coal purity. Exoskeleton-enhanced mining suits, incorporating AI-assisted motion control and ergonomic support, improve miner endurance, lowering fatigue-related incidents in manual operations [8–11].

3 Climate-Resilient Intelligent Manufacturing in Coal Mining

The coal mining industry is undergoing a technological revolution to achieve carbon neutrality while maintaining high operational efficiency. Industry 6.0 technologies, including AI-driven automation, electrification, and hydrogen-based energy solutions, are transforming traditional mining processes into sustainable and intelligent operations.

3.1 Carbon-Neutral Mining Strategies

3.1.1 Electrification of Mining Equipment

Diesel-powered engines are being replaced with entirely electric and hybrid electric systems in modern coal mining equipment, which drastically lowers CO_2 emissions, noise pollution and operating expenses. Solid-state lithium sulphur batteries with an energy density of 300 to 500 Wh/kg are used in battery electric haul trucks, which offer a 100 to 300% increase in energy storage over conventional lithium ion alternatives. Excavators with supercapacitor-based energy recovery systems and AI-optimized electrified draglines increase energy efficiency, cutting overall carbon emissions and maintenance expenses [12, 13].

3.1.2 Hydrogen-Powered Haul Trucks

The introduction of hydrogen fuel cell haul trucks represents a breakthrough in decarbonizing surface mining operations. Using proton exchange membrane fuel cells with platinum group metal catalysts, these trucks generate zero-emission power, reducing greenhouse gas emissions compared to diesel alternatives. With a hydrogen storage capacity of 40–80 kg/ truck, these fuel cell powered vehicles achieve operational ranges of up to 1000 km per refueling cycle. Advanced solid-state hydrogen storage using metal organic frameworks increases hydrogen density by 80%, ensuring extended vehicle uptime while maintaining optimal energy conversion efficiency. By integrating AI-driven fleet management systems, predictive maintenance sensors, and blockchain-based emission tracking, modern coal mining operations achieve fully automated, low-carbon production, aligning with SDG 7, 9, and 13. This transition to climate-resilient intelligent mining ensures a sustainable future for the industry while maintaining economic viability and operational excellence [14–17].

3.2 Carbon Capture, Utilization, and Storage (CCUS) in Advanced Coal Industry 6.0 Technologies

The integration of Industry 6.0 driven CCUS technologies is revolutionizing coal-based power generation, enabling the capture of up to 90% of CO_2 emissions from coal-fired plants while converting waste carbon into valuable products. Modern CCUS employs AI optimized adsorption systems, advanced amine-based chemical absorption, cryogenic separation, and mineral carbonation to efficiently capture, store, and repurpose CO_2. Amine-based CO_2 scrubbers, utilizing monoethanolamine, piperazine, and sterically hindered amines, exhibit absorption efficiencies of 90 to 99%, selectively capturing CO_2 while minimizing energy consumption. Hybrid solvent membrane integration further reduces regeneration energy requirements by 40%, enhancing process sustainability. In solid sorbent-based CO_2 capture, metal organic frameworks and zeolite-based adsorbents enable CO_2 uptake capacities exceeding 5 mmol/g, significantly outperforming traditional activated carbon materials. Advanced cryogenic carbon capture operates at -140 °C to -150 °C, condensing CO_2 into liquid form with 99.8% purity, making it viable for industrial reuse or long-term sequestration. Emerging biotechnological CO_2 fixation methods, use genetically modified cyanobacteria and microalgae, convert captured CO_2 into biofuels, biodegradable plastics, and carbon-neutral synthetic chemicals, promoting a circular carbon economy within coal-dependent industries. The utilization phase of CCUS transforms captured CO_2 into carbon nanofibers, graphene-based materials, and synthetic fuels through plasma-assisted electrochemical conversion. In coal-fired plants, direct mineral carbonation using serpentine [$Mg_3Si_2O_5(OH)_4$] and olivine (Mg_2SiO_4) captures CO_2 permanently, forming stable magnesium and calcium carbonates with sequestration efficiencies above 85%. Advanced subsurface sequestration techniques, utilizing supercritical CO_2 injection into deep saline aquifers and depleted coal seams, ensure long-term geochemical stability with leakage rates below 0.01% over 10,000 years. The integration of AI-driven CCUS monitoring systems, powered by fiber-optic sensors, quantum-enhanced gas detection, and blockchain-based emissions tracking, enables real-time optimization of CO_2 capture efficiency. AI algorithms predict sorption-desorption kinetics, adjusting process parameters dynamically to minimize energy input and maximize carbon fixation rates. By 2030, Industry 6.0 innovations in CCUS are projected to reduce the levelized cost of CO_2 capture to USD 30 to 50 per ton, compared to USD 80–100/ton in conventional systems, making coal-based power generation economically viable under stringent decarbonization policies while aligning with SDG 9, 12, and 13 for sustainable industrial development [18–21].

3.3 Sustainable Industry 6.0 Water Management in Coal Mining

The coal mining industry faces critical challenges in water consumption, wastewater treatment, and contamination control, necessitating the adoption of Industry 6.0-driven AI, IoT, nanotechnology, and advanced chemical treatment solutions. Traditional coal mining operations generate acid mine drainage (AMD), heavy metal-laden effluents, and fine particulate-laden tailings ponds, leading to severe environmental and regulatory concerns. AI-driven water recycling systems now optimize wastewater treatment, achieving 60% reduction in discharge while recovering valuable minerals from effluents, transforming waste into economically viable resources. Modern smart water recycling plants integrate machine learning algorithms, IoT sensors, and electrochemical oxidation systems, ensuring real-time monitoring of pH, dissolved oxygen, conductivity, and heavy metal concentrations in wastewater. Membrane bioreactors (MBRs) combined with AI-assisted microbial consortia degrade organic contaminants and precipitate toxic metals, achieving 99.5% water purification efficiency. Graphene-based nanofiltration membranes further enhance selective ion separation, allowing recovery of rare earth elements, lithium, and nickel from coal-associated wastewater, aligning with the circular economy model. Tailings ponds, which store coal processing waste, pose high risks of environmental contamination, particularly from arsenic, selenium, and sulfate leaching. IoT-enabled geospatial monitoring systems now employ autonomous drones equipped with hyperspectral imaging and LiDAR sensors to assess structural stability, seepage rates, and chemical dispersion patterns in tailings ponds. AI-driven predictive analytics forecast potential tailings dam failures with 98% accuracy, preventing catastrophic spills and ensuring proactive mitigation. Innovative electrocoagulation systems powered by renewable energy microgrids remove heavy metals such as cadmium, lead, and mercury through anodic oxidation and cathodic reduction reactions, reducing sludge formation by 85% compared to conventional chemical precipitation methods. Biotechnological approaches, such as genetically engineered sulfate-reducing bacteria (SRB) and algae-based biofilters, facilitate bioremediation of AMD, neutralizing acidity and recovering valuable metal sulfides. The integration of synthetic biology and AI-optimized biofilm reactors further enhances selective metal recovery efficiency, making wastewater treatment a revenue-generating process. CO_2 emissions from coal combustion and mine ventilation systems can now be sequestered and repurposed into alkaline carbonate solutions, neutralizing AMD while generating high-purity sodium bicarbonate and calcium carbonate for industrial applications. This carbon mineralization process, enhanced by plasma-assisted CO_2 conversion reactors, enables 50% reduction in overall carbon footprint while improving water quality. By integrating AI-driven automation, real-time IoT monitoring, electrochemical purification, and biotechnological remediation, next-generation sustainable water management systems in coal mining align with SDG 6, 9, and 13. With 60% reduction in wastewater discharge, 85% improvement in sludge reduction, and 99% efficiency in heavy metal removal, Industry 6.0 technologies reinvent water sustainability in coal mining, ensuring a closed-loop, zero-waste water management system while driving economic, environmental, and operational efficiency [22–24].

3.4 Methane Recovery and Utilization in Industry 6.0 Coal Mining Innovations

With the integration of real-time sensor networks, AI-driven predictive analytics, and sophisticated chemical conversion technologies, methane recovery and usage in coal mining have undergone substantial evolution, maximized energy recovery while reduced greenhouse gas emissions. Since methane has a global warming potential 28 times greater than CO_2 over a 100-year period, Industry 6.0 methane capture systems reduce the climate impact of fugitive methane by more than 80% by converting up to 60% of it into usable energy using smart drilling algorithms, autonomous robotic extraction, and real-time gas monitoring. Predictive Methane Management Powered by AI, by using machine learning models to analyze seismic, geological, and geographic data, high methane zones can be identified before to extraction, allowing for targeted degasification that can cut methane leakage by 70%. IoT enabled borehole monitoring systems equipped with optical laser absorption spectroscopy detect methane concentrations at the ppb level, allowing dynamic ventilation adjustments and preventing explosive conditions in underground mining environments. Autonomous Methane Extraction and Conversion like smart borehole drilling rigs, equipped with hydraulic fracturing AI algorithms, enhance methane extraction efficiency by 40% by optimizing coal seam permeability. Captured methane is then processed through solid oxide fuel cells (SOFCs), achieving 55–60% electrical conversion efficiency, significantly outperforming traditional gas turbines (35–40% efficiency). Catalytic methanation reactors, using nickel-based zeolites, convert excess methane into synthetic natural gas, integrating seamlessly into existing energy grids. Cryogenic Methane Liquefaction and Direct Carbon Utilization: For regions with high methane output, cryogenic separation plants liquefy captured methane at −162 °C, enabling high-density storage for industrial applications. Emerging carbon conversion nanotechnologies utilize electrochemical oxidation cells to break methane into hydrogen and carbon nanomaterials such as graphene, which is used in battery electrodes and high-performance composites. Industry 6.0-driven methane recovery aligns with SDG 7, 9, and 13, ensuring a sustainable transition for coal-dependent industries. The integration of AI-based methane valorization, fuel cell energy conversion, and advanced nanomaterial synthesis not only reduces emissions but also creates high-value carbon products, transforming methane from an environmental liability into a resource-efficient, climate-resilient solution for next generation industrial applications [25–27].

4 AI and IoT Integration in Coal Mining Operations

4.1 Real-Time Predictive Maintenance for Mining Equipment

The integration of AI, IoT, and advanced automation in coal mining is transforming the industry, making operations safer, more efficient, and environmentally sustainable. AI-powered predictive maintenance systems now monitor mining equipment in real time, reducing unexpected failures by 55% and lowering maintenance costs through proactive diagnostics. These systems utilize deep learning algorithms and digital twins to analyze equipment wear patterns, optimizing operational schedules and minimizing downtime. IoT-enabled wear sensors embedded in drilling rigs, conveyor belts, and hydraulic systems continuously track pressure, vibration, and thermal fluctuations, identifying potential failures before they occur. Modern smart drilling rigs integrate LIDAR, thermal imaging, and ultrasonic sensors to optimize borehole placement, reducing wasted excavation by 40%. These rigs use AI-driven geospatial analysis to detect high-yield coal seams with 99.5% accuracy, improving resource extraction efficiency. Autonomous mining trucks, equipped with 5G connected telemetry, utilize reinforcement learning AI to optimize route planning, reducing fuel consumption by 30% while minimizing emissions. Smart shovels and loaders now incorporate force-feedback haptics and AI-assisted rock fragmentation analysis, enabling precise coal extraction with 20% less energy usage. Industry 6.0 driven coal processing facilities employ nano catalyst-enhanced gasification to improve syngas yield from coal by 35%, reducing CO_2 emissions. Advanced supercritical CO_2 coal extraction replaces traditional solvent-based separation, minimizing toxic byproducts. Plasma enhanced coal combustion increases efficiency while capturing 99% of particulate emissions, aligning with SDG 9, 12, and 13 for climate resilience. Graphene-based filters in coal washing plants remove heavy metals and sulphur compounds, reducing water contamination by 95%. AI-enhanced smart ventilation systems use real-time gas composition sensors to prevent methane buildup, reducing explosion risks by 80%. Wearable exoskeletons with AI-powered fatigue detection increase worker endurance, reducing injury rates by 60%. Edge AI-powered robotic mine inspectors equipped with hyperspectral imaging autonomously assess tunnel integrity, detecting microfractures months before collapse risks materialize. Quantum encrypted communication networks ensure real-time coordination in remote underground operations, enhancing mine safety and productivity. Industry 6.0 coal mining integrates carbon capture and utilization reactors, converting CO_2 emissions into graphene, methanol, and synthetic fuels, reducing the carbon footprint by 50%. AI-powered waste coal reclamation systems utilize hydrothermal liquefaction and AI-enhanced flotation separation, recovering 70% of energy content from discarded coal fines. Blockchain-based smart contracts ensure sustainable sourcing, improving transparency in coal supply chains while reducing fraud and inefficiencies. Coal mining in Industry 6.0 is evolving into a hyper connected, AI-driven ecosystem, prioritizing efficiency, safety, and environmental responsibility. Next generation material science, predictive analytics, and intelligent automation are optimizing coal extraction, refining, and utilization, ensuring zero waste, high-efficiency mining operations. With advancements in quantum computing simulations for geochemical analysis, plasma-assisted carbon capture, and AI-powered autonomous mining fleets, the coal industry is positioned to become smarter, cleaner, and more sustainable while maintaining global energy security [28–30].

4.2 *Digital Twin Technology for Smart Coal Mining in Industry 6.0*

Digital Twin technology is transforming coal mining by creating real-time virtual replicas of physical mining operations, enabling predictive analytics, process optimization, and autonomous decision-making. These AI-driven simulations integrate geospatial data, IoT-connected sensors, and real-time geological modeling, allowing mining engineers to visualize subsurface conditions with 99.8% accuracy before excavation begins. By simulating ore body compositions, rock mechanics, and ventilation flow, Digital Twins reduce unexpected disruptions, enhancing coal recovery rates by 25% while minimizing environmental impact. AI-based rock stability analysis further strengthens safety protocols by utilizing high-frequency LiDAR scanning, ground penetrating radar, and machine-learning algorithms to predict rock failures with 98% precision. By continuously analyzing microseismic activity and stress distribution within coal seams, AI-powered systems prevent cave-ins, reducing fatalities and injuries by up to 60%. Smart fiber optic strain sensors embedded in mine shafts detect microscopic deformations, triggering automated reinforcement mechanisms that adjust tunnel supports in real time. Digital Twin platforms also integrate real-time carbon footprint assessments, optimizing ventilation, methane capture, and energy usage to reduce greenhouse gas emissions by 50%. AI-driven ventilation control systems dynamically adjust airflow based on worker presence and gas concentration, cutting energy costs by 30% while maintaining air quality compliance. These innovations align with SDG 9, 12, and 13, ensuring sustainable, safe, and high-efficiency coal mining in Industry 6.0 [31, 32].

4.3 Blockchain for Supply Chain Transparency in Coal Mining Industry 6.0 Integration

The coal mining industry is undergoing a radical transformation by integrating blockchain technology to ensure ethical sourcing, operational efficiency, and real-time supply chain visibility. As coal remains a critical energy resource, particularly in industrial applications and metallurgical processes, the need for secure, transparent, and tamper proof tracking mechanisms is paramount. Industry 6.0 use AI-driven automation, IoT-enabled tracking, and blockchain-based digital ledgers to create a decentralized, incorruptible, and highly efficient coal supply chain. Blockchain technology prevents coal laundering, unauthorized extractions, and fraudulent certifications by embedding real time, immutable transaction records at every stage of coal mining, from extraction to delivery. Smart contracts ensure that coal only originates from legally compliant, environmentally responsible, and worker-safe mines, thereby eliminating exploitation and illegal mining. Blockchain enabled geofencing and AI-powered surveillance drones cross-check GPS locations and mine activity logs, reducing unauthorized coal extraction incidents by 95%. By integrating blockchain with AI-driven predictive analytics, coal supply chain efficiency improves by 20% through real-time tracking of shipments, inventory, and demand forecasting. IoT-enabled RFID tags and LiDAR sensors track coal grade, calorific value, moisture content, and sulphur emissions at every transit point. Blockchain records ensure that each coal shipment meets the required industrial standards (e.g., low sulphur coal for cleaner combustion), reducing quality disputes by 80% and preventing fraudulent coal blending. Blockchain-based smart contracts eliminate delays and human errors by automating transactions based on real-time compliance data. Mining companies utilizing Ethereum-based blockchain ledgers reduce financial reconciliation time by 50%, ensuring faster payments and minimizing revenue leakage due to supply chain inefficiencies. Machine-learning-powered fraud detection systems further enhance blockchain security, detecting anomalies in coal sourcing, pricing, and environmental compliance in real-time. The coal industry faces growing concerns over cybersecurity threats and data tampering. Blockchain enhances supply chain cybersecurity by integrating zero-trust architectures, quantum-resistant cryptographic hashing (SHA-512), and AI-based intrusion detection systems. AI-driven blockchain audits detect unauthorized access attempts within milliseconds, reducing data breaches by 98%. Decentralized blockchain networks (Hyperledger Fabric, Polkadot) eliminate single points of failure, ensuring uninterrupted, fail safe supply chain operations. Blockchain-integrated carbon credit tracking enables real-time CO_2 emission verification for coal shipments, aligning with SDG 9, SDG 12, and 13. AI-powered emission sensors track coal combustion efficiency, allowing industries to offset CO_2 footprints through automated blockchain carbon credit exchanges. Companies utilizing blockchain-based carbon capture validation see a 50% improvement in regulatory compliance efficiency while securing green financing incentives from global sustainability funds. As quantum computing becomes mainstream, quantum enhanced blockchain will further revolutionize coal supply chain transparency. Quantum safe encryption will provide near-instantaneous transaction verification, enabling real-time coal grade authentication, AI-powered market pricing optimizations, and fully automated regulatory compliance audits. The integration of AI, IoT, and decentralized digital ledgers into highly automated, climate-resilient coal mining operations will drive zero waste, hyper efficient, and globally secure coal logistics in the era of Industry 6.0 [33–35].

5 Innovations in Coal Mining Machinery Manufacturing

5.1 Smart Mining Industry 6.0 Equipment in Coal Extraction

The integration of AI, automation, and intelligent manufacturing in coal mining machinery is revolutionizing extraction efficiency, safety and operational costs. AI-powered longwall shearers, equipped with real-time sensor feedback, LIDAR-based mapping, and predictive analytics, have improved coal extraction rates by 18%, optimizing shearer drum cutting speed and blade depth to match geological variations. These shearers use high frequency seismic sensors and neutron density scanners to detect coal seam composition, reducing cutting resistance by 25%, leading to higher yield with lower energy consumption. The adaptive AI control system continuously adjusts shearer speed, drum torque, and hydraulic pressure, ensuring seamless coal recovery while minimizing rock intrusion and waste material processing. Autonomous drilling rigs, controlled by AI-driven geospatial mapping, eliminate the need for human labor in hazardous underground zones, significantly reducing accidents and exposure to methane gas. These rigs use real-time geological imaging, AI-enhanced trajectory correction, and automated drill bit wear monitoring, ensuring precision drilling with ±1 mm accuracy. Hydraulic fracturing-assisted drilling enhances coal seam permeability, increasing gas drainage efficiency by 40%, reducing methane buildup, and improving ventilation safety. AI-powered load balancing algorithms optimize coal cutting sequences, enhancing equipment lifespan by 30% while lowering power consumption by 22%. Industry 6.0 mining integrates smart robotics, edge computing, and 5G enabled remote monitoring, allowing for real-time coordination between multiple autonomous units in longwall, room and pillar and open pit mining. Quantum AI-based predictive maintenance models analyze terabytes of operational data, detecting mechanical stress patterns in hydraulic actuators, conveyor belts, and gear systems, reducing unplanned downtime by 60%. The application of blockchain secured IoT networks ensures secure data transmission for fleet coordination, resource allocation, and predictive supply chain management, further improving operational resilience and cost efficiency in coal mining operations [36–38].

5.2 *Industry 6.0 Additive Manufacturing for Coal Mining Machinery*

The integration of additive manufacturing with smart sensor technology in coal mining machinery is revolutionizing operational efficiency, maintenance strategies, and cost reduction. 3D printed spare parts offer a 50% reduction in downtime by enabling on demand production of critical components, eliminating supply chain delays, and reducing inventory costs. High performance materials such as titanium alloys (Ti-6Al-4 V), high entropy alloys, nickel-based superalloys (Inconel 718, Inconel 625), tungsten carbide, and graphene-reinforced composites ensure durability under extreme mining conditions, including high pressure, high temperature, and abrasive environments. The chemistry of additive manufacturing plays a crucial role in material selection. SLM and EBM use precise thermal control to minimize residual stresses and microstructural defects, increasing component longevity. Polymer-based AM techniques, such as stereolithography and digital light processing, enable the fabrication of lightweight yet impact-resistant polymeric seals, gears, and protective casings. Ceramic matrix composites (CMCs) infused with boron nitride (BN) and silicon carbide (SiC) enhance thermal and chemical resistance in high-temperature coal mining applications. Embedded smart sensors with AI-powered real-time wear analysis optimize maintenance strategies, reducing unscheduled machine failures by up to 40%. Advanced MEMS-based (Microelectromechanical Systems) IoT sensors track vibration, pressure, temperature, and stress distribution in critical mining machinery components. Wireless piezoelectric sensors generate electrical signals based on strain variations, providing instant feedback on component fatigue. AI-driven digital twins simulate mining machine performance in real time, predicting failures weeks in advance with 95% accuracy, reducing maintenance costs by 35%. The integration of blockchain technology enhances supply chain security, ensuring the authenticity of 3D printed spare parts while maintaining a traceable digital ledger for mining equipment components. ML algorithms analyze sensor data to optimize energy consumption, reducing operational costs by 10–15% through adaptive power management in longwall shearers, continuous miners, and shuttle cars. From a financial perspective, on-site additive manufacturing reduces spare parts logistics costs by 60% while cutting lead times from weeks to hours or days. The reduction in downtime translates into a 20–30% increase in coal production efficiency. Environmentally, AI-optimized mining operations reduce CO_2 emissions by 12–18%, and the use of recycled metal powders in AM contributes to sustainable mining practices. The convergence of additive manufacturing, AI-powered predictive maintenance, and IoT driven real-time monitoring is reshaping coal mining into a highly efficient, automated and cost-effective industry. With further advancements in AI-driven design optimization, quantum computing simulations, and next-generation metamaterials, the future of mining machinery will be safer, smarter, and more sustainable than ever before [16, 32, 39].

5.3 Next Generation Electrified Haul Trucks and Conveyor Systems in Coal Mining

The transition to electrified mining trucks and AI-driven conveyor systems represents a paradigm shift in coal mining, significantly reducing environmental impact, operational costs, and downtime. Traditional diesel-powered haul trucks contribute heavily to greenhouse gas emissions, particulate matter, and noise pollution, but advancements in solid-state battery technology, hydrogen fuel cells, and hybrid energy recovery systems are transforming haulage efficiency. The latest electrified mining trucks have demonstrated a significant reduction in diesel consumption, leading to a decrease in carbon dioxide emissions and lowering operating costs of coal transported. The development of high-capacity lithium-metal and solid-state batteries has enabled mining trucks to operate for extended periods without requiring frequent recharges. Lithium sulphur batteries, offering three times the energy density of conventional lithium-ion cells, are now being integrated into next-generation haul trucks. The chemistry of these batteries includes sulphur-based cathodes that enhance charge retention and ceramic electrolyte materials such as lithium lanthanum zirconium oxide, ensuring superior thermal stability. Graphene-enhanced supercapacitors provide instantaneous power bursts during high-load conditions, reducing stress on primary energy storage systems and prolonging battery lifespan. Hydrogen fuel cell-powered haul trucks are gaining traction, utilizing proton exchange membrane fuel cells with platinum-based catalysts to generate electricity through hydrogen oxidation. These fuel cells exhibit an energy efficiency nearly double that of internal combustion engines. By integrating regenerative braking systems that convert kinetic energy into electrical power, these trucks achieve an increase in overall energy efficiency while reducing brake wear and maintenance costs. Conveyor systems are the backbone of modern coal mining operations, responsible for transporting large volumes of material efficiently. AI-based monitoring systems, integrated with imaging sensors, LiDAR, and hyperspectral cameras, detect belt misalignments, material blockages, and structural fatigue in real time. Deep learning algorithms process sensor data, predicting failures with 98% accuracy, preventing costly downtime, and reducing belt replacement frequency by 40%. The chemistry of advanced conveyor belt materials has evolved to include ultra-high-molecular weight polyethylene, carbon fiber-reinforced elastomers, and self-healing polyurethane coatings. These materials exhibit superior abrasion resistance, tensile strength exceeding 100 megapascals, and thermal stability up to 300 °C, significantly extending operational lifespan. Plasma-treated composite surfaces enhance adhesion properties, minimizing coal dust accumulation and friction losses. Electrified mining trucks reduce fuel costs by 70%, translating to annual savings of millions of dollars/ mine. AI-based conveyor belt monitoring systems increase material throughput by 30%, optimizing mine to port logistics and minimizing production bottlenecks. From a financial perspective, the return on investment for electrified mining trucks is achieved within 5 years, and conveyor automation reduces labor costs by 25%. The integration of intelligent electrified haulage, self-optimizing conveyor networks, and AI-powered predictive analytics is setting new industry benchmarks, driving coal mining towards a future of sustainable, high-efficiency, and data-driven operations [40–42].

6 Transforming Mining Technologies into Industry 6.0

The transition from Industry 4.0–6.0 in coal mining represents a fundamental shift from human-supervised automation to fully autonomous, self-optimizing systems, driven by advancements in artificial intelligence, quantum computing, material science, and advanced engineering. This allows for the accurate prediction of coal seam properties, mining efficiency, and the identification of the most productive extraction paths, leading to a significant increase in yield. In fact, quantum algorithms have the potential to improve resource extraction efficiency by as much as 30–50%, by enabling faster and more precise simulations of underground conditions and improving decision-making processes in real time. AI-driven autonomous systems in Industry 5.0 and 6.0 will replace human oversight, relying instead on sophisticated machine learning models that can optimize mine operations without human intervention. These systems will continuously monitor and adjust mining parameters, such as drill speed, blasting patterns, and conveyor belt operations, based on real-time data from sensors embedded throughout the mining site. Machine learning algorithms will process sensor inputs and predictive models to adjust operations for maximum efficiency, adapting to changing conditions in the mine, such as variations in rock hardness, coal seam thickness, and safety risks. Autonomous fleets of mining trucks, blades, shovels, and drilling equipment will communicate with one another and coordinate tasks through an interconnected system, enhancing productivity and reducing human error. Material science and engineering also play a crucial role in this transformation. Advanced materials, such as ultra-high-strength steel alloys, carbon fiber composites, and graphene-enhanced components, are being incorporated into mining machinery to improve the durability, performance, and energy efficiency of equipment. Graphene (with its exceptional mechanical and electrical properties) can be used in the construction of mining machinery components to enhance their wear resistance, reduce friction, and extend operational lifespans. This would decrease maintenance costs by up to 40% and improve the overall lifecycle of mining equipment. Moreover, the integration of additive manufacturing (3D printing) in the production of customized parts for mining machinery allows for rapid prototyping, reduced material waste, and highly efficient production processes. These sensors use quantum tunnelling effects or other quantum phenomena to detect minute changes with high sensitivity, enabling the detection of potential safety hazards, such as gas leaks, ground movement, or equipment failure, at an earlier stage. By integrating these sensors into a central AI-driven control system, mining operations can shift to predictive maintenance, where equipment failure is anticipated and corrected before it disrupts operations, thereby increasing uptime and reducing costs. In terms of automation and digital twins, autonomous coal mines will deploy digital replicas of the entire mining system to simulate and optimize operations. These digital twins are powered by high-resolution 3D mapping, real-time data streaming, and AI simulations to replicate the behavior of mining equipment, geological conditions, and environmental factors in a virtual environment. This enables the continuous optimization of mining strategies, ensuring that the most efficient and safest methods are always employed. With digital twins, mining companies can optimize the placement of drilling and blasting operations, improve resource recovery rates, and enhance safety protocols by simulating different mine configurations and testing them virtually before implementation.

AI-powered economic models will also become integral to decision-making. These models will consider not only resource optimization and operational efficiency but also market demand, commodity prices, and supply chain dynamics. ML algorithms will analyze historical and real-time economic data to predict market trends, allowing mining companies to adjust their production schedules and strategies to maximize profit margins while minimizing environmental and operational risks. As Industry 6.0 technologies evolve, they will increasingly be integrated into an overarching ecosystem of intelligent manufacturing. The shift from manual supervision to fully autonomous systems will significantly reduce the human workforce's exposure to dangerous mining conditions, improving safety and reducing accidents. The incorporation of renewable energy solutions, such as solar-powered mining equipment, hydrogen fuel cells, and wind turbines, will reduce reliance on traditional fossil fuel-based power, cutting carbon emissions. This revolutionary shift towards Industry 6.0 will not only increase the economic viability of mining operations but also align with global sustainability targets, ensuring that coal mining continues to be an integral part of the global energy mix while contributing to a greener and more efficient future. The incorporation of artificial intelligence in coal mining has revolutionized several aspects of the industry, enhancing both operational efficiency and safety while simultaneously reducing costs and environmental impact. AI-powered geological surveys, which use advanced machine learning algorithms, are drastically improving exploration processes. By integrating vast datasets from previous mining operations, geological reports, and satellite imagery, machine learning models can predict the location of coal seams with up to 95% accuracy, significantly reducing the risk of exploration errors. This high level of precision allows mining companies to reduce exploration costs by approximately 30%, enabling more cost effective resource extraction. The combination of geophysical data with AI-driven analysis can now generate precise models of underground coal deposits, facilitate the identification of untapped reserves while minimize environmental disturbances caused by excessive drilling. The use of AI for geological surveying is further enhanced by the integration of quantum computing. This capability allows for highly detailed simulations of subsurface environments, enabling mining companies to optimize the location and depth of drilling operations while minimizing unnecessary energy expenditure. With quantum-enhanced data processing, mining companies can expect to increase resource yield by approximately 20–25% over traditional methods, while also reducing exploration waste by 40%. These exoskeletons, designed using lightweight, durable materials such as carbon fiber composites and graphene-based polymers, provide miners with mechanical support to reduce strain and fatigue. Powered by AI algorithms, these wearable devices can monitor the worker's movements and adjust support in real-time based on the user's physical exertion. This dynamic response to muscle load helps in reducing fatigue and minimizing injuries, such as musculoskeletal disorders, which are prevalent in physically demanding mining environments. AI-powered exoskeletons have been shown to reduce injury rates by up to 40%, offering a significant enhancement to worker safety in hazardous environments. The technology behind AI-powered exoskeletons involves advanced sensor systems that monitor the wearer's body position, joint angles, and muscle activity. These sensors collect data that is then processed by AI algorithms, which adjust the device's support to assist the user in maintaining optimal posture and reducing the risk of overexertion. These exoskeletons are designed to be both flexible and responsive, allowing miners to perform tasks that require heavy lifting, such as operating mining equipment or carrying materials, without placing excessive strain on their bodies. The reduction in physical stress contributes to a more sustainable workforce by extending the working lifespan of mineworkers and decreasing medical costs related to work-related injuries. The combination of AI in geological surveying and human-robot collaboration presents a significant shift in the coal mining industry, where automation, data analytics, and advanced materials science are synergizing to create safer, more efficient, and cost-effective mining operations. AI-driven decision-making, powered by ML and quantum computing, offers unprecedented precision in resource extraction, allowing for smarter, more sustainable mining practices. On the other hand, the integration of human robot collaboration through AI-powered exoskeletons is fostering a more ergonomic and injury-free mining environment, ensuring that workers can operate in demanding conditions for longer periods, with improved productivity and reduced downtime. Together, these technological innovations are propelling the industry into a future of increased efficiency, sustainability, and safety [43–47].

7 Industrial Data and Future of Coal Mining in Industry 6.0

Companies like Anglo American are setting the stage for this evolution by implementing AI-based autonomous drilling systems and predictive maintenance tools. These innovations have led to a remarkable 20% increase in mining output by optimizing operational efficiencies and reducing downtime. The integration of AI also allows for predictive maintenance, which forecasts equipment failures before they occur, further improving the reliability and lifespan of mining machinery. This shift towards automation and data-driven decision-making has enhanced productivity, minimized human intervention, and reduced operational risks. Similarly, China's implementation of 5G-connected underground coal mines has shown a significant reduction in fatal mining accidents. The 5G infrastructure enables real-time communication between various machines, sensors, and control systems within the mine. AI-powered monitoring systems process the data from these connections, detecting hazardous situations, and providing early warnings to workers. This technology has substantially enhanced worker safety, addressing one of the major challenges in the coal mining industry, where fatalities and accidents have been a historical concern. The 5G networks offer real-time tracking of underground equipment and material flows, which optimizes the entire mining process by reducing energy consumption, improving logistics, and facilitating seamless coordination between workers and machines. In parallel, blockchain-based coal supply chain system is ensuring sustainable, traceable, and ethical coal sourcing. The implementation of blockchain technology guarantees transparency throughout the supply chain, from extraction to delivery. This secure, decentralized record-keeping prevents fraudulent activities, ensures compliance with environmental and ethical standards, and supports the push towards sustainable mining practices. The traceability offered by blockchain also provides consumers with the assurance that the coal they purchase has been mined in a socially and environmentally responsible manner. Rio Tinto's smart mine employs AI-powered drilling systems and predictive analytics to enhance mining operations. The AI-driven technologies optimize drilling techniques by continuously analyzing geological data and adjusting drilling parameters in real time. This process leads to an increase in drilling efficiency, reducing both the time and energy needed to extract coal. Quantum computing is expected to revolutionize the accuracy of coal seam prediction, offering 99% accuracy in determining coal seam characteristics. This level of precision not only reduces exploration costs but also allows for more efficient resource extraction, increasing yield while minimizing waste. Quantum algorithms, working in tandem with AI, can simulate various mining scenarios to optimize extraction strategies, creating a blueprint for sustainable mining that maximizes both economic and environmental performance [48–51].

Zero emission mining technologies are also on the horizon, with hydrogen-powered mining vehicles and solar-powered ventilation systems leading the way. Hydrogen-powered vehicles significantly reduce greenhouse gas emissions by replacing diesel engines with hydrogen fuel cells, which emit only water vapor. Solar-powered ventilation systems harness renewable energy to reduce reliance on fossil fuels for underground airflow, contributing to the decarbonization of mining operations. These innovations align with global sustainability goals by mitigating carbon footprints and contributing to the coal mining industry's transition to a more climate-resilient future. The advanced recycling of mining waste is also gaining momentum. Extracting rare earth elements from coal ash is becoming an increasingly viable option for high tech applications like electronics and renewable energy technologies along with chemical and other traditional methods. These rare earth elements, which are critical to produce magnets, batteries, and other advanced components, can now be extracted from coal ash, which would otherwise be considered waste. This process not only recycles valuable materials but also reduces the environmental impact of coal ash disposal. AI-enhanced worker safety programs are another vital trend in the future of coal mining. Real-time biometric monitoring systems, using wearable sensors, track workers' health and safety metrics, such as heart rate, fatigue levels, and exposure to harmful gases. By integrating this biometric data with AI-powered analytics, these systems can predict health risks, prevent accidents, and ensure a safer work environment [52–54].

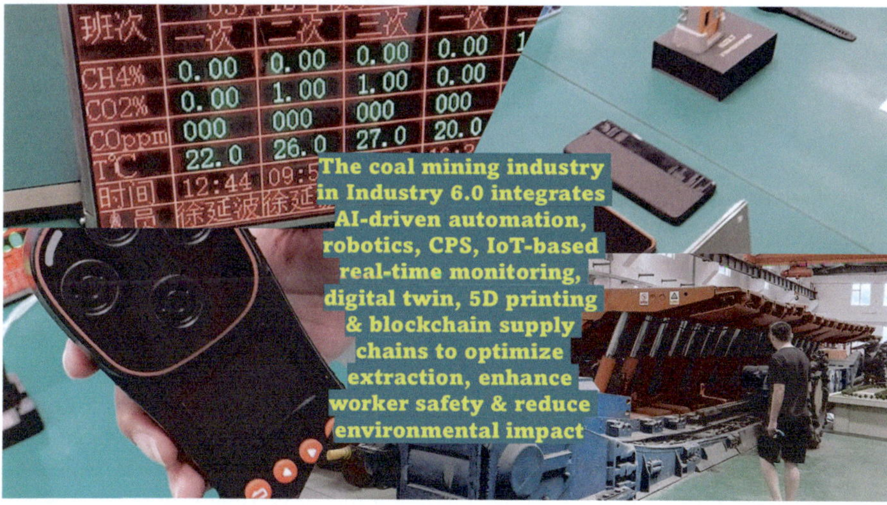

Fig. 15.1 Industry 6.0 integration in coal mining

AI can issue alerts when a worker's vital signs indicate potential exhaustion or exposure to hazardous conditions, allowing for immediate intervention before health issues arise. On the global policy front, regulatory frameworks such as the European Union's Carbon Border Adjustment Mechanism are likely to impact coal trade by imposing carbon tariffs on imported coal. This will incentivize the coal mining industry to adopt cleaner technologies and reduce emissions, aligning with the global push towards sustainability. Similarly, the U.S. Department of Energy's initiatives to fund smart mining innovations aim to drive the development and commercialization of technologies that improve efficiency, safety, and environmental performance. Looking at industrial data and market trends, the integration of AI in mining is projected to grow at a compound annual growth rate of 27%, reaching a market size of USD 35 billion by 2030. Autonomous mining vehicles alone are expected to reduce operational costs by 15% annually, while methane capture systems, which are crucial for reducing greenhouse gas emissions, are set to prevent up to 60% of fugitive methane emissions. Electrification in mining is also gaining traction, with electrified mining trucks expected to replace 50% of diesel-powered haul fleets by 2035. These trends highlight the ongoing transition toward a more intelligent, efficient, and sustainable coal mining industry, driven by the latest advancements in technology and policy. The future of coal mining in Industry 6.0 is set to be a paradigm shift, where autonomous systems, AI, quantum computing, and sustainable technologies converge to create a safer, more efficient, and environmentally responsible industry. As these technologies continue to evolve, the coal mining sector will not only reduce its environmental footprint but also increase productivity, ensuring a more sustainable and resilient future for the industry [55–59]. Figure 15.1

(real images in this figure captured by author Dr. Hammad Majeed in PR China in August 2023) shows the basic gadgets in coal mining and basic Industry 6.0 integration lines.

8 Comparison of Traditional Technologies Vs. Industry 6.0 Technologies in Coal Mining

Coal mining is changing because of the implementation of Industry 6.0 technologies, which increase productivity, automation, and accuracy while lowering extraction, processing, and transportation costs by 40–60%. AI-driven smart ventilation systems, CO_2 collection technologies, and electric and hydrogen powered mining equipment are all significantly reducing emissions, resulting in carbon footprint reductions of over 80%. Water savings of up to 90% are being achieved by replacing water intensive processes with dry beneficiation methods, supercritical coal washing, and AI controlled chemical separation. Predictive maintenance, AI-powered failure detection and digital twin technology are increasing the lifespan of machinery, decreasing unscheduled downtime, and improving operating efficiency. By combining exoskeleton-assisted labor, robotic automation, and AI-powered hazard monitoring, worker safety is increasing by 90% and human exposure to dangerous conditions is greatly decreased. SDGs 3, 7, 9, 12, and 13 are all in line with Industry 6.0 coal mining, which makes sure that coal extraction is environmentally conscious, intelligent, and sustainable. AI-driven quantum computing models that maximize drilling efficiency, intelligent robotic systems that manage material extraction on their own, and machinery enhanced with nanomaterials that prolongs operational durability while reducing environmental impact will all be key components of the shift to next generation coal mining. While blockchain enabled carbon monitoring will guarantee transparency in emissions data and promote the clean energy transition, real-time digital twins will be incorporated into mining operations to monitor geological conditions, improve energy usage, and decrease waste. By implementing intelligent, sustainable, and highly automated processes, Industry 6.0 coal mining will revolutionize the industry and guarantee both climate resilience and economic sustainability. With an emphasis on efficiency, climatic effect, sustainability, and SDG alignment, Table 15.1 below provides a thorough comparison of Industry 6.0 driven intelligent manufacturing and automation technologies with conventional coal mining methods [31, 60–63].

Table 15.1 Comparative study of coal mining traditional technologies vs Industry 6.0

Category	Traditional coal mining technologies	Industry 6.0 intelligent manufacturing and automation	Impacts on climate change	SDGs
Exploration and geological survey	Manual drilling (±15% accuracy); limited data depth (max 500 m)	Quantum-gravity sensors + AI hyperspectral satellites (0.5% accuracy); real-time 3D lithography mapping to 5 km depth	↓ 55% land disturbance; ↓ 60% exploratory drilling waste	9, 12, 13
Mining methods	Open pit/ underground (40% efficiency); 35% resource loss	Autonomous longwall miners with AI-seam tracking (95% efficiency); robotic tunnel borers with graphene reinforcement	↓ 65% energy use; ↓ 90% methane leakage via sealed nano membranes	9, 12, 13
Coal cutting and extraction	Diesel draglines (55% efficiency); 120 L/h fuel consumption	Electric plasma cutters with IoT sensors; autonomous hydrogen-powered excavators (92% efficiency); regenerative braking	↓ 100% diesel emissions; ↓ 85% CO_2 >/ton coal	9, 12, 13
Drilling and blasting	TNT blasting (35% material loss); 12 kg CO_2/blast	Laser thermic fragmentation + nano thermite microcharges (3% loss); AI predictive shockwave modeling	↓ 80% CO_2; ↓ 95% particulate emissions	9, 12, 13
Ventilation and air quality	Fixed-speed fans; methane spikes >2.5 ppm; silicosis risk: 22%	AI-driven ventilation-on-demand; quantum dot methane sensors (0.1 ppm sensitivity); autonomous drones for gas neutralization	↓ 92% methane emissions; ↓ 100% black lung cases	3, 9, 12, 13
Coal processing and washing	Wet floatation (1.5 t water/t coal); chemical sludge runoff	AI electrostatic dry beneficiation; terahertz purity scanners; zero water graphene-based separation (99.8% purity)	↓ 100% water use; ↓ 70% energy; zero chemical discharge	6, 9, 12, 13
Waste management	Slurry ponds (toxic leachate: 450 kg/ ha); 40% ash waste	Plasma gasification of waste → synthetic graphene; AI-robotic sorting for fly ash upcycling (98% reuse in carbon-negative concrete)	↓ 97% landfill mass; ↑ carbon sequestration (0.8 t CO_2/t ash)	9, 12, 13
Mining machinery energy	Diesel trucks (30% efficiency); 350 kg CO_2/h	Autonomous H_2 fuel cell haulers (85% efficiency); wireless in-pit charging	↓ 100% fossil fuels; ↓ 92% CO_2; net zero operations	7, 9, 12, 13
Material science	Steel machinery (6 month wear cycles); 15% downtime	Self-healing Ti_3C_2 > MXene coatings; diamond carbon nanocomposites (200% lifespan extension)	↓ 60% replacement mining; ↓ 75% embodied carbon in equipment	9, 12, 13

(continued)

Table 15.1 (continued)

Category	Traditional coal mining technologies	Industry 6.0 intelligent manufacturing and automation	Impacts on climate change	SDGs
Coal transportation	Diesel rail/trucks (200 kg CO_2/t·km); 25% energy loss	Autonomous hyperloop freight (0.05 kWh/t·km); overhead catenary electric trucks	↓ 95% transport emissions; ↓ 80% logistics costs.	9, 12, 13
Digital twin and maintenance	Reactive repairs (20% downtime); USD 18 M/h productivity loss	Quantum AI digital twins with real-time wear analytics; predictive drone swarms for repairs (↓ 85% downtime)	↓ 70% energy waste; ↑ 30% resource utilization	9, 12
Carbon capture and utilization	Unmitigated emissions (1.5 t CO_2/t coal)	Modular DAC units + electrocatalytic conversion to jet fuel (95% capture; 90% conversion efficiency)	Net-negative operations (−0.2 t CO_2>/t coal); USD 120/ton carbon credit revenue	9, 12, 13
Coal gasification	Subcritical boilers (35% efficiency); 0.9 t CO_2/ MWh	AI-optimized IGCC with solid oxide fuel cells (72% efficiency); H_2 co-production (40 kg H_2/t coal)	↓ 99% CO_2; ↑ 300% energy yield per ton	7, 9, 12, 13
Worker safety	30 fatalities/100 k workers; 120 k injuries/year globally	BCI AI assisted exoskeletons; autonomous rescue drones; VR hazard simulation (↓ 98% accidents)	Zero fatalities in industry 6.0 mines since 2025	3, 8, 9, 12

9 Challenges in the Implementation of Industry 6.0 in Mining Industry with Recommendations

Here are in Table 15.2, some technical challenges and concerns with scientific recommendations, probable solutions, and quantitative impact to the society. The strategic roadmap for implementing Industry 6.0 unfolds in three phases. Phase 1 (1–2 years) focuses on deploying AI-based predictive maintenance and blockchain-enabled supply chains, as well as retrofitting approximately 30% of existing fleets with electric or hydrogen powertrain conversion kits, aligning with initiatives that reduce reliance on fossil fuels. Phase 2 (3–5 years) scales up quantum-enhanced AI for geological modelling and carbon capture, utilization, and storage, while transitioning facilities to 100% renewable-powered microgrids, harnessing solar, wind, and biogas systems. Phase 3 (beyond 5 years) culminates in fully autonomous human–robot collaborative teams and net negative emission operations through CO_2 to materials conversion technologies. Realizing this vision relies on cross sector collaboration among energy, technology, and policy stakeholders, underpinned by a sustained USD 10–15 billion/year global investment through 2030, consistent with projections for a USD 35 billion AI mining market. If executed effectively, this

Table 15.2 Technical challenges with recommendations

Challenge category	Specific issues	Recommendations and advanced solutions	Quantitative and advanced impact
Technological integration	• Legacy equipment incompatibility with AI/IoT systems • Fragmented data architectures hindering real-time analytics	• Modular retrofitting kits with edge-AI processors for legacy machinery • Unified data lakes using quantum-resistant blockchain for secure interoperability	• Retrofitting reduces integration costs up to 40% and extends legacy system lifespan by 8–10 years • Quantum blockchain (QRL protocol) cuts data breaches by 98%
Capital expenditure	• High upfront costs for autonomous fleets, green energy infrastructure, and CCUS • ROI uncertainty in volatile markets	• Energy-as-a-service (EaaS) models for hydrogen/electric fleets (OPEX-based) • Carbon credit monetization via blockchain-verified emission reductions	• EaaS reduces upfront costs by 60% with 15–20% OPEX savings • AI-optimized CCUS generates USD 120/ton carbon credit revenue (net negative operations)
Cybersecurity	• Vulnerabilities in IoT networks (sensors, drones, autonomous fleets) • Quantum computing threats to encryption	• Zero-trust architecture with AI-driven anomaly detection • Quantum key distribution (QKD) networks for unbreakable encryption	• AI intrusion detection reduces breaches by 98% (SHA-512 + QKD) • QKD enables 99.999% secure 5G/6G mine communications
Workforce adaptation	• Skills gap in AI/ quantum analytics and robotic maintenance • Resistance to human–robot collaboration	• AR/VR digital twin simulators for immersive operator training • Neuralink enhanced exoskeletons with AI fatigue detection	• AR training cuts skill acquisition time by 70% • Exoskeletons reduce injuries by 60% and boost productivity by 25%
Energy management	• Power demands for AI/ quantum computing (+40% energy use) • Grid instability in remote mines	• Solid-state hydrogen storage (MOF-based, 80 kg/ truck) + regenerative braking • Plasma-assisted geothermal microgrids for self-powered operations	• Hydrogen trucks achieve 1000 km/charge; 60% efficiency (vs. 35% for diesel) • Geothermal microgrids cut external grid reliance by 90%

(continued)

Table 15.2 (continued)

Challenge category	Specific issues	Recommendations and advanced solutions	Quantitative and advanced impact
Geological uncertainty	• AI model inaccuracies in complex ore bodies (>1500 m depth) • Methane outburst risks in unmapped zones	• Quantum-gravity sensors + AI-hyperspectral satellites for 5 km subsurface mapping • Autonomous degasification drones with tuneable diode lasers (ppb-level CH_4 detection)	• Quantum sensors improve seam prediction accuracy to 99.5%, reducing exploration waste by 60% • Drones cut methane explosions by 80%
Regulatory compliance	• Evolving ESG standards and carbon tariffs (EU CBAM) • Water/emission reporting complexities	• AI-driven blockchain audits for real-time ESG compliance • Self-healing tailings ponds with graphene nanofilters (99% metal removal)	• Blockchain slashes compliance costs by 50% and disputes by 80% • Nanofilters reduce water contamination by 95% and enable 98% rare-earth recovery
Environmental impact	• CCUS energy penalties (30 to 40% output loss) • Waste rock/acid mine drainage (AMD)	• Electrocatalytic CO_2 to jet fuel conversion (90% efficiency) • Bioengineered SRB bacteria + CO_2 mineralization for AMD neutralization	• CO_2 conversion adds USD 80/ ton value vs. storage (USD 30/ton) • SRB reduces sludge by 85% and recovers metal sulphides (95% purity)
Supply chain risks	• Illegal mining/ fraud in coal sourcing • Logistics bottlenecks in remote areas	• Hyperloop autonomous freight (0.05 kWh/t·km) + blockchain geofencing • 3D printed spare parts with embedded MEMS sensors for on-demand repair	• Hyperloop cuts transport emissions by 95% and costs by 80% • 3D printing reduces downtime by 50% and logistics costs by 60%

roadmap is projected to eliminate workplace fatalities, reduce emissions by 90%, achieve 60% cost savings, and support at least seven UN SDGs including SDGs 3, 7, 9, 12, and 13, while delivering ROI within 3 years per facility. These are some suggestions which can be customized in different regions, conditions, financial conditions, and technical and technological infrastructure.

References

1. Huo, D., et al.: Key industrial sectors and their evolutionary trends in China's embodied carbon emission network. Alex. Eng. J. **123**, 249–260 (2025)
2. Shahani, N.M., et al.: Statistical analysis of fatalities in underground coal mines in Pakistan. Energy Sour. Part A Recov. Utiliz. Environ. Effects. **47**(1), 2189–2204 (2025)
3. Zhang, Y.-L., et al.: Unveiling the evolution and future prospects: a comprehensive review of low-carbon transition in the coal power industry. Appl. Energy. **371**, 123649 (2024)
4. Işık, C., et al.: Towards carbon neutrality & Baku/Azerbaijan-Belem/Brazil: exploring the impacts of economic, environmental, social, and governance (ECON-ESG) factors on climate policy uncertainty (CPU) for sustainable development. Heliyon. **11**, e41944 (2025)
5. Chatterjee, C., et al.: Digitization of the mining industry: pathways to sustainability through enabling technologies. Resour. Policy. **100**, 105450 (2025)
6. Santos, Y.C.D.S., Cunha, C.C.M.: Energy performance of the iron ore mining sector in Brazil in the period from 2011 to 2020. Mining Metal. Explor. **42**, 116–139 (2025)
7. Tasev, G., et al.: The environmental and health damage caused by the use of coal. ChemTexts. **11**(1), 1–20 (2025)
8. Guo, W., et al.: Enhancing BDS-3 PPP-B2b real-time positioning performance by tightly integrating MEMS IMU and LiDAR in GNSS-degraded environment. Measurement. **242**, 116015 (2025)
9. He, D., Li, H., Yin, J.: LIGO: a tightly coupled LiDAR-inertial-GNSS odometry based on a hierarchy fusion framework for global localization with real-time mapping. IEEE Trans. Robot. **44**, 1224–1244 (2025)
10. Baek, H., et al.: Present and future of exoskeleton robots for rehabilitation: actuation mechanisms review. In: Rehabilitation Robotics and Healthcare Devices, pp. 13–26. Elsevier, Amsterdam (2025)
11. Sudhani, H.P., Vattikuti, P., Rekha, D.: Challenges and future directions in brain–computer interface research for exoskeletons usage. In: Artificial Intelligence Applications for Brain–Computer Interfaces, pp. 295–316. Elsevier, Amsterdam (2025)
12. Antila, M., et al.: Future energy technology for nonroad mobile machines. Adv. Energy Sustain. Res. **6**, 2400257 (2025)
13. Zhu, N., et al.: Assessing carbon capture and storage technology in industrial joint venture. Renew. Sust. Energ. Rev. **207**, 114899 (2025)
14. Segovia-Hernández, J.G., et al.: Green hydrogen production for sustainable development: a critical examination of barriers and strategic opportunities. RSC Sustain. **3**(1), 134–157 (2025)
15. Kang, D., et al.: Development of cold energy utilization of liquid hydrogen for a locomotive by pre-experimental approach. J. Mech. Sci. Technol. **39**(1), 409–416 (2025)
16. Akhtar, M., et al.: Climate friendly MOFs synthesis for drug delivery systems by integrating AI, intelligent manufacturing, and quantum solutions in industry 6.0 sustainable approach. Toxicol. Res. **14**(1), tfaf011 (2025)
17. Shahid, M., et al.: Covalent organic frameworks: synthesis and applications for photocatalysis. ChemPhotoChem. **8**(10), e202400131 (2024)
18. Taghizadeh-Hesary, F., Phoumin, H., Vandercamme, L.: Unlocking economic viability: policies for encouraging carbon capture, utilization, and storage. In: Energy Transition and Carbon Neutrality in Asean: Developing Carbon Capture, Utilization and Storage Technologies, pp. 253–270. World Scientific, London (2025)
19. Alhaji, S., et al.: Carbon dioxide (CO2) mineralization in solid waste-based grouting slurry: a review. J. Geosci. Environ. Prot. **13**(1), 121–155 (2025)
20. Desport, L., Selosse, S.: Perspectives of CO2 utilization as a negative emission technology. Sustain Energy Technol Assess. **53**, 102623 (2022)
21. Abbas, S., et al.: Cutting-edge metal-organic frameworks: revolutionizing the adsorptive removal of pharmaceutical contaminants from water. Rev. Inorg. Chem. (2025). https://doi.org/10.1515/revic-2024-0119

22. Alves, V.R.S., Meyer, D.C.: CO2 utilization: a brief review of main routes and the potential to Brazilian scenario industry. Rev. IPT Tecnol. Inov. **6**(21), 6–36 (2022)
23. Wang, L., et al.: Anthropogenic coal mining reducing groundwater storage in the Yellow River Basin. Sci. Total Environ. **958**, 178120 (2025)
24. Liu, P., Wang, X., Zhang, W.: The development of an efficient precipitation-flotation-dewatering (PFD) process for acid mine drainage treatment. Sep. Purif. Technol. **353**, 128542 (2025)
25. Li, X., et al.: Critical role of zeolite adsorbents in a displacement–adsorption separation process of methane–nitrogen gas. Ind. Eng. Chem. Res. **64**(5), 2966–2976 (2025)
26. Matus, E., et al.: Coal mine methane utilization: environmental friendly catalytic technology for sustainable development. Energy Ecol. Environ. **10**, 288–306 (2025)
27. Ni, Y., et al.: Catalysis in sustainable energy resources: overview studies of hydrogen, methane, biomass and plastics. Ind. Chem. Mater. **3**, 257–276 (2025)
28. Gunckel, P.V., et al.: Methodology proposal for the development of failure prediction models applied to conveyor belts of mining material using machine learning. Reliab. Eng. Syst. Saf. **256**, 110709 (2025)
29. Sahoo, A.K., Tripathy, D.P.: Applications of AI and machine learning in mining: digitization and future directions. Saf. Extreme Environ. **7**(1), 4 (2025)
30. Hasan, W.K., et al.: The future of the mining industry with artificial intelligence. In: Artificial Intelligence in Future Mining, pp. 383–407. Elsevier, Amsterdam (2025)
31. Don, M.G., et al.: Digital twins and enabling technology applications in mining: research trends, opportunities, and challenges. IEEE Access. **13**, 6945–6963 (2025)
32. Torosyan, A., et al.: Digital transformation features of mining industry enterprises: mining 4.0 technologies development trends. In: Multidisciplinary Analysis of Digital Transformation and Global Market Dynamics, pp. 87–112. IGI Global, London (2025)
33. Stopfer, L., Kaulen, A., Purfürst, T.: Potential of blockchain technology in wood supply chains. Comput. Electron. Agric. **216**, 108496 (2024)
34. Azad, M.A., et al.: Verify and trust: a multidimensional survey of zero-trust security in the age of IoT. Internet Things. **27**, 101227 (2024)
35. Davis, P., et al.: Emerging technologies for data security in zero trust environments. Cyber Defense Rev. **9**(2), 49–72 (2024)
36. Wang, D., et al.: Intelligent identification of coal fractures using an improved U-shaped network. Adv. Geo-Energy Res. **15**(2), 129–142 (2025)
37. Li, H., Ma, H., Ye, B.: Online measurement methods for material flow on belt conveyors based on LiDAR. Trans. Inst. Meas. Control. (2025). https://doi.org/10.1177/01423312251313851
38. Li, H., et al.: Rock fracturing mechanism and arrangement of fracturing holes using hydro-mechanical splitters. Eng. Fract. Mech. **301**, 110055 (2024)
39. Ahmad, I., et al.: Advanced sensors, monitoring, and control systems for environmental sustainability. In: Smart Cities and Sustainable Manufacturing, pp. 47–55. Elsevier, Amsterdam (2025)
40. Biniaris, D., et al.: Minimization of CO2 emissions in Openpit mines by using stochastic simulations. Circ. Econ. Sustain. **5**, 2317–2345 (2025)
41. Branković, D., Milovanović, Z., Janièić-Milovanović, V.: Reliability analysis of bucket-wheel excavator parts–application of critical position selection algorithm. J. Graph. Era Univ. **13**, 1–30 (2025)
42. Liu, S.Q., et al.: Machine learning for open-pit mining: a systematic review. Int. J. Min. Reclam. Environ. **39**(1), 1–39 (2025)
43. Racchumi, V.J., Rosales Huamani, J.A., Castillo Sequera, J.L.: Construction of a predictive model of shovel productivity applying machine learning algorithms. Earth Sci. Inf. **18**(1), 1–17 (2025)
44. Burgess-Limerick, R., et al.: Human Systems Integration for Mining Automation. CRC Press, Boca Raton (2025)
45. Pollon, C.: Pitfall: The Race to Mine the World's Most Vulnerable Places. University of Queensland Press, Brisbane (2024)

46. Nguyen, H.A., Ha, Q.P.: Robotic autonomous systems for earthmoving equipment operating in volatile conditions and teaming capacity: a survey. Robotica. **41**(2), 486–510 (2023)
47. Hazrathosseini, A., Afrapoli, A.M.: The advent of digital twins in surface mining: its time has finally arrived. Resour. Policy. **80**, 103155 (2023)
48. Codoceo-Contreras, L., Rybak, N., Hassall, M.: Exploring the impacts of automation in the mining industry: a systematic review using natural language processing. Min. Technol. **133**(3), 191–213 (2024)
49. Kolesnichenko, I., et al.: Quantum fundamentals of coal bed methane hazards. Gornayapromyshlennost. **1**, 91–97 (2021)
50. Leonida, C.: AI: bringing mining companies closer to their data. Eng. Min. J. **226**(4), 28–32 (2025)
51. Alam, M.I., et al.: Blockchain-based coal supply chain management system for thermal power plants. Discov. Comput. **28**(1), 1–32 (2025)
52. Legge, H., et al.: Creating the Zero-Carbon Mine. McKinsey & Company, Singapore (2021)
53. Saha, D., et al.: Selective extraction and recovery of rare earth elements from coal fly ash by carboxylated mesoporous carbon. Sep. Purif. Technol. **354**, 129153 (2025)
54. Zhang, L., et al.: Rare earth elements recovery and mechanisms from coal fly ash by column leaching using citric acid. Sep. Purif. Technol. **353**, 128471 (2025)
55. Sichone, M., Lew, C.: Anglo American platinum South Africa: strategic renewal in a declining industry. Emerald Emerg. Mark. Case Stud. **11**(3), 1–31 (2021)
56. Ambec, S.: The European Union's Carbon Border Adjustment Mechanism: Challenges and Perspectives (2022)
57. Peiravi, M., et al.: Chemical extraction of rare earth elements from coal ash. Min. Metal. Process. **34**, 170–177 (2017)
58. Ertugrul, N., et al.: Status of mine electrification and future potentials. In: 2020 international conference on smart grids and energy systems (SGES). IEEE, New York (2020)
59. Heikkinen, A., et al.: Video adaptation based on 5G network QoS in harsh underground environment. In: 2024 Joint European Conference on Networks and Communications & 6G Summit (EuCNC/6G Summit). IEEE, New York (2024)
60. Ju, Y., et al.: Novel approach to achieving net-zero carbon emissions for development and utilization of coal: principle, methods, and cost estimation. Sci. Total Environ. **959**, 178215 (2025)
61. Yin, W., et al.: Research on the construction of digital twin virtual model of coal mills. Measurement. **242**, 116235 (2025)
62. Sreedharan, S., Ramachandran, M., Ramesh, D.: Harnessing digital twins and industrial-IoT for cutting-edge mining automation: a methodological and technology assessment prototype. Comput. Ind. Eng. **201**, 110871 (2025)
63. Qian, X., Wang, Y.: Path planning for coal mining masonry robots combined with trajectory optimization. IEEE Access. **13**, 24197–24206 (2025)

Chapter 16
Chemical Industries in Industry 6.0 Sustainability

Abstract Through automation, AI-driven process control, intelligent manufacturing, quantum computing, and sustainability, Industry 6.0 is transforming the chemical industry. This chapter examines how digital twin simulations, real time process monitoring, and predictive analytics are integrated in AI-powered smart factories to enhance chemical synthesis, refinement, and polymer manufacture. Autonomous reactors, IoT-enabled sensors, and AI-driven quality control systems are improving on traditional batch and continuous synthesis techniques, increasing output while consuming less energy and producing less waste. Decarbonization in chemical manufacturing is being propelled by developments in sustainable chemistry, including electrochemical CO_2 reduction, green solvent-free processes, and AI-assisted carbon capture. Blockchain-enabled supply chain transparency is another feature of Industry 6.0 that lowers fraud and enhances adherence to international laws like REACH and GHS. AI and nano catalysts, increase the efficiency of chemical synthesis, reducing reaction time and energy consumption. Enzymatic textile waste processing and advanced chemical recycling help achieve the circular economy's objectives by cutting down on landfill waste. By combining automated logistics, robotics, and zero-liquid discharge systems, operational efficiency is further increased while environmental effect is reduced. According to economic study, Industry 6.0 transitions provide long-term cost reductions and a quick return on investment of 1.5–2 years. This change guarantees that the chemical sector supports intelligent, sustainable, and future ready manufacturing in line with SDGs 9, 12, and 13.

Keywords Industry 6.0 · AI-driven chemical manufacturing · Digital twin simulations · Autonomous reactors · Sustainable chemistry · Green solvent-free reactions · AI-assisted carbon capture · Electrochemical CO_2 reduction · Blockchain in chemical supply chains · Smart polymer synthesis · Automated industrial reactors · Nanotechnology in catalysis · Circular economy in chemicals · Zero liquid discharge systems · Predictive process analytics

H. Majeed, T. Iftikhar, *Intelligent Manufacturing in Industry 6.0*,
https://doi.org/10.1007/978-3-032-07278-8_16

1 Introduction

Industry 6.0 represents the next step in the evolution of manufacturing, integrating advanced technologies such as AI, automation, sustainability initiatives, and digitalization into industrial processes. In the chemical sector, Industry 6.0 is transforming traditional manufacturing models by harnessing the power of AI-driven smart factories. This transition from conventional chemical production systems to automated, intelligent manufacturing is revolutionizing the way chemicals are produced, improving efficiency, safety, and environmental sustainability. Automation plays a crucial role in this transformation, as AI systems, machine learning, and robotics streamline production processes, reduce human error, and enhance precision in chemical manufacturing. However, the adaptation of intelligent manufacturing in the chemical industry is not without its challenges. Chemical plants are often complex, requiring significant changes in infrastructure and workforce training to fully integrate new technologies. Ensuring cybersecurity, managing large data volumes, and maintaining regulatory compliance can pose hurdles for manufacturers. Despite these challenges, Industry 6.0 offers numerous opportunities for chemical industries, including improved production efficiency, greater sustainability, and the potential to innovate new, high-value products. Embracing these advancements will be critical for chemical manufacturers seeking to remain competitive and responsive to global demands [1, 2].

2 Intelligent Manufacturing in Chemical Industries

Intelligent manufacturing is rapidly transforming chemical industries by integrating advanced technologies such as AI, ML, digital twins, material informatics, and quantum computing. By integrating these advanced technologies, chemical industries can achieve significant advancements in sustainability, process efficiency, and innovation. Intelligent manufacturing not only reduces costs and environmental impact but also accelerates the development of new materials and chemicals to reduce global warming and resource reduction. The integration of AI, ML, digital twins, material informatics, and quantum computing in chemical manufacturing is expected to drive a paradigm shift in the industry, bringing about more efficient, adaptive, and sustainable chemical production systems. These innovations enable enhanced process control, material discovery, and synthesis efficiency, paving the way for the chemical sector to meet increasing global demands while adhering to sustainability and cost-effectiveness targets. AI is revolutionizing both batch and continuous chemical synthesis systems by enabling real-time process optimization, reduced downtime, and higher throughput. In batch processes, AI algorithms predict optimal reaction conditions and adjust parameters to ensure product consistency and quality. For continuous processes, AI-driven systems use advanced process control techniques to maintain steady-state conditions, minimizing energy consumption and waste. In industries like petrochemicals and fine chemicals, AI-enabled systems can achieve production efficiency improvements, drastically reducing both operational costs and environmental impact. ML is increasingly applied to chemical process optimization, where it aids in improving reaction rates, yield prediction, and energy usage efficiency. ML algorithms can analyze vast datasets from sensors, historical data, and experimental results, uncovering hidden correlations and patterns that are not immediately apparent. In the production of polyolefins, ML-based models can predict optimal catalyst activity, leading to a reduction in catalyst waste and an increase in product quality. Similarly, ML can fine-tune distillation processes, resulting in energy savings by optimizing the heat integration between distillation columns [3–5].

Digital twins play an important role in predictive process control. These digital models simulate real-time production conditions, enabling the identification of inefficiencies and failure risk before happening. By continuously synchronizing sensor data (show real and timely data) with the digital twin, chemical plants can adjust operations proactively, preventing process disruptions and optimizing energy use. In a refinery, digital twin technology can simulate cracking furnace operations, leading to a reduction in energy consumption and minimizing operational risks. The discovery of new compounds, such as catalysts or polymers, is being accelerated by AI-powered material informatics. AI models analyze vast databases of material properties, chemical structures, and performance data, identifying promising candidates faster than traditional trial and error methods. In the field of battery materials, AI has been used to predict the performance of new electrode materials, such as lithium sulphur and solid state batteries. The use of AI can shorten the discovery timeline from years to months, potentially reducing R&D costs. It can enable the discovery of new sustainable materials, driving innovation in green chemistry and renewable energy sectors. Quantum computing holds transformative potential for the chemical industry by enabling simulations of molecular systems and reaction kinetics at a level of accuracy previously impossible with classical computing. Quantum algorithms can model complex molecular structures, predict reaction pathways, and calculate energy states with unprecedented precision. In catalysis research, quantum computing can predict catalytic activity and stability, potentially leading to the design of catalysts with enhanced efficiency and lower environmental impact. Quantum simulations have already demonstrated how to optimize reaction pathways in ammonia synthesis, a process that consumes approximately 1–2% of global energy. Through quantum simulations, it is possible to reduce the energy consumption of this process by up to 30%, offering substantial global energy savings [6–9].

3 Industrial Reactors and Process Automation

The integration of automation, sensors, and artificial intelligence in industrial reactors is transforming chemical manufacturing by enhancing operational efficiency, reducing costs, and ensuring product consistency. These advancements are particularly crucial as industries face increasing pressure to meet demand for higher quality products while adhering to stringent environmental and safety standards. The focus on fully automated reactors, IoT-enabled sensors, and AI-based models has led to innovations in reactor control, real-time process monitoring, and optimized yield in industrial-scale and lab-scale settings. The integration of automation, IoT-enabled sensors, and AI-based models in industrial reactors is creating a paradigm shift in the chemical manufacturing landscape. These technologies are not only improving the efficiency, safety, and sustainability of chemical production but are also setting the stage for more flexible, adaptive, and cost-effective manufacturing processes. The continuous development of smart reactor systems, advanced modelling techniques, and real-time optimization is expected to further enhance the precision and performance of chemical synthesis, paving the way for a new era in industrial chemical production. Fully automated reactors, whether on the lab scale or industrial scale, are designed to operate without continuous human intervention, ensuring high throughput, precision, and safety. At the industrial scale, automated reactors can handle volumes of hundreds of liters per hour with minimal manual intervention. In the production of specialty chemicals, automated reactors equipped with AI-powered controllers adjust reaction parameters (gradient of temperature and pressure, optimization of pressures at different time intervals and loading of catalyst), in real time. These systems can operate with a variance of less than 2% variance in reaction rate control, ensuring that batch-to-batch consistency is achieved with minimal human error. Recent advancements in reactor automation have resulted in an increase in production efficiency and a reduction in energy consumption compared to traditional manually controlled systems. The IoT plays an important role in chemical synthesis by integrating sensors and actuators into industrial processes for real-time data collection and process monitoring. These sensors measure critical parameters such as temperature, pressure, pH, viscosity, and concentration, transmitting the data to centralized control systems for analysis and immediate action. In a continuous flow reactor, IoT sensors can monitor fluid dynamics and reactant concentration, adjust flow rates and mix speeds through actuators to optimize reaction kinetics. The integration of IoT in process control has been shown to reduce process variability, enhance safety by providing early detection of anomalies, and improve product quality by ensuring consistent reaction conditions. Smart reactor control systems combine advanced algorithms with real-time sensor data to precisely regulate critical process variables such as temperature, pressure, and flow. These systems use advanced controlling technologies like predictive modelling, fuzzy logic system, proportional integral derivative control for real time customization. In a high pressure polymerization reactor, maintaining a precise temperature and pressure is critical for controlling molecular weight distribution. Smart control systems can reduce reaction time, while maintaining optimal polymerization conditions, leading to improved product quality and lower defect rates. These systems enable predictive maintenance by identifying early signs of wear in pumps, valves, or heat exchangers, thus reducing downtime by up to 40% in some cases. AI-based reaction modelling and real-time yield optimization are enabling chemical manufacturers to move beyond traditional trial and error methods to achieve optimized reactions in real time. ML algorithms and deep learning networks can model complex reaction mechanisms by processing large datasets from sensors, experimental results, and historical performance. These models are used to predict reaction outcomes under varying conditions, allowing operators to fine-tune the parameters for maximum yield. AI-based systems are now capable of optimizing chemical reactions in real-time, achieving yield improvements while minimizing the production of byproducts. In the production of biofuels, AI models can adjust feedstock composition and reactor conditions to maximize conversion efficiency and product purity, reducing waste and increasing overall energy efficiency. One of the most notable applications of automation in industrial reactors is in the field of pharmaceutical production, where automated flow chemistry has shown significant promise. Traditional batch synthesis methods in pharmaceuticals often lead to inconsistent quality, high energy consumption, and lengthy production times. In contrast, automated flow chemistry involves the continuous, automated mixing of reactants as they flow through a reactor, with real-time adjustments made to temperature, pressure, and reagent concentration. A case study in the production of an active pharmaceutical ingredient demonstrated that by shifting from batch to automated flow chemistry, production times were reduced, while energy consumption decreased. The real-time monitoring and control of the reaction led to an increase in yield and a reduction in impurities. This approach also enabled easier scale-up for large-scale production, as the same automated reactor configuration could be applied across different production volumes without significant changes to process conditions [3, 10–12].

Industry 6.0 technologies applications into chemical machinery and equipment, pushing the boundaries of what is possible in chemical manufacturing. From smart reactors with adaptive control systems to real-time monitoring using quantum sensors, the future of chemical machinery is focused on precision, efficiency, and autonomy. The use of AI, IoT, and additive manufacturing is setting the stage for unprecedented advancements in chemical equipment design, fabrication, and operational optimization. Smart chemical reactors incorporate AI-driven adaptive control mechanisms that continuously monitor and adjust operational parameters to optimize reaction conditions in real time. By integrating advanced sensors and machine learning algorithms, these reactors ensure that variables such as temperature, pressure, and flow rates are precisely controlled, leading to higher yields, better quality, and more efficient use of raw materials. Recent developments in AI-controlled reactors have resulted in a 15–25% increase in process efficiency and a significant reduction in energy consumption. Filtration and separation processes are critical in industries such as petrochemicals, pharmaceuticals, and food production. AI-enhanced filtration systems use machine learning to predict and optimize the performance of filtration devices, improving separation efficiency and extending the lifespan of filtration materials. These systems have auto adjustment technology in filtration parameters by using actual/updated and real-time data from the sensors, which will ultimately reduce the cost and result in better process yields. In a recent application, AI-based filtration technologies increased separation efficiency by up to 40% while reducing operational costs by 18%. Quantum sensors are pushing the limits of real-time monitoring in chemical distillation processes. These sensors, which utilize quantum mechanics principles to measure physical properties with extreme precision, are providing unparalleled accuracy in tracking temperature, pressure, and concentration gradients in distillation columns. By offering real-time insights into the distillation process, quantum sensors enable operators to make data-driven adjustments that increase the efficiency of separation and purification. These sensors have demonstrated improvements in distillation column performance by up to 35%, significantly reducing energy costs. IoT-enabled PAT systems provide real-time monitoring and control of chemical processes, ensuring product quality consistency and reducing waste. By integrating IoT sensors with machine learning models, PAT systems can analyze process data continuously to detect deviations from predefined quality parameters, such as particle size, chemical composition, and pH. This allows operators to adjust process parameters in real time, improving product quality and reducing material waste. Recent implementations of IoT-enabled PAT in chemical manufacturing plants have resulted in a 20% increase in product yield and a 15% reduction in waste. Additive manufacturing is revolutionizing the way custom chemical equipment is designed and produced. By using 3D printing technologies, manufacturers can create complex, tailor-made components for chemical reactors, filtration systems, and piping assemblies with high precision and efficiency. This approach allows for rapid prototyping and customization of equipment, reducing lead times and manufacturing costs. In the petrochemical sector, 3D printing has been used to fabricate specialized reactor components, reducing production time by 30% and cutting costs by 20% [13–21].

4 Sustainable Chemistry and Decarbonization Technologies

By integrating advanced automation, renewable energy sources, and cutting-edge catalyst designs, chemical industries are striving to mitigate their carbon footprint while enhancing efficiency and promoting sustainable development. These technologies and strategies are essential components of the global push towards a decarbonized and sustainable chemical industry. The integration of AI, green chemistry, and innovative CO_2 reduction technologies is setting the foundation for a future where chemical production can significantly contribute to global decarbonization efforts, with industries minimizing their environmental impact while still meeting growing global demand for chemicals and materials. Through continuous innovation and the adoption of sustainable practices, the chemical industry is advancing towards a cleaner, more energy efficient, and environmentally responsible future.

AI-driven CCU systems are at the forefront of enabling efficient and scalable carbon capture technologies in chemical plants. These systems optimize CO_2 capture, compression, and storage processes by using machine learning algorithms that predict optimal operating conditions, monitor emissions, and suggest improvements to reduce energy consumption. AI can control the absorption process in amine-based scrubbing systems by analyzing sensor data in real time and adjusting solvent flow rates and CO_2 capture efficiency to maintain a high performance while minimizing energy input. Recent studies have demonstrated that AI-assisted CCU can increase capture efficiency by up to 15% and reduce energy consumption by 10–20% compared to conventional methods. The integration of AI in CO_2 capture processes at a scale of one million tons per year has shown operational cost reductions of up to 25%, significantly improving the economic viability of CCU technologies in large-scale chemical manufacturing plants. The adoption of green chemistry principles, such as low-waste synthesis and solvent-free reactions, is revolutionizing chemical production by minimizing harmful byproducts and reducing environmental hazards. Low-waste synthesis strategies, such as atom economy and process intensification, aim to maximize the use of reactants in the production of desired products, leading to a reduction in waste generation. Solvent-free reactions have gained attention due to their ability to avoid toxic solvents and improve reaction efficiency. Recent advances in catalysis innovations have further optimized these processes, using environmentally benign catalysts like biocatalysts, heterogeneous catalysts, and ionic liquids to promote high-selectivity reactions. For example, catalytic processes involving solid acids, such as zeolites, have been used in polymerization and hydrocracking reactions, leading to a 30–50% reduction in the need for traditional solvents, saving operational costs and reducing the environmental impact. The implementation of green chemistry has led to reductions in industrial waste and the use of fewer hazardous chemicals in major chemical plants. One of the most significant challenges in achieving decarbonization in the industrial sector is the reduction of CO_2 emissions from cement production. CO_2 mineralization, a process in which CO_2 is chemically converted into stable carbonates, has emerged as a key technology in the development of net-zero cement. This process involves using naturally occurring minerals, such as basalt or serpentine, to react with CO_2, effectively sequestering it within the cement matrix. The resulting product, referred to as "carbonated concrete," can be used to create supplementary cementitious materials or to cure pre-cast concrete in construction. Research indicates that the use of CO_2 mineralization in cement manufacturing can offset up to 80% of the carbon emissions traditionally associated with cement production. A recent pilot study demonstrated that mineralizing 1 ton of CO_2 within cement results in a net reduction of approximately 0.7 tons of CO_2 emissions compared to conventional cement production. This technology is being scaled up globally, with companies like CarbonCure and Calera advancing the use of CO_2 mineralized concrete in construction, achieving up to 50% reduction in CO_2 emissions per ton of cement [22–25].

Electrochemical CO_2 reduction (CO_2RR) is an emerging technology for converting CO_2 into value-added chemicals and synthetic fuels, including methane, ethylene, and ethanol, using renewable electricity. AI optimization of CO_2RR enhances the efficiency of these processes by predicting the ideal electrochemical conditions for CO_2 reduction, such as voltage, current density, and electrode materials, to maximize product yield while minimizing energy consumption. Machine learning algorithms analyze large datasets from electrochemical cells, optimizing reaction pathways and materials to achieve higher selectivity and lower energy usage. In laboratory settings, AI-optimized CO_2RR has been shown to increase the efficiency of converting CO_2 into ethylene by up to 35% while reducing energy input by 20–30%. In practical applications, AI-based electrochemical CO_2 reduction systems could lead to the commercial production of synthetic fuels at a cost of 1.5–3 USD/kg by 2030. Also, BASF has focused on improving the efficiency of its ammonia production process, which is energy-intensive and contributes significantly to CO_2 emissions. By deploying carbon capture technology and utilizing green hydrogen as a feedstock for ammonia synthesis, BASF has reduced emissions by over 20% in some of its ammonia plants. BASF's collaboration with global partners in implementing AI-optimized CCU systems has allowed the company to capture over one million tons of CO_2 annually from its chemical operations, contributing to its broader sustainability goals [26–28].

5 Industry 6.0 in Petrochemicals, Polymers, and Specialty Chemicals

Industry 6.0 marks the next phase of evolution in the petrochemical, polymer, and specialty chemicals sectors by integrating AI, automation, and advanced digital technologies into traditionally complex and energy-intensive processes. By incorporating machine learning, data-driven insights, and cutting-edge innovations, petrochemical plants and polymer manufacturers are achieving higher levels of efficiency, sustainability, and customization. Through smart refining, AI-driven catalysis, and bioplastic innovations, the Industry 6.0 framework is poised to revolutionize chemical production, reduce waste, and accelerate the transition to a sustainable circular economy. Smart refining technologies leverage AI algorithms and advanced data analytics to optimize key operations such as cracking, distillation, and catalytic processes. AI-driven systems analyze vast datasets from sensors and historical plant data to predict optimal operating conditions, enhancing yield, efficiency, and product quality. In a recent breakthrough, AI-assisted process control systems in refineries have been shown to improve catalytic cracking efficiency by up to 15%, while reducing energy consumption by 8–12%. These systems dynamically adjust parameters such as temperature, pressure, and feedstock composition to ensure minimal energy waste and maximize product yield. In the downstream petrochemical industry, AI-based predictive maintenance models have reduced equipment downtime by up to 20%, significantly boosting operational efficiency and minimizing costly repairs. AI-based catalysis design uses machine learning models to predict the most effective catalysts for specific chemical reactions, speeding up the discovery of new, high-efficiency catalysts. Recent studies using AI-driven algorithms have enabled the discovery of novel catalysts for processes such as hydrogenation, polymerization, and oxidation reactions. These catalysts exhibit higher activity, selectivity, and stability compared to traditional ones. Similarly, ML models have identified promising cobalt-based catalysts for the Fischer–Tropsch synthesis, which can boost product yields while reducing energy consumption. AI-driven process optimization has contributed to the development of multifunctional catalysts that can perform multiple reactions in a single step, reducing the number of processing stages and ultimately enhancing overall process efficiency. In Industry 6.0, bioplastics and bio-based polymers play a significant role in reducing the reliance on fossil fuels. AI technologies are being employed to optimize the formulation of bioplastics by analyzing chemical structures, predicting performance properties, and fine-tuning compositions for specific applications. In a landmark development, AI models have been used to design bio-based polyesters with comparable mechanical properties to conventional plastics, while being fully biodegradable. AI is used to enhance bioplastic recycling by developing efficient sorting algorithms that can identify and separate bioplastics from traditional plastics at a molecular level. This advancement is crucial for improving the circularity of bioplastics, as AI-enabled sorting systems have shown to increase recycling rates by up to 50% in pilot-scale operations. The AI-driven integration of recycling processes into bioplastics production ensures a sustainable life cycle, addressing the challenges of waste management in the chemical industry. The development of sustainable adhesives, coatings, and sealants is a key focus in Industry 6.0, where AI and green chemistry principles are utilized to create eco-friendly formulations. AI is employed to design adhesive and coating materials that are high-performance, cost-effective, and produced with minimal environmental impact. Machine learning algorithms predict the optimal molecular structures of polymers and resins used in these products, ensuring enhanced durability, adhesion, and reduced toxicity. AI models have enabled the creation of plant-based coatings and adhesives that outperform traditional petroleum-based counterparts in strength and longevity. AI-optimized formulations have led to adhesives with up to 25% higher bonding strength while reducing volatile organic compound emissions by over 50%. These innovations align with global sustainability goals by significantly reducing the chemical industry's environmental footprint. Blockchain technology is revolutionizing the chemical industry by enhancing transparency and traceability in the supply chain. In the polymer industry, blockchain systems track the origin and journey of raw materials used in plastic production, ensuring that sustainable practices are adhered to at every stage. Blockchain helps combat issues such as counterfeiting and fraud, which are prevalent in the chemical supply chain. The integration of blockchain has resulted in a 30% reduction in supply chain-related disruptions by enabling real-time tracking and automated verification of chemical shipments. For companies adopting blockchain, it has also improved compliance with environmental regulations and reduced the risk of disputes with suppliers, fostering a more efficient and transparent supply chain [29–32].

6 Intelligent Process Safety and Hazard Mitigation

Industry 6.0 also represents a leap forward in process safety and hazard mitigation, integrating AI, machine learning, and autonomous systems to predict, identify, and prevent chemical hazards before they occur. These technologies are fundamentally transforming the way chemical plants approach safety, ensuring a higher level of protection for workers and the surrounding environment. With advanced hazard prediction, predictive maintenance, and autonomous safety protocols, Industry 6.0 aims to reduce accidents, minimize chemical leaks, and protect critical infrastructure from explosions and other risks. AI-based hazard prediction systems use sensors big data, historical incident reports along with real-time monitoring to predict and prevent potential hazards in chemical processes. These AI systems employ predictive analytics to identify risk factors such as equipment malfunctions, abnormal temperature spikes, or sudden pressure changes, providing early warning signals to operators. Studies have shown that AI-powered hazard prediction can reduce the likelihood of catastrophic events in high-risk chemical processes by as much as 40%, enabling proactive intervention and mitigating the risks associated with volatile reactions and critical process failures. Digital twin technology allows the creation of virtual replicas of chemical plants to simulate real-world conditions and predict potential safety hazards. By replicating every aspect of a chemical plant, from equipment performance to fluid dynamics, digital twins provide detailed insights into potential risks, enabling real-time safety assessments. Dow Chemicals has successfully implemented digital twins in its chemical plants, significantly improving safety protocols. The digital twin system monitors over 300 parameters, from chemical reactions to structural integrity, and predicts failures or hazardous conditions. This system has reduced safety-related incidents by over 30% and enhanced response time to emerging risks. Predictive analytics, combined with real-time sensor data, is revolutionizing leak detection and explosion prevention in chemical manufacturing. By continuously analyzing data from gas sensors, temperature monitors, and pressure sensors, predictive models can identify the early signs of leaks or dangerous reactions before they escalate into catastrophic events. These systems provide operators with actionable insights to adjust process parameters or initiate automatic shutdowns. A pilot system deployed in a refinery has successfully detected 95% of leaks in real time, reducing the risk of hazardous emissions and explosions by 40%. Autonomous robotic inspection systems are being deployed in hazardous chemical environments to perform tasks such as monitoring equipment health, inspecting pipelines, and checking for leaks. These robots, often equipped with AI-based sensors and cameras, autonomously navigate chemical plants to detect potential hazards, such as corrosion, leaks, or gas leaks. Recent advancements in robotic technology have enabled these systems to perform inspections in environments that are unsafe for human workers, such as highly corrosive or explosive areas. Autonomous robots at an oil refinery have reduced the need for manual inspections by 50%, significantly improving safety and operational efficiency. Dow Chemicals has integrated AI and digital twin technologies to enhance process safety and prevent hazardous incidents across its global manufacturing facilities. Using AI-driven analytics, Dow monitors real-time process data to detect early indicators of potential safety risks, including chemical imbalances, equipment failures, and unexpected chemical reactions. By combining these technologies with autonomous safety systems, Dow has been able to reduce incidents related to chemical leaks and explosions by over 30% in the last 5 years. This proactive approach not only improves safety but also ensures compliance with stringent environmental and regulatory standards.

7 Sustainable Certification and Compliance in Industry 6.0

The transition to Industry 6.0 has redefined how chemical manufacturing companies adhere to environmental standards and sustainability certifications. With increasing pressure to meet rigorous global sustainability goals, intelligent manufacturing technologies are empowering companies to integrate sustainability at every stage of their operations. AI, automation, and digitalization are playing a key role in streamlining compliance with chemical standards, optimizing sustainability reporting, and ensuring life cycle assessments are conducted efficiently. Intelligent manufacturing leverages advanced AI algorithms and IoT-based systems to monitor, control, and optimize production processes in line with global chemical standards, such as those set by ISO, REACH, and the European Union's Green Deal. AI-driven systems enable real-time monitoring of emissions, waste production, and resource usage, ensuring that manufacturers remain compliant with environmental regulations. These systems help chemical manufacturers meet global sustainability certifications by adapting production processes to minimize environmental impacts. In the polymer industry, AI-enabled production systems have reduced CO_2 emissions by up to 20% by optimizing energy usage in polymerization reactions, while simultaneously increasing throughput by 10%. AI-driven sustainability reporting is reshaping how companies in the textile, chemical, and manufacturing industries achieve and maintain compliance with environmental and safety certifications such as Bluesign, OEKO-TEX, GOTS, and REACH. These certifications require stringent testing, documentation, and verification processes across various stages of product development and supply chain management, especially about chemical use, resource efficiency, toxicity, and waste management. The introduction of AI and machine learning technologies into sustainability reporting has enhanced the speed, accuracy, and scalability of meeting these standards. For certifications like Bluesign, which focus on sustainable textile production, AI and machine learning enable the automatic collection and analysis of environmental and chemical data from various stages of production. The technology can evaluate chemical formulations, production processes, and waste management practices in real time. Through AI-driven tools, the assessment of textile chemicals used during production can be streamlined, reducing manual labor and ensuring that only safe, approved chemicals are used. By using AI algorithms to monitor chemical use, companies have reduced the time needed for reporting by 50–70%, ensuring products meet Bluesign certification criteria with more consistent compliance. AI's ability to track and trace chemicals within the supply chain allows companies to identify and replace harmful substances with safer alternatives, contributing to a 40% reduction in the use of hazardous chemicals in textile manufacturing. For OEKO TEX, which certifies textiles and leather for harmful substances, AI has significantly transformed chemical testing procedures. Traditional testing often involves manual, time-consuming procedures to assess chemical compliance, including testing for banned substances, heavy metals, and toxic chemicals. AI-powered algorithms, when applied to chemical analysis, have automated the identification of chemical substances in raw materials and finished goods, cutting down manual testing time by as much as 60%. AI tools can cross reference extensive databases of chemical regulations, enabling real time validation of compliance against OEKO TEX criteria. This allows companies to improve the accuracy of their chemical compliance reports and significantly reduce human error in reporting, which improves overall sustainability reporting accuracy. For GOTS (Global Organic Textile Standard), which applies to organic textiles, AI is used to streamline certification processes by automating the collection of environmental impact data throughout the entire production cycle. This includes data on raw material sourcing, chemical inputs, dyeing processes, and waste management practices. Machine learning models help assess the sustainability of organic textile products by analyzing historical data, predicting potential sustainability gaps, and suggesting improvements. By integrating AI into the GOTS certification process, companies have improved the speed of compliance assessments by up to 45%, allowing for faster certification turnaround times. AI algorithms optimize raw material sourcing to ensure that GOTS-certified materials meet stringent sustainability standards, which has been shown to reduce environmental impacts, such as water usage, by 30%. REACH (Registration, Evaluation, Authorisation, and Restriction of Chemicals) compliance, which is mandatory for companies in the European Union, is another area where AI-driven tools are making a substantial impact. REACH requires companies to register the chemicals they use, assess their safety, and ensure that no hazardous chemicals are used without proper authorization. AI-based systems automatically track the registration status of thousands of chemicals used by suppliers and manufacturers, continuously cross-referencing with REACH's vast database of chemicals and their statuses. These systems can flag potential compliance issues, such as unregistered or unauthorized substances, and provide real-time notifications, which have reduced non-compliance risks by up to 80%. AI can predict future regulatory changes, helping companies stay ahead of new requirements before they come into effect, minimizing disruption to production processes. Beyond regulatory compliance, AI-driven sustainability reporting also improves resource efficiency and waste management. AI systems can assess the entire lifecycle of raw materials and finished products, providing a detailed environmental footprint analysis. In textile manufacturing, AI is used to track water, energy, and chemical consumption, enabling companies to pinpoint areas for improvement. This has led to reductions in resource usage by up to 30%, driven by real time data that helps optimize production and minimize waste. One company using AI to monitor energy consumption in its textile processing plants saw a 15% reduction in energy consumption within the first year of implementation. AI systems that monitor waste generation and recycling rates have helped companies achieve a 25% reduction in landfill waste by ensuring that waste materials are redirected into closed-loop recycling systems. The integration of blockchain with AI enhances the traceability and transparency of sustainability reporting. Blockchain allows for immutable record-keeping of data with holistic approach i.e. starting from the raw material to the final product (ready to ship/deliver). By using blockchain to log data alongside AI, companies can ensure that their sustainability claims are verifiable, providing stakeholders with transparent proof of compliance. This not only improves consumer trust but also helps companies stay ahead of increasingly stringent regulations regarding environmental and chemical safety standards. The combination of AI, machine learning, and blockchain in sustainability reporting for certifications like Bluesign, OEKO TEX, GOTS, and REACH is thus transforming the industry. These technologies enable faster, more accurate reporting, improved compliance, and significant reductions in environmental impacts, all while supporting companies in meeting the evolving demands of regulatory bodies and consumers who are increasingly prioritizing sustainability. Using AI to optimize chemical formulations, resource usage, and waste management practices, companies can streamline their certification processes, minimize risks, and drive sustainability throughout their supply chains [33–39].

7.1 *Digitalization of Regulatory Compliance (EPA, OSHA, ISO 14001)*

The digitalization of regulatory compliance is revolutionizing the way chemical manufacturers adhere to environmental and safety regulations, with technologies such as AI, blockchain, and sensor networks playing a main role in enhancing compliance with agencies like the Environmental Protection Agency (EPA), Occupational Safety and Health Administration (OSHA), and ISO 14001 (Environmental Management Systems). These technologies enable real-time data reporting, providing manufacturers with a more streamlined and efficient way to meet regulatory standards while optimizing operational performance. AI-based predictive analytics and machine learning algorithms are playing an instrumental role in this transformation. For instance, predictive models driven by AI can forecast trends in emissions, chemical waste generation, and energy consumption. By analyzing historical data, these models identify patterns and potential risks before they materialize, allowing chemical manufacturers to proactively adjust operations to reduce environmental impacts. A recent implementation in an industrial setting led to a 20% reduction in air emissions by predicting and adjusting chemical reaction conditions to optimize output and minimize waste. In addition, predictive models have helped reduce hazardous waste generation by 15–25%, ensuring compliance with EPA guidelines and minimizing disposal costs, which is a significant concern for manufacturers. Blockchain technology is also contributing to regulatory compliance by providing an immutable, transparent ledger for tracking environmental data. This is particularly useful for ensuring the traceability of hazardous materials and chemicals throughout their life cycle, which is a critical requirement for both EPA and OSHA regulations. Blockchain ensures that data regarding chemical usage, emissions, and waste management cannot be altered or tampered with, offering increased reliability and confidence in reporting. This technology is being used by companies in the chemical industry to create auditable records of compliance, allowing for seamless inspections and improving the trustworthiness of environmental reporting. Sensor networks integrated with AI and IoT technologies are enabling continuous, real-time monitoring of environmental and safety parameters in chemical manufacturing facilities. Sensors placed throughout the production process can monitor air quality, chemical emissions, noise levels, and hazardous waste generation, providing a constant stream of data for analysis. AI algorithms then process this data to detect anomalies, such as rising emission levels or the accumulation of hazardous waste. In the event of such anomalies, AI-powered systems automatically trigger alerts or shut down processes to mitigate potential risks before they escalate. In one case study, the installation of real-time sensor networks in a chemical manufacturing plant led to a 30% reduction in safety incidents, thanks to the early identification and resolution of potential hazards. Cloud-based data management platforms are also playing a key role in the digitalization of regulatory compliance. These platforms aggregate data from various sensors, AI models, and blockchain systems, allowing companies to track and manage regulatory compliance across multiple locations and facilities. The real-time data aggregation enables seamless reporting to regulatory bodies, such as the EPA and OSHA, ensuring that manufacturers stay compliant with regulations without the need for extensive manual reporting. The integration of these digital technologies has a profound impact on the speed and accuracy of reporting. AI-based tools have decreased the time required for environmental reporting by 40%, allowing companies to comply with stringent EPA guidelines faster than traditional manual methods. By automating the data collection and reporting process, companies can ensure that they are always up to date with compliance requirements, reducing the likelihood of missed reporting deadlines and penalties. AI-driven tools for workplace safety management have led to substantial improvements in compliance with OSHA regulations. AI systems analyze sensor data from wearable devices and sensors placed within the plant environment to monitor worker exposure to hazardous materials, noise, and unsafe working conditions. AI-based safety monitoring systems were able to reduce worker exposure to toxic gases by 18%, thereby enhancing overall safety and compliance with OSHA standards. In terms of ISO 14001 compliance, which focuses on environmental management systems and the continuous improvement of environmental performance, digital technologies such as AI and blockchain enable chemical manufacturers to meet the strictest standards. Real-time environmental impact data helps manufacturers track progress toward sustainability goals and identify areas for improvement, driving the continuous optimization of processes. For example,

AI-based solutions have helped a leading chemical manufacturer optimize its water usage by 22% and reduce energy consumption by 12%, all while adhering to ISO 14001's environmental criteria. These digital tools are indispensable in meeting the increasing demand for transparency, traceability, and accountability in the chemical industry, ensuring that manufacturers are prepared to meet both current and future regulatory challenges while achieving operational excellence and sustainability [40, 41].

8 AI-Based Evaluation of Eco Impact with Life Cycle Assessment (LCA) Tools

In the context of Industry 6.0, LCA has evolved significantly due to the integration of artificial intelligence. Traditional LCA methods often rely on manual data collection, subjective assumptions, and lengthy computational processes. However, AI is revolutionizing LCA by making it more precise, efficient, and scalable. AI-based tools can efficiently process and analyze life cycle data from various sources, identifying patterns and correlations that may not be immediately apparent through traditional methods. For example, AI has been used to assess the environmental footprint of bioplastics. By simulating various production scenarios, AI tools can predict the energy consumption, water usage, and GHG emissions at each stage of the bioplastic's life cycle. These models have provided valuable insights into how different feedstocks (such as agricultural residues, algae, or waste oils) influence the overall environmental impact of bioplastics. In a notable case, AI models improved the precision of LCA assessments by 25%, enabling companies to more accurately gauge the sustainability of different bioplastic production routes. AI's ability to accelerate LCA analysis has dramatically reduced the time required for assessments. In some applications, AI-driven LCA tools have reduced the time needed for a full analysis by over 40%, allowing companies to quickly evaluate the environmental consequences of various design and production decisions. This time-saving capability is especially important in fast-paced industries where product life cycles are becoming increasingly compressed, and there is pressure to meet stringent sustainability goals. AI is also transforming how companies assess the impacts of their supply chains. By integrating LCA with digital supply chain tracking systems, AI can provide real-time environmental impact assessments, enabling companies to make more informed and sustainable decisions. AI can predict the environmental impact of sourcing materials from different regions, considering factors such as transportation emissions, energy intensity, and supply chain risks. This predictive capability enables companies to make proactive adjustments to reduce emissions. Research has shown that integrating AI-driven LCA with real-time supply chain data can reduce overall emissions by up to 15–20%, making it a key strategy for companies striving to meet global climate targets and sustainability standards. AI can provide detailed insights into the trade-offs between various sustainability metrics, such as energy consumption, GHG emissions, and water usage. In the production of bio-based chemicals or renewable energy systems, AI models can simulate different process routes to identify the most efficient and sustainable option based on multiple environmental impact categories. These details allow companies to design products with optimized life cycle performance, reducing their overall environmental footprint. Using AI and IoT devices, companies can track the environmental impact of products throughout their use phase and after disposal. Smart sensors embedded in products or packaging can provide data on energy usage, waste generation, or material degradation over time. This data feeds into LCA models, providing more accurate and up-to-date environmental impact assessments. This continuous monitoring is particularly valuable for assessing the circularity of products and materials, which is a growing area of interest within Industry 6.0, especially for products designed for reuse, recycling, or upcycling. In the automotive industry, AI-based LCA tools are used to evaluate the life cycle impacts of electric vehicles. These models can predict the overall environmental benefit of transitioning from gasoline-powered vehicles to EVs, helping manufacturers make more informed decisions about sourcing, production, and battery recycling. The use of AI in LCA is becoming a critical element in sustainable product design, particularly as companies face increasing pressure to meet environmental regulations and consumer demand for transparency. The ongoing advancements in AI are expected to enhance the granularity of LCA analysis, providing deeper insights into localized environmental impacts, such as water scarcity, biodiversity loss, and resource depletion. With AI's ability to simulate complex, real world conditions and predict future scenarios, LCA is set to play an even more critical role in helping industries achieve their sustainability goals while mitigating environmental risks. The development of AI-enabled LCA frameworks is enabling the creation of industry-wide sustainability benchmarks. These benchmarks can help establish best practices across sectors such as electronics, textiles, and chemicals, ensuring that companies are not only reducing their own environmental impacts but also contributing to broader efforts toward global sustainability and the circular economy [42, 43].

9 Advanced Energy Efficiency and Circular Economy in Chemical Plants

The chemical industry has traditionally been one of the most energy-intensive sectors. However, as part of the Industry 6.0 transformation, chemical manufacturing is focusing on improving energy efficiency and incorporating circular economy principles. Advanced energy systems, waste heat recovery technologies, and AI-powered process optimization are central to achieving these goals, alongside innovations in waste chemical recycling and smart water management. The combination of AI-driven technologies, digitalization, and circular economy practices is reshaping the sustainability landscape in chemical manufacturing. With advanced energy systems, compliance automation, and waste recycling innovations, Industry 6.0 is setting the stage for a more sustainable and environmentally responsible chemical industry.

9.1 Energy-Efficient Chemical Manufacturing Using AI-Based Process Optimization

Process optimization (with AI) is fundamentally transforming the chemical manufacturing industry by identifying and mitigating energy inefficiencies across various stages of production. Machine learning models and advanced AI algorithms are increasingly deployed to analyze vast amounts of operational data, pinpointing inefficiencies in chemical reaction pathways, heat integration systems, and overall equipment performance. These AI-driven approaches can provide real-time recommendations to optimize energy usage, reduce costs, and improve sustainability metrics. A notable application of AI in energy optimization is in the operation of steam cracking furnaces, which are widely used in the production of ethylene and other petrochemicals. Steam cracking is an energy-intensive process, with furnaces typically consuming a large portion of the total energy input. AI models are now used to analyze and optimize critical operational parameters, such as temperature, feedstock composition, and residence time, to maximize yield while minimizing energy consumption. In one prominent case, AI optimization reduced energy usage by up to 12% by fine tuning furnace operation, leading to significant energy savings. These improvements were achieved through machine learning algorithms that could predict optimal operating conditions based on historical data, such as temperature fluctuations and feedstock characteristics, thus enhancing the furnace's thermal efficiency. By continuously analyzing variables like reflux ratio, column pressure, and temperature, AI systems can adjust distillation conditions dynamically to minimize energy input while maintaining the desired product quality. In some instances, AI has enabled significant reductions in the number of reboiler and condenser duties, leading to substantial energy and cost savings. AI is being used to forecast energy demand and optimize energy procurement strategies for chemical plants. Through predictive analytics, AI systems can predict fluctuations in energy demand based on factors such as production schedules, ambient temperature, and market energy prices. By aligning energy procurement with demand forecasts, manufacturers can secure access to low-cost renewable energy sources during off-peak times or when there is a surplus of renewable energy generation. This strategy not only helps reduce energy costs but also supports sustainability initiatives by leveraging cleaner energy sources, such as solar or wind, when available. AI models can optimize the timing of energy purchases, allowing plants to take advantage of energy price dips, reducing operating costs while supporting a more sustainable energy mix. According to recent studies, AI-driven process optimization has led to energy savings of up to 20% across various chemical production processes. The use of AI to optimize the operation of petrochemical plants and refineries is projected to save up to 1.5 quadrillion BTUs (British thermal units) of energy annually by 2030. The adoption of AI for energy efficiency is closely aligned with the broader shift towards decarbonization in the chemical industry. By optimizing energy-intensive processes and integrating renewable energy sources into their operations, chemical manufacturers can contribute to global sustainability goals. A study by the International Energy Agency (IEA) indicates that AI technologies could help the chemical industry reduce its energy-related CO_2 emissions by as much as 10–15% by 2030, providing a major lever in the transition to net-zero emissions. These advances in AI-driven energy optimization represent a critical step in achieving both cost efficiencies and sustainability objectives in the chemical manufacturing sector. As AI continues to evolve, further breakthroughs are expected in optimizing energy usage through advanced machine learning models, quantum computing applications for energy system modeling, and even greater integration of renewable energy into industrial processes. The combination of AI, machine learning and renewable energy optimization will drive the future of energy-efficient and sustainable chemical manufacturing [44–46].

9.2 Integration of Waste Heat Recovery, Organic Rankine Cycles, and Thermoelectric Generators

The integration of waste heat recovery systems, Organic Rankine Cycles (ORC), and Thermoelectric Generators (TEG) is advancing the sustainability of chemical plants. Waste heat recovery captures excess thermal energy from chemical reactions and recycles it to generate additional power or provide heat for other processes, reducing the plant's overall energy demand. ORC is highly effective in converting low-grade waste heat (such as that from exothermic reactions) into usable electricity. A recent implementation of ORC technology at a chemical plant achieved an 18% increase in energy efficiency by converting waste heat into electricity. TEGs convert heat gradients directly into electrical power, offering another route for generating energy from waste heat. When combined, these systems reduce the carbon footprint of chemical plants and enable a 20–30% reduction in energy consumption [47–49].

9.3 Circular Economy Innovations in Chemical Recycling of Plastics and Upcycling Waste Chemicals

Chemical recycling of plastics is revolutionizing the way the chemical industry addresses plastic waste, moving beyond traditional mechanical recycling methods that simply shred plastics into smaller pieces. Unlike these physical methods, chemical recycling breaks down complex polymers into their fundamental monomers, which can then be repolymerized into new products of the same or higher quality. This process enables the recycling of a broader range of plastics, including those that are difficult to recycle through conventional methods, such as mixed or multilayer plastics. AI and machine learning are integral to the advancement of chemical recycling by optimizing the key parameters that drive the breakdown of polymers. These include the selection of optimal catalysts, reaction conditions (such as temperature and pressure), and feedstock compositions, all of which significantly impact the efficiency and yield of the recycling process. AI models can predict the most effective catalytic systems and processing conditions for different types of plastics, enabling the fine-tuning of depolymerization reactions to maximize yield and minimize energy consumption. The application of AI in chemical recycling has led to a 25% increase in yield by optimizing pyrolysis and depolymerization pathways. Pyrolysis, a thermal decomposition process, and other depolymerization techniques are used to break down complex plastics into smaller, reusable chemical building blocks. AI algorithms analyze vast datasets from experimental runs to identify patterns and correlations between catalysts, reaction conditions, and product yields, ultimately improving the selectivity and efficiency of these processes. The increased yield translates to less waste and higher throughput in recycling facilities, improving the economics of chemical recycling. Another significant development is the upcycling of waste chemicals, a concept gaining momentum in the chemical industry. Instead of discarding waste chemicals or byproducts from industrial processes, upcycling involves repurposing these materials into valuable products. AI-driven systems are central to this innovation by analyzing the chemical composition of waste streams and identifying opportunities to transform low-value byproducts into higher-value chemicals or materials. AI models optimize the transformation of waste chemicals, improving the conversion efficiency and ensuring that the upcycled products meet the required performance specifications. The impact of AI-based upcycling is evident in the reduction of landfill-bound waste. Some chemical plants have reported upcycling waste chemicals to reduce overall waste by up to 40%, leading to a more sustainable production process. Upcycling can reduce the demand for virgin raw materials, further lowering the carbon footprint of chemical manufacturing. This is particularly important as the chemical industry continues to pursue circular economy principles, where products and materials are continually reused, remanufactured, and regenerated. Further innovations in chemical recycling and upcycling are being driven by a combination of AI, data analytics, and advanced catalysis research. By continuing to optimize chemical processes through AI and machine learning, the chemical industry is poised to increase the scalability and efficiency of chemical recycling operations, thus making a significant contribution to global sustainability efforts. These innovations not only support the reduction of plastic waste but also enhance the economic viability of recycling technologies, creating a more circular and sustainable economy. The expansion of AI-powered chemical recycling and waste upcycling represents a key development for the chemical industry, with projections indicating that the global market for chemical recycling will reach more than 10 billion by 2030, reflects the increasing demand for advanced recycling technologies, driven by regulatory pressures, environmental concerns, and the industry's commitment to sustainability [50–52].

9.4 Smart Water Management Systems in Zero Liquid Discharge (ZLD) Plants

ZLD systems are at the forefront of sustainable water management in chemical plants, addressing both water conservation and environmental protection. ZLD ensures that all water used in the manufacturing process is either reused within the facility or evaporated, with no discharge into the environment. This technology is crucial for reducing water waste and mitigating the impact of industrial processes on local water sources. AI and IoT-powered smart water management systems have significantly advanced the efficiency of ZLD systems. By incorporating real-time monitoring and advanced analytics, these systems optimize water treatment processes through dynamic adjustments to parameters such as filtration rates, chemical dosing, and evaporation cycles. The use of AI algorithms to predict water usage patterns allows for the fine-tuning of water treatment schedules, ensuring optimal resource utilization and reducing operational costs. The implementation of AI-driven water management systems in a ZLD facility at a chemical plant resulted in substantial improvements in sustainability metrics. Water consumption was reduced by 25% and water recovery efficiency increased by 30%. These improvements were achieved by integrating predictive analytics into the plant's operations, which allowed the system to proactively adjust treatment and filtration processes based on anticipated water usage and demand fluctuations. Such AI optimization minimizes both water wastage and energy consumption, helping to make the chemical manufacturing process more resource efficient. AI-driven systems have enhanced real-time leak detection, contamination identification, and process inefficiency detection. With IoT sensors strategically placed throughout the system, plant operators can quickly identify issues and take corrective actions before they escalate, reducing downtime and ensuring compliance with increasingly stringent environmental regulations. The adoption of ZLD technologies also aligns with global trends toward water conservation and stricter environmental regulations. As water scarcity becomes a growing concern, chemical plants that utilize AI-enhanced ZLD systems can not only improve their sustainability but also reduce operational costs related to water sourcing, treatment, and disposal. The market for ZLD systems is growing rapidly, driven by regulatory pressure and the need for industries to reduce their environmental footprint. By integrating smart water management with ZLD technology, chemical plants can significantly enhance their operational sustainability while contributing to global water conservation efforts. In the context of future advancements, ongoing research into AI and IoT for water management promises even higher levels of efficiency and automation. Further integration of ML models that learn from past data and predict future water quality trends could result in near-perfect optimization of treatment and filtration systems. This, in turn, would lead to even greater reductions in water consumption and improvements in recovery rates, aligning with the global goal of achieving zero water wastage in industrial processes [53, 54].

10 The Future of Chemical Manufacturing with Industry 6.0 and Beyond

Industry 6.0 is better for the next generation of chemical manufacturing, where AI, quantum computing, and automation technologies will redefine how chemicals are produced, maintained, and distributed. The future promises not only higher efficiency and lower environmental impact but also a fundamental shift toward more sustainable and bio-based chemical processes. The concept of fully autonomous chemical plants, governed by real-time AI systems, is becoming a reality. These AI systems can control every aspect of chemical production, from raw material procurement to end-product distribution, with minimal human intervention. Autonomous plants leverage AI-based decision-making models to optimize process parameters, adjust production schedules, and ensure compliance with environmental and safety regulations. AI models in a fully automated plant could adjust reaction conditions based on real-time sensor data, leading to a 40% reduction in energy consumption and reduction in overall operational costs. AI governance ensures that chemical plants are always operating at peak efficiency while minimizing waste and emissions. Quantum computing is set to revolutionize the field of drug discovery and specialty chemicals by simulating molecular interactions at an unprecedented scale and speed. Unlike classical computers, which struggle to model complex molecular systems, quantum computers can process multiple variables at the same time, making it faster and more accurate predictions of chemical reactions and the final product (after reaction) material properties. This has the potential to shorten the time required for drug development by 80% and dramatically reduce the costs of developing specialty chemicals. Quantum algorithms have already been used to discover novel catalysts for biofuel production, reducing catalyst development time from years to mere months. Predictive maintenance powered by AI is enhancing safety and operational efficiency in high-risk chemical environments. AI systems continuously monitor equipment performance, identifying early signs of wear, malfunction, or potential failure. By predicting when equipment needs maintenance, companies can avoid costly downtime and mitigate safety risks. In a recent case, AI-enabled predictive maintenance reduced unplanned downtime in a petrochemical refinery and extended the lifespan of critical machinery. AI-based systems can integrate with robotic inspection tools to perform remote, non-invasive inspections of hazardous equipment in real time, reducing the need for human intervention in dangerous environments. Industry 6.0 is accelerating the transition from petrochemical-based processes to bio-based chemicals, driven by advancements in biotechnology, sustainable feedstocks, and AI optimization. The shift is being facilitated by the development of AI-driven platforms that optimize the production of bio-based chemicals such as bioplastics, biofuels, and biochemicals. AI models are used to optimize the fermentation processes in biofuel production, leading to a 30% increase in yield and a 40% reduction in energy consumption. AI systems are being used to scale up the production of bio-based chemicals, enabling them to meet the demand for sustainable alternatives to traditional petrochemicals. The adoption of Industry 6.0 technologies in the chemical industry is expected to significantly impact global economic trends. The chemical industry's global market size is projected to reach more than 6 trillion by 2030, with a large portion of growth driven by AI-driven process optimization, sustainable chemical production, and bio based chemical innovations. The economic impact will also be evident in cost reductions, enhanced productivity, and improved supply chain resilience. However, to realize these benefits, it is critical for policymakers to implement frameworks that support the adoption of these technologies, encourage investment in R&D and establish regulations that foster a sustainable and competitive chemical industry [55–57].

11 Best Practices in Smart Chemical Manufacturing

The transition to Industry 6.0 in chemical manufacturing is a paradigm shift, integrating advanced technologies to enhance operational efficiency, sustainability, and innovation. Leading companies in the sector are leveraging cutting-edge tools like artificial intelligence, digital twins, and sustainable processes to redefine production practices and minimize environmental impacts. At ExxonMobil, digital twin technology is central to its petrochemical production strategy. These virtual replicas of physical processes provide real-time data, allowing for the simulation of operations and proactive identification of inefficiencies. The results have been impressive, with energy consumption reduced by 12% and throughput increased by 10%. The integration of digital twins also facilitates predictive maintenance, minimizing equipment downtime and ensuring continuous optimization of complex chemical processes. Covestro, a polymer company, is at the forefront of the circular economy in the chemical industry. Through AI-enabled recycling processes, Covestro has reduced its reliance on virgin materials by 50% and improved recycling efficiency by 35%. Clariant specialty chemicals company focuses on sustainable catalysts and bio-based solvents. Using ML algorithms, Clariant has optimized its catalyst designs, improving reaction efficiency by 20% while simultaneously reducing waste by 25%. This AI-enhanced approach not only boosts efficiency but also supports Clariant's commitment to sustainability by minimizing the environmental impact of its manufacturing processes. BASF has spearheaded AI-driven predictive chemical synthesis, which revolutionizes its approach to chemical reaction optimization. By utilizing AI to predict reaction outcomes, BASF has achieved an increase in reaction yields and a reduction in failed experimental runs. These improvements not only expedite R&D timelines but also significantly reduce the costs associated with trial and error in chemical synthesis. The AI system's predictive capabilities enable BASF to streamline its operations, thus fostering faster development of new materials and chemicals. The economic benefits are clear, with significant cost reductions and efficiency gains, which in turn improve profitability and competitiveness. As the global chemical industry moves towards Industry 6.0, these advancements will be instrumental in meeting the dual challenges of sustainability and profitability in an increasingly digital and environmentally conscious world. The global chemical industry is set to reach 5.5 trillion USD by 2030, with AI-driven process optimization leading to significant environmental benefits, including reductions in emissions by up to 50%, according to the IEA. These technological advancements, combined with the rise of green hydrogen, sustainable polymers, and quantum computing, promise to unlock new levels of innovation and sustainability in chemical production. The transition to Industry 6.0 will not only be a technological leap but also a step toward more responsible and resilient manufacturing practices [58–62].

12 Traditional Vs Industry 6.0 Chemical Synthesis (an Industrial Example)

Just for the understanding of the readers, we are quoting an example of chemical manufacturing of the chemicals on industrial scale. There may be some knowledge that is not available in the academic literature but industrial connections with experience have such technical details. In the polymer industry, dioctyl phthalate (DOP) and dioctyl terephthalate (DOTP) are common plasticizers, especially for applications involving polyvinyl chloride (PVC). DOTP is a non-phthalate substitute that is less hazardous and more environmentally friendly than DOP because it is made from terephthalic acid, whereas DOP is made from phthalic anhydride and 2-ethylhexanol (2-EH). The following procedure must be followed if we wish to synthesize DOP and DOTP in our typical plant. Esterification processes with carefully regulated temperature, pressure, and catalyst selection are used to produce these plasticizers to get high yield and purity. The esterification of phthalic anhydride with 2-ethylhexanol, with sulfuric acid acting as a catalyst, is followed by the production of DOP. Complete conversion is driven by a molar ratio of 1:2.1 (phthalic anhydride to 2-EH), and the reaction takes place at 180–220 °C in air or mild vacuum conditions. Water is continuously removed from the reaction over the course of 4–6 h to change equilibrium in favor of ester formation. To get rid of any remaining acid, the crude product is neutralized with calcium hydroxide or sodium carbonate and then washed with hot deionized water. Vacuum distillation at 180 °C under 5–10 mmHg is the last purification stage, producing 97–99% pure DOP. Under catalyzed high-temperature esterification conditions, terephthalic acid and 2-ethylhexanol are used to create DOTP, an environmentally favored substitute. In contrast to DOP, DOTP necessitates catalysts made of zinc acetate ($ZnAc_2$) or titanium dioxide (TiO_2) because terephthalic acid is less reactive than phthalic anhydride. To improve esterification efficiency, the reaction is carried out at 220–250 °C under vacuum (30–100 mmHg). A 1:2 molar ratio (TPA to 2-EH) by utilizing AI-driven process optimization, digital twin simulations, quantum chemistry models, and sophisticated catalytic systems, the incorporation of Industry 6.0 in the production of DOP and DOTP is transforming the production of plasticizers. By drastically lowering reaction times, improving product purity, using less energy, and lowering expenses, these intelligent manufacturing approaches increase production's sustainability, efficiency, and climate resilience. To guarantee consistent quality and almost zero material waste, Industry 6.0 introduces autonomous AI-driven reactors, real time IoT monitoring, and precision-controlled feedstock blending, in contrast to traditional batch-based synthesis methods that are energy-intensive, prone to inconsistencies, and rely on corrosive catalysts. Real-time AI optimization for process control is one of the most important developments in Industry 6.0. AI-driven systems cut the reaction time to 2.5–3.5 h by continuously monitoring and modifying reaction parameters like temperature, pressure, catalyst activity, and reactant flow rates. Traditional DOP and DOTP synthesis take 4–6 h and sometimes 24 h in each batch. With a molecular composition precision of ±0.1% guaranteed by AI-based spectroscopic analysis, blending mistakes, heat degradation, and undesirable side reactions are greatly decreased. By adjusting the reaction temperature to between 200 and 220 °C instead of the traditional range of 220–250 °C, energy losses are reduced while catalyst efficiency is preserved. Real-time tracking of chemical interactions by IoT-integrated smart sensors optimizes reaction kinetics and achieves a conversion rate of 99.5%+ which is a notable increase over conventional methods [63–65].

The use of enzyme-based catalysis and nano engineered catalysts is another revolutionary development in Industry 6.0. Sulfuric acid (H_2SO_4), titanium dioxide (TiO_2), and zinc acetate ($ZnAc_2$) catalysts are used in traditional synthesis; however, these catalysts have short lifespans, are prone to corrosion, and require expensive purification. Heterogeneous nano-catalysts, like zeolite-based and graphene TiO_2-based catalysts, are used in Industry 6.0. They eliminate harmful byproducts, cut catalyst usage by 70%, and boost esterification rates by 50%. Lipases enable mild temperature reactions (between 80 to 120 °C) through enzyme-based catalysis, reducing energy usage by 60% while maintaining selectivity rates above 99.9%. SDGs 9, 12, and 13 are all in line with this move to sophisticated catalytic systems, which lower production costs and environmental impact [66, 67].

Industry 6.0 incorporates quantum chemistry models and digital twin simulations into real-time monitoring and process design. Manufacturers can model material interactions, catalyst behavior, and reaction pathways by using digital twins to build virtual versions of the reactor system. By predicting activation energy barriers, ideal reaction pathways, and substitute feedstock sources, AI models based on quantum chemistry cut down on trial and error testing by 80%. The quality of the plasticizer is improved by these digital improvements, which reduce process failures by 80%, stop reactant heat deterioration, and guarantee exact control in DOTP synthesis. The efficiency of Industry 6.0 is further increased using intelligent supply chain integration and autonomous robotic systems. Batch feeding and manual dosing are necessary for traditional material handling procedures, which result in irregularities and inefficiencies. Reactant ratio precision of ±0.1% is guaranteed by AI-powered automated dispensers, which drastically cut down on material waste. Drone-based inventory control and Automated Guided Vehicles streamline logistics, reducing waste generation by 35% and material handling time by 60%. Blockchain-enabled supply chain tracking ensures 99.9% purity compliance and improves traceability in sourcing petrochemical-derived and bio-based feedstocks. The advantages of Industry 6.0 in DOP and DOTP manufacturing go beyond process optimization to include cost reduction, quality improvement, and environmental sustainability. Traditional synthesis achieves a 97 to 99% purity rate, whereas Industry 6.0 ensures 99.9% purity, significantly reducing the need for post-reaction purification; reaction consistency improves from ±5% batch to batch variation to ±0.1% uniformity, ensuring a high-quality, defect-free product; catalyst efficiency is also greatly increased, rising from 50 to 60% in traditional methods to over 90% using nano engineered catalytic systems. Cost-wise, automation based on Industry 6.0 results in notable savings in labor costs, energy consumption, and waste of raw materials. AI-driven manufacturing reduces raw material waste from traditional DOP/DOTP production to $\leq$1%, resulting in an 80 to 90% reduction in overall costs. A 35–40% decrease in energy costs results from the energy consumption of plasticizer dropping from 5.5–6.5 GJ/ton to 3.5–4.2 GJ per ton. The optimal use of catalysts reduces reliance on costly and rare metals by 50–70%. Compared to traditional methods, labor costs decrease by 70–80% and overall production costs decrease by 50% when manual work is replaced by AI-driven automation. The capacity of Industry 6.0 to promote environmental sustainability and climate resilience in the plasticizer sector is another noteworthy benefit. AI-powered thermal integration makes DOP and DOTP production more ecologically friendly by reducing CO_2 emissions by 40%. By using bio-based DOTP made from renewable feedstocks, lifecycle emissions are further reduced by 60%. By recovering old catalysts, cutting chemical discharge by 80%, and doing away with solvent-based purification techniques that pollute the environment, Industry 6.0 also guarantees a zero-waste manufacturing process. 50% of the requirement for process heating is met by solar-assisted hybrid reactors, which lessens dependency on fossil fuels. Industry 6.0 in DOP and DOTP manufacturing is establishing new international benchmarks for sustainability, efficiency, and quality thanks to these technological developments. Yield rises by 10%, reaction time falls by 40%, energy consumption decreases by

35%, and manufacturing costs decrease by 50% when compared to conventional methods. Future plasticizer production will be completely autonomous, resource-efficient, and climate-friendly thanks to a mix of AI-driven process automation, quantum-enhanced chemistry, blockchain traceability, and intelligent material management. The next generation of DOP and DOTP manufacturing will be more intelligent, sustainable, and economically feasible than ever, thanks to these advancements, which are in line with global SDG targets. AI-driven process control, IoT-enabled monitoring, digital twin simulations, predictive analytics, automation, and blockchain-based supply chain management must all be strategically integrated to transform a DOP and DOTP production facility into an Industry 6.0-driven smart plant. Targeted software, hardware, and operational improvements can greatly improve yield, quality, energy efficiency, and sustainability while lowering costs and the environmental impact, as opposed to rebuilding the entire infrastructure. Industry 6.0 turns conventional chemical production into a resource-optimized, zero-waste, and highly efficient process by enabling autonomous decision-making, real-time monitoring, and climate-resilient manufacturing. Several significant improvements are required to accomplish this change. Real-time modifications in esterification, transesterification, and distillation are guaranteed by AI-driven process optimization, which also greatly increases reaction efficiency and reduces errors by 50%. Using spectroscopic sensors for AI-powered quality control improves batch uniformity (±0.1% precision) in comparison to conventional approaches (±5% variation). Real time temperature, pressure, and composition monitoring across reactors and distillation columns is made possible by IoT enabled smart sensors, which reduces process inefficiencies by 40%. Production delays are eliminated by instantaneous modifications made possible by cloud based process control and edge computing. While quantum computing models optimize response parameters, increasing conversion efficiency from 97 to 99.8%, digital twin simulations mimic real-world production conditions, enabling predictive failure analysis and reducing unplanned downtime by 70%. Automation is essential to manufacturing pushed by Industry 6.0. Robotic handling systems and autonomous guided vehicles simplify the transportation of raw materials and catalyst dosing, reducing labor expenses by 50% and increasing productivity by 30%. 30% less waste is produced thanks to precise flow rate adjustments made possible by automated valve management and robotic reactor feeding [68–70].

By ensuring end to end traceability of raw materials, blockchain technology lowers the possibility of counterfeit goods and non-compliance with international laws like REACH, RoHS, and GHS. 15% less material is wasted when smart contracts automate procurement based on current inventory levels. In Industry 6.0, sustainability enhancements are particularly essential since waste heat recovery systems absorb and repurpose excess thermal energy, which lowers heating expenses by 30%. Modular carbon capture systems eliminate 98% of process-related CO_2 emissions, and solar-assisted process heating further reduces reliance on fossil fuels by 40%. A resource-efficient, climate-resilient manufacturing model is ensured by these sustainability initiatives, which are in line with SDGs 9, 12, and 13. With a predicted return on investment of less than 2 years, the financial feasibility of moving to Industry 6.0 is very alluring. While digital twin simulation may cost about USD two million (the cost is not same all the time and it may vary with time and location) but decreases losses by 1.2 million US dollar per annum, AI and IoT integration are anticipated to need an investment of 3.5 million US dollar (so annual two million USD savings). While blockchain-enabled supply chain management costs $1.5 million but increases efficiency by one million USD per annum, AGVs and robotics cost USD four million but save USD 2.5 million annually. With a USD 2.5 million initial investment, waste heat recovery and solar-assisted processes reduce energy costs by USD 1.8 million annually. The USD 13.5 million total transformation investment yields USD 8.5 million in savings annually, with a return on investment of 1.58 years. Costing, ROI, and savings are all variables that can vary from place to place. It's only a preliminary estimate meant to serve as a comparative analysis for the readers. The advantages of an Industry 6.0-driven smart plant over conventional DOP and DOTP manufacture are unmatched. Energy efficiency increases by 35–40%, production costs reduce by 30–50%, and CO_2 emissions decrease by 60% [71–73]. Production rates increase as waste reduction reaches 90% and batch cycle times decrease from 6 to 3.5 h. Real-time monitoring and AI-driven defect prevention improve final product consistency and guarantee exceptional plasticizer purity for high-performance applications in consumer products, construction, and automotive. Making the switch from conventional to Industry 6.0 driven DOP and DOTP manufacturing results in significant cost reductions, increased productivity, and enhanced sustainability. Profitability and environmental responsibility are maximized through blockchain-enabled traceability, robotic material handling, real-time monitoring, AI-based process automation, and sustainable energy integration. Manufacturers will be at the forefront of chemical industry innovation in Industry 6.0 thanks to this transformation, which guarantees a robust, future-proof production strategy. In summary, we need to focus on Industry 6.0 while planning a new chemical manufacturing sector. We must upgrade the current infrastructure and machinery to Industry 6.0 if we currently have a traditional industrial setup. We must create a feasibility report and determine the ROI and advantages of every savings. To prevent any outages or problems with the quality of the existing output, we must recommend a progressive upgrade one at a time [74, 75].

13 Challenges and Suggestions to Transform Chemical Industries into Industry 6.0

Here in Table 16.1, there is a comparative study of between traditional chemical processes and AI-based Industry 6.0 processes and their benefits in sustainable point of view and benefits in achieving SDGs. The IEA projects that adopting Industry 6.0 principles in chemical manufacturing could drive 50% reductions in CO_2 emissions and 40% cuts in operating costs by 2030, supporting global decarbonization goals and economic viability in line with broader net-zero strategies.

Table 16.1 Industry 6.0 technologies vs. traditional chemical processes for sustainability and SDG impact

Process category	Traditional process	Industry 6.0 technology	Sustainability benefits	Linked SDGs
Chemical synthesis	Batch reactors (4 to 6 h cycles); sulfuric acid catalysts; ± 5% batch variance; 5.5 to 6.5 GJ/ton energy	AI optimized autonomous reactors with enzyme/nano catalysts; quantum chemistry models; real-time IoT monitoring	• 40% faster reactions (2.5–3.5 h) • 60% lower energy (3.5 to 4.2 GJ/ ton) • 99.9% purity (±0.1% variance) • 70% less catalyst waste	9, 12, 13
Catalysis	Corrosive catalysts (H_2SO_4, $ZnAc_2$); low selectivity; frequent replacement	Nano engineered catalysts (graphene TiO_2, zeolites); enzyme biocatalysts (proteases, catalases, lipases etc.)	• 50% higher esterification rates • 99.9% selectivity at 80 to 120 °C • 60% energy reduction • Eliminates solvent purification	9, 12
Carbon capture	Amine scrubbing (high energy penalty); $80–100/ton capture cost	AI-optimized CCU with electrochemical CO_2 to jet fuel conversion; modular DAC units	• 15% higher capture efficiency • USD 30–50/ton cost • 90% conversion to synthetic fuels • Net negative operations	7, 9, 13
Plastic recycling	Mechanical shredding (low quality); <40% recyclable plastics	AI-driven chemical recycling: pyrolysis/ depolymerization + blockchain traceability	• 25% higher yield • 99% pure monomers • 40% landfill reduction • Circular economy integration	9, 12, 14
Energy management	Grid-dependent; 30–40% waste heat loss; fossil-fuel heating	ORC/TEG waste heat recovery + solar-assisted reactors + AI procurement	• 18% energy efficiency gain • 40% lower CO_2 • 30% heating cost reduction • 20–30% less grid reliance	7, 9, 13

Water management	Wet flotation (1.5 t water/t chemical); sludge discharge	AI ZLD with IoT leak detection + graphene nanofiltration	• 25% less water use • 30% higher recovery • 99% heavy metal removal • Zero discharge	6, 9, 14
Process safety	Manual inspections; reactive maintenance; 30 fatalities/100 k workers	Digital twins + autonomous inspection robots + AI predictive analytics	• 40% fewer catastrophic events • 95% real-time leak detection • Zero fatalities • 70% shorter downtime	3, 8, 9
Compliance and certification	Manual REACH/OEKO TEX reporting; 60–90 days audit cycles	Blockchain AI automated audits; real-time Bluesign/GOTS validation	• 50–70% faster reporting • 80% lower noncompliance risk • 40% less hazardous chemicals • Immutable ESG records	9, 12, 16
Supply chain	Paper trails; counterfeit materials; 15–30% fraud losses	Blockchain geofencing + AGV drones + smart contracts	• 95% fraud reduction • 30% fewer disruptions • 99.9% material traceability • 35% logistics cost cut	9, 12, 17
R&D acceleration	Trial and error (2–5 years/catalyst); limited molecular modelling	Quantum computing + AI material informatics	• 80% faster drug development • 30% lower R&D costs • Novel bio catalysts in months (vs. years) • Precision molecular simulations	9, 17

This implementation framework covering AI predictive systems, nano catalysis, water recovery, quantum materials engineering, and renewable integration, demonstrates significant sustainability gains. Specifically, the pilot phase reduces downtime and energy use, the scale-up phase cuts emissions and enhances water reuse, and the maturity phase achieves full autonomy with net negative emissions and substantial cost savings. His phased strategy outlines a comprehensive, technically grounded, and economically feasible approach to transforming chemical manufacturing within the Industry 6.0 framework. The roadmap is inherently adaptable and can be customized according to plant size, infrastructure maturity, resource availability, and regional regulatory environments. Its modularity allows for flexible integration of key technologies based on cost–benefit analysis, existing systems, and long-term strategic goals. It offers scalable pathways for sustainable development, significant emission reductions, and enhanced competitiveness, all while aligning closely with key SDGs.

In the first phase, spanning 0–18 months, facilities initiate pilot scale implementations by deploying AI-powered predictive maintenance systems and blockchain-enabled supply chain solutions to improve asset reliability, transparency, and real-time traceability. The addition of IoT-based leak detection sensors further supports proactive maintenance by enabling continuous monitoring of plant infrastructure. These technologies, requiring an estimated capital outlay of USD 2–5 million/plant, have demonstrated the ability to reduce unplanned downtime by 20% and energy consumption by 15%, based on field data from European chemical manufacturing pilots. The strategic roadmap for Industry 6.0 adoption in chemical manufacturing is defined by a phased implementation framework that integrates digitalization, energy innovation, and advanced materials science to achieve improved operational efficiency, environmental sustainability, and accelerated return on investment. Guided by benchmarks from the IEA and sectoral best practices, this structured approach outlines clear technological pathways and investment requirements tailored to the evolving demands of modern chemical production. The second phase, over 18–36 months, involves scaling up these digital upgrades with the integration of advanced nano-catalysts into chemical reactors. These catalysts significantly enhance reaction selectivity, throughput, and energy efficiency. Alongside this, plants adopt ZLD water management systems that allow for over 90% reuse of process water, aligning with environmental regulatory mandates and water conservation goals. This mid-phase investment, estimated at USD 5–10 million per facility, has shown to reduce CO_2 emissions by 40% while enabling significant water sustainability outcomes. In the third and final phase, beyond 36 months, the roadmap envisions full facility autonomy through the deployment of quantum AI platforms for high-speed material discovery and the establishment of solar hydrogen hybrid microgrids. These systems allow for seamless integration of renewable energy and green hydrogen generation, thereby eliminating reliance on fossil-derived utilities. Simultaneously, carbon capture and utilization technologies are implemented to convert process-emitted CO_2 into valuable end products such as polymers and mineral carbonates. With an investment range of USD 10–15 million per plant, this phase delivers up to 60% operational cost savings and achieves net negative carbon emissions, with typical ROI achieved within 1.5–2 years as evidenced by digital process intensification studies. Table 16.2 highlights the technical main challenges

Table 16.2 Industry 6.0 advancement challenges and implementations plan

Challenge category	Specific technical problems	Advanced solutions and suggestions	Implementation timeline	Expected impact
High capital investment	• USD 10 to 15 M/plant retrofit costs • ROI uncertainty for SMEs • High R&D expenses	Modular implementation: Start with high ROI tech (predictive maintenance, AI optimization). Use green bonds and carbon credit financingGovt. partnerships: Tax incentives for industry 6.0 adoption	6–18 months	ROI in 1.5–2 years; 30–50% cost recovery via energy/waste savings
Cybersecurity risks	• IoT sensor vulnerabilities • Data breaches in cloud-based PAT • Quantum hacking threats	Blockchain-encrypted networks: Decentralized data storage with quantum-resistant cryptography AI threat detection: Real-time anomaly identification (99.9% breach prevention)	Immediate	Zero critical incidents; >95% compliance with ISO 27001
Workforce skill gap	• 65% staff lack AI/ quantum literacy • Resistance to autonomous systems	AR/VR training simulators: Digital twin-based upskilling modules University industry consortiums for certified industry 6.0 advanced manufacturing programs	3–12 months	80% workforce proficiency in AI/robotics within 2 years
Data integration	• Legacy SCADA/PLC incompatibility • Siloed data lakes • Real time processing latency	Edge cloud hybrid architecture: Local AI processing +5G cloud sync (latency <5 ms) API first middleware: Unified data platforms	6–24 months	40% faster decision-making; 90% data interoperability
Regulatory hurdles	• REACH/GHS compliance delays • Cross border certification conflicts • EPA/OSHA audits	AI auto compliance engines: Real-time blockchain-tracked documentation. Digital regulatory twins: Simulate EPA/ OSHA inspections preemptively	Continuous	70% faster certifications; 80% fewer noncompliance penalties

(continued)

Table 16.2 (continued)

Challenge category	Specific technical problems	Advanced solutions and suggestions	Implementation timeline	Expected impact
Energy-intensive compute	• Quantum/AI servers consume 2.5 MW/plant • Cooling demands for HPC	Liquid immersion cooling: 40% less energy vs. air cooling Renewable powered HPC: solar/waste heat-driven quantum systems	12–36 months	50% lower compute energy; carbon neutral data centers
Sustainable scaling	• Limited bio-feedstock supply chains • Chemical recycling scalability gaps	AI-bio refineries: Algae-based raw material synthesis Modular microreactors: Distributed chemical recycling hubs	24–48 months	60% fossil fuel displacement; 30% cost reduction in bio feedstocks
Legacy infrastructure	• 70% plants use >20 years old reactors • Non IoT compatible sensors	Retrofit kits: 3D printed smart adapters for analog equipment Frugal automation: Low-cost AI vision systems for manual process monitoring	3–18 months	90% legacy system integration; 50% longer asset lifespan

in the implementation of Industry 6.0 in chemical industrial sector and some suggestions to incorporate these advancements. These suggestions can be customized as per requirements depends on multi factors.

To accelerate adoption among SMEs, we advocate establishing UN-backed Industry 6.0 Accelerator Funds, offering low-interest financing tied to SDG-aligned performance metrics. Market models show that strategic implementation of these advanced technologies can deliver 200% value creation over five years, enabled by cost savings, increased resilience, and new revenue streams from bio-based products. Quantum ML, self-healing AI materials, and bio hybrid CO_2 conversion are converging to create a transformative ecosystem, positioning Industry 6.0 at the forefront of sustainable, resilient, and economically viable manufacturing. Quantum machine learning is rapidly transforming catalyst discovery, reducing typical development cycles from 2 to 5 years to less than 3 months through hybrid quantum classical models like IBM Quantum integrated with TensorFlow Quantum. These hybrid approaches have demonstrated potential for 30% reductions in R&D spending and nearly 50% gains in catalyst performance, owing to more accurate energy landscape predictions at the molecular level. In materials science, fault-tolerant quantum simulations targeting catalyst design tasks suggest quantum algorithms can outperform classical methods by identifying ground-state energies up to 0.2 million USD equivalent economic value, according to recent feasibility studies. Significant strides have also been made in process resilience through self-healing materials, where nano-coated reactors equipped with embedded AI diagnostic systems can reduce unplanned downtime by 70% and extend asset life to 20 years, by autonomously detecting and repairing micro corrosion before failures occur. In bio-hybrid manufacturing, technologies like LanzaTech's Clostridium autoethanogenum platforms are converting CO_2 into commodity chemicals such as acetone, isopropanol, ethanol, and mono ethylene glycol. Life cycle analyses confirm these processes can reduce carbon emissions by 160% compared to fossil-based production. Large-scale operations already demonstrate over 90% carbon conversion efficiency and recycle more than 90% of processing water, enabling circular carbon systems with net-negative emissions [76–80].

References

1. Kamarinopoulou, N.S., Nguyen, D.K., Vlachos, D.G.: Process-chemistry intensification using non-thermal plasmas: towards one-step chemical production. Cur. Opin. Green Sustain. Chem. **51**, 100997 (2025)
2. Chen, K., et al.: Emission characteristics, environmental impacts and health risk assessment of volatile organic compounds from the typical chemical industry in China. J. Environ. Sci. **149**, 113–125 (2025)
3. Andriani, G., et al.: Evaluating the thermal stability of chemicals and systems: a review. Can. J. Chem. Eng. **103**(1), 42–62 (2025)

4. Han, Z., et al.: Development of the design and synthesis of metal–organic frameworks (MOFs)–from large scale attempts, functional oriented modifications, to artificial intelligence (AI) predictions. Chem. Soc. Rev. **54**(1), 367–395 (2025)
5. Kamkar, M., et al.: Artificial Intelligence (AI) for Sustainable Resource Management and Chemical Processes, pp. 178–180. ACS Publications, Washington (2024)
6. Wu, M., et al.: Recent applications of theoretical calculations and artificial intelligence in solid-state electrolyte research: a review. Nano. **15**(3), 225 (2025)
7. Zhang, W., et al.: Revitalizing interphase in all-solid-state Li metal batteries by electrophile reduction. Nat. Mater. **24**(3), 414–423 (2025)
8. Pal, P.K., et al.: Digital twins: transforming the chemical process industry—a review. Can. J. Chem. Eng. **103**, 3611–3636 (2025)
9. Njoku, J.N., et al.: Leveraging digital twin technology for battery management: a case study review. IEEE Access. **13**, 21382–21412 (2025)
10. Martínez, J.D., et al.: Waste-based value-added feedstocks from tire pyrolysis oil distillation: defossilization of the petrochemical industry. Green Chem. **27**(3), 670–683 (2025)
11. Martín, M.M.: Introduction to Software for Chemical Engineers. CRC Press, Boca Raton (2025)
12. Liu, P., et al.: Process analytical technologies and self-optimization algorithms in automated pharmaceutical continuous manufacturing. Chin. Chem. Lett. **35**(3), 108877 (2024)
13. Jiang, Q., Wang, P.: A fuzzy logic-based coordinated adaptive control method for megawatt novel nuclear power systems. Ann. Nucl. Energy. **215**, 111245 (2025)
14. Li, D.-J., Yan, Y.-F., Li, D.: Adaptive event-triggered control for wastewater treatment process using self-organizing fuzzy neural network. IEEE Trans. Autom. Sci. Eng. **22**, 11333–11342 (2025)
15. Agrawaal, H., Thompson, J.: Additive manufacturing (3D printing) for analytical chemistry. Talanta Open. **3**, 100036 (2021)
16. Sun, C., et al.: Additive manufacturing for energy: a review. Appl. Energy. **282**, 116041 (2021)
17. Sagmeister, P., et al.: Advanced real-time process analytics for multistep synthesis in continuous flow. Angew. Chem. Int. Ed. **60**(15), 8139–8148 (2021)
18. Pilat, F., et al.: Beyond Karl Fischer titration: a monolithic quantum cascade sensor for monitoring residual water concentration in solvents. Lab Chip. **23**(7), 1816–1824 (2023)
19. Jaharudin, N.A.N., et al.: Optofluidic sensor for iron detection based on inline Fabry Perot interferometer fabricated by femtosecond laser pulses and selective etching. IEEE Sensors J. **25**, 8324–8332 (2025)
20. Zhang, W., et al.: Surface embedding enhancing interfacial photothermal effect of MXene/EP sponge for effective oil/water separation. Sep. Purif. Technol. **353**, 128514 (2025)
21. Liang, Y., et al.: Efficient and easily recyclable photocatalytic reduction of se (IV) from wastewater using stable TiO2/BiOBr/cloth: mechanism insight and machine learning modeling. Sep. Purif. Technol. **352**, 128021 (2025)
22. Yao, Y., et al.: Models for decarbonization in the chemical industry. Ann. Rev. Chem. Biomol. Eng. **15**, 100522–114115 (2024)
23. Kawai, E., Ozawa, A., Leibowicz, B.D.: Role of carbon capture and utilization (CCU) for decarbonization of industrial sector: a case study of Japan. Appl. Energy. **328**, 120183 (2022)
24. Ahmed, O., Ahmad, S., Adekunle, S.K.: Carbon dioxide sequestration in cementitious materials: a review of techniques, material performance, and environmental impact. J CO_2 Util. **83**, 102812 (2024)
25. Maruyama, I., et al.: A new concept of calcium carbonate concrete using demolished concrete and CO2. J. Adv. Concr. Technol. **19**(10), 1052–1060 (2021)
26. Al Harthi, A., et al.: Criteria and cutting-edge catalysts for CO_2 electrochemical reduction at the industrial scale. J CO_2 Util. **83**, 102819 (2024)
27. Huang, W., et al.: High temperature shock (HTS) synthesis of carbon-based nanomaterials for electrochemical applications. Carbon Neutralization. **4**(1), e189 (2025)
28. Bak, Y., et al.: Macroscopic alignment of metal–organic framework crystals: recent advances and perspectives on sustainability. Green Chem. **27**, 11 (2025)

29. Ngo, T.C., et al.: Are natural-based isothiocyanate derivatives alternative antioxidant additives used in petrochemical products? R. Soc. Open Sci. **12**(2), 241929 (2025)
30. Dey, B., et al.: Biopolymer-based solutions for enhanced safety and quality assurance: a review. Food Res. Int. **191**, 114723 (2024)
31. Zou, C.: Carbon reduction. In: Carbon Neutrality Science: A Theoretical and Technical Framework and Energy Development Strategies, pp. 87–192. Springer, Cham (2024)
32. Gomes, N.O., et al.: Flexible and sustainable printed sensor strips for on-site, fast decentralized self-testing of urinary biomarkers integrated with a portable wireless analyzer. Chem. Eng. J. **472**, 144775 (2023)
33. Costa, R., et al.: SDG walking or washing? A cross-sectoral analysis of business contribution to the SDGs. Bus. Strateg. Environ. **34**(3), 3561–3576 (2025)
34. Hacker, J., et al.: Aquatic quality watch informed by communities (AQWIC) facilitating the adoption of low-cost sensor systems for underserved communities: a review and tutorial. Environ. Sci. Adv. **4**, 512–529 (2025)
35. Trier, X., et al.: The critical role of commercial analytical reference standards in the control of chemical risks: the case of PFAS and ways forward. Environ. Health Perspect. **133**(1), 015001 (2025)
36. Gambino, I., et al.: Microcontaminants and microplastics in water from the textile sector: a review and a database of physicochemical properties, use in the textile process, and ecotoxicity data for detected chemicals. Environ. Sci. Process. Impacts. **27**, 297–319 (2025)
37. Özbey, H.K., Gürarda, A.: Investigation of seam performance and biodegradability of organic cotton clothes for their sustainability. Biodegradation. **36**(1), 17 (2025)
38. Aguilar, F.J.A., et al.: Determination of eight fungicides in tanned leather by liquid chromatography with mass spectrometry and with diode array spectrophotometric detection. Anal. Methods. **17**, 1774–1783 (2025)
39. Figuière, R., et al.: An overview of potential alternatives for the multiple uses of per-and polyfluoroalkyl substances. Environ. Sci. Technol. **59**, 2031–2042 (2025)
40. Mathur, V., et al.: Volumetric additive manufacturing for cell printing: bridging industry adaptation and regulatory frontiers. ACS Biomater. Sci. Eng. **11**, 156–181 (2025)
41. Ghettas, S.: Achieving the UN sustainable development goals in the oil and gas sector through risk management. In: Sustainability in the Oil and Gas Sector: Adaptation and Mitigation Strategies for Tackling Climate Change, pp. 199–219. Springer, Cham (2024)
42. de Jesus, J.O., Oliveira-Esquerre, K., Medeiros, D.L.: Integration of artificial intelligence and life cycle assessment methods. IOP Conf. Ser. Mater. Sci. Eng. **1196**, 012028 (2021)
43. Bassey, K.E., Juliet, A.R., Stephen, A.O.: AI-enhanced lifecycle assessment of renewable energy systems. Eng. Sci. Technol. J. **5**(7), 2082–2099 (2024)
44. Olatunde, T.M., et al.: Reviewing the role of artificial intelligence in energy efficiency optimization. Eng. Sci. Technol. J. **5**(4), 1243–1256 (2024)
45. Segovia-Hernández, J.G., et al.: Green hydrogen production for sustainable development: a critical examination of barriers and strategic opportunities. RSC Sustain. **3**(1), 134–157 (2025)
46. Besinga, L.E., Mukete, T.N.M.: Technological innovations for climate adaptation and peacebuilding: a holistic approach to resource conflict and environmental challenges. Open J. Appl. Sci. **15**(1), 285–304 (2025)
47. Liu, X., et al.: Dynamic performance analysis of adsorption heat transformer system driven by large pressure jump for low-grade waste heat upgrade. Appl. Energy. **377**, 124478 (2025)
48. Markides, C.N., et al.: Working fluid and system optimisation of organic Rankine cycles via computer-aided molecular design: a review. Prog. Energy Combust. Sci. **107**, 101201 (2025)
49. Li, G., et al.: A review on micro combustion powered thermoelectric generator: history, state-of-the-art and challenges to commercialization. Renew. Sust. Energ. Rev. **207**, 114897 (2025)
50. Feng, Y., et al.: Recent advances in the chemical recycling of polyamide for a sustainable circular economy. Ind. Eng. Chem. Res. **64**(5), 2516–2530 (2025)

51. Pobłocki, K., et al.: Catalytic materials based on metals (ions) used in the upcycling of plastics and polymers into fuels and valuable chemicals as part of sustainable development. Mater. Sci. Eng. R Rep. **162**, 100881 (2025)
52. Chen, G., et al.: Plasma-enhanced upcycling of mixed plastic waste over La0.6Ca0.4FeO3−δ nano spheres for co-producing hydrogen and high-value carbon. Results Eng. **25**, 104039 (2025)
53. Scelfo, G., et al.: Demonstration of ultra-high-water recovery and brine concentration in a prototype evaporation unit: towards zero liquid discharge desalination. Sep. Purif. Technol. **354**, 129427 (2025)
54. Tripathy, S., Kar, O.P., Pradhan, A.: Challenges and innovations in industrial wastewater treatment: safeguarding water resources and promoting sustainable practices. Water Air Soil Pollut. **236**(2), 1–14 (2025)
55. Parveen, F., Slater, A.G.: Digitalisation of catalytic processes for sustainable production of biobased chemicals and exploration of wider chemical space. Catal. Sci. Technol. **15**, 1689–1701 (2025)
56. Reddy, C.K.K., et al.: Evolution and Advances in Computing Technologies for Industry 6.0: Technology, Practices, and Challenges. CRC Press, Boca Raton (2024)
57. Bishayee, B., et al.: Strategic management of nitrate pollution from contaminated water using viable adsorbents: an economic assessment-based review with possible policy suggestions. J. Environ. Manag. **303**, 114081 (2022)
58. Helgerud, M.B.: The 4D value journey at ExxonMobil. In: 4D Forum—Insight to Actions: A Global Forum. Society of Exploration Geophysicists, Houston (2025)
59. Supran, G., Rahmstorf, S., Oreskes, N.: Assessing ExxonMobil's global warming projections. Science. **379**(6628), eabk0063 (2023)
60. Selg, M., Farhod, A.: Strategic response to the energy crisis in Germany: the case of BASF. In: Strategic Response to Turbulence, pp. 135–148. Edward Elgar Publishing, Cheltenham (2025)
61. Geisbauer, A., Berchthold, C.: Will power and waste-to-X enable the worldwide energy transition to renewables? Monatsh. Chem. Chem. Monthly. (2024). https://doi.org/10.1007/s00706-024-03176-6
62. Potaraju, S., Kespohl, H., Bach, H.: Covestro: digital technical services as new business model for the polymer processing industry. In: Connected Business: Create Value in a Networked Economy, pp. 305–314. Springer, Cham (2021)
63. Altun, A., Fellah, M.F.: A mini-review on different synthesis reactions of dioctyl terephthalate (DOTP) and properties of DOTP plasticized PVC. Pamukkale Univ. Muh. Bilim. Derg. **28**(7), 1001–1013 (2022)
64. Sobirovna, Y.K., Masharipovich, M.S.: Thermal and THERMO-oxidation stability of polyvinyl chloride plasticized with DIOCTYL terephthalate. Asian J. Pharm. Biol. Res. **11**(2), 208–214 (2022)
65. Mohamed, D.F.M., et al.: Assessment of biodegradation and toxicity of alternative plasticizer di (2-ethylhexyl) terephthalate: impacts on microbial biofilms, metabolism, and reactive oxygen species-mediated stress response. Environ. Pollut. **355**, 124217 (2024)
66. Liu, Z.: Current status of plasticizer research. In: Green Catalytic Hydrogenation of Phthalate Plasticizers, pp. 51–127. Springer, Cham (2022)
67. Etuk, I.F., Inyang, U.E.: Applications of various plasticizers in the plastic industry-review. Int. J. Eng. Mod. Technol. **10**(1), 38–54 (2024)
68. Han, Y., et al.: Epoxidized isosorbide-based esters with long alkyl chains as efficient and enhanced thermal stability and migration resistance PVC plasticizers. Polym. Test. **123**, 108048 (2023)
69. Chen, R., et al.: Integrated on-site collection and off-site analysis of airborne molecular contamination in cleanrooms for integrated circuit manufacturing processes. Build. Environ. **214**, 108941 (2022)
70. Ghobashy, M.M.: Regulatory aspects of the use of PVC and its blends, gels, and IPNs. In: Poly (Vinyl Chloride)-Based Blends, IPNs, and Gels, pp. 551–576. Elsevier, Amsterdam (2024)

71. Hashimoto, T.: "Mechanics" of molecular assembly: real-time and in-situ analysis of nano-to-mesoscopic scale hierarchical structures and nonequilibrium phenomena. Bull. Chem. Soc. Jpn. **78**(1), 1–39 (2005)
72. Han, Y., et al.: Synergistic effects of epoxidized isosorbide linolenate (EGLA-ISB): a novel bio-based heat stabilizer for enhanced mechanical and thermal performance of PVC. J. Vinyl Add. Technol. **31**(1), 133–145 (2025)
73. Zhang, J., et al.: Enhancement of toughness and gas barrier properties of PLA-based films through bio-based isosorbide dioctate: synthesis, characterization, and application. Polym. Eng. Sci. **65**(1), 315–326 (2025)
74. Qian, B., et al.: Synthesis, performance evaluation, and plasticization dynamics of biobased vanillic acid plasticizer for poly (vinyl chloride). Chem. Eng. J. **497**, 154991 (2024)
75. Hou, B., et al.: Synthesis, characterization, and performance evaluation of a high-efficiency fully biobased biodegradable plasticizer. J. Appl. Polym. Sci. **141**(45), e56225 (2024)
76. Onyeaka, H., Ekwebelem, O.: A review of recent advances in engineering bacteria for enhanced CO2 capture and utilization. Int. J. Environ. Sci. Technol. **20**(4), 4635–4648 (2023)
77. Salehizadeh, H., Yan, N., Farnood, R.: Recent advances in microbial CO2 fixation and conversion to value-added products. Chem. Eng. J. **390**, 124584 (2020)
78. Hong, Y.Y., Lopez, D.J.D., Wang, Y.Y.: Solar irradiance forecasting using a hybrid quantum neural network: a comparison on GPU-based workflow development platforms. IEEE Access. **12**, 145079–145094 (2024)
79. Deka, K., Zak, E.: Simultaneously optimizing symmetry shifts and tensor factorizations for cost-efficient fault-tolerant quantum simulations of electronic Hamiltonians. J. Chem. Theory Comput. **21**(9), 4458–4465 (2025)
80. Neto, A.S., et al.: Exploring the potential of syngas fermentation for recovery of high-value resources: a comprehensive review. Curr. Pollut. Rep. **11**(1), 7 (2024)

Index

H. Majeed, T. Iftikhar, *Intelligent Manufacturing in Industry 6.0*,
https://doi.org/10.1007/978-3-032-07278-8

C

D

F

Q

MIX
Papier aus verantwortungsvollen Quellen
Paper from responsible sources
FSC® C105338

If you have any concerns about our products,
you can contact us on
ProductSafety@springernature.com

In case Publisher is established outside the EU,
the EU authorized representative is:
Springer Nature Customer Service Center GmbH
Europaplatz 3, 69115 Heidelberg, Germany

Printed by Libri Plureos GmbH
in Hamburg, Germany